Structural Equation Modeling

WILEY SERIES IN PROBABILITY AND STATISTICS

Established by WALTER A. SHEWHART and SAMUEL S. WILKS

A complete list of the titles in this series appears at the end of this volume.

Structural Equation Modeling

A Bayesian Approach

Sik-Yum Lee
Department of Statistics
Chinese University of Hong Kong

BICENTENNIAL 1807 WILEY 2007 BICENTENNIAL

John Wiley & Sons, Ltd

Telephone (+44) 1243 779777

Email (for orders and customer service enquiries): cs-books@wiley.co.uk
Visit our Home Page on www.wiley.com

Other Wiley Editorial Offices

John Wiley & Sons Inc., 111 River Street, Hoboken, NJ 07030, USA

Jossey-Bass, 989 Market Street, San Francisco, CA 94103-1741, USA

Wiley-VCH Verlag GmbH, Boschstr. 12, D-69469 Weinheim, Germany

John Wiley & Sons Australia Ltd, 42 McDougall Street, Milton, Queensland 4064, Australia

John Wiley & Sons (Asia) Pte Ltd, 2 Clementi Loop #02-01, Jin Xing Distripark, Singapore 129809

John Wiley & Sons Canada Ltd, 22 Worcester Road, Etobicoke, Ontario, Canada M9W 1L1

Wiley also publishes its books in a variety of electronic formats. Some content that appears in print may not be available in electronic books.

Library of Congress Cataloging in Publication Data

British Library Cataloguing in Publication Data

A catalogue record for this book is available from the British Library

ISBN-13 978-0-470-02423-2(HB)

Typeset in 10/12pt Galliard by Integra Software Services Pvt. Ltd, Pondicherry, India
Printed and bound in Great Britain by TJ International, Padstow, Cornwall
This book is printed on acid-free paper responsibly manufactured from sustainable forestry in which at least two trees are planted for each one used for paper production.

For Mable Lee and Timothy Lee

Contents

About the Author

Sik-Yum Lee is a professor of statistics at the Chinese University of Hong Kong. He earned his Ph.D. in biostatistics at the University of California, Los Angeles, USA. He received a distinguished service award from the International Chinese Statistical Association, is a former president of the Hong Kong Statistical Society, and is an elected member of the International Statistical Institute and a Fellow of the American Statistical Association. He serves as Associate Editor for *Psychometrika* and *Computational Statistics & Data Analysis*, and as a member of the Editorial Board of *British Journal of Mathematical and Statistical Psychology*, *Structural Equation Modeling*, *Handbook of Computing and Statistics with Applications* and *Chinese Journal of Medicine*. His research interests are in structural equation models, latent variable models, Bayesian methods and statistical diagnostics. He is editor of *Handbook of Latent Variable and Related Models* and author of over 140 papers.

Preface

Substantive theory usually involves observed and latent variables. An observed variable can be directly measured through a single measurement, such as income or systolic blood pressure. Latent variables are those that cannot be directly measured. Usually, one has to use several observed variables to assess the characteristic of a latent variable. It is very easy to give examples of latent variables in behavioral, biological, educational, medical, psychological and social sciences. For instance, in medical science, obesity is a latent variable that should be assessed by body mass, waist and hip indexes; lipid is a latent variable that is better assessed by non-high density lipoprotein, low density lipoprotein and triglyceride; blood pressure is another latent variable that should be measured by both systolic and diastolic blood pressures. Basically, structural equation models (SEMs) are regression models with observed and latent variables. For example, the influential LISREL model is composed of two simple regression equations: the measurement equation that relates the observed variables to the latent variables, and the structural equation that relates the endogenous latent variables to other latent variables. Due to the contributions of many psychometricians, including but not limited to Karl Jöreskog and Peter Bentler, and their LISREL and EQS6 programs, SEMs have been extensively applied not only to behavioral, educational and social science, but also to biological and medical sciences in the last quarter of a century.

The excellent book of Bollen (1989) on standard SEMs was published more than 15 years ago. Despite the widespread use of the standard SEMs, and the rapid growth in the new developments of nonlinear SEMs, two-level SEMs, and mixtures of SEMs, as well as SEMs for more complex data structures, such as dichotomous, ordered categorical, binary and missing data, there are very few new reference/textbooks in the field and there have been very few practical applications of the aforementioned recent developments to substantive research. This unexpected phenomenon is the motivation for writing this book, to provide a reference for researchers in various disciplines, and a textbook for graduate students in statistics, biostatistics and psychometrics. My main purpose is to introduce a Bayesian approach for developing efficient and rigorous statistical methodologies in SEMs and applying them to practical problems. Given the importance of latent variables and the popularity of the regression models, I hope that this book will help to promote SEMs to become a mainstream in statistics and psychometrics, and stimulate more applications.

The theme of this book is on the Bayesian analysis of SEMs. An introduction is given in Chapter 1, and some standard models are briefly discussed in Chapter 2. Chapter 3 presents the basic asymptotic theory for the traditional maximum likelihood (ML) and generalized least squares (GLS) approaches in analyzing the covariance structure of the model. The reason for including these nonBayesian materials is to provide a rigorous technical background related to the statistical results that are given in a large number of commercial software packages. For example, I present detailed proof on the consistency and asymptotic normality of the estimators, as well as the asymptotic chi-square distribution of the goodness-of-fit statistic. As far as I know, these kind of technical materials cannot be found in existing books on SEMs. Chapter 4 gives a detailed introduction to Bayesian estimation, whereas Chapter 5 treats Bayesian model comparison through the Bayes factor. Materials provided in these two core chapters can be applied to standard SEMs and their generalizations. Chapters 6 to 13 present the descriptions of the Bayesian approach as applied to SEMs with ordered categorical variables, SEMs with dichotomous variables, nonlinear SEMs, two-level nonlinear SEMs, multisample SEMs, mixtures of SEMs, SEMs with missing data and SEMs with variables from an exponential family of distributions. The Bayesian methodologies are illustrated by analyses of real examples in various fields via our tailor-made programs and/or the general software WinBUGS. The fundamental material in these chapters is self-contained. However, for generality and for enhancing wider applications, certain sections in the chapters discuss combinations of models. These sections depend on some of the material presented in previous chapters.

This book requires an understanding of some fundamental concepts in statistics. In particular, the concept of conditional distributions is necessary to understand the key ideas of the posterior distribution and the sampling-based computational methods. It does not need a knowledge of factor analysis or SEMs, but such a knowledge will enhance the appreciation of the Bayesian methodologies. It requires some basic knowledge of convergence in probability and convergence in distribution to understand the materials in Chapter 3. However, other chapters are completely independent of this chapter. Readers who do not have an interest in the asymptotic theory of the traditional ML and GLS approaches may skip Chapter 3. I am pleased to receive any comments about the book via email at sylee@sta.cuhk.edu.hk.

I owe a great debt to organizations and individuals for the use of their data sets. These include: World Value Study Group, World Values Survey, 1981–1984 and 1990–1993, for allowing the use of the Inter-university Consortium for Political and Social Research (ICPSR) data set; the Faculty of Education and Hong Kong Institute of Educational Research, the Chinese University of Hong Kong for providing part of the data set in the Accelerated Schools for Quality Education (ASQE) Project; D. E. Morisky, J. A. Stein and their colleagues for providing the AIDS data set; and Juliana C. N. Chan for providing the dataset about patient's non-adherence. I am deeply grateful to the Medical Research Council Biostatistics Unit (Cambridge, UK), and the Department of Epidemiology and Public Health of the Imperial College School of Medicine at St. Mary's Hospital (London, UK) for providing the powerful software WinBUGS as an alternative program in obtaining the Bayesian solutions. Without the kind support from the above mentioned organizations and individuals, this book might not have existed. The Research Grants Council of the Hong Kong Special Administration Region and the Chinese University of Hong Kong have provided financial support for my research and the writing of this book.

This book owes much to many people. It is certainly a great pleasure to thank them individually for their help and influence in producing this book, which represents most of my recent work in SEMs. My advisor, R. I. Jennrich, taught me computational methods and deeply inspired me with his serious attitude toward research. P. M. Bentler led me to the exciting field of SEM; he has had a distinctive influence on me during my early work in this field. I am grateful to colleagues in the Department of Statistics, the Chinese University of Hong Kong. In particular, W.Y Poon has given me generous support in research and administration work; W. H. Wong introduced the idea of data augmentation and Markov chain Monte Carlo methods during his 3 year stay in our department; and X. Y. Song gave numerous constructive suggestions in many chapters and essentially wrote all the programs for analyzing the artificial and real examples. I am very fortunate to have excellent students and research assistants. Those to whom I am greatly indebted include S. J. Wang, J. Q. Shi, W. Zhang, H. T. Zhu, X. Y. Song, J. S. Fu, L. Xu, B. Lu, Y. M. Xia, N. S. Tang, Y. Li, F. Chen, J. H. Cai, C. T. Poon and Y. Zhou. Most of them read the manuscript and gave helpful comments. WinBUGS results were mostly obtained by B. Lu. A number of people tackled the tedious task of typing the manuscript and drawing the path diagrams. For their excellent work and their patience with my handwriting, I would like to thank K. H. Leung and E. L. S. Tam. I am grateful to all the wonderful people on the John Wiley editorial staff, particularly Kelly Board, Lucy Bryan, Wendy Hunter, Simon Lightfoot, Kathryn Sharples, and Vidya Vijayan, and their design team for their continued assistance, encouragement and support of my work. Last and most important, I owe deepest thanks to Mable Lee and Timothy Lee for their constant understanding, support, encouragement and love that greatly release me from the pressure of the observed and latent variables.

SIK-YUM LEE
Department of Statistics
Chinese University of Hong Kong
Shatin, NT,
Hong Kong

1

Introduction

1.1 STANDARD STRUCTURAL EQUATION MODELS

In behavioural, educational, medical, and social sciences, substantive theory usually involves two kinds of variables, namely manifest (observed) and latent variables. Manifest variables are those that can be measured directly, such as income, test scores, systolic blood pressure, diastolic blood pressure, weight or heart rate. All data are records of measurements from observed data. Very often, it is necessary to deal with latent variables that cannot be directly measured by a single manifest variable. Examples are intelligence, personality, quantitative ability, anxiety, buying behavior, blood pressure and health condition. In practice, the characteristics of a latent variable can be partially measured by a linear combination of some manifest variables. For example, the quantitative ability of secondary school students can be reflected by their test scores in mathematics, physics and chemistry; the blood pressure of a patient can be measured by systolic and diastolic blood pressures. In most substantive research, it is important to establish an appropriate model to evaluate a series of simultaneous hypotheses about the impacts of latent variables and manifest variables on the other variables, and take the measurement errors into account. Structural equation models (SEMs) are well recognized as the most important statistical method to serve the above purpose. SEMs can be applied to many fields. For example, they can be applied to market research for establishing interrelationships between demand and supply, and the attitude and behaviour of the customers; to environmental science for investigating how health is affected by air and water pollution; to education for measuring the growth of intelligence and its relationship to personality and school environment; and to medicine for analyzing quality of life data and/or examining the impacts of

Structural Equation Modeling: A Bayesian Approach S-Y. Lee

physicians' concern, social influence and cognition on patients' adherence to medication.

The standard SEM, in particular the LISREL model (Jöreskog and Sörbom, 1996), is composed of two components. The first component is a confirmatory factor analysis model which relates the latent variables to all their corresponding manifest variables (indicators) and takes the measurement errors into account. This component can be regarded as a regression model which regresses the manifest variables with a small number of latent variables. The second component is again a regression type structural equation which regresses the endogenous (dependent) latent variables with the linear terms of some endogenous and exogenous (independent) latent variables. As latent variables are random, they cannot be directly analyzed by techniques in ordinary regression that are based on raw observations. However, conceptually, SEMs are formulated by the familiar regression type model, hence they are easy to apply in practice.

1.2 COVARIANCE STRUCTURE ANALYSIS

For standard SEMs, the covariance matrix of the manifest random vector $\mathbf{y}$ contains all the unknown parameters in the model. Hence, the classical methods for analyzing standard SEMs focused on the sample covariance matrix $\mathbf{S}$ and not the raw individual random vectors $\mathbf{y}_i$. This involves the formulation of the covariance structure $\boldsymbol{\Sigma}(\boldsymbol{\theta})$, which is a matrix function of the unknown parameters vector $\boldsymbol{\theta}$; estimation of $\boldsymbol{\theta}$ by minimizing (or maximizing) some objective functions that measure the discrepancy between $\mathbf{S}$ and $\boldsymbol{\Sigma}(\boldsymbol{\theta})$, such as the maximum likelihood (ML) function or the generalized least square (GLS) function; and the derivation of asymptotic goodness-of-fit statistics for assessing whether $\boldsymbol{\Sigma}(\boldsymbol{\theta})$ fits $\mathbf{S}$. As this kind of analysis emphasizes the population covariance matrix and the sample covariance matrix, it is often called covariance structure analysis. Today, more than a dozen user-friendly SEM software packages have been developed on the basis of covariance structure analysis approach with the sample covariance matrix. Typical examples are LISREL, EQS6 and AMOS. The covariance structure analysis approach depends heavily on the asymptotic normality of $\mathbf{S}$, either in defining the objective function or in deriving the asymptotic properties for statistical inferences. When the distribution of the manifest random vector $\mathbf{y}_i$ is multivariate normal and the sample size is reasonably large, the asymptotic distribution of $\mathbf{S}$ accurately approximates to the claimed multivariate normal distribution, and as a result this approach works fine. However, under slightly more complex situations that are common in substantive research, the covariance structure analysis approach on the basis of $\mathbf{S}$ is not effective and may encounter theoretical and computational problems.

It is well-recognized that estimating nonlinear terms (particularly the interaction term) among latent variables in the structural equation is an important issue in behavioral, social and psychological science (see Kenny and Judd, 1984; Bagozzi, Baumgartner and Yi, 1992). Due to the presence of the nonlinear terms of latent variables, the endogenous latent variables and the related manifest variables in $\mathbf{y}_i$ are not normally distributed. Hence, the sample covariance matrix of the raw sample observations is inadequate for modeling the nonlinear relationships. The product indicator approach (Jöreskog and Yang, 1996; Marsh, Wen and Hau, 2004) that artifically added products of indicators to $\mathbf{y}_i$ and used the sample covariance matrix of the enlarged manifest random vector for analysis, cannot provide a satisfactory method to cope with the problem (see Lee, Song, and Poon, 2004). For dichotomous or ordered categorical data, the sample covariance matrix of the raw sample data cannot be used. The multistage estimation procedures in LISREL or EQS produced estimates that are less optimal than the exact ML estimates and cannot be applied to analyze nonlinear terms of latent variables. For missing data that have a small number of observations within some missing patterns, the covariance structure analysis approach would also encounter serious difficulties because the sample covariance matrices corresponding to those patterns could be singular. The degree of difficulty is further compounded with a large number of missing patterns in the data set, or the missing data are missing with a nonignorable missing mechanism. For hierarchical data, the individual observations are correlated; this induces a problem in the covariance structure analysis with the sample covariance matrix.

In view of the above discussion, it is clear that although the covariance structure analysis approach based on the sample covariance matrix works well for the standard SEMs under the normality assumption, it cannot be applied to more complex models or data structures that are commonly encountered in substantive research. To develop sound statistical methods for those complex situations, it is necessary to develop better statistical methods which are based on the individual observations and their basic model, rather than the sample covariance matrix.

1.3 WHY A NEW BOOK?

In the past few years, the growth of SEM has been very rapid. New models and statistical methods have been developed for better analyses of more complex data structures in substantive research. These include but are not limited to: (i) SEMs with dichotomous and/or ordered categorical data (Shi and Lee, 1998, 2000; Moustaki, 2003; Rabe-Hesketh, Skrondal and Pickles, 2004; among others); (ii) nonlinear SEMs (Kenny and Judd, 1984; Klein and Moosbrugger, 2000; Lee and Song, 2003a; Wall and Amemiya, 2000; among others); (iii) linear or nonlinear SEMs with covariates (Lee and Shi, 2000; Moustaki, 2003; Song and Lee, 2006a; among others); (iv) two-level or multilevel SEMs (Lee and Shi, 2001; Ansari and Jedidi, 2000; Rabe-Hesketh, Skrondal and Pickles,

2004; Song and Lee, 2004; among others); (v) multisample SEMs (Song and Lee, 2001, 2002b); (vi) mixtures of SEMs (Jedidi, Jagpal and DeSarbo, 1997; Arminger, Stein and Wittenberg, 1999; Dolan and van der Maas, 1998; Zhu and Lee, 2001; Lee and Song, 2003b; among others); (vii) SEMs with missing data that are missing at random or with a nonignorable mechanism (Jamshidian and Bentler, 1999; Song and Lee, 2002a; Lee and Song, 2004a; Lee and Tang, 2006a; Song and Lee, 2006b; among others); and (viii) SEMs with variables from the exponential family distributions (Wedel and Kamakura, 2001; Rabe-Hesketh, Skrondal and Pickles, 2004; Lee and Tang, 2006b; Song and Lee, 2006b; among others). The above articles not only provide theoretical results, but also have significant practical value. For instance, it is very common in practice to encounter ordered categorical data with missing data, hierarchical data and/or heterogeneous data, and hence developments of sound statistical methods to cope with such practical situations are useful.

The primary goal of all the existing commercial software packages in SEMs is for analyzing the standard SEMs under the normal assumption. They cannot effectively and efficiently analyze the more complex models and/or data structures mentioned above. At the moment, there are only a limited number of reference/textbooks in SEM. Moreover, the emphasis of all the existing books, for example Bollen (1989), was devoted to the standard SEMs, and focused on the covariance structure analysis approach. Hence, despite the widespread use of SEMs, and the importance of the aforementioned complex models for sound and rigorous analyses of real data sets, there have been very few practical applications of the recent developments of SEMs to substantive research. The limited applications are not due to a lack of relevant substantive applications that required such models and their associative statistical methods. Rather, the main reasons are that the applied researchers are not familiar with these models and methods, and the existing commercial SEM software cannot produce satisfactory solutions for coping with the complex situations.

Therefore, there is a need for a new reference/textbook for the second generation of SEM which involves a much wider class of SEMs that include the standard SEMs and their useful generations. This book should provide a more appropriate approach than the covariance structure analysis approach in analyzing the general class of SEMs, together with a dependable software for obtaining reliable and rigorous results for statistical inference.

1.4 OBJECTIVES OF THE BOOK

One of the basic objectives of this book is to propose a Bayesian approach for analyzing some useful structural equation models and/or data structures that are commonly encountered in substantive research. This includes the treatments of the standard SEMs and their useful generalizations in various ways. More specifically, the generalizations are SEMs with ordered categorical variables, SEMs with dichotomous variables, nonlinear SEMs, two-level SEMs, multisample SEMs, mixtures of SEMs, SEMs with ignorable and/or nonignorable missing data, SEMs with variables from the exponential family distributions, and some of their combinations. In formulating various SEMs, and in developing the Bayesian methods, the emphasis is placed on the raw individual random observations rather than on the sample covariance matrix. This formulation has several advantages. First, the development of statistical methods is based on the first moment properties of the raw individual observations which is simpler than the second moment properties of the sample covariance matrix. Hence, it is easier to apply in more complex situations. Second, it leads to a direct estimation of the latent variables which is better than the classical regression method or the Bartlett's method for obtaining the factor score estimates. Third, as it directly models manifest variables with their latent variables through the familiar regression equations, it gives a more direct interpretation and can utilize the common techniques in regression such as outlier and residual analyses in conducting statistical analysis. The advantages of a Bayesian approach are that it allows the use of genuine prior information in addition to the information that is available in the observed data for producing better results, provides useful statistics such as the mean and percentiles of the posterior distribution of the unknown parameters, and gives more reliable results for small samples (see Dunson, 2000; Lee and Song; 2004b; Scheines, Hoijtink and Boomsma, 1999).

The next aim is to describe the technique of data augmentation (Tanner and Wong, 1987), and introduce some efficient tools in statistical computing for analyzing SEMs. The key idea of data augmentation is to augment the observed data with the latent quantities, which could be the latent variables or the unobservable data (missing data or the unobservable continuous measurements that underlie the dichotomous and/or ordered categorical data), so that the Bayesian analysis is feasible with the complete-data set. The introduced tools in statistical computing include some Markov chain Monte Carlo (MCMC) methods, such as the Gibbs sampler (Geman and Geman, 1984) and the Metropolis – Hastings algorithm (Metropolis *et al.*, 1953; Hastings, 1970) and path sampling (Gelman and Meng, 1998). The strategy of data augmentation followed by MCMC methods will be used repeatedly to analyze complex SEMs and data structures throughout this book. The Bayesian methodologies will be demonstrated through real examples in the fields of education, management, medicine, psychology and sociology.

One of the main goals is the introduction of the freely available software WinBUGS (Spiegelhalter, Thomas, Best and Lunn, 2003) to the field of SEM. This software is able to produce reliable Bayesian statistics including the Bayesian estimates and their standard error estimates for a wide range of statistical models (Congdon, 2003). For most SEMs, it also provides the Deviance Information Criterion (DIC, see Speigelhalter, Thomas, Best and Lunn, 2003) for model comparison or goodness-of-fit assessment of the hypothesized model. The applications of WinBUGS to various SEMs are illustrated by real and/or artifical examples in Chapter 4 and Chapters 6 to 13. With the availability of the dependable software WinBUGS, researchers can apply the Bayesian approach, which has the aforementioned advantages and the same statistical optimality as the maximum likelihood, without implementing their own computer programs. Given its potential in analyzing various kinds of models, readers are recommended to make some effort to get to know more about WinBUGS and its power in solving practical problems.

1.5 DATA SETS AND NOTATIONS

Real examples require real data sets. The author owes a great debt to many academic organizations and colleagues for allowing him to use their valuable data in the real examples of this book. As it may not be appropriate to describe every real data set, here is a brief description of the Inter-university Consortium for Political and Social Research (ICPSR) data set which was collected by the World Value Study Group (1994) in the project *World Value Survey 1981–1984 and 1990–1993* (World Value Study Group, ICPSR Version). The ICPSR data set has been used in many of the illustrative examples. It was collected in 45 societies around the world on very broad topics related to family life, the meaning and purpose of life, work and contemporary political and social issues (see the Summary of the ICPSR data set). It provides a rich source of real data in relation to management, psychology and sociology. We have used the data obtained in several countries, such as Canada, UK and USA. The variables (or items in the questionnaire) that have been used in subsequent chapters are collectively presented in Appendix 1.1.

There are not enough symbols for different types of observations in relation to observable manifest continuous and discrete variables, or covariates, unobservable measurements in relation to missing data or continuous measurements underlie the discrete data, latent variables, as well as different types of parameters, such as thresholds, structural parameters in the model, and hyperparameters in the prior distributions. Hence, if the context is clear, some letters of the Greek alphabet may be used to serve different purposes. For example, α has been used to denote (i) an unknown threshold in defining an ordered categorical variable, (ii) an intercept in the measurement equation, and (iii) a hyperparameter in some prior distributions. Nevertheless, some general notation is given as in Table 1.1.

Table 1.1 Typical notation.

Symbols	Meaning
$\mathbf{u}, \mathbf{v}, \mathbf{x}, \mathbf{y}$	Manifest random vectors
$\boldsymbol{\omega}$	Latent random vector in the measurement equation
$\boldsymbol{\eta}$	Endogenous latent vector in the structural equation
$\boldsymbol{\xi}$	Exogenous latent vector in the structural equation
$\boldsymbol{\varepsilon}, \boldsymbol{\delta}$	Random vectors of measurement errors
$\boldsymbol{\Lambda}$	Factor loading matrix in the measurement equation
$\boldsymbol{\Pi}, \mathbf{B}, \boldsymbol{\Gamma}$	Matrices of regression coefficients in the structural equation
$\boldsymbol{\Phi}$	Covariance matrix of latent variables
$\boldsymbol{\Psi}_\varepsilon, \boldsymbol{\Psi}_\delta$	Diagonal covariance matrices of measurement errors with diagonal elements $\psi_{\varepsilon k}$ and $\psi_{\delta k}$, respectively
$\alpha_{o\varepsilon k}, \beta_{o\varepsilon k}, \alpha_{o\delta k}, \beta_{o\delta k}$	Hyperparameters in the Gamma distributions of $\psi_{\varepsilon k}$ and $\psi_{\delta k}$, respectively
$\mathbf{R}_o, \rho_o$	Hyperparameters in the Wishart distribution related to the prior distribution of $\boldsymbol{\Phi}$
$\boldsymbol{\Lambda}_{ok}, \mathbf{H}_{oyk}$	Hyperparameters in the multivariate normal distribution related to the prior distribution of the kth row of $\boldsymbol{\Lambda}$ in the measurement equation with $\mathbf{y}$
$\boldsymbol{\Lambda}_{o\omega k}, \mathbf{H}_{o\omega k}$	Hyperparameters in the multivariate normal distribution related to the prior distribution of the kth row of $\boldsymbol{\Lambda}_{\omega k}$ in the structural equation
$\mathbf{I}_q$	A q by q identity matrix. Sometimes just $\mathbf{I}$ is used to denote an identity matrix if its dimension is clear

APPENDIX 1.1

This appendix provides the questions/items in the Data Collection Description, World Values Study Group (1994), World Values Survey, 1981–1984 and 1990–1993 that have been used as manifest variables in the illustrative examples of this book.

Thinking about your reasons for doing voluntary work, please use the following five-point scale to indicate how important each of the reasons below have been in your own case. (WHERE 1 IS UNIMPORTANT AND 5 IS VERY IMPORTANT)

V 62 Religious beliefs 1 2 3 4 5

During the past few weeks, did you ever feel . . . (Yes: 1 No: 2)

V 89 Bored 1 2

V 91 Depressed or very unhappy 1 2

V 93 Upset because somebody criticized you 1 2

V 96 All things considered, how satisfied are you with your life as a whole these days?

1 2 3 4 5 6 7 8 9 10

Dissatisfied Satisfied

Here are some aspects of a job that people say are important. Please look at them and tell me which ones you personally think are important in a job. (Mentioned: 1; Not Mentioned: 2)

V 99	Good pay	1	2
V 100	Pleasant people to work with	1	2
V 102	Good job security	1	2
V 103	Good chances for promotion	1	2
V 111	A responsible job	1	2

V 115 How much pride, if any, do you take in the work that you do?

1 A great deal, 2 Some, 3 Little, 4 None

V 116 Overall, how satisfied or dissatisfied are you with your job?

1 2 3 4 5 6 7 8 9 10

Dissatisfied Satisfied

V 117 How free are you to make decisions in your job?

1 2 3 4 5 6 7 8 9 10

Not at all A great deal

V 129 When jobs are scarce, people should be forced to retire early,

Agree 1; Neither 2; Disagree 3

V 132 How satisfied are you with the financial situation of your household?

1 2 3 4 5 6 7 8 9 10

Dissatisfied Satisfied

V 176 How important is God in your life? 10 means very important and 1 means not at all important.

1 2 3 4 5 6 7 8 9 10

V 179 How often do you pray to God outside of religious services? Would you say . . .

1 Often
2 Sometimes
3 Hardly ever
4 Only in times of crisis
5 Never

V 180 Overall, how satisfied or dissatisfied are you with your home life?

1	2	3	4	5	6	7	8	9	10
Dissatisfied								Satisfied	

V 241 How interested would you say you are in politics?

1 Very interested

2 Somewhat interested

3 Not very interested

4 Not at all interested

Now I'd like you to tell me your views on various issues. How would you place your views on this scale? 1 means you agree completely with the statement on the left, 10 means you agree completely with the statement on the right, or you can choose any number in between.

V 252

1 2 3 4 5 6 7 8 9 10

Individual should take more responsibility for providing for themselves	The state should take more responsibility to ensure that everyone is provided for

V 253

1 2 3 4 5 6 7 8 9 10

People who are unemployed should have to take any job available or lose their unemployment benefits	People who are unemployed should have the right to refuse a job they do not want

V 254

1 2 3 4 5 6 7 8 9 10

Competition is good. It stimulates people to work hard and develop new ideas	Competition is harmful. It brings out the worst in perople

V 255

1 2 3 4 5 6 7 8 9 10

In the long run, hard work usually brings a better life	Hard work doesn't generally brings success — it's more a matter of luck and connections

Please tell me for each of the following statements whether you think it can always be justified, never be justified, or something in between.

V 296 Claiming government benefits which you are not entitled to

1 2 3 4 5 6 7 8 9 10
Never Always

V 297 Avoiding a fare on public transport

1 2 3 4 5 6 7 8 9 10
Never Always

V 298 Cheating on tax if you have the chance

1 2 3 4 5 6 7 8 9 10
Never Always

V 314 Failing to report damage you've done accidentally to a parked vehicle

1 2 3 4 5 6 7 8 9 10
Never Always

I am going to read out some statements about the government and the economy. For each one, could you tell me how much you agree or disagree?

V 336 Our government should be made much more open to the public

1 2 3 4 5 6
Agree Completely Disagree Completely

V 337 We are more likely to have a healthy economy if the government allows more freedom for individuals to do as they wish

1 2 3 4 5 6
Agree Completely Disagree Completely

V 339 Political reform in this country is moving too rapidly

1 2 3 4 5 6
Agree Completely Disagree Completely

REFERENCES

Ansari, A. and Jedidi, K. (2000) Bayesian factor analysis for multilevel binary observations. *Psychometrika*, **65**, 475–497.

Arminger, G., Stein, P. and Wittenberg, J. (1999) Mixtures of conditional mean- and covariance-structure models. *Psychometrika*, **64**, 475–494.

Bagozzi, R. P., Baumgartner, H. and Yi, Y. (1992) State versus action orientation and the theory of reasoned action: an application to coupon usage. *Journal of Consumer Research*, **18**, 505–517.

Bollen, K. A. (1989) *Structural Equation Models with Latent Variables*, NJ: John Wiley & Sons, Inc.

Congdon, P. (2003) *Applied Bayesian Modeling*, Hoboken, NJ: John Wiley & Sons, Inc.

Dolan, C. V. and van der Maas, J. J. L. (1998) Fitting multivariate normal mixtures subject to structural equation modeling. *Psychometrika*, **63**, 227–254.

Dunson, D. B. (2000) Bayesian latent variable models for clustered mixed outcomes. *Journal of the Royal Statistical Society, Series B*, **62**, 355–366.

Gelman, A. and Meng, X. L. (1998) Simulating normalizing constants: from importance sampling to bridge sampling to path sampling. *Statistical Science*, **13**, 163–185.

Geman, S. and Geman, D. (1984) Stochastic relaxation, Gibbs distribution, and the Bayesian restoration of images. *IEEE Transactions on Pattern Analysis and Machine Intelligence*, **6**, 721–741.

Hastings, W. K. (1970) Monte Carlo sampling methods using Markov chains and their application. *Biometrika*, **57**, 97–109.

Jamshidian, M. and Bentler, P. M. (1999) ML estimation of mean and covariance structures with missing data using complete data routines. *Journal of Educational and Behavioral Statistics*, **24**, 21–41.

Jedidi, K., Jagpal, H. S. and DeSarbo, W. S. (1997) STEMM: A general finite mixture structural equation model. *Journal of Classification*, **14**, 23–50.

Jöreskog, K. G. and Sörbom, D. (1996) LISREL 8: *Structural Equation Modeling with the SIMPLIS Command Language*. Scientific Software International.

Jöreskog, K. G. and Yang, F. (1996) Nonlinear structural equation models: the Kenny–Judd model with interaction effects. In G. A. Marcoulides and R. E. Schumacker (eds), *Advanced Structural Equation Modeling Techniques* (pp. 57–88). Hillsdale, NJ: LEA.

Kenny, D. A. and Judd, C. M. (1984) Estimating the nonlinear and interactive effects of latent variables. *Psychological Bulletin*, **96**, 201–210.

Klein, A. and Moosbrugger, M. (2000) Maximum likelihood estimation of latent interaction effects with the LMS method. *Psychometrika*, **65**, 457–474.

Lee, S. Y. and Shi, J. Q. (2000) Bayesian analysis of structural equation model with fixed covariates. *Structural Equation Modeling*, **7**, 411–430.

Lee, S. Y. and Shi, J. Q. (2001) Maximum likelihood estimation of two-level latent variable models with mixed continuous and polytomous data. *Biometrics*, **57**, 787–794.

Lee, S. Y. and Song, X. Y. (2003a) Model comparison of nonlinear structural equation models with fixed covariates. *Psychometrika*, **68**, 27–47.

Lee, S. Y. and Song, X. Y. (2003b) Bayesian model selection for mixtures of structural equation models with an unknown number of components. *British Journal of Mathematical and Statistical Psychology*, **56**, 145–165.

Lee, S. Y. and Song, X. Y. (2004a) Bayesian model comparison of nonlinear structural equation models with missing continuous and ordinal categorical data. *British Journal of Mathematical and Statistical Psychology*, **57**, 131–150.

Lee, S. Y. and Song, X. Y. (2004b) Evaluation of Bayesian and maximum likelihood approaches in analyzing structural equation models with small sample sizes. *Multivariate Behavioral Research*, **39**, 653–686.

Lee, S. Y. and Tang, N. S. (2006a) Bayesian analysis of nonlinear structural equation models with nonignorable missing data. *Psychometrika*, in press.

Lee, S. Y. and Tang, N. S. (2006b) Bayesian analysis of structural equation models with mixed exponential family and ordered categorical data. *British Journal of Mathematical and Statistical Psychology*, **59**, 151–172.

Lee, S. Y., Song, X. Y. and Poon, W. Y. (2004) Comparison of approaches in estimating interaction and quadratic effects of latent variables. *Multivariate Behavioral Research*, **39**, 37–67.

Marsh, H. W., Wen, Z. and Hau, K. T. (2004) Structural equation models of latent interaction: evaluation of alternative estimation strategies and indicator construction. *Psychological Methods*, **9**, 275–300.

Metropolis, N. *et al.* (1953) Equations of state calculations by fast computing machine. *Journal of Chemical Physics*, **21**, 1087–1091.

Moustaki, I. (2003) A general class of latent variable methods for ordinal manifest variables with covariate effects on the manifest and latent variables. *British Journal of Mathematical and Statistical Psychology*, **56**, 337–357.

Rabe-Hesketh S., Skrondal, A. and Pickles, A. (2004) Generalized multilevel structural equation modeling. *Psychometrika*, **69**, 167–190.

Scheines, R., Hoijtink, H. and Boomsma, A. (1999) Bayesian estimation and testing of structural equation models. *Psychometrika*, **64**, 37–52.

Shi, J. Q. and Lee, S. Y. (1998) Bayesian sampling-based approach for factor analysis model with continuous and polytomous data. *British Journal of Mathematical and Statistical Psychology*, **51**, 233–252.

Shi, J. Q. and Lee, S. Y. (2000) Latent variable models with mixed continuous and polytomous data. *Journal of the Royal Statistical Society, Ser B*, **62**, 77–87.

Song, X. Y. and Lee, S. Y. (2001) Bayesian estimation and test for factor analysis model with continuous and polytomous data in several populations. *British Journal of Mathematical and Statistical Psychology*, **54**, 237–263.

Song, X. Y. and Lee, S. Y. (2002a) Analysis of structural equation model with ignorable missing continuous and polytomous data. *Psychometrika*, **67**, 261–288.

Song, X. Y. and Lee, S. Y. (2002b) Bayesian estimation and testing for nonlinear factor analysis in several populations. *Structural Equation Modeling*, **9**, 523–553.

Song, X. Y. and Lee, S. Y. (2004) Bayesian analyses of two-level nonlinear structural equation models with continuous and polytomous data. *British Journal of Mathematical and Statistical Psychology*, **57**, 29–52.

Song, X. Y. and Lee, S. Y. (2006a) Bayesian analysis of structural equation models with nonlinear covariates and latent variables. *Multivariate Behavioral Research*, **41**, 337–365.

Song, X. Y. and Lee, S. Y. (2006b) Bayesian analysis of latent variable models with nonignorable missing outcomes from exponential family. *Statistics in Medicine*, in press.

Spiegelhalter, D. J., Thomas, A., Best, N. G. and Lunn, D. (2003) *WinBugs User Manual. Version 1.4.* Cambridge, UK: MRC Biostatistics Unit.

Tanner, M. A. and Wong, W. H. (1987) The calculation of posterior distributions by data augmentation(with discussion). *Journal of the American Statistical Association*, **82**, 528–550.

Wall, M. M. and Amemiya, Y. (2000) Estimation for polynomial structural equation models. *Journal of the American Statistical Association*, **95**, 929–940.

Wedel, M. and Kamakura, W. (2001) Factor analysis with (mixed) observed and latent variables in the exponential family. *Psychometrika*, **55**, 515–530.

World Values, Study Group (1994) World Values Survey, 1981–1984 and 1990–1993 (Computer file). ICPSR version. Ann Arbor, MI: Institute for Social Research (producer), Ann Arbor, MI: Inter-university Consortium for Political and Social Research (distributor).

Zhu, H. T. and Lee, S. Y. (2001) A Bayesian analysis of finite mixtures in the LISREL model. *Psychometrika*, **66**, 133–152.

2

Some Basic Structural Equation Models

2.1 INTRODUCTION

The main objective of this chapter is to introduce some basic structural equation models (SEMs). Definitions of the models will be given, their basic features discussed and the possible improvements that can be achieved through the Bayesian approach will be pointed out. The covariance structure approach for analyzing these models will be discussed in Chapter 3. The Bayesian approach, which has several key advantages over the covariance structure approach, will be discussed in Chapters 4 to 13.

Historically, factor analysis is the most basic SEM which was developed by psychometricians (e.g. Spearman, 1904; Thurstone, 1944) to study internal relationships of a set of variables. Its primary concern was on hypotheses testing about the organization of mental ability. Nowadays, the model has been applied to a much wider range of situations, for example, analyzing sets of economics quantities, sets of tests of attitudes and behaviors, and sets of physical measurements, etc..

The basic motivation for developing exploratory factor analysis is that for a given set of response variables one wants to search a fewer number of uncorrelated latent factors that will account for the intercorrelations of the response variables so that when the latent factors are partialled out from the response variables there no longer remain any correlations between them. It can be regarded as an exploratory data analysis method at the initial stage of the development of SEM. Starting with a given data set, the relating statistical problems

Structural Equation Modeling: A Bayesian Approach S-Y. Lee

are determining the number of factors for the structure, estimating the parameters and achieving meaningful interpretations of the factors via appropriate rotations of the factor loading matrix. All these can be viewed as exploratory data analysis for a better understanding of the given data. Development of the confirmatory factor analysis was made by Jöreskog (1969). In this model, the latent factors are allowed to be correlated. Moreover, any values may be preassigned in advance for any number of specified factor loadings, factor correlations and unique variances. Typical application of the model is on confirmatory studies, where the experimenter has already obtained a certain amount of knowledge about the model from the substantive theory or exploratory analysis, and is in a position to formulate a more precise confirmatory factor model with a given number of latent factors. Usually, the model is identified by the fixed parameters, the resulting solution will be directly interpretable, and the subsequent rotation of the factor loading matrix is not necessary. In the analysis, the unspecified parameters are estimated conditionally on the preassigned values of the specified parameters. Goodness-of-fit of the proposed model to the data is usually assessed via a discrepancy function that measures the difference between the estimated and the observed covariance matrices. It can be used to 'confirm' the results obtained from exploratory analysis of the model. Moreover, the flexibility of fixing and freeing of parameters at will provides a powerful way for establishing plausible models for substantive theory in the real world, and inspires the later stage development of SEMs. Shortly after solving the fundamental problems of the confirmatory factor analysis model, Jöreskog (1970) developed the second-order factor model. This model was further generalized to higher-order factor models, and higher-order moment structures by Bentler (1976, 1983).

As pointed out by Bentler (1983), the most exciting development in structural equation modeling has been the integration of the confirmatory factor analysis model and the simultaneous equation model, achieved by Ward Keesling, David Wiley and Karl Jöreskog (Jöreskog, 1977; Wiley, 1973). The several generations of the computer program LISREL (Jöreskog and Sörbom, 1983–1996) have had a tremendous impact on the early stage development of SEMs. The LISREL model is defined by a structural equation which is essentially a set of simultaneous linear equations with latent variables, and two measurement models that relate the latent variables to the manifest variables via confirmatory factor analysis models. One of its key features is the distinction between latent and manifest variables. A model on the basis of a generic linear structural matrix equation with no such distinction has been developed by Bentler and Weeks (1980). The well-known EQS6 (Bentler and Wu, 2002) program is based on this model. Nowadays, there are a number of other programs in the field such as AMOS, CALIS, COSAN, LINCS, MECOSA, MPLUS, PLS, RAM, RAMONA, SEPATH, STREAMS and TETRAD II, among others. In order to conserve space, the underlying models of these programs, and other classical models in the field such as the multi-mode models (Tucker, 1966; Bentler and Lee, 1979) and moment structure model (Bentler, 1983), will not be discussed. Also the controversy criticisms about SEMs, such as those related to causality, the use of latent variables, etc., are not specifically discussed because they have been well addressed by Bollen (1989). Moreover, some of the criticisms, for example about the falsifiability of the models and the distributional assumptions, would be alleviated by the Bayesian model comparison, and the generality of the more subtle SEMs that are less reliant on the normality assumption of the manifest variables.

Owing to the strong demand for more general models to cope with the complicated real life practical problems, a number of useful generalizations of the basic SEMs have been proposed. The most important representatives are models with ordered categorical variables, dichotomous variables, nonlinear models, multilevel models, multisample models, mixtures models, models with missing data and models with the variables from the exponential family of distribution. These models will be discussed in subsequent chapters of this book.

In this chapter, the standard exploratory and confirmatory factor analysis models are discussed, as are the higher-order factor analysis models, the LISREL model, and the Bentler and Weeks (1980) model that are related to the well-known LISREL program and the EQS program, respectively.

2.2 EXPLORATORY FACTOR ANALYSIS

Exploratory factor analysis (EFA) is a basic model and has received a lot of attention in the field for many years. This section makes no attempt to give a complete treatment of the subject. Comprehensive discussions may be found in Lawley and Maxwell (1971) or Mulaik (1972).

2.2.1 Model Definition

The exploratory factor analysis model is defined by a $p \times 1$ random vector $\mathbf{x}$ that satisfies the following linear equation:

$$\mathbf{x} = \boldsymbol{\Lambda}\boldsymbol{\xi} + \boldsymbol{\epsilon}, \qquad (2.1)$$

where $\boldsymbol{\Lambda}(p \times q)$ is a matrix of factor loadings, $\boldsymbol{\xi}(q \times 1)$ is a random vector of latent common factors and $\boldsymbol{\epsilon}(p \times 1)$ is a random vector of error measurements (sometimes called latent unique factors or residuals). It is assumed that $\boldsymbol{\xi}$ is distributed as $N[\mathbf{0}, \mathbf{I}]$, and $\boldsymbol{\epsilon}$ is distributed as $N[\mathbf{0}, \boldsymbol{\Psi}_\epsilon]$, where $\boldsymbol{\Psi}_\epsilon$ is a diagonal matrix, and $\boldsymbol{\xi}$ is uncorrelated with $\boldsymbol{\epsilon}$. Generally, q is much smaller than p. The manifest random vector $\mathbf{x}$ is distributed as $N[\mathbf{0}, \boldsymbol{\Sigma}]$, where

$$\boldsymbol{\Sigma} = \boldsymbol{\Lambda}\boldsymbol{\Lambda}^T + \boldsymbol{\Psi}_\epsilon. \tag{2.2}$$

Clearly, $\text{cov}(\mathbf{x}, \boldsymbol{\xi}) = \boldsymbol{\Lambda}$. Hence, the correlations between the latent factors and the manifest variables are given by the elements in the factor loading matrix $\boldsymbol{\Lambda}$. The variance of the kth manifest variable is equal to

$$\sigma_{kk} = \lambda_{k1}^2 + \cdots + \lambda_{kq}^2 + \psi_{\epsilon k},$$

where λ_{kh} and $\psi_{\epsilon k}$ are the (k,h)th element of $\boldsymbol{\Lambda}$ and the kth elements of $\boldsymbol{\Psi}_\epsilon$, respectively. The quantity $\lambda_{k1}^2 + \cdots + \lambda_{kq}^2$ is called the communality which represents the variance contributed by the latent factors.

In principal component analysis, a set of p variables is transformed linearly and orthogonally to an equal number of new uncorrelated variables, such that the total variance is unchanged. In contrast to principal component analysis, the EFA model explains most of the dependence structure by a much smaller number of common factors with the residuals accounted for by the unique factors. It can also be viewed as a linear regression model in which a vector of manifest variables is regressed on a smaller dimensional random vector of latent factors. In practical applications, EFA has been widely used to group together manifest variables that are related to a particular latent factor for the purpose of data reduction.

2.2.2 Identification and Analysis of the Model

Suppose that a random sample $\{\mathbf{x}_1, \ldots, \mathbf{x}_n\}$ of $\mathbf{x}$ has been obtained. We define the sample covariance matrix by

$$\mathbf{S} = (n-1)^{-1}\sum_{i=1}^{n}(\mathbf{x}_i - \bar{\mathbf{x}})(\mathbf{x}_i - \bar{\mathbf{x}})^T,$$

where $\bar{\mathbf{x}}$ is the sample mean. The sample covariance matrix is an unbiased estimator of $\boldsymbol{\Sigma}$, and $(n-1)\mathbf{S}$ is distributed as $W_p(\boldsymbol{\Sigma}, n-1)$, a p-dimensional Wishart distribution with $n-1$ degrees of freedom (see Anderson, 1984). In practice, unknown elements in the parameter matrices $\boldsymbol{\Lambda}$ and $\boldsymbol{\Psi}_\epsilon$ are estimated from $\mathbf{S}$. Before estimation, we have to consider the identification of the model.

If $q = 1$, then $\boldsymbol{\Lambda}$ is unique apart from a plausible change of sign of all its elements, which corresponds to changing the sign of the factors. When $q > 1$, there is an infinity of choices for $\boldsymbol{\Lambda}$, because Equations (2.1) and (2.2) are still satisfied if $\boldsymbol{\xi}$ is replaced by $\mathbf{A}\boldsymbol{\xi}$ and $\boldsymbol{\Lambda}$ by $\boldsymbol{\Lambda}\mathbf{A}^T$, where $\mathbf{A}$ is any orthogonal matrix of rank q. This factor indeterminacy problem may be solved by imposing identification conditions on the parameters (see Anderson and Rubin, 1956; Jöreskog, 1967; and Jennrich and Robinson, 1969). Some examples of the identification conditions are: (i) $\boldsymbol{\Lambda}^T\Sigma^{-1}\Lambda$ is diagonal; (ii) $\boldsymbol{\Lambda}^T\boldsymbol{\Psi}_\epsilon^{-1}\Lambda$ is diagonal; (iii) $\boldsymbol{\Lambda}^T\mathbf{S}^{-1}\Lambda$ is diagonal; or (iv) $\lambda_{kh} = 0$ for all $h = 2, \ldots, q$ and $k < h$. An example of $q = 3$ corresponding to (iv) is given by

$$\begin{bmatrix} \lambda_{11} & 0 & 0 \\ \lambda_{21} & \lambda_{22} & 0 \\ \lambda_{31} & \lambda_{32} & \lambda_{33} \\ \vdots & \vdots & \vdots \\ \lambda_{p1} & \lambda_{p2} & \lambda_{p3} \end{bmatrix}.$$

See Anderson and Rubin (1956) for more theoretical discussion and examples on the identification conditions.

For a proposed model with a specified fixed number of factors, estimates of the parameters subject to an appropriate identification condition are obtained by minimizing an objective function. Since the solution is usually not in closed form, estimates are obtained via some iterative procedures. Procedures for obtaining the ML estimates of $\boldsymbol{\Lambda}$ and $\boldsymbol{\psi}_\epsilon$ that subject to (ii) and (iii) were developed by Jöreskog (1967) and Jennrich and Robinson (1969) respectively. ML estimates subject to (iv) can be obtained by the scoring algorithm or the Gauss Newton algorithm that will be discussed in Chapter 3. It should be remembered that ML estimates of the parameters that are obtained on the basis of different identification conditions are equivalent. In practice, the estimate of $\boldsymbol{\Lambda}$ is usually rotated by some methods such as the varimax rotation (see e.g. Lawley and Maxwell, 1971) for better interpretations.

To apply the exploratory factor analysis model for finding the dependence structure of the manifest variables, selecting an appropriate number of latent factors is an important issue. Hence, a basic statistical problem is on testing the hypothesis concerning q. Consider the following negative log-likelihood function on the basis of the information provided by $\mathbf{S}$:

$$L(\tilde{\boldsymbol{\Lambda}}, \tilde{\boldsymbol{\Psi}}_\epsilon) = \log|(\tilde{\boldsymbol{\Lambda}}\tilde{\boldsymbol{\Lambda}}^T + \tilde{\boldsymbol{\Psi}}_\epsilon)| + \text{tr}\{\mathbf{S}(\tilde{\boldsymbol{\Lambda}}\tilde{\boldsymbol{\Lambda}}^T + \tilde{\boldsymbol{\Psi}}_\epsilon)^{-1}\} - \log|\mathbf{S}| - p,$$

where $\{\tilde{\boldsymbol{\Lambda}}, \tilde{\boldsymbol{\Psi}}_\epsilon\}$ are the ML estimates of $\{\boldsymbol{\Lambda}, \boldsymbol{\Psi}_\epsilon\}$. The likelihood ratio criterion is given by $nL(\tilde{\boldsymbol{\Lambda}}, \tilde{\boldsymbol{\Psi}}_\epsilon)$. Under the null hypothesis that the true model has q latent factors, the asymptotic distribution of this statistic is χ^2_{df}, where the 'degrees of freedom' df is equal to the number of parameters in a general symmetric covariance matrix minus the number of unknown parameters in the proposed model (see Chapter 3, Section 3.2). Hence,

$$df = \frac{p(p+1)}{2} - \left[(pq+p) - \frac{q(q-1)}{2}\right] = \frac{1}{2}\left[(p-q)^2 - (p+q)\right].$$

The proposed model with q latent factors is rejected at a chosen type I error level if $nL(\tilde{\Lambda}, \tilde{\Psi}_\epsilon)$ is larger than the corresponding critical value that is obtained from the chi-square table.

In practice, a sequential procedure is usually applied to decide the number of latent factors in fitting the data. Starting with a small q_1, say $q_1 = 1$, one can estimate the parameters and perform the above goodness-of-fit test. If the test criterion is not significant at the chosen level, the model with q_1 latent factors can be regarded as a plausible model for the data; otherwise the procedure continues with $q_1 + 1$. However, the above procedure is open to the following objections: (i) the significance level for the test criterion has not been adjusted to take into account that a sequence of hypotheses is being tested, with each one dependent on the rejection of all the previous tests; (ii) even if the null hypothesis is not rejected, it does not provide supportive evidence for the model specified by the null hypothesis; (iii) the null hypothesis is usually rejected in situations with huge sample sizes. A better Bayesian approach for selecting q is via the Bayes factor (see Berger, 1985), which will be discussed in detail in Chapter 5, Section 5.2, and Chapter 6, Section 6.5.

More detailed discussions on the estimation, such as the asymptotic distributions of the estimators and the iterative procedures for achieving the solution, are left until Chapter 3 under the general framework of the traditional covariance structure analysis approach.

2.3 CONFIRMATORY AND HIGHER-ORDER FACTOR ANALYSIS MODELS

2.3.1 Confirmatory Factor Analysis Model

The confirmatory factor analysis (CFA) model is a natural extension of the EFA model. It is defined by

$$\mathbf{x} = \Lambda\boldsymbol{\xi} + \boldsymbol{\epsilon}, \tag{2.3}$$

where the definitions of $\mathbf{x}$, $\boldsymbol{\xi}$, $\boldsymbol{\epsilon}$, and Λ, and the distributional assumption of $\boldsymbol{\epsilon}$ are the same as in an EFA model. In a CFA model, the latent common factors are allowed to be correlated so that $\boldsymbol{\xi}$ is distributed as $N[\mathbf{0}, \boldsymbol{\Phi}]$ with a positive definite covariance matrix $\mathbf{\Phi}$. Under the assumption that $\boldsymbol{\xi}$ is independent with $\boldsymbol{\epsilon}$, the covariance matrix of $\mathbf{x}$ is

$$\mathbf{\Sigma} = \mathbf{\Lambda}\mathbf{\Phi}\mathbf{\Lambda}^T + \mathbf{\Psi}_\epsilon. \tag{2.4}$$

In this model, elements of $\mathbf{\Lambda}$, $\mathbf{\Phi}$ and $\mathbf{\Psi}_\epsilon$ are allowed to be fixed at preassigned values. The positions of the fixed parameters and their preassigned values represent part of the hypotheses that the investigator wishes to test about the model. For example, $\lambda_{kh} = 0$ means that the hth common factor does not enter into the kth manifest (response) variable, and $\phi_{jh} = 0$ means the jth common factor is uncorrelated with the hth factor. The proposed model involves specifications of the number of latent factors, as well as the positions and preassigned values of the fixed parameters. In almost all practical applications, the proposed CFA model is identified by the fixed parameters in $\mathbf{\Lambda}$, $\mathbf{\Phi}$ and $\mathbf{\Psi}_\epsilon$ at appropriate preassigned values. Estimates of the unknown parameters are unique, and hence the factor rotation is not relevant in this situation.

In EFA analysis, we are given a data set corresponding to a number of manifest variables, and we wish to group together some correlated manifest variables to a latent factor which hopefully can be interpreted as a meaningful latent construct (via factor rotation). This is an exploratory data-driven procedure that does not depend on any prior knowledge of the manifest variables. On the other hand, most CFA applications are confirmatory. From the subject-matter knowledge, we already have in mind some correlated manifest variables that would be grouped to a latent construct (factor) of interest. Typical examples of the latent constructs are the job attitude and satisfaction in organization and management research, consumer satisfaction and emotion in marketing research, physical and mental health in quality of life research, etc.. More specifically, in the WHOQOL quality of life instrument (Power, Bullingen, Hazper and WHOQOL Group, 1999), investigators used items such as 'pain', 'medication', 'energy and fatigue', 'mobility', 'sleep and rest', 'daily activities' and 'work capacity' to reflect the latent construct about 'physical health'. In general, as the latent constructs and the corresponding items can be naturally regarded as latent factors and the related manifest variables, the CFA is a logical confirmatory statistical model for substantive research that involves manifest and latent constructs.

The CFA model has been applied widely to many substantive problems. As an example to illustrate the flexibility of the CFA model, we now present its application to analyze the multitrait–multimethod model that is useful for assessing the convergent and discriminant validity of latent constructs. Consider a situation with $t(\geq 3)$ traits which are measured by $m(\geq 3)$ methods. Let x_{jk} represent an observed variable that corresponds to the jth trait measured by the kth method. The multitrait – multimethod model is given as

$$x_{jk} = \alpha_{jk}\xi_{Tj} + \beta_{jk}\xi_{Mk} + \epsilon_{jk}, \quad j = 1, \ldots, t; \; k = 1, \ldots, m$$

where $\boldsymbol{\xi}_{Tj}$ and $\boldsymbol{\xi}_{Mk}$ represent the jth trait and the kth method, respectively, α_{jk} and β_{jk} are unknown coefficients and ϵ_{jk} is the residual. This model can be expressed as a CFA model as follows:

$$\begin{bmatrix} x_{11} \\ \vdots \\ x_{1m} \\ x_{21} \\ \vdots \\ x_{2m} \\ \vdots \\ x_{t1} \\ \vdots \\ x_{tm} \end{bmatrix} = \begin{bmatrix} \alpha_{11} & 0 & \cdots & 0 & \beta_{11} & \ddots & 0 \\ \vdots & \vdots & \vdots & \vdots & \ddots & \ddots & \vdots \\ \alpha_{1m} & 0 & \cdots & 0 & 0 & \cdots & \beta_{1m} \\ 0 & \alpha_{21} & \cdots & 0 & \beta_{21} & \cdots & 0 \\ \vdots & \vdots & \vdots & \vdots & \ddots & \ddots & \vdots \\ 0 & \alpha_{2m} & \cdots & 0 & 0 & \cdots & \beta_{2m} \\ \vdots & \vdots & \vdots & \vdots & \vdots & \ddots & \vdots \\ 0 & 0 & \cdots & \alpha_{t1} & \beta_{t1} & \cdots & 0 \\ \vdots & \vdots & \vdots & \vdots & \vdots & \ddots & \vdots \\ 0 & 0 & \cdots & \alpha_{tm} & 0 & \cdots & \beta_{tm} \end{bmatrix} \begin{bmatrix} \xi_{T1} \\ \vdots \\ \xi_{Tt} \\ \xi_{M1} \\ \vdots \\ \xi_{Mm} \end{bmatrix} + \begin{bmatrix} \epsilon_{11} \\ \vdots \\ \epsilon_{1m} \\ \epsilon_{21} \\ \vdots \\ \epsilon_{2m} \\ \vdots \\ \epsilon_{t1} \\ \vdots \\ \epsilon_{tm} \end{bmatrix},$$

where the zeros are fixed parameters. The unknown parameters in the loading matrix are those unknown coefficients α_{jk} and β_{jk}. Correlations among traits, methods, and traits and methods are represented by the elements in Φ, the correlation matrix of the vector of latent factors $(\xi_{T1}, \ldots, \xi_{Tt}, \xi_{M1}, \ldots, \xi_{Mm})^T$. Similar to many other applications of the CFA model, the clear associations of the latent factors and their indicators suggest a specific form of the loading matrix with appropriate elements fixed at zero.

In general, it can be seen from Equation (2.4) that all unknown parameters in the model are involved in the covariance matrix $\boldsymbol{\Sigma}$. Hence, an approach for developing statistical theory is based on the framework of covariance structure analysis. The covariance matrix $\boldsymbol{\Sigma}(\boldsymbol{\theta})$ is regarded as a matrix function of the parameter vector $\boldsymbol{\theta}$ which contain the unknown parameters $\boldsymbol{\theta}$, an objective function $f(\boldsymbol{\Sigma}(\boldsymbol{\theta}), \mathbf{S})$ which measures the discrepancy between the proposed covariance matrix and the sample covariance matrix $\mathbf{S}$ is minimized. Commonly used objective functions are the negative log-likelihood function and the generalized least square function. The goodness-of-fit of the proposed model is assessed by $nf(\boldsymbol{\Sigma}(\tilde{\boldsymbol{\theta}}), \mathbf{S})$ at the estimate $\tilde{\boldsymbol{\theta}}$ of $\boldsymbol{\theta}$ (see Chapter 3 for more detailed discussions in the framework of covariance structure analysis).

2.3.2 Estimation of Factor Scores

In addition to the estimation of the structural parameters in the underlying covariance matrix, a primary interest is on estimating the random vector $\boldsymbol{\xi}$ of latent factor scores.

The regression method is based on the following joint distribution of $\mathbf{x}$ and $\boldsymbol{\xi}$:

$$\begin{pmatrix}\mathbf{x}\\ \boldsymbol{\xi}\end{pmatrix} \stackrel{D}{=} N\left[\begin{pmatrix}0\\0\end{pmatrix}, \begin{pmatrix}\boldsymbol{\Lambda}\boldsymbol{\Phi}\boldsymbol{\Lambda}^T + \boldsymbol{\Psi}_\epsilon & \boldsymbol{\Lambda}\boldsymbol{\Phi}\\ \boldsymbol{\Phi}\boldsymbol{\Lambda}^T & \boldsymbol{\Phi}\end{pmatrix}\right].$$

Regressing $\boldsymbol{\xi}$ on $\mathbf{x}$, we have

$$\boldsymbol{\xi} = \boldsymbol{\Phi}\boldsymbol{\Lambda}^T(\boldsymbol{\Lambda}\boldsymbol{\Phi}\boldsymbol{\Lambda}^T + \boldsymbol{\Psi}_\epsilon)^{-1}\mathbf{x} = \boldsymbol{\Phi}\boldsymbol{\Lambda}^T\boldsymbol{\Sigma}^{-1}\mathbf{x}. \tag{2.5}$$

For EFA, $\boldsymbol{\Phi}$ in Equation (2.5) is taken to be the identity matrix. Another method is to minimize the sum of squares of the standardized residuals, that is

$$(\mathbf{x} - \boldsymbol{\Lambda}\boldsymbol{\xi})^T\boldsymbol{\Psi}_\epsilon^{-1}(\mathbf{x} - \boldsymbol{\Lambda}\boldsymbol{\xi}).$$

For both the CFA and EFA models, the solution is

$$\boldsymbol{\xi} = (\boldsymbol{\Lambda}^T\boldsymbol{\Psi}_\epsilon^{-1}\boldsymbol{\Lambda})^{-1}\boldsymbol{\Lambda}^T\boldsymbol{\Psi}_\epsilon^{-1}\mathbf{x}. \tag{2.6}$$

Estimates of $\boldsymbol{\xi}$ that are obtained by the above methods depend on the true parameter matrices. Since these matrices are unknown in practice, they are replaced by estimates that are obtained from some estimation procedures. From Equations (2.5) and (2.6), the estimates of $\boldsymbol{\xi}$ corresponding to an individual $\mathbf{x}$ are

$$\tilde{\boldsymbol{\xi}} = \hat{\boldsymbol{\Phi}}\hat{\boldsymbol{\Lambda}}^T\hat{\boldsymbol{\Sigma}}^{-1}\mathbf{x}, \tag{2.7}$$

or

$$\tilde{\boldsymbol{\xi}} = (\hat{\boldsymbol{\Lambda}}^T\hat{\boldsymbol{\Psi}}_\epsilon^{-1}\hat{\boldsymbol{\Lambda}})^{-1}\hat{\boldsymbol{\Lambda}}^T\hat{\boldsymbol{\Psi}}_\epsilon^{-1}\mathbf{x}. \tag{2.8}$$

These methods are easy to apply. However, the factor score estimates obtained are rarely used in practice, probably due to the following deficiencies of these methods: (i) the sampling errors are ignored, and (ii) as $\tilde{\boldsymbol{\xi}}$ is a nonlinear function of the parameter estimates, its distribution could be very complicated. Hence, it is rather difficult to use these estimates for further rigorous statistical analyses. A better Bayesian method that does not have these deficiencies will be given in Chapter 4.

2.3.3 Higher-order Factor Analysis Model

The CFA model can be generalized to a higher-order factor analysis model in a natural way. The second-order model is given by

$$\mathbf{x} = \mathbf{B}(\boldsymbol{\Lambda}\boldsymbol{\xi} + \boldsymbol{\epsilon}) + \boldsymbol{\delta} = \mathbf{B}\boldsymbol{\Lambda}\boldsymbol{\xi} + \mathbf{B}\boldsymbol{\epsilon} + \boldsymbol{\delta}, \tag{2.9}$$

where $\mathbf{B}$ and $\boldsymbol{\Lambda}$ are factor loading matrices, $\boldsymbol{\xi}$ is a random vector of latent common factors, and $\boldsymbol{\epsilon}$ and $\boldsymbol{\delta}$ are residuals. Random vectors $\boldsymbol{\xi}, \boldsymbol{\epsilon}$ and $\boldsymbol{\delta}$ are independently distributed as $N[\mathbf{0}, \boldsymbol{\Phi}]$, $N[\mathbf{0}, \boldsymbol{\Psi}_\epsilon]$ and $N[\mathbf{0}, \boldsymbol{\Psi}_\delta]$, respectively, where $\boldsymbol{\Psi}_\epsilon$ and $\boldsymbol{\Psi}_\delta$ are diagonal matrices. The covariance matrix of $\mathbf{x}$ is equal to

$$\boldsymbol{\Sigma}(\boldsymbol{\theta}) = \mathbf{B}(\boldsymbol{\Lambda}\boldsymbol{\Phi}\boldsymbol{\Lambda}^T + \boldsymbol{\Psi}_\epsilon)\mathbf{B}^T + \boldsymbol{\Psi}_\delta. \tag{2.10}$$

According to Jöreskog (1970), elements of the parameter vector $\boldsymbol{\theta}$ are of three kinds: (i) fixed parameters that have been assigned given values, (ii) constrained parameters that are unknown but equal to one or more other parameters, and (iii) free unknown parameters. By utilizing the flexibility of fixing arbitrary parameters at given values at will, Jöreskog (1970) showed this model can be applied to a number of statistical models, such as test theory models, simplex models and models for several sets of congeneric test scores.

Higher-order factor analysis model can be defined as:

$$\mathbf{x} = \boldsymbol{\Lambda}_{(1)}\boldsymbol{\Lambda}_{(2)} \cdots \boldsymbol{\Lambda}_{(k)}\boldsymbol{\xi}_k + \cdots + \boldsymbol{\Lambda}_{(1)}\boldsymbol{\Lambda}_{(2)}\boldsymbol{\xi}_2 + \boldsymbol{\Lambda}_{(1)}\boldsymbol{\xi}_1, \tag{2.11}$$

where $\boldsymbol{\Lambda}_{(1)}, \ldots, \boldsymbol{\Lambda}_{(k)}$ are loading matrices, and $\boldsymbol{\xi}_1, \ldots, \boldsymbol{\xi}_k$ are independently distributed random vectors which may be latent factors or error measurements. The covariance matrix of $\mathbf{x}$ is given by

$$\boldsymbol{\Sigma} = \boldsymbol{\Lambda}_{(1)} \cdots \boldsymbol{\Lambda}_{(k)}\boldsymbol{\Phi}_k\boldsymbol{\Lambda}_{(k)}^T \cdots \boldsymbol{\Lambda}_{(1)}^T + \cdots + \boldsymbol{\Lambda}_{(1)}\boldsymbol{\Lambda}_{(2)}\boldsymbol{\Phi}_2\boldsymbol{\Lambda}_{(2)}^T\boldsymbol{\Lambda}_{(1)}^T + \boldsymbol{\Lambda}_{(1)}\boldsymbol{\Phi}_1\boldsymbol{\Lambda}_{(1)}^T, \tag{2.12}$$

where $\boldsymbol{\Phi}_1, \ldots, \boldsymbol{\Phi}_k$ are covariance matrices of $\boldsymbol{\xi}_1, \ldots, \boldsymbol{\xi}_k$, respectively. Parameters in $\boldsymbol{\Lambda}_{(1)}, \ldots, \boldsymbol{\Lambda}_{(k)}$ and $\boldsymbol{\Phi}_1, \ldots, \boldsymbol{\Phi}_k$ are allowed to be fixed at any given values.

2.4 THE LISREL MODEL

In the CFA model, correlations among latent variables can be assessed by their covariance matrix; however, latent variables are never regressed on the other variables. The basic goal of the development of SEMs is to generalize the CFA model for assessing how latent variables affect each other in various ways. A typical representation of SEM is the LISREL model, which consists of two major components, namely the measurement equations and the structural equation.

The measurement equations are defined by the following CFA models:

$$\mathbf{x}_1 = \boldsymbol{\Lambda}_1\boldsymbol{\eta} + \boldsymbol{\epsilon}_1 \tag{2.13}$$

$$\mathbf{x}_2 = \boldsymbol{\Lambda}_2\boldsymbol{\xi} + \boldsymbol{\epsilon}_2, \tag{2.14}$$

where $\mathbf{x}_1(r \times 1)$ and $\mathbf{x}_2(s \times 1)$ are random vectors of the manifest variables which are the respective indicators for $\boldsymbol{\eta}$ and $\boldsymbol{\xi}$, $\boldsymbol{\Lambda}_1(r \times q_1)$ and $\boldsymbol{\Lambda}_2(s \times q_2)$ are loading matrices, $\boldsymbol{\epsilon}_1(r \times 1)$ and $\boldsymbol{\epsilon}_2(s \times 1)$ are random vectors of error measurements. It is assumed that $\boldsymbol{\epsilon}_1$ and $\boldsymbol{\epsilon}_2$ are uncorrelated with $\boldsymbol{\eta}$, $\boldsymbol{\xi}$ and $\boldsymbol{\delta}$, and the distributions of these random vectors are normal with zero means. Given the observed data in manifest random vectors $\mathbf{x}_1$ and $\mathbf{x}_2$, the measurement equations appropriately group together the correlated manifest variables to form latent variables in $\boldsymbol{\eta}$ and $\boldsymbol{\xi}$. This is done by assigning fixed parameters and defining unknown parameters in $\boldsymbol{\Lambda}_1$ and $\boldsymbol{\Lambda}_2$.

The structural equation, which specifies relationships among the identified latent variables, is defined by the equation:

$$\boldsymbol{\eta} = \boldsymbol{\Pi}\boldsymbol{\eta} + \boldsymbol{\Gamma}\boldsymbol{\xi} + \boldsymbol{\delta}, \tag{2.15}$$

where $\boldsymbol{\eta}(q_1 \times 1)$ is an endogenous random vector of latent variables and $\boldsymbol{\xi}(q_2 \times 1)$ is an exogenous random vector of latent variables, $\boldsymbol{\Pi}(q_1 \times q_1)$ and $\boldsymbol{\Gamma}(q_1 \times q_2)$ are unknown matrices of regression coefficients that represent the causal effects among $\boldsymbol{\eta}$ and $\boldsymbol{\xi}$, and $\boldsymbol{\delta}(q_1 \times 1)$ is a random vector of error measurements or residuals. It is assumed that $(\mathbf{I} - \boldsymbol{\Pi})$ is nonsingular, $\boldsymbol{\xi}$ is uncorrelated with $\boldsymbol{\delta}$, and the means of these random vectors are zero.

Let $\boldsymbol{\Phi}$, $\boldsymbol{\Psi}_\delta$, $\boldsymbol{\Psi}_{\epsilon 1}$ and $\boldsymbol{\Psi}_{\epsilon 2}$ be the covariance matrices of $\boldsymbol{\xi}$, $\boldsymbol{\delta}$, $\boldsymbol{\epsilon}_1$ and $\boldsymbol{\epsilon}_2$, respectively, the covariance matrix of $(\mathbf{x}_1^T, \mathbf{x}_2^T)$ is

$$\boldsymbol{\Sigma} = \begin{bmatrix} \boldsymbol{\Lambda}_1(\mathbf{I}-\boldsymbol{\Pi})^{-1}\{\boldsymbol{\Gamma}\boldsymbol{\Phi}\boldsymbol{\Gamma}^T + \boldsymbol{\Psi}_\delta\}(\mathbf{I}-\boldsymbol{\Pi})^{-T}\boldsymbol{\Lambda}_1^T + \boldsymbol{\Psi}_{\epsilon 1} & \boldsymbol{\Lambda}_1(\mathbf{I}-\boldsymbol{\Pi})^{-1}\boldsymbol{\Gamma}\boldsymbol{\Phi}\boldsymbol{\Lambda}_2^T \\ \boldsymbol{\Lambda}_2\boldsymbol{\Phi}\boldsymbol{\Gamma}^T(\mathbf{I}-\boldsymbol{\Pi})^{-T}\boldsymbol{\Lambda}_1^T & \boldsymbol{\Lambda}_2\boldsymbol{\Phi}\boldsymbol{\Lambda}_2^T + \boldsymbol{\Psi}_{\epsilon 2} \end{bmatrix}. \tag{2.16}$$

The elements of $\boldsymbol{\Sigma}$ are functions of the parameter matrices $\boldsymbol{\Lambda}_1, \boldsymbol{\Lambda}_2, \boldsymbol{\Pi}, \boldsymbol{\Gamma}, \boldsymbol{\Phi}$, $\boldsymbol{\Psi}_\delta$, $\boldsymbol{\Psi}_{\epsilon 1}$, and $\boldsymbol{\Psi}_{\epsilon 2}$. The elements of these parameter matrices are of three kinds: (i) fixed parameters that have been assigned at given values, (ii) constrained unknown parameters that are equal to one or more other parameters, and (iii) free unconstrained parameters. The flexibility of allowing arbitrary parameters at any given values gives high power for the LISREL model to subsume many useful models. Clearly, the EFA and CFA models are its special cases. Let $\mathbf{\Pi} = \mathbf{0}$ and substitute the resulting Equation (2.15) to (2.13), we see that the second-order factor analysis model is also its special case. If we assume $\mathbf{\Lambda}_1 = \mathbf{I}$, $\mathbf{\Lambda}_2 = \mathbf{I}$, $\boldsymbol{\epsilon}_1 = \mathbf{0}$, and $\boldsymbol{\epsilon}_2 = \mathbf{0}$, then Equation (2.15) becomes

$$\mathbf{x}_1 = \mathbf{\Gamma}\mathbf{x}_2 + \boldsymbol{\delta},$$

which is the simultaneous equation model in econometrics.

Like applications of many statistical models, applications of the LISREL model or SEMs rely on substantive knowledge to build a model. It is a confirmatory tool rather than an exploratory tool. In practical applications, we usually have a clear objective of the study, and some basic background about the key structure of the model that is obtained either from the subject knowledge or from preliminary data analysis. This basic background will be used in both model specification and interpretation.

The LISREL as defined in Equations (2.13), (2.14) and (2.15) specified the key number q_1, and q_2. The choice of these numbers is important in formulating the model and in interpretation of the statistical results. In substantive applications, the basic background of the given p manifest variables in $\mathbf{y}$ together with the sample correlation matrix of these variables usually give good choices of q_1 and q_2. To give a simple illustrative example in medical research, suppose we are interested in studying the effects of blood pressure and obesity on kidney failure which is assessed by urinary albumin creatinine ratio (ACR) and plasma creatinine (PCr). In addition to ACR and PCr, suppose the other manifest variables are: stystolic blood pressure (SBP), diastolic blood pressure (DBP), body mass index (BMI), hip index (HIP), and waist index (WST). From the medical knowledge of these variables, the following groupings are clear. (i) The manifest variables ACR and PCr provide key information about kidney disease severity; hence, they are grouped together to form the endogenous latent variable that can be interpreted as 'kidney disease severity, η'. (ii) Based on the medical knowledge and/or information from sample correlation matrix, {SBP, DBP} are taken to be the manifest variables of the latent variable that can be interpreted as 'blood pressure, ξ_1'. (iii) Similarly, {BMI, HIP, WST} are grouped together in a latent variable that can be interpreted as 'obesity, ξ_2'. Hence, it is natural to group the seven manifest variables into three latent variables, and take $q_1 = 1$ and $q_2 = 2$. Let

$$\mathbf{\Lambda}_1 = \begin{bmatrix} 1.0^* \\ \lambda_{1,21} \end{bmatrix} \text{ and } \mathbf{\Lambda}_2 = \begin{bmatrix} 1.0^* & 0^* \\ \lambda_{2,21} & 0^* \\ 0^* & 1.0^* \\ 0^* & \lambda_{2,42} \\ 0^* & \lambda_{2,52} \end{bmatrix},$$

where parameters with asterisks are known parameters which are fixed at the preassigned values. This notation will be used throughout this book. According to the usual practice of factor analysis, the fixed value of 1.0 is used to specify the 'scale' of the unknown parameters (factor loadings) in its column. The non-overlapping factor loading matrix $\mathbf{\Lambda}_2$ is taken because SBP and DBP are clear indicators of 'blood pressure, ξ_1', whilst BMI, HIP and WST are not; and BMI, HIP and WST are clear indicators of 'obesity, ξ_2', whilst SBP and DBP are not. Obviously, ACR and PCr are the manifest variables related to η. Hence the measurement equations are equal to:

$$\mathbf{x}_1 = \begin{bmatrix} 1.0^* \\ \lambda_{1,21} \end{bmatrix} \eta + \begin{bmatrix} \epsilon_{11} \\ \epsilon_{12} \end{bmatrix} \tag{2.17}$$

$$\mathbf{x}_2 = \begin{bmatrix} 1.0^* & 0^* \\ \lambda_{2,21} & 0^* \\ 0^* & 1.0^* \\ 0^* & \lambda_{2,42} \\ 0^* & \lambda_{2,52} \end{bmatrix} \begin{bmatrix} \xi_1 \\ \xi_2 \end{bmatrix} + \begin{bmatrix} \epsilon_{21} \\ \epsilon_{22} \\ \epsilon_{23} \\ \epsilon_{24} \\ \epsilon_{25} \end{bmatrix}. \tag{2.18}$$

The latent variables are related by the following structural equation:

$$\eta = \gamma_1 \xi_1 + \gamma_2 \xi_2 + \delta, \tag{2.19}$$

hence, $\mathbf{\Pi} = \mathbf{0}$ and $\mathbf{\Gamma} = (\gamma_1 \ \ \gamma_2)$. The covariance matrices of $(\xi_1, \xi_2)^T$, $(\epsilon_{11}, \epsilon_{12})^T$ and $(\epsilon_{21}, \ldots, \epsilon_{25})^T$ are respectively equal to

$$\mathbf{\Phi} = \begin{bmatrix} \phi_{11} & \phi_{12} \\ \phi_{21} & \phi_{22} \end{bmatrix}, \ \mathbf{\Psi}_{\epsilon 1} = \text{diag}(\psi_{11}, \psi_{12}), \text{and } \mathbf{\Psi}_{\epsilon 2} = \text{diag}(\psi_{21}, \ldots, \psi_{25}).$$

To simplify the notation, we consider the following equivalent formulation of the LISREL model. Let $\mathbf{v} = (\mathbf{x}_1^T, \mathbf{x}_2^T)^T$, $\boldsymbol{\omega} = (\boldsymbol{\eta}^T, \boldsymbol{\xi}^T)^T$ and $\boldsymbol{\epsilon} = (\boldsymbol{\epsilon}_1^T, \boldsymbol{\epsilon}_2^T)^T$, so that Equations (2.13) and (2.14) can be expressed as

$$\begin{aligned} \mathbf{v} = \begin{pmatrix} \mathbf{x}_1 \\ \mathbf{x}_2 \end{pmatrix} &= \begin{pmatrix} \mathbf{\Lambda}_1 & \mathbf{0} \\ \mathbf{0} & \mathbf{\Lambda}_2 \end{pmatrix} \begin{pmatrix} \boldsymbol{\eta} \\ \boldsymbol{\xi} \end{pmatrix} + \begin{pmatrix} \boldsymbol{\epsilon}_1 \\ \boldsymbol{\epsilon}_2 \end{pmatrix} \\ &= \mathbf{\Lambda}\boldsymbol{\omega} + \boldsymbol{\epsilon}. \end{aligned} \tag{2.20}$$

In all applications of SEMs, the form of $\mathbf{\Lambda}$ is given by Equation (2.20) with the zero matrices at the specified positions. The parameter matrix $\mathbf{\Lambda}_2$ could be non-overlapping. Clearly, Equation (2.20) is equivalent to Equations (2.13) and (2.14). Hence, the LISREL model can be formulated as a CFA model as defined in Equation (2.20), and the following linear structural equation that is equivalent to Equation (2.15):

$$\boldsymbol{\eta} = (\boldsymbol{\Pi} \ \ \boldsymbol{\Gamma}) \begin{pmatrix} \boldsymbol{\eta} \\ \boldsymbol{\xi} \end{pmatrix} + \boldsymbol{\delta} = \boldsymbol{\Lambda}_{\omega}\boldsymbol{\omega} + \boldsymbol{\delta}, \tag{2.21}$$

where $\boldsymbol{\Lambda}_{\omega} = (\boldsymbol{\Pi}, \boldsymbol{\Gamma})$.

2.5 THE BENTLER–WEEKS MODEL

The Bentler–Weeks model (Bentler and Weeks, 1980) is a combination of a structural equation model and a selection model. The structural equation model is defined as

$$\boldsymbol{\eta} = \boldsymbol{\Pi}\boldsymbol{\eta} + \boldsymbol{\Gamma}\boldsymbol{\xi} \tag{2.22}$$

where $\boldsymbol{\eta}(q_1 \times 1)$ is a random vector of dependent variables, $\boldsymbol{\xi}(q_2 \times 1)$ is a random vector of independent variables, $\boldsymbol{\Pi}(q_1 \times q_1)$ and $\boldsymbol{\Gamma}(q_1 \times q_2)$ are matrices of regression coefficients. It is assumed that $(\mathbf{I} - \boldsymbol{\Pi})$ is nonsingular. In general, the parameters involved in Equation (2.22) are of the three kinds that are similar to those described in the LISREL model. This equation relates all the manifest and latent variables under consideration. A variable is included in $\boldsymbol{\eta}$ if it is ever considered to be a dependent variable in any structural equation; otherwise it is included in $\boldsymbol{\xi}$ and considered as an independent variable in the model. Thus, a key concept involves the designation of each and every variable in the system as either a dependent or an independent variable. The random vector $\boldsymbol{\eta}$ consists of all manifest dependent variables and certain latent variables such as common factors of any level and residuals. The random vector $\boldsymbol{\xi}$ contains the manifest and latent variables that are not structural functions of other manifest or latent variables. Manifest variables in $\boldsymbol{\eta}$ and $\boldsymbol{\xi}$ are represented by $\mathbf{x}$ and $\mathbf{y}$, which are extracted via the selection model:

$$\mathbf{x} = \mathbf{G}_x\boldsymbol{\eta}, \quad \mathbf{y} = \mathbf{G}_y\boldsymbol{\xi}, \tag{2.23}$$

where $\mathbf{G}_x$ and $\mathbf{G}_y$ are selection matrices with entries 1.0 or 0.0. Let $\boldsymbol{\Phi}$ be the covariance matrix of $\boldsymbol{\xi}$, the covariance matrix of $(\mathbf{x}^T, \mathbf{y}^T)^T$ is equal to

$$\boldsymbol{\Sigma} = \begin{bmatrix} \mathbf{G}_x(\mathbf{I}-\boldsymbol{\Pi})^{-1}\boldsymbol{\Gamma}\boldsymbol{\Phi}\boldsymbol{\Gamma}^T(\mathbf{I}-\boldsymbol{\Pi})^{-T}\mathbf{G}_x{}^T & \mathbf{G}_x(\mathbf{I}-\boldsymbol{\Pi})^{-1}\boldsymbol{\Gamma}\boldsymbol{\Phi}\mathbf{G}_y^T \\ \mathbf{G}_y\boldsymbol{\Phi}\boldsymbol{\Gamma}^T(\mathbf{I}-\boldsymbol{\Pi})^{-T}\mathbf{G}_x^T & \mathbf{G}_y\boldsymbol{\Phi}\mathbf{G}_y^T \end{bmatrix}. \tag{2.24}$$

This covariance matrix can be more simply represented as

$$\mathbf{\Sigma} = \mathbf{G}_o(\mathbf{I}-\mathbf{\Pi}_o)^{-1}\mathbf{\Gamma}_o\mathbf{\Phi}\mathbf{\Gamma}_o^T(\mathbf{I}-\mathbf{\Pi}_o)^{-T}\mathbf{G}_o^T, \tag{2.25}$$

where

$$\mathbf{G}_o = \begin{bmatrix} \mathbf{G}_x & \mathbf{0} \\ \mathbf{0} & \mathbf{G}_y \end{bmatrix}, \mathbf{\Pi}_o = \begin{bmatrix} \mathbf{\Pi} & \mathbf{0} \\ \mathbf{0} & \mathbf{0} \end{bmatrix} \text{ and } \mathbf{\Gamma}_o = \begin{bmatrix} \mathbf{\Gamma} \\ \mathbf{I} \end{bmatrix}$$

There are only three parameter matrices in this simple representation, namely $\mathbf{G}_o$, $\mathbf{\Pi}_o$ and $\mathbf{\Gamma}_o$.

Comparing covariance matrices given in Equations (2.16) and (2.24), it can be seen that the Bentler–Weeks model is a special case of the LISREL model. Conversely, let Equation (2.22) be expressed by the following block vectors and block matrices.

$$\begin{bmatrix} \mathbf{x}_1 \\ \mathbf{x}_2 \\ \boldsymbol{\eta} \end{bmatrix} = \begin{bmatrix} \mathbf{0} & \mathbf{0} & \mathbf{\Lambda}_1 \\ \mathbf{0} & \mathbf{0} & \mathbf{0} \\ \mathbf{0} & \mathbf{0} & \mathbf{\Pi} \end{bmatrix} \begin{bmatrix} \mathbf{x}_1 \\ \mathbf{x}_2 \\ \boldsymbol{\eta} \end{bmatrix} + \begin{bmatrix} \mathbf{0} & \mathbf{I} & \mathbf{0} & \mathbf{0} \\ \mathbf{\Lambda}_2 & \mathbf{0} & \mathbf{I} & \mathbf{0} \\ \mathbf{\Gamma} & \mathbf{0} & \mathbf{0} & \mathbf{I} \end{bmatrix} \begin{bmatrix} \boldsymbol{\xi} \\ \boldsymbol{\epsilon}_1 \\ \boldsymbol{\epsilon}_2 \\ \boldsymbol{\delta} \end{bmatrix}.$$

Note that the covariance matrix of $(\boldsymbol{\xi}^T, \boldsymbol{\epsilon}_1^T, \boldsymbol{\epsilon}_2^T, \boldsymbol{\delta}^T)$ is

$$\begin{bmatrix} \mathbf{\Phi} & \mathbf{0} & \mathbf{0} & \mathbf{0} \\ \mathbf{0} & \mathbf{\Psi}_{\epsilon 1} & \mathbf{0} & \mathbf{0} \\ \mathbf{0} & \mathbf{0} & \mathbf{\Psi}_{\epsilon 2} & \mathbf{0} \\ \mathbf{0} & \mathbf{0} & \mathbf{0} & \mathbf{\Psi}_{\delta} \end{bmatrix}.$$

Using the LISREL notation as in Equations (2.13), (2.14) and (2.15), the LISREL model can be regarded as a special case of the Bentler–Weeks model. This shows that the Bentler–Weeks model is equivalent to the LISREL model. Thus, it has the same power of practical applicability as the LISREL model. See Bentler and Weeks (1980) for more discussion on the comparison with other models on various theoretical and practical aspects. The EQS6 program (Bentler and Wu, 2002) is written on the basis of this model.

2.6 DISCUSSION

This chapter presents the most basic SEMs, including the EFA, CFA, LISREL and the Bentler–Weeks models. These SEMs have one common characteristic, namely the covariance matrices contain the underlying unknown parameters of the models. Under the normality assumption, the ML or the generalized least squares theory for analyzing these models can be derived through a covariance structure analysis approach, on the basis of the sample covariance matrix. As outputs of almost all common software in SEMs provide the ML and the GLS solutions, the asymptotic properties of these two approaches are important and will be discussed on Chapter 3.

In the real world, there are a large number of practical problems that cannot be handled by these standard SEMs. Hence, there is a strong demand for generalizations of the standard SEMs to cope with more complicated substantive theory and complex data sets. A number of important generalizations will be discussed in subsequent chapters, based on the Bayesian approach. For easier understanding of the model, better interpretation of the results and conveniences in developing the Bayesian methodologies, these generalizations are basically developed on the basis of LISREL type models with integral components that are basically defined by Equations (2.20) and (2.21).

REFERENCES

Anderson, T. W. (1984) *An Introduction to Multivariate Statistical Analysis* (2nd edn). New York: John Wiley & Sons, Inc.

Anderson, T. W. and Rubin, H. (1956) Statistical inference in factor analysis. *Proceedings of the Third Berkeley Symposium on Mathematical Statistics and Probability*. 110–150. Berkeley: University of California Press.

Bentler, P. M. (1976) Multistructural statistical models applied to factor analysis. *Multivariate Behavioral Research*, **11**, 3–25.

Bentler, P. M. (1983) Some contributions to efficient statistics for structural models: specification and estimation of moment structures. *Psychometrika*, **48**, 493–517.

Bentler, P. M. and Lee, S. Y. (1979) A statistical development of three-mode factor analysis. *British Journal of Mathematical and Statistical Psychology*, **32**, 87–104.

Bentler, P. M. and Weeks, D. G. (1980) Linear structural equations with latent variables. *Psychometrika*, **45**, 289–308.

Bentler, P. M. and Wu, E. J. C (2002) *EQS6 for Windows User Guide*. Enciuo, CA.: Multivariate Software, Inc.

Berger, J. O. (1985) *Statistical Decision Theory and Bayesian Analysis*. New York: Springer-Verlag.

Bollen, K. A. (1989) *Structural Equations with Latent Variables*. New York: John Wiley & Sons, Inc.

Jennrich R. I. and Robinson, S. M. (1969) A Newton–Raphson algorithm for maximum likelihood. factor analysis. *Psychometrika*, **34**, 111–123.

Jöreskog, K. G. (1967) Some contributions to maximum likelihood factor analysis. *Psychometrika*, **32**, 443–482.

Jöreskog, K. G. (1969) A general approach to confirmatory maximum likelihood factor analysis. *Psychometrika*, **34**, 183–202.

Jöreskog, K. G. (1970) A general method for analysis of covariance structures. *Biometrika*, **57**, 239–251.

Jöreskog, K. G. (1977) Factor analysis by least-squares and maximum likelihood methods. In K. Enslein, A. Ralston and H. S. Wilf (eds), Statistical Methods for Digital Computers. New York: John Wiley & Sons, Inc..

Jöreskog, K. G. and Sörbom, D. (1994) LISREL VIII: *A Guide to the Program and Applications*, Chicago, IL: SPSS, Inc..

Jöreskog, K. G. and Sörbom, D. (1996) *LISREL 8: Structural Equation Modeling with the SIMPLIS Command Language*. Hove and London: Scientific Software International.

Lawley, D. N. and Maxwell, A. E. (1971) *Factor Analysis as a Statistical Method* (2nd edn). New York: Elsevier.

Mulaik, S. A. (1972). *The Foundations of Factor Analysis*. New York: McGraw-Hill Book Company.

Power, M., Bullingen, M., Hazper, A. and WHOQOL Group (1999) The World Health Organization WHOQOL-100: tests of the universality of quality of life in 15 different cultural groups worldwide. *Health Psychology*, **18**, 495–505.

Spearman, C. (1904) General intelligence objectively determined and measured. *American Journal of Psychology*, **15**, 201–293.

Thurstone, L. L. (1944) A multiple group method of factoring the correlation matrix. *Psychometrika*, **10**, 73–38.

Tucker, L. R. (1966) Some mathematical notes on three-mode factor analysis.*Psychometrika*, **31**, 279–311.

Wiley, D. (1973) The identification problem for structural equation models with unmeasured variables. In A. S. Goldberger and O. D. Duncan (eds), *Structural Equation Models in the Social Sciences* pp. 69–83. New York: Academic Press.

3

Covariance Structure Analysis

3.1 INTRODUCTION

Structural equation models (SEMs) are typically formulated with random observed variables, latent variables and error measurements. Although the model equations are very similar to regression equations, common statistical methods in analyzing the regression model that are derived on the basis of the raw observations cannot be applied to SEMs, mainly because of the presence of the latent variables. For example, consider the following simple confirmatory factor analysis (CFA) model:

$$\mathbf{x}_i = \boldsymbol{\Lambda}\boldsymbol{\xi}_i + \boldsymbol{\epsilon}_i, \qquad i = 1, \ldots, n$$

where $\boldsymbol{\xi}_i$ is a vector of latent variables (factor scores) with a distribution $N[\mathbf{0}, \ \boldsymbol{\Phi}]$, and $\boldsymbol{\Lambda}$ and $\boldsymbol{\epsilon}_i$ are similarly defined as in Chapter 2, Section 2.3. In this model, $\boldsymbol{\xi}_i$ is not observed as in a regression model. Consequently, the generalized least square or maximum likelihood methods in regression cannot be applied. However, if $\boldsymbol{\xi}_i$ is given, the model becomes a regression model and the resulting analysis is much simpler.

It is noted that for all the standard SEMs presented in Chapter 2, the parameters involved in the model are contained in the covariance matrix of the observed variables. For example, in CFA, $\boldsymbol{\Sigma} = \boldsymbol{\Lambda}\boldsymbol{\Phi}\boldsymbol{\Lambda}^T + \boldsymbol{\Psi}_\epsilon$ and the unknown parameters are those free elements in $\boldsymbol{\Lambda}, \boldsymbol{\Phi}$ and $\boldsymbol{\Psi}_\epsilon$. Hence, the covariance matrix plays an important role, and the basic statistical analysis can be regarded as the covariance structure analysis (CSA).

In the general CSA, the covariance matrix of the observed random vector of the hypothesized model is formulated as a matrix function of the unknown parameters vector $\boldsymbol{\theta}$, say $\boldsymbol{\Sigma}(\boldsymbol{\theta})$. Since the sample covariance matrix $\mathbf{S}$ is an unbiased estimator of the true covariance matrix, one approach in developing statistical methods to analyze $\boldsymbol{\Sigma}(\boldsymbol{\theta})$ is based on $\mathbf{S}$. Let $f[\boldsymbol{\Sigma}(\boldsymbol{\theta}), \mathbf{S}]$ be an objective function which measures the discrepancy between $\boldsymbol{\Sigma}(\boldsymbol{\theta})$ and $\mathbf{S}$. A method for estimating $\boldsymbol{\theta}$ is by finding the value of $\boldsymbol{\theta}$ that minimizes $f[\boldsymbol{\Sigma}(\boldsymbol{\theta}), \mathbf{S}]$. Moreover, if the final function value evaluated at the estimate of $\boldsymbol{\theta}$ is large, then one may conclude that $\boldsymbol{\Sigma}(\boldsymbol{\theta})$ does not agree with the data, and hence the proposed model is rejected. Otherwise, $\boldsymbol{\Sigma}(\boldsymbol{\theta})$ may be regarded as a plausible model for the given data. The degree of the 'largeness' for rejecting a hypothesized model is assessed by a goodness-of-fit statistic. After achieving a plausible model, further statistical analyses are conducted through the asymptotic distribution of the estimator.

The existing software in the field of SEM are mainly developed on the basis of the LISREL model, the Bentler and Weeks model, or their equivalent forms. Usually, the outputs give same basic statistics, such as the estimates of the unknown parameters, their standard error estimates, and the p-value of a chi-square goodness-of-fit test. The asymptotic properties of these statistics are important in conducting statistical inferences and obtaining correct interpretations of the results. One objective of this chapter is to derive the asymptotic properties that are associated with the special functions of $f[\boldsymbol{\Sigma}(\boldsymbol{\theta}), \mathbf{S}]$ that give the well-known maximum likelihood (ML) and generalized least squares (GLS) approaches.

Except for very special cases, the minimum of $f[\boldsymbol{\Sigma}(\boldsymbol{\theta}), \mathbf{S}]$ cannot be obtained in closed form, hence some iterative procedure is required. Historically, the search for an efficient algorithm for obtaining the ML estimate of parameters in the EFA model was the focus of attention in the 1960s. The major breakthrough came from the effort of Karl Jöreskog who successfully implemented the Fletcher–Powell (FP) algorithm to get the ML solution. Eventually, this algorithm became the major computing tool in the LISREL (Jöreskog and Sörbom, 1996) program. Another program EQS6 (Bentler and Wu, 2002) applies the Gauss–Newton (GN) algorithm for getting the generalized least squares (GLS) estimate, and a GN-type algorithm on an iteratively reweighted least squares mode for obtaining the ML estimate (see Lee and Jennrich, 1979). These algorithms are discussed in Section 3.6.

Inspired by the nice work of Browne (1974), we will start with a GLS approach. Asymptotic properties of the GLS estimate, such as consistency, asymptotic normality and the asymptotic distribution of the goodness-of-fit test statistics, will be derived on the basis of the normal assumption and some mild conditions on the model. Other reasons for using a GLS approach are:

(i) it extends naturally and conveniently to the asymptotically distribution-free method, and (ii) the best GLS estimator is asymptotically equivalent to the ML estimator. Because of (ii), the ML estimator has the same asymptotic properties as the GLS estimator. These properties provide the foundation for statistical inferences of the standard SEMs.

To derive expressions for implementing the iterative procedures and achieving the asymptotic properties of the estimators, we need to differentiate some matrix functions with respect to some matrix of variables. A brief outline of a matrix calculus method is given in Appendix 3.1. Preliminary results concerning the convergence of sequences of random variables and distribution functions are given in Appendix 3.2. Proofs of some theoretical results are presented in Appendix 3.3.

3.2 DEFINITIONS, NOTATIONS AND PRELIMINARY RESULTS

Let $\mathbf{A}(p \times q) = (a_{kh})$ and $\mathbf{B}(r \times s)$ be any matrices; we define the right Kronecker product of $\mathbf{A}$ and $\mathbf{B}$ by

$$\mathbf{A} \otimes \mathbf{B} = \begin{bmatrix} a_{11}\mathbf{B} & a_{12}\mathbf{B} & \cdots & a_{1q}\mathbf{B} \\ a_{21}\mathbf{B} & a_{22}\mathbf{B} & \cdots & a_{2q}\mathbf{B} \\ \vdots & \vdots & & \vdots \\ a_{p1}\mathbf{B} & a_{p2}\mathbf{B} & \cdots & a_{pq}\mathbf{B} \end{bmatrix}.$$

Therefore, $\mathbf{A} \otimes \mathbf{B}$ is a pr by qs block matrix with a typical (k, h)th block $a_{kh}\mathbf{B}$. When $\mathbf{A}$ is a scalar, this product reduces to the ordinary product of a scalar and a matrix. This Kronecker product has the following properties.

Lemma 1

(i) $(\mathbf{A}_1 \otimes \mathbf{A}_2)(\mathbf{A}_3 \otimes \mathbf{A}_4) = (\mathbf{A}_1\mathbf{A}_3) \otimes (\mathbf{A}_2\mathbf{A}_4)$, where $\mathbf{A}_i, i = 1, \ldots, 4$ are matrices such that $\mathbf{A}_1\mathbf{A}_3$ and $\mathbf{A}_2\mathbf{A}_4$ are well defined.

(ii) $(\mathbf{A}_1 \otimes \mathbf{A}_2)^T = \mathbf{A}_1^T \otimes \mathbf{A}_2^T$.

(iii) $(\mathbf{A}_1 \otimes \mathbf{A}_2)^{-1} = \mathbf{A}_1^{-1} \otimes \mathbf{A}_2^{-1}$, where $\mathbf{A}_1$ and $\mathbf{A}_2$ are nonsingular matrices.

(iv) If $\mathbf{A}$ is a positive definite matrix, then $\mathbf{A} \otimes \mathbf{A}$ is positive definite.

For any matrix $\mathbf{A}(p \times q)$ and any symmetric matrix $\mathbf{B}(p \times p)$, we define

$$\text{vec}\mathbf{A} = (a_{11}, \ldots, a_{1q}, a_{21}, \ldots, a_{2q}, \ldots, a_{p1}, \ldots, a_{pq})^T$$
$$\text{vecs}\mathbf{B} = (b_{11}, b_{21}, b_{22}, b_{31}, b_{32}, b_{33}, \ldots, b_{p1}, \ldots, b_{pp})^T.$$

Hence, vec$\mathbf{A}$ is a $(pq \times 1)$ column vector which is formed by stacking rows of $\mathbf{A}$ sequentially, and vecs$\mathbf{B}$ is a $p^* = 2^{-1}p(p+1)$ by 1 column vector which is formed by stacking the lower triangular elements of $\mathbf{B}$, row by row sequentially. We have the following additional properties of Kronecker product and the vec operator.

Lemma 2

(i) $(\mathbf{A}_1 \otimes \mathbf{A}_2)\text{vec}\mathbf{C} = \text{vec}\mathbf{A}_1\mathbf{C}\mathbf{A}_2^T$, where $\mathbf{A}_1, \mathbf{A}_2$ and $\mathbf{C}$ are matrices such that $\mathbf{A}_1\mathbf{C}\mathbf{A}_2^T$ is well defined.

(ii) $(\text{vec}\mathbf{C}_1)^T(\mathbf{A}_1 \otimes \mathbf{A}_2)(\text{vec}\mathbf{C}_2) = \text{tr}\ \mathbf{C}_1^T\mathbf{A}_1\mathbf{C}_2\mathbf{A}_2^T$, where $\mathbf{A}_1, \mathbf{A}_2, \mathbf{C}_1$ and $\mathbf{C}_2$ are matrices such that $\mathbf{C}_1^T\mathbf{A}_1\mathbf{C}_2\mathbf{A}_2^T$ is well defined.

It can be shown that vecs$\mathbf{B}$ and vec$\mathbf{B}$ are related by (see Browne, 1974)

$$\text{vecs}\ \mathbf{B} = \mathbf{K}_p^T\ \text{vec}\ \mathbf{B}, \tag{3.1}$$

where the typical element of the matrix $\mathbf{K}_p(p^2 \times p^*)$ is

$$\mathbf{K}_p(ij, kh) = 2^{-1}(\delta_{ik}\delta_{jh} + \delta_{ih}\delta_{jk}),\ i \le p,\ j \le p,\ k \le h \le p$$

and δ_{ik} represents the Kronecker's delta such that $\delta_{ik} = 1$ if $i = k$, and $\delta_{ik} = 0$ if $i \neq k$. Hence, $\mathbf{K}_p^T$ reduces vec$\mathbf{B}$ to vecs$\mathbf{B}$ which only contains the non-redundant elements in the lower-triangular part of the synmetirc matrix. An example $\mathbf{K}_p$ with $p = 3$ is

$$K_3 = \left[\begin{array}{c|cc|ccc} 1 & 0 & 0 & 0 & 0 & 0 \\ 0 & \frac{1}{2} & 0 & 0 & 0 & 0 \\ 0 & 0 & 0 & \frac{1}{2} & 0 & 0 \\ \hline 0 & \frac{1}{2} & 0 & 0 & 0 & 0 \\ 0 & 0 & 1 & 0 & 0 & 0 \\ 0 & 0 & 0 & 0 & \frac{1}{2} & 0 \\ \hline 0 & 0 & 0 & \frac{1}{2} & 0 & 0 \\ 0 & 0 & 0 & 0 & \frac{1}{2} & 0 \\ 0 & 0 & 0 & 0 & 0 & 1 \end{array}\right].$$

The rank of $\mathbf{K}_p$ is p^* and its left inverse is

$$\mathbf{K}_p^- = (\mathbf{K}_p^T\ \mathbf{K}_p)^{-1}\mathbf{K}_p^T. \tag{3.2}$$

This left inverse can be used to express vec**B** in terms of vecs**B** as follows,

$$\text{vec}\mathbf{B} = (\mathbf{K}_p^-)^T \text{ vecs}\mathbf{B}. \tag{3.3}$$

Another important matrix $\mathbf{M}_p(p^2 \times p^2)$ is defined by

$$\mathbf{M}_p = \mathbf{K}_p\mathbf{K}_p^- = \mathbf{K}_p(\mathbf{K}_p^T\mathbf{K}_p)^{-1}\mathbf{K}_p^T.$$

An example of $\mathbf{M}_p$ with $p = 3$ is given by

$$\left[\begin{array}{ccc|ccc|ccc} 1 & 0 & 0 & 0 & 0 & 0 & 0 & 0 & 0 \\ 0 & \frac{1}{2} & 0 & \frac{1}{2} & 0 & 0 & 0 & 0 & 0 \\ 0 & 0 & \frac{1}{2} & 0 & 0 & 0 & \frac{1}{2} & 0 & 0 \\ \hline 0 & \frac{1}{2} & 0 & \frac{1}{2} & 0 & 0 & 0 & 0 & 0 \\ 0 & 0 & 0 & 0 & 1 & 0 & 0 & 0 & 0 \\ 0 & 0 & 0 & 0 & 0 & \frac{1}{2} & 0 & \frac{1}{2} & 0 \\ \hline 0 & 0 & \frac{1}{2} & 0 & 0 & 0 & \frac{1}{2} & 0 & 0 \\ 0 & 0 & 0 & 0 & 0 & \frac{1}{2} & 0 & \frac{1}{2} & 0 \\ 0 & 0 & 0 & 0 & 0 & 0 & 0 & 0 & 1 \end{array}\right].$$

The matrix $\mathbf{M}_p$ has the following properties.

Lemma 3

(i) $\mathbf{M}_p$ is a symmetric idempotent matrix of rank p^*, so $\mathbf{M}_p\mathbf{M}_p = \mathbf{M}_p$.

(ii) $\mathbf{M}_p\mathbf{K}_p = \mathbf{K}_p$.

(iii) $\mathbf{M}_p$ vec**B** = vec**B**, for any symmetric matrix **B**.

(iv) $\mathbf{M}_p(\mathbf{A} \otimes \mathbf{A}) = (\mathbf{A} \otimes \mathbf{A})\mathbf{M}_q$, for any matrix $\mathbf{A}(p \times q)$.

More properties of the matrices $\mathbf{K}_p$, $\mathbf{M}_p$ and the Kronecker product can be found in Browne (1974), and Bentler and Lee (1975).

For simplicity, we use '·' over a function to denote its derivatives. Let $\mathbf{\Sigma}(\boldsymbol{\theta})$ be a symmetric matrix of differentiable functions of $\boldsymbol{\theta}$, we have $\dot{\mathbf{\Sigma}} = \dot{\mathbf{\Sigma}}(\boldsymbol{\theta}) = \partial\mathbf{\Sigma}(\boldsymbol{\theta})/\partial\boldsymbol{\theta}$, and $\ddot{\mathbf{\Sigma}} = \ddot{\mathbf{\Sigma}}(\boldsymbol{\theta}) = \partial^2\mathbf{\Sigma}(\boldsymbol{\theta})/\partial\boldsymbol{\theta}\partial\boldsymbol{\theta}$. From the definition of $\dot{\mathbf{\Sigma}}$, its jth row is

$$\dot{\mathbf{\Sigma}}_j = (\partial\sigma_{11}/\partial\theta_j, \ldots, \partial\sigma_{1p}/\partial\theta_j, \ldots, \partial\sigma_{p1}/\partial\theta_j, \ldots, \partial\sigma_{pp}/\partial\theta_j),$$

which is equal to

$$\dot{\boldsymbol{\Sigma}}_j = \text{vec}\begin{bmatrix} \frac{\partial \sigma_{11}}{\partial \theta_j} & \cdots & \frac{\partial \sigma_{1p}}{\partial \theta_j} \\ \vdots & \vdots & \vdots \\ \frac{\partial \sigma_{p1}}{\partial \theta_j} & \cdots & \frac{\partial \sigma_{pp}}{\partial \theta_j} \end{bmatrix}.$$

Hence, it follows from Lemma 3 (iii) that

$$\mathbf{M}_p \dot{\boldsymbol{\Sigma}}_j^T = \dot{\boldsymbol{\Sigma}}_j^T \quad \text{and} \quad \dot{\boldsymbol{\Sigma}}_j = \dot{\boldsymbol{\Sigma}}_j \mathbf{M}_p. \tag{3.4}$$

This is a useful result for deriving the asymptotic distributions.

3.3 GLS ANALYSIS OF COVARIANCE STRUCTURE

3.3.1 The GLS Approach

We consider a random vector $\mathbf{x}(p \times 1)$ which has a mean vector $\mathbf{0}$ and a covariance matrix $\boldsymbol{\Sigma}_0$. Suppose $\boldsymbol{\Sigma}_0$ is a matrix function of a true though unknown parameter vector $\boldsymbol{\theta}_0(q \times 1)$, so that $\boldsymbol{\Sigma}_0 = \boldsymbol{\Sigma}(\boldsymbol{\theta}_0)$. To develop statistical theory for analyzing the covariance structure, we regard $\boldsymbol{\theta}(q \times 1)$ as a vector of mathematical variables which can take the value $\boldsymbol{\theta}_0$ and/or any estimate of $\boldsymbol{\theta}_0$. Moreover, $\boldsymbol{\Sigma}(\boldsymbol{\theta})$ will be regarded as a matrix function of $\boldsymbol{\theta}$. In the following text, if the context is clear, we will denote $\boldsymbol{\Sigma}(\boldsymbol{\theta})$ by $\boldsymbol{\Sigma}$ to simplify notation.

The following mild regularity conditions will be assumed throughout this chapter.

(a) The matrix $\boldsymbol{\Sigma}_0$ is positive definite.

(b) The vector $\boldsymbol{\theta}_0$ is an interior point in the parameter space, that is there exists a neighborhood of $\boldsymbol{\theta}_0$ that completely lies inside the parameter space.

(c) The model is identified, that is $\boldsymbol{\Sigma}(\boldsymbol{\theta}_0) = \boldsymbol{\Sigma}(\boldsymbol{\theta}^*)$ implies $\boldsymbol{\theta}_0 = \boldsymbol{\theta}^*$.

(d) All partial derivatives of the first three orders of $\boldsymbol{\Sigma}$ with respect to elements of $\boldsymbol{\theta}$ are continuous and bounded in a neighborhood of $\boldsymbol{\theta}_0$.

(e) The q by p^2 matrix of partial derivatives $\partial \boldsymbol{\Sigma}/\partial \boldsymbol{\theta}$ is of full rank in a neighborhood of $\boldsymbol{\theta}_0$.

Condition (a) implies that the distribution of $\mathbf{x}$ is not degenerate. Condition (d) implies that all elements of $\boldsymbol{\Sigma}$ are continuous functions of $\boldsymbol{\theta}$; this condition and condition (b) are necessary in applying the 'mean-value' theorem (see Bartle,

1964, p. 210) in the derivation of the asymptotic properties. Condition (c) is essential for obtaining a unique estimator of $\boldsymbol{\theta}_0$; iterative estimation procedures will diverge without this condition. Condition (e) is a natural assumption which eliminates collinearity among the elements of $\boldsymbol{\theta}$. Hence for each of the models discussed in Chapter 2, appropriate identification conditions have to be imposed to guarantee that condition (c) is satisfied. In the following, we assume that all the models and $\boldsymbol{\theta}_0$ in the discussion satisfy the above regularity conditions.

Let $\{\mathbf{x}_1, \ldots, \mathbf{x}_n\}$ be a random sample of $\mathbf{x}$, such that all $\mathbf{x}_i$ are identically and independently distributed (iid) according to $N[\mathbf{0}, \boldsymbol{\Sigma}_0]$. Throughout this chapter, n is assumed to be significantly larger than p^*. The sample covariance matrix is

$$\mathbf{S} = (n-1)^{-1} \sum_{i=1}^{n} (\mathbf{x}_i - \overline{\mathbf{x}})(\mathbf{x}_i - \overline{\mathbf{x}})^T,$$

where $\overline{\mathbf{x}} = n^{-1}(\mathbf{x}_1 + \cdots + \mathbf{x}_n)$ is the sample mean. (Rigorously, some asymptotic results should be presented with $n-1$; however, as n is large, the difference is very minor.) This matrix is positive definite with probability 1, it is an unbiased estimate of $\boldsymbol{\Sigma}_0$, and it also converges to $\boldsymbol{\Sigma}_0$ in probability. Moreover, it follows from the 'Multivariate Central Limit Theorem' that (see Anderson, 1984 Chapter 4)

$$n^{1/2}\{\mathrm{vecs}(\mathbf{S} - \boldsymbol{\Sigma}_0)\} \xrightarrow{L} N[\mathbf{0},\ 2\mathbf{K}_p^T(\boldsymbol{\Sigma}_0 \otimes \boldsymbol{\Sigma}_0)\mathbf{K}_p], \tag{3.5}$$

where '$\xrightarrow{L}$' denotes convergence in distribution (see Appendix 3.2). A key feature of Equation (3.5) is that the covariances/variances of elements of $\mathbf{S}$, which are quadratic functions of $\{\mathbf{x}_1, \ldots, \mathbf{x}_n\}$, can be expressed by $\boldsymbol{\Sigma}_0$, the central second moment of $\mathbf{x}$. This is the most basic result for developing the theory of CSA under the multivariate normal assumption. An analogous result that is obtained via Equation (3.5) is

$$\begin{aligned} n^{1/2}\, \mathrm{vec}(\mathbf{S} - \boldsymbol{\Sigma}_0) &= n^{1/2}\, \mathbf{K}_p^{-T}\mathrm{vecs}(\mathbf{S} - \boldsymbol{\Sigma}_0) \\ &\xrightarrow{L} N[\mathbf{0},\ 2\mathbf{K}_p^{-T}\mathbf{K}_p^T(\boldsymbol{\Sigma}_0 \otimes \boldsymbol{\Sigma}_0)\mathbf{K}_p\mathbf{K}_p^{-}] \\ &= N[\mathbf{0},\ 2\mathbf{M}_p(\boldsymbol{\Sigma}_0 \otimes \boldsymbol{\Sigma}_0)\mathbf{M}_p]. \end{aligned} \tag{3.6}$$

The asymptotic distribution of $\mathbf{S}$ as given by Equation (3.5) motivates the consideration of the non-linear regression model

$$\mathrm{vecs}\mathbf{S} = \mathrm{vecs}\boldsymbol{\Sigma}(\theta) + \boldsymbol{\epsilon}. \tag{3.7}$$

where $\boldsymbol{\epsilon}$ is the residual vector which has an asymptotic distribution $N[\mathbf{0}, 2n^{-1}\mathbf{K}_p^T(\boldsymbol{\Sigma}_0 \otimes \boldsymbol{\Sigma}_0)\mathbf{K}_p]$. The definition of the GLS function is motivated from the residual quadratic form:

$$\{\text{vecs}(\mathbf{S}-\boldsymbol{\Sigma})\}^T\{2n^{-1}\mathbf{K}_p^T(\boldsymbol{\Sigma}_0 \otimes \boldsymbol{\Sigma}_0)\mathbf{K}_p\}^{-1}\{\text{vecs}(\mathbf{S}-\boldsymbol{\Sigma})\}. \tag{3.8}$$

The inverse of $\mathbf{K}_p^T(\boldsymbol{\Sigma}_0 \otimes \boldsymbol{\Sigma}_0)\mathbf{K}_p$ is $\mathbf{K}_p^-(\boldsymbol{\Sigma}_0 \otimes \boldsymbol{\Sigma}_0)^{-1}\mathbf{K}_p^{-T}$, because from the definition and properties of $\mathbf{M}_p$,

$$\begin{aligned}
&\{\mathbf{K}_p^T(\boldsymbol{\Sigma}_0 \otimes \boldsymbol{\Sigma}_0)\mathbf{K}_p\}\{\mathbf{K}_p^-(\boldsymbol{\Sigma}_0 \otimes \boldsymbol{\Sigma}_0)^{-1}\mathbf{K}_p^{-T}\}\\
=\;& \mathbf{K}_p^T(\boldsymbol{\Sigma}_0 \otimes \boldsymbol{\Sigma}_0)\mathbf{M}_p(\boldsymbol{\Sigma}_0 \otimes \boldsymbol{\Sigma}_0)^{-1}\mathbf{K}_p^{-T}\\
=\;& \mathbf{K}_p^T\mathbf{M}_p(\boldsymbol{\Sigma}_0 \otimes \boldsymbol{\Sigma}_0)(\boldsymbol{\Sigma}_0 \otimes \boldsymbol{\Sigma}_0)^{-1}\mathbf{K}_p^{-T}\\
=\;& \mathbf{K}_p^T\,\mathbf{K}_p^{-T} = (\mathbf{K}_p^-\,\mathbf{K}_p)^T = \mathbf{I}.
\end{aligned}$$

Substituting this result in (3.8), it follows from Equation (3.3) and Lemma 1 that

$$\begin{aligned}
&n2^{-1}\{\text{vecs}(\mathbf{S}-\boldsymbol{\Sigma})\}^T\,\mathbf{K}_p^-(\boldsymbol{\Sigma}_0 \otimes \boldsymbol{\Sigma}_0)^{-1}\mathbf{K}_p^{-T}\{\text{vecs}(\mathbf{S}-\boldsymbol{\Sigma})\}\\
=\;& n2^{-1}\{\text{vec}(\mathbf{S}-\boldsymbol{\Sigma})\}^T\,(\boldsymbol{\Sigma}_0^{-1} \otimes \boldsymbol{\Sigma}_0^{-1})\{\text{vec}(\mathbf{S}-\boldsymbol{\Sigma})\}.
\end{aligned}$$

This residual quadratic form is a function of $\boldsymbol{\Sigma}_0^{-1}$. As $\boldsymbol{\Sigma}_0$ is unknown, we shall replace $\boldsymbol{\Sigma}_0^{-1}$ by a p by p positive definite matrix $\mathbf{V}$, and consider the GLS function:

$$\begin{aligned}
G(\boldsymbol{\theta}) &= 2^{-1}\{\text{vec}(\mathbf{S}-\boldsymbol{\Sigma})\}^T(\mathbf{V} \otimes \mathbf{V})\{\text{vec}(\mathbf{S}-\boldsymbol{\Sigma})\}\\
&= 2^{-1}\,\text{tr}\{(\mathbf{S}-\boldsymbol{\Sigma})\mathbf{V}\}^2.
\end{aligned} \tag{3.9}$$

The matrix $\mathbf{V}$ can be a constant positive definite matrix or a stochastic matrix that converges to a positive definite matrix $\mathbf{V}^*$. When $\mathbf{V}$ equals to the identity matrix, the GLS function reduces to a least squares function.

Definition 3.1 The GLS estimator $\tilde{\boldsymbol{\theta}}$ of $\boldsymbol{\theta}_0$ is the vector that minimizes $G(\boldsymbol{\theta})$.

On the basis of the techniques in matrix calculus (see Appendix 3.1) the gradient vector $\dot{\mathbf{G}}(\boldsymbol{\theta})$ and the Hessian matrix $\ddot{\mathbf{G}}(\boldsymbol{\theta})$ of $G(\boldsymbol{\theta})$ can be derived. The results are given by the following lemma.

Lemma 4

$$\text{(i)}\ \ \dot{\mathbf{G}}(\boldsymbol{\theta}) = \frac{\partial G(\boldsymbol{\theta})}{\partial \boldsymbol{\theta}} = -\dot{\boldsymbol{\Sigma}}(\boldsymbol{\theta})(\mathbf{V}\otimes\mathbf{V})\text{vec}(\mathbf{S}-\boldsymbol{\Sigma}). \tag{3.10}$$

$$\text{(ii)}\ \ \ddot{\mathbf{G}}(\boldsymbol{\theta}) = \frac{\partial^2 G(\boldsymbol{\theta})}{\partial \boldsymbol{\theta}\partial \boldsymbol{\theta}} = \dot{\boldsymbol{\Sigma}}(\boldsymbol{\theta})(\mathbf{V}\otimes\mathbf{V})\dot{\boldsymbol{\Sigma}}(\boldsymbol{\theta})^T - \ddot{\boldsymbol{\Sigma}}(\boldsymbol{\theta})\{\mathbf{I}_q\otimes((\mathbf{V}\otimes\mathbf{V})\text{vec}(\mathbf{S}-\boldsymbol{\Sigma}))\}, \tag{3.11}$$

where $\mathbf{I}_q$ is a q by q identity matrix.

The proof of Lemma 4 is given in Appendix 3.3. Since $G(\tilde{\boldsymbol{\theta}})$ is the minimum of $G(\boldsymbol{\theta})$, $\dot{\mathbf{G}}(\tilde{\boldsymbol{\theta}}) = \mathbf{0}$, and $\ddot{\mathbf{G}}(\tilde{\boldsymbol{\theta}})$ is positive definite. As $\ddot{\mathbf{G}}(\boldsymbol{\theta})$ is a continuous function, $\ddot{\mathbf{G}}(\boldsymbol{\theta})$ is also positive definite within a sufficiently small neighborhood of $\tilde{\boldsymbol{\theta}}$. Moreover, since $\mathbf{S} \xrightarrow{p} \boldsymbol{\Sigma}_0$, where '$\xrightarrow{p}$' denotes convergence in probability, it follows from Equation (3.11) that

$$\ddot{\mathbf{G}}(\boldsymbol{\theta}_0) \xrightarrow{p} \boldsymbol{\Omega}(\mathbf{V}^*) = \dot{\boldsymbol{\Sigma}}(\boldsymbol{\theta}_0)(\mathbf{V}^*\otimes\mathbf{V}^*)\dot{\boldsymbol{\Sigma}}(\boldsymbol{\theta}_0)^T. \tag{3.12}$$

Since $\dot{\boldsymbol{\Sigma}}(\boldsymbol{\theta}_0)$ is full rank and $(\mathbf{V}^*\otimes\mathbf{V}^*)$ is a positive definite; $\boldsymbol{\Omega}(\mathbf{V}^*)$, and the Hessian matrix $\ddot{\mathbf{G}}(\boldsymbol{\theta}_0)$ are positive definite. These matrices are important for deriving the asymptotic distribution of $\tilde{\boldsymbol{\theta}}$.

3.3.2 Asymptotic Properties of the GLS Estimator

Asymptotic properties presented in this section are useful for analyzing covariance structures. The proof of these asymptotic results is given in Appendix 3.3. Conditions (a) – (e) are assumed throughout.

Theorem 3.1 The GLS estimator is consistent, that is $\tilde{\boldsymbol{\theta}} \xrightarrow{p} \boldsymbol{\theta}_0$.
It follows from Theorem 3.1 that if n is sufficient large, the GLS estimator $\tilde{\boldsymbol{\theta}}$ would be very close to the true parameter vector $\boldsymbol{\theta}_0$. This property gives confidence in the GLS estimator. The asymptotic distribution of $\tilde{\boldsymbol{\theta}}$ is given by the next theorem.

Theorem 3.2 Let $\mathbf{C}(\boldsymbol{\theta}_0) = 2\boldsymbol{\Omega}(\mathbf{V}^*)^{-1}\boldsymbol{\Omega}(\mathbf{V}^*\boldsymbol{\Sigma}_0\mathbf{V}^*)\boldsymbol{\Omega}(\mathbf{V}^*)^{-1}$, where $\boldsymbol{\Omega}(\mathbf{V}^*) = \dot{\boldsymbol{\Sigma}}(\boldsymbol{\theta}_0)(\mathbf{V}^*\otimes\mathbf{V}^*)\dot{\boldsymbol{\Sigma}}(\boldsymbol{\theta}_0)^T$, then

$$n^{1/2}(\tilde{\boldsymbol{\theta}} - \boldsymbol{\theta}_0) \xrightarrow{L} N[\mathbf{0}, \mathbf{C}(\boldsymbol{\theta}_0)]. \tag{3.13}$$

The asymptotic covariance matrix of $\tilde{\boldsymbol{\theta}}$ is rather complicated. However, if $\mathbf{V}$ is chosen such that $\mathbf{V}^* = \boldsymbol{\Sigma}_0^{-1}$, the asymptotic covariance matrix of $\tilde{\boldsymbol{\theta}}$ is significantly simplified. Moreover, the resulting GLS estimator will have the minimum asymptotic variance, in the sense as given by the following corollary.

Corollary 3.1 If $\mathbf{V} \xrightarrow{P} \mathbf{V}^* = \boldsymbol{\Sigma}_0^{-1}$, then

$$n^{1/2}(\tilde{\boldsymbol{\theta}} - \boldsymbol{\theta}_0) \xrightarrow{L} N[\mathbf{0}, 2\boldsymbol{\Omega}(\boldsymbol{\Sigma}_0^{-1})^{-1}]. \tag{3.14}$$

Moreover, $\mathbf{C}(\boldsymbol{\theta}_0) - 2\boldsymbol{\Omega}(\boldsymbol{\Sigma}_0^{-1})^{-1}$ is a positive definite matrix.

Browne (1974) called the GLS estimator with $\mathbf{V}^* = \boldsymbol{\Sigma}_0^{-1}$ the 'best' GLS estimator. Clearly, if $\mathbf{V} = \mathbf{S}^{-1}$, then $\mathbf{V}^* = \boldsymbol{\Sigma}_0^{-1}$. Hence, $\mathbf{S}^{-1}$ is a logical choice for the weight matrix $\mathbf{V}$ in $G(\boldsymbol{\theta})$.

Let $\tilde{\boldsymbol{\Sigma}} = \boldsymbol{\Sigma}(\tilde{\boldsymbol{\theta}})$ and $\tilde{\boldsymbol{\Omega}}(\tilde{\boldsymbol{\Sigma}}^{-1}) = \dot{\boldsymbol{\Sigma}}(\tilde{\boldsymbol{\theta}})(\tilde{\boldsymbol{\Sigma}}^{-1} \otimes \tilde{\boldsymbol{\Sigma}}^{-1})\dot{\boldsymbol{\Sigma}}(\tilde{\boldsymbol{\theta}})^T$. Since $\tilde{\boldsymbol{\Omega}}(\tilde{\boldsymbol{\Sigma}}^{-1})$ converges in probability to $\boldsymbol{\Omega}(\boldsymbol{\Sigma}_0^{-1})$, standard error estimates of elements in $\tilde{\boldsymbol{\theta}}$ can be obtained via the corresponding diagonal elements of $2n^{-1}\tilde{\boldsymbol{\Omega}}(\tilde{\boldsymbol{\Sigma}}^{-1})^{-1}$. Other statistical inference of $\boldsymbol{\theta}_0$, for example the asymptotic t-test or confidence interval of an individual parameter, can be obtained from standard multivariate methods.

The following theorem provides an asymptotic test statistic for assessing the goodness-of-fit of the hypothesized model.

Theorem 3.3 If $\mathbf{V}^* = \boldsymbol{\Sigma}_0^{-1}$, then the asymptotic distribution of $nG(\tilde{\boldsymbol{\theta}})$ is chi-square with $p^* - q$ degrees of freedom, that is,

$$nG(\tilde{\boldsymbol{\theta}}) \xrightarrow{L} \chi^2_{p^*-q}. \tag{3.15}$$

On the basis of Theorem 3.3, the proposed model $\boldsymbol{\Sigma}(\boldsymbol{\theta})$ is rejected if $nG(\tilde{\boldsymbol{\theta}})$ is larger than $\chi^2_{p^*-q}(\alpha)$, the critical value given by the chi-square table with $p^* - q$ degrees of freedom, at the given type I error level α. The proposed model $\boldsymbol{\Sigma}(\boldsymbol{\theta})$ is not rejected if $nG(\tilde{\boldsymbol{\theta}})$ is less than $\chi^2_{p^*-q}(\alpha)$. However, this does not mean that $\boldsymbol{\Sigma}(\boldsymbol{\theta})$ can be accepted as the correct model by the theory of hypothesis testing. In the common practice of SEM, the proposed model is regarded as a plausible model if $\boldsymbol{\Sigma}(\boldsymbol{\theta})$ is not rejected.

3.4 ML ANALYSIS OF COVARIANCE STRUCTURE

Given that the random sample $\{\mathbf{x}_1, \ldots, \mathbf{x}_n\}$ is from a multivariate normal distribution $N[\boldsymbol{\mu}_0, \boldsymbol{\Sigma}_0]$, the exact distribution of the sample covariance matrix $\mathbf{S}$ is a Wishart distribution with a probability density function (see Anderson, 1984),

$$f(\mathbf{S}|\boldsymbol{\Sigma}_0) = \frac{C\exp\{-(n-1)/2\}\text{tr}\,\mathbf{S}\boldsymbol{\Sigma}_0^{-1}}{|\boldsymbol{\Sigma}_0|^{(n-1)/2}},$$

where C is an appropriate normalizing constant. Hence, the negative log-likelihood function is equal to

$$-\log C + \frac{(n-1)}{2}[\log|\boldsymbol{\Sigma}(\boldsymbol{\theta})| + \text{tr}\,\mathbf{S}\boldsymbol{\Sigma}(\boldsymbol{\theta})^{-1}],$$

where $\boldsymbol{\theta}$ is treated as a vector of mathematical variables as in the GLS estimation. In CSA, it is common to work with the following discrepancy function (see Jöreskog, 1978; Browne, 1984),

$$F(\boldsymbol{\theta}) = \log|\boldsymbol{\Sigma}(\boldsymbol{\theta})| + \text{tr}\,\mathbf{S}\boldsymbol{\Sigma}(\boldsymbol{\theta})^{-1} - \log|\mathbf{S}| - p. \tag{3.16}$$

Definition 3.2 The vector $\tilde{\boldsymbol{\theta}}_M$ that minimizes $F(\boldsymbol{\theta})$ is defined as the ML estimate of $\boldsymbol{\theta}_0$.

It follows from standard statistical theory that $\tilde{\boldsymbol{\theta}}_M$ is a consistent estimator of $\boldsymbol{\theta}_0$, that is, $\tilde{\boldsymbol{\theta}}_M \xrightarrow{p} \boldsymbol{\theta}_0$. The goodness-of-fit of the proposed model can be tested via the asymptotic distribution of the likelihood ratio criterion.

Let the null hypothesis be H_0: The covariance matrix of $\mathbf{x}$ is given by the hypothesized model $\boldsymbol{\Sigma}(\boldsymbol{\theta})$, and let the general hypothesis be H^*: The covariance matrix of $\mathbf{x}$ is any arbitrary positive definite matrix. The likelihood ratio criterion is

$$LR = \frac{f(\mathbf{S}|\boldsymbol{\Sigma}(\tilde{\boldsymbol{\theta}}_M))}{f(\mathbf{S}|(n-1)n^{-1}\mathbf{S})}, \tag{3.17}$$

where $(n-1)n^{-1}\mathbf{S}$ is the ML estimate of the covariance matrix under the general hypothesis. It follows from the definition of $f(\cdot|\cdot)$ that

$$-2\log(LR) = -2\log\left[\frac{|\boldsymbol{\Sigma}(\tilde{\boldsymbol{\theta}}_M)|^{-(n-1)/2}\exp\{-(n-1)\,\text{tr}\,\mathbf{S}\boldsymbol{\Sigma}(\tilde{\boldsymbol{\theta}}_M)^{-1}/2\}}{|(n-1)n^{-1}\mathbf{S}|^{-(n-1)/2}\exp[-(n-1)\,\text{tr}\,\mathbf{S}\{(n-1)n^{-1}\mathbf{S}\}^{-1}/2]}\right]$$

$$\xrightarrow{p} nF(\tilde{\boldsymbol{\theta}}_M).$$

From the standard ML theory, $-2\log(LR) \xrightarrow{L} \chi^2_{p^*-q}$, hence the following theorem is valid.

Theorem 3.4 The asymptotic distribution of $nF(\tilde{\boldsymbol{\theta}}_M)$ is chi-square with $p^* - q$ degrees of freedom, that is

$$nF(\tilde{\boldsymbol{\theta}}_M) \xrightarrow{L} \chi^2_{p*-q}.$$

On the basis of the techniques in matrix calculus (see Appendix 3.1) the gradient vector $\dot{\mathbf{F}}(\boldsymbol{\theta})$ and the Hessian matrix $\ddot{\mathbf{F}}(\boldsymbol{\theta})$ can be derived. Results are given by the following lemma.

Lemma 5

$$\dot{\mathbf{F}}(\boldsymbol{\theta}) = -\dot{\boldsymbol{\Sigma}}(\boldsymbol{\theta})(\boldsymbol{\Sigma}^{-1}\otimes\boldsymbol{\Sigma}^{-1})\text{vec}(\mathbf{S}-\boldsymbol{\Sigma}). \tag{3.18}$$

$$\ddot{\mathbf{F}}(\boldsymbol{\theta}) = \frac{\partial^2 F(\boldsymbol{\theta})}{\partial\boldsymbol{\theta}\partial\boldsymbol{\theta}} = \dot{\boldsymbol{\Sigma}}(\boldsymbol{\theta})(\boldsymbol{\Sigma}^{-1}\otimes\boldsymbol{\Sigma}^{-1})\dot{\boldsymbol{\Sigma}}(\boldsymbol{\theta})^T - 2\dot{\boldsymbol{\Sigma}}(\boldsymbol{\theta})\{\boldsymbol{\Sigma}^{-1}\otimes\boldsymbol{\Sigma}^{-1}(\mathbf{S}-\boldsymbol{\Sigma})\boldsymbol{\Sigma}^{-1}\}\dot{\boldsymbol{\Sigma}}(\boldsymbol{\theta})^T$$
$$-\ddot{\boldsymbol{\Sigma}}(\boldsymbol{\theta})\{\mathbf{I}_q\otimes((\boldsymbol{\Sigma}^{-1}\otimes\boldsymbol{\Sigma}^{-1})\text{vec}(\mathbf{S}-\boldsymbol{\Sigma}))\}. \tag{3.19}$$

Similar to $\ddot{\mathbf{G}}(\boldsymbol{\theta})$ in the GLS estimation, $\ddot{\mathbf{F}}(\boldsymbol{\theta})$ is also positive definite within a sufficiently small neighborhood of $\tilde{\boldsymbol{\theta}}_M$. Moreover, it follows from Equation (3.19) that

$$\ddot{\mathbf{F}}(\tilde{\boldsymbol{\theta}}_M) \xrightarrow{p} \dot{\boldsymbol{\Sigma}}(\boldsymbol{\theta}_0)(\boldsymbol{\Sigma}_0^{-1}\otimes\boldsymbol{\Sigma}_0^{-1})\dot{\boldsymbol{\Sigma}}(\boldsymbol{\theta}_0)^T = \boldsymbol{\Omega}(\boldsymbol{\Sigma}_0^{-1}) \tag{3.20}$$

and

$$E\{\ddot{\mathbf{F}}(\boldsymbol{\theta}_0)\} = \boldsymbol{\Omega}(\boldsymbol{\Sigma}_0^{-1}). \tag{3.21}$$

Hence, $\boldsymbol{\Omega}(\boldsymbol{\Sigma}_0^{-1})$ is the information matrix of $F(\boldsymbol{\theta})$ evaluated at $\boldsymbol{\theta}_0$.

It follows from Equation (3.18) and (3.20), and similar arguments to those in the proof of Theorem 3.2 that

$$n^{1/2}(\tilde{\boldsymbol{\theta}}_M - \theta_0) \xrightarrow{p} \boldsymbol{\Omega}(\boldsymbol{\Sigma}_0^{-1})^{-1}\dot{\boldsymbol{\Sigma}}(\boldsymbol{\theta}_0)(\boldsymbol{\Sigma}_0^{-1}\otimes\boldsymbol{\Sigma}_0^{-1})\{n^{1/2}\text{vec}(\mathbf{S}-\boldsymbol{\Sigma}_0)\}. \tag{3.22}$$

From (A3.5) in Appendix 3.3, and (3.22), $n^{1/2}(\tilde{\boldsymbol{\theta}} - \tilde{\boldsymbol{\theta}}_M)$ converges in probability to

$$\{\boldsymbol{\Omega}(V^*)^{-1}\dot{\boldsymbol{\Sigma}}(\boldsymbol{\theta}_0)(\mathbf{V}^* \otimes \mathbf{V}^*) - \boldsymbol{\Omega}(\boldsymbol{\Sigma}_0^{-1})^{-1}\dot{\boldsymbol{\Sigma}}(\boldsymbol{\theta}_0)(\boldsymbol{\Sigma}_0^{-1} \otimes \boldsymbol{\Sigma}_0^{-1})\}n^{1/2}\,\mathrm{vec}\,(\mathbf{S} - \boldsymbol{\Sigma}_0).$$

Since $\{\boldsymbol{\Omega}(V^*)^{-1}\dot{\boldsymbol{\Sigma}}(\boldsymbol{\theta}_0)(\mathbf{V}^* \otimes \mathbf{V}^*) - \boldsymbol{\Omega}(\boldsymbol{\Sigma}_0^{-1})^{-1}\dot{\boldsymbol{\Sigma}}(\boldsymbol{\theta}_0)(\boldsymbol{\Sigma}_0^{-1} \otimes \boldsymbol{\Sigma}_0^{-1})\}$ converges to zero in probability and $n^{1/2}\mathrm{vec}(\mathbf{S} - \boldsymbol{\Sigma}_0)$ converges in distribution, $n^{1/2}(\tilde{\boldsymbol{\theta}} - \tilde{\boldsymbol{\theta}}_M)$ converges to zero in probability. Therefore for the 'best' GLS estimator $\tilde{\boldsymbol{\theta}}$ with $\mathbf{V}^* = \boldsymbol{\Sigma}_0^{-1}$, the following result is established.

Theorem 3.5 The 'best' GLS estimator and the ML estimator are asymptotically equivalent in the sense that

$$n^{1/2}(\tilde{\boldsymbol{\theta}} - \tilde{\boldsymbol{\theta}}_M) \xrightarrow{P} \mathbf{0}. \tag{3.23}$$

Corollary 3.2

(i) $n^{1/2}(\tilde{\boldsymbol{\theta}}_M - \boldsymbol{\theta}_0) \xrightarrow{L} N[\mathbf{0}, 2\boldsymbol{\Omega}(\boldsymbol{\Sigma}_0^{-1})^{-1}]$.

(ii) $nG(\tilde{\boldsymbol{\theta}}_M) \xrightarrow{L} \chi^2_{p^*-q}$.

(iii) $nF(\tilde{\boldsymbol{\theta}}) \xrightarrow{L} \chi^2_{p^*-q}$.

The proof of Corollary 3.2 is given in Appendix 3.3. Results in this corollary indicate that the ML estimator and the GLS estimator are closely related. Theoretically, we may use either the ML or the GLS estimator and its asymptotic properties in analyzing the model.

The derivations of the asymptotic properties of the GLS and ML approaches heavily depend on the validity of the result about the asymptotic distribution of the sample covariance matrix as given in (3.5). Thus, to apply these asymptotic properties in practice, we have to check whether the following related conditions are satisfied: (i) the data are iid, (ii) the data are coming from a normal distribution, and (iii) the sample size is large enough. Clearly, these properties are invalid for data that are not identically or not independently distributed. For some non-normal distributions, the asymptotic covariance matrix of S may heavily depend on the fourth-order cumulant (see Section 3.5), and hence may be very different from $2\mathbf{K}_p^T(\boldsymbol{\Sigma}_0 \otimes \boldsymbol{\Sigma}_0)\mathbf{K}_p$. Hence, as expected, the ML and GLS approaches are not robust to the normal assumption. Moreover, the ML approach (or GLS approach) is not robust to small sample sizes, as demonstrated by many simulation studies (see for example, Lee and Song, 2004). The demand for rigorous and efficient statistical methods for analyzing the complex non-normal data motivates the rapid growth of SEM.

3.5 ASYMPTOTICALLY DISTRIBUTION-FREE METHODS

To address the problem on the violation of the assumption about multivariate normality, Browne (1984) developed an asymptotically distribution-free (ADF) approach that does not require this assumption. The main ideas of this approach are briefly discussed in this section.

Suppose $\mathbf{x}_1, \ldots, \mathbf{x}_n$ are iid observations that are sampled from a distribution, not necessarily multivariate normal, with a mean vector $\boldsymbol{\mu}$, a covariance matrix $\boldsymbol{\Sigma}_0 = \boldsymbol{\Sigma}(\boldsymbol{\theta}_0)$, and finite eighth-order moments. Analysis of the covariance structure is again on the basis of the sample covariance matrix $\mathbf{S}$. The finite sample distribution of $n^{1/2}\text{vecs}(\mathbf{S} - \boldsymbol{\Sigma}_0)$ has a zero mean vector and a covariance matrix with typical element (see Kendall and Stuart, 1969, Section 13.11),

$$\begin{aligned} &\text{Cov}[(n-1)^{1/2}\{\mathbf{S}(i,j) - \boldsymbol{\Sigma}_0(i,j)\}, (n-1)^{1/2}\{\mathbf{S}(k,h) - \boldsymbol{\Sigma}_0(k,h)\}] \\ = \;& \boldsymbol{\Sigma}_0(i,k)\boldsymbol{\Sigma}_0(j,h) + \boldsymbol{\Sigma}_0(i,h)\boldsymbol{\Sigma}_0(j,k) + (n-1)n^{-1}\kappa_{ijkh}, \end{aligned}$$

where κ_{ijkh} is a fourth-order cumulant which is given by

$$\kappa_{ijkh} = \delta_{ijkh} - \boldsymbol{\Sigma}_0(i,j)\boldsymbol{\Sigma}_0(k,h) - \boldsymbol{\Sigma}_0(i,k)\boldsymbol{\Sigma}_0(j,h) - \boldsymbol{\Sigma}_0(i,h)\boldsymbol{\Sigma}_0(j,k)$$

with

$$\delta_{ijkh} = E\{(x(i) - \mu(i)\}\{x(j) - \mu(j)\}\{x(k) - \mu(k)\}\{x(h) - \mu(h)\}.$$

According to the multivariate central limit theorem (see Anderson, 1984, Chapter 3),

$$n^{1/2}\text{vecs}(\mathbf{S} - \boldsymbol{\Sigma}_0) \xrightarrow{L} N[\mathbf{0}, \boldsymbol{\Sigma}^*], \tag{3.24}$$

where

$$\begin{aligned} \boldsymbol{\Sigma}^*(ij, kh) &= \boldsymbol{\Sigma}_0(i,k)\boldsymbol{\Sigma}_0(j,h) + \boldsymbol{\Sigma}_0(i,h)\boldsymbol{\Sigma}_0(j,k) + \kappa_{ijkh} \\ &= \delta_{ijkh} - \boldsymbol{\Sigma}_0(i,j)\boldsymbol{\Sigma}_0(k,h). \end{aligned} \tag{3.25}$$

The asymptotic distribution of $\mathbf{S}$ as given in (3.24) is similar to that given in (3.5). The difference is on their asymptotic covariance matrices. Under the multivariate normality assumption, the fourth-order cumulants κ_{ijkh} are equal to zero, hence $\boldsymbol{\Sigma}^*(ij, kh)$ reduces to $\boldsymbol{\Sigma}_0(i,k)\boldsymbol{\Sigma}_0(j,h) + \boldsymbol{\Sigma}_0(i,h)\boldsymbol{\Sigma}_0(j,k)$ (in matrix form, $\boldsymbol{\Sigma}^* = 2\mathbf{K}_p^T(\boldsymbol{\Sigma}_0 \otimes \boldsymbol{\Sigma}_0)\mathbf{K}_p)$ which has the same form as given by (3.5). In general, the convergence in distribution that is based on the general case given in (3.24) is much slower than that in (3.5) under the multivariate normal distributions. Similar to the normal theory, the asymptotic distribution of $\mathbf{S}$ as given by (3.24) and (3.25) is very important in the development of the statistical properties of the ADF method.

The ADF estimator $\tilde{\boldsymbol{\theta}}_A$ of $\boldsymbol{\theta}_0$ under the general situation is defined as the vector that minimizes the following GLS function

$$G_A(\boldsymbol{\theta}) = 2^{-1}[\text{vecs}\{\mathbf{S} - \boldsymbol{\Sigma}(\boldsymbol{\theta})\}]^T\mathbf{W}^{-1}[\text{vecs}\{\mathbf{S} - \boldsymbol{\Sigma}(\boldsymbol{\theta})\}], \tag{3.26}$$

where $\mathbf{W}$ is a positive definite stochastic weight matrix. Here, we just consider the $\mathbf{W}$ which converges to $\boldsymbol{\Sigma}^*$ in probability. A natural candidate of $\mathbf{W}$ is the matrix with typical element

$$\mathbf{W}(ij, kh) = s_{ijkh} - \mathbf{S}(i, j)\mathbf{S}(k, h), \tag{3.27}$$

where

$$s_{ijkh} = n^{-1}\sum_{t=1}^{n}\{x_t(i) - \bar{x}(i)\}\{x_t(j) - \bar{x}(j)\}\{x_t(k) - \bar{x}(k)\}\{x_t(h) - \bar{x}(h)\},$$

which is the sample fourth-order moment about the mean. This matrix is positive definite with probability one.

Using similar arguments to those used in Section 3.3 and Appendix 3.3, it can be shown that the following asymptotic properties relating to $\tilde{\boldsymbol{\theta}}_A$ are valid.

Theorem 3.6 The ADF estimator $\tilde{\boldsymbol{\theta}}_A$ of $\boldsymbol{\theta}_0$ has the following asymptotic properties:

(i) $\tilde{\boldsymbol{\theta}}_A$ is consistent, that is, $\tilde{\boldsymbol{\theta}}_A \xrightarrow{P} \boldsymbol{\theta}_0$.

(ii) $n^{1/2}(\tilde{\boldsymbol{\theta}}_A - \boldsymbol{\theta}_0) \xrightarrow{L} N[\mathbf{0}, \mathbf{C}_A(\boldsymbol{\theta}_0)]$, where $\mathbf{C}_A(\boldsymbol{\theta}_0) = \{\dot{\boldsymbol{\Sigma}}(\boldsymbol{\theta}_0)\boldsymbol{\Sigma}^{*-1}\dot{\boldsymbol{\Sigma}}(\boldsymbol{\theta}_0)^T\}^{-1}$.

(iii) $nG_A(\tilde{\boldsymbol{\theta}}_A) \xrightarrow{L} \chi^2_{p*-q}$.

The goodness-of-fit of the hypothesized model can be tested through (iii) of Theorem 3.6 in a standard manner as before. Standard error estimates of elements of $\tilde{\boldsymbol{\theta}}_A$ may be obtained from the square roots of the corresponding diagonal elements of $n^{-1}\dot{\boldsymbol{\Sigma}}(\tilde{\boldsymbol{\theta}}_A)\,\tilde{\boldsymbol{\Sigma}}^{*-1}\dot{\boldsymbol{\Sigma}}(\tilde{\boldsymbol{\theta}}_A)^T$ with a consistent estimator $\tilde{\boldsymbol{\Sigma}}^*$ of $\boldsymbol{\Sigma}^*$. Both the LISREL and EQS6 programs have options for analyzing standard

SEMs via the ADF method. Practically, as pointed out by Hu, Bentler and Kano (1992), and Bentler and Dudgeon (1996), the ADF theory requires extremely large sample sizes to attain its claimed asymptotic results.

The dimension of the weight matrix $\mathbf{W}$ in Equation (3.26) is p^* by p^*, hence the computational burden for obtaining $\tilde{\boldsymbol{\theta}}_A$ is rather heavy for large size problems. Bentler (1983) and Browne (1984) noted that this computational burden can be substantially alleviated if the sample observations are coming from an elliptical distribution. Under an elliptical distribution, the general element of the asymptotic covariance matrix $\boldsymbol{\Sigma}^*$ is

$$\boldsymbol{\Sigma}^*(ij, kh) = (\kappa+1)\{\boldsymbol{\Sigma}_0(i, k)\boldsymbol{\Sigma}_0(j, h)\} + \kappa\{\boldsymbol{\Sigma}_0(i, j)\boldsymbol{\Sigma}_0(k, h)\},$$

which depends on a relative kurtosis coefficient κ and elements of $\boldsymbol{\Sigma}_0$. In matrix form,

$$\boldsymbol{\Sigma}^* = \mathbf{K}_p^T\{2(\kappa+1)(\boldsymbol{\Sigma}_0 \otimes \boldsymbol{\Sigma}_0) + \kappa \operatorname{vec}(\boldsymbol{\Sigma}_0)\operatorname{vec}(\boldsymbol{\Sigma}_0)^T\}\mathbf{K}_p.$$

Let $\tilde{\kappa}$ and $\tilde{\boldsymbol{\Sigma}}$ be consistent estimates of κ and $\boldsymbol{\Sigma}_0$, respectively, a consistent estimate of $\boldsymbol{\Sigma}^*$ is

$$\mathbf{W} = \mathbf{K}_p^T\{2(\tilde{\kappa}+1)(\tilde{\boldsymbol{\Sigma}} \otimes \tilde{\boldsymbol{\Sigma}}) + \tilde{\kappa} \operatorname{vec}(\tilde{\boldsymbol{\Sigma}}) \operatorname{vec}(\tilde{\boldsymbol{\Sigma}})^T\}\mathbf{K}_p.$$

The corresponding GLS function is (see Bentler, 1983; Browne, 1984)

$$G_A(\boldsymbol{\theta}) = 2^{-1}(\tilde{\kappa}+1)^{-1} \operatorname{tr}[\{\mathbf{S} - \boldsymbol{\Sigma}(\boldsymbol{\theta})\}\tilde{\boldsymbol{\Sigma}}^{-1}]^2 - \frac{\tilde{\kappa}[\operatorname{tr}\{\mathbf{S} - \boldsymbol{\Sigma}(\boldsymbol{\theta})\}\tilde{\boldsymbol{\Sigma}}^{-1}]^2}{4(\tilde{\kappa}+1) + 2p\tilde{\kappa}(\tilde{\kappa}+1)}.$$

Clearly, a natural choice of $\tilde{\boldsymbol{\Sigma}}$ is $\mathbf{S}$. An estimate of κ that is proposed by Browne (1984) is

$$\hat{\kappa} = \{np(p+2)\}^{-1}\sum_{i=1}^{n}\{(\mathbf{x}_i - \bar{\mathbf{x}})^T\mathbf{S}^{-1}(\mathbf{x}_i - \bar{\mathbf{x}})\}^2 - 1.$$

Other estimates of κ are available in Bentler (1992). The above $G_A(\boldsymbol{\theta})$ just involves scalars and p by p symmetric matrices. Computational efforts and computer storage are greatly reduced. The EQS6 (Bentler and Wu, 2002) program has an option to compute GLS estimates under elliptical distributions. As the elliptical distribution is symmetrical and has longer tails, the option developed under this distribution is effective in handling data with heavy tails. However, it may not be effective in handling skewed data. Compared with the general ADF method, the method based on an elliptical distribution requires a relatively smaller sample size to achieve the asymptotic results.

3.6 SOME ITERATIVE PROCEDURES

In general, there are many iterative procedures which could be considered for minimizing an objective function. Here, we shall restrict our attention to those related to the LISREL (Jöreskog and Sörbom, 1996) and EQS6 (Bentler and Wu, 2002); namely the Newton–Raphson, Scoring, Fletcher–Powell, and Gauss–Newton algorithms. The Newton–Raphson and the Fletcher–Powell algorithms can be used for minimizing any arbitrary objective functions, the Scoring algorithm is most suitable for minimizing the likelihood function $F(\boldsymbol{\theta})$, while the Gauss–Newton algorithm is most suitable for minimizing the GLS function $G(\boldsymbol{\theta})$.

In general, suppose $\boldsymbol{\theta}$ is a vector of unknown parameters and we want to estimate $\boldsymbol{\theta}$ by minimizing a real-valued objective function $Q(\boldsymbol{\theta})$. The main purpose of nonlinear programming is to locate $\tilde{\boldsymbol{\theta}}$, the value of $\boldsymbol{\theta}$ for which $Q(\boldsymbol{\theta})$ is smallest. We start with an initial guess $\boldsymbol{\theta}^{(1)}$ and iteratively generate a sequence $\boldsymbol{\theta}^{(2)}, \boldsymbol{\theta}^{(3)}, \ldots$ which hopefully will converge to $\tilde{\boldsymbol{\theta}}$. We will call the computation of $\boldsymbol{\theta}^{(i+1)}$ the ith iteration, the vector $\boldsymbol{\theta}^{(i)}$ the ith iterate, and

$$\Delta\boldsymbol{\theta}^{(i)} = \boldsymbol{\theta}^{(i+1)} - \boldsymbol{\theta}^{(i)} \tag{3.28}$$

the ith step. If $Q(\boldsymbol{\theta}^{(i+1)})$ is less than $Q(\boldsymbol{\theta}^{(i)})$, then the ith step is said to be acceptable.

All the algorithms considered in this chapter will be of the following form:

(i) Set $i = 1$, guess $\boldsymbol{\theta}^{(1)}$.

(ii) At the ith iteration, determine a scalar ρ_i and a positive definite matrix $\mathbf{R}^{(i)}$ such that $\Delta\boldsymbol{\theta}^{(i)} = -\rho^{(i)}\mathbf{R}^{(i)}\dot{\mathbf{Q}}(\boldsymbol{\theta}^{(i)})$ is an acceptable step, where $\dot{\mathbf{Q}}(\boldsymbol{\theta}^{(i)})$ is the gradient vector of $Q(\boldsymbol{\theta})$ at $\boldsymbol{\theta}^{(i)}$.

(iii) Check convergence. If the algorithm is not yet converged, go to step (ii) and continue.

The Newton–Raphson algorithm is to choose the inverse of the Hessian matrix, $\ddot{\mathbf{Q}}(\boldsymbol{\theta}^{(i)})^{-1}$, for $\mathbf{R}^{(i)}$; hence its ith step is defined by

$$\Delta\boldsymbol{\theta}^{(i)} = -\rho^{(i)}\ddot{\mathbf{Q}}(\boldsymbol{\theta}^{(i)})^{-1}\dot{\mathbf{Q}}(\boldsymbol{\theta}^{(i)}). \tag{3.29}$$

If $Q(\boldsymbol{\theta})$ is quadratic, this algorithm converges in one iteration from any starting value. For general functions, this method gives quadratic convergence near the minimum. However, for poor starting values, $\ddot{\mathbf{Q}}(\boldsymbol{\theta})$ may not be positive definite. Greenstadt (1967) modified this algorithm by replacing the negative definitive Hessian matrix with some positive definite matrix. In practice, we can use a positive definite matrix for $\mathbf{R}^{(i)}$, such as the inverse of $E(\ddot{\mathbf{Q}}(\boldsymbol{\theta}))$ or the identity matrix, at the first few iterations and then switch to $\ddot{\mathbf{Q}}(\boldsymbol{\theta})^{-1}$ near the minimum, for achieving quicker convergence.

Another major disadvantage of the Newton–Raphson algorithm is that it requires the second-order partial derivatives of $Q(\boldsymbol{\theta})$. Sometimes these derivatives are difficult to obtain and take a long time to compute. Some algorithms, which use only first-order derivatives, will be considered below.

The well-known LISREL program uses the Fletcher–Powell algorithm (Fletcher and Powell, 1963) to minimize the objective function. Steps of this algorithm are:

(i) Start with $\boldsymbol{\theta}^{(1)}$ and any positive definite matrix $\mathbf{A}^{(1)}$. Set $i = 1$.

(ii) At the ith iteration, choose $\rho^{(i)}$, the value of ρ such that $Q(\boldsymbol{\theta}^{(i)} - \rho\mathbf{A}^{(i)}\dot{\mathbf{Q}}(\boldsymbol{\theta}^{(i)}))$ is minimized with respect to ρ.

(iii) Obtain $\boldsymbol{\theta}^{(i+1)} = \boldsymbol{\theta}^{(i)} - \rho^{(i)}\mathbf{A}^{(i)}\dot{\mathbf{Q}}(\boldsymbol{\theta}^{(i)}), \Delta\boldsymbol{\theta}^{(i)}, \dot{\mathbf{Q}}(\boldsymbol{\theta}^{(i+1)})$, and $\Delta\dot{\mathbf{Q}}^{(i)} = \dot{\mathbf{Q}}(\boldsymbol{\theta}^{(i+1)}) - \dot{\mathbf{Q}}(\boldsymbol{\theta}^{(i)})$.

(iv) Set $\mathbf{A}^{(i+1)} = \mathbf{A}^{(i)} + \Delta\mathbf{A}_i^{(i)}$ and $i = i + 1$, where

$$\Delta\mathbf{A}^{(i)} = \frac{1}{\Delta\boldsymbol{\theta}^{(i)^T}\Delta\dot{\mathbf{Q}}^{(i)}}\Delta\boldsymbol{\theta}^{(i)}\Delta\boldsymbol{\theta}^{(i)^T} - \frac{1}{\Delta\dot{\mathbf{Q}}^{(i)^T}\mathbf{A}^{(i)}\Delta\dot{\mathbf{Q}}^{(i)}}\mathbf{A}^{(i)}\Delta\dot{\mathbf{Q}}^{(i)}\Delta\dot{\mathbf{Q}}^{(i)^T}\mathbf{A}^{(i)}.$$

(v) Check convergence. If the algorithm is not yet converged, go to step (ii) and continue.

The initial matrix $\mathbf{A}^{(1)}$ can be chosen to be any positive definite symmetric matrix that approximates the inverse of the Hessian matrix. If we do not have such an approximation, the identity matrix may be used. Fletcher and Powell (1963) showed that if $\mathbf{A}^{(i)}$ is positive definite, then $\mathbf{A}^{(i+1)}$ is positive definite. Hence, starting with a positive definite matrix $\mathbf{A}^{(i)}$ always gives an acceptable step. Moreover, if $Q(\boldsymbol{\theta})$ is a quadratic function, then

$$\mathbf{A}^{(q+1)} = \ddot{\mathbf{Q}}(\boldsymbol{\theta})^{-1}. \tag{3.30}$$

It should be noted that the above results depend on the optional choice of $\rho^{(i)}$. For minimizing $Q(\boldsymbol{\theta})$, a highly accurate of $\rho^{(i)}$ is not important. For example, Jöreskog (1967) used the 'extrapolation and interpolation' technique of Davidon (1959) for getting $\rho^{(i)}$.

Equation (3.30) motivates using $2n^{-1}\tilde{\mathbf{A}}$ to estimate the asymptotic covariance matrix and standard errors of $\tilde{\boldsymbol{\theta}}$ or $\tilde{\boldsymbol{\theta}}_M$, where $\tilde{\mathbf{A}}$ denotes the converged value of the matrix $\mathbf{A}$ in the Fletcher–Powell algorithm. However, Lee and Jennrich

(1979) showed that this method for approximating the asymptotic covariance matrix or the standard errors fails to give answers consistent with the appropriate Hessian matrix or the information matrix. In the LISREL program, standard errors are computed via the inverse of the information matrix.

The scoring algorithm can be applied when $Q(\boldsymbol{\theta})$ depends on observed values of random variables. In this algorithm, the Hessian matrix is replaced by its expectation. Hence

$$\Delta\boldsymbol{\theta}^{(i)} = -\rho^{(i)}[E\{\ddot{\mathbf{Q}}(\boldsymbol{\theta}^{(i)})\}]^{-1}\dot{\mathbf{Q}}(\boldsymbol{\theta}^{(i)}). \tag{3.31}$$

This algorithm is usually more robust to bad starting values, because $E\{\ddot{\mathbf{Q}}(\boldsymbol{\theta}^{(i)})\}$ is often positive definite. Moreover, as $E\{\ddot{\mathbf{Q}}(\boldsymbol{\theta}^{(i)})\}$ usually just involves first partial derivatives, it requires less computing time per iteration than the Newton–Raphson algorithm.

The Gauss–Newton algorithm is particularly attractive for minimizing objective functions of the form

$$Q(\boldsymbol{\theta}) = 2^{-1}\mathbf{e}(\boldsymbol{\theta})^T\mathbf{W}\mathbf{e}(\boldsymbol{\theta}),$$

where $\mathbf{e}(\boldsymbol{\theta})$ is a column vector of residuals and $\mathbf{W}$ is a weight matrix. From the product rule of matrix calculus,

$$\begin{aligned}
\dot{\mathbf{Q}}(\boldsymbol{\theta}) &= 2^{-1}[\dot{\mathbf{e}}(\boldsymbol{\theta})\{1 \otimes \mathbf{W}\mathbf{e}(\boldsymbol{\theta})\} + \dot{\mathbf{e}}(\boldsymbol{\theta})\{\mathbf{W}\mathbf{e}(\boldsymbol{\theta}) \otimes 1\}] = \dot{\mathbf{e}}(\boldsymbol{\theta})\mathbf{W}\mathbf{e}(\boldsymbol{\theta}) \\
\ddot{\mathbf{Q}}(\boldsymbol{\theta}) &= 2^{-1}\frac{\partial}{\partial\boldsymbol{\theta}}[\dot{\mathbf{e}}(\boldsymbol{\theta})\{1 \otimes \mathbf{W}\mathbf{e}(\boldsymbol{\theta})\} + \dot{\mathbf{e}}(\boldsymbol{\theta})\{\mathbf{W}\mathbf{e}(\boldsymbol{\theta}) \otimes 1\}] \\
&= \dot{\mathbf{e}}(\boldsymbol{\theta})\mathbf{W}\dot{\mathbf{e}}(\boldsymbol{\theta})^T + \ddot{\mathbf{e}}(\boldsymbol{\theta})\mathbf{W}\mathbf{e}(\boldsymbol{\theta}).
\end{aligned}$$

Since $\mathbf{e}(\boldsymbol{\theta})$ is usually small, particularly near the minimum, the last term in $\ddot{\mathbf{Q}}(\boldsymbol{\theta})$ is neglected in the Gauss–Newton algorithm. Let $\mathbf{U}(\boldsymbol{\theta}) = \dot{\mathbf{e}}(\boldsymbol{\theta})\mathbf{W}\dot{\mathbf{e}}(\boldsymbol{\theta})^T$, the ith step of the Gauss–Newton algorithm is defined by

$$\Delta\boldsymbol{\theta}^{(i)} = -\rho^{(i)}\mathbf{U}(\boldsymbol{\theta}^{(i)})^{-1}\dot{\mathbf{Q}}(\boldsymbol{\theta}^{(i)}). \tag{3.32}$$

In the scoring algorithm or the Gauss–Newton algorithm, the step size $\rho^{(i)}$ is commonly chosen via step-halving, that is, using the first value in the sequence $\{1, 2^{-1}, 2^{-2}, \ldots\}$ that provides an acceptable step.

When applying the Fletcher–Powell algorithm to the GLS and ML approaches of the covariance structure analysis under the multivariate normal assumption, $\dot{\mathbf{Q}}(\boldsymbol{\theta}) = \dot{\mathbf{G}}(\boldsymbol{\theta})$ or $\dot{\mathbf{Q}}(\boldsymbol{\theta}) = \dot{\mathbf{F}}(\boldsymbol{\theta})$. The scoring algorithm for the ML estimation becomes (see Equations (3.18) and (3.19))

$$\triangle\boldsymbol{\theta}^{(i)} = \rho^{(i)}\{\dot{\boldsymbol{\Sigma}}(\boldsymbol{\theta})(\boldsymbol{\Sigma}^{-1}\otimes\boldsymbol{\Sigma}^{-1})\dot{\boldsymbol{\Sigma}}(\boldsymbol{\theta})^T\}^{-1}\dot{\boldsymbol{\Sigma}}(\boldsymbol{\theta})(\boldsymbol{\Sigma}^{-1}\otimes\boldsymbol{\Sigma}^{-1})\mathrm{vec}(\mathbf{S}-\boldsymbol{\Sigma})\Big|_{\boldsymbol{\theta}=\boldsymbol{\theta}^{(i)}}. \tag{3.33}$$

For the Gauss–Newton algorithm in the GLS estimation, $\mathbf{e}(\boldsymbol{\theta}) = \mathrm{vecs}(\mathbf{S}-\boldsymbol{\Sigma})$ and $\mathbf{W} = (\mathbf{V}\otimes\mathbf{V})$, see Equation (3.9). Hence, from Equations (3.10) and (3.11),

$$\triangle\boldsymbol{\theta}^{(i)} = \rho^{(i)}\{\dot{\boldsymbol{\Sigma}}(\boldsymbol{\theta})(\mathbf{V}\otimes\mathbf{V})\dot{\boldsymbol{\Sigma}}(\boldsymbol{\theta})^T\}^{-1}\dot{\boldsymbol{\Sigma}}(\boldsymbol{\theta})(\mathbf{V}\otimes\mathbf{V})\mathrm{vec}(\mathbf{S}-\boldsymbol{\Sigma})\Big|_{\boldsymbol{\theta}=\boldsymbol{\theta}^{(i)}}. \tag{3.34}$$

Comparing Equations (3.33) and (3.34), we see that the scoring algorithm is an iteratively reweighted Gauss–Newton algorithm, that is, one in which the weight matrix $\mathbf{V} = \boldsymbol{\Sigma}^{-1}$ changes with $\boldsymbol{\theta}$ from iteration to iteration. This means that the Gauss–Newton algorithm may be used for both GLS and ML estimations. Standard errors for the estimates are obtained from $E\{\ddot{\mathbf{F}}(\tilde{\boldsymbol{\theta}})\}^{-1}$ or $\mathbf{U}(\tilde{\boldsymbol{\theta}})^{-1}$, which are by-products of the scoring algorithm or the Gauss–Newton algorithm. The well-known EQS program uses the Gauss–Newton algorithm for obtaining the GLS and ML solutions.

To apply the Fletcher–Powell algorithm to the GLS approach under ADF theory with objective function $G_A(\boldsymbol{\theta})$, $\dot{\mathbf{Q}}(\boldsymbol{\theta}) = \dot{\mathbf{G}}_A(\boldsymbol{\theta}) = -\dot{\boldsymbol{\Sigma}}(\boldsymbol{\theta})\mathbf{W}^{-1}\mathrm{vec}(\mathbf{S}-\boldsymbol{\Sigma})$. When applying the Gauss–Newton algorithm to ADF estimation, Equation (3.32) becomes

$$\triangle\boldsymbol{\theta}^{(i)} = -\rho^{(i)}\{\dot{\boldsymbol{\Sigma}}(\boldsymbol{\theta})\mathbf{W}^{-1}\dot{\boldsymbol{\Sigma}}(\boldsymbol{\theta})^T\}^{-1}\dot{\mathbf{G}}_A(\boldsymbol{\theta})\Big|_{\boldsymbol{\theta}=\boldsymbol{\theta}^{(i)}}. \tag{3.35}$$

3.6.1 Numerical Examples

In this section, we give an example to illustrate the Fletcher–Powell (FP) algorithm that is used in the LISREL program, and the Gauss–Newton (GN) algorithm and its iteratively reweighted mode that are used in the EQS program for obtaining the GLS and ML estimates. Although these algorithms can be applied to more general models, for simplicity, only the exploratory factor analysis defined by Equation (2.2) is considered. Hence the parameter vector $\boldsymbol{\theta}$ contains the unknown parameters in $\boldsymbol{\Lambda}$ and diag $\boldsymbol{\Psi}_\epsilon$ (the vector that contains the diagonal elements in $\boldsymbol{\Psi}_\epsilon$).

Based on the matrix calculus given in Appendix 3.1, the gradient vector of the objective functions are given by

$$\dot{\mathbf{F}}(\boldsymbol{\theta}) = \begin{bmatrix} \dfrac{\partial F(\boldsymbol{\theta})}{\partial \boldsymbol{\Lambda}} \\ \dfrac{\partial F(\boldsymbol{\theta})}{\partial \text{diag}\boldsymbol{\Psi}_\epsilon} \end{bmatrix} = -\begin{bmatrix} 2\ \text{vec}\{\boldsymbol{\Sigma}^{-1}(\mathbf{S}-\boldsymbol{\Sigma})\boldsymbol{\Sigma}^{-1}\boldsymbol{\Lambda}\} \\ \mathbf{C}_p\{\boldsymbol{\Sigma}^{-1}(\mathbf{S}-\boldsymbol{\Sigma})\boldsymbol{\Sigma}^{-1}\} \end{bmatrix},$$

$$\dot{\mathbf{Q}}(\boldsymbol{\theta}) = \begin{bmatrix} \dfrac{\partial Q(\boldsymbol{\theta})}{\partial \boldsymbol{\Lambda}} \\ \dfrac{\partial Q(\boldsymbol{\theta})}{\partial \text{diag}\boldsymbol{\Psi}_\epsilon} \end{bmatrix} = -\begin{bmatrix} 2\ \text{vec}\{\mathbf{V}(\mathbf{S}-\boldsymbol{\Sigma})\mathbf{V}\boldsymbol{\Lambda}\} \\ \mathbf{C}_p(\mathbf{V}(\mathbf{S}-\boldsymbol{\Sigma})\mathbf{V}\} \end{bmatrix}.$$

The implementation of the FP algorithm and the GN algorithm requires these first derivatives. It can be shown that the matrix $\mathbf{U}(\boldsymbol{\theta})$ required by the GN algorithm (see Equation (3.32)) is given by

$$\mathbf{U}(\boldsymbol{\theta}) = \begin{bmatrix} 4(\mathbf{V}\otimes\boldsymbol{\Lambda}^T\mathbf{V})\mathbf{M}_p(\mathbf{I}\otimes\boldsymbol{\Lambda}) & 2(\mathbf{V}\otimes\boldsymbol{\Lambda}^T\mathbf{V})\mathbf{C}_p^T \\ 2\mathbf{C}_p(\mathbf{V}\otimes\mathbf{V}\boldsymbol{\Lambda}^T) & \mathbf{V}*\mathbf{V} \end{bmatrix}$$

where $\mathbf{V}*\mathbf{V}$ is the Hadamard product of $\mathbf{V}$ which is a $p \times p$ matrix with the general (i, j)th entry v_{ij}^2. For the fixed parameters in $\boldsymbol{\Lambda}$, the corresponding rows in $\dot{\mathbf{F}}(\boldsymbol{\theta})$ and $\dot{\mathbf{Q}}(\boldsymbol{\theta})$, and the corresponding rows and columns of $U(\boldsymbol{\theta})$ are removed in the implemention. In the illustration, we use $\mathbf{V} = \mathbf{S}^{-1}$ in analyzing the Emmett (1949) data which consist of nine variables on a sample of 211 subjects. Starting values are obtained by the principal component factor analysis (see e.g. Afifi and Azen, 1974). The data set is fitted with a three-factor model, and the upper triangle elements of $\boldsymbol{\Lambda}$ are fixed at the starting values for identifying the model. We say that an algorithm has been converged if the root mean squares of the gradient vector or the root mean squares of $\Delta\boldsymbol{\theta}$ is less than 0.0001.

In obtaining the GLS estimates, the GN algorithm and the FP algorithm converge quickly; the root mean square of the gradient and the root mean square of $\boldsymbol{\Delta\theta}$ approach zero, and the function values converge to the same

Table 3.1 Convergence of the Gauss–Newton algorithm; on Emmett's (1949) data in the GLS estimation.

Iteration	Function value	RMS $\dot{\mathbf{Q}}(\boldsymbol{\theta})$	RMS $\Delta\boldsymbol{\theta}$	$\lambda_{1,1}$	$\lambda_{9,2}$	$\psi_{\epsilon 2}$
0	0.88830	0.4293	—	0.749	−0.285	0.337
1	0.63940	0.5467	0.2822	0.499	0.238	0.417
2	0.07278	0.0919	0.1229	0.595	0.414	0.413
3	0.03414	0.0123	0.0499	0.656	0.339	0.414
4	0.03324	0.0010	0.0145	0.661	0.340	0.416
5	0.03322	0.0003	0.0023	0.661	0.339	0.416
6	0.03322	0.0001	0.0012	0.662	0.339	0.416
7	0.03322	0.0000	0.0003	0.662	0.339	0.416

Taken from Lee and Jennrich (1979).

Table 3.2 Convergence of the Fletcher–Powell algorithm; on Emmett's (1949) data in the GLS estimation.

Iteration	Function value	RMS $\dot{\mathbf{Q}}(\boldsymbol{\theta})$	RMS $\Delta\boldsymbol{\theta}$	$\lambda_{1,1}$	$\lambda_{9,2}$	$\psi_{\epsilon 2}$
0	0.88830	0.4293	—	0.749	−0.285	0.337
5	0.20870	0.1028	0.0522	0.673	0.037	0.413
10	0.05665	0.0550	0.0143	0.697	0.260	0.407
15	0.03641	0.0207	0.0123	0.669	0.335	0.411
20	0.03363	0.0059	0.0032	0.660	0.342	0.414
25	0.03324	0.0021	0.0008	0.661	0.341	0.416
30	0.03322	0.0004	0.0002	0.662	0.341	0.416
35	0.03322	0.0001	0.0000	0.662	0.341	0.416

Taken from Lee and Jennrich (1979).

minimum value. To give some idea of the convergence, summaries of the algorithms are presented in Tables 3.1 and 3.2, respectively. The final parameter estimates obtained by these two algorithms are very close to each other; almost all of them are exactly the same in the first three decimal places. Hence, only the estimates and their standard errors estimates that are obtained via the GN algorithm are presented in Table 3.3. In obtaining the ML estimates, the iteratively reweighted GN algorithm and the FP algorithm converged in six and 32 iterations. The similar convergence summaries are not presented. The ML estimates obtained by the two algorithms are very close, hence only those obtained via the GN algorithm are presented in Table 3.4, together with the standard errors estimates. As expected, the asymptotically equivalent GLS and ML results are very close.

Table 3.3 The GLS solution by the GN algorithm; standard error estimates are in parentheses and parameters with an asterisk are fixed.

$\hat{\Lambda}$			$\hat{\Psi}_\epsilon$
0.66(0.07)	0.32*	−0.08*	0.45(0.05)
0.62(0.20)	0.39(0.34)	0.19*	0.42(0.05)
0.55(0.20)	0.20(0.37)	0.22(0.16)	0.60(0.07)
0.29(0.18)	0.84(0.11)	0.04(0.83)	0.21(0.04)
0.19(0.24)	0.75(0.23)	0.16(0.79)	0.37(0.05)
0.21(0.12)	0.88(0.12)	0.11(0.94)	0.17(0.05)
0.72(0.11)	0.27(0.21)	0.07(0.17)	0.39(0.05)
0.51(0.30)	0.18(0.52)	−0.48(0.22)	0.47(0.15)
0.81(0.09)	0.34(0.14)	0.00(0.15)	0.23(0.04)

Table 3.4 The ML solution obtained by iteratively reweighted GN algorithm; standard error estimates are in parentheses and parameters with an asterisk are fixed.

$\hat{\Lambda}$			$\hat{\Psi}_\epsilon$
0.66(0.07)	0.32*	−0.08*	0.45(0.05)
0.62(0.19)	0.39(0.31)	0.19*	0.43(0.05)
0.54(0.18)	0.20(0.33)	0.22(0.15)	0.62(0.07)
0.29(0.16)	0.84(0.09)	0.03(0.78)	0.21(0.04)
0.19(0.21)	0.75(0.20)	0.15(0.74)	0.38(0.05)
0.23(0.11)	0.88(0.12)	−0.11(0.88)	0.18(0.05)
0.72(0.11)	0.27(0.20)	0.08(0.16)	0.40(0.05)
0.52(0.29)	0.17(0.50)	−0.49(0.24)	0.46(0.17)
0.81(0.09)	0.34(0.14)	0.01(0.15)	0.23(0.04)

APPENDIX 3.1: MATRIX CALCULUS

There are many methods for obtaining derivatives of a matrix of function with respect to a matrix of their variables. The method described here is an adaptation of McDonald and Swaminathan (1973).

Definition 3.1.1 Given a p by q matrix $\mathbf{Y} = (y_{ij})$ whose elements are differentiable functions of elements of an m by n matrix $\mathbf{X} = (x_{ij})$, the matrix derivatives of $\mathbf{Y}$ with respect to $\mathbf{X}$ is

$$\frac{\partial \mathbf{Y}}{\partial \mathbf{X}} = \begin{bmatrix} \frac{\partial y_{11}}{\partial x_{11}} \cdots \frac{\partial y_{1q}}{\partial x_{11}} \cdots \frac{\partial y_{p1}}{\partial x_{11}} \cdots \frac{\partial y_{pq}}{\partial x_{11}} \\ \vdots \quad \vdots \quad \vdots \quad \vdots \\ \frac{\partial y_{11}}{\partial x_{1n}} \cdots \frac{\partial y_{1q}}{\partial x_{1n}} \cdots \frac{\partial y_{p1}}{\partial x_{1n}} \cdots \frac{\partial y_{pq}}{\partial x_{1n}} \\ \vdots \quad \vdots \quad \vdots \quad \vdots \\ \frac{\partial y_{11}}{\partial x_{m1}} \cdots \frac{\partial y_{1q}}{\partial x_{m1}} \cdots \frac{\partial y_{p1}}{\partial x_{m1}} \cdots \frac{\partial y_{pq}}{\partial x_{m1}} \\ \vdots \quad \vdots \quad \vdots \quad \vdots \\ \frac{\partial y_{11}}{\partial x_{mn}} \cdots \frac{\partial y_{1q}}{\partial x_{mn}} \cdots \frac{\partial y_{p1}}{\partial x_{mn}} \cdots \frac{\partial y_{pq}}{\partial x_{mn}} \end{bmatrix}.$$

The order of $\partial \mathbf{Y}/\partial \mathbf{X}$ is mn by pq. It is obtained by differentiation of all elements of $\mathbf{Y}$, row by row sequentially, with respect to all elements of $\mathbf{X}$, also following the manner of row by row sequentially. Recall

$$\text{vec}\mathbf{X} = (x_{11}, \ldots, x_{1n}, x_{21}, \ldots, x_{2n}, \ldots x_{m1}, \ldots, x_{mn})^T,$$

it follows from the definition of $\partial \mathbf{Y}/\partial \mathbf{X}$ that

$$\frac{\partial \mathbf{Y}}{\partial \mathbf{X}} = \frac{\partial \mathbf{Y}}{\partial \text{vec}\mathbf{X}} = \frac{\partial \mathbf{Y}}{\partial (\text{vec}\mathbf{X})^T} = \frac{\partial \text{vec}\mathbf{Y}}{\partial \mathbf{X}} = \frac{\partial (\text{vec}\mathbf{Y})^T}{\partial \mathbf{X}}.$$

Examples. (1) Let $\mathbf{X}$ be a 3 by 2 matrix with distinct variables,

$$\mathbf{E}_{32} = \frac{\partial \mathbf{X}^T}{\partial \mathbf{X}} = \left[\begin{array}{ccc|ccc} 1 & 0 & 0 & 0 & 0 & 0 \\ 0 & 0 & 0 & 1 & 0 & 0 \\ \hline 0 & 1 & 0 & 0 & 0 & 0 \\ 0 & 0 & 0 & 0 & 1 & 0 \\ \hline 0 & 0 & 1 & 0 & 0 & 0 \\ 0 & 0 & 0 & 0 & 0 & 1 \end{array}\right]$$

The typical element of $\mathbf{E}_{mn}$ corresponding to an m by n matrix $\mathbf{X}$ is

$$\begin{aligned} \mathbf{E}_{mn}[n(j-1)+k,\, m(i-1)+h] &= 1 \text{ if } j = h, k = i, \\ &= 0 \quad \text{otherwise,} \end{aligned}$$

where $0 < j, h \leq m$, and $0 < i, k \leq n$.

(2) Let $\mathbf{X}$ be a 3 by 3 symmetric matrix with distinct lower-triangular variables,

$$\mathbf{T}_3 = \frac{\partial \mathbf{X}}{\partial \text{vecs}\mathbf{X}} = \left[\begin{array}{ccc|ccc|ccc} 1 & 0 & 0 & 0 & 0 & 0 & 0 & 0 & 0 \\ \hline 0 & 1 & 0 & 1 & 0 & 0 & 0 & 0 & 0 \\ 0 & 0 & 0 & 0 & 1 & 0 & 0 & 0 & 0 \\ \hline 0 & 0 & 1 & 0 & 0 & 0 & 1 & 0 & 0 \\ 0 & 0 & 0 & 0 & 0 & 1 & 0 & 1 & 0 \\ 0 & 0 & 0 & 0 & 0 & 0 & 0 & 0 & 1 \end{array}\right].$$

The typical element of $\mathbf{T}_n$ corresponding to an n by n symmetric matrix $\mathbf{X}$ is

$$\begin{aligned} \mathbf{T}_n[j(j-1)+k,\, n(i-1)+h] &= 1 \text{ if } j=i,\, k=h \text{ or } j=h,\, i=k, \\ &= 0 \quad \text{otherwise.} \end{aligned}$$

(3) Let $\mathbf{X}$ be a 3 by 3 diagonal matrix with distinct diagonal variables,

$$\mathbf{C}_3 = \frac{\partial \mathbf{X}}{\partial \text{diag}\mathbf{X}} = \left[\begin{array}{ccc|ccc|ccc} 1 & 0 & 0 & 0 & 0 & 0 & 0 & 0 & 0 \\ 0 & 0 & 0 & 0 & 1 & 0 & 0 & 0 & 0 \\ 0 & 0 & 0 & 0 & 0 & 0 & 0 & 0 & 1 \end{array}\right]$$

The general typical element of $\mathbf{C}_p$ corresponding to a p by p diagonal matrix $\mathbf{X}$ is

$$\begin{aligned} \mathbf{C}_p(i,j) &= 1 \text{ if } j = p(i-1)+i, \\ &= 0 \quad \text{otherwise.} \end{aligned}$$

Matrices $\mathbf{E}_{mn}$, $\mathbf{T}_n$ and $\mathbf{C}_n$ are common and useful for implementing iterative procedures for estimation of CSA. See Bentler and Lee (1975) for more properties of these matrices among themselves, with the Kronecker product, $\mathbf{K}_p$ and $\mathbf{M}_p$.

Now, let us recall the 'chain rule' in scalar calculus. Let $z = g(y_1, \ldots, y_p)$ be a differentiable function of $y_1, \ldots, y_p$, and for each $k = 1, \ldots, p$, let $y_k = f_k(x_1, \ldots, x_q)$ be a differentiable function of $x_1, \ldots, x_q$. From the theory of partial differentiation,

$$\frac{\partial z}{\partial x_i} = \frac{\partial g(y_1, \ldots, y_p)}{\partial x_i}$$

$$= \sum_{k=1}^{p} \frac{\partial f_k(x_1, \ldots, x_q)}{\partial x_i} \frac{\partial g(y_1, \ldots, y_p)}{\partial y_k}$$

$$= \sum_{k=1}^{p} \frac{\partial y_k}{\partial x_i} \frac{\partial z}{\partial y_k}. \tag{A3.1}$$

This is the 'chain rule' in scalar calculus. To apply it correctly, $y_1, \ldots, y_p$ are considered as mathematically independent variables in $\partial g(y_1, \ldots, y_p)/\partial y_k$. With this fact in mind, we now present the 'chain rule' in matrix calculus.

Theorem 3.1.1 (*The Chain Rule*). If elements of $\mathbf{Z}$ are differentiable functions of elements of $\mathbf{Y}$, and elements of $\mathbf{Y}$ are differentiable functions of elements of $\mathbf{X}$, then

$$\frac{\partial \mathbf{Z}}{\partial \mathbf{X}} = \frac{\partial \mathbf{Y}}{\partial \mathbf{X}} \frac{\partial \mathbf{Z}}{\partial \mathbf{Y}}. \tag{A3.2}$$

The proof is straightforward. It should be noted that in applying this rule, $\partial \mathbf{Z}/\partial \mathbf{Y}$ in Equation (A3.2) is calculated with elements of $\mathbf{Y}$ being treated as mathematically independent variables.

Theorem 3.1.2 (*The Product Rule*). If elements of $\mathbf{Y}(p \times r)$ and $\mathbf{Z}(r \times q)$ are differentiable functions of elements of $\mathbf{X}$, then

$$\frac{\partial \mathbf{YZ}}{\partial \mathbf{X}} = \frac{\partial \mathbf{Y}}{\partial \mathbf{X}}(\mathbf{I}_p \otimes \mathbf{Z}) + \frac{\partial \mathbf{Z}}{\partial \mathbf{X}}(\mathbf{Y}^T \otimes \mathbf{I}_q). \tag{A3.3}$$

The proof is straightforward. Some useful results are given below.

Theorem 3.1.3 Suppose elements of $\mathbf{Y}(p \times p)$ and $\mathbf{Z}(r \times m)$ are differentiable functions of elements of $\mathbf{X}$, then:

(i) $\dfrac{\partial \mathrm{tr}\mathbf{Y}}{\partial \mathbf{X}} = \dfrac{\partial \mathbf{Y}}{\partial \mathbf{X}} \mathrm{vec}\, \mathbf{I}_p.$

(ii) $\dfrac{\partial |\mathbf{Y}|}{\partial \mathbf{X}} = \dfrac{\partial \mathbf{Y}}{\partial \mathbf{X}} |\mathbf{Y}| \mathrm{vec}(\mathbf{Y}^{T-1})$, if $\mathbf{Y}$ is non-singular.

(iii) $\dfrac{\partial \log |\mathbf{Y}|}{\partial \mathbf{X}} = \dfrac{\partial \mathbf{Y}}{\partial \mathbf{X}} \mathrm{vec}(\mathbf{Y}^{T-1})$, if $\mathbf{Y}$ is non-singular.

(iv) $\dfrac{\partial \mathbf{Y}^{-1}}{\partial \mathbf{X}} = -\dfrac{\partial \mathbf{Y}}{\partial \mathbf{X}}(\mathbf{Y}^{T-1} \otimes \mathbf{Y}^{-1})$, if $\mathbf{Y}$ is non-singular.

(v) $\dfrac{\partial \mathbf{AZB}}{\partial \mathbf{X}} = \dfrac{\partial \mathbf{Z}}{\partial \mathbf{X}}(\mathbf{A}^T \otimes \mathbf{B})$, if $\mathbf{A}$ and $\mathbf{B}$ are constant matrices such that $\mathbf{AZB}$ is well defined.

(vi) $\dfrac{\partial \mathbf{ZAZ}^T}{\partial \mathbf{X}} = \dfrac{\partial \mathbf{Z}}{\partial \mathbf{X}}(\mathbf{I}_r \otimes \mathbf{AZ}^T) + \dfrac{\partial \mathbf{Z}^T}{\partial \mathbf{X}}(\mathbf{AZ}^T \otimes \mathbf{I}_r)$, if $\mathbf{A}(m \times m)$ is a constant symmetric matrix.

Proof. The proof of (i) is straightforward. To prove (ii) note that $|\mathbf{Y}| = \Sigma_{j=1}^p y_{ij} \Upsilon_{ij}$, where Υ_{ij} is the cofactor of y_{ij}, which is also equal to the (i, j)th entry of $|\mathbf{Y}|\mathbf{Y}^{T-1}$. Hence $\partial |\mathbf{Y}| / \partial y_{ij} = \Upsilon_{ij}$ and

$$\frac{\partial |\mathbf{Y}|}{\partial \mathbf{X}} = \frac{\partial \mathbf{Y}}{\partial \mathbf{X}} \mathrm{vec}(|\mathbf{Y}|\mathbf{Y}^{T-1}) = \frac{\partial \mathbf{Y}}{\partial \mathbf{X}} |\mathbf{Y}| \mathrm{vec}(\mathbf{Y}^{T-1}).$$

Assertion (iii) follows immediately from (ii). To prove (iv), it follows from $\mathbf{YY}^{-1} = \mathbf{I}$ that

$$\mathbf{0} = \frac{\partial \mathbf{YY}^{-1}}{\partial \mathbf{X}} = \frac{\partial \mathbf{Y}}{\partial \mathbf{X}}(\mathbf{I}_p \otimes \mathbf{Y}^{-1}) + \frac{\partial \mathbf{Y}^{-1}}{\partial \mathbf{X}}(\mathbf{Y}^T \otimes \mathbf{I}_p).$$

Hence,

$$\frac{\partial \mathbf{Y}^{-1}}{\partial \mathbf{X}} = -\frac{\partial \mathbf{Y}}{\partial \mathbf{X}}(\mathbf{I}_p \otimes \mathbf{Y}^{-1})(\mathbf{Y}^T \otimes \mathbf{I}_p)^{-1} = -\frac{\partial \mathbf{Y}}{\partial \mathbf{X}}(\mathbf{Y}^{T-1} \otimes \mathbf{Y}^{-1}).$$

Assertions (v) and (vi) follow directly from the product rule.

APPENDIX 3.2: SOME BASIC RESULTS IN PROBABILITY THEORY

This appendix contains the basic results in probability theory needed for deriving the asymptotic results of the GLS, ML and ADF estimators. Proofs are not presented, they can be found in other textbooks in probability theory or mathematical statistics such as Rao (1973, Chapter 2), among others.

Definition 3.2.1 Let $\{X_n\} = \{X_n; n = 1, 2, \ldots\}$ be a sequence of random variables.

(i) $\{X_n\}$ is said to converge to a constant c *in probability* if, for every $\varepsilon > 0$,

$$\lim_{n \to \infty} P(|X_n - c| > \varepsilon) = 0.$$

Such convergence is denoted by $X_n \xrightarrow{p} c$.

(ii) $\{X_n\}$ is said to converge to a random variable X *in probability* if the sequence of random variables $\{(X_n - X);\, n = 1, 2, \ldots\}$ converges to zero in probability. Such convergence is denoted by $X_n \xrightarrow{p} X$.

(iii) Let $\{F_n;\, n = 1, 2, \ldots\}$ be the sequence of distribution functions corresponding to $\{X_n\}$. Then $\{X_n\}$ is said to converge *in distribution* to a random variable X with distribution function F if

$$F_n \longrightarrow F \text{ as } n \longrightarrow \infty,$$

at all continuity points of F. Such convergence is denoted by $X_n \xrightarrow{L} X$. The approximating distribution is called the asymptotic distribution of X_n.

In most of our applications, the random variable X_n stands for a statistic computed from a sample of size n. Since the actual distribution of X_n is difficult to find, it is necessary to obtain its asymptotic distribution for statistical inferences. Some basic properties of the above defined convergences are presented via the following theorem. These properties are essential for obtaining the asymptotic distribution.

Theorem 3.2.1 Let $(X_n, Y_n);\, n = 1, 2, \ldots$ be a sequence of pairs of variables. Then

(i) $X_n \xrightarrow{L} X,\ Y_n \xrightarrow{p} 0 \Longrightarrow X_n Y_n \xrightarrow{p} 0.$

(ii) $X_n \xrightarrow{L} X,\ Y_n \xrightarrow{p} c \Longrightarrow X_n + Y_n \xrightarrow{L} X + c.$

(iii) $X_n \xrightarrow{p} X,\ Y_n \xrightarrow{p} Y \Longrightarrow X_n Y_n \xrightarrow{p} XY.$

(iv) $X_n - Y_n \xrightarrow{p} 0,\ X_n \xrightarrow{L} X \Longrightarrow Y_n \xrightarrow{L} X.$

(v) If $X_n \xrightarrow{p} X$ and $g(\cdot)$ is a continuous function, then

$$g(X_n) \xrightarrow{p} g(X).$$

Since $X \xrightarrow{L} X$ is always true, so if $X_n - X \xrightarrow{p} 0$, then it follows from (iv) that $X_n \xrightarrow{L} X$. Hence, convergence in probability implies convergence in distribution.

For simplicity, the above definitions and theorem are given in terms of random variables, but they are also valid for random vectors.

APPENDIX 3.3: PROOFS OF SOME RESULTS

PROOF OF LEMMA 4

$$
\begin{aligned}
\dot{\mathbf{G}}(\boldsymbol{\theta}) = \frac{\partial G(\boldsymbol{\theta})}{\partial \boldsymbol{\theta}} &= \frac{1}{2}\frac{\partial\, \mathrm{tr}[(\mathbf{S}-\boldsymbol{\Sigma})\mathbf{V}]^2}{\partial \boldsymbol{\theta}} \\
&= \frac{1}{2}\frac{\partial(\mathbf{S}-\boldsymbol{\Sigma})}{\partial \boldsymbol{\theta}}\frac{\partial(\mathbf{S}-\boldsymbol{\Sigma})\mathbf{V}}{\partial(\mathbf{S}-\boldsymbol{\Sigma})}\frac{\partial[(\mathbf{S}-\boldsymbol{\Sigma})\mathbf{V}]^2}{\partial(\mathbf{S}-\boldsymbol{\Sigma})\mathbf{V}}\frac{\partial \mathrm{tr}[(\mathbf{S}-\boldsymbol{\Sigma})\mathbf{V}]^2}{\partial[(\mathbf{S}-\boldsymbol{\Sigma})\mathbf{V}]^2} \\
&= -\frac{1}{2}\dot{\boldsymbol{\Sigma}}(\boldsymbol{\theta})\{\mathbf{I}_p \otimes \mathbf{V})\{[\mathbf{I}_p \otimes (\mathbf{S}-\boldsymbol{\Sigma})\mathbf{V}] + [\mathbf{V}(\mathbf{S}-\boldsymbol{\Sigma}) \otimes \mathbf{I}_p]\}\mathrm{vec}\, \mathbf{I}_p \\
&= -\frac{1}{2}\dot{\boldsymbol{\Sigma}}(\boldsymbol{\theta})\{\mathbf{I}_p \otimes \mathbf{V}(\mathbf{S}-\boldsymbol{\Sigma})\mathbf{V} + \mathbf{V}(\mathbf{S}-\boldsymbol{\Sigma}) \otimes \mathbf{V}\}\mathrm{vec}\, \mathbf{I}_p \\
&= -\frac{1}{2}\dot{\boldsymbol{\Sigma}}(\boldsymbol{\theta})\{2\mathrm{vec}[\mathbf{V}(\mathbf{S}-\boldsymbol{\Sigma})\mathbf{V}]\} \\
&= -\dot{\boldsymbol{\Sigma}}(\boldsymbol{\theta})\mathrm{vec}[\mathbf{V}(\mathbf{S}-\boldsymbol{\Sigma})\mathbf{V}]
\end{aligned}
$$

$$
\begin{aligned}
\ddot{\mathbf{G}}(\boldsymbol{\theta}) = \frac{\partial^2 G(\boldsymbol{\theta})}{\partial \boldsymbol{\theta}\partial \boldsymbol{\theta}} &= -\frac{\partial}{\partial \boldsymbol{\theta}}[\dot{\boldsymbol{\Sigma}}(\boldsymbol{\theta})(\mathbf{V}\otimes\mathbf{V})\mathrm{vec}(\mathbf{S}-\boldsymbol{\Sigma})] \\
&= -\ddot{\boldsymbol{\Sigma}}(\boldsymbol{\theta})[\mathbf{I}_q \otimes ((\mathbf{V}\otimes\mathbf{V})\mathrm{vec}(\mathbf{S}-\boldsymbol{\Sigma}))] - \frac{\partial(\mathbf{V}\otimes\mathbf{V})\mathrm{vec}(\mathbf{S}-\boldsymbol{\Sigma})}{\partial \boldsymbol{\theta}}[\dot{\boldsymbol{\Sigma}}(\boldsymbol{\theta})^T \otimes 1] \\
&= -\ddot{\boldsymbol{\Sigma}}(\boldsymbol{\theta})[\mathbf{I}_q \otimes ((\mathbf{V}\otimes\mathbf{V})\mathrm{vec}(\mathbf{S}-\boldsymbol{\Sigma}))] - \frac{\partial(\mathbf{S}-\boldsymbol{\Sigma})}{\partial \boldsymbol{\theta}}(\mathbf{V}\otimes\mathbf{V})\dot{\boldsymbol{\Sigma}}(\boldsymbol{\theta})^T.
\end{aligned}
$$

PROOF OF THEOREM 3.1

Consider the quadratic form

$$G^*(\boldsymbol{\theta}) = 2^{-1}[\mathrm{vec}(\boldsymbol{\Sigma}_0 - \boldsymbol{\Sigma})]^T(\mathbf{V}^* \otimes \mathbf{V}^*)[\mathrm{vec}(\boldsymbol{\Sigma}_0 - \boldsymbol{\Sigma})].$$

Since the model is identified and $(\mathbf{V}^* \otimes \mathbf{V}^*)$ is positive definite, this quadratic form has its unique minimum which is equal to zero at $\boldsymbol{\theta} = \boldsymbol{\theta}_0$. Compare it with

$$G(\boldsymbol{\theta}) = 2^{-1}[\mathrm{vec}(\mathbf{S} - \boldsymbol{\Sigma})]^T(\mathbf{V} \otimes \mathbf{V})[\mathrm{vec}(\mathbf{S} - \boldsymbol{\Sigma})].$$

Since $\mathbf{S}$ and $\mathbf{V}$ converge respectively to $\boldsymbol{\Sigma}_0$ and $\mathbf{V}^*$ in probability, and $\boldsymbol{\Sigma}$ is bounded in a neighborhood of $\boldsymbol{\theta}_0$, $G(\boldsymbol{\theta})$ converges in probability to $G^*(\boldsymbol{\theta})$ uniformly in a neighborhood of $\boldsymbol{\theta}_0$. Since $G(\boldsymbol{\theta})$ is continuous, its unique minimum $\tilde{\boldsymbol{\theta}}$ converges to the unique minimum $\boldsymbol{\theta}_0$ of $G^*(\boldsymbol{\theta})$ in probability. This proof is an adoption of a proof of Browne (1974).

PROOF OF THEOREM 3.2

It follows from the definition of $\tilde{\boldsymbol{\theta}}$ that $\dot{\mathbf{G}}(\tilde{\boldsymbol{\theta}}) = \mathbf{0}$. Applying the 'mean-value' theorem (see Bartle, 1964, p. 210) to $\dot{\mathbf{G}}(\tilde{\boldsymbol{\theta}})$, we have

$$\mathbf{0} = \dot{\mathbf{G}}(\tilde{\boldsymbol{\theta}}) = \dot{\mathbf{G}}(\boldsymbol{\theta}_0) + \ddot{\mathbf{G}}(\boldsymbol{\theta}^*)(\tilde{\boldsymbol{\theta}} - \boldsymbol{\theta}_0), \tag{A3.4}$$

where $\boldsymbol{\theta}^*$ is a vector which lies between $\tilde{\boldsymbol{\theta}}$ and $\boldsymbol{\theta}_0$. From Equations (3.10) and (A3.4),

$$\ddot{\mathbf{G}}(\boldsymbol{\theta}^*)[n^{1/2}(\tilde{\boldsymbol{\theta}} - \boldsymbol{\theta}_0)] = -n^{1/2}\dot{\mathbf{G}}(\boldsymbol{\theta}_0) = \dot{\boldsymbol{\Sigma}}(\boldsymbol{\theta}_0)(\mathbf{V} \otimes \mathbf{V})[n^{1/2}\text{vec}(\mathbf{S} - \boldsymbol{\Sigma}_0)].$$

Thus,

$$n^{1/2}(\tilde{\boldsymbol{\theta}} - \boldsymbol{\theta}_0) = \ddot{\mathbf{G}}(\boldsymbol{\theta}^*)^{-1}\dot{\boldsymbol{\Sigma}}(\boldsymbol{\theta}_0)(\mathbf{V} \otimes \mathbf{V})[n^{1/2}\,\text{vec}(\mathbf{S} - \boldsymbol{\Sigma}_0)].$$

From Equations (3.6) and (3.12), and knowing that $\mathbf{V}$ converges to $\mathbf{V}^*$ in probability, we have

$$\begin{aligned} n^{1/2}(\tilde{\boldsymbol{\theta}} - \boldsymbol{\theta}_0) &\xrightarrow{p} \boldsymbol{\Omega}(\mathbf{V}^*)^{-1}\dot{\boldsymbol{\Sigma}}(\boldsymbol{\theta}_0)(\mathbf{V}^* \otimes \mathbf{V}^*)[n^{1/2}\,\text{vec}(\mathbf{S} - \boldsymbol{\Sigma}_0)] \\ &\xrightarrow{L} N[\mathbf{0}, \mathbf{C}(\boldsymbol{\theta}_0)], \end{aligned} \tag{A3.5}$$

where

$$\begin{aligned} \mathbf{C}(\boldsymbol{\theta}_0) &= \boldsymbol{\Omega}(\mathbf{V}^*)^{-1}\dot{\boldsymbol{\Sigma}}(\boldsymbol{\theta}_0)(\mathbf{V}^* \otimes \mathbf{V}^*)\{2\mathbf{M}_p(\boldsymbol{\Sigma}_0 \otimes \boldsymbol{\Sigma}_0)\mathbf{M}_p\}(\mathbf{V}^* \otimes \mathbf{V}^*)\dot{\boldsymbol{\Sigma}}(\boldsymbol{\theta}_0)^T\boldsymbol{\Omega}(\mathbf{V}^*)^{-1} \\ &= 2\boldsymbol{\Omega}(\mathbf{V}^*)^{-1}\dot{\boldsymbol{\Sigma}}(\boldsymbol{\theta}_0)(\mathbf{V}^* \otimes \mathbf{V}^*)(\boldsymbol{\Sigma}_0 \otimes \boldsymbol{\Sigma}_0)(\mathbf{V}^* \otimes \mathbf{V}^*)\dot{\boldsymbol{\Sigma}}(\boldsymbol{\theta}_0)^T\boldsymbol{\Omega}(\mathbf{V}^*)^{-1} \\ &= 2\boldsymbol{\Omega}(\mathbf{V}^*)^{-1}\boldsymbol{\Omega}(\mathbf{V}^*\boldsymbol{\Sigma}_0\mathbf{V}^*)\boldsymbol{\Omega}(\mathbf{V}^*)^{-1}. \end{aligned}$$

This completes the proof.

The idea of the above proof can be applied to other objective functions. Roughly, asymptotic normality of the corresponding estimator can be achieved if the gradient vector of the objective function is asymptotic normal.

PROOF OF COROLLARY 3.1

If $\mathbf{V}^* = \mathbf{\Sigma}_0^{-1}$, then

$$\mathbf{C}(\theta_0) = 2\mathbf{\Omega}(\mathbf{\Sigma}_0^{-1})^{-1}\mathbf{\Omega}(\mathbf{\Sigma}_0^{-1}\mathbf{\Sigma}_0\mathbf{\Sigma}_0^{-1})\mathbf{\Omega}(\mathbf{\Sigma}_0^{-1})^{-1} = 2\mathbf{\Omega}(\mathbf{\Sigma}_0^{-1})^{-1}.$$

Now, consider the quadratic form

$$\begin{aligned}
&2[\mathbf{\Omega}(\mathbf{V}^*)^{-1}\dot{\mathbf{\Sigma}}(\boldsymbol{\theta}_0)(\mathbf{V}^*\otimes\mathbf{V}^*) - \mathbf{\Omega}(\mathbf{\Sigma}_0^{-1})^{-1}\dot{\mathbf{\Sigma}}(\boldsymbol{\theta}_0)(\mathbf{\Sigma}_0^{-1}\otimes\mathbf{\Sigma}_0^{-1})](\mathbf{\Sigma}_0\otimes\mathbf{\Sigma}_0)\\
&[\mathbf{\Omega}(\mathbf{V}^*)^{-1}\dot{\mathbf{\Sigma}}(\boldsymbol{\theta}_0)(\mathbf{V}^*\otimes\mathbf{V}^*) - \mathbf{\Omega}(\mathbf{\Sigma}_0^{-1})^{-1}\dot{\mathbf{\Sigma}}(\boldsymbol{\theta}_0)(\mathbf{\Sigma}_0^{-1}\otimes\mathbf{\Sigma}_0^{-1})]^T\\
=\,&2[\mathbf{\Omega}(\mathbf{V}^*)^{-1}\dot{\mathbf{\Sigma}}(\boldsymbol{\theta}_0)(\mathbf{V}^*\otimes\mathbf{V}^*)(\mathbf{\Sigma}_0\otimes\mathbf{\Sigma}_0)(\mathbf{V}^*\otimes\mathbf{V}^*)\dot{\mathbf{\Sigma}}(\boldsymbol{\theta}_0)^T\mathbf{\Omega}(\mathbf{V}^*)^{-1}\\
&-\mathbf{\Omega}(\mathbf{V}^*)^{-1}\dot{\mathbf{\Sigma}}(\boldsymbol{\theta}_0)(\mathbf{V}^*\otimes\mathbf{V}^*)(\mathbf{\Sigma}_0\otimes\mathbf{\Sigma}_0)(\mathbf{\Sigma}_0^{-1}\otimes\mathbf{\Sigma}_0^{-1})\dot{\mathbf{\Sigma}}(\boldsymbol{\theta}_0)^T\mathbf{\Omega}(\mathbf{\Sigma}_0^{-1})^{-1}\\
&-\mathbf{\Omega}(\mathbf{\Sigma}_0^{-1})^{-1}\dot{\mathbf{\Sigma}}(\boldsymbol{\theta}_0)(\mathbf{\Sigma}_0^{-1}\otimes\mathbf{\Sigma}_0^{-1})(\mathbf{\Sigma}_0\otimes\mathbf{\Sigma}_0)(\mathbf{V}^*\otimes\mathbf{V}^*)\dot{\mathbf{\Sigma}}(\boldsymbol{\theta}_0)^T\mathbf{\Omega}(\mathbf{V}^*)^{-1}\\
&+\mathbf{\Omega}(\mathbf{\Sigma}_0^{-1})^{-1}\dot{\mathbf{\Sigma}}(\boldsymbol{\theta}_0)(\mathbf{\Sigma}_0^{-1}\otimes\mathbf{\Sigma}_0^{-1})(\mathbf{\Sigma}_0\otimes\mathbf{\Sigma}_0)(\mathbf{\Sigma}_0^{-1}\otimes\mathbf{\Sigma}_0^{-1})\dot{\mathbf{\Sigma}}(\boldsymbol{\theta}_0)^T\mathbf{\Omega}(\mathbf{\Sigma}_0^{-1})^{-1}]\\
=\,&2[\mathbf{\Omega}(V^*)^{-1}\dot{\mathbf{\Sigma}}(\boldsymbol{\theta}_0)(\mathbf{V}^*\mathbf{\Sigma}_0\mathbf{V}^*\otimes\mathbf{V}^*\mathbf{\Sigma}_0\mathbf{V}^*)\dot{\mathbf{\Sigma}}(\boldsymbol{\theta}_0)^T\mathbf{\Omega}(\mathbf{V}^*)^{-1}\\
&-\mathbf{\Omega}(\mathbf{\Sigma}_0^{-1})^{-1} - \mathbf{\Omega}(\mathbf{\Sigma}_0^{-1})^{-1} + \mathbf{\Omega}(\mathbf{\Sigma}_0^{-1})^{-1}]\\
=\,&\mathbf{C}(\boldsymbol{\theta}_0) - 2\mathbf{\Omega}(\mathbf{\Sigma}_0^{-1})^{-1}.
\end{aligned}$$

Since $(\mathbf{\Sigma}_0\otimes\mathbf{\Sigma}_0)$ is positive definite, the above quadratic form and hence $\mathbf{C}(\boldsymbol{\theta}_0) - 2\mathbf{\Omega}(\mathbf{\Sigma}_0^{-1})^{-1}$ is positive definite. This completes the proof of the corollary.

PROOF OF THEOREM 3.3

On the basis of the 'mean-value' theorem, there exists a $\boldsymbol{\theta}^*$ between $\tilde{\boldsymbol{\theta}}$ and $\boldsymbol{\theta}_0$ such that

$$\begin{aligned}
n^{1/2}\text{vec}(\mathbf{S}-\tilde{\mathbf{\Sigma}}) &= n^{1/2}\text{vec}[(\mathbf{S}-\mathbf{\Sigma}_0) - (\tilde{\mathbf{\Sigma}}-\mathbf{\Sigma}_0)]\\
&= n^{1/2}\text{vec}(\mathbf{S}-\mathbf{\Sigma}_0) - \dot{\mathbf{\Sigma}}(\boldsymbol{\theta}^*)^T[n^{1/2}(\tilde{\boldsymbol{\theta}}-\boldsymbol{\theta}_0)].
\end{aligned}$$

From above and (A3.5)

$$\begin{aligned}
&n^{1/2}\text{vec}(\mathbf{S}-\tilde{\mathbf{\Sigma}})\\
&\xrightarrow{p} n^{1/2}\text{vec}(\mathbf{S}-\mathbf{\Sigma}_0) - \dot{\mathbf{\Sigma}}(\boldsymbol{\theta}^*)^T\mathbf{\Omega}(\mathbf{\Sigma}_0^{-1})^{-1}\dot{\mathbf{\Sigma}}(\boldsymbol{\theta}_0)(\mathbf{\Sigma}_0^{-1}\otimes\mathbf{\Sigma}_0^{-1})[n^{1/2}\text{vec}(\mathbf{S}-\mathbf{\Sigma}_0)]\\
&\xrightarrow{p} \mathbf{\Pi}_0[n^{1/2}\text{vec}(\mathbf{S}-\mathbf{\Sigma}_0)],
\end{aligned}\qquad\text{(A3.6)}$$

where $\mathbf{\Pi}_0 = [\mathbf{I}_{p^2} - \dot{\mathbf{\Sigma}}(\boldsymbol{\theta}_0)^T \mathbf{\Omega}(\mathbf{\Sigma}_0^{-1})^{-1} \dot{\mathbf{\Sigma}}(\boldsymbol{\theta}_0)(\mathbf{\Sigma}_0^{-1} \otimes \mathbf{\Sigma}_0^{-1})]$, $\mathbf{I}_{p^2}$ is a $p^2 \times p^2$ identity matrix.

Substitute (A3.6) to $nG(\tilde{\boldsymbol{\theta}})$,

$$nG(\tilde{\boldsymbol{\theta}}) = 2^{-1}[n^{1/2}\text{vec}(\mathbf{S} - \tilde{\mathbf{\Sigma}})]^T(\mathbf{V} \otimes \mathbf{V})[n^{1/2}\text{vec}(\mathbf{S} - \tilde{\mathbf{\Sigma}})]$$
$$\xrightarrow{p} [(2^{-1}n)^{1/2}\text{vec}(\mathbf{S} - \mathbf{\Sigma}_0)]^T \mathbf{\Pi}_0^T(\mathbf{\Sigma}_0^{-1} \otimes \mathbf{\Sigma}_0^{-1})\mathbf{\Pi}_0[(2^{-1}n)^{1/2}\text{vec}(\mathbf{S} - \mathbf{\Sigma}_0)].$$

Now, $(2^{-1}n)^{1/2}\text{vec}(\mathbf{S} - \mathbf{\Sigma}_0) \xrightarrow{L} N[\mathbf{0}, \mathbf{M}_p(\mathbf{\Sigma}_0 \otimes \mathbf{\Sigma}_0)\mathbf{M}_p]$. Hence, from a standard theorem on quadratic form (see Graybill, 1961 p. 83) it suffices to show $\mathbf{\Pi}_0^T(\mathbf{\Sigma}_0^{-1} \otimes \mathbf{\Sigma}_0^{-1})\mathbf{\Pi}_0\ \mathbf{M}_p(\mathbf{\Sigma}_0 \otimes \mathbf{\Sigma}_0)\mathbf{M}_p$ is an idempotent matrix with rank $p^* - q$.

First, we note that

$$\begin{aligned}
&\mathbf{\Pi}_0^T(\mathbf{\Sigma}_0^{-1} \otimes \mathbf{\Sigma}_0^{-1})\mathbf{\Pi}_0 \\
&= \{(\mathbf{\Sigma}_0^{-1} \otimes \mathbf{\Sigma}_0^{-1}) - (\mathbf{\Sigma}_0^{-1} \otimes \mathbf{\Sigma}_0^{-1})\dot{\mathbf{\Sigma}}(\boldsymbol{\theta}_0)^T\mathbf{\Omega}(\mathbf{\Sigma}_0^{-1})^{-1}\dot{\mathbf{\Sigma}}(\boldsymbol{\theta}_0)(\mathbf{\Sigma}_0^{-1} \otimes \mathbf{\Sigma}_0^{-1})\}\mathbf{\Pi}_0 \\
&= (\mathbf{\Sigma}_0^{-1} \otimes \mathbf{\Sigma}_0^{-1}) - (\mathbf{\Sigma}_0^{-1} \otimes \mathbf{\Sigma}_0^{-1})\dot{\mathbf{\Sigma}}(\boldsymbol{\theta}_0)^T\mathbf{\Omega}(\mathbf{\Sigma}_0^{-1})\dot{\mathbf{\Sigma}}(\boldsymbol{\theta}_0)(\mathbf{\Sigma}_0^{-1} \otimes \mathbf{\Sigma}_0^{-1}) \\
&\quad - (\mathbf{\Sigma}_0^{-1} \otimes \mathbf{\Sigma}_0^{-1})\dot{\mathbf{\Sigma}}(\boldsymbol{\theta}_0)^T\mathbf{\Omega}(\mathbf{\Sigma}_0^{-1})^{-1}\dot{\mathbf{\Sigma}}(\boldsymbol{\theta}_0)(\mathbf{\Sigma}_0^{-1} \otimes \mathbf{\Sigma}_0^{-1}) \\
&\quad + (\mathbf{\Sigma}_0^{-1} \otimes \mathbf{\Sigma}_0^{-1})\dot{\mathbf{\Sigma}}(\boldsymbol{\theta}_0)^T\mathbf{\Omega}(\mathbf{\Sigma}_0^{-1})^{-1}\dot{\mathbf{\Sigma}}(\boldsymbol{\theta}_0)(\mathbf{\Sigma}_0^{-1} \otimes \mathbf{\Sigma}_0^{-1})^{-1} \\
&\quad \times \dot{\mathbf{\Sigma}}(\boldsymbol{\theta}_0)^T\mathbf{\Omega}(\mathbf{\Sigma}_0^{-1})^{-1}\dot{\mathbf{\Sigma}}(\boldsymbol{\theta}_0)(\mathbf{\Sigma}_0^{-1} \otimes \mathbf{\Sigma}_0^{-1}) \\
&= [\mathbf{I}_{p^2} - (\mathbf{\Sigma}_0^{-1} \otimes \mathbf{\Sigma}_0^{-1})\dot{\mathbf{\Sigma}}(\boldsymbol{\theta}_0)^T\mathbf{\Omega}(\mathbf{\Sigma}_0^{-1})\dot{\mathbf{\Sigma}}(\boldsymbol{\theta}_0)](\mathbf{\Sigma}_0^{-1} \otimes \mathbf{\Sigma}_0^{-1}).
\end{aligned}$$

Hence, it follows from Equation (3.4) and Lemma 3 that

$$\begin{aligned}
&\mathbf{\Pi}_0^T(\mathbf{\Sigma}_0^{-1} \otimes \mathbf{\Sigma}_0^{-1})\mathbf{\Pi}_0\mathbf{M}_p(\mathbf{\Sigma}_0 \otimes \mathbf{\Sigma}_0)\mathbf{M}_p \\
&= [\mathbf{I}_{p^2} - (\mathbf{\Sigma}_0^{-1} \otimes \mathbf{\Sigma}_0^{-1})\dot{\mathbf{\Sigma}}(\boldsymbol{\theta}_0)^T\mathbf{\Omega}(\mathbf{\Sigma}_0^{-1})^{-1}\dot{\mathbf{\Sigma}}(\boldsymbol{\theta}_0)](\mathbf{\Sigma}_0^{-1} \otimes \mathbf{\Sigma}_0^{-1})(\mathbf{\Sigma}_0 \otimes \mathbf{\Sigma}_0)\mathbf{M}_p \\
&= \mathbf{M}_p - (\mathbf{\Sigma}_0^{-1} \otimes \mathbf{\Sigma}_0^{-1})\dot{\mathbf{\Sigma}}(\boldsymbol{\theta}_0)^T\mathbf{\Omega}(\mathbf{\Sigma}_0^{-1})^{-1}\dot{\mathbf{\Sigma}}(\boldsymbol{\theta}_0).
\end{aligned}$$

To show this is an idempotent matrix, we note that

$$\begin{aligned}
&[\mathbf{M}_p - (\mathbf{\Sigma}_0^{-1} \otimes \mathbf{\Sigma}_0^{-1})\dot{\mathbf{\Sigma}}(\boldsymbol{\theta}_0)^T\mathbf{\Omega}(\mathbf{\Sigma}_0^{-1})^{-1}\dot{\mathbf{\Sigma}}(\boldsymbol{\theta}_0)] \\
&\quad \times [\mathbf{M}_p - (\mathbf{\Sigma}_0^{-1} \otimes \mathbf{\Sigma}_0^{-1})\dot{\mathbf{\Sigma}}(\boldsymbol{\theta}_0)^T\mathbf{\Omega}(\mathbf{\Sigma}_0^{-1})^{-1}\dot{\mathbf{\Sigma}}(\boldsymbol{\theta}_0)] \\
&= \mathbf{M}_p - (\mathbf{\Sigma}_0^{-1} \otimes \mathbf{\Sigma}_0^{-1})\dot{\mathbf{\Sigma}}(\boldsymbol{\theta}_0)^T\mathbf{\Omega}(\mathbf{\Sigma}_0^{-1})^{-1}\dot{\mathbf{\Sigma}}(\boldsymbol{\theta}_0) \\
&\quad - (\mathbf{\Sigma}_0^{-1} \otimes \mathbf{\Sigma}_0^{-1})\dot{\mathbf{\Sigma}}(\boldsymbol{\theta}_0)^T\mathbf{\Omega}(\mathbf{\Sigma}_0^{-1})^{-1}\dot{\mathbf{\Sigma}}(\boldsymbol{\theta}_0)
\end{aligned}$$

$$+(\boldsymbol{\Sigma}_0^{-1}\otimes\boldsymbol{\Sigma}_0^{-1})\dot{\boldsymbol{\Sigma}}(\boldsymbol{\theta}_0)^T\boldsymbol{\Omega}(\boldsymbol{\Sigma}_0^{-1})^{-1}[\dot{\boldsymbol{\Sigma}}(\boldsymbol{\theta}_0)(\boldsymbol{\Sigma}_0^{-1}\otimes\boldsymbol{\Sigma}_0^{-1})\dot{\boldsymbol{\Sigma}}(\boldsymbol{\theta}_0)^T]\boldsymbol{\Omega}(\boldsymbol{\Sigma}_0^{-1})^{-1}\dot{\boldsymbol{\Sigma}}(\boldsymbol{\theta}_0)$$
$$=\mathbf{M}_p-(\boldsymbol{\Sigma}_0^{-1}\otimes\boldsymbol{\Sigma}_0^{-1})\dot{\boldsymbol{\Sigma}}(\boldsymbol{\theta}_0)^T\boldsymbol{\Omega}(\boldsymbol{\Sigma}_0^{-1})^{-1}\dot{\boldsymbol{\Sigma}}(\boldsymbol{\theta}_0).$$

Hence, $\boldsymbol{\Pi}_0^T(\boldsymbol{\Sigma}_0^{-1}\otimes\boldsymbol{\Sigma}_0^{-1})\boldsymbol{\Pi}_0\mathbf{M}_p(\boldsymbol{\Sigma}_0\otimes\boldsymbol{\Sigma}_0)\mathbf{M}_p$ is idempotent. Its rank is equal to

$$\begin{aligned}
&\operatorname{tr}(\mathbf{M}_p)-\operatorname{tr}[(\boldsymbol{\Sigma}_0^{-1}\otimes\boldsymbol{\Sigma}_0^{-1})\dot{\boldsymbol{\Sigma}}(\boldsymbol{\theta}_0)^T\boldsymbol{\Omega}(\boldsymbol{\Sigma}_0^{-1})^{-1}\dot{\boldsymbol{\Sigma}}(\boldsymbol{\theta}_0)]\\
&=p(p+1)/2-\operatorname{tr}[\boldsymbol{\Omega}(\boldsymbol{\Sigma}_0^{-1})^{-1}\dot{\boldsymbol{\Sigma}}(\boldsymbol{\theta}_0)(\boldsymbol{\Sigma}_0^{-1}\otimes\boldsymbol{\Sigma}_0^{-1})\dot{\boldsymbol{\Sigma}}(\boldsymbol{\theta}_0)^T]\\
&=p^*-\operatorname{tr}\,\mathbf{I}_q=p^*-q.
\end{aligned}$$

This completes the proof.

PROOF OF LEMMA 5

$$\begin{aligned}
\dot{\mathbf{F}}(\boldsymbol{\theta})&=\frac{\partial F(\boldsymbol{\theta})}{\partial\boldsymbol{\theta}}=\frac{\partial}{\partial\boldsymbol{\theta}}[\log|\boldsymbol{\Sigma}|+\operatorname{tr}(S\boldsymbol{\Sigma}^{-1})]\\
&=\frac{\partial\log|\boldsymbol{\Sigma}|}{\partial\boldsymbol{\theta}}+\dot{\boldsymbol{\Sigma}}(\boldsymbol{\theta})\frac{\partial\boldsymbol{\Sigma}^{-1}}{\partial\boldsymbol{\Sigma}}\;\frac{\partial\mathbf{S}\boldsymbol{\Sigma}^{-1}}{\partial\boldsymbol{\Sigma}^{-1}}\;\frac{\partial\operatorname{tr}(\mathbf{S}\boldsymbol{\Sigma}^{-1})}{\partial\mathbf{S}\boldsymbol{\Sigma}^{-1}}\\
&=\dot{\boldsymbol{\Sigma}}(\theta)\operatorname{vec}(\boldsymbol{\Sigma}^{-1})+\dot{\boldsymbol{\Sigma}}(\boldsymbol{\theta})(-\boldsymbol{\Sigma}^{-1}\otimes\boldsymbol{\Sigma}^{-1})(\mathbf{S}\otimes\mathbf{I}_p)\operatorname{vec}\,\mathbf{I}_p\\
&=\dot{\boldsymbol{\Sigma}}(\boldsymbol{\theta})\operatorname{vec}(\boldsymbol{\Sigma}^{-1})-\dot{\boldsymbol{\Sigma}}(\boldsymbol{\theta})\operatorname{vec}(\boldsymbol{\Sigma}^{-1}\mathbf{S}\boldsymbol{\Sigma}^{-1})\\
&=-\dot{\boldsymbol{\Sigma}}(\boldsymbol{\theta})\operatorname{vec}[\boldsymbol{\Sigma}^{-1}(\mathbf{S}-\boldsymbol{\Sigma})\boldsymbol{\Sigma}^{-1}]\\
&=-\dot{\boldsymbol{\Sigma}}(\boldsymbol{\theta})(\boldsymbol{\Sigma}^{-1}\otimes\boldsymbol{\Sigma}^{-1})\operatorname{vec}(\mathbf{S}-\boldsymbol{\Sigma}).
\end{aligned}\tag{A3.7}$$

Further,

$$\begin{aligned}
&\frac{\partial\boldsymbol{\Sigma}^{-1}(\mathbf{S}-\boldsymbol{\Sigma})\boldsymbol{\Sigma}^{-1}}{\partial\boldsymbol{\theta}}=\frac{\partial\boldsymbol{\Sigma}^{-1}}{\partial\boldsymbol{\theta}}[\mathbf{I}_p\otimes(\mathbf{S}-\boldsymbol{\Sigma})\boldsymbol{\Sigma}^{-1}]+\frac{\partial(\mathbf{S}-\boldsymbol{\Sigma})\boldsymbol{\Sigma}^{-1}}{\partial\boldsymbol{\theta}}(\boldsymbol{\Sigma}^{-1}\otimes\mathbf{I}_p)\\
&=-\dot{\boldsymbol{\Sigma}}(\theta)(\boldsymbol{\Sigma}^{-1}\otimes\boldsymbol{\Sigma}^{-1})[\mathbf{I}_p\otimes(\mathbf{S}-\boldsymbol{\Sigma})\boldsymbol{\Sigma}^{-1}]+[\frac{\partial(\mathbf{S}-\boldsymbol{\Sigma})}{\partial\boldsymbol{\theta}}(\mathbf{I}_p\otimes\boldsymbol{\Sigma}^{-1})\\
&\quad+\frac{\partial\boldsymbol{\Sigma}^{-1}}{\partial\boldsymbol{\theta}}(\mathbf{S}-\boldsymbol{\Sigma})\otimes\mathbf{I}_p)](\boldsymbol{\Sigma}^{-1}\otimes\mathbf{I}_p)\\
&=-\dot{\boldsymbol{\Sigma}}(\boldsymbol{\theta})[\boldsymbol{\Sigma}^{-1}\otimes\boldsymbol{\Sigma}^{-1}(\mathbf{S}-\boldsymbol{\Sigma})\boldsymbol{\Sigma}^{-1}]+\{-\dot{\boldsymbol{\Sigma}}(\boldsymbol{\theta})(\mathbf{I}_p\otimes\boldsymbol{\Sigma}^{-1})\\
&\quad-\dot{\boldsymbol{\Sigma}}(\boldsymbol{\theta})(\boldsymbol{\Sigma}^{-1}\otimes\boldsymbol{\Sigma}^{-1})[(\mathbf{S}-\boldsymbol{\Sigma})\otimes\mathbf{I}_p]\}(\boldsymbol{\Sigma}^{-1}\otimes\mathbf{I}_p)\\
&=-\dot{\boldsymbol{\Sigma}}(\boldsymbol{\theta})[\boldsymbol{\Sigma}^{-1}\otimes\boldsymbol{\Sigma}^{-1}(\mathbf{S}-\boldsymbol{\Sigma})\boldsymbol{\Sigma}^{-1}+\boldsymbol{\Sigma}^{-1}(\mathbf{S}-\boldsymbol{\Sigma})\boldsymbol{\Sigma}^{-1}\otimes\boldsymbol{\Sigma}^{-1}]-\dot{\boldsymbol{\Sigma}}(\boldsymbol{\theta})(\boldsymbol{\Sigma}^{-1}\otimes\boldsymbol{\Sigma}^{-1}).
\end{aligned}\tag{A3.8}$$

From Equations (3.21) and (3.22),

$$
\begin{aligned}
\ddot{\mathbf{F}}(\boldsymbol{\theta}) &= \frac{\partial^2 F}{\partial\boldsymbol{\theta}\partial\boldsymbol{\theta}} = \frac{\partial}{\partial\boldsymbol{\theta}}\left\{-\dot{\boldsymbol{\Sigma}}(\boldsymbol{\theta})\mathrm{vec}[\boldsymbol{\Sigma}^{-1}(\mathbf{S}-\boldsymbol{\Sigma})\boldsymbol{\Sigma}^{-1}]\right\}, \\
&= -\ddot{\boldsymbol{\Sigma}}(\boldsymbol{\theta})\{\mathbf{I}_q\otimes\mathrm{vec}[\boldsymbol{\Sigma}^{-1}(\mathbf{S}-\boldsymbol{\Sigma})\boldsymbol{\Sigma}^{-1}]\} - \frac{\partial\boldsymbol{\Sigma}^{-1}(\mathbf{S}-\boldsymbol{\Sigma})\boldsymbol{\Sigma}^{-1}}{\partial\boldsymbol{\theta}}[\dot{\boldsymbol{\Sigma}}(\boldsymbol{\theta})^T\otimes 1] \\
&= \dot{\boldsymbol{\Sigma}}(\boldsymbol{\theta})(\boldsymbol{\Sigma}^{-1}\otimes\boldsymbol{\Sigma}^{-1})\dot{\boldsymbol{\Sigma}}(\boldsymbol{\theta})^T - \ddot{\boldsymbol{\Sigma}}(\boldsymbol{\theta})[\mathbf{I}_q\otimes(\boldsymbol{\Sigma}^{-1}\otimes\boldsymbol{\Sigma}^{-1})\mathrm{vec}(\mathbf{S}-\boldsymbol{\Sigma})] \\
&\quad - \dot{\boldsymbol{\Sigma}}(\boldsymbol{\theta})\{[\boldsymbol{\Sigma}^{-1}\otimes\boldsymbol{\Sigma}^{-1}(\mathbf{S}-\boldsymbol{\Sigma})\boldsymbol{\Sigma}^{-1}] + [\boldsymbol{\Sigma}^{-1}(\mathbf{S}-\boldsymbol{\Sigma})\boldsymbol{\Sigma}^{-1}\otimes\boldsymbol{\Sigma}^{-1}]\}\dot{\boldsymbol{\Sigma}}(\boldsymbol{\theta})^T \\
&= \dot{\boldsymbol{\Sigma}}(\boldsymbol{\theta})(\boldsymbol{\Sigma}^{-1}\otimes\boldsymbol{\Sigma}^{-1})\dot{\boldsymbol{\Sigma}}(\boldsymbol{\theta})^T - 2\dot{\boldsymbol{\Sigma}}(\boldsymbol{\theta})[\boldsymbol{\Sigma}^{-1}\otimes\boldsymbol{\Sigma}^{-1}(\mathbf{S}-\boldsymbol{\Sigma})\boldsymbol{\Sigma}^{-1}]\dot{\boldsymbol{\Sigma}}(\boldsymbol{\theta})^T \\
&\quad - \ddot{\boldsymbol{\Sigma}}(\boldsymbol{\theta})[\mathbf{I}_q\otimes(\boldsymbol{\Sigma}^{-1}\otimes\boldsymbol{\Sigma}^{-1})\mathrm{vec}(\mathbf{S}-\boldsymbol{\Sigma})]. \qquad \text{(A3.9)}
\end{aligned}
$$

PROOF OF COROLLARY 3.2

(i) It follows from Corollary 3.1 and Equation (3.23) that

$$
\begin{aligned}
n^{1/2}(\tilde{\boldsymbol{\theta}}_M - \boldsymbol{\theta}_0) &= n^{1/2}(\tilde{\boldsymbol{\theta}}_M - \tilde{\boldsymbol{\theta}}) + n^{1/2}(\tilde{\boldsymbol{\theta}} - \boldsymbol{\theta}_0) \\
&\xrightarrow{L} N[\mathbf{0}, 2\boldsymbol{\Omega}(\boldsymbol{\Sigma}_0^{-1})^{-1}].
\end{aligned}
$$

(ii) From the 'mean-value' theorem, there exists a $\boldsymbol{\theta}^*$ in between $\tilde{\boldsymbol{\theta}}_M$ and $\tilde{\boldsymbol{\theta}}$ such that

$$
\begin{aligned}
nG(\tilde{\boldsymbol{\theta}}_M) - nG(\tilde{\boldsymbol{\theta}}) &= n^{1/2}\dot{\mathbf{G}}(\boldsymbol{\theta}^*)[n^{1/2}(\tilde{\boldsymbol{\theta}}_M - \tilde{\boldsymbol{\theta}})] \\
&= -\{\dot{\boldsymbol{\Sigma}}(\boldsymbol{\theta}^*)(\mathbf{V}\otimes\mathbf{V})n^{1/2}\mathrm{vec}[\mathbf{S}-\boldsymbol{\Sigma}(\boldsymbol{\theta}^*)]\}[n^{1/2}(\tilde{\boldsymbol{\theta}}_M - \tilde{\boldsymbol{\theta}})]. \qquad \text{(A3.10)}
\end{aligned}
$$

Similarly, there exists a $\boldsymbol{\theta}^{**}$ between $\boldsymbol{\theta}^*$ and $\tilde{\boldsymbol{\theta}}$ that

$$
\begin{aligned}
n^{1/2}\,\mathrm{vec}[\mathbf{S}-\boldsymbol{\Sigma}(\boldsymbol{\theta}^*)] - n^{1/2}\,\mathrm{vec}[\mathbf{S}-\boldsymbol{\Sigma}(\tilde{\boldsymbol{\theta}})] &= -n^{1/2}\,\mathrm{vec}[\boldsymbol{\Sigma}(\boldsymbol{\theta}^*) - \boldsymbol{\Sigma}(\tilde{\boldsymbol{\theta}})] \\
&= -\dot{\boldsymbol{\Sigma}}(\boldsymbol{\theta}^{**})^T n^{1/2}(\boldsymbol{\theta}^* - \tilde{\boldsymbol{\theta}}),
\end{aligned}
$$

which converges in probability to zero because $n^{1/2}(\boldsymbol{\theta}^* - \tilde{\boldsymbol{\theta}})$ converges in probability to zero and $\dot{\boldsymbol{\Sigma}}(\boldsymbol{\theta}^{**})$ is bounded. Hence, $n^{1/2}\mathrm{vec}[\mathbf{S}-\boldsymbol{\Sigma}(\boldsymbol{\theta}^*)]$ converges in distribution to the same one as $n^{1/2}\mathrm{vec}[\mathbf{S}-\boldsymbol{\Sigma}(\tilde{\boldsymbol{\theta}})]$. It follows from Equation (A3.10) that $nG(\tilde{\boldsymbol{\theta}}_M)$ converges to $nG(\tilde{\boldsymbol{\theta}})$ in probability and hence they have the same asymptotic distribution, which is χ^2_{p*-q}. This completes the proof of (ii). The proof of (iii) is very similar.

REFERENCES

Afifi, A. A. and Azen, S. P. (1974) *Statistical Analysis:A Computer Oriented Approach*. New York: Academic Press.

Anderson, T. W. (1984) *An Introduction to Multivariate Statistical Analysis* (2nd edn). New York: John Wiley & Sons, Inc..

Bartle, R. G. (1964) *The Elements of Real Analysis*. New York: John Wiley & Sons, Inc..

Bentler, P. M. (1983) Some contributions to efficient statistics for structural models: specification and estimation of moment structures. *Psychometrika*, **48**, 493–517.

Bentler, P. M. (1992) *EQS: Structural Equation Program Manual*, Los Angeles, CA: BMDP Statistical Software.

Bentler, P. M. and Dudgeon, P. (1996) Covariance structure analysis: statistical practice, theory and directions. *Annual Review of Psychology*, **47**, 563–592.

Bentler, P. M. and Lee, S. Y. (1975) Some extensions of matrix calculus. *General Systems*, **20**, 145–150.

Bentler, P. M. and Wu, E. J. C. (2002) *EQS 6 for Windows User's Guide*. Encino, CA: Multivariate Software.

Browne, M. W. (1974) Generalized least squares estimators in the analysis of covariance structures. *South African Statistical Journal*, **8**, 1–24.

Browne, M. W. (1984). Asymptotic distribution free methods in analysis of covariance structure. *British Journal of Mathematical and Statistical Psychology*, **37**, 62–83.

Davidon, W. C. (1959). Variable metric method for minimization. *U.S. Atomic Energy Commission, Argonne National Laboratories, Research and Development Report ANL-5990*, p. 27.

Emmett, W. G. (1949) Factor analysis by Lawley's method of maximum likelihood. *British Journal of Mathematical and Statistical Psychology*, **2**, 90–97.

Fletcher, R. and Powell, M. (1963) A rapidly convergent descent method for minimization. *Computer Journal*, **6**, 163–168.

Greenstadt, J. (1967) On the relative efficiencies of gradient method. *Mathematical Computing*, **21**, 360–367.

Graybill, F. A. (1961). *An Introduction to Linear Statistical Models*. New York: McGraw-Hill.

Hu, L., Bentler, P. M. and Kano, Y. (1992) Can test statistics in covariance structure analysis be trusted? *Psychological Bulletin*, **112**, 351–362.

Jöreskog, K. G. (1967) Some contributions to maximum likelihood factor analysis. *Psychometrika*, **32**, 443–482.

Jöreskog, K. G. (1978) Structural analysis of covariance and correlation matrices. *Psychometrika*, **43**, 443–477.

Jöreskog, K. G. and Sörbom, D. (1996) *LISREL 8: Structural Equation Modeling with the SIMPLIS Command Language*. Hove and London: Scientific Software International.

Kendall, M. G. and Stuart, A. (1969) *The Advanced Theory of Statistics*, vol. 1. London: Charles Griffin.

Lee, S. Y. and Jennrich, R. I. (1979) A study of algorithms for covariance structure analysis with specific comparisons using factor analysis. *Psychometrika*, **44**, 99–113.

Lee, S. Y. and Song, X. Y. (2004) Evaluation of the Bayesian and maximum likelihood approaches in analyzing structural equation models with small sample sizes. *Multivariate Behavioral Research*, **39**, 653–686.

McDonald, R. P. and Swaminathan, H. (1973) A simple matrix calculus with applications to multivariate analysis. *General System*, **XVIII**, 37–54.

Rao, CR. (1973) *Linear Statistical Inference and its Applications* (2nd edn). New York: John Wiley & Sons, Inc..

4

Bayesian Estimation of Structural Equation Models

4.1 INTRODUCTION

In Chapter 3, we discussed the GLS and ML approaches in a covariance structure analysis framework for analyzing the standard structural equation model. The statistical theory that is associated with the GLS and ML approaches as well as the computational algorithms are developed on the basis of the sample covariance matrix $\mathbf{S}$. For example, the form of the GLS function (see Equation (3.9)) and the derivation of the asymptotic properties of the GLS estimator heavily depend on the asymptotic distribution of $\mathbf{S}$ (see Equation (3.5)). Hence, these approaches work well under certain assumptions that ensure the validity of Equation (3.5). Typically, random observations are assumed to be identically and independently distributed according to a multivariate normal distribution. If some of the assumptions are violated, $\mathbf{S}$ and/or its asymptotic properties may be difficult to derive. Unfortunately, as the real world is complicated, the required assumptions cannot be satisfied by a large number of substantive problems. Hence, there is a strong demand for new developments of new statistical methods for handling more general models and complex data structures. This strong demand produces the recent growth of SEMs.

The basic objective of this book is to introduce a Bayesian approach for analyzing not only the standard SEMs but also their useful generalizations which have been developed in recent years. In contrast to the existing covariance structure analysis approach, we focus on the use of the raw observations rather than the sample covariance matrix. To solve the difficulties that are induced

Structural Equation Modeling: A Bayesian Approach S-Y. Lee

by the complexities of the model and the data, the following general strategy has been emphasized and used repeatedly throughout this book. First, we treat the latent variables in the model, and the latent measurements (such as the real missing data, or the continuous measurements that are associated with the discrete data) as missing data then we analyze the model on the basis of the complete data set that contains the observed data and all the missing data by applying some powerful tools in statistical computing. As the complete data set is much easier to handle, the difficulties that are induced by the complexities of the model and the data are alleviated. The choice of the Bayesian approach is justified by the following points.

The basic attractive feature of a Bayesian approach is its flexibility to utilize useful prior information for achieving better results. In many practical problems, statisticians may have good prior information from some sources, for example the knowledge of experts and analyses of similar data and/or past data. For example, in research relating to organization and management that involves latent variables about job performance and job satisfaction, we may have some prior information about the correlation of these latent variables, say a relatively large value is one that is larger than 0.4; we also may have some prior information on the values of the factor loadings, say a relatively large loading that corresponds to 'salary' and 'job satisfaction'. For situations without accurate prior information, some type of non-informative prior distributions can be used in a Bayesian approach. In these cases, the accuracy of the Bayesian estimates is close to that of the ML estimates.

It is well known that the statistical properties of the ML approach are asymptotic. Hence, they are valid for situations with large sample sizes. In the context of some basic SEMs, many studies (see, for example, Boomsma, 1982; Chou, Bentler and Satorra, 1991; Hu, Bentler and Kano, 1992; Hoogland and Boomsma, 1998) have been devoted to study the behaviors of the ML asymptotic properties with small sample sizes. It was concluded by such research that the properties of the statistics are not robust for small sample sizes. The reason for this phenomenon is that the derivation of the statistics heavily depends on the important result that the sample covariance matrix $\mathbf{S}$ is asymptotically normal. However, even if the given data are normal, the distribution of $\mathbf{S}$ approaches normal only if the corresponding sample size is large. On the contrary, as pointed out by many important articles in Bayesian analyses of structural equation models (Ansari and Jedidi, 2000; Ansari, Jedidi and Dube, 2002; Ansari, Jedidi and Jagpal, 2000; Dunson, 2000; Scheines, Hoijtink and Boomsma, 1999; Lee and Song, 2004), the sampling-based Bayesian methods depend less on asymptotic theory, and hence have the potential to produce reliable results even with small samples.

The posterior distributions of parameters and latent variables can be estimated by using a sufficiently large number of observations that are simulated from the posterior distribution of the unknown parameters through efficient tools in statistical computing such as the various Markov chain Monte Carlo

(MCMC) methods. Means as well as quantiles of this posterior distribution can be estimated from the simulated observations. These quantities are useful in making statistical inferences. For example, the Bayesian estimates of the unknown parameters and the latent variables can be obtained from the corresponding sample means of the posterior distribution. From these estimates, estimated residuals can be obtained. In some complex situations, these estimated residuals are useful in assessing the goodness-of-fit of the proposed model and detecting outliers. Finally, the Bayes factor that is closely related with the Bayesian approach gives a more flexible and natural statistic for model comparison than the classical likelihood ratio test (see Kass and Raftery, 1995). We will give a detailed discussion on the model comparison with the Bayes factor in Chapter 5.

Before the 20th century, the Bayesian approach received little attention in SEM. Contributions are only limited to factor analysis (see, for example, Martin and McDonald, 1975; Lee, 1981; Bartholomew, 1981). More recently, the idea of data augmentation (Tanner and Wong, 1987) and the powerful tools in statistical computing for simulating observations from posterior distributions have greatly enhanced the applicability of the general Bayesian approach. A number of generalizations of the standard SEM have been separately developed by this approach. These include the developments of models with fixed covariates, nonlinear models, multilevel models, multisample models, mixture models, models with mixed continuous, dichotomous and/or ordered categorical variables, models with missing data, and models with data that are coming from an exponential family of distributions. These developments will be described in subsequent chapters. More importantly, we show that the freely available software WinBUGS has the potential to produce various Bayesian statistics, such as the Bayesian estimates, their standard error estimates and the estimates of the latent variables. Given the availability of WinBUGS, it is convenient for the applied researchers to use the Bayesian approach for applying either the standard SEMs or the complex SEMs to substantive problems.

The objective of this chapter is to provide an introduction of the Bayesian approach to SEMs. It is not intended to present a full coverage of the general Bayesian theory. Readers may refer to other excellent books, for example Box and Tiao (1973), and Gelman, Carlin, Stern and Rubin (1995) for more details of this general statistical method. This chapter begins with a section on the basic ideas of the Bayesian approach in estimation, including the discussion of the prior distribution and the posterior analysis by some MCMC methods. A Bayesian estimation of the CFA model is presented in Section 4.3, with a real example to illustrate the concept, and some simulation results on the accuracy of the Bayesian estimates in small samples. Section 4.4 considers Bayesian estimation of a LISREL type model. The final section demonstrates the use of WinBUGS to obtain Bayesian estimates of the parameters, standard error estimates and latent variable estimates. Some technical details are given in the Appendices to the chapter.

4.2 BASIC PRINCIPLES AND CONCEPTS OF BAYESIAN ANALYSIS OF SEMs

4.2.1 Bayesian Estimation

The Bayesian approach is well recognized in the statistics literature as an attractive approach to analyze a wide variety of models (Berger, 1985; Congdon, 2003). To introduce this approach for analyzing SEMs, we let M be an arbitrary SEM with a vector of unknown parameters $\boldsymbol{\theta}$, and let $\mathbf{Y}$ be the observed data set of raw observations with a sample size n. In a non-Bayesian approach, for example in an ML approach, $\boldsymbol{\theta}$ is not considered as random. In a Bayesian approach, $\boldsymbol{\theta}$ is considered to be random with a distribution (called prior distribution) and an associative (prior) density function, say, $p(\boldsymbol{\theta}|M)$ (see Berger, 1985, and the references therein for the theoretical and practical rationales for treating $\boldsymbol{\theta}$ as random). For simplicity, we use $p(\boldsymbol{\theta})$ to denote $p(\boldsymbol{\theta}|M)$. Bayesian inference is based on the observed data $\mathbf{Y}$ and the prior distribution of $\boldsymbol{\theta}$.

Let $p(\mathbf{Y}, \boldsymbol{\theta}|M)$ be the probability density function of the joint distribution of $\mathbf{Y}$ and $\boldsymbol{\theta}$ under M. The behavior of $\boldsymbol{\theta}$ under the given data $\mathbf{Y}$ is fully described by the conditional distribution of $\boldsymbol{\theta}$ given $\mathbf{Y}$. This conditional distribution is called the posterior distribution of $\boldsymbol{\theta}$. Let $p(\boldsymbol{\theta}|\mathbf{Y}, M)$ be the density function of the posterior distribution, which is called the posterior density function. The posterior distribution of $\boldsymbol{\theta}$ or its density plays the most important role in the Bayesian analysis of the model. Based on a well-known identity in probability, we have $p(\mathbf{Y}, \boldsymbol{\theta}|M) = p(\mathbf{Y}|\boldsymbol{\theta}, M)p(\boldsymbol{\theta}) = p(\boldsymbol{\theta}|\mathbf{Y}, M)p(\mathbf{Y}|M)$. As $p(\mathbf{Y}|M)$ does not depend on $\boldsymbol{\theta}$, and can be regarded as a constant with fixed $\mathbf{Y}$, we have

$$\log p(\boldsymbol{\theta}|\mathbf{Y}, M) \propto \log p(\mathbf{Y}|\boldsymbol{\theta}, M) + \log p(\boldsymbol{\theta}). \qquad (4.1)$$

Note that $p(\mathbf{Y}|\boldsymbol{\theta}, M)$ can be regarded as the likelihood function, because it is the probability density of $\mathbf{y}_1, \ldots, \mathbf{y}_n$ conditional on the parameter vector $\boldsymbol{\theta}$. It follows from Equation (4.1) that the posterior density function incorporates the sample information and the prior information through the likelihood function $p(\mathbf{Y}|\boldsymbol{\theta}, M)$ and the prior density function $p(\boldsymbol{\theta})$. Note also that $p(\mathbf{Y}|\boldsymbol{\theta}, M)$ depends on the sample size, whereas $p(\boldsymbol{\theta})$ does not. When the sample size becomes arbitrarily large, $\log(\mathbf{Y}|\boldsymbol{\theta}, M)$ could be very large and hence $\log p(\mathbf{Y}|\boldsymbol{\theta}, M)$ dominates $\log p(\boldsymbol{\theta})$. In this situation, the prior distribution of $\boldsymbol{\theta}$ plays a less important role, and the posterior density function $\log p(\boldsymbol{\theta}|\mathbf{Y}, M)$ is close to the log-likelihood function $\log p(\mathbf{Y}|\boldsymbol{\theta}, M)$. Hence, Bayesian and ML approaches are asymptotically equivalent, and the Baysian estimates have the same optimal properties as the ML estimates. When the sample sizes are small or moderate, the prior distribution of $\boldsymbol{\theta}$ plays a significant role in the Bayesian approach. Hence, in substantive research problems where the sample sizes are small or moderate, prior information about the parameter vector $\boldsymbol{\theta}$ can be incorporated into the Bayesian analysis through the prior distribution of $\boldsymbol{\theta}$ in order to achieve better results (see below for the utilization of useful prior information in the analysis). For many practical problems, researchers may have good prior information from experts, from analyses of similar or past data or from some other sources. More accurate results can be achieved by incorporating the appropriate prior information in the analysis through the prior distribution of $\boldsymbol{\theta}$. Thus, the selection of the prior density is an important issue in Bayesian analysis. In the following sections and chapters, the symbol M will be omitted if the context is clear. Moreover, '$p(\boldsymbol{\theta}|\mathbf{y})\stackrel{D}{=}$' will also be used to denote that 'the conditional distribution of $[\boldsymbol{\theta}|\mathbf{y}]$ is distributed as', if the context is clear.

4.2.2 Prior Distributions

Prior distribution of $\boldsymbol{\theta}$ represents the distribution of possible parameter values, from which the parameter $\boldsymbol{\theta}$ has been drawn. Basically, there are two kinds of prior distributions, namely the non-informative prior distributions and the informative prior distributions. Non-informative prior distributions associate with situations where the prior distributions have no population basis. They are used when we have little prior information, and hence the prior distributions play a minimal role in the posterior distribution. The associated prior density is regarded as vague, diffuse, flat or non-informative, for example density that is proportional to a constant or has an extremely huge variance. In this case, the Bayesian estimation is unaffected by information external to the observed data. For informative prior distribution, we may have prior knowledge about this distribution, either from closed related data or from the subjective knowledge of experts. Usually, an informative prior distribution has its own parameters, which are called hyperparameters.

A commonly used informative prior distribution in the general Bayesian approach to statistical problems is the conjugate prior distribution. Let us consider an example with the univariate binomial model. Considered as a function of θ, the likelihood of an observation y is of the form

$$p(y|\theta) = \binom{n}{y} \theta^y (1-\theta)^{n-y}.$$

If the prior density of θ is of the same form, then it can be seen from Equation (4.1) that the posterior density will also be of this form. More specifically, consider the prior density of θ:

$$p(\theta) \propto \theta^{\alpha-1}(1-\theta)^{\beta-1}, \tag{4.2}$$

which is a beta distribution with hyperparameters parameters α and β. Then,

$$\begin{aligned} p(\theta|y) &\propto p(y|\theta)p(\theta) \\ &\propto \theta^y(1-\theta)^{n-y}\theta^{\alpha-1}(1-\theta)^{\beta-1} \\ &= \theta^{y+\alpha-1}(1-\theta)^{n-y+\beta-1} \end{aligned} \tag{4.3}$$

which is a beta distribution with parameters $y+\alpha$ and $n-y+\beta$. We see that $p(\theta)$ and $p(\theta|y)$ are of the same form. The property that the posterior distribution follows the same parametric form as the prior distribution is called conjugacy, and the prior distribution is called a conjugate prior distribution (Gelman, Carlin, Stern and Rubin, 1995). In the above example, the beta prior distribution is a conjugate family for the binomial likelihood.

Consider another example with a sample of independently and identically distributed (iid) observations $y_1, \ldots, y_n$ from $N[\theta, \sigma^2]$, where σ^2 is known. Let $\mathbf{Y} = (y_1, \ldots, y_n)$, the likelihood function is

$$p(\mathbf{Y}|\theta) = \frac{1}{(2\pi)^{\frac{n}{2}}\ \sigma^n}\ \exp\left[-\frac{1}{2\sigma^2}\sum_{i=1}^{n}(y_i-\theta)^2\right].$$

Considered as a function of θ, the likelihood is an exponential of a quadratic form in θ. Thus, a conjugate prior distribution of θ can be parameterized as

$$p(\theta) \propto \exp\left[-\frac{1}{2\tau_0^2}(\theta-\mu_0)^2\right], \tag{4.4}$$

that is, $\theta \stackrel{D}{=} N[\mu_0, \tau_0^2]$, where μ_0 and τ_0^2 are hyperparameters. The posterior density is

$$\begin{aligned} p(\theta|\mathbf{Y}) &\propto p(\theta)p(\mathbf{Y}|\theta) \\ &\propto \exp\left[-\frac{1}{2\tau_0^2}\ (\theta-\mu_0)^2\right]p(\mathbf{Y}|\theta) \\ &\propto \exp\left\{-\frac{1}{2}\left[\frac{1}{\tau_0^2}\ (\theta-\mu_0)^2+\frac{1}{\sigma^2}\sum_{i=1}^{n}(y_i-\theta)^2\right]\right\} \end{aligned} \tag{4.5}$$

which can be shown to belong to the same distribution family as the distribution of θ.

If $\mathbf{Y}$ is from $N[\theta, \sigma^2]$ with θ known and σ^2 unknown, then

$$p(\mathbf{Y}|\sigma^2) \propto (\sigma^2)^{-n/2}\prod_{i=1}^{n}\exp\left\{\frac{1}{2\sigma^2}(y_i-\theta)^2\right\}$$

$$= (\sigma^2)^{-n/2} \exp\left(-\frac{n}{2\sigma^2} \, v\right),$$

where $v = n^{-1}\Sigma_{i=1}^{n}(y_i - \theta)^2$. The corresponding conjugate prior density of σ^2 is

$$p(\sigma^2) \propto (\sigma^2)^{-(\alpha_0+1)} \exp(-\beta_0/\sigma^2). \tag{4.6}$$

This is an inverted-gamma density function with hyperparameters α_0 and β_0. The posterior distribution of σ^2 is

$$\begin{aligned} p(\sigma^2|\mathbf{Y}) &\propto p(\sigma^2)p(\mathbf{Y}|\sigma^2) \\ &\propto (\sigma^2)^{-[n/2+(\alpha_0+1)]} \exp\left[-\frac{1}{2\sigma^2}(nv+2\beta_0)\right], \end{aligned} \tag{4.7}$$

which belong to the inverted-gamma distribution family. Equivalently, we may consider the conjugate prior distribution for σ^{-2} as gamma (α_0^*, β_0^*) with hyperparameters α_0^* and β_0^*. That is

$$p(\sigma^{-2}) \;\propto\; (\sigma^{-2})^{\alpha_0^*-1} \exp(-\beta_0^*\sigma^{-2}). \tag{4.8}$$

Again, the resulting posterior distribution belongs to the gamma distribution family.

Now consider the situation with a sample of i.i.d observations $y_1, \ldots, y_n$ from $N[\mu, \sigma^2]$, where μ and σ^2 are unknown. To obtain some idea for finding the conjugate prior distribution of μ and σ^2, we note that

$$p(\mu, \sigma^2|\mathbf{Y}) \propto p(\mathbf{Y}|\mu, \sigma^2)p(\mu, \sigma^2). \tag{4.9}$$

The likelihood $p(\mathbf{Y}|\mu, \sigma^2)$ is proportional to

$$\begin{aligned} &(\sigma^2)^{-n/2} \exp\left[-\frac{1}{2\sigma^2}\sum_{i=1}^{n}(y_i - \mu)^2\right] \\ &= (\sigma^2)^{-n/2} \exp\left\{-\frac{1}{2\sigma^2}[(n-1)s^2 + n(\bar{y} - \mu)^2]\right\}, \end{aligned} \tag{4.10}$$

where $s^2 = (n-1)^{-1}\sum_{i=1}^{n}(y_i - \bar{y})^2$ is the sample variance. Hence, if the prior distribution of (μ, σ^2) has the same form as given in Equation (4.10), we expect that the posterior density $p(\mu, \sigma^2|\mathbf{Y})$ will also have the same form. On the basis of the following equality,

$$p(\mu, \sigma^2) = p(\sigma^2)p(\mu|\sigma^2),$$

we need to specify $p(\mu|\sigma^2)$ and $p(\sigma^2)$. Motivated from the above discussion on treating σ^2 in a normal distribution, see Equation (4.7), a natural choice for the prior distribution of σ^2 is an inverted-gamma distribution, with hyperparameters α_0 and β_0, that is,

$$p(\sigma^2) \propto (\sigma^2)^{-(\alpha_0+1)} \exp(-\beta_0/\sigma^2). \tag{4.11}$$

A natural choice for the prior distribution of μ given σ^2 is a normal distribution, say $N[\mu_0, \sigma^2\tau_0^2]$, with hyperparameters μ_0 and τ_0^2, that is

$$p(\mu|\sigma^2) \propto (\sigma^2)^{-1/2} \exp[-(\mu-\mu_0)^2/(2\sigma^2\tau_0^2)]. \tag{4.12}$$

Combining Equations (4.11) and (4.12), the prior distribution of (μ, σ^2) is of the form

$$\begin{aligned} p(\mu, \sigma^2) &\propto (\sigma^2)^{-(\alpha_0+1)}(\sigma^2)^{-1/2} \exp(-\beta_0/\sigma^2)\exp[-(\mu-\mu_0)^2/(2\sigma^2\tau_0^2)], \\ &\propto (\sigma^2)^{-(\alpha_0+1)}(\sigma^2)^{-1/2} \exp\left[-\frac{1}{2\sigma^2}\left(2\beta_0 + \frac{(\mu-\mu_0)^2}{\tau_0^2}\right)\right], \end{aligned}$$

which is equivalent to the so-called normal-inverted gamma distribution. On the basis of this prior distribution, it follows from Equations (4.9) and (4.10) that the posterior density $p(\mu, \sigma^2|\mathbf{Y})$ is proportional to

$$\begin{aligned} &(\sigma^2)^{-(\alpha_0+1)} \exp(-\beta_0/\sigma^2) \times (\sigma^2)^{-1/2} \exp[-(\mu-\mu_0)^2/(2\sigma^2\tau_0^2)] \\ &\times (\sigma^2)^{-n/2} \exp\{-[(n-1)s^2 + n(\bar{y}-\mu_0)^2]/(2\sigma^2)\}, \end{aligned}$$

which can be shown to be in the form of a normal-inverted gamma distribution. Hence, a conjugate type prior distribution of (μ, σ^2) is given by Equations (4.11) and (4.12).

The above discussion on scalar parameters motivates the selection of conjugate type prior distributions for the parameters in Bayesian analyses of SEMs. For instance, considering the following factor analysis model corresponding to a measurement equation:

$$\mathbf{y}_i = \mathbf{\Lambda}\boldsymbol{\omega}_i + \boldsymbol{\epsilon}_i,$$

where $\boldsymbol{\omega}_i(q \times 1)$ is distributed as $N[\mathbf{0}, \mathbf{\Phi}]$, and $\boldsymbol{\epsilon}_i$ is distributed as $N[\mathbf{0}, \mathbf{\Psi}_\epsilon]$, where $\mathbf{\Psi}_\epsilon$ is diagonal with diagonal elements $\psi_{\epsilon k}$, and $\boldsymbol{\epsilon}_i$ and $\boldsymbol{\omega}_i$ are independent.

Let Λ_k^T be the kth row of Λ. A conjugate type prior distribution of $(\Lambda_k, \psi_{\epsilon k})$ is

$$\psi_{\epsilon k} \stackrel{D}{=} \text{Inverted Gamma } (\alpha_{0\epsilon k}^*, \beta_{0\epsilon k}^*) \text{ or equivalently} \psi_{\epsilon k}^{-1} \stackrel{D}{=} \text{Gamma } (\alpha_{0\epsilon k}, \beta_{0\epsilon k});$$
$$\text{and } [\Lambda_k | \psi_{\epsilon k}] \stackrel{D}{=} N\left[\Lambda_{0k}, \psi_{\epsilon k}\mathbf{H}_{0yk}\right], \tag{4.13}$$

where $\alpha_{0\epsilon k}, \beta_{0\epsilon k}, \alpha_{0\epsilon k}^*, \beta_{0\epsilon k}^*$, and elements in Λ_{0k} and $\mathbf{H}_{0yk}$ are hyperparameters, and $\mathbf{H}_{0yk}$ is a positive definite matrix. The conjugate prior distribution of Φ^{-1} is a q-dimensional Wishart distribution,

$$\Phi^{-1} \stackrel{D}{=} W_q[\mathbf{R}_0, \rho_0], \text{ or equivalently } \Phi \stackrel{D}{=} IW_q(\mathbf{R}_0^*, \rho_0) \tag{4.14}$$

where $W_q[\mathbf{R}_0, \rho_0]$ is a q-dimensional Wishart distribution with hyperparameters ρ_0 and a positive definite matrix $\mathbf{R}_0$, and $IW_q[\mathbf{R}_0^*, \rho_0]$ is an inverted Wishart distribution with hyperparameters ρ_0 and a positive definite matrix $\mathbf{R}_0^*$. This is a multivariate extension of the prior distribution of $\psi_{\epsilon k}^{-1}$ in Equation (4.13). As the forms of the structural equation and the factor analysis model are similar, analogous conjugate prior distributions as given above are also used for the parameters involved.

If the hyperparameters in the conjugate prior distributions are not known, then they may be treated as unknown parameters and thus have their own prior distributions in a full Bayesian analysis. These hyperprior distributions again have their own hyperparameters. As a result, the problem will become tedious. Hence, for convenience, we usually assign fixed known values to the hyperparameters in the conjugate prior distributions. In fact, existing works in Bayesian analysis of SEMs, such as Shi and Lee (1998), Lee and Zhu (1999), and Song and Lee (2001), used conjugate type prior distributions with given hyperparameters values. It has been shown that these distributions work well for many SEMs. Therefore, in this book, we will use the conjugate prior distributions in most of our Bayesian analyses.

In specifying conjugate prior distributions, we assign values to their hyperparameters. In practice, these preassigned values (prior inputs) represent the available prior knowledge. In general, if we have confidence to have good prior information about a parameter, then it is desirable to take a small variance in the corresponding prior distribution; otherwise a larger variance should be selected. For example, if we have confidence that the true Λ_k is not too far away from the preassigned hyperparameter value Λ_{0k}, then $\mathbf{H}_{0yk}$ should be taken as a matrix with small variances (such as $0.5\mathbf{I}$). The choice of $\alpha_{0\epsilon k}$ and $\beta_{0\epsilon k}$ is based on the same general rationale and the nature of the parameter $\psi_{\epsilon k}$ in the model. First we note that the distribution of the error measurement ϵ_k is $N[0, \psi_{\epsilon k}]$. Hence, if we think that the variation of ϵ_k with its mean value 0 is small (that is, $\Lambda_k^T \boldsymbol{\omega}_i$ is a good predictor for y_{ik}), then the prior distribution of

$\psi_{\epsilon k}$ should have a small mean value as well as a small variance. Otherwise, the prior distribution $\psi_{\epsilon k}$ should have a large mean value and/or a large variance as well. This gives some idea in choosing the hyperparameters $\alpha_{0\epsilon k}$ and $\beta_{0\epsilon k}$ in the inverted Gamma distribution (see Equation (4.13)). Note that for the inverted Gamma distribution, the mean is equal to $\beta_{0k}/(\alpha_{0k}-1)$, and the variance is equal to $\beta_{0k}^2/[(\alpha_{0k}-1)^2(\alpha_{0k}-2)]$. Hence, we may take $\alpha_{0k}=9$ and $\beta_{0k}=4$ for a situation where we have confidence that $\mathbf{\Lambda}_k^T\boldsymbol{\omega}_i$ is a good predictor of y_{ik} in the measurement equation. With this choice, the mean of $\psi_{\epsilon k}$ is $4/8=0.5$, and the variance of $\psi_{\epsilon k}$ is $4^2/(9-1)^2(9-2)=1/28$. For a situation with little confidence, we may take $\alpha_{0k}=6$ and $\beta_{0k}=10$, then the mean of $\psi_{\epsilon k}$ is 2.0 and the variance is 1.0. Now, consider the choice of $\mathbf{R}_0$ and ρ_0 in the prior distribution of $\mathbf{\Phi}$, see Equation (4.14). It follows from Muirhead (1982, p. 97) that the mean of $\mathbf{\Phi}$ is $\mathbf{R}_0^{-1}/\{\rho_0-q-1\}$. Hence, if we have confidence that $\mathbf{\Phi}$ is not too far away from $\mathbf{\Phi}_0$, we can choose $\mathbf{R}_0^{-1}$ and ρ_0 such that $\mathbf{R}_0^{-1}=(\rho_0-q-1)\mathbf{\Phi}_0$. Other values of $\mathbf{R}_0^{-1}$ and ρ_0 may be considered for situations without good prior information. Based on our experience in handling situations with nonaccurate prior inputs, large values of β_{0k} or ρ_0 should be taken so that the variance of the corresponding prior distribution is large enough.

Now, we discuss some methods to obtain $\mathbf{\Lambda}_{0k}$ and $\mathbf{R}_0$. As we mentioned before, these hyperparameter values may be obtained from the subjective knowledge of field experts and/or analysis of past or closely related data. If this kind of information is not available and the sample size is small, we may consider using non-informative prior distributions. Based on the general Bayesian framework in Zellner (1971), some common non-informative prior distributions in SEMs are

$$p(\mathbf{\Lambda},\mathbf{\Psi}_\epsilon)\propto p(\psi_{\epsilon 1},\ldots,\psi_{\epsilon p})\propto\prod_{k=1}^{p}\psi_{\epsilon k}^{-1}\text{ and }p(\mathbf{\Phi})\propto|\mathbf{\Phi}|^{-(q+1)/2}. \tag{4.15}$$

In Equation (4.15), the prior distribution of the unknown parameters in $\mathbf{\Lambda}$ is implicitly taken to be proportional to a constant. Note that no hyperparameters are involved in these non-informative prior distribution. Bayesian analysis on the basis of the above non-informative prior distribution is basically equivalent to the Bayesian analysis with conjugate prior distributions given by Equations (4.13) and (4.14) that have very large variances. If the sample is large, one possible method is to use a portion of the data, say one-half or less, to conduct an auxiliary Bayesian estimation with non-informative priors to obtain initial Bayesian estimates. Then, use the remaining data to do the actual Bayesian analysis, by utilizing the initial Bayesian estimates as hyperparameter values. For situations with moderate sample sizes, the Bayesian analysis may be done by applying data-dependent prior inputs that are obtained from the initial estimation of the whole data set. Although the above methods are reasonable, we emphasize that we are not routinely recommending them for every practical application. In general, the issue of selecting prior inputs should be carefully approached on a problem-by-problem basis. Moreover, it is desirable to conduct a sensitivity study to see whether the results are robust for prior inputs. This can be done by perturbing the given hyperparameter values or considering some ad hoc prior inputs.

4.2.3 Posterior Analysis

The Bayesian estimate of $\boldsymbol{\theta}$ is usually defined as the mean or the mode of $p(\boldsymbol{\theta}|\mathbf{Y})$. Martin and McDonald (1975) and Lee (1981) respectively obtained the Bayesian estimates of $\boldsymbol{\theta}$ in EFA and CFA models from the mode of $\log p(\boldsymbol{\theta}|\mathbf{Y})$. They worked directly with Equation (4.1) by finding the maximum of $\log p(\mathbf{Y}|\boldsymbol{\theta}) + \log\ p(\boldsymbol{\theta})$, or its equivalent form by using iterative procedures as described in Section 3.6. This procedure will encounter difficulties if $\log\ p(\mathbf{Y}|\boldsymbol{\theta})$ is intractable. A way of handling these difficulties is by applying the EM algorithm (Dempster, Laird and Rubin, 1977) for finding the maximum of $\log p(\mathbf{Y}|\theta) + \log p(\boldsymbol{\theta})$. In this book, we are mainly interested in estimating the mean of the posterior distribution of the unknown parameters.

Theoretically, the mean of the posterior distribution $[\boldsymbol{\theta}|\mathbf{Y}]$ could be obtained via integration. For most situations, the integration does not have a closed form. However, if we can simulate a sufficiently large number of observations from $[\boldsymbol{\theta}|\mathbf{Y}]$ (or from $p(\boldsymbol{\theta}|\mathbf{Y})$), we can approximate the mean, and/or other useful statistics, through the simulated observations. Hence, to solve the problem, it is important to develop efficient methods for drawing observations from the posterior distribution (or the posterior density). For most nonstandard SEMs, the posterior distribution $[\boldsymbol{\theta}|\mathbf{Y}]$ is complicated. It is difficult to derive this distribution and simulate observations from it. A major break through for posterior simulation is the idea of data augmentation that was proposed by Tanner and Wong (1987). The strategy is to treat latent quantities as hypothetical missing data and augment the observed data with them so that the posterior distribution (or the posterior density) based on the complete data set is relatively easy to analyze. This strategy has been found to be extremely useful in statistics; see, for example, Rubin (1991), Zeger and Karim (1991), Albert and Chib (1993), among many others. It is particularly useful for SEMs which involve latent variables. The feature that makes SEMs different from the common regression model and the simultaneous equation model is the existence of latent random variables. In fact, for many situations, the presence of latent variables causes the difficulties in analyzing the model. However, if the latent random variables are observed, the SEM will become the familiar regression model or simultaneous equation model that can be handled without much difficulty. For example, if the factor scores in a factor analytic model are not random but observed data, then the model becomes the regression model.

Hence, the above-mentioned strategy based on data augmentation provides an useful approach to solve the problem that is induced by latent variables. Augmenting the observed variables in complicated SEMs with the latent variables that are treated as hypothetical missing data, we usually can obtain the Bayesian solution on the basis of the complete-data set. More specifically, instead of working with the intractable posterior density $p(\boldsymbol{\theta}|\mathbf{Y})$, we will work with $p(\boldsymbol{\theta}, \boldsymbol{\Omega}|\mathbf{Y})$, where $\boldsymbol{\Omega}$ is the set of latent variables in the model. For most cases, $p(\boldsymbol{\theta}, \boldsymbol{\Omega}|\mathbf{Y})$ is still not in closed form and it is difficult to deal with it directly. However, on the basis of the complete-data set $(\mathbf{\Omega}, \mathbf{Y})$, the conditional distribution $[\boldsymbol{\theta}|\mathbf{\Omega}, \mathbf{Y}]$ is usually standard; moreover, the conditional distribution $[\mathbf{\Omega}|\boldsymbol{\theta}, \mathbf{Y}]$ can be derived from the definition of the model without much difficulty. Consequently, we can apply some MCMC methods to simulate observations from $p(\boldsymbol{\theta}, \mathbf{\Omega}|\mathbf{Y})$ by drawing observations iteratively from their full conditional densities $p(\boldsymbol{\theta}|\mathbf{\Omega}, \mathbf{Y})$ and $p(\mathbf{\Omega}|\boldsymbol{\theta}, \mathbf{Y})$. (Following the terminology in MCMC methods, we may call $p(\boldsymbol{\theta}|\mathbf{\Omega}, \mathbf{Y})$ and $p(\mathbf{\Omega}|\boldsymbol{\theta}, \mathbf{Y})$ conditional distributions, if the context is clear.) A useful algorithm to do this is the following Gibbs sampler (Geman and Geman, 1984).

In the model M, suppose the parameter vector $\boldsymbol{\theta}$ and the latent matrix $\mathbf{\Omega}$ are respectively decomposed into the following components or subvectors $\boldsymbol{\theta} = (\boldsymbol{\theta}_1, \ldots, \boldsymbol{\theta}_a)$ and $\mathbf{\Omega} = (\mathbf{\Omega}_1, \ldots, \mathbf{\Omega}_b)$. The Gibbs sampler is a Markov chain algorithm which performs an alternating conditional sampling at each of its iterations. It cycles through the components of $\boldsymbol{\theta}$ and $\mathbf{\Omega}$, drawing each component conditional on the values of all the other components. More specifically, at the jth iteration with current values $\boldsymbol{\theta}^{(j)} = (\boldsymbol{\theta}_1^{(j)}, \ldots, \boldsymbol{\theta}_a^{(j)})$ and $\mathbf{\Omega}^{(j)} = (\mathbf{\Omega}_1^{(j)}, \ldots, \mathbf{\Omega}_b^{(j)})$, it simulates in turn,

$$
\begin{array}{lll}
\boldsymbol{\theta}_1^{(j+1)} & \text{from} & p(\boldsymbol{\theta}_1|\boldsymbol{\theta}_2^{(j)}, \ldots, \boldsymbol{\theta}_a^{(j)}, \mathbf{\Omega}^{(j)}, \mathbf{Y}), \\
\boldsymbol{\theta}_2^{(j+1)} & \text{from} & p(\boldsymbol{\theta}_2|\boldsymbol{\theta}_1^{(j+1)}, \ldots, \boldsymbol{\theta}_a^{(j)}, \mathbf{\Omega}^{(j)}, \mathbf{Y}), \\
\vdots & & \vdots \\
\boldsymbol{\theta}_a^{(j+1)} & \text{from} & p(\boldsymbol{\theta}_a|\boldsymbol{\theta}_1^{(j+1)}, \ldots, \boldsymbol{\theta}_{a-1}^{(j+1)}, \mathbf{\Omega}^{(j)}, \mathbf{Y}), \\
\mathbf{\Omega}_1^{(j+1)} & \text{from} & p(\mathbf{\Omega}_1|\boldsymbol{\theta}^{(j+1)}, \mathbf{\Omega}_2^{(j)}, \cdots, \mathbf{\Omega}_b^{(j)}, \mathbf{Y}), \\
\mathbf{\Omega}_2^{(j+1)} & \text{from} & p(\mathbf{\Omega}_2|\boldsymbol{\theta}^{(j+1)}, \mathbf{\Omega}_1^{(j+1)}, \ldots, \mathbf{\Omega}_b^{(j)}, \mathbf{Y}), \\
\vdots & & \vdots \\
\mathbf{\Omega}_b^{(j+1)} & \text{from} & p(\mathbf{\Omega}_b|\boldsymbol{\theta}^{(j+1)}, \mathbf{\Omega}_1^{(j+1)}, \ldots, \mathbf{\Omega}_{b-1}^{(j+1)}, \mathbf{Y}).
\end{array}
\tag{4.16}
$$

There are $a+b$ steps in the jth iteration of the Gibbs sampler. At each step, each component in $\boldsymbol{\theta}$ and $\mathbf{\Omega}$ is updated conditional on the latest values of the other components. We may simulate the components in $\mathbf{\Omega}$ first, then the components in $\boldsymbol{\theta}$, or vice versa. Most of the full conditional distributions in (4.16) are the standard normal, gamma or Wishart distributions. Simulating observations from them is rather straightforward. For nonstandard conditional distributions, the Metropolis–Hastings (MH) algorithm (Metropolis *et al.*, 1953; Hastings, 1970) may have to be used for efficient simulation. A brief description of the MH algorithm is given in Appendix 4.1.

It has been shown (Geman & Geman, 1984; Geyer, 1992) that under mild regularity conditions, the joint distribution of $(\boldsymbol{\theta}^{(j)}, \mathbf{\Omega}^{(j)})$ converges at an exponential rate to the desired posterior distribution $[\boldsymbol{\theta}, \mathbf{\Omega}|\mathbf{Y}]$, after a sufficiently large number of iterations, say J. If the iterations have not proceeded long enough, the simulated observations may not be representative of the posterior distribution. Moreover, after the algorithm has reached approximate convergence, observations obtained at the early iterations should be discarded because they are still not part of the target distribution. The required number of iterations for achieving convergence of the Gibbs sampler, that is the burn-in iteration J, can be determined by plots of the simulated sequences of the individual parameters. At convergence, parallel sequences generated with different starting values should be mixed well together. Examples of sequences for which convergence looks reasonable, and sequences that have not reached convergence are presented in Figure 4.1. Another method for monitoring convergence of the Gibbs sampler is the following method as described in Gelman (1996). Based on different starting values of the structural parameters and latent variables, parallel sequences of observations are generated and the 'estimated potential scale reduction (EPSR)' values corresponding to the parameters are calculated sequentially as the runs proceed. As suggested by Gelman (1996), convergence of these sequences has been achieved where the EPSR values are all less than 1.2. The computation of the EPSR values is presented in Appendix 4.2. A minor problem with iterative simulation draws is their within-sequence correlation. In general, inference from correlated observations is less precise than from the same number of independent observations. To obtain a less correlated sample,

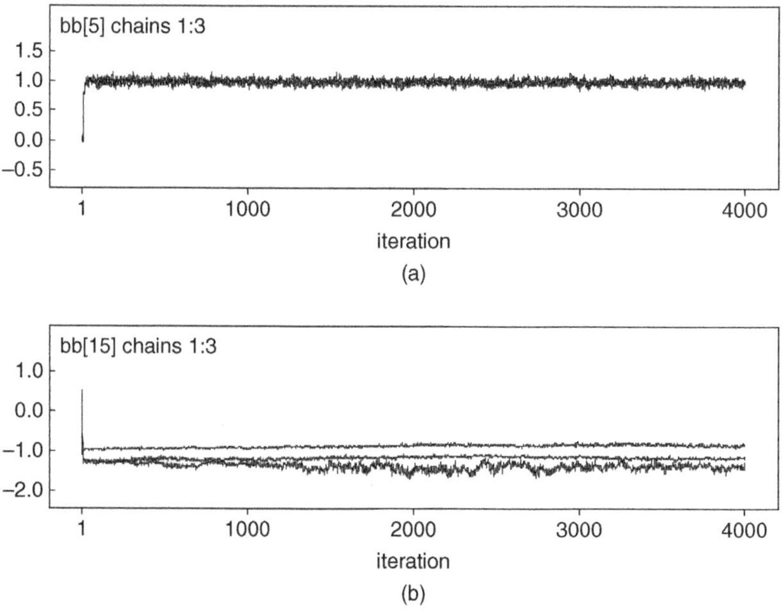

Figure 4.1 Sample traces of chains: (a) for which convergence looks reasonable; (b) which have not reached convergence.

observations may be collected in cycles with indices $J+c, J+2c, \ldots, J+T^*c$ for some spacing c (see Gelfand & Smith , 1990). However, in most practical applications a small c will suffice for many statistical analyses such as obtaining estimates of the parameters and standard errors (see Zeger and Karim (1991); Albert and Chib (1993)). In the numerical illustrations of the remaining chapters, we will use $c = 1$.

Statistical inference of the model can then be conducted on the basis of a simulated sample of observations from $p(\boldsymbol{\theta}, \boldsymbol{\Omega}|\mathbf{Y})$, namely $\{(\boldsymbol{\theta}^{(t)}, \boldsymbol{\Omega}^{(t)}) : t = 1, \ldots, T^*\}$. The Bayesian estimate of $\boldsymbol{\theta}$ as well as the numerical standard error estimates can be obtained from

$$\hat{\boldsymbol{\theta}} = T^{*-1} \sum_{t=1}^{T^*} \boldsymbol{\theta}^{(t)}, \tag{4.17}$$

$$\widehat{\mathrm{Var}}(\boldsymbol{\theta}|\mathbf{Y}) = (T^* - 1)^{-1} \sum_{t=1}^{T^*} (\boldsymbol{\theta}^{(t)} - \hat{\boldsymbol{\theta}})(\boldsymbol{\theta}^{(t)} - \hat{\boldsymbol{\theta}})^T. \tag{4.18}$$

It can be shown that (Geyer, 1992) $\hat{\boldsymbol{\theta}}$ tends to $E(\boldsymbol{\theta}|\mathbf{Y})$ as T^* tends to infinity. It can be seen from Equation (4.1) that $\hat{\boldsymbol{\theta}}$ has the same large sample properties as the ML estimate. Other statistical inference on $\boldsymbol{\theta}$ can be carried out based on the simulated sample, $\{\boldsymbol{\theta}^{(t)} : t = 1, \ldots, T^*\}$. For instance, the 2.5 %, 50 % and 97.5 % points of the sampled distribution of an individual parameter give a 95 % posterior interval and convey skewness in its marginal posterior density. The construction of the posterior interval does not depend on any asymptotic results. The total number of draws, T^*, that is required for statistical analysis depends on the form of the posterior distribution. Clearly, different choices of sufficiently large T^* would produce close estimates, although they may not be exactly equal.

As the posterior distribution of $\boldsymbol{\theta}$ given $\mathbf{Y}$ describes the distributional behaviors of $\boldsymbol{\theta}$ with the given data, the dispersion of $\boldsymbol{\theta}$ can be assessed through $\mathrm{var}(\boldsymbol{\theta}|\mathbf{Y})$, with an estimate given by Equation (4.18), based on the sample covariance matrix of the simulated observations. Let θ_k be the kth element of $\boldsymbol{\theta}$. The positive square root of the kth diagonal element in $\widehat{\mathrm{Var}}(\boldsymbol{\theta}|\mathbf{Y})$ is commonly taken as the standard error estimate of the Bayesian estimate $\hat{\theta}_k$. While this estimate provides some information about the variation of $\hat{\theta}_k$, it may not be appropriate to construct a 'z-score' in hypothesis testing for some complex SEMs with non-standard data. In general Bayesian analysis, the issue of hypothesis testing is formulated as a model comparison problem, and is handled by some model comparison statistics such as the Bayes factor. (see Chapter 5 for more detailed discussions).

For any individual $\mathbf{y}_i$, let $\boldsymbol{\omega}_i$ be the vector of latent variables, $E(\boldsymbol{\omega}_i|\mathbf{y}_i)$ and $\mathrm{Var}(\boldsymbol{\omega}_i|\mathbf{y}_i)$ be the posterior mean and the posterior covariance matrix.

A Bayesian estimate $\hat{\boldsymbol{\omega}}_i$ can be obtained through $\{\boldsymbol{\Omega}^{(t)}, t = 1, \ldots, T^*\}$ as follows:

$$\hat{\boldsymbol{\omega}}_i = T^{*-1} \sum_{t=1}^{T^*} \boldsymbol{\omega}_i^{(t)}, \tag{4.19}$$

where $\boldsymbol{\omega}_i^{(t)}$ is the ith column of $\boldsymbol{\Omega}^{(t)}$. This gives a direct Bayesian estimate which is not expressed in terms of the structural parameter estimates. Hence, in contrast to the classical methods in estimating latent variables, no sampling errors of the estimates are involved in the Bayesian method. It can be shown (Geyer, 1992) that $\hat{\boldsymbol{\omega}}_i$ is a consistent estimate of $E(\boldsymbol{\omega}_i|\mathbf{y}_i)$. A consistent estimate of Var $(\boldsymbol{\omega}_i|\mathbf{y}_i)$ can be obtained as

$$\widehat{\mathrm{Var}}(\boldsymbol{\omega}_i|\mathbf{y}_i) = (T^* - 1)^{-1} \sum_{t=1}^{T^*} (\boldsymbol{\omega}_i^{(t)} - \hat{\boldsymbol{\omega}}_i)(\boldsymbol{\omega}_i^{(t)} - \hat{\boldsymbol{\omega}}_i)^T. \tag{4.20}$$

In practice, numerical standard error estimates of elements in $\hat{\boldsymbol{\omega}}_i$ can be obtained via square roots of diagonal elements in $\widehat{\mathrm{Var}}(\boldsymbol{\omega}_i|\mathbf{y}_i)$. Therefore, the proposed estimation procedure based on the Gibbs sampler also produces Bayesian estimates of the latent variables as well as their numerical standard errors estimates. These estimates $\hat{\boldsymbol{\omega}}_i$ can be used for outlier and residual analysis, and for assessing the goodness-of-fit of the measurement equation or the structural equation, particularly in analyzing complicated SEMs (see more discussions on this issue in Chapter 8, Section 8.4.3). It should be noted that as the data information for estimating $\hat{\boldsymbol{\omega}}_i$ is only given by the single observation $\mathbf{y}_i$, we cannot expect $\hat{\boldsymbol{\omega}}_i$ to be an accurate estimate of the true latent variable $\boldsymbol{\omega}_{i_0}$ (see the simulation study reported in Lee and Shi (2000) on the estimation of factor scores in a factor analysis model). However, the empirical distribution of the Bayesian estimates $\{\hat{\boldsymbol{\omega}}_1, \ldots, \hat{\boldsymbol{\omega}}_n\}$ is close to the distribution of the true factor scores $\{\hat{\boldsymbol{\omega}}_{1_0}, \ldots, \hat{\boldsymbol{\omega}}_{n0}\}$ (Shi and Lee (1998)).

Before ending this section, we emphasize the importance of the following strategy in analyzing complex SEMs: (i) Apply the idea of data augmentation to augment the observed data $\mathbf{Y}$ with the unknown quantities, which could be the latent variables, missing data, etc., that caused the difficulties in the problem, and then work with the joint posterior distribution in the posterior analysis. (ii) Apply MCMC tools in statistical computing to draw observations from the full conditional distributions of the joint posterior distribution. This strategy will be repeatedly applied in subsequent chapters.

4.3 BAYESIAN ESTIMATION OF THE CFA MODEL

To illustrate the Bayesian estimation and the associative MCMC method described in Section 4.2, we present a detailed application in the context of

CFA model. Consider the following CFA model which is equivalent to that as defined in Equation (2.3) of Chapter 2. For $i = 1, \ldots, n$

$$\mathbf{y}_i = \mathbf{\Lambda}\boldsymbol{\omega}_i + \boldsymbol{\epsilon}_i, \tag{4.21}$$

where $\mathbf{y}_i$ is a $p \times 1$ observed random vector, $\mathbf{\Lambda}$ is a $p \times q$ factor loading matrix, $\boldsymbol{\omega}_i$ is a $q \times 1$ vector of factor scores and $\boldsymbol{\epsilon}_i$ is a $p \times 1$ random vector of error measurements which is independent of $\boldsymbol{\omega}_i$. Suppose $\boldsymbol{\epsilon}_i$ is distributed as $N[\mathbf{0}, \mathbf{\Psi}_\epsilon]$, where $\mathbf{\Psi}_\epsilon$ is a diagonal matrix, and $\boldsymbol{\omega}_i$ is distributed as $N[\mathbf{0}, \mathbf{\Phi}]$ with some positive definite covariance matrix $\mathbf{\Phi}$.

Let $\mathbf{Y} = (\mathbf{y}_1, \ldots, \mathbf{y}_n)$ be the observed data matrix, $\mathbf{\Omega} = (\boldsymbol{\omega}_1, \ldots, \boldsymbol{\omega}_n)$ be the matrix of latent factor scores, and $\boldsymbol{\theta}$ be the structural parameter vector that contains the unknown elements of $\mathbf{\Lambda}$, $\mathbf{\Phi}$ and $\mathbf{\Psi}_\epsilon$ in the model. It is assumed that this CFA model is identified. One common method to achieve this assumption is to set some appropriate elements in $\mathbf{\Lambda}$ to fixed known values. From a Bayesian point of view, this is equivalent to assigning the fixed values to the corresponding parameters with probability one. In the analysis, fixed parameters are not estimated. In the Bayesian analysis, we will treat the latent factor scores in $\mathbf{\Omega}$ as hypothetical missing data, and augment the observed data set $\mathbf{Y}$ with $\mathbf{\Omega}$ in the posterior analysis. A sufficiently large sample of $(\boldsymbol{\theta}, \mathbf{\Omega})$ from the joint posterior distribution $[\boldsymbol{\theta}, \mathbf{\Omega}|\mathbf{Y}]$ is generated by the following Gibbs sampler algorithm. At the $(j+1)$th iteration with current values of $\mathbf{\Omega}^{(j)}$, $\mathbf{\Psi}_\epsilon^{(j)}$, $\mathbf{\Lambda}^{(j)}$ and $\mathbf{\Phi}^{(j)}$:

$$\begin{aligned}
&\text{(i)} \quad \text{Generate } \mathbf{\Omega}^{(j+1)} \text{ from } p(\mathbf{\Omega}|\mathbf{\Psi}_\epsilon^{(j)}, \mathbf{\Lambda}^{(j)}, \mathbf{\Phi}^{(j)}, \mathbf{Y}).\\
&\text{(ii)} \quad \text{Generate } \mathbf{\Psi}_\epsilon^{(j+1)} \text{ from } p(\mathbf{\Psi}_\epsilon|\mathbf{\Omega}^{(j+1)}, \mathbf{\Lambda}^{(j)}, \mathbf{\Phi}^{(j)}, \mathbf{Y}).\\
&\text{(iii)} \quad \text{Generate } \mathbf{\Lambda}^{(j+1)} \text{ from } p(\mathbf{\Lambda}|\mathbf{\Omega}^{(j+1)}, \mathbf{\Psi}_\epsilon^{(j+1)}, \mathbf{\Phi}^{(j)}, \mathbf{Y}).\\
&\text{(iv)} \quad \text{Generate } \mathbf{\Phi}^{(j+1)} \text{ from } p(\mathbf{\Phi}|\mathbf{\Omega}^{(j+1)}, \mathbf{\Psi}_\epsilon^{(j+1)}, \mathbf{\Lambda}^{(j+1)}, \mathbf{Y}).
\end{aligned} \tag{4.22}$$

The conditional distributions involved in steps (4.22) are required in the implementation of the Gibbs sampler.

4.3.1 Conditional Distributions

The derivation of $p(\mathbf{\Omega}|\mathbf{\Psi}_\epsilon, \mathbf{\Lambda}, \mathbf{\Phi}, \mathbf{Y}) = p(\mathbf{\Omega}|\boldsymbol{\theta}, \mathbf{Y})$ is based on the definition of the model and the distributional properties of the random vectors $\mathbf{y}_i$ and $\boldsymbol{\omega}_i$. Note that for $i = 1, \ldots, n$, $\boldsymbol{\omega}_i$ are mutually independent, and $\mathbf{y}_i$ are also mutually independent given $(\boldsymbol{\omega}_i, \boldsymbol{\theta})$. Hence, we have

$$p(\mathbf{\Omega}|\mathbf{Y}, \boldsymbol{\theta}) = \prod_{i=1}^{n} p(\boldsymbol{\omega}_i|\mathbf{y}_i, \boldsymbol{\theta}) \propto \prod_{i=1}^{n} p(\boldsymbol{\omega}_i|\boldsymbol{\theta})\, p(\mathbf{y}_i|\boldsymbol{\omega}_i, \boldsymbol{\theta}). \tag{4.23}$$

Moreover, since the conditional distributions of $\boldsymbol{\omega}_i$ given $\boldsymbol{\theta}$, and $\mathbf{y}_i$ given $(\boldsymbol{\omega}_i, \boldsymbol{\theta})$ are $N(\mathbf{0}, \boldsymbol{\Phi})$, and $N(\boldsymbol{\Lambda}\boldsymbol{\omega}_i, \boldsymbol{\Psi}_\epsilon)$ respectively, it can be shown (Lindley and Smith (1972), pp. 4–5) that the conditional distribution of $\boldsymbol{\omega}_i$ given $(\mathbf{y}_i, \boldsymbol{\theta})$ is equal to

$$[\boldsymbol{\omega}_i|\mathbf{y}_i, \boldsymbol{\theta}] \stackrel{D}{=} N\left[(\boldsymbol{\Phi}^{-1}+\boldsymbol{\Lambda}^T\boldsymbol{\Psi}_\epsilon^{-1}\boldsymbol{\Lambda})^{-1}\boldsymbol{\Lambda}^T\boldsymbol{\Psi}_\epsilon^{-1}\mathbf{y}_i, (\boldsymbol{\Phi}^{-1}+\boldsymbol{\Lambda}^T\boldsymbol{\Psi}_\epsilon^{-1}\boldsymbol{\Lambda})^{-1}\right]. \tag{4.24}$$

Hence, the conditional distribution of $\boldsymbol{\Omega}$ given $(\mathbf{Y}, \boldsymbol{\theta})$ can be obtained from Equations (4.23) and (4.24).

The conditional distribution of $\boldsymbol{\theta}$ given $(\mathbf{Y}, \boldsymbol{\Omega})$ is proportional to $p(\boldsymbol{\theta})p(\mathbf{Y}, \boldsymbol{\Omega}|\boldsymbol{\theta})$. Hence, it is necessary to select the prior density function $p(\boldsymbol{\theta})$ that represents the prior information of $\boldsymbol{\theta}$. Based on the factor analysis model as defined in Equation (4.21), we first note that with given $\boldsymbol{\Omega}$ the underlying CFA model becomes a regression model with parameters $\boldsymbol{\Lambda}$ and $\boldsymbol{\Psi}_\epsilon$ only. On the other hand, the parameter matrix $\boldsymbol{\Phi}$ is only involved in the distribution of $\boldsymbol{\omega}_i$. Hence, it is reasonable to assume that the prior distributions of $(\boldsymbol{\Lambda}, \boldsymbol{\Psi}_\epsilon)$ and $\boldsymbol{\Phi}$ are independent. As a result, we specify the prior distribution of $\boldsymbol{\theta}$ as follows:

$$p(\boldsymbol{\theta}) = p(\boldsymbol{\Lambda}, \boldsymbol{\Phi}, \boldsymbol{\Psi}_\epsilon) = p(\boldsymbol{\Lambda}, \boldsymbol{\Psi}_\epsilon)\, p(\boldsymbol{\Phi}). \tag{4.25}$$

Moreover, the conditional distribution of $\mathbf{Y}$ given $\boldsymbol{\Omega}$ only depends on $\boldsymbol{\Lambda}$ and $\boldsymbol{\Psi}_\epsilon$, and the distribution of $\boldsymbol{\Omega}$ only involves $\boldsymbol{\Phi}$. Consequently, we assume that

$$\begin{aligned} p(\boldsymbol{\Lambda}, \boldsymbol{\Psi}_\epsilon, \boldsymbol{\Phi}|\mathbf{Y}, \boldsymbol{\Omega}) &= p(\boldsymbol{\theta}|\mathbf{Y}, \boldsymbol{\Omega}) \propto p(\mathbf{Y}, \boldsymbol{\Omega}|\boldsymbol{\theta})p(\boldsymbol{\theta}) = p(\mathbf{Y}|\boldsymbol{\theta}, \boldsymbol{\Omega})p(\boldsymbol{\Omega}|\boldsymbol{\theta})p(\boldsymbol{\theta}) \\ &= p(\mathbf{Y}|\boldsymbol{\theta}, \boldsymbol{\Omega})p(\boldsymbol{\Omega}|\boldsymbol{\theta})p(\boldsymbol{\Lambda}, \boldsymbol{\Psi}_\epsilon)p(\boldsymbol{\Phi}) \\ &= [p(\boldsymbol{\Lambda}, \boldsymbol{\Psi}_\epsilon)p(\mathbf{Y}|\boldsymbol{\Lambda}, \boldsymbol{\Psi}_\epsilon, \boldsymbol{\Omega})][p(\boldsymbol{\Omega}|\boldsymbol{\Phi})p(\boldsymbol{\Phi})]. \end{aligned} \tag{4.26}$$

Since the first term of the product on the right-hand side of Equation (4.26) depends only on $(\boldsymbol{\Lambda}, \boldsymbol{\Psi}_\epsilon)$ while the second term depends only on $\boldsymbol{\Phi}$, the marginal conditional densities $p(\boldsymbol{\Lambda}, \boldsymbol{\Psi}_\epsilon|\mathbf{Y}, \boldsymbol{\Omega})$ and $p(\boldsymbol{\Phi}|\mathbf{Y}, \boldsymbol{\Omega})$ are proportional to $p(\boldsymbol{\Lambda}, \boldsymbol{\Psi}_\epsilon)p(\mathbf{Y}|\boldsymbol{\Lambda}, \boldsymbol{\Psi}_\epsilon, \boldsymbol{\Omega})$ and $p(\boldsymbol{\Omega}|\boldsymbol{\Phi})p(\boldsymbol{\Phi})$, respectively. As a result, these densities can be treated separately.

The following conjugate type prior distributions are considered as prior distributions for $(\boldsymbol{\Lambda}, \boldsymbol{\Psi}_\epsilon)$ and $\boldsymbol{\Phi}$. Let $\psi_{\epsilon k}$ and $\boldsymbol{\Lambda}_k^T$ be the kth diagonal elements of $\boldsymbol{\Psi}_\epsilon$ and the kth row of $\boldsymbol{\Lambda}$, respectively. For any $k \neq h$, we assume that the prior distribution of $(\psi_{\epsilon k}, \boldsymbol{\Lambda}_k)$ is independent of $(\psi_{\epsilon h}, \boldsymbol{\Lambda}_h)$. Moreover, we take

$$\begin{aligned} &\psi_{\epsilon k}^{-1} \stackrel{D}{=} Gamma[\alpha_{0\epsilon k}, \beta_{0\epsilon k}], \\ &[\boldsymbol{\Lambda}_k|\psi_{\epsilon k}] \stackrel{D}{=} N[\boldsymbol{\Lambda}_{0k}, \psi_{\epsilon k}\mathbf{H}_{0yk}], \text{ and } \boldsymbol{\Phi}^{-1} \stackrel{D}{=} W_q[\mathbf{R}_0, \rho_0], \end{aligned} \tag{4.27}$$

where $Gamma(\alpha, \beta)$ represents the gamma distribution with shape parameter $\alpha > 0$ and inverse scale parameter $\beta > 0$, $W_q[\cdot, \cdot]$ denotes an q-dimensional Wishart distribution, $\alpha_{0\epsilon k}$, $\beta_{0\epsilon k}$, $\mathbf{\Lambda}_{0k}$, ρ_0 and the positive definite matrices $\mathbf{H}_{0yk}$ and $\mathbf{R}_0$ are hyperparameters whose values are assumed to be given from the prior information of previous studies or other sources.

Let $\mathbf{Y}_k^T$ be the kth row of $\mathbf{Y}$, $\mathbf{A}_k = (\mathbf{H}_{0yk}^{-1} + \mathbf{\Omega}\mathbf{\Omega}^T)^{-1}$, $\mathbf{a}_k = \mathbf{A}_k(\mathbf{H}_{0yk}^{-1}\mathbf{\Lambda}_{0k} + \mathbf{\Omega}\mathbf{Y}_k)$, and $\beta_{\epsilon k} = \beta_{0\epsilon k} + 2^{-1}(\mathbf{Y}_k^T\mathbf{Y}_k - \mathbf{a}_k^T\mathbf{A}_k^{-1}\mathbf{a}_k + \mathbf{\Lambda}_{0k}^T\mathbf{H}_{0yk}^{-1}\mathbf{\Lambda}_{0k})$, it can be shown as in Appendix 4.3 that for $k = 1, \ldots, p$, the conditional distribution of $(\mathbf{\Lambda}_k, \psi_{\epsilon k}^{-1})$ given $\mathbf{Y}$ and $\mathbf{\Omega}$ is independently distributed as the following normal–gamma distribution (Broemeling, 1985):

$$[\psi_{\epsilon k}^{-1}|\mathbf{Y}, \mathbf{\Omega}] \stackrel{D}{=} Gamma\,[n/2 + \alpha_{0\epsilon k},\ \beta_{\epsilon k}], \quad \text{and} \quad [\mathbf{\Lambda}_k|\mathbf{Y}, \mathbf{\Omega}, \psi_{\epsilon k}^{-1}] \stackrel{D}{=} N\,[\mathbf{a}_k, \psi_{\epsilon k}\mathbf{A}_k]. \tag{4.28}$$

Since $p(\mathbf{\Lambda}_k, \psi_{\epsilon k}^{-1}|\mathbf{Y}, \mathbf{\Omega}) = p(\psi_{\epsilon k}^{-1}|\mathbf{Y}, \mathbf{\Omega})p(\mathbf{\Lambda}_k|\mathbf{Y}, \mathbf{\Omega}, \psi_{\epsilon k}^{-1})$, the conditional distribution of $(\mathbf{\Lambda}_k, \psi_{\epsilon k}^{-1})$ given $(\mathbf{Y}, \mathbf{\Omega})$ can be obtained via Equations (4.28). In the implementation of the Gibbs sampler, we just simulate an observation $\psi_{\epsilon k}^{-1}$ from Equations (4.28), then we can obtain an observation $\psi_{\epsilon k}$ by taking the inverse. This $\psi_{\epsilon k}$ will be used in the other parts of the program, for example, for simulating $\mathbf{\Lambda}_k$ in Equation (4.28). This method will be used in simulating the variances of the error measurements throughout this book.

To derive $p(\mathbf{\Phi}|\mathbf{Y}, \mathbf{\Omega})$, we first note from Equation (4.26) that it is proportional to $p(\mathbf{\Phi})p(\mathbf{\Omega}|\mathbf{\Phi})$. As $\boldsymbol{\omega}_i$ are independent, we have

$$p(\mathbf{\Phi}|\mathbf{Y}, \mathbf{\Omega}) \propto p(\mathbf{\Phi}) \prod_{i=1}^{n} p(\boldsymbol{\omega}_i|\mathbf{\Phi}). \tag{4.29}$$

From the prior distribution of $\mathbf{\Phi}^{-1}$ given in Equation (4.27), it implies that $\mathbf{\Phi} \stackrel{D}{=} IW_q[\mathbf{R}_0^{-1}, \rho_0]$, where $IW_q[\mathbf{R}_0^{-1}, \rho_0]$ denotes an q-dimensional inverted Wishart distribution. Moreover, since the distribution of $\boldsymbol{\omega}_i$ given $\mathbf{\Phi}$ is $N(\mathbf{0}, \mathbf{\Phi})$, we have

$$\begin{aligned} p(\mathbf{\Phi}|\mathbf{Y}, \mathbf{\Omega}) &\propto \left[|\mathbf{\Phi}|^{-(\rho_0+q+1)/2} \exp\left(-\frac{1}{2}\mathrm{tr}[\mathbf{R}_0^{-1}\mathbf{\Phi}^{-1}]\right)\right] \\ &\quad \times \left\{|\mathbf{\Phi}|^{-n/2} \exp\left[-\frac{1}{2}\sum_{i=1}^{n} \boldsymbol{\omega}_i^T\mathbf{\Phi}^{-1}\boldsymbol{\omega}_i\right]\right\} \\ &= |\mathbf{\Phi}|^{-(n+\rho_0+q+1)/2} \exp\left\{-\frac{1}{2}\ \mathrm{tr}[\mathbf{\Phi}^{-1}(\mathbf{\Omega}\mathbf{\Omega}^T + \mathbf{R}_0^{-1})]\right\}. \end{aligned} \tag{4.30}$$

Since the right-hand side of Equation (4.30) is proportional to the density function of an inverted Wishart distribution (see Zellner, 1971), we have

$$[\mathbf{\Phi}|\mathbf{Y}, \mathbf{\Omega}] \stackrel{D}{=} IW_q[(\mathbf{\Omega}\mathbf{\Omega}^T + \mathbf{R}_0^{-1}),\ n + \rho_0]. \tag{4.31}$$

The above derivation can be extended to handle the general situation with fixed known elements in $\mathbf{\Lambda}$ as follows. Consider $\mathbf{\Lambda}_k^T$, the kth row of $\mathbf{\Lambda}$ with certain fixed parameters. Let $\mathbf{c}_k$ be the corresponding $1 \times q$ row vector such that $c_{kj} = 0$ if λ_{kj} is a fixed parameter; and $c_{kj} = 1$ if λ_{kj} is an unknown parameter, for $k = 1, \ldots, p, j = 1, \ldots, q$, and $r_k = c_{k1} + \ldots + c_{kq}$. Moreover, let $\mathbf{\Lambda}_k^{*T}$ be the 1 by r_k row vector that contains the unknown parameters in $\mathbf{\Lambda}_k$; and let $\mathbf{\Omega}_k^*$ be the r_k by n submatrix of $\mathbf{\Omega}$ such that for $j = 1, \ldots, r_k$, all the rows corresponding to $c_{kj} = 0$ are deleted. Let $\mathbf{Y}_k^{*T} = (y_{1k}^*, \ldots, y_{nk}^*)$ with

$$y_{ik}^* = y_{ik} - \sum_{j=1}^{q} \lambda_{kj} \omega_{ij} (1 - c_{kj}).$$

As an example, let $\Lambda_k^T = (1, \lambda_{k2}, \lambda_{k3}, 0)$, where 1 and 0 are fixed. Then $\mathbf{c}_k = (0, 1, 1, 0)$, $r_k = 2$, $\mathbf{\Lambda}_k^{*T} = (\lambda_{k2}, \lambda_{k3})$, $y_{ik}^* = y_{ik} - \omega_{i1}$, and

$$\mathbf{\Omega}_k^* = \begin{bmatrix} \omega_{21} & \ldots & \omega_{2n} \\ \omega_{31} & \ldots & \omega_{3n} \end{bmatrix},$$

which is obtained by deleting the first and the fourth rows of $\mathbf{\Omega}$.

Suppose the conjugate prior distribution defined in Equation (4.27) about the loading matrix is

$$[\mathbf{\Lambda}_k^* | \psi_{\epsilon k}] \stackrel{D}{=} N[\mathbf{\Lambda}_{0k}^*, \ \psi_{\epsilon k} \mathbf{H}_{0yk}^*],$$

for some hyperparameters $\mathbf{\Lambda}_{0k}^*$ and $\mathbf{H}_{0yk}^*$. Let $\mathbf{A}_k^* = (\mathbf{H}_{0yk}^{*-1} + \mathbf{\Omega}_k^* \mathbf{\Omega}_k^{*T})^{-1}$, $\mathbf{a}_k^* = \mathbf{A}_k^* (\mathbf{H}_{0yk}^{*-1} \mathbf{\Lambda}_{0k}^* + \mathbf{\Omega}_k^* \mathbf{Y}_k^*)$, and $\beta_{\epsilon k} = \beta_{0\epsilon k} + \frac{1}{2}(\mathbf{Y}_k^{*T} \mathbf{Y}_k^* - \mathbf{a}_k^{*T} \mathbf{A}_k^{*-1} \mathbf{a}_k^* + \mathbf{\Lambda}_{0k}^{*T} \mathbf{H}_{0yk}^{*-1} \mathbf{\Lambda}_{0k}^*)$. Then, for $k = 1, \ldots, p$, it can be shown from exactly the same reasoning as given above and in Appendix 4.3 that the posterior distributions of $(\mathbf{\Lambda}_k^*, \psi_{\epsilon k}^{-1})$ and $\mathbf{\Phi}$ corresponding to the conjugate priors are respectively given by:

$$[\psi_{\epsilon k}^{-1} | \mathbf{Y}, \mathbf{\Omega}] \stackrel{D}{=} Gamma[n/2 + \alpha_{0\epsilon k}, \beta_{\epsilon k}], \ [\mathbf{\Lambda}_k^* | \mathbf{Y}, \mathbf{\Omega}, \psi_{\epsilon k}^{-1}] \stackrel{D}{=} N[\mathbf{a}_k^*, \psi_{\epsilon k} \mathbf{A}_k^*],$$
$$\text{and } [\mathbf{\Phi} | \mathbf{Y}, \mathbf{\Omega}] \stackrel{D}{=} IW_q[(\mathbf{\Omega}\mathbf{\Omega}^T + \mathbf{R}_0^{-1}), \ n + \rho_0]. \tag{4.32}$$

It can be shown by a similar derivation, as in Appendix 4.3, that for $k = 1, \ldots, p$ the posterior distributions of $(\mathbf{\Lambda}_k^*, \mathbf{\Omega}, \psi_{\epsilon k})$ obtaining from the non-informative prior distributions given in (4.15) are:

$$[\psi_{\epsilon k}^{-1} | \mathbf{Y}, \mathbf{\Omega}] \stackrel{D}{=} Gamma[(n - r_k)/2, 2^{-1} \mathbf{Y}_k^{*T} \{\mathbf{I}_n - \mathbf{\Omega}_k^{*T} (\mathbf{\Omega}_k^* \mathbf{\Omega}_k^{*T})^{-1} \mathbf{\Omega}_k^*\} \mathbf{Y}_k^*] \text{ and}$$
$$[\mathbf{\Lambda}_k^* | \mathbf{Y}, \mathbf{\Omega}, \psi_{\epsilon k}^{-1}] \stackrel{D}{=} N[(\mathbf{\Omega}_k^* \mathbf{\Omega}_k^{*T})^{-1} \mathbf{\Omega}_k^* \mathbf{Y}_k^*, \psi_{\epsilon k} (\mathbf{\Omega}_k^* \mathbf{\Omega}_k^{*T})^{-1}], \tag{4.33}$$

where $\mathbf{I}_n$ denotes the identity matrix of order n. Moreover,

$$[\mathbf{\Phi}|\mathbf{Y}, \mathbf{\Omega}] \stackrel{D}{=} IW_q[\mathbf{\Omega}\mathbf{\Omega}^T, n]. \tag{4.34}$$

It can be shown that if $\mathbf{H}_{0yk}^{*-1}$ tends to the zero matrix, then $\mathbf{A}_k^* = (\mathbf{\Omega}_k^*\mathbf{\Omega}_k^{*T})^{-1}$, $\mathbf{a}_k^* = (\mathbf{\Omega}_k^*\mathbf{\Omega}_k^{*T})^{-1}\mathbf{\Omega}_k^*\mathbf{Y}_k^*$, and $\beta_{\epsilon k} = \beta_{0\epsilon k} + 2^{-1}\mathbf{Y}_k^{*T}[\mathbf{I}_n - \mathbf{\Omega}_k^{*T}(\mathbf{\Omega}_k^*\mathbf{\Omega}_k^{*T})^{-1}\mathbf{\Omega}_k^*]\mathbf{Y}_k^*$. Under this situation, if $\alpha_{0\epsilon k}$ tends to $-r_k/2$ and $\beta_{0\epsilon k}$ tends to zero, the conditional distributions $[\psi_{\epsilon k}^{-1}|\mathbf{Y}, \mathbf{\Omega}]$ and $[\mathbf{\Lambda}_k^*|\mathbf{Y}, \mathbf{\Omega}, \psi_{\epsilon k}^{-1}]$ given in Equations (4.32) reduce to those given in Equations (4.33) with non-informative priors. Moreover, if ρ_0 tends to zero, and $\mathbf{R}_0^{-1}$ tends to the zero matrix, the conditional distribution $[\mathbf{\Phi}|\mathbf{Y}, \mathbf{\Omega}]$ given in Equation (4.32) reduces to that given in Equation (4.34). Note that the above hyperparameters are associated with prior distributions having extremely huge variances. Hence, conditional distributions that are obtained on the basis of non-informative priors are special cases of the conditional distributions that are obtained from conjugate prior distributions with very large variances.

Let $\{(\boldsymbol{\theta}^{(t)}, \mathbf{\Omega}^{(t)}), t = 1, \ldots, T^*\}$ be the observations of $(\boldsymbol{\theta}, \mathbf{\Omega})$ generated by the Gibbs sampler from the joint posterior distribution of $\boldsymbol{\theta}$ and $\mathbf{\Omega}$ given $\mathbf{Y}$, $E(\boldsymbol{\theta}|\mathbf{Y})$ and $\mathrm{Var}(\boldsymbol{\theta}|\mathbf{Y})$ are the posterior mean vector and the posterior covariance matrix, respectively. The Bayesian estimate $\hat{\boldsymbol{\theta}}$ can easily be obtained from the simulated random observations as in Equation (4.17). An estimate of the posterior covariance matrix can be obtained easily via Equation (4.18).

Now consider the posterior analysis about the factor scores. For any given individual $\mathbf{y}_i$, let $E(\boldsymbol{\omega}_i|\mathbf{y}_i)$ and $\mathrm{Var}(\boldsymbol{\omega}_i|\mathbf{y}_i)$ be the posterior mean and the posterior covariance matrix, respectively, and let $\boldsymbol{\omega}_{io}$ be the true factor scores of $\mathbf{y}_i$. A Bayesian estimate $\hat{\boldsymbol{\omega}}_i$ of $\boldsymbol{\omega}_{io}$ can be obtained on the basis of the simulated sample from the posterior distribution as in Equation (4.19). This gives a direct Bayesian estimate that does not express in terms of the structural parameter estimates, and is a consistent estimate of $E(\boldsymbol{\omega}_i|\mathbf{y}_i)$. A consistent estimate of $\mathrm{Var}(\boldsymbol{\omega}_i|\mathbf{y}_i)$ can be similarly obtained as in Equation (4.20). In practice, standard error estimates of elements in $\hat{\boldsymbol{\omega}}_i$ can be obtained via $\widehat{\mathrm{Var}}(\boldsymbol{\omega}_i|\mathbf{y}_i)$. It should be noted that both $E(\boldsymbol{\omega}_i|\mathbf{y}_i)$ and $\mathrm{Var}(\boldsymbol{\omega}_i|\mathbf{y}_i)$ are difficult to assess using the classical theory of factor analysis (see Bartholomew (1981)).

4.3.2 A Numerical Example

In this example, a real data set (see Fuller, 1987, p. 154) from a study about the writing skill of non-native speakers of English is considered. One hundred faculty members were asked to read and give scores to two essays using 11 items. The information on each item in the data set is the sum of scores on that item for the two essays. A part of the data set that involved six items was analyzed by Fuller (1987). To show aspects of the Gibbs sampler and the Bayesian estimates, this part of the data set was analyzed based on the assumption that the random observations are coming from a multivariate normal population with a factor analysis model. Following the suggestion in Fuller (1987), the following structure of $\mathbf{\Lambda}$ is considered in the analysis:

$$\mathbf{\Lambda}^T = \begin{bmatrix} \lambda_{11} & \lambda_{21} & 0 & 0 & 1 & 0 \\ \lambda_{12} & \lambda_{22} & \lambda_{32} & \lambda_{42} & 0 & 1 \end{bmatrix},$$

where elements with '0' and '1' are treated as fixed known values. Hence, the structural parameter vector $\boldsymbol{\theta}$ contains the unknown elements of $\mathbf{\Lambda}$, the upper triangular elements of $\mathbf{\Phi}$ and diagonal elements of $\mathbf{\Psi}_\epsilon$. The total number of unknown structural parameters is 15.

Bayesian estimates with conjugate prior distributions are first obtained. In this example, we do not have any historical information about the values of the hyperparameters. For illustration, ML estimates that are obtained from the first 40 observations are used to provide values of the hyperparameters in $\mathbf{\Lambda}^*_{0k}$, $\mathbf{R}_0$, $\alpha_{0\epsilon k}$ and $\beta_{0\epsilon k}$, $k = 1, \ldots, p$. Let $\tilde{\mathbf{\Lambda}}^*_k$, $\tilde{\psi}_{\epsilon k}$ and $\tilde{\mathbf{\Phi}}$ be the corresponding ML estimates. We select $\mathbf{\Lambda}^*_{0k} = \tilde{\mathbf{\Lambda}}^*_k$; and since $E(\psi_{\epsilon k}) = \beta_{0\epsilon k}/(\alpha_{0\epsilon k} - 1)$ and $E(\mathbf{\Phi}) = \mathbf{R}_0^{-1}/(\rho_0 - q - 1)$, we take $\alpha_{0\epsilon k} = 3$, $\rho_0 = r + 4$, $\beta_{0\epsilon k} = (\alpha_{0\epsilon k} - 1)\tilde{\psi}_{\epsilon k}$ and $\mathbf{R}_0^{-1} = (\rho_0 - q - 1)\tilde{\mathbf{\Phi}}$. Finally, for convenience, we take $\mathbf{H}_{0yk}$ to be a r_k by r_k identity matrix. The Bayesian estimates were obtained with the remaining 60 observations. The convergence of the Gibbs sampler is monitored by the 'estimated potential scale reduction (EPSR)' values obtained from three parallel sequences of the 15 structural parameters generated with different starting values. As suggested by Gelman (1996), convergence of these sequences has been achieved if the EPSR values are all less than 1.2. Figure 4.2 shows the plots of the EPSR values against the iteration numbers. We observe from this figure that the sequences converged rapidly. The values of EPSR for all parameters are less than 1.1 after about 250 iterations.

A total of $T^* = 4000$ observations are collected after 250 iterations with the spacing $c = 1$. Then, the Bayesian estimates and their numerical standard errors estimates are obtained via Equations (4.17) and (4.18), respectively. Moreover, ML estimates of the structural parameters are also obtained from the given data set $\mathbf{Y}$ and LISREL VIII (Jöreskog and Sörbom, 1996). Bayesian estimates (BAY), ML estimates and estimates of the corresponding standard errors are presented in Table 4.1. We observe that the ML estimate of $\psi_{\epsilon 5}$, the unique variance corresponding to the fifth item, is equal to 0.06. Hence, this estimate is very close to a Heywood case. As pointed out by Lee (1980), Heywood cases in the ML estimation can be avoided by imposing inequality constraints on $\psi_{\epsilon k}$ with a penalty function. In the Bayesian approach, the conjugate prior distribution of $\psi_{\epsilon k}^{-1}$ specified $\psi_{\epsilon k}$ in a region of positive values and hence has a similar effect as adding a penalty function. Hence, no Heywood cases are found in the Bayesian solution because of the penalty function induced by the prior distribution on $\psi_{\epsilon k}^{-1}$. This phenomenon agrees with the discussion in Martin and McDonald (1975).

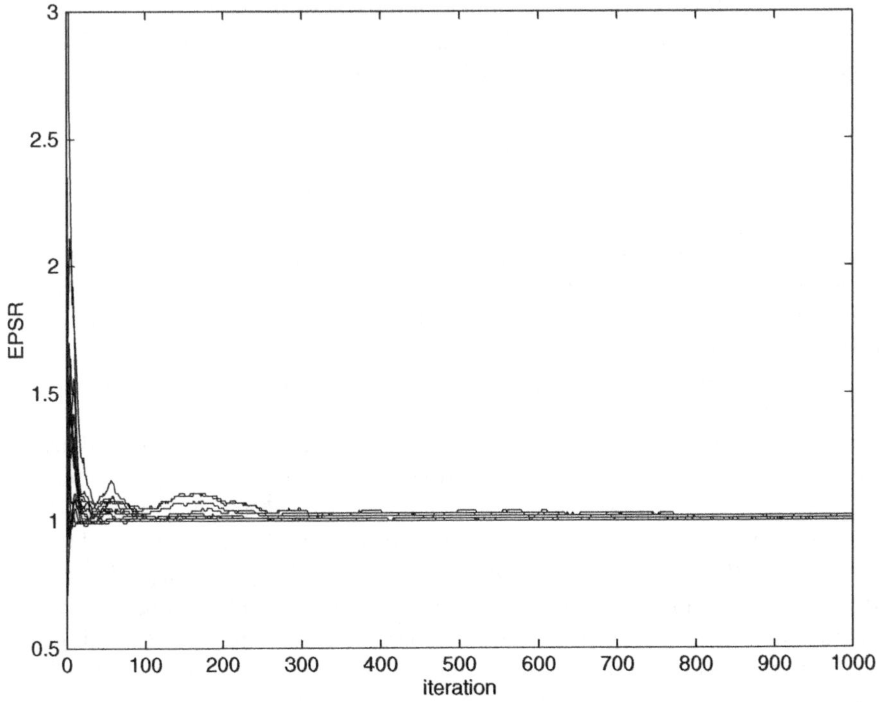

Figure 4.2 EPSR values in the analysis of the language data (this figure is taken from Lee and Shi (2000)).

Table 4.1 Bayesian estimates (BAY) and ML estimates and their standard errors in the language example

	Estimates		Standard errors			Estimates		Standard errors	
Parameters	ML	BAY	ML	BAY	Parameters	ML	BAY	ML	BAY
λ_{11}	0.77	1.04	0.18	0.18	$\psi_{\epsilon 1}$	0.97	0.79	0.21	0.18
λ_{12}	0.28	0.06	0.18	0.20	$\psi_{\epsilon 2}$	1.03	1.02	0.17	0.16
λ_{21}	0.53	0.65	0.14	0.14	$\psi_{\epsilon 3}$	0.97	1.03	0.20	0.21
λ_{22}	0.50	0.42	0.16	0.17	$\psi_{\epsilon 4}$	0.97	1.01	0.21	0.21
λ_{32}	1.19	1.25	0.14	0.19	$\psi_{\epsilon 5}$	0.06	0.40	0.26	0.13
λ_{42}	1.27	1.34	0.15	0.19	$\psi_{\epsilon 6}$	0.89	0.93	0.17	0.17
ϕ_{11}	2.35	1.92	0.43	0.35					
ϕ_{12}	1.15	1.09	0.25	0.24					
ϕ_{22}	1.41	1.31	0.32	0.31					

Note: Table 4.1 is taken from Lee and Shi (2000).

Bayesian estimates of the parameters in this example have also been obtained with spacings $c = 50$ and 100. We find that the estimates obtained with different c are very close to each other. Using $c = 1$, Bayesian estimates have also been obtained with $T^* = 2000$ and 8000. We found that these estimates are very close to the previous estimates with $T^* = 4000$ (difference only at the third decimal place). Hence, it seems that the choices of c and T^* are not important in this analysis.

4.3.3 Robustness for Small Sample Sizes

Statistical properties of the estimates and the goodness-of-fit test that are obtained from the ML and GLS methods in the CSA approach or the ADF approach are asymptotically true only. Hence, theoretically, large sample sizes are required for making valid statistical inferences. In past years, a lot of studies (see for example Boomsma, 1982; Chou, Bentler and Satorra, 1991; Hu, Bentler and Kano, 1992; Yung and Bentler, 1994; Hoogland and Boomsma, 1998) have been devoted to study the behaviors of these approaches with small sample sizes. From such research, it is concluded that the properties of the ML approach are not robust for small sample sizes. In the Bayesian approach, the sampling-based procedures generate a sufficiently large number of observations, say T^*, from the augmented joint posterior distribution $p(\boldsymbol{\theta}, \boldsymbol{\Omega}|\mathbf{Y})$ by some MCMC methods, and use the empirical distribution to approximate the posterior distribution. Now, an interesting question is: for a sufficiently large T^*, is the empirical distribution of the generated sample able to give an accurate approximation of the underlying posterior distribution, regardless of the sample size? For instance, the more specific question in estimation is: is the Bayesian estimate obtained via the sample mean of a sufficiently large sample of observations drawn from the joint posterior distribution $p(\boldsymbol{\theta}, \boldsymbol{\Omega}|\mathbf{Y})$ close to the parameter vector $\boldsymbol{\theta}_0$ even if the information provided by the given data $\mathbf{Y}$ is limited with a small sample size? Recently, Lee and Song (2004) addressed this question via a simulation study. Some of their basic results are discussed below. We will present the results of their simulation study that is based on a two-factor CFA model defined by Equation (4.21) with the following specifications:

$$\boldsymbol{\Lambda}^T = \begin{bmatrix} 1.0^* & \lambda_{21} & \lambda_{31} & \lambda_{41} & 0^* & 0^* & 0^* \\ 0^* & 0^* & 0^* & \lambda_{42} & 1.0^* & \lambda_{62} & \lambda_{72} \end{bmatrix},$$

$$\boldsymbol{\Phi} = \begin{bmatrix} \phi_{11} & \phi_{12} \\ \phi_{21} & \phi_{22} \end{bmatrix}, \quad \text{and} \quad \boldsymbol{\Psi}_\epsilon = \text{diag}(\psi_{\epsilon 1}, \ldots, \psi_{\epsilon 7}),$$

where parameters with asterisks are treated as fixed parameters. The number of unknown parameters is 16. These parameters are classified into three groups: the unknown coefficients λ_{21}, λ_{31}, λ_{41}, λ_{42}, λ_{62} and λ_{72}; the covariance of the latent factor ϕ_{21}; and the variances of the error measurements, $\psi_{\epsilon 1}, \ldots, \psi_{\epsilon 7}$. We

Table 4.2 Design with respect to parameters' magnitudes.

	Regression λ_{kh}	Covariance of latent variables ϕ_{21}	Variance of errors $\psi_{\epsilon k}$
V_1	0.8	0.6	0.36
V_2	0.8	0.6	0.72
V_3	0.8	0.2	0.36
V_4	0.8	0.2	0.72
V_5	0.4	0.6	0.36
V_6	0.4	0.6	0.72
V_7	0.4	0.2	0.36
V_8	0.4	0.2	0.72

Note: Tables 4.2 to 4.6 are taken from Lee and Song (2004).

consider comparatively small and large values for parameter(s) in each group. More specifically, the combinations of different true parameter values are given by $V_1, \ldots, V_8$ in Table 4.2. For example, V_1 and V_8 corresponding to situations with the smallest (relative) and the largest error measurement variances, respectively. The true values of ϕ_{11} and ϕ_{22} are 1.0. The most important factor in the simulation study is on the sample sizes. We consider sample sizes given by $n = da$, where a is the number of parameters in the model and $d = 2, 3, 4$ and 5. Hence, $n = 32$, 48, 64 and 80 are considered for each $V_1, \ldots, V_8$.

For each of the $32(8 \times 4)$ combinations, the accuracy of the Bayesian estimates is assessed by means of 200 replications, with data-dependent prior inputs that are obtained via an auxiliary estimation with non-informative priors. The bias (BIAS) of the estimates and the following root mean squares (RMS) between the true values and the corresponding estimates are computed:

$$\text{RMS of } \hat{\theta}(h) = \left\{ \frac{1}{200} \sum_{r=1}^{200} [\hat{\theta}_r(h) - \theta_0(h)]^2 \right\}^{1/2}, \tag{4.35}$$

where $\hat{\theta}(h)$ and $\theta_0(h)$ are the hth elements of $\hat{\theta}$ and its true value, respectively. To study the behavior of the numerical standard error estimates, let $\text{SD}(\theta(h))$ be the sample standard deviation obtained from $\{\hat{\theta}_r(h) : r = 1, \ldots, 200\}$, and $\text{SE}(\theta(h))$ be the mean of the numerical standard errors estimates of $\hat{\theta}(h)$ obtained via Equation (4.18). If the standard errors estimates obtained from Equation (4.18) are close to the sample standard deviations, $\text{SE}(\theta(h))$ should be close to $\text{SD}(\theta(h))$, and $\text{SE}(\theta(h))/\text{SD}(\theta(h))$ should be close to 1.0. Hence the ratio $\text{SE}(\theta(h))/\text{SD}(\theta(h))$ is used to assess the behavior of the numerical standard error estimates.

We only present the results of some randomly selected parameters corresponding to regression coefficients and variances of the errors in the measurement and the covariance of the latent variables. The 'BIAS', 'RMS' and 'SE/SD'

Table 4.3 Performance of the Bayesian approach under $V_1 : \lambda_{kh} = 0.8, \phi_{21} = 0.6, \psi_{\epsilon k} = 0.36$, and $V_2 : \lambda_{kh} = 0.8, \phi_{21} = 0.6, \psi_{\epsilon k} = 0.72$.

	V_1					V_2				
	Par	32	48	64	80	Par	32	48	64	80
BIAS	λ_{21}	0.032	0.040	0.025	0.004	λ_{21}	0.039	0.033	0.038	0.029
	λ_{41}	0.032	0.020	0.006	0.021	λ_{41}	0.054	0.062	0.061	0.041
	λ_{62}	0.050	0.020	0.004	0.019	λ_{62}	0.035	0.019	0.012	0.006
	λ_{72}	0.048	0.036	0.018	0.025	λ_{72}	0.009	0.006	0.012	0.003
	ϕ_{11}	0.016	0.018	0.003	0.006	ϕ_{11}	0.057	0.003	0.002	0.018
	ϕ_{21}	0.098	0.061	0.067	0.039	ϕ_{21}	0.123	0.108	0.068	0.096
	ϕ_{22}	0.094	0.077	0.028	0.063	ϕ_{22}	0.093	0.068	0.108	0.012
	$\psi_{\epsilon 1}$	0.036	0.028	0.011	0.007	$\psi_{\epsilon 1}$	0.009	0.004	0.004	0.010
	$\psi_{\epsilon 3}$	0.013	0.014	0.005	0.003	$\psi_{\epsilon 3}$	0.029	0.026	0.011	0.025
	$\psi_{\epsilon 7}$	0.033	0.023	0.018	0.011	$\psi_{\epsilon 7}$	0.004	0.015	0.003	0.011
RMS	λ_{21}	0.145	0.121	0.097	0.090	λ_{21}	0.203	0.159	0.159	0.149
	λ_{41}	0.188	0.157	0.123	0.122	λ_{41}	0.232	0.236	0.223	0.203
	λ_{62}	0.143	0.102	0.090	0.093	λ_{62}	0.185	0.147	0.141	0.141
	λ_{72}	0.120	0.108	0.096	0.101	λ_{72}	0.170	0.166	0.135	0.133
	ϕ_{11}	0.271	0.233	0.206	0.195	ϕ_{11}	0.303	0.237	0.228	0.215
	ϕ_{21}	0.231	0.189	0.165	0.157	ϕ_{21}	0.268	0.211	0.186	0.179
	ϕ_{22}	0.284	0.238	0.211	0.202	ϕ_{22}	0.314	0.274	0.291	0.225
	$\psi_{\epsilon 1}$	0.074	0.071	0.059	0.059	$\psi_{\epsilon 1}$	0.127	0.127	0.133	0.117
	$\psi_{\epsilon 3}$	0.065	0.065	0.057	0.048	$\psi_{\epsilon 3}$	0.134	0.124	0.117	0.113
	$\psi_{\epsilon 7}$	0.066	0.066	0.059	0.054	$\psi_{\epsilon 7}$	0.121	0.114	0.116	0.110
SE/SD	λ_{21}	1.097	1.102	1.159	1.063	λ_{21}	1.078	1.175	1.037	0.998
	λ_{41}	1.104	1.095	1.183	1.090	λ_{41}	1.272	1.111	1.047	1.029
	λ_{62}	1.046	1.156	1.148	1.002	λ_{62}	1.119	1.167	1.050	0.989
	λ_{72}	1.276	1.132	1.095	0.926	λ_{72}	1.218	1.019	1.090	1.036
	ϕ_{11}	1.157	1.133	1.101	1.051	ϕ_{11}	1.226	1.257	1.166	1.126
	ϕ_{21}	1.094	1.069	1.082	0.986	ϕ_{21}	1.050	1.130	1.072	1.067
	ϕ_{22}	1.233	1.204	1.102	1.116	ϕ_{22}	1.233	1.171	1.056	1.079
	$\psi_{\epsilon 1}$	1.599	1.384	1.372	1.236	$\psi_{\epsilon 1}$	1.465	1.297	1.141	1.213
	$\psi_{\epsilon 3}$	1.422	1.257	1.241	1.308	$\psi_{\epsilon 3}$	1.296	1.223	1.170	1.119
	$\psi_{\epsilon 7}$	1.640	1.303	1.281	1.216	$\psi_{\epsilon 7}$	1.456	1.319	1.178	1.134

under $V_j, j = 1, \ldots, 8$ with different true parameter values are reported in Tables 4.3 to 4.6, respectively. The following phenomena are observed from the results on 'BIAS' and 'RMS' which reveal the accuracy of the estimates: (i) Even under situations with small sample sizes ($n = 2a$ or $3a$, where a is the number of unknown parameters), the 'BIAS' values are acceptable. For instance, over 70 % of them are less than 0.05, and about 90 % of them are less than 0.1. These results imply that the Bayesian estimates are close to the true values of the parameters. (ii) The overall empirical performances in terms of RMS are

Table 4.4 Performance of the Bayesian approach under $V_3 : \lambda_{kh} = 0.8, \phi_{21} = 0.2, \psi_{\epsilon k} = 0.36$, and $V_4 : \lambda_{kh} = 0.8, \phi_{21} = 0.2, \psi_{\epsilon k} = 0.72$.

	V_3					V_4				
	Par	32	48	64	80	Par	32	48	64	80
	λ_{21}	0.036	0.034	0.028	0.037	λ_{21}	0.087	0.073	0.065	0.069
	λ_{41}	0.061	0.053	0.041	0.036	λ_{41}	0.086	0.076	0.044	0.051
	λ_{62}	0.015	0.009	0.001	0.003	λ_{62}	0.006	0.017	0.003	0.001
	λ_{72}	0.031	0.012	0.003	0.003	λ_{72}	0.003	0.012	0.004	0.008
BIAS	ϕ_{11}	0.004	0.039	0.002	0.016	ϕ_{11}	0.051	0.072	0.024	0.044
	ϕ_{21}	0.019	0.029	0.025	0.022	ϕ_{21}	0.040	0.035	0.016	0.018
	ϕ_{22}	0.083	0.049	0.018	0.017	ϕ_{22}	0.060	0.026	0.012	0.030
	$\psi_{\epsilon 1}$	0.039	0.031	0.023	0.024	$\psi_{\epsilon 1}$	0.003	0.015	0.014	0.018
	$\psi_{\epsilon 3}$	0.014	0.007	0.001	0.004	$\psi_{\epsilon 3}$	0.021	0.032	0.010	0.032
	$\psi_{\epsilon 7}$	0.030	0.026	0.016	0.015	$\psi_{\epsilon 7}$	0.008	0.010	0.005	0.010
	λ_{21}	0.146	0.121	0.101	0.098	λ_{21}	0.223	0.193	0.152	0.158
	λ_{41}	0.161	0.136	0.127	0.115	λ_{41}	0.242	0.225	0.195	0.172
	λ_{62}	0.130	0.116	0.100	0.078	λ_{62}	0.169	0.160	0.135	0.131
	λ_{72}	0.131	0.108	0.099	0.093	λ_{72}	0.179	0.165	0.160	0.137
RMS	ϕ_{11}	0.260	0.219	0.200	0.199	ϕ_{11}	0.258	0.238	0.217	0.227
	ϕ_{21}	0.182	0.141	0.138	0.123	ϕ_{21}	0.193	0.157	0.150	0.116
	ϕ_{22}	0.247	0.259	0.209	0.185	ϕ_{22}	0.287	0.258	0.228	0.214
	$\psi_{\epsilon 1}$	0.073	0.075	0.064	0.061	$\psi_{\epsilon 1}$	0.127	0.136	0.125	0.120
	$\psi_{\epsilon 3}$	0.066	0.061	0.051	0.055	$\psi_{\epsilon 3}$	0.136	0.126	0.124	0.116
	$\psi_{\epsilon 7}$	0.067	0.067	0.060	0.055	$\psi_{\epsilon 7}$	0.120	0.137	0.119	0.121
	λ_{21}	1.111	1.133	1.150	1.110	λ_{21}	1.145	1.109	1.241	1.106
	λ_{41}	1.197	1.160	1.025	1.012	λ_{41}	1.182	1.049	1.007	1.031
	λ_{62}	1.122	1.043	1.066	1.218	λ_{62}	1.261	1.134	1.186	1.098
	λ_{72}	1.119	1.112	1.071	1.021	λ_{72}	1.179	1.102	1.008	1.053
SE/SD	ϕ_{11}	1.187	1.177	1.155	1.034	ϕ_{11}	1.326	1.262	1.261	1.090
	ϕ_{21}	1.134	1.185	1.056	1.058	ϕ_{21}	1.142	1.148	1.063	1.244
	ϕ_{22}	1.412	1.053	1.108	1.127	ϕ_{22}	1.308	1.191	1.207	1.202
	$\psi_{\epsilon 1}$	1.717	1.355	1.391	1.385	$\psi_{\epsilon 1}$	1.494	1.269	1.286	1.244
	$\psi_{\epsilon 3}$	1.423	1.319	1.370	1.188	$\psi_{\epsilon 3}$	1.294	1.236	1.125	1.122
	$\psi_{\epsilon 7}$	1.578	1.322	1.243	1.238	$\psi_{\epsilon 7}$	1.448	1.144	1.180	1.073

acceptable. Together with the findings in (i), we can conclude that the Bayesian estimates obtained from small samples are fairly accurate. (iii) Comparing V_1 with V_2, V_3 with V_4, V_5 with V_6 and V_7 with V_8, the overall performances under V_1, V_3, V_5 and V_7 are better than those under V_2, V_4, V_6 and V_8, respectively. That is, performances with models having small error measurement variances are better. Based on the nature of a CFA model, small error variances implies a better fit of the dependent variables in $\mathbf{y}_i$ by the independent variables in $\boldsymbol{\omega}_i$. Hence, it is fairly logical to have better performances when working with better models. (iv) Comparing V_1 with V_3, V_2 with V_4, V_5 with V_7 and V_6 with V_8,

Table 4.5 Performance of the Bayesian approach under $V_5 : \lambda_{kh} = 0.4, \phi_{21} = 0.6, \psi_{\epsilon k} = 0.36$, and $V_6 : \lambda_{kh} = 0.4, \phi_{21} = 0.6, \psi_{\epsilon k} = 0.72$.

	V_5					V_6				
	Par	32	48	64	80	Par	32	48	64	80
BIAS	λ_{21}	0.014	0.037	0.013	0.022	λ_{21}	0.024	0.048	0.040	0.031
	λ_{41}	0.021	0.025	0.007	0.021	λ_{41}	0.044	0.077	0.056	0.068
	λ_{62}	0.001	0.006	0.002	0.008	λ_{62}	0.021	0.001	0.020	0.003
	λ_{72}	0.012	0.014	0.008	0.000	λ_{72}	0.002	0.001	0.000	0.014
	ϕ_{11}	0.011	0.005	0.012	0.002	ϕ_{11}	0.033	0.033	0.008	0.015
	ϕ_{21}	0.151	0.100	0.086	0.070	ϕ_{21}	0.175	0.185	0.132	0.126
	ϕ_{22}	0.020	0.008	0.007	0.016	ϕ_{22}	0.074	0.020	0.011	0.026
	$\psi_{\epsilon 1}$	0.053	0.047	0.039	0.032	$\psi_{\epsilon 1}$	0.028	0.034	0.017	0.034
	$\psi_{\epsilon 3}$	0.007	0.009	0.004	0.006	$\psi_{\epsilon 3}$	0.029	0.029	0.021	0.028
	$\psi_{\epsilon 7}$	0.012	0.009	0.006	0.007	$\psi_{\epsilon 7}$	0.021	0.010	0.002	0.026
RMS	λ_{21}	0.133	0.108	0.085	0.084	λ_{21}	0.204	0.165	0.156	0.138
	λ_{41}	0.150	0.145	0.121	0.121	λ_{41}	0.237	0.212	0.208	0.192
	λ_{62}	0.129	0.096	0.096	0.074	λ_{62}	0.192	0.155	0.147	0.131
	λ_{72}	0.136	0.104	0.102	0.091	λ_{72}	0.174	0.165	0.154	0.117
	ϕ_{11}	0.262	0.214	0.211	0.178	ϕ_{11}	0.293	0.248	0.235	0.222
	ϕ_{21}	0.255	0.197	0.177	0.168	ϕ_{21}	0.284	0.272	0.228	0.209
	ϕ_{22}	0.245	0.216	0.194	0.189	ϕ_{22}	0.294	0.244	0.257	0.211
	$\psi_{\epsilon 1}$	0.077	0.074	0.071	0.068	$\psi_{\epsilon 1}$	0.115	0.142	0.129	0.130
	$\psi_{\epsilon 3}$	0.055	0.055	0.054	0.049	$\psi_{\epsilon 3}$	0.119	0.107	0.113	0.100
	$\psi_{\epsilon 7}$	0.064	0.053	0.052	0.048	$\psi_{\epsilon 7}$	0.119	0.114	0.115	0.099
SE/SD	λ_{21}	1.101	1.141	1.174	1.084	λ_{21}	1.054	1.168	1.034	1.045
	λ_{41}	1.323	1.140	1.140	1.053	λ_{41}	1.243	1.271	1.099	1.142
	λ_{62}	1.065	1.166	1.014	1.180	λ_{62}	1.080	1.134	1.038	1.024
	λ_{72}	1.016	1.084	0.955	0.949	λ_{72}	1.199	1.061	0.993	1.150
	ϕ_{11}	1.213	1.282	1.158	1.221	ϕ_{11}	1.318	1.289	1.236	1.218
	ϕ_{21}	1.118	1.143	1.107	1.015	ϕ_{21}	1.193	1.083	1.042	1.066
	ϕ_{22}	1.330	1.243	1.245	1.163	ϕ_{22}	1.390	1.351	1.131	1.311
	$\psi_{\epsilon 1}$	2.296	2.011	1.817	1.684	$\psi_{\epsilon 1}$	2.036	1.509	1.510	1.490
	$\psi_{\epsilon 3}$	1.527	1.339	1.199	1.213	$\psi_{\epsilon 3}$	1.381	1.348	1.123	1.163
	$\psi_{\epsilon 7}$	1.339	1.383	1.244	1.245	$\psi_{\epsilon 7}$	1.365	1.228	1.096	1.173

the general performances of V_3, V_4, V_7 and V_8 are slightly better than those of V_1, V_2, V_5 and V_6 respectively. Hence, performances of models having small latent variables correlation are better. This is logical, as the performances are expected to be better for models with small multicollinearity among independent variables. (v) Consider two kinds of indicators (manifest variables) within each V_j. The first kind involves indicators for just one latent factor ξ_1 or ξ_2, such as the first three and the last three indicators in the model. The second kind involves indicators for both ξ_1 and ξ_2, such as the fourth indicator in the model. The empirical performances of the estimates associated with loadings

Table 4.6 Performance of the Bayesian approach under $V_7 : \lambda_{kh} = 0.4, \phi_{21} = 0.2, \psi_{\epsilon k} = 0.36$, and $V_8 : \lambda_{kh} = 0.4, \phi_{21} = 0.2, \psi_{\epsilon k} = 0.72$.

	V_7					V_8				
	Par	32	48	64	80	Par	32	48	64	80
BIAS	λ_{21}	0.054	0.040	0.038	0.039	λ_{21}	0.052	0.063	0.071	0.057
	λ_{41}	0.051	0.029	0.034	0.043	λ_{41}	0.081	0.112	0.079	0.073
	λ_{62}	0.002	0.027	0.022	0.028	λ_{62}	0.007	0.034	0.054	0.018
	λ_{72}	0.006	0.016	0.014	0.003	λ_{72}	0.010	0.037	0.045	0.033
	ϕ_{11}	0.053	0.057	0.079	0.068	ϕ_{11}	0.067	0.090	0.112	0.088
	ϕ_{21}	0.045	0.036	0.049	0.016	ϕ_{21}	0.074	0.056	0.053	0.042
	ϕ_{22}	0.011	0.055	0.035	0.020	ϕ_{22}	0.051	0.084	0.057	0.053
	$\psi_{\epsilon 1}$	0.072	0.066	0.057	0.058	$\psi_{\epsilon 1}$	0.054	0.067	0.081	0.078
	$\psi_{\epsilon 3}$	0.011	0.001	0.002	0.000	$\psi_{\epsilon 3}$	0.050	0.025	0.016	0.033
	$\psi_{\epsilon 7}$	0.013	0.005	0.014	0.008	$\psi_{\epsilon 7}$	0.034	0.030	0.029	0.015
RMS	λ_{21}	0.146	0.121	0.117	0.103	λ_{21}	0.200	0.191	0.175	0.161
	λ_{41}	0.161	0.112	0.110	0.103	λ_{41}	0.253	0.212	0.175	0.171
	λ_{62}	0.116	0.108	0.095	0.083	λ_{62}	0.202	0.168	0.159	0.139
	λ_{72}	0.132	0.107	0.098	0.078	λ_{72}	0.190	0.171	0.163	0.137
	ϕ_{11}	0.243	0.235	0.214	0.200	ϕ_{11}	0.261	0.246	0.226	0.242
	ϕ_{21}	0.181	0.160	0.149	0.136	ϕ_{21}	0.186	0.171	0.166	0.158
	ϕ_{22}	0.257	0.215	0.198	0.201	ϕ_{22}	0.225	0.247	0.231	0.225
	$\psi_{\epsilon 1}$	0.099	0.094	0.088	0.090	$\psi_{\epsilon 1}$	0.141	0.142	0.162	0.143
	$\psi_{\epsilon 3}$	0.067	0.055	0.049	0.051	$\psi_{\epsilon 3}$	0.126	0.129	0.106	0.107
	$\psi_{\epsilon 7}$	0.061	0.059	0.056	0.052	$\psi_{\epsilon 7}$	0.123	0.124	0.106	0.104
SE/SD	λ_{21}	1.120	1.093	0.989	1.025	λ_{21}	1.213	1.094	1.115	1.061
	λ_{41}	1.115	1.256	1.133	1.128	λ_{41}	1.098	1.251	1.269	1.167
	λ_{62}	1.248	1.169	1.138	1.196	λ_{62}	1.125	1.199	1.115	1.080
	λ_{72}	1.106	1.138	1.083	1.181	λ_{72}	1.224	1.170	1.063	1.148
	ϕ_{11}	1.317	1.171	1.173	1.142	ϕ_{11}	1.419	1.372	1.450	1.198
	ϕ_{21}	1.178	1.086	1.053	1.012	ϕ_{21}	1.327	1.165	1.074	1.020
	ϕ_{22}	1.251	1.274	1.247	1.113	ϕ_{22}	1.643	1.342	1.315	1.262
	$\psi_{\epsilon 1}$	2.013	1.872	1.715	1.594	$\psi_{\epsilon 1}$	1.799	1.760	1.512	1.689
	$\psi_{\epsilon 3}$	1.292	1.312	1.324	1.143	$\psi_{\epsilon 3}$	1.359	1.112	1.225	1.120
	$\psi_{\epsilon 7}$	1.436	1.225	1.221	1.157	$\psi_{\epsilon 7}$	1.333	1.157	1.223	1.122

(here, λ_{21}, λ_{62} and λ_{72}) that belong to the first kind indicators are better than those belonging to the second kind (λ_{41} in this case). This indicates the expected result that the complexity of the factor loading structure has some effect on the accuracy of the estimates. (vi) The empirical performances of the estimates for the variance and covariance (ϕ_{11}, ϕ_{12} and ϕ_{22}) of the latent factors are not as good as the others.

The ratios SE/SD corresponding to most of the parameters having different magnitudes in $V_1, \ldots, V_8$ are over 1.0. Based on the definitions of SE($\theta(h)$) and SD($\theta(h)$), the sample standard deviation of $\{\hat{\theta}_r(h), r = 1, \ldots, 200\}$ is smaller than the mean of the numerical standard error estimates computed from the simulated observations via Equation (4.18). That is, the variability of the Bayesian estimates is relatively small. This may be regarded as a good property of the Bayesian estimates from an estimation point of view. On the other hand, the numerical standard error estimates of the Bayesian approach produced by Equation (4.18) are overestimated. However, as there are not too many ratios that are over 1.5, particularly with $n = 4a$ or $5a$, the induced impact is not very substantial. Moreover, it should be noted that the use of the numerical standard error estimates in a Bayesian approach is not as important as in the ML approach. For instance, we do not use them in model comparison or hypothesis testing.

Lee and Song (2004) also provided simulation results that were obtained by the ML method in the context of the CSA approach. Based on their results, they arrived at the same conclusion given by many previous studies in the literature that the ML–CSA approach is not recommended for situations with small sample sizes.

4.4 BAYESIAN ESTIMATION OF STANDARD SEMs

In this section, we further illustrate the Bayesian estimation by considering a standard SEM that is equivalent to the most commonly used LISREL model. It is composed of a measurement equation and a structural equation. The measurement equation is basically equivalent to the following CFA model as given in Equation (4.21):

$$\mathbf{y}_i = \mathbf{\Lambda}\boldsymbol{\omega}_i + \boldsymbol{\epsilon}_i, \tag{4.36}$$

with the same definition and assumption for $\mathbf{\Lambda}$, $\boldsymbol{\omega}_i$ and $\boldsymbol{\epsilon}_i$, except that the covariance matrix of $\boldsymbol{\omega}_i$ is no longer $\mathbf{\Phi}$. Let $\boldsymbol{\omega}_i = (\boldsymbol{\eta}_i^T, \boldsymbol{\xi}_i^T)^T$ be a partition of $\boldsymbol{\omega}_i$ into an $q_1 \times 1$ dependent latent vector $\boldsymbol{\eta}_i$, and an $q_2 \times 1$ independent latent vector $\boldsymbol{\xi}_i$. The structural equation for assessing the relationship between $\boldsymbol{\eta}_i$ and $\boldsymbol{\xi}_i$ is given by

$$\boldsymbol{\eta}_i = \mathbf{\Pi}\boldsymbol{\eta}_i + \mathbf{\Gamma}\boldsymbol{\xi}_i + \boldsymbol{\delta}_i, \tag{4.37}$$

where $\mathbf{\Pi}(q_1 \times q_1)$ and $\mathbf{\Gamma}(q_1 \times q_2)$ are unknown parameter matrices of regression coefficients, and $\boldsymbol{\delta}_i$ is a $q_1 \times 1$ random vector of error measurements. It is assumed that $\boldsymbol{\xi}_i$ is distributed as $N[\mathbf{0}, \mathbf{\Phi}]$, and $\boldsymbol{\delta}_i$ is distributed as $N[\mathbf{0}, \mathbf{\Psi}_\delta]$, where $\mathbf{\Psi}_\delta$ is a diagonal matrix and $\boldsymbol{\xi}_i$ and $\boldsymbol{\delta}_i$ are independent. For simplicity, it is further assumed that $\mathbf{\Pi}_0 = |\mathbf{I} - \mathbf{\Pi}|$ is a positive constant independent with elements in $\mathbf{\Pi}$. This model is not identified, but it can be identified by restricting appropriate elements in $\mathbf{\Lambda}$, $\mathbf{\Pi}$ and/or $\mathbf{\Gamma}$ at fixed known values.

Let $\mathbf{Y} = (\mathbf{y}_1, \ldots, \mathbf{y}_n)$ and $\boldsymbol{\Omega} = (\boldsymbol{\omega}_1, \ldots, \boldsymbol{\omega}_n)$, and let $\boldsymbol{\theta}$ be the vector of unknown parameters in $\boldsymbol{\Lambda}$, $\boldsymbol{\Psi}_\epsilon$, $\boldsymbol{\Pi}$, $\boldsymbol{\Gamma}$, $\boldsymbol{\Phi}$ and $\boldsymbol{\Psi}_\delta$. In the posterior analysis, we augment the observed data $\mathbf{Y}$ with the matrix of latent variable $\boldsymbol{\Omega}$, and consider the joint posterior distribution $[\boldsymbol{\theta}, \boldsymbol{\Omega}|\mathbf{Y}]$. Again, a sufficiently large number of observations are generated from this posterior distribution by the Gibbs sampler, which is implemented as follows. At the $(j+1)$th iteration with current values of $\boldsymbol{\Omega}^{(j)}$ and $\boldsymbol{\theta}^{(j)}$:

$$\begin{aligned} &\text{(i)} \quad \text{Generate } \boldsymbol{\Omega}^{(j+1)} \text{ from } p(\boldsymbol{\Omega}|\boldsymbol{\theta}^{(j)}, \mathbf{Y}), \\ &\text{(ii)} \quad \text{Generate } \boldsymbol{\theta}^{(j+1)} \text{ from } p(\boldsymbol{\theta}|\boldsymbol{\Omega}^{(j+1)}, \mathbf{Y}). \end{aligned} \tag{4.38}$$

Note that $\boldsymbol{\theta}$ involves components that correspond to $\boldsymbol{\Lambda}$, $\boldsymbol{\Psi}_\epsilon$, $\boldsymbol{\Pi}$, $\boldsymbol{\Gamma}$, $\boldsymbol{\Phi}$ and $\boldsymbol{\Psi}_\delta$. This Gibbs sampler is similar to the one for the CFA model, except that more components corresponding to the additional parameters are involved.

The SEM defined by Equations (4.36) and (4.37) is a straightforward generalization of the CFA model through an additional structural Equation (4.37). The conditional distributions involved in the Gibbs sampler that correspond to the measurement equation are very similar to those that are associated with the CFA model. This is also true for the structural equation, because it is essentially a regression model or a factor analysis model with latent variables. Hence, the generalization of the conditional distributions in the Gibbs sampler for a CFA model to an SEM does not involve too much difficulty.

Under the similar definition and assumption, $p(\boldsymbol{\Omega}|\mathbf{Y}, \boldsymbol{\theta})$ can be expressed as in Equation (4.23), with the conditional distribution of $\boldsymbol{\omega}_i$ given $(\mathbf{y}_i, \boldsymbol{\theta})$ is similarly given as in Equation (4.24). However, in Equation (4.24) $\boldsymbol{\Phi}$ should be replaced by the following covariance matrix of $\boldsymbol{\omega}$, $\boldsymbol{\Sigma}_\omega$, which is derived on the basis of the SEM defined in Equations (4.36) and (4.37):

$$\boldsymbol{\Sigma}_\omega = \begin{bmatrix} \boldsymbol{\Pi}_0^{-1}(\boldsymbol{\Gamma}\boldsymbol{\Phi}\boldsymbol{\Gamma}^T + \boldsymbol{\Psi}_\delta)\boldsymbol{\Pi}_0^{-T} & \boldsymbol{\Pi}_0^{-1}\boldsymbol{\Gamma}\boldsymbol{\Phi} \\ \boldsymbol{\Phi}\boldsymbol{\Gamma}^T\boldsymbol{\Pi}_0^{-T} & \boldsymbol{\Phi} \end{bmatrix}. \tag{4.39}$$

The conditional distribution of $\boldsymbol{\theta}$ given $(\mathbf{Y}, \boldsymbol{\Omega})$ is proportional to $p(\boldsymbol{\theta})p(\mathbf{Y}, \boldsymbol{\Omega}|\boldsymbol{\theta})$. We note that when $\boldsymbol{\Omega}$ is given, Equations (4.36) and (4.37) are regression models. Let $\boldsymbol{\theta}_y$ be the unknown parameters in $\boldsymbol{\Lambda}$ and $\boldsymbol{\Psi}_\epsilon$ associated with the measurement equation; and $\boldsymbol{\theta}_\omega$ be the unknown parameters in $\boldsymbol{\Pi}$, $\boldsymbol{\Gamma}$, $\boldsymbol{\Phi}$ and $\boldsymbol{\Psi}_\delta$ associated with the structural model with the latent variables. It is natural to assume that the prior distribution of $\boldsymbol{\theta}_y$ is independent of the prior distribution of $\boldsymbol{\theta}_\omega$, that is, $p(\boldsymbol{\theta}) = p(\boldsymbol{\theta}_y)p(\boldsymbol{\theta}_\omega)$. Moreover, $p(\mathbf{Y}|\boldsymbol{\Omega}, \boldsymbol{\theta}) = p(\mathbf{Y}|\boldsymbol{\Omega}, \boldsymbol{\theta}_y)$ and $p(\boldsymbol{\Omega}|\boldsymbol{\theta}) = p(\boldsymbol{\Omega}|\boldsymbol{\theta}_\omega)$. Hence,

$$p(\boldsymbol{\theta}_y, \boldsymbol{\theta}_\omega|\mathbf{Y}, \boldsymbol{\Omega}) \propto [p(\mathbf{Y}|\boldsymbol{\Omega}, \boldsymbol{\theta}_y)p(\boldsymbol{\theta}_y)][p(\boldsymbol{\Omega}|\boldsymbol{\theta}_\omega)p(\boldsymbol{\theta}_\omega)].$$

Since the first term of the product on the right-hand side depends only on $\boldsymbol{\theta}_y$, whereas the second term depends only on $\boldsymbol{\theta}_\omega$, the marginal conditional densities $\boldsymbol{\theta}_y$ and $\boldsymbol{\theta}_\omega$ are proportional to $p(\mathbf{Y}|\boldsymbol{\Omega}, \boldsymbol{\theta}_y)p(\boldsymbol{\theta}_y)$ and $p(\boldsymbol{\Omega}|\boldsymbol{\theta}_\omega)p(\boldsymbol{\theta}_\omega)$, respectively. Consequently, these conditional densities can be treated separately.

The marginal conditional distribution of $\boldsymbol{\theta}_y$ is $p(\boldsymbol{\Lambda}, \boldsymbol{\Psi}_\epsilon|\mathbf{Y}, \boldsymbol{\Omega})$. Under the conjugate prior distributions given in Equation (4.27), this conditional distribution can be obtained from Equation (4.28).

Now, consider the conditional distribution of $\boldsymbol{\theta}_\omega$ that is proportional to $p(\boldsymbol{\Omega}|\boldsymbol{\theta}_\omega)p(\boldsymbol{\theta}_\omega)$. Let $\boldsymbol{\Omega}_1 = (\boldsymbol{\eta}_1, \ldots, \boldsymbol{\eta}_n)$ and $\boldsymbol{\Omega}_2 = (\boldsymbol{\xi}_1, \ldots, \boldsymbol{\xi}_n)$. Since the distribution of $\boldsymbol{\xi}_i$ only involves $\boldsymbol{\Phi}$, $p(\boldsymbol{\Omega}_2|\boldsymbol{\theta}_\omega) = p(\boldsymbol{\Omega}_2|\boldsymbol{\Phi})$. Moreover, it is natural to assume that the prior distribution of $\boldsymbol{\Phi}$ is independent of the prior distributions of $\boldsymbol{\Pi}$, $\boldsymbol{\Gamma}$ and $\boldsymbol{\Phi}_\delta$. Hence,

$$p(\boldsymbol{\Omega}|\boldsymbol{\theta}_\omega)p(\boldsymbol{\theta}_\omega) = [p(\boldsymbol{\Omega}_1|\boldsymbol{\Omega}_2, \boldsymbol{\Pi}, \boldsymbol{\Gamma}, \boldsymbol{\Psi}_\delta)p(\boldsymbol{\Pi}, \boldsymbol{\Gamma}, \boldsymbol{\Psi}_\delta)][p(\boldsymbol{\Omega}_2|\boldsymbol{\Phi})p(\boldsymbol{\Phi})],$$

and the marginal conditional densities of $(\boldsymbol{\Pi}, \boldsymbol{\Gamma}, \boldsymbol{\Psi}_\delta)$ and $\boldsymbol{\Phi}$ can be treated separately.

Consider a conjugate type prior distribution for $\boldsymbol{\Phi}$ with $\boldsymbol{\Phi}^{-1} \stackrel{D}{=} \mathbf{W}_{q_2}[\mathbf{R}_0, \rho_0]$, a Wishart distribution with hyperparameters ρ_0 and a positive definite matrix $\mathbf{R}_0$. It can be shown by reasoning similar to that used in Section 4.3.1 that

$$p(\boldsymbol{\Omega}_2|\boldsymbol{\Phi})p(\boldsymbol{\Phi}) \propto |\boldsymbol{\Phi}|^{-(n+\rho_0+q_2+1)/2} \exp\left\{-\frac{1}{2}\operatorname{tr}[\boldsymbol{\Phi}^{-1}(\boldsymbol{\Omega}_2\boldsymbol{\Omega}_2^T + \mathbf{R}_0^{-1})]\right\}.$$

Thus, the conditional distribution of $\boldsymbol{\Phi}$ given $\boldsymbol{\Omega}_2$ is given by

$$[\boldsymbol{\Phi}|\boldsymbol{\Omega}_2] \stackrel{D}{=} IW_{q_2}[(\boldsymbol{\Omega}_2\boldsymbol{\Omega}_2^T + \mathbf{R}_0^{-1}), n + \rho_0]. \tag{4.40}$$

Rewrite Equation (4.37) as $\boldsymbol{\eta}_i = \boldsymbol{\Lambda}_\omega\boldsymbol{\omega}_i + \boldsymbol{\delta}_i$, where $\boldsymbol{\Lambda}_\omega = (\boldsymbol{\Pi}, \boldsymbol{\Gamma})$. This is very similar to a factor analysis model, and when $\boldsymbol{\Omega}$ is given this is a regression model. Let $\psi_{\delta k}$ be the kth diagonal element of $\boldsymbol{\Psi}_\delta$, and $\boldsymbol{\Lambda}_{\omega k}^T$ be the kth row of $\boldsymbol{\Lambda}_\omega$. The prior distributions of $(\psi_{\delta k}, \boldsymbol{\Lambda}_{\omega k})$ are similarly selected via the following conjugate type distributions:

$$\psi_{\delta k}^{-1} \stackrel{D}{=} Gamma[\alpha_{0\delta k}, \beta_{0\delta k}], \text{ and}$$

$$[\boldsymbol{\Lambda}_{\omega k}|\psi_{\delta k}] \stackrel{D}{=} N[\boldsymbol{\Lambda}_{0\omega k}, \psi_{\delta k}\mathbf{H}_{0\omega k}],\ k = 1, 2, \ldots, k_1, \tag{4.41}$$

where $\alpha_{0\delta k}$, $\beta_{0\delta k}$ and $\mathbf{H}_{0\omega k}$ are given hyperparameters. Moreover, it is assumed that for $h \neq k$, $(\psi_{\delta k}, \mathbf{\Lambda}_{\omega k})$ and $(\psi_{\delta h}, \mathbf{\Lambda}_{\omega h})$ are independent. Then, following the same reasoning as before, it can be shown that:

$$
\begin{aligned}
[\psi_{\delta k}^{-1}|\mathbf{\Omega}] &\stackrel{D}{=} Gamma\,[n/2 + \alpha_{0\delta k}, \beta_{\delta k}], \text{ and} \\
[\mathbf{\Lambda}_{\omega k}|\mathbf{\Omega}, \psi_{\delta k}^{-1}] &\stackrel{D}{=} N\,[\mathbf{a}_{\omega k}, \psi_{\delta k}\mathbf{A}_{\omega k}],
\end{aligned} \tag{4.42}
$$

where $\mathbf{A}_{\omega k} = (\mathbf{H}_{0\omega k}^{-1} + \mathbf{\Omega}\mathbf{\Omega}^T)^{-1}$, $\mathbf{a}_{\omega k} = \mathbf{A}_{\omega k}(\mathbf{H}_{0\omega k}^{-1}\mathbf{\Lambda}_{0\omega k} + \mathbf{\Omega}\mathbf{\Omega}_{1k})$ and $\beta_{\delta k} = \beta_{0\delta k} + \frac{1}{2}(\mathbf{\Omega}_{1k}^T\mathbf{\Omega}_{1k} - \mathbf{a}_{\omega k}^T\mathbf{A}_{\omega k}^{-1}\mathbf{a}_{\omega k} + \mathbf{\Lambda}_{0\omega k}^T\mathbf{H}_{0\omega k}^{-1}\mathbf{\Lambda}_{0\omega k})$, in which $\mathbf{\Omega}_{1k}^T$ is the kth row of $\mathbf{\Omega}_1$.

The situation with fixed parameters can be handled in a similar way to Section 4.3.1. See also Appendix 5.1 for the treatment of fixed parameters in an SEM with fixed covariates.

It should be emphasized that the Bayesian analysis of the CFA model and the SEM are based on the basic model of the raw individual observations, rather than the sample covariance matrix. Again, estimates of the latent variables can be easily obtained through the observations of $\mathbf{\Omega}$ simulated by the Gibbs sampler.

4.5 BAYESIAN ESTIMATION VIA WINBUGS

The freely available software WinBUGS (**W**indows version of **B**ayesian inference **U**sing **G**ibbs **S**ampling) is useful to produce reliable Bayesian statistics for a very wide range of statistical models. The algorithm used in WinBUGS is mainly developed using MCMC techniques, such as the Gibbs sampler (Geman and Geman, 1984) and the MH algorithm (Metropolis *et al.*, 1953; Hastings, 1970). It has been shown that under broad conditions, this software can provide simulated samples from the joint posterior distribution of the unknown quantities, such as parameters and latent variables in the model. As discussed in previous sections, Bayesian estimates of the unknown parameters and their standard error estimates in the model can be obtained from these samples for conducting statistical inferences.

The advanced version of the program is WinBUGS 1.4 running under Windows, which is developed by the Medical Research Council (MRC) Biostatistics Unit (Cambridge, UK) and the Department of Epidemiology and Public Health of the Imperial College School of Medicine at St Mary's Hospital (London). It can be downloaded from the website http://www.mrc-bsu.cam.ac.uk/bugs/. The free version of WinBUGS is a restricted version; one needs to email the BUGS project for a key that will let the user use the full version. The WinBUGS Manual (Spiegelhalter, Thomas, Best and Lunn, 2003), which gives brief instructions on WinBUGS, is available online. See also Lawson, Browne and Vidal Rodeiro (2003, Chapter 4) for supplementary descriptions.

The following LISREL type model is used to illustrate the use of WinBUGS in conducting Bayesian analysis of SEMs. The measurement equation of the model is defined by nine manifest variables in $\mathbf{y}_i = (y_{i1}, \ldots, y_{i9})^T$ and three latent variables in $\boldsymbol{\omega}_i = (\eta_i, \xi_{i1}, \xi_{i2})^T$ as follows:

$$\begin{aligned} y_{i1} &= \alpha_1 + \eta_i + \epsilon_{i1}, \; y_{ik} = \alpha_k + \lambda_{k1}\eta_i + \epsilon_{ik}, \quad k = 2, 3 \\ y_{i4} &= \alpha_4 + \xi_{i1} + \epsilon_{i4}, \; y_{ik} = \alpha_k + \lambda_{k2}\xi_{i1} + \epsilon_{ik}, \; k = 5, 6 \\ y_{i7} &= \alpha_7 + \xi_{i2} + \epsilon_{i7}, \; y_{ik} = \alpha_k + \lambda_{k3}\xi_{i2} + \epsilon_{ik}, \; k = 8, 9 \end{aligned} \tag{4.43}$$

where ϵ_{ik}, $k = 1, \ldots, p$, is independently distributed as $N[0, \psi_{\epsilon k}]$, and is also independent with $\boldsymbol{\omega}_i$. The structural equation is defined by

$$\eta_i = \gamma_1\xi_{i1} + \gamma_2\xi_{i2} + \delta_i = \boldsymbol{\Gamma}\boldsymbol{\xi}_i + \delta_i \tag{4.44}$$

where $\boldsymbol{\xi}_i = (\xi_{i1}, \xi_{i2})^T$ is distributed as $N[\mathbf{0}, \boldsymbol{\Phi}]$, and δ_i is distributed as $N[0, \psi_\delta]$, $\boldsymbol{\xi}_i$ and δ_i are independent. An artificial random sample of size $n = 300$ is simulated via the following true population values of the free parameters:

$$\begin{aligned} &\alpha_1 = \cdots = \alpha_9 = 0.0, \; \lambda_{21} = \lambda_{52} = \lambda_{83} = 0.9, \\ &\lambda_{31} = \lambda_{62} = \lambda_{93} = 0.7, \\ &\psi_{\epsilon1} = \psi_{\epsilon2} = \psi_{\epsilon3} = 0.3, \; \psi_{\epsilon4} = \cdots = \psi_{\epsilon7} = 0.5, \psi_{\epsilon8} = \psi_{\epsilon9} = 0.4, \\ &\gamma_1 = 0.6, \gamma_2 = 0.4, \\ &\phi_{11} = \phi_{22} = 1.0, \phi_{12} = 0.3, \text{ and } \psi_\delta = 0.36. \end{aligned} \tag{4.45}$$

where ϕ_{11}, ϕ_{21} and ϕ_{22} are distinct elements in $\boldsymbol{\Phi}$.

The following conjugate prior distributions are used in the Bayesian estimation. For $k = 1, \ldots, p$,

$$\begin{aligned} &\boldsymbol{\alpha} \stackrel{D}{=} N[\mathbf{0}, \mathbf{I}], \boldsymbol{\Phi}^{-1} \stackrel{D}{=} W_2[\mathbf{I}, 5], \\ &\psi_{\epsilon k}^{-1} \stackrel{D}{=} Gamma\,(9, 4), \quad \boldsymbol{\Lambda}_k \stackrel{D}{=} N[\boldsymbol{\Lambda}_{0k}, \psi_{\epsilon k}\mathbf{I}], \\ &\psi_\delta^{-1} \stackrel{D}{=} Gamma\,(9, 4), \quad \boldsymbol{\Gamma} \stackrel{D}{=} N[\boldsymbol{\Gamma}_0, \psi_\delta\mathbf{I}], \end{aligned} \tag{4.46}$$

where elements in $\boldsymbol{\Lambda}_{0k}$ and $\boldsymbol{\Gamma}_0$ are taken to be 0.8, and 0.5, respectively, and $\mathbf{I}'s$ are the identity matrices of appropriate orders.

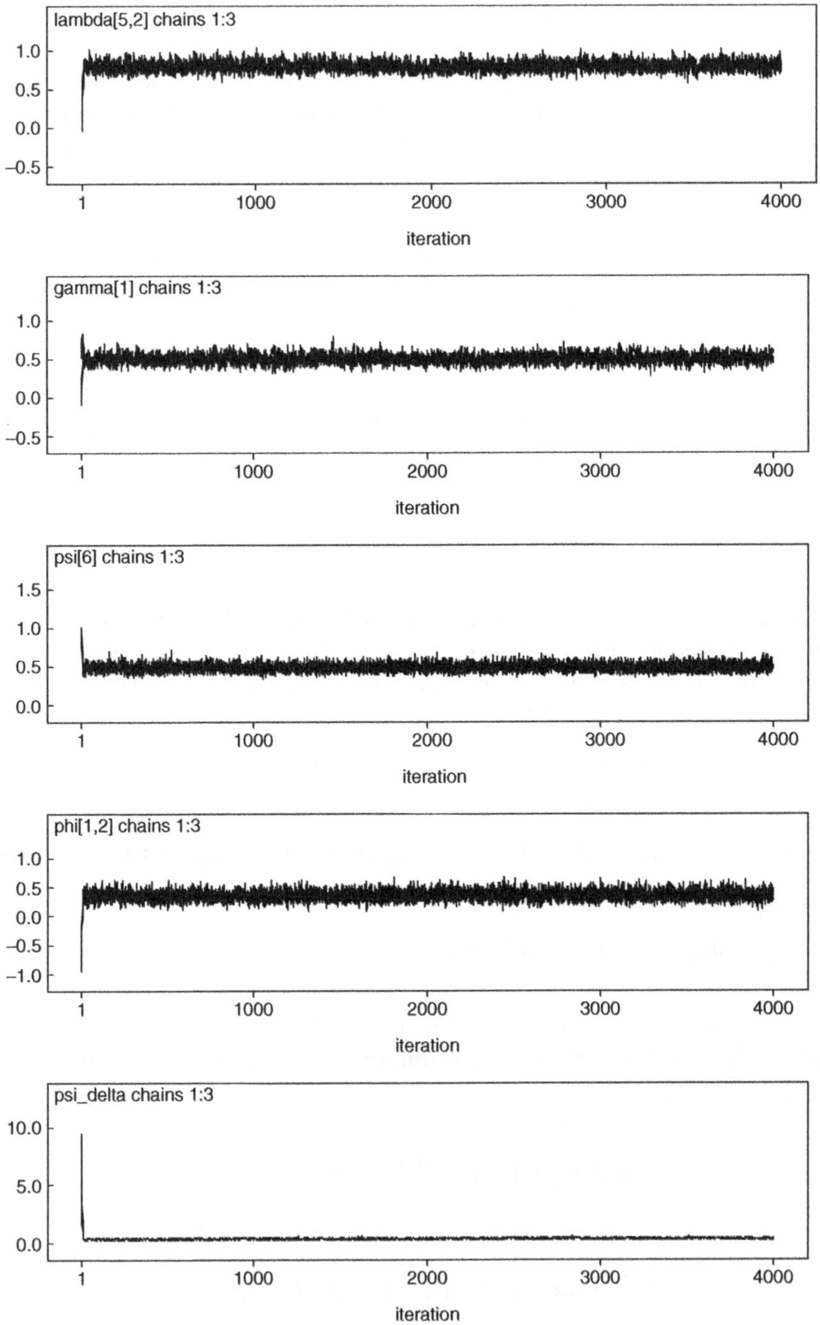

Figure 4.3 Three chains of observation corresponding to λ_{52}, γ_1, $\psi_{\epsilon 6}$, ϕ_{12} and ψ_δ generated by different initial values.

To specify the model in the WinBUGS language, the measurement Equation (4.43) is reformulated as: $y_{ik} \stackrel{D}{=} N[\mu_{ik}, \psi_k]$, where

$$\mu_{i1} = \alpha_1 + \eta_i, \quad \mu_{ik} = \alpha_k + \lambda_{k1}\eta_i, \quad k = 2, 3$$
$$\mu_{i4} = \alpha_4 + \xi_{i1}, \quad \mu_{ik} = \alpha_k + \lambda_{k2}\xi_{i1}, \quad k = 5, 6$$
$$\mu_{i7} = \alpha_7 + \xi_{i2}, \quad \mu_{ik} = \alpha_k + \lambda_{k3}\xi_{i2}, \quad k = 8, 9;$$

and the structural Equation (4.44) is reformulated by defining the conditional distribution of η_i given ξ_{i1} and ξ_{i2} as $N[\nu_i, \psi_\delta]$, where

$$\nu_i = \gamma_1\xi_{i1} + \gamma_2\xi_{i2}. \tag{4.47}$$

WinBUGS was used to analyze the artificial data with $n = 300$ which were simulated according to the model defined above, with true parameter values given in Equation (4.46). The WinBUGS codes together with some instructions are given in the following website: http://www.wiley.com /go/lee_structural. The artificial data set is also given in this website. The hyperparameter values are given in Equation (4.45). Plots of sequences of observations corresponding to some parameters are displayed in Figure 4.3. These plots indicate that the algorithm converged in less than 1000 iterations. We discard 2000 burn-in

Table 4.7 Bayesian estimates (EST) and their standard error (SE) of the artificial example obtained through WinBUGS.

Par	True value	EST	SE	Par	True value	EST	SE
α_1	0.0	−0.005	0.068	$\psi_{\epsilon 1}$	0.3	0.376	0.041
α_2	0.0	0.007	0.069	$\psi_{\epsilon 2}$	0.3	0.331	0.040
α_3	0.0	−0.020	0.052	$\psi_{\epsilon 3}$	0.3	0.279	0.028
α_4	0.0	0.120	0.072	$\psi_{\epsilon 4}$	0.5	0.462	0.063
α_5	0.0	0.106	0.063	$\psi_{\epsilon 5}$	0.5	0.496	0.053
α_6	0.0	0.019	0.057	$\psi_{\epsilon 6}$	0.5	0.479	0.047
α_7	0.0	0.120	0.070	$\psi_{\epsilon 7}$	0.5	0.514	0.067
α_8	0.0	0.047	0.061	$\psi_{\epsilon 8}$	0.4	0.471	0.053
α_9	0.0	0.040	0.052	$\psi_{\epsilon 9}$	0.4	0.371	0.039
λ_{21}	0.9	1.042	0.055	γ_1	0.6	0.501	0.058
λ_{31}	0.7	0.741	0.045	γ_2	0.4	0.440	0.060
λ_{52}	0.9	0.794	0.061	ϕ_{11}	1.0	1.089	0.134
λ_{62}	0.7	0.684	0.055	ϕ_{21}	0.3	0.346	0.077
λ_{83}	0.9	0.841	0.070	ϕ_{22}	1.0	0.964	0.127
λ_{93}	0.7	0.702	0.056	ψ_δ	0.36	0.357	0.047

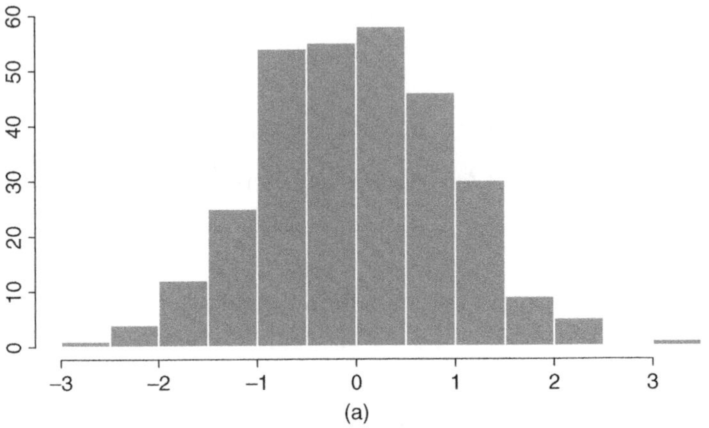

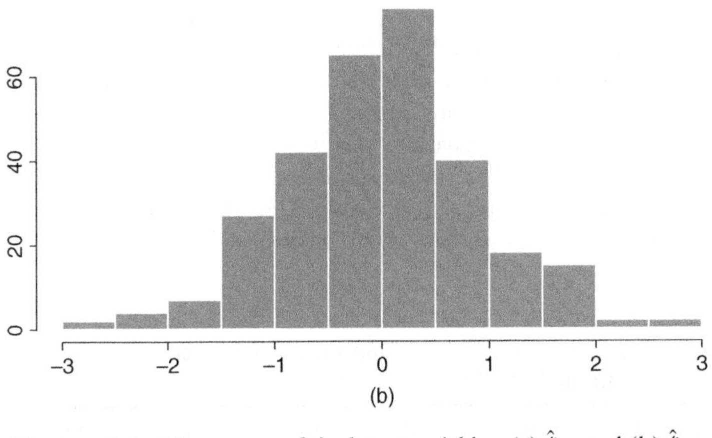

Figure 4.4 Histograms of the latent variables: (a) $\hat{\xi}_{i1}$, and (b) $\hat{\xi}_{i2}$.

iterations, and use $T^* = 2000$ observations to obtain the Bayesian results, see Equation (4.17) to (4.20). The Bayesian estimates and their standard error estimates are given in Table 4.7. We observe that the Bayesian estimates are reasonably close to their true values and the standard error estimates are reasonable. The sample variances and covariance of $\hat{\xi}_{i1}$ and $\hat{\xi}_{i2}$ are equal to 0.932, 0.818 and 0.354, respectively; they are reasonably close to the true values of $\mathbf{\Phi}(\phi_{11} = \phi_{22} = 1.0, \phi_{12} = 0.3)$. Histograms that correspond to the set of latent variable estimates $\hat{\xi}_{i1}$ and $\hat{\xi}_{i2}$ are shown in Figure 4.4. These histograms indicate that the distributions of $\hat{\xi}_{i1}$, and $\hat{\xi}_{i2}$ are close to normal. The estimated residual plots of $\hat{\epsilon}_{i1}$, $\hat{\epsilon}_{i5}$ and $\hat{\delta}_i$ against the case number are displayed in Figure 4.5, and plots of estimated residual $\hat{\epsilon}_{i1}$ versus $\hat{\xi}_{i1}$, $\hat{\xi}_{i2}$ and $\hat{\eta}_i$, $\hat{\delta}_i$ versus $\hat{\xi}_{i1}$ and $\hat{\xi}_{i2}$ are

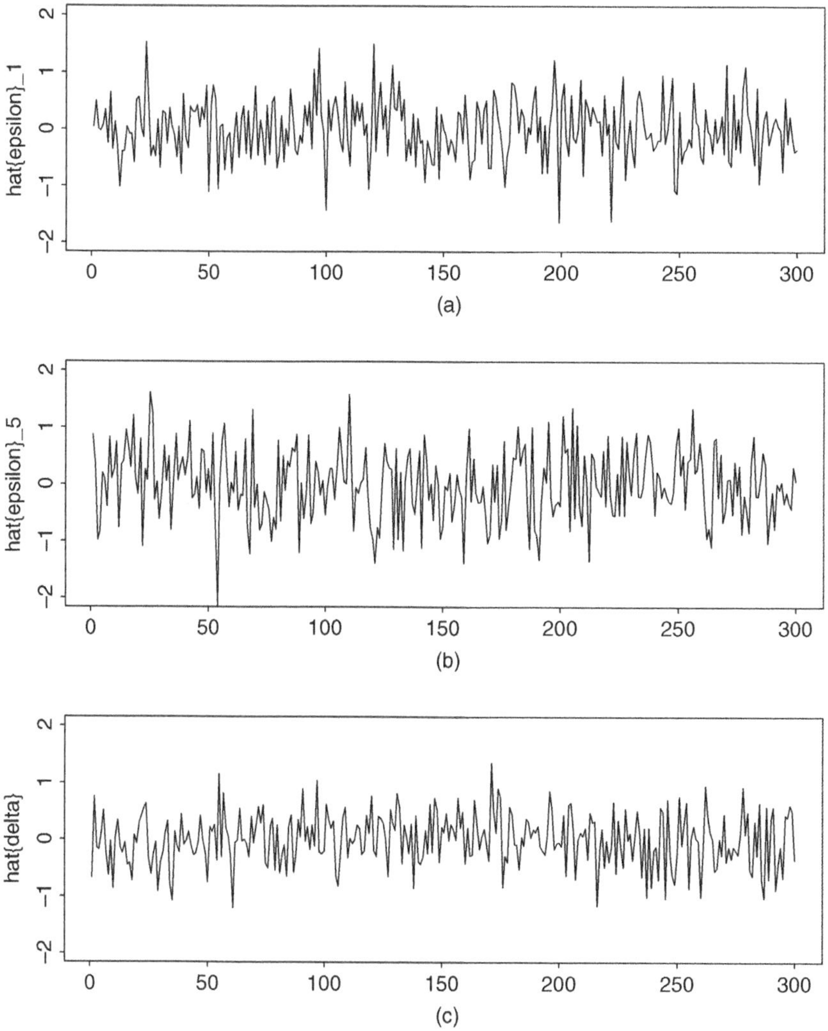

Figure 4.5 Estimated residual plots: (a) $\hat{\epsilon}_{i1}$, (b) $\hat{\epsilon}_{i5}$ and (c) $\hat{\delta}_i$.

shown in Figures 4.6 and 4.7 respectively. Other residual plots are similar but they are not presented to save space. These plots lie within two parallel horizontal lines that are centered at zero, and no linear or quadratic trends are detected. This indicates that the measurement model and the structural equation of the proposed model adequately fit the data. Basically, these residual plots are interpreted in a similar way to a regression model and may be used for outlier analysis.

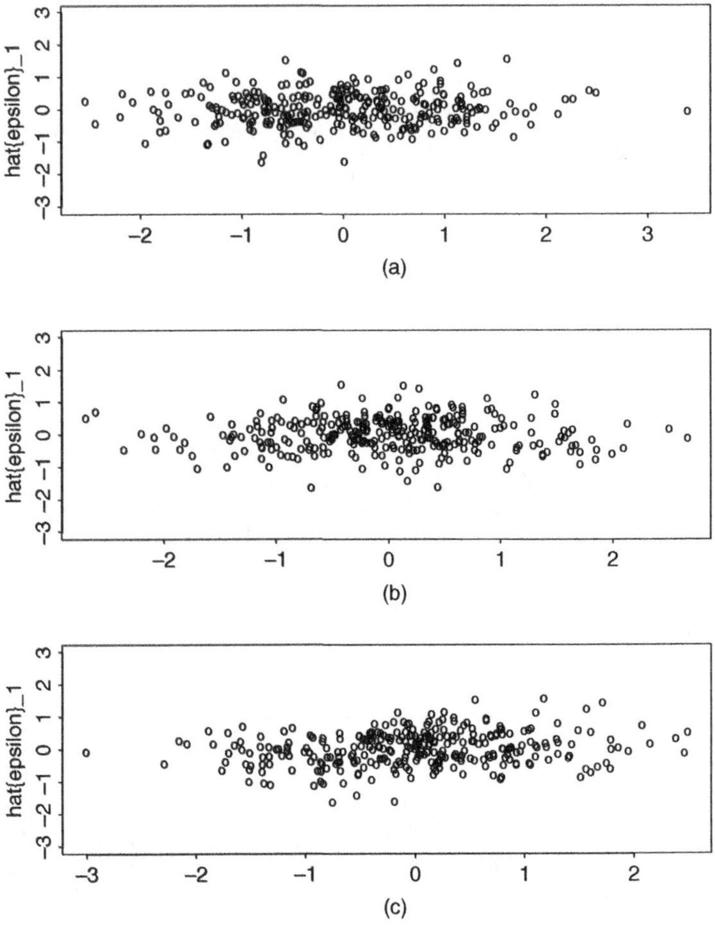

Figure 4.6 Plots of estimated residual $\hat{\epsilon}_{i1}$ versus: (a) $\hat{\xi}_{i1}$, (b) $\hat{\xi}_{i2}$ and (c) $\hat{\eta}_i$.

APPENDIX 4.1: THE METROPOLIS–HASTINGS ALGORITHM

Suppose we wish to simulate observations say $\{X_j, j = 1, 2, \ldots\}$ from a conditional distribution with target density $\pi(\cdot)$. At the jth iteration of the Metropolis–Hastings (MH) algorithm with a current X_j, the next X_{j+1} is chosen by first sampling a candidate point Υ from a proposal distribution $q(\cdot|X_j)$ which is easy to sample. This candidate point Υ is accepted as X_{j+1} with probability

$$\min\left(1, \frac{\pi(\Upsilon)q(X_j|\Upsilon)}{\pi(X_j)q(\Upsilon|X_j)}\right).$$

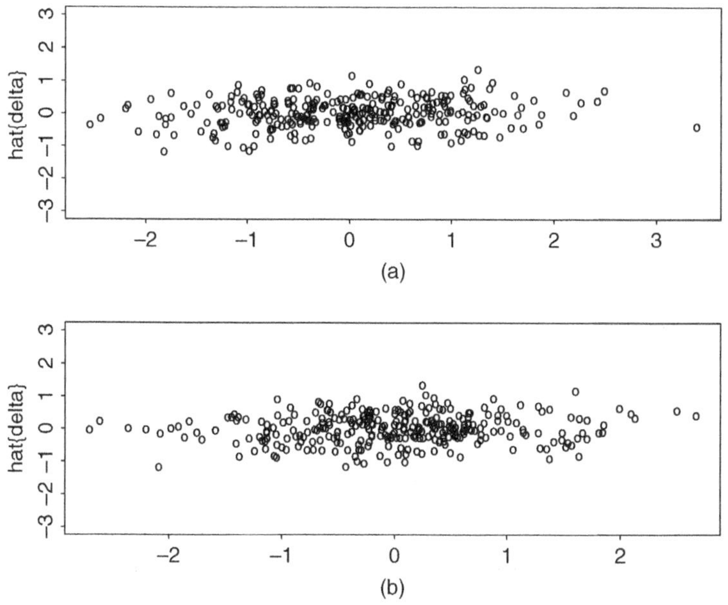

Figure 4.7 Plots of estimated residual $\hat{\delta}_i$ versus: (a) $\hat{\xi}_{i1}$ and (b) $\hat{\xi}_{i2}$.

If the candidate point Y is rejected, then $X_{j+1} = X_j$ and the chain does not move.

The proposal distribution $q(\cdot|\cdot)$ can have any form and the stationary distribution of the Markov chain will be the target distribution with density $\pi(\cdot)$. In most analyses of SEMs considered in this book, we will take $q(\cdot|X)$ to be a normal distribution with mean X and some covariance matrix.

APPENDIX 4.2: EPSR VALUE

Assessing convergence of a Monte Carlo simulation procedure should be on the basis of several simulation sequences generated independently via different starting values. The following approach (Gelman, 1996) involves monitoring each scalar estimate (e.g. parameter) of interest separately. Let n be the length of each sequence, after discarding the first part of the simulations. For each scalar estimate, say ψ, let $\psi_{jk}(j = 1, \ldots, n; k = 1, \ldots, K)$ be the draws from K parallel sequences of length n. The between- and within-sequences variances are computed as

$$B = \frac{n}{K-1}\sum_{k=1}^{K}(\psi_{\cdot k} - \psi_{\cdot\cdot})^2, \text{ where } \psi_{\cdot k} = n^{-1}\sum_{j=1}^{n}\psi_{jk},\ \psi_{\cdot\cdot} = K^{-1}\sum_{k=1}^{K}\psi_{\cdot k}$$

$$W = \frac{1}{K}\sum_{k=1}^{K}s_k^2, \text{ where } s_k^2 = (n-1)^{-1}\sum_{j=1}^{n}(\psi_{jk} - \psi_{\cdot k})^2.$$

The estimate of $\mathrm{var}(\psi|\mathbf{Y})$, the marginal posterior variance of the estimate, is then obtained by a weighted average of B and W as follows

$$\widehat{\mathrm{Var}}(\psi) = \frac{n-1}{n}W + \frac{1}{n}B.$$

The 'estimated potential scale reduction (EPSR)' is defined as

$$\hat{R}^{1/2} = [\widehat{\mathrm{Var}}(\psi)/W]^{1/2}.$$

As the simulation converges, $\hat{R}^{1/2}$ should be close to 1.0. In monitoring convergence, all EPSR values for all scalar estimates are computed. The whole simulation procedure is said to be converged if all the EPSR values are less than 1.2.

APPENDIX 4.3: DERIVATIONS OF CONDITIONAL DISTRIBUTIONS

To simplify notation in the derivation of $p(\mathbf{\Lambda}_k, \psi_{\epsilon k}|\mathbf{Y}, \mathbf{\Omega})$, we let $\nu_k = \psi_{\epsilon k}^{-1}$. From Equations (4.28), the conjugate prior density of ν_k, and the conjugate prior density of $\mathbf{\Lambda}_k$ given ν_k, are proportional to $\nu_k^{\alpha_{0k}-1}\exp(-\beta_{0k}\nu_k)$ and $\nu_k^{q/2}\exp[-\frac{1}{2}(\mathbf{\Lambda}_k - \mathbf{\Lambda}_{0k})^T\mathbf{H}_{0yk}^{-1}(\mathbf{\Lambda}_k - \mathbf{\Lambda}_{0yk})\nu_k]$, respectively. Also, from Equation (4.21), it can be seen that the likelihood of $\mathbf{Y}$ is given by

$$p(\mathbf{Y}|\mathbf{\Lambda}, \mathbf{\Psi}_\epsilon, \mathbf{\Omega}) \propto |\mathbf{\Psi}_\epsilon|^{-n/2}\exp\left[-\frac{1}{2}\sum_{i=1}^{n}(\mathbf{y}_i - \mathbf{\Lambda}\boldsymbol{\omega}_i)^T\mathbf{\Psi}_\epsilon^{-1}(\mathbf{y}_i - \mathbf{\Lambda}\boldsymbol{\omega}_i)\right].$$

Let $\mathbf{Y}_k^T$ be the kth row of $\mathbf{Y}, y_{ik}$ be the ith component of $\mathbf{Y}_k^T, \mathbf{A}_k^* = (\mathbf{\Omega}\mathbf{\Omega}^T)^{-1}\mathbf{\Omega}\mathbf{Y}_k$, and $b_k = \mathbf{Y}_k^T\mathbf{Y}_k - \mathbf{Y}_k^T\mathbf{\Omega}^T(\mathbf{\Omega}\mathbf{\Omega}^T)^{-1}\mathbf{\Omega}\mathbf{Y}_k = \mathbf{Y}_k^T\mathbf{Y}_k - \mathbf{A}_k^{*T}(\mathbf{\Omega}\mathbf{\Omega}^T)\mathbf{A}_k^*$, the exponential term in $p(\mathbf{Y}|\mathbf{\Lambda}, \mathbf{\Psi}_\epsilon, \mathbf{\Omega})$ can be expressed as

$$-\frac{1}{2}\sum_{i=1}^{n}(\mathbf{y}_i - \mathbf{\Lambda}\boldsymbol{\omega}_i)^T\mathbf{\Psi}_\epsilon^{-1}(\mathbf{y}_i - \mathbf{\Lambda}\boldsymbol{\omega}_i) = -\frac{1}{2}\sum_{i=1}^{n}\sum_{k=1}^{p}\psi_{\epsilon k}^{-1}(y_{ik} - \mathbf{\Lambda}_k^T\boldsymbol{\omega}_i)^2$$

$$= -\frac{1}{2}\sum_{k=1}^{p}\left\{\nu_k\left[\sum_{i=1}^{n}y_{ik}^2 - 2\mathbf{\Lambda}_k^T\sum_{i=1}^{n}y_{ki}\boldsymbol{\omega}_i + \mathrm{tr}(\mathbf{\Lambda}_k\mathbf{\Lambda}_k^T\sum_{i=1}^{n}\boldsymbol{\omega}_i\boldsymbol{\omega}_i^T)\right]\right\}$$

$$= -\frac{1}{2}\sum_{k=1}^{p}\{\nu_k[\mathbf{Y}_k^T\mathbf{Y}_k - 2\mathbf{\Lambda}_k^T\mathbf{\Omega}\mathbf{Y}_k + \mathbf{\Lambda}_k^T(\mathbf{\Omega}\mathbf{\Omega}^T)\mathbf{\Lambda}_k]\}$$

$$= -\frac{1}{2}\sum_{k=1}^{p}\{\nu_k[\mathbf{Y}_k^T\mathbf{Y}_k - \mathbf{Y}_k^T\mathbf{\Omega}^T(\mathbf{\Omega}\mathbf{\Omega}^T)^{-1}\mathbf{\Omega}\mathbf{Y}_k]$$

$$+ \nu_k[\mathbf{\Lambda}_k - (\mathbf{\Omega}\mathbf{\Omega}^T)^{-1}\mathbf{\Omega}\mathbf{Y}_k]^T(\mathbf{\Omega}\mathbf{\Omega}^T)[\mathbf{\Lambda}_k - (\mathbf{\Omega}\mathbf{\Omega}^T)^{-1}\mathbf{\Omega}\mathbf{Y}_k]\Big\}.$$

$$= -\frac{1}{2}\sum_{k=1}^{p}\{\nu_k[b_k + (\mathbf{\Lambda}_k - \mathbf{A}_k^*)^T(\mathbf{\Omega}\mathbf{\Omega}^T)(\mathbf{\Lambda}_k - \mathbf{A}_k^*)]\}.$$

Therefore, it follows from the likelihood of $\mathbf{Y}$ and the conjugate density of $\mathbf{\Lambda}_k$ and ν_k that:

$$p(\mathbf{\Lambda}, \nu_1, \ldots, \nu_p|\mathbf{Y}, \mathbf{\Omega}) \propto \prod_{k=1}^{p}\Big[\nu_k^{n/2+q/2+\alpha_{0\epsilon k}-1}$$

$$\times \exp\Big\{-\frac{1}{2}\nu_k[(\mathbf{\Lambda}_k - \mathbf{A}_k^*)^T(\mathbf{\Omega}\mathbf{\Omega}^T)(\mathbf{\Lambda}_k - \mathbf{A}_k^*)$$

$$+(\mathbf{\Lambda}_k - \mathbf{\Lambda}_{0k})^T\mathbf{H}_{0yk}^{-1}(\mathbf{\Lambda}_k - \mathbf{\Lambda}_{0k})] - \nu_k(\beta_{0\epsilon k} + b_k/2)\Big\}\Big]$$

$$= \prod_{k=1}^{p} p(\mathbf{\Lambda}_k, \nu_k|\mathbf{Y}, \mathbf{\Omega}).$$

From the above equation, it can be seen that the conditional distributions of $(\mathbf{\Lambda}_k, \nu_k)$ given $(\mathbf{Y}, \Omega)$ are mutually independent for $k = 1, \cdots, p$. Hence, it is sufficient to derive $p(\mathbf{\Lambda}_k, \nu_k|\mathbf{Y}, \mathbf{\Omega})$.

Let $\mathbf{A}_k = (\mathbf{H}_{0yk}^{-1} + \mathbf{\Omega}\mathbf{\Omega}^T)^{-1}$ and $\mathbf{a}_k = \mathbf{A}_k(\mathbf{H}_{0yk}^{-1}\mathbf{\Lambda}_{0k} + \mathbf{\Omega}\mathbf{Y}_k)$, it follows that

$$(\mathbf{\Lambda}_k - \mathbf{A}_k^*)^T(\mathbf{\Omega}\mathbf{\Omega}^T)(\mathbf{\Lambda}_k - \mathbf{A}_k^*) + (\mathbf{\Lambda}_k - \mathbf{\Lambda}_{0k})^T\mathbf{H}_{0yk}^{-1}(\mathbf{\Lambda}_k - \mathbf{\Lambda}_{0k})$$
$$= (\mathbf{\Lambda}_k - \mathbf{a}_k)^T\mathbf{A}_k^{-1}(\mathbf{\Lambda}_k - \mathbf{a}_k) - \mathbf{a}_k^T\mathbf{A}_k^{-1}\mathbf{a}_k + \mathbf{A}_k^{*T}\mathbf{\Omega}\mathbf{\Omega}^T\mathbf{A}_k^* + \mathbf{\Lambda}_{0k}^T\mathbf{H}_{0yk}^{-1}\mathbf{\Lambda}_{0k}.$$

Hence

$$p(\mathbf{\Lambda}_k, \nu_k|\mathbf{Y}, \mathbf{\Omega}) = p(\nu_k|\mathbf{Y}, \mathbf{\Omega})\, p(\mathbf{\Lambda}_k|\mathbf{Y}, \mathbf{\Omega}, \nu_k)$$

$$\propto \Big[\nu_k^{n/2+\alpha_{0\epsilon k}-1}\exp(-\beta_{\epsilon k}\nu_k)\Big]\cdot\Big\{\nu_k^{q/2}\exp\,[-\frac{1}{2}(\mathbf{\Lambda}_k - \mathbf{a}_k)^T\mathbf{A}_k^{-1}(\mathbf{\Lambda}_k - \mathbf{a}_k)\nu_k]\Big\}$$

where $\beta_{\epsilon k} = \beta_{0\epsilon k} + 2^{-1}(\mathbf{Y}_k^T\mathbf{Y}_k - \mathbf{a}_k^T\mathbf{A}_k^{-1}\mathbf{a}_k + \mathbf{\Lambda}_{0k}^T\mathbf{H}_{0yk}^{-1}\mathbf{\Lambda}_{0k})$. Thus, the posterior distribution of $(\mathbf{\Lambda}_k, \nu_k)$ given $\mathbf{Y}$ and $\mathbf{\Omega}$ is the following normal–gamma distribution (Broemeling, 1985):

$$[\nu_k|\mathbf{Y}, \mathbf{\Omega}] \stackrel{D}{=} \text{Gamma}[n/2 + \alpha_{0\epsilon k}, \ \beta_{\epsilon k}], \text{ and } [\mathbf{\Lambda}_k|\mathbf{Y}, \mathbf{\Omega}, \nu_k] \stackrel{D}{=} N[\mathbf{a}_k, \ \nu_k^{-1}\mathbf{A}_k].$$

REFERENCES

Albert, J. H. and Chib, S. (1993) Bayesian analysis of binary and polychotomous response data. *Journal of the American Statistical Association*, **88**, 669–679.

Ansari, A. and Jedidi, K. (2000) Bayesian factor analysis for multilevel binary observations. *Psychometrika*, **65**, 475–498.

Ansari, A., Jedidi, K. and Dube, L. (2002) Heterogeneous factor analysis models: A Bayesian approach. *Psychometrika*, **67**, 49–78.

Ansari, A., Jedidi, K. and Jagpal, S. (2000) A hierarchical Bayesian methodology for treating heterogeneity in structural equation models. *Marketing Science*, **19**, 328–347.

Bartholomew, D. J. (1981) Posterior analysis of the factor model. *British Journal of Mathematics and Statistical Psychology*, **34**, 93–99.

Berger, J. O. (1985) *Statistical Decision Theory and Bayesian Analysis.* New York: Springer-Verlag.

Boomsma, A. (1982) The robustness of LISREL against small sample sizes in factor analysis model. In K. G. Jörkog and H. Wold (eds), *System under Indirect Observation: Causality, Structure, Prediction* pp. 149–173. Amsterdam: North-Holland.

Box, G. E. P. and Tiao, G. C. (1973) *Bayesian Inference in Statistical Analysis.* Reading, MA: Addison-Wesley.

Broemeling, L. D. (1985) *Bayesian Analysis of Linear Models.* New York: Marcel Dekker Inc..

Chou, C. P., Bentler, P. M. and Satorra, A. (1991) Scaled test statistics and robust standard errors for non-normal data in covariance structure analysis: A Monte Carlo study. *British Journal of Mathematical and Statistical Psychology*, **44**, 347–357.

Congdon, P. (2003) *Applied Bayesian Modeling.* Hoboken, New York: John Wiley & Sons, Inc..

Dempster, A. P., Laird, N. M. and Rubin, D. B. (1977) Maximum likelihood from incomplete data via the EM algorithm (with discussion). *Journal of the Royal Statistical Society, Series B*, **39**, 1–38.

Dunson, D. B. (2000) Bayesian latent variable models for clustered mixed outcomes. *Journal of the Royal Statistical Society, Series B*, **62**, 355–366.

Fuller, W. A. (1987) *Measurement Error Models.* New York: John Wiley & Sons, Inc..

Gelfand, A. E. and Smith, A. F. M. (1990) Sampling-based approaches to calculating marginal densities. *Journal of the American Statistical Association*, **85**, 398–409.

Gelman, A. (1996) Inference and monitoring convergence. In W. R. Gilks, S. Richardson, and D. J. Spiegelhalter (eds), *Markov Chain Monte Carlo in Practice*, pp. 131–144. London: Chapman and Hall.

Gelman, A., Carlin, J. B., Stern, H. S. and Rubin, D. B. (1995) *Bayesian Data Analysis.* London: Chapman & Hall Ltd.

Geman, S. and Geman, D. (1984) Stochastic relaxation, Gibbs distribution and the Bayesian restoration of images. *IEEE Transactions on Pattern Analysis and Machine Intelligence*, **6**, 721–741.

Geyer, C. J. (1992) Practical Markov chain Monte Carlo. *Statistical Science*, **7**, 473–511.

Hastings, W. K. (1970) Monte Carlo sampling methods using Markov chains and their application. *Biometrika*, **57**, 97–109.

Hoogland, J. J. and Boomsma, A. (1998) Robustness studies in covariance structure modeling: an overview and a meta analysis. *Sociological Methods & Research*, **26**, 329–368.

Hu, L., Bentler, P. M. and Kano, Y. (1992) Can test statistics in covariance structure analysis be trusted. *Psychological Bulletin*, **112**, 351–362.

Jöreskog, K. G. and Sörbom, D. (1996) *LISREL 8: Structural Equation Modeling with the SIMPLIS Command Language*. Hove and London: Scientific Software International.

Kass, R. E. and Raftery, A. E. (1995) Bayes factors. *Journal of the American Statistical Association*, **90**, 773–795.

Lawson, A. B., Browne, W. J. and Vidal Rodeiro, C. L. (2003) *Disease Mapping with WinBUGS and MLWIN*. Cluchester: John Wiley & Sons, Ltd.

Lee, S. Y. (1980) Estimation of covariance structure models with parameters subject to functional restraints. *Psychometrika*, **45**, 309–324.

Lee, S. Y. (1981) A Bayesian approach to confirmatory factor analysis. *Psychometrika*, **46**, 153–160.

Lee, S. Y. and Shi, J. Q. (2000) Joint Bayesian analysis of factor scores and structural parameters in the factor analysis model. *Annals of the Institute of Statistical mathematics*, **52**, 722–736.

Lee, S. Y. and Song, X. Y. (2004) Evaluation of the Bayesian and maximum likelihood approaches in analyzing structural equation models with small sample sizes. *Multivariate Behavioral Research*. **39**, 653–686.

Lee, S. Y. and Zhu, H. T. (1999) Statistical analysis of nonlinear factor analysis models. *British Journal of Mathematical and Statistical Psychology*, **52**, 225–242.

Lindley, D. V. and Smith, A. F. M. (1972) Bayes estimates for the linear model (with discussion). *Journal of the Royal Statistical Society, Series B*, 1–42.

Martin, J. K. and McDonald, R. P. (1975) Bayesian estimation in unrestricted factor analysis: a treatment for Heyword cases. *Psychometrika*, **40**, 505–577.

Metropolis, N. *et al.* (1953) Equations of state calculations by fast computing machine. *Journal of Chemical Physics*, **21**, 1087–1091.

Muirhead, R. J. (1982) *Aspects of Multivariate Statistical Theory*. New York: John Wiley & Sons, Inc..

Rubin, D. B. (1991) EM and beyond. *Psychometrika*, **56**, 241–254.

Scheines, R., Hoijtink, H. and Boomsma, A. (1999) Bayesian estimation and testing of structural equation models. *Psychometrika*, **64**, 37–52.

Shi, J. Q. and Lee, S. Y. (1998) Bayesian sampling-based approach for factor analysis model with continuous and polytomous data. *British Journal of Mathematical and Statistical Psychology*, **51**, 233–252.

Song, X. Y. and Lee, S. Y. (2001) Bayesian estimation and test for factor analysis model with continuous and polytomous data in several populations. *British Journal of Mathematical and Statistical Psychology*, **54**, 237–263.

Spiegelhalter, D. J., Thomas, A., Best, N. G. and Lunn, D. (2003) *WinBUGS User Manual. Version 1.4.* Cambridge, England: MRC Biostatistics Unit.

Tanner, M. A. and Wong, W. H. (1987) The calculation of posterior distributions by data augmentation(with discussion). *Journal of the American Statistical Association*, **82**, 528–550.

Yung, Y. F. and Bentler, P. M. (1994) Bootstrap-corrected ADF test statistics in covariance structure analysis. *British Journal of Mathematical and Statistical Psychology*, **47**, 63–84.

Zeger, S. L. and Karim, M. R. (1991) Generalized linear models with random effects: A Gibbs sampling approach. *Journal of the American Statistical Association*, **86**, 79–86.

Zellner, A. (1971) *An Introduction to Bayesian Inference in Econometrics*. New York: John Wiley & Sons, Inc..

5

Model Comparison and Model Checking

5.1 INTRODUCTION

One important statistical inference beyond estimation is on testing of various hypotheses about the model. In the field of structural equation modeling, a common approach for hypothesis testing is to use the significance tests on the basis of p-values that are determined by some asymptotic distributions of the test statistics. As pointed out in the statistics literature (see e.g. Berger and Sellke, 1987; Berger and Dalampady, 1987; Kass and Raftery, 1995) there are problems associated with such an approach. Those which are related to SEMs are discussed as follows:

(i) Tests on the basis of p-values tend to reject the null hypothesis too frequently with large sample sizes. A dramatic example with a sample size 113 556 was given by Raftery (1986), where a substantively meaningful model (associated with the null hypothesis) that explained 99.7 % of the deviance was rejected by a standard chi-squared test with an extremely small p-value. In the traditional analysis of SEMs, various descriptive fit indexes, such as the well-known normed or non-normed fit indexes (Bentler and Bonett, 1980) and the comparative fit index (Bentler, 1992) have been proposed as complementary measures for the goodness-of-fit of the model. Very often, the values of the fit indexes are over 0.95, but the p-values of the χ^2-test are less than 0.01. Under these situations, the conclusions drawn from these two testing methods seem contradictory.

(ii) The p-value of a significance test in hypothesis testing is a measure of evidence against the null model, not a means of supporting/proving the model. Hence, the conclusion of a significance test can only be used to reject the null hypothesis and cannot offer an assessment of the strength of the evidence in favor of the null hypothesis. As a result, even the chi-square goodness-of-fit test does not reject the null hypothesis, nor can it be used to conclude that the posited model is better than the alternative model, or to conclude that the given data support the posited model. On the other hand, rejection of the null hypothesis by such a test does not indicate the alternative model is better.

(iii) The significance tests, as well as the descriptive fit indexes mentioned above, cannot be applied to test non-nested hypotheses or to compare non-nested models. Therefore only a hierarchy of nested hypotheses can be assessed see, for example, Bollen (1989). However, we are very often interested in assessing non-nested SEMs in practical applications.

In this chapter, we consider a Bayesian approach for hypothesis testing that does not have the above problems. As we can associate a hypothesis with a model, testing the null hypothesis H_0 against its alternative H_1 can be regarded as comparing two models corresponding to H_0 and H_1. Hence we will use the general term 'model comparison' to represent hypothesis testing, model comparison and/or model selection. In the field of SEM, we are often interested in comparing a discrete set of competing models. Typical examples are comparing an FA model with three factors with an FA model with four factors, comparing an SEM with an interaction term of exogenous latent variables with one without the interaction term, comparing a mixture SEM with two components with one with three components, and so on. The well-known statistic in Bayesian model comparison, namely the Bayes factor (Berger, 1985; Kass and Raftery, 1995), will be emphasized and applied to the problem of model comparison. Other methods that emphasize for comparing continuous families of models (see Gelman, Carlin, Stern and Rubin 2003, and the references therein) are not considered.

In general, the computation of the Bayes factor is difficult. Various computational methods have been proposed (Kass and Raftery, 1995). A simple but rough approximation, namely the Bayesian Information Criterion (BIC) has been used for model comparison of some SEMs. For example, Raftery (1993) applied it to the LISREL model, Lee and Song (2001) applied it to a two-level SEM, and Jedidi, Jagpal, and DeSarbo (1997) applied it to finite mixtures of SEMs with a fixed number of components, among others. Other useful methods for computing the Bayes factor have been established on the basis of posterior simulation, using recently developed MCMC methods. DiCiccio, Kass, Raftery and Wasserman (1997) provided a comparative study on a variety of methods, from Laplace approximation to importance sampling and bridge sampling, and concluded that bridge sampling is an attractive method. Gelman and Meng (1998) showed that path sampling is a direct extension of the bridge sampling. Naturally, it is expected that path sampling is even better. In this book, we emphasize the application of path sampling (Gelman and Meng, 1998) to compute Bayes factors for model comparisons of various SEMs.

Differing from estimation, Bayesian model comparison using the Bayes factor may be sensitive to prior distributions of the parameters. Hence, these distributions should be selected with care, and sensitivity analysis of the prior inputs should be conducted.

An introduction of the Bayes factor will be presented in Section 5.2, followed by the discussion of path sampling for computing this statistic in Section 5.3. Section 5.4 provides an application of the methodology to SEMs with fixed covariates. Some other methods for model comparison and model checking are given in Section 5.5, and a discussion is given in Section 5.6.

5.2 BAYES FACTOR

In this section, we introduce an important Bayesian statistic, the Bayes factor (Berger, 1985; Kass and Raftery, 1995), for model comparison/selection. This statistic has a solid logical foundation that offers great flexibility. It has been extensively applied to a lot of statistical models, see the references given in Kass and Raftery (1995). Its applicability is further enhanced by the powerful MCMC methods that have been recently developed in statistical computing.

Suppose the given data $\mathbf{Y}$ with a sample size n have arisen under one of the two competing models M_1 and M_0 according to a probability density $p(\mathbf{Y}|M_1)$ or $p(\mathbf{Y}|M_0)$, respectively. Let $p(M_0)$ be the prior probability of M_0 and $p(M_1) = 1 - p(M_0)$, and let $p(M_k|\mathbf{Y})$ be the posterior probability, for $k = 0, 1$. From the Bayes theorem, we obtain

$$p(M_k|\mathbf{Y}) = \frac{p(\mathbf{Y}|M_k)p(M_k)}{p(\mathbf{Y}|M_1)p(M_1) + p(\mathbf{Y}|M_0)p(M_0)}, \quad k = 0, 1.$$

Hence,

$$\frac{p(M_1|\mathbf{Y})}{p(M_0|\mathbf{Y})} = \frac{p(\mathbf{Y}|M_1)\ p(M_1)}{p(\mathbf{Y}|M_0)\ p(M_0)}. \tag{5.1}$$

The Bayes factor for comparing M_1 and M_0 is defined as

$$B_{10} = \frac{p(\mathbf{Y}|M_1)}{p(\mathbf{Y}|M_0)}. \tag{5.2}$$

From Equation (5.1), we see that

$$\text{posterior odds} = \text{Bayes factor} \times \text{prior odds}.$$

In the special case where the competitive models M_1 and M_0 are equally probable a priori so that $p(M_1) = p(M_0) = 0.5$, the Bayes factor is equal to the posterior odds in favor of M_1. In general, it is a summary of evidence provided by the data in favor of M_1 as opposed to M_0, or in favor of M_0 as opposed to M_1. It may reject a null hypothesis associated with M_0, or may equally provide evidence in favor of the null hypothesis or the alternative hypothesis associated with M_1. Moreover, unlike the significance test approach that is based on the likelihood ratio criterion and its asymptotic chi-square statistic, the comparison does not depend on the assumption that either model is 'true'. Moreover, it can be seen from Equation (5.2) that the same data set is used in the comparison, hence, it does not favor the alternative hypothesis (or M_1) in extremely large samples. Finally, it can be applied to compare non-nested models M_0 and M_1.

According to the suggestion given in Kass and Raftery (1995), the criterion that is given in Table 5.1 is used for interpreting B_{10} and $2 \log B_{10}$. Kass and Raftery (1995) pointed out that these categories furnish appropriate guidelines for practical applications of the Bayes factor. Depending on the competing models M_0 and M_1 for fitting a given data set, if the Bayes factor (or 2 log Bayes factor) rejects the null hypothesis H_0 that is associated with M_0, we can conclude that the data give evidence to support the alternative hypothesis H_1 that is associated with M_1. Similarly if the Bayes factor rejects H_1, a more definite conclusion of supporting H_0 can be attained.

The interpretation of evidence provided by Table 5.1 depends on the specific context. For two non-nested competitive models, say M_1 and M_0, we should select M_0 if $2 \log B_{10}$ is negative. If $2 \log B_{10}$ is in (0, 2), we may interpret M_1 is slightly better than M_0 and hence it may be better to select M_1. The choice of M_1 is more definite if $2 \log B_{10}$ is larger than 6. For two nested competitive models, say M_0 is nested in the more complicated model M_1, if M_1 is significantly better than M_0, it can be much larger than 6. Then the above criterion will suggest a decisive conclusion to select M_1. However, if $2 \log B_{10}$ is in (0, 2), then the difference between M_0 and M_1 is 'not worth more than a bare mention'.

Table 5.1 Interpretation of Bayes factor.

B_{10}	$2 \log B_{10}$	Evidence against H_0 (M_0)
< 1	< 0	Supports $H_0(M_0)$
1 to 3	0 to 2	Not worth more than a bare mention
3 to 20	2 to 6	Positive
20 to 150	6 to 10	Strong
> 150	> 10	Decisive

Under this situation, great caution should be taken in drawing a conclusion. According to the 'parsimonious' guideline in practical applications, it may be desirable to select M_0 if it is simpler than M_1. The criterion given in Table 5.1 is a suggestion, it is not necessary to regard it as a strict rule. Similarly in frequentist hypothesis testing, one may take the type I error to be 0.05 or 0.10, and the choice is decided with other factors in the substantive situation. Similar to other data analyses, for conclusions drawn from the marginal cases, it is always helpful to conduct other analysis, for example residual analysis, to cross-validate the results. Generally speaking, model selection should be approached on a problem-by-problem basis. It is also desirable to take the opinions from experts into account if no clear conclusion can be drawn.

From Equation (5.2), we see that the density $p(\mathbf{Y}|M_k)$ is involved in the Bayes factor. This function is obtained by integrating $p(\mathbf{Y}|\boldsymbol{\theta}_k, M_k)p(\boldsymbol{\theta}_k|M_k)$ over the parameter space, that is

$$p(\mathbf{Y}|M_k) = \int p(\mathbf{Y}|\boldsymbol{\theta}_k, M_k)p(\boldsymbol{\theta}_k|M_k)\,d\boldsymbol{\theta}_k, \tag{5.3}$$

where $\boldsymbol{\theta}_k$ is the parameter vector in M_k, $p(\boldsymbol{\theta}_k|M_k)$ is its prior density and $p(\mathbf{Y}|\boldsymbol{\theta}_k, M_k)$ is the probability density of $\mathbf{Y}$ given $\boldsymbol{\theta}_k$. The dimension of this integral is equal to the dimension of $\boldsymbol{\theta}_k$. This quantity can be interpreted as the marginal likelihood of the data, obtained by integrating the joint density of $(\mathbf{Y}, \boldsymbol{\theta}_k)$ over $\boldsymbol{\theta}_k$. It can also be interpreted as the predictive probability of the data; that is, the probability of seeing the data that actually were observed, calculated before any data became available. Sometimes, it is also called an integrated likelihood. Note that, as in the computation of the likelihood ratio statistic but unlike in some other applications of likelihood, all constants appearing in the definition of the likelihood $p(\mathbf{Y}|\boldsymbol{\theta}_k, M_k)$ must be retained when computing B_{10}. In fact, B_{10} is closely related to the likelihood ratio statistic, in which the parameters $\boldsymbol{\theta}_k$ are eliminated by maximization rather than by integration. Very often, it is extremely difficult to obtain B_{10} analytically, and various analytic and numerical approximations have been proposed in the literature. For example, Chib (1995), and Chib and Jeliazkov (2001) respectively developed efficient algorithms for computing the marginal likelihood through MCMC chains produced by the Gibbs sampler and the MH algorithm. Based on the results of DiCiccio, Kass, Raftery and Wasserman (1997) and the recommendation of Gelman and Meng (1998), we will apply path sampling to compute the Bayes factor for model comparison.

5.3 PATH SAMPLING

A procedure based on path sampling (Gelman and Meng, 1998) is introduced in this section for computing the Bayes factor. The key idea of path sampling is to compute the ratio of normalizing constants of probability densities (or equivalently difference of the logarithm of them). Hence, it can be applied to compute the Bayes factor. Following Gelman and Meng (1998), we motivate this computing tool from importance sampling (see for example Gelfand and Dey, 1994) and bridge sampling (Meng and Wong, 1996).

In the context of SEMs, we consider two competitive models M_0 and M_1 with the matrix $\boldsymbol{\Omega}$ of latent variables. Let $\boldsymbol{\theta}$ be the vector of unknown parameters in both M_1 and M_0. Usually, it is quite difficult to apply path sampling to evaluate $p(\mathbf{Y}|M_k)$ under complicated situations. Hence, we first use the idea of data augmentation to include the latent variables in the computation, by considering the complete-data set $(\mathbf{Y}, \boldsymbol{\Omega})$. Basically, we compute

$$z(k) = p(\mathbf{Y}|M_k) = \int p(\mathbf{Y}, \boldsymbol{\Omega}, \boldsymbol{\theta}|M_k)\, d\boldsymbol{\Omega}\, d\boldsymbol{\theta}, \quad k = 0, 1$$

and obtain B_{10} as the ratio of $z(1)/z(0)$.

The key idea of importance sampling is the following equality:

$$z(1) = p(\mathbf{Y}|M_1) = \int \frac{p(\mathbf{Y}, \boldsymbol{\Omega}, \boldsymbol{\theta}|M_1)}{p(\boldsymbol{\Omega}, \boldsymbol{\theta}|\mathbf{Y}, M_1)} p(\boldsymbol{\Omega}, \boldsymbol{\theta}|\mathbf{Y}, M_1)\, d\boldsymbol{\Omega}\, d\boldsymbol{\theta}. \tag{5.4}$$

Viewing $p(\boldsymbol{\Omega}, \boldsymbol{\theta}|\mathbf{Y}, M_1)$ as a probability density function, $z(1)$ is the expectation of $p(\mathbf{Y}, \boldsymbol{\Omega}, \boldsymbol{\theta}|M_1)/p(\boldsymbol{\Omega}, \boldsymbol{\theta}|\mathbf{Y}, M_1)$ with respect to the joint distribution of $\boldsymbol{\Omega}$ and $\boldsymbol{\theta}$ under M_1. On the basis of the simple idea that approximating the expectation by the sample mean of observations in a sufficiently large sample, it can be approximated as below,

$$z(1) \cong \frac{1}{J} \sum_{j=1}^{J} \frac{p(\mathbf{Y}, \boldsymbol{\Omega}^{(j)}, \boldsymbol{\theta}^{(j)}|M_1)}{p(\boldsymbol{\Omega}^{(j)}, \boldsymbol{\theta}^{(j)}|\mathbf{Y}, M_1)}, \tag{5.5}$$

where $\{(\boldsymbol{\Omega}^{(j)}, \boldsymbol{\theta}^{(j)}), j = 1, \cdots, J\}$ is a sample of observations drawn from the target distribution $p(\boldsymbol{\Omega}, \boldsymbol{\theta}|\mathbf{Y}, M_1)$ via some posterior simulation methods. Similarly we can obtain $z(0)$. Finally, the Bayes factor can be estimated via $z(1)/z(0)$. In this method, we need to simulate observations from both $p(\boldsymbol{\Omega}, \boldsymbol{\theta}|\mathbf{Y}, M_0)$ and $p(\boldsymbol{\Omega}, \boldsymbol{\theta}|\mathbf{Y}, M_1)$. The following slight modification may be preferable. It follows from an identity in Meng and Wong (1996, Equation 1.4) that the ratio of normalizing constants can be expressed as

$$\frac{z(1)}{z(0)} = E_0\left[\frac{p(\mathbf{Y}, \boldsymbol{\Omega}, \boldsymbol{\theta}|M_1)}{p(\mathbf{Y}, \boldsymbol{\Omega}, \boldsymbol{\theta}|M_0)}\right], \quad \text{when} \quad \Theta_1 \subset \Theta_0, \tag{5.6}$$

where Θ_k is the set containing $(\boldsymbol{\Omega}, \boldsymbol{\theta})$ such that $p(\boldsymbol{\Omega}, \boldsymbol{\theta}|\mathbf{Y}, M_k) > 0,\ k = 0, 1$, and E_k is the expectation taken with respect to $p(\boldsymbol{\Omega}, \boldsymbol{\theta}|\mathbf{Y}, M_k)$. Then,

$$B_{10} = \frac{z(1)}{z(0)} \cong \frac{1}{J}\sum_{j=1}^{J} \frac{p(\mathbf{Y}, \boldsymbol{\Omega}^{(j)}, \boldsymbol{\theta}^{(j)}|M_1)}{p(\mathbf{Y}, \boldsymbol{\Omega}^{(j)}, \boldsymbol{\theta}^{(j)}|M_0)}, \tag{5.7}$$

where $\{(\boldsymbol{\Omega}^{(j)}, \boldsymbol{\theta}^{(j)}), j = 1, \cdots, J\}$ are sampled from $p(\boldsymbol{\Omega}, \boldsymbol{\theta}|\mathbf{Y}, M_0)$. Hence, this application of the importance sampling only requires posterior simulation of $p(\boldsymbol{\Omega}, \boldsymbol{\theta}|\mathbf{Y}, M_0)$. The application of importance sampling is fairly straightforward. However, this method is only effective if M_0 and M_1 are close to each other.

Bridge sampling is a better method for computing the Bayes factor in the case where M_0 and M_1 may not be close to each other. Its development is based on following important identity whose general form has been studied in detail by Meng and Wong (1996):

$$\frac{z(1)}{z(0)} = \frac{E_0[p(\mathbf{Y}, \boldsymbol{\Omega}, \boldsymbol{\theta}|M_1)\alpha(\boldsymbol{\Omega}, \boldsymbol{\theta})]}{E_1[p(\mathbf{Y}, \boldsymbol{\Omega}, \boldsymbol{\theta}|M_0)\alpha(\boldsymbol{\Omega}, \boldsymbol{\theta})]}, \tag{5.8}$$

where $\alpha(\boldsymbol{\Omega}, \boldsymbol{\theta})$ is an arbitrary function satisfying

$$0 < \left| \int_{\Theta_0 \cap \Theta_1} \alpha(\boldsymbol{\Omega}, \boldsymbol{\theta}) p(\mathbf{Y}, \boldsymbol{\Omega}, \boldsymbol{\theta}|M_1) p(\mathbf{Y}, \boldsymbol{\Omega}, \boldsymbol{\theta}|M_0) d\boldsymbol{\Omega} d\boldsymbol{\theta} \right| < \infty.$$

The existence of such $\alpha(\boldsymbol{\Omega}, \boldsymbol{\theta})$ is guaranteed if the set of $(\boldsymbol{\Omega}, \boldsymbol{\theta})$ that satisfies $p(\mathbf{Y}, \boldsymbol{\Omega}, \boldsymbol{\theta}|M_1) > 0$ and $p(\mathbf{Y}, \boldsymbol{\Omega}, \boldsymbol{\theta}|M_0) > 0$ is not empty; that is, there is some overlap between M_0 and M_1. Taking $\alpha(\boldsymbol{\Omega}, \boldsymbol{\theta}) = 1/p(\mathbf{Y}, \boldsymbol{\Omega}, \boldsymbol{\theta}|M_0)$, Equation (5.8) reduces to Equation (5.6), which is the method based on the importance sampling. Now, we define

$$\alpha(\boldsymbol{\Omega}, \boldsymbol{\theta}) = \frac{p(\mathbf{Y}, \boldsymbol{\Omega}, \boldsymbol{\theta}|M_{\frac{1}{2}})}{p(\mathbf{Y}, \boldsymbol{\Omega}, \boldsymbol{\theta}|M_0) p(\mathbf{Y}, \boldsymbol{\Omega}, \boldsymbol{\theta}|M_1)},$$

where $p(\mathbf{Y}, \boldsymbol{\Omega}, \boldsymbol{\theta}|M_{\frac{1}{2}})$ is a density in between $p(\mathbf{Y}, \boldsymbol{\Omega}, \boldsymbol{\theta}|M_0)$ and $p(\mathbf{Y}, \boldsymbol{\Omega}, \boldsymbol{\theta}|M_1)$ having support $\Theta_0 \cap \Theta_1$ with a model $M_{\frac{1}{2}}$ in between M_0 and M_1. Substitute this $\alpha(\boldsymbol{\Omega}, \boldsymbol{\theta})$ into Equation (5.8), we have

$$\begin{aligned} \frac{z(1)}{z(0)} &= \frac{E_0[p(\mathbf{Y}, \boldsymbol{\Omega}, \boldsymbol{\theta}|M_{\frac{1}{2}})/p(\mathbf{Y}, \boldsymbol{\Omega}, \boldsymbol{\theta}|M_0)]}{E_1[p(\mathbf{Y}, \boldsymbol{\Omega}, \boldsymbol{\theta}|M_{\frac{1}{2}})/p(\mathbf{Y}, \boldsymbol{\Omega}, \boldsymbol{\theta}|M_1)]} \\ &\cong \frac{J_0^{-1}\sum_{j=1}^{J_0}[p(\mathbf{Y}, \boldsymbol{\Omega}_0^{(j)}, \boldsymbol{\theta}_0^{(j)}|M_{\frac{1}{2}})/p(\mathbf{Y}, \boldsymbol{\Omega}_0^{(j)}, \boldsymbol{\theta}_0^{(j)}|M_0)]}{J_1^{-1}\sum_{j=1}^{J_1}[p(\mathbf{Y}, \boldsymbol{\Omega}_1^{(j)}, \boldsymbol{\theta}_1^{(j)}|M_{\frac{1}{2}})/p(\mathbf{Y}, \boldsymbol{\Omega}_1^{(j)}, \boldsymbol{\theta}_1^{(j)}|M_1)]}, \end{aligned} \tag{5.9}$$

where $\{(\boldsymbol{\Omega}_0^{(j)}, \boldsymbol{\theta}_0^{(j)})\}, j = 1, \cdots, J_0\}$ and $\{(\boldsymbol{\Omega}_1^{(j)}, \boldsymbol{\theta}_1^{(j)}), j = 1, \cdots, J_1\}$ are drawn from $p(\boldsymbol{\Omega}, \boldsymbol{\theta}|\mathbf{Y}, M_0)$ and $p(\boldsymbol{\Omega}, \boldsymbol{\theta}|\mathbf{Y}, M_1)$, respectively. Hence, in contrast to the importance sampling which directly approximates $z(1)/z(0)$ via M_0 and M_1, the bridge sampling makes use of a sensible choice of a bridge model $M_{\frac{1}{2}}$ so that M_0 is closer to $M_{\frac{1}{2}}$ and $M_{\frac{1}{2}}$ is closer to M_1, for improving the effectiveness and accuracy.

The efficiency of bridge sampling depends on the overlap of M_0 and $M_{\frac{1}{2}}$, as well as $M_{\frac{1}{2}}$ and M_1. The more they overlap, the more efficient is the algorithm. If $p(\mathbf{Y}, \boldsymbol{\Omega}, \boldsymbol{\theta}|M_0)$ and $p(\mathbf{Y}, \boldsymbol{\Omega}, \boldsymbol{\theta}|M_1)$ are very far apart, then the algorithm will be unstable even with the optimal bridge. Hence, it is useful to construct a series of $L-1$ intermediate models, from which we can make draws. Let $p(\mathbf{Y}, \boldsymbol{\Omega}, \boldsymbol{\theta}|t_\ell)$, where $\ell = 0, \cdots, L$ and t_ℓ in $[0, 1]$, represent a class of densities corresponding to models M_ℓ in between M_0 and M_1. For example, $p(\mathbf{Y}, \boldsymbol{\Omega}, \boldsymbol{\theta}|t_0 = 0)$ and $p(\mathbf{Y}, \boldsymbol{\Omega}, \boldsymbol{\theta}|t_L = 1)$ represent M_0 and M_1, respectively. For each pair of consecutive functions $p(\mathbf{Y}, \boldsymbol{\Omega}, \boldsymbol{\theta}|t_\ell)$ and $p(\mathbf{Y}, \boldsymbol{\Omega}, \boldsymbol{\theta}|t_{\ell-\frac{1}{2}})$, it follows from Equations (5.8) and (5.9) that

$$\frac{z(1)}{z(0)} = \prod_{\ell=1}^{L} \frac{E_{\ell-1}[p(\mathbf{Y}, \boldsymbol{\Omega}, \boldsymbol{\theta}|t_{\ell-\frac{1}{2}})/p(\mathbf{Y}, \boldsymbol{\Omega}, \boldsymbol{\theta}|t_{\ell-1})]}{E_\ell[p(\mathbf{Y}, \boldsymbol{\Omega}, \boldsymbol{\theta}|t_{\ell-\frac{1}{2}})/p(\mathbf{Y}, \boldsymbol{\Omega}, \boldsymbol{\theta}|t_\ell)]}.$$

Let L tend to infinity and consider the index t_ℓ as a parameter in $[0, 1]$. Gelman and Meng (1998) proved that

$$\begin{aligned}\log \frac{z(1)}{z(0)} &= \lim_{L\to\infty} \log \prod_{\ell=1}^{L} \frac{E_{\ell-1}[p(\mathbf{Y}, \boldsymbol{\Omega}, \boldsymbol{\theta}|t_{\ell-\frac{1}{2}})/p(\mathbf{Y}, \boldsymbol{\Omega}, \boldsymbol{\theta}|t_{\ell-1})]}{E_\ell[p(\mathbf{Y}, \boldsymbol{\Omega}, \boldsymbol{\theta}|t_{\ell-\frac{1}{2}})/p(\mathbf{Y}, \boldsymbol{\Omega}, \boldsymbol{\theta}|t_\ell)]} \\ &= \int_0^1 E_{\boldsymbol{\Omega},\boldsymbol{\theta}}[U(\mathbf{Y}, \boldsymbol{\Omega}, \boldsymbol{\theta}, t)]\mathrm{d}t,\end{aligned} \tag{5.10}$$

where $E_{\boldsymbol{\Omega},\boldsymbol{\theta}}$ denotes the expectation with respect to the distribution $p(\boldsymbol{\Omega}, \boldsymbol{\theta}|\mathbf{Y}, t)$, and

$$U(\mathbf{Y}, \boldsymbol{\Omega}, \boldsymbol{\theta}, t) = \frac{\mathrm{d}}{\mathrm{d}t} \log p(\mathbf{Y}, \boldsymbol{\Omega}|\boldsymbol{\theta}, t). \tag{5.11}$$

Following a procedure in Gelman and Meng (1998) to evaluate numerically the above integral over t via the method in Ogata (1989, 1990), an estimate of $\log B_{10}$ can be obtained. Specifically, we first order the values of S fixed grids $\{t_{(s)}\}_{s=0}^S$ such that $t_{(0)} = 0 < t_{(1)} < t_{(2)} < \cdots < t_{(S)} < t_{(S+1)} = 1$, and estimate $\log B_{10}$ by

$$\widehat{\log B_{10}} = \frac{1}{2}\sum_{s=0}^{S}(t_{(s+1)} - t_{(s)})(\overline{U}_{(s+1)} + \overline{U}_{(s)}), \tag{5.12}$$

where $\overline{U}_{(s)}$ is the average of the values of $U(\mathbf{Y}, \boldsymbol{\Omega}, \boldsymbol{\theta}, t)$ on the basis of all simulation draws for which $t = t_{(s)}$, that is,

$$\overline{U}_{(s)} = J^{-1} \sum_{j=1}^{J} U(\mathbf{Y}, \boldsymbol{\Omega}^{(j)}, \boldsymbol{\theta}^{(j)}, t_{(s)}), \tag{5.13}$$

in which $\{(\boldsymbol{\Omega}^{(j)}, \boldsymbol{\theta}^{(j)}), j = 1, \cdots, J\}$ are simulated observations drawn from $p(\boldsymbol{\Omega}, \boldsymbol{\theta}|\mathbf{Y}, t_{(s)})$. This is the path sampling approach for estimating $\log B_{10}$. We see that it is an extension of the bridge sampling from one bridge to an infinite number of bridges. Based on the equality $p(\boldsymbol{\Omega}, \boldsymbol{\theta}|\mathbf{Y}) = p(\mathbf{Y}, \boldsymbol{\Omega}, \boldsymbol{\theta})/p(\mathbf{Y})$ and treating $p(\mathbf{Y})$ as a normalizing constant, we have another derivation of Equation (5.10). The details are presented in Appendix 5.1.

To apply the path sampling procedure, we need to define a linked model M_t to link M_0 and M_1, such that when $t = 0$, $M_t = M_0$ and when $t = 1$, $M_t = M_1$. Then, we obtain $U(\mathbf{Y}, \boldsymbol{\Omega}, \boldsymbol{\theta}, t)$ by differentiating the logarithm of the complete-data likelihood function under M_t with respect to t, and finally estimate $\log B_{10}$ via Equation (5.12). The main computation is on simulating the sample of observations $\{(\boldsymbol{\Omega}^{(j)}, \boldsymbol{\theta}^{(j)}), j = 1, \cdots, J\}$ from $p(\boldsymbol{\Omega}, \boldsymbol{\theta}|\mathbf{Y}, t_{(s)})$, for $s = 0, \ldots, S$. This task can be done via some efficient MCMC methods, such as the Gibbs sampler and the MH algorithm as described in the Chapter 4. See also the illustrative examples given in the next section and other chapters in this book. For most SEMs, $S = 20$ and $J = 1000$ would provide results that are accurate enough for most practical applications. Experience indicates that $S = 10$ is also acceptable.

The path sampling approach has several attractive features. Its implementation is simple, the main programming task is on simulating observations from $p(\boldsymbol{\Omega}, \boldsymbol{\theta}|\mathbf{Y}, t_{(s)})$. In general, as pointed out by Gelman and Meng (1998), we can always construct a continuous path to link two competing models. Hence the method can be applied to a wide variety of models. Bayesian estimates of the unknown parameters and other interesting statistics under M_0 and M_1 can be obtained via the simulated observations at $t = 0$ and $t = 1$. Distinct from most existing approaches in computing the Bayes factor, the prior density is not directly involved in the evaluation. Furthermore, the logarithm scale of Bayes factor is computed, which is generally more stable than the ratio scale. In a comparative study on a variety of methods for computing the Bayes factor, DiCiccio, Kass, Raftery and Wasserman (1997) concluded that bridge sampling typically provides an order of magnitude of improvement on other methods in computing the Bayes factor. As path sampling is a generalization of bridge sampling, it has the potential to have even more improvement.

5.4 AN APPLICATION: BAYESIAN ANALYSIS OF SEMs WITH FIXED COVARIATES

As noted by Sammel and Ryan (1996), very often it is useful to accommodate fixed covariates in a model so as to give more ingredients and flexibility for developing better models. Clearly, like other statistical models, residuals errors in the measurement and structural equations of the model can be reduced by additionally incorporating fixed covariates. For the measurement equation, fixed covariates provide a more subtle structure, and hence enable us to assess a more precise relationship between the latent variables and their manifest variables. For the structural equation, fixed covariates provide more ingredients for accounting the endogenous latent variables, in addition to the exogenous latent variables. For a possible exogenous variable that can be accounted by a single manifest variable, we can directly assess its causal effect on the endogenous variables by treating it as a fixed covariate in the structural equation. In this way, it is not necessary to include this manifest variable in the measurement equation, and the formulation of the model is simplified. The main focus of this subsection is to develop a Bayesian procedure for estimation and model comparison of SEMs with fixed covariates on the measurement and structural equations.

Consider the following measurement equation for the $p \times 1$ manifest random vector $\mathbf{y}_i$ measured on an individual i:

$$\mathbf{y}_i = \mathbf{A}\mathbf{x}_i + \boldsymbol{\Lambda}\boldsymbol{\omega}_i + \boldsymbol{\epsilon}_i; \quad i = 1, \cdots, n, \tag{5.14}$$

in which $\mathbf{A}$ and $\boldsymbol{\Lambda}$ are unknown parameter matrices, $\mathbf{x}_i$ is a $r_1 \times 1$ vector of fixed covariates, $\boldsymbol{\omega}_i$ is an $q \times 1$ latent random vector and $\boldsymbol{\epsilon}_i$ is a $p \times 1$ random vector of error measurements with distribution $N[\mathbf{0}, \boldsymbol{\Psi}_\epsilon]$, where $\boldsymbol{\Psi}_\epsilon$ is diagonal and $\boldsymbol{\epsilon}_i$ is independent with $\boldsymbol{\omega}_i$. Furthermore, we model the latent subvectors $\boldsymbol{\eta}_i$ and $\boldsymbol{\xi}_i$ of $\boldsymbol{\omega}_i$ further with an additional $r_2 \times 1$ vector of fixed covariates $\mathbf{z}_i$ via the following structural equation:

$$\boldsymbol{\eta}_i = \mathbf{B}\mathbf{z}_i + \boldsymbol{\Pi}\boldsymbol{\eta}_i + \boldsymbol{\Gamma}\boldsymbol{\xi}_i + \boldsymbol{\delta}_i, \tag{5.15}$$

where $\boldsymbol{\omega}_i = (\boldsymbol{\eta}_i^T, \boldsymbol{\xi}_i^T)^T$, $\boldsymbol{\eta}_i$ and $\boldsymbol{\xi}_i$ are respectively $q_1 \times 1$ and $q_2 \times 1$ latent vectors; $\mathbf{B}, \boldsymbol{\Pi}$ and $\boldsymbol{\Gamma}$ are unknown parameter matrices; $\boldsymbol{\xi}_i$ and $\boldsymbol{\delta}_i$ are independently distributed as $N[\mathbf{0}, \boldsymbol{\Phi}]$ and $N[\mathbf{0}, \boldsymbol{\Psi}_\delta]$, respectively, where $\boldsymbol{\Psi}_\delta$ is a diagonal covariance matrix. The measurement and structural equations defined in Equations (5.14) and (5.15) reduces to the ordinary simultaneous equation model if $\boldsymbol{\Pi} = \mathbf{0}$, $\boldsymbol{\Gamma} = 0$ and $\boldsymbol{\Lambda} = \mathbf{0}$. Let $\boldsymbol{\Lambda} = (\boldsymbol{\Lambda}_1 \ \boldsymbol{\Lambda}_2)$ be a partition of $\boldsymbol{\Lambda}$, where $\boldsymbol{\Lambda}_1$ and $\boldsymbol{\Lambda}_2$ are $p \times q_1$ and $p \times q_2$ matrices, respectively; and $\boldsymbol{\Pi}_0 = \mathbf{I} - \boldsymbol{\Pi}$. We further assume that $|\mathbf{\Pi}_0|$ is a constant independent with elements in $\mathbf{\Pi}$. The covariance structure of $\boldsymbol{\omega}_i$ is given by

$$\mathbf{\Sigma}_\omega = \begin{bmatrix} \mathbf{\Pi}_0^{-1}(\mathbf{\Gamma\Phi\Gamma}^T + \mathbf{\Psi}_\delta)\mathbf{\Pi}_0^{-T} & \mathbf{\Pi}_0^{-1}\mathbf{\Gamma\Phi} \\ \mathbf{\Phi\Gamma}^T\mathbf{\Pi}_0^{-T} & \mathbf{\Phi} \end{bmatrix}.$$

The marginal model for the manifest random vector $\mathbf{y}_i$ is given by

$$(\mathbf{y}_i|\mathbf{A}, \mathbf{B}, \mathbf{\Lambda}, \mathbf{\Pi}, \mathbf{\Gamma}, \mathbf{\Phi}, \mathbf{\Psi}_\delta, \mathbf{\Psi}_\epsilon) \stackrel{D}{=} N[\mathbf{A}\mathbf{x}_i + \mathbf{\Lambda}_1\mathbf{\Pi}_0^{-1}\mathbf{B}\mathbf{z}_i,\ \mathbf{\Lambda\Sigma}_\omega\mathbf{\Lambda}^T + \mathbf{\Psi}_\epsilon],$$

where the mean structure is defined as a linear combination of various fixed covariates, while the covariance structure is equal to the LISREL model (Jöreskog and Sörbom, 1996). It should also be noted that the fixed covariates $\mathbf{x}_i$ and $\mathbf{z}_i$ can be discrete, ordered categorical or continuous measurements. The mean vector can be represented by a column in $\mathbf{A}$ corresponding to covariates that are fixed at 1.0. Following a common practice in structural equation modeling to identify the model, some appropriate elements in $\mathbf{\Lambda}$, $\mathbf{\Phi}$, $\mathbf{\Pi}$ and $\mathbf{\Gamma}$ will be restricted at known values and not be estimated. Below, we assume that the covariance structure of $\mathbf{y}_i$ is identified.

Let $\mathbf{Y} = (\mathbf{y}_1, \cdots, \mathbf{y}_n)$, $\mathbf{X} = (\mathbf{x}_1, \cdots, \mathbf{x}_n)$ and $\mathbf{Z} = (\mathbf{z}_1, \cdots, \mathbf{z}_n)$ be the data matrices; and let $\mathbf{\Omega} = (\boldsymbol{\omega}_1, \cdots, \boldsymbol{\omega}_n)$ be the matrix of latent vectors. Moreover, let $\boldsymbol{\theta}$ be the structural parameter vector that contains all the unknown parameters in $\mathbf{A}, \mathbf{B}, \mathbf{\Lambda}, \mathbf{\Pi}, \mathbf{\Gamma}, \mathbf{\Phi}, \mathbf{\Psi}_\delta$ and $\mathbf{\Psi}_\epsilon$. One main objective is to illustrate the path sampling procedure in computing the log Bayes factor for model comparison. A sequence of random observations from the joint posterior distribution $[\boldsymbol{\theta}, \mathbf{\Omega}|\mathbf{Y}, t]$ will be generated by the Gibbs sampler which is implemented as below:

At the $(j+1)$th iteration with current value $\boldsymbol{\theta}^{(j)}$ and $\mathbf{\Omega}^{(j)}$.

Step (a): Generate a random variate $\mathbf{\Omega}^{(j+1)}$ from the conditional distribution $[\mathbf{\Omega}|\mathbf{Y}, \boldsymbol{\theta}^{(j)}, t]$.

Step (b): Generate a random variate $\boldsymbol{\theta}^{(j+1)}$ from the conditional distribution $[\boldsymbol{\theta}|\mathbf{Y}, \mathbf{\Omega}^{(j+1)}, t]$, and return to 'Step (a)' if necessary.

As the model defined by Equations (5.14) and (5.15) are essentially regression models with latent variables and similar to the standard SEM defined by Equations (4.36) and (4.37), the conditional distributions $[\mathbf{\Omega}|\mathbf{Y}, \boldsymbol{\theta}]$ and $[\boldsymbol{\theta}|\mathbf{Y}, \mathbf{\Omega}]$ under the conjugate prior distribution are very similar to those presented in Section 4.4.1. For completeness, expressions of these conditional distributions are discussed in Appendix 5.2. For a given t, corresponding conditional distributions can be obtained by incorporating these expressions with t.

To illustrate the application of path sampling for computing the Bayes factor with a real example, a small portion of the Inter-university Consortium for Political and Social Research (ICPSR) data set collected in the project World Value Survey 1981–1984 and 1990–1993 (World Value Study Group, (1994)) is analyzed. In our illustration, only the data obtained from Canada are used. Six variables in the full data set (variables 180, 96, 62, 176, 116 and 117; see Appendix 1.1) that are related to the respondents' job, religious belief and homelife are taken as manifest variables in $\mathbf{y}$. Intercepts with $x_{i1} = 1.0$ and a variable x_{i2} (variable 353) about gender are taken as fixed covariates for the manifest variables in the measurement equation. An additional variable (variable 364) about socio-economic status is taken as a fixed covariate z_i in the structural equation. For simplicity, cases with missing data are not used, and the remaining sample size is 454.

Parameter matrices $\mathbf{A}$ and $\mathbf{\Lambda}$ in the measurement equation (5.14) are specified as follows:

$$\mathbf{A}^T = \begin{pmatrix} a_{11} \cdots a_{61} \\ a_{12} \cdots a_{62} \end{pmatrix}, \text{ and } \mathbf{\Lambda}^T = \begin{bmatrix} 1.0 & \lambda_{21} & 0 & 0 & 0 & 0 \\ 0 & 0 & 1.0 & \lambda_{42} & 0 & 0 \\ 0 & 0 & 0 & 0 & 1.0 & \lambda_{63} \end{bmatrix},$$

where ones's and zero's in $\mathbf{\Lambda}$ were treated as fixed parameters to identify the model. In this model, there are three latent factors: η, ξ_1 and ξ_2, which can be roughly interpreted as 'life', 'religious belief' and 'job satisfaction', respectively. We first conduct an auxiliary Bayesian estimation with non-informative priors to obtain some prior inputs for the conjugate prior distributions. Then, we do an actual simulation from the conditional distributions to obtain some idea about the convergence of the Gibbs sampler with the hyperparameter values that are obtained from the auxiliary estimation. We observe that the Gibbs sampler converged in less than 800 iterations. To be conservative, we take a burn-in phase of 1000 iterations and for each $t_{(s)}$ an additional $J = 2000$ observations are collected after convergence for computing the Bayes factor. The number of grids is taken to be 20.

The following three models are compared:

$$M_0 : \mathbf{y} = \mathbf{A}\mathbf{x} + \mathbf{\Lambda}\boldsymbol{\omega} + \boldsymbol{\epsilon}, \text{ and } \eta = bz + \gamma_1\xi_1 + \gamma_2\xi_2 + \delta;$$

$$M_1 : \mathbf{y} = \mathbf{\Lambda}\boldsymbol{\omega} + \boldsymbol{\epsilon}, \text{ and } \eta = bz + \gamma_1\xi_1 + \gamma_2\xi_2 + \delta;$$

$$M_2 : \mathbf{y} = \mathbf{A}\mathbf{x} + \mathbf{\Lambda}\boldsymbol{\omega} + \boldsymbol{\epsilon}, \text{ and } \eta = \gamma_1\xi_1 + \gamma_2\xi_2 + \delta.$$

Note that M_1 and M_2 are non-nested, while both of them are nested in M_0. Models M_0 and M_1 only differ in the measurement equations, hence they can be linked up by a parameter t in [0, 1] as

$$M_{t01} : y = (1-t)\mathbf{A}\mathbf{x} + \mathbf{\Lambda}\boldsymbol{\omega} + \boldsymbol{\epsilon}$$

with the same structural equation. Clearly when $t = 0$, $M_{t01} = M_0$, and when $t = 1$, $M_{t01} = M_1$. The parameter vector $\boldsymbol{\theta}$ in the linked model M_{t01} contains the unknown matrix $\mathbf{A}$ that is not involved in M_1. Models M_0 and M_2 have the same measurement equation. Their structural equations are linked up by a parameter t in $[0, 1]$ as

$$M_{t02} : \eta = (1 - t)bz + \gamma_1\xi_1 + \gamma_2\xi_2 + \delta.$$

Clearly when $t = 0$, $M_{t02} = M_0$, and when $t = 1$, $M_{t02} = M_2$. Note that the parameter vector $\boldsymbol{\theta}$ in M_{t02} contains b that is not involved in M_2. Finally, models M_1 and M_2 can be linked up as

$$\begin{aligned} M_{t12} : \mathbf{y} &= t\mathbf{A}\mathbf{x} + \boldsymbol{\Lambda}\boldsymbol{\omega} + \boldsymbol{\epsilon}, \\ \eta &= (1 - t)bz + \gamma_1\xi_1 + \gamma_2\xi_2 + \delta, \end{aligned}$$

where when $t = 0$, $M_{t12} = M_1$, and when $t = 1$, $M_{t12} = M_2$.

The logarithm complete-data likelihood functions corresponding to linked models M_{t01}, M_{t02} and M_{t12} are respectively given by:

$$\begin{aligned} \log p_{01}(\mathbf{Y}, \boldsymbol{\Omega}|\boldsymbol{\theta}, t) = -\frac{1}{2}\Bigg\{ & c^* + \sum_{i=1}^{n} \boldsymbol{\xi}_i^T \boldsymbol{\Phi}^{-1} \boldsymbol{\xi}_i \\ & + \sum_{i=1}^{n} [\mathbf{y}_i - (1-t)\mathbf{A}\mathbf{x}_i - \boldsymbol{\Lambda}\boldsymbol{\omega}_i]^T \boldsymbol{\Psi}_\epsilon^{-1} [\mathbf{y}_i - (1-t)\mathbf{A}\mathbf{x}_i - \boldsymbol{\Lambda}\boldsymbol{\omega}_i] \\ & + \sum_{i=1}^{n} (\eta_i - bz_i - \gamma_1\xi_{i1} - \gamma_2\xi_{i2}) \psi_\delta^{-1} (\eta_i - bz_i - \gamma_1\xi_{i1} - \gamma_2\xi_{i2}) \Bigg\}, \\ \log p_{02}(\mathbf{Y}, \boldsymbol{\Omega}|\boldsymbol{\theta}, t) = -\frac{1}{2}\Bigg\{ & c^* + \sum_{i=1}^{n} \boldsymbol{\xi}_i^T \boldsymbol{\Phi}^{-1} \boldsymbol{\xi}_i \\ & + \sum_{i=1}^{n} (\mathbf{y}_i - \mathbf{A}\mathbf{x}_i - \boldsymbol{\Lambda}\boldsymbol{\omega}_i)^T \boldsymbol{\Psi}_\epsilon^{-1} (\mathbf{y}_i - \mathbf{A}\mathbf{x}_i - \boldsymbol{\Lambda}\boldsymbol{\omega}_i) \\ & + \sum_{i=1}^{n} [\eta_i - (1-t)bz_i - \gamma_1\xi_{i1} - \gamma_2\xi_{i2}] \psi_\delta^{-1} \\ & \times [\eta_i - (1-t)bz_i - \gamma_1\xi_{i1} - \gamma_2\xi_{i2}] \Bigg\} \\ \log p_{12}(\mathbf{Y}, \boldsymbol{\Omega}|\boldsymbol{\theta}, t) = -\frac{1}{2}\Bigg\{ & c^* + \sum_{i=1}^{n} \boldsymbol{\xi}_i^T \boldsymbol{\Phi}^{-1} \boldsymbol{\xi}_i \\ & + \sum_{i=1}^{n} (\mathbf{y}_i - t\mathbf{A}\mathbf{x}_i - \boldsymbol{\Lambda}\boldsymbol{\omega}_i)^T \boldsymbol{\Psi}_\epsilon^{-1} (\mathbf{y}_i - t\mathbf{A}\mathbf{x}_i - \boldsymbol{\Lambda}\boldsymbol{\omega}_i) \end{aligned}$$

$$+\sum_{i=1}^{n}[\eta_i-(1-t)bz_i-\gamma_1\xi_{i1}-\gamma_2\xi_{i2}]\psi_\delta^{-1}$$
$$\times[\eta_i-(1-t)bz_i-\gamma_1\xi_{i1}-\gamma_2\xi_{i2}]\},$$

where c^* is a constant that is equal to $n(p+q)\log(2\pi)+n\log|\boldsymbol{\Psi}_\epsilon|+n\log|\psi_\delta|+n\log|\boldsymbol{\Phi}|$, and $\boldsymbol{\theta}$ is the parameter vector in the linked model. Derivatives of these functions with respect to t are given by

$$\frac{\mathrm{d}\,\log p_{01}(\mathbf{Y},\boldsymbol{\Omega}|\boldsymbol{\theta},t)}{\mathrm{d}t}=-\sum_{i=1}^{n}\{\mathbf{y}_i-(1-t)\mathbf{A}\mathbf{x}_i-\boldsymbol{\Lambda}\boldsymbol{\omega}_i\}^T\boldsymbol{\Psi}_\epsilon^{-1}\mathbf{A}\mathbf{x}_i,$$
$$\frac{\mathrm{d}\,\log p_{02}(\mathbf{Y},\boldsymbol{\Omega}|\boldsymbol{\theta},t)}{\mathrm{d}t}=-\sum_{i=1}^{n}\{\eta_i-(1-t)bz_i-\gamma_1\xi_{i1}-\gamma_2\xi_{i2}\}\psi_\delta^{-1}bz_i,$$

and

$$\frac{\mathrm{d}\,\log p_{12}(\mathbf{Y},\boldsymbol{\Omega}|\boldsymbol{\theta},t)}{\mathrm{d}t}=\sum_{i=1}^{n}\{(\mathbf{y}_i-t\mathbf{A}\mathbf{x}_i-\boldsymbol{\Lambda}\boldsymbol{\omega}_i)^T\boldsymbol{\Psi}_\epsilon^{-1}\mathbf{A}\mathbf{x}_i$$
$$-(\eta_i-(1-t)bz_i-\gamma_1\xi_{i1}-\gamma_2\xi_{i2})\psi_\delta^{-1}bz_i\}.$$

These derivatives give $U(\mathbf{Y},\boldsymbol{\Omega},\boldsymbol{\theta},t)$ under various situations. The log Bayes factors are estimated via Equations (5.12) and (5.13).

To give some idea about the sensitivity of the Bayesian analysis with respect to priors, we consider the model comparison with three types of prior inputs: prior inputs obtained from an auxiliary estimation (type I), those equal to half (type II) and twice (type III) of the type I prior inputs. The estimated twice log Bayes factors are reported in Table 5.2. Clearly, the conclusions drawn from these results are not very sensitive to these prior inputs. From Table 5.2, we see that all $2\widehat{\log B_{01}}$ and $2\widehat{\log B_{21}}$ are significantly larger than 10, and hence M_1 is rejected. All $2\widehat{\log B_{02}}$ are less than 2, that means the more complicated model M_0 is not significantly better than the simpler model M_2. Hence, M_2 may be selected. It can be concluded from the model comparison results that the effect of the fixed covariate 'gender' is important in relating the latent variables with their manifest variables, whilst the

Table 5.2 Estimated twice log Bayes factors.

	Prior inputs		
	Type I	Type II	Type III
$2\log B_{01}$	162.2	158.8	163.4
$2\log B_{02}$	1.058	0.668	0.534
$2\log B_{21}$	165.0	166.8	171.4

fixed covariate about socio-economic status has little impact on the endogenous latent variable on 'life'.

To give some information about the convergence of the Gibbs sampler in the context of the selected model M_2, plots of several sequences of some randomly selected parameters values generated from different starting points against the iteration numbers are displayed in Figure 5.1. We observe that the sequences mixed well in less than 800 iterations. The EPSR values against the iteration numbers are plotted in Figure 5.2. We observe that these values are less than 1.2 after 800 iterations. For completeness, the Bayesian estimates and the standard errors estimates of the unknown parameters in M_2 under prior inputs type I are presented in Table 5.3. Interpretations of the results given in this table are straightforward.

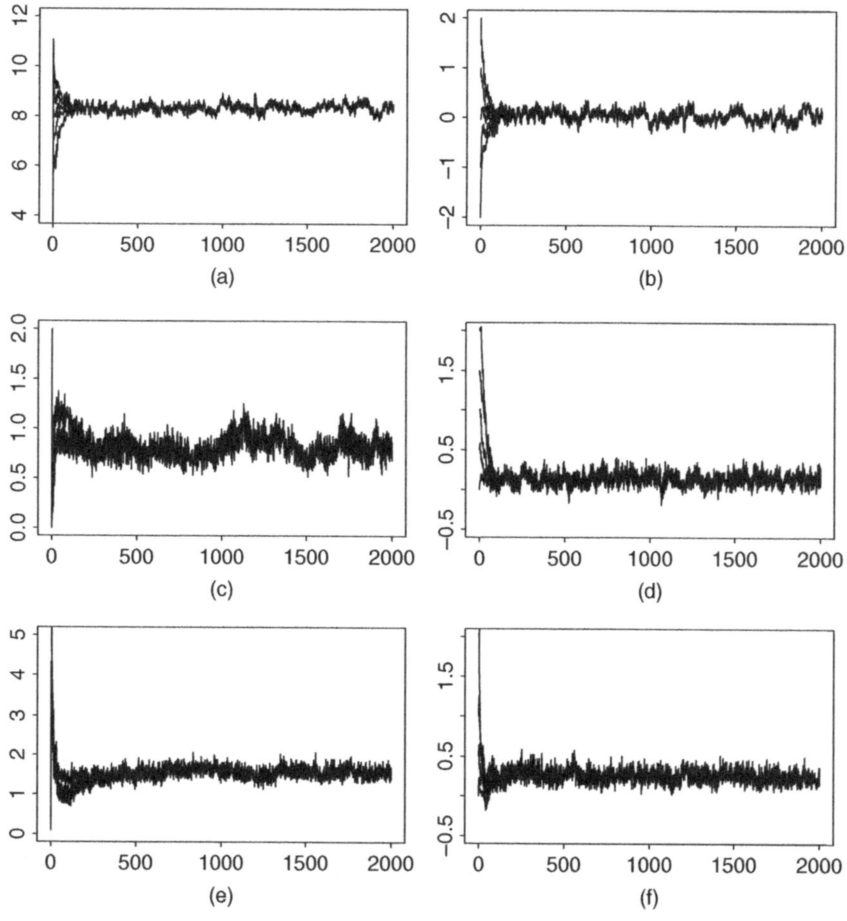

Figure 5.1 (a)–(f) are plots of parallel sequences corresponding to different starting values of a_{11}, a_{12}, λ_{63}, γ_1, $\psi_{\epsilon 3}$ and ϕ_{12} against iteration numbers with different starting values.

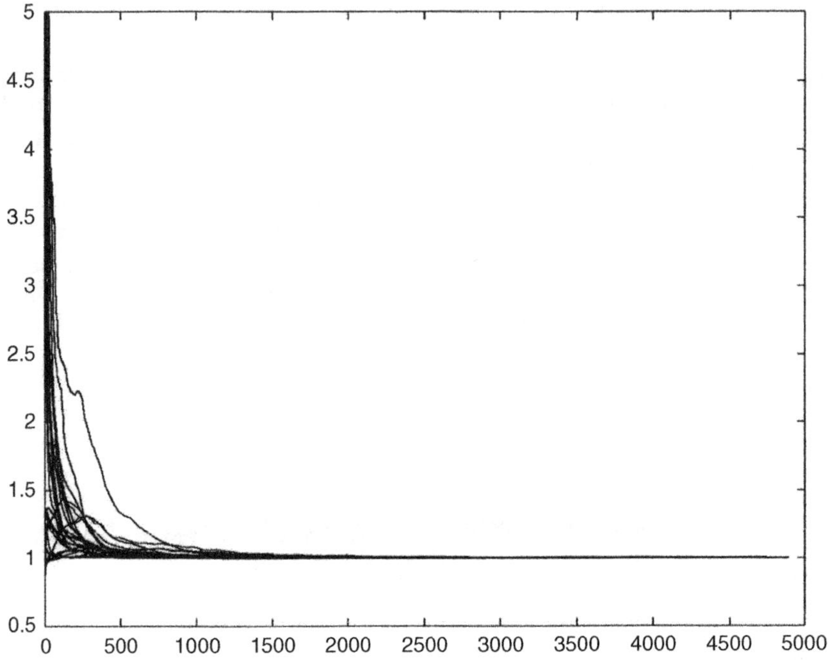

Figure 5.2 EPSR values against the number of iterations.

Table 5.3 Bayesian estimates and their standard errors.

Parameters	Est.	SE	Parameters	Est.	SE
a_{11}	8.34	0.21	λ_{21}	1.07	0.09
a_{21}	7.74	0.21	λ_{42}	2.78	0.26
a_{31}	2.67	0.21	λ_{63}	0.82	0.09
a_{41}	5.66	0.33	ϕ_{11}	0.74	0.13
a_{51}	7.76	0.24	ϕ_{12}	0.25	0.08
a_{61}	7.57	0.32	ϕ_{22}	1.91	0.19
a_{12}	0.04	0.14	$\psi_{\epsilon 1}$	1.26	0.13
a_{22}	0.16	0.13	$\psi_{\epsilon 2}$	0.74	0.12
a_{32}	0.07	0.13	$\psi_{\epsilon 3}$	1.57	0.13
a_{42}	0.88	0.20	$\psi_{\epsilon 4}$	1.56	0.34
a_{52}	0.19	0.15	$\psi_{\epsilon 5}$	0.87	0.14
a_{62}	−.12	0.21	$\psi_{\epsilon 6}$	3.22	0.25
γ_1	0.16	0.08	ψ_{δ}	0.87	0.12
γ_2	0.44	0.06			

5.5 OTHER METHODS

5.5.1 Bayesian Information Criterion and Akaike Information Criterion

A simple approximation of $2\log B_{10}$ which does not depend on the prior density is the following Schwarz criterion S^* (Schwarz, 1978):

$$2\log B_{10} \cong 2S^* = 2[\log p(\mathbf{Y}|\tilde{\boldsymbol{\theta}}_1, M_1) - \log p(\mathbf{Y}|\tilde{\boldsymbol{\theta}}_0, M_0)] - (d_1 - d_0)\log n, \tag{5.16}$$

where $\tilde{\boldsymbol{\theta}}_1$ and $\tilde{\boldsymbol{\theta}}_0$ are ML estimates of $\boldsymbol{\theta}_1$ and $\boldsymbol{\theta}_0$ under M_1 and M_0, respectively, d_1 and d_0 are the dimensions of $\boldsymbol{\theta}_1$ and $\boldsymbol{\theta}_0$ and n is the sample size. Minus $2S^*$ is the following well-known Bayesian information criterion (BIC) for comparing M_1 and M_0:

$$\text{BIC}_{10} = -2S^* \cong -2\log B_{10} = 2\log B_{01}. \tag{5.17}$$

The interpretation of BIC_{10} can be based on Table 5.1. For each M_k, $k = 0, 1$, we define

$$\text{BIC}_k = -2\log p(\mathbf{Y}|\tilde{\boldsymbol{\theta}}_k, M_k) + d_k \log n. \tag{5.18}$$

Then $2\log B_{10} = \text{BIC}_0 - \text{BIC}_1$. Hence, it follows from Table 5.1 that the model M_k with the smaller BIC_k value is selected.

As n tends to infinity, it has been shown (Schwarz, 1978) that

$$\frac{S^* - \log B_{10}}{\log B_{10}} \to 0,$$

thus S^* may be viewed as an approximation to $\log B_{10}$. This approximation is of the order O(1), hence it does not give the exact $\log B_{10}$ even for large samples. However, as pointed out by Kass and Raftery (1995), it can be used for scientific reporting as long as the number of degrees of freedom $(d_1 - d_0)$ involved in the comparison is small relative to the sample size n. The BIC is appealing in that it is relatively simple and can be applied even when the priors $p(\boldsymbol{\theta}_k|M_k)$ are hard to set precisely. The ML estimates of $\boldsymbol{\theta}_1$ and $\boldsymbol{\theta}_0$ are involved in the computation of BIC. Sometimes obtaining the ML estimates is difficult for some complex SEMs. In practice, since the Bayesian estimates and the ML estimates are close to each other in large samples, they can be used to replace the ML estimates in computing the BIC. The order of approximation is not changed and the BIC obtained can be interpreted using the rough criteria given in Table 5.1.

The Akaike information criterion (AIC, Akaike, 1973) associated with a competing model M_k is given by

$$\mathrm{AIC}_k = -2\log p(\mathbf{Y}|\tilde{\boldsymbol{\theta}}_k, M_k) + 2d_k \tag{5.19}$$

which does not involve the sample size n. The interpretation of AIC_k is similar to BIC_k. Hence, M_k is selected if its AIC_k is smaller. Comparing Equations (5.18) and (5.19), we see that BIC tends to favor simpler models more than those selected by AIC.

5.5.2 Deviance Information Criterion

Another goodness-of-fit or model comparison statistic that takes into account the number of unknown parameters in the model is the deviance information criterion (DIC), see Spiegelhalter, Best, Carlin and van der Linde (2002). This statistic can be viewed as a generalization of AIC. Under a competing model M_k with a vector of unknown parameter $\boldsymbol{\theta}_k$ of dimension d_k, let $\{\boldsymbol{\theta}_k^{(j)} : j = 1, \ldots, J\}$ be a sample of observations simulated from the posterior distribution. We define

$$\mathrm{DIC}_k = -\frac{2}{J}\sum_{j=1}^{J}\log p(\mathbf{Y}|\boldsymbol{\theta}_k^{(j)}, M_k) + 2d_k. \tag{5.20}$$

In model comparison, the model with the smaller DIC value is selected.

The computational burden of DIC is similar to the computational burden of BIC or AIC, and is less heavy than that of the Bayes factor. In analyzing a hypothesized model, WinBUGS (Spiegelhalter, Best and Lunn, 2003) also produces a DIC value. This DIC value can be used for model comparison. However, as pointed out in the WinBUGS User Manual in practical applications of DIC it is important to note the following points: (i) DIC assumes the posterior mean to be a good estimate of the parameter. There are circumstances, such as with mixture models, in which WinBUGS will not give the DIC values. (ii) If the difference in DIC is small, for example less than 5, and the models make very different inferences, then just reporting the model with the lowest DIC could be misleading. (iii) DIC can be applied to non-nested models. Moreover, similar to the Bayes factor, BIC and AIC, DIC gives clear conclusion to support the null hypothesis or the alternative hypothesis. Detailed discussions of DIC can be found in Spiegelhalter, Best, Carlin and van der Linde (2002), and Celeux, Forbes, Roberts and Titterington (2003).

5.5.3 Posterior Predictive p-value

The Bayes factor can be used to assess the goodness-of-fit of the hypothesized model by taking M_0 or M_1 to be the saturated model. However, when analyzing some complex SEMs it is rather difficult to define a saturated model. For example, in analyzing nonlinear SEMs, the distribution associated with the hypothesized model is not normal. Hence, the model having a normal distribution with a general unstructured covariance matrix cannot be regarded as a saturated model. Under these situations, it is not appropriate to apply model comparison to access goodness-of-fit of the hypothesized model. A simple and more convenient alternative without involving a basic saturated model is the posterior predictive p-values (PP p-values) introduced by Meng (1994) on the basis of the posterior assessment in Rubin (1984). It has been shown (see Gelman, Meng and Stern, 1996 and the references therein) that this approach is conceptually and computationally simple, and is useful in model-checking for a wide variety of complicated situations. Moreover, the required computation is a by-product of the common Bayesian simulation procedures such as the Gibbs sampler or its related algorithms. A brief outline of this method is given in Appendix 5.3. The proposed model may be considered as plausible if the PP p-value estimate is not far from 0.5. As pointed out by Meng (1994), and Carlin and Louis (1996), the PP p-value is only useful for the goodness-of-fit assessment of a single model, it is not suitable for comparing different models. Moreover, it may show good fit for some inappropriate models, due to the problem of using the data twice. Hence, the PP p-value should only be used as a complementary statistic for the Bayes factor. Bayarri and Berger (2000) proposed a partial posterior p-value to overcome the problem of using the data twice. Recently, this statistic has been applied to a nonlinear SEM with nonignorable missing data (see Lee and Tang, 2006).

5.5.4 Residual and Outliers Analysis

Many common model checking methods in data analysis, including residual analysis and test for outliers, can be incorporated in the Bayesian analysis. An advantage of the sampling-based Bayesian approach to SEMs is that we can obtain the estimates of the latent variables through the posterior simulation so that reliable estimates of the residuals in the measurement equation and the structural equation can be obtained. The graphical interpretation of these residuals is similar to those in other statistical models, for example, regression.

Consider the SEMs with fixed covariates as described in Section 5.4. Estimates of the residuals in the measurement equation can be obtained from Equation (5.14) as:

$$\hat{\boldsymbol{\epsilon}}_i = \mathbf{y}_i - \hat{\mathbf{A}}\mathbf{x}_i - \hat{\boldsymbol{\Lambda}}\hat{\boldsymbol{\omega}}_i \quad i = 1, \ldots, n, \tag{5.21}$$

where $\hat{\mathbf{A}}$, $\hat{\boldsymbol{\Lambda}}$ and $\hat{\boldsymbol{\omega}}_i$ are Bayesian estimates that are obtained from the corresponding simulated observations through the MCMC methods. Plots of $\hat{\boldsymbol{\epsilon}}_i$ with $\hat{\boldsymbol{\omega}}_i$ give useful information for the fit of the measurement equation. For a reasonable good fit, the plots should lie within two parallel horizontal lines that are not widely separated apart and centered at zero. Estimates of residuals in the structural equation can be obtained from Equation (5.15) as:

$$\hat{\boldsymbol{\delta}}_i = (\mathbf{I} - \hat{\mathbf{\Pi}})\hat{\boldsymbol{\eta}}_i - \hat{\mathbf{B}}\mathbf{z}_i - \hat{\mathbf{\Gamma}}\hat{\boldsymbol{\xi}}_i, \quad i = 1, \cdots, n, \tag{5.22}$$

where $\hat{\mathbf{\Pi}}, \hat{\mathbf{B}}, \hat{\mathbf{\Gamma}}, \hat{\boldsymbol{\eta}}_i$ and $\hat{\boldsymbol{\xi}}_i$ are Bayesian estimates obtained via the MCMC methods. The interpretation and the use of plots of $\hat{\boldsymbol{\delta}}_i$ are similar.

The residual estimates $\hat{\boldsymbol{\epsilon}}_i$ can also be used for outliers analysis. A particular observation $\mathbf{y}_i$ whose residual is far from the expected value $(\bar{\boldsymbol{\epsilon}})$ may be informally regarded as an outlier. Moreover, the QQ plots (Johnson and Wichern, 1992) of $\hat{\epsilon}_{ij}, j = 1, \cdots, p$, and/or $\hat{\delta}_{ik}, k = 1, \cdots, r_2$, can be used to check the assumption of normality.

5.6 DISCUSSION

The main objective of this chapter is to consider Bayesian model comparison of SEMs, with emphasis on the computation of the well-known Bayes factor through path sampling. In applying path sampling, it is necessary to find a path t in $[0, 1]$ to join the competing models M_1 and M_2. In most cases, it is fairly straightforward to find such a path or the corresponding linked model. However, for some complicated situations that involve M_1 and M_2 being far apart, it is difficult to find a path that directly links M_1 and M_2. In most situations, this difficulty can be solved by using appropriate auxiliary models $M_a, M_b, \cdots$, in between M_1 and M_2. For example, suppose that M_a and M_b are appropriate auxiliary models such that M_a can be linked with M_1, and M_b, and M_b can be linked with M_2. Then

$$\frac{p(\mathbf{Y}|M_1)}{p(\mathbf{Y}|M_2)} = \frac{p(\mathbf{Y}|M_1)/p(\mathbf{Y}|M_a)}{p(\mathbf{Y}|M_2)/p(\mathbf{Y}|M_a)} \quad \text{and} \quad \frac{p(\mathbf{Y}|M_2)}{p(\mathbf{Y}|M_a)} = \frac{p(\mathbf{Y}|M_2)/p(\mathbf{Y}|M_b)}{p(\mathbf{Y}|M_a)/p(\mathbf{Y}|M_b)}.$$

Hence, $\log B_{12} = \log B_{1a} + \log B_{ab} - \log B_{2b}$. Each logarithm Bayes factor can be computed through path sampling (see an example in Chapter 10).

The prior distributions of $\boldsymbol{\theta}$ are either directly or indirectly involved in the computation of the Bayes factor. The BIC defined in Equation (5.17) does not depend directly on prior densities. However, if the ML estimates are replaced by the Bayesian estimates in the computation, then the posterior simulations in obtaining the Bayesian estimates will depend on the prior distributions of $\boldsymbol{\theta}$. The importance sampling, bridge sampling and path sampling approaches are indirectly affected by the priors during the simulation of observations from the posterior distributions that depend on the prior distributions. Thus, a concern is to choose prior distributions to represent the available information. Once this is done, an important issue is the sensitivity of the Bayes factor to the choices of priors.

Consider the problem of comparing M_0 with M_1, as pointed out by Kass and Raftery (1995), using a prior with a very large spread on the parameters under M_1 so as to make it 'noninformative'; this will force the Bayes factor to favor the competitive model M_0. This is known as the 'Bartlett's paradox'. For example, in the context of linear and log-linear models, Spiegelhalter and Smith (1982) showed that the Bayes factor may be problematic if the prior distribution of the parameters involved in the comparison (or in the hypothesis) are noninformative. To avoid this difficulty, priors on parameters under the model comparison are generally taken to be proper and do not have too big a spread. The conjugate families suggested in the previous section with reasonable spread are good choices. Parameters not under the model comparison are regarded as nuisance parameters. As pointed out by Kass and Raftery (1995), under mild regularity conditions, choices of priors of the nuisance parameters do not greatly affect the comparison results. Hence, noninformative priors can be used for the nuisance parameters. However, more care should be taken for model comparison involving parameters with non-informative prior. One simple method suggested in Kass and Raftery (1995) on the basis of the idea in Lempers (1971) is to set aside part of the data to use as a training sample which is combined with the noninformative prior distribution to produce an informative prior distribution. The Bayes factor is then computed from the remainder of the data. More advanced methods have been suggested (see, for example, O'Hagan (1995) and Berger and Pericchi (1996), among others.

To study the sensitivity of the Bayes factor to the choices of the prior inputs in terms of the hyperparameters' values, a common method (see Kass and Raftery, 1995; Lee and Song, 2003, among others) is to perturb the prior inputs. For example, if the prior distribution is $N[\mu_0, \sigma_0^2]$ in which the given hyperparameters are μ_0 and σ_0^2, the hyperparameters may be perturbed by changing μ_0 to $\mu_0 \pm c$ and halving and doubling σ_0^2, and the Bayes factor recomputed.

For most SEMs, the freely available software WinBUGS produces the DIC value of the hypothesized model. In practice, this value conveniently gives an alternative statistic for model comparison. The residual plots are also helpful in assessing the goodness-of-fit of the measurement equation and structural equation.

APPENDIX 5.1: ANOTHER PROOF OF EQUATION (5.10)

This approach derives Equation (5.10) on the basis of the equality $p(\mathbf{\Omega}, \boldsymbol{\theta}|\mathbf{Y}) = p(\mathbf{Y}, \mathbf{\Omega}, \boldsymbol{\theta})/p(\mathbf{Y})$. The marginal density of $p(\mathbf{Y})$ can be treated as the normalizing constant of $p(\mathbf{\Omega}, \boldsymbol{\theta}|\mathbf{Y})$. Consider the following class of densities with a continuous parameter $t \in [0, 1]$:

$$p(\mathbf{\Omega}, \boldsymbol{\theta}|\mathbf{Y}, t) = \frac{1}{z(t)} p(\mathbf{Y}, \mathbf{\Omega}, \boldsymbol{\theta}|t), \tag{A5.1}$$

where

$$z(t) = p(\mathbf{Y}|t) = \int p(\mathbf{Y}, \mathbf{\Omega}, \boldsymbol{\theta}|t) d\mathbf{\Omega} d\boldsymbol{\theta} = \int p(\mathbf{Y}, \mathbf{\Omega}|\boldsymbol{\theta}, t) p(\boldsymbol{\theta}) d\mathbf{\Omega} d\boldsymbol{\theta}, \tag{A5.2}$$

with $p(\boldsymbol{\theta})$ be the prior density of $\boldsymbol{\theta}$ which is independent of t. In computing the Bayes factor, we need to construct a path using a parameter $t \in [0, 1]$ to link the two competing models M_0 and M_1 together, so that $z(1) = p(\mathbf{Y}|1) = p(\mathbf{Y}|M_1)$, $z(0) = p(\mathbf{Y}|0) = p(\mathbf{Y}|M_0)$, and $B_{10} = z(1)/z(0)$. From Equations (A5.1) and (A5.2), we have

$$\begin{aligned}
\frac{d \log(z(t))}{dt} &= \int \frac{1}{z(t)} \frac{d}{dt} p(\mathbf{Y}, \mathbf{\Omega}, \boldsymbol{\theta}|t) d\mathbf{\Omega} d\boldsymbol{\theta} \\
&= \int \frac{p(\mathbf{\Omega}, \boldsymbol{\theta}|\mathbf{Y}, t)}{p(\mathbf{Y}, \mathbf{\Omega}, \boldsymbol{\theta}|t)} \frac{d}{dt} p(\mathbf{Y}, \mathbf{\Omega}, \boldsymbol{\theta}|t) d\mathbf{\Omega} d\boldsymbol{\theta} \\
&= \int \left[\frac{d}{dt} \log p(\mathbf{Y}, \mathbf{\Omega}, \boldsymbol{\theta}|t) \right] \cdot p(\mathbf{\Omega}, \boldsymbol{\theta}|\mathbf{Y}, t) d\mathbf{\Omega} d\boldsymbol{\theta} \\
&= E_{\Omega,\theta} \left[\frac{d}{dt} \log p(\mathbf{Y}, \mathbf{\Omega}, \boldsymbol{\theta}|t) \right],
\end{aligned} \tag{A5.3}$$

where $E_{\Omega,\theta}$ is the expectation with respect to the distribution $p(\mathbf{\Omega}, \boldsymbol{\theta}|\mathbf{Y}, t)$. Define:

$$U(\mathbf{Y}, \mathbf{\Omega}, \boldsymbol{\theta}, t) = \frac{d}{dt} \log p(\mathbf{Y}, \mathbf{\Omega}, \boldsymbol{\theta}|t) = \frac{d}{dt} \log p(\mathbf{Y}, \mathbf{\Omega}|\boldsymbol{\theta}, t). \tag{A5.4}$$

As $p(\boldsymbol{\theta})$ is independent of t, $U(\mathbf{Y}, \mathbf{\Omega}, \boldsymbol{\theta}, t)$ does not involve the prior density $p(\boldsymbol{\theta})$. It follows from Equations (A5.3) and (A5.4) and integrating from 0 to 1, that

$$\log \frac{z(1)}{z(0)} = \int_0^1 E_{\Omega,\theta}[U(\mathbf{Y}, \mathbf{\Omega}, \boldsymbol{\theta}, t)] dt.$$

APPENDIX 5.2: CONDITIONAL DISTRIBUTIONS FOR SIMULATING $(\theta, \Omega | Y, T)$

We first note that for $i = 1, \cdots, n$, $\boldsymbol{\omega}_i$ are conditionally independent given $\boldsymbol{\theta}$ and $\mathbf{y}_i$ are also conditionally independent with given $(\boldsymbol{\omega}_i, \boldsymbol{\theta})$. Hence,

$$p(\boldsymbol{\Omega}|\mathbf{Y}, \boldsymbol{\theta}) \propto \prod_{i=1}^{n} p(\boldsymbol{\omega}_i|\boldsymbol{\theta})\, p(\mathbf{y}_i|\boldsymbol{\omega}_i, \boldsymbol{\theta}). \tag{A5.5}$$

It implies that the conditional distribution of $\boldsymbol{\omega}_i$ given $(\mathbf{y}_i, \boldsymbol{\theta})$ are mutually independent for different i, and $p(\boldsymbol{\omega}_i|\mathbf{y}_i, \boldsymbol{\theta}) \propto p(\boldsymbol{\omega}_i|\boldsymbol{\theta})p(\mathbf{y}_i|\boldsymbol{\omega}_i, \boldsymbol{\theta})$. From the fact that

$$[\boldsymbol{\omega}_i|\boldsymbol{\theta}] \stackrel{D}{=} N\left[\begin{pmatrix}\boldsymbol{\Pi}_0^{-1}\mathbf{B}\mathbf{z}_i \\ \mathbf{0}\end{pmatrix}, \boldsymbol{\Sigma}_\omega\right],$$

and $[\mathbf{y}_i|\boldsymbol{\omega}_i, \boldsymbol{\theta}] \stackrel{D}{=} N[\mathbf{A}\mathbf{x}_i + \boldsymbol{\Lambda}\boldsymbol{\omega}_i, \boldsymbol{\Psi}_\epsilon]$, we have

$$[\boldsymbol{\omega}_i|\mathbf{y}_i, \boldsymbol{\theta}] \stackrel{D}{=} N\left[\boldsymbol{\Sigma}^{*-1}\boldsymbol{\Lambda}^T\boldsymbol{\Psi}_\epsilon^{-1}(\mathbf{y}_i - \mathbf{A}\mathbf{x}_i) + \boldsymbol{\Sigma}^{*-1}\boldsymbol{\Sigma}_\omega^{-1}\begin{pmatrix}\boldsymbol{\Pi}_0^{-1}\mathbf{B}\mathbf{z}_i \\ \mathbf{0}\end{pmatrix}, \boldsymbol{\Sigma}^{*-1}\right] \tag{A5.6}$$

where $\boldsymbol{\Sigma}^* = \boldsymbol{\Sigma}_\omega^{-1} + \boldsymbol{\Lambda}^T\boldsymbol{\Psi}_\epsilon^{-1}\boldsymbol{\Lambda}$. We see that the conditional distribution $[\boldsymbol{\omega}_i|\mathbf{y}_i, \boldsymbol{\theta}]$ is a normal distribution, and it is similar to that given in Equation (4.25).

CONDITIONAL DISTRIBUTION $(\theta | Y, \Omega)$

The conditional distribution of $\boldsymbol{\theta}$ given $(\mathbf{Y}, \boldsymbol{\Omega})$ is proportional to $p(\boldsymbol{\theta})p(\mathbf{Y}, \boldsymbol{\Omega}|\boldsymbol{\theta})$. We note that $\boldsymbol{\Omega}$ is given, and Equations (5.14) and (5.15) are linear models with fixed covariates. Let $\boldsymbol{\theta}_y$ be the unknown parameters in $\mathbf{A}, \boldsymbol{\Lambda}$ and $\boldsymbol{\Psi}_\epsilon$ associated with the measurement equation and $\boldsymbol{\theta}_\omega$ be the unknown parameters in $\mathbf{B}, \boldsymbol{\Pi}, \boldsymbol{\Gamma}, \boldsymbol{\Phi}$ and $\boldsymbol{\Psi}_\delta$ associated with the structural model. Again, it is assumed that the prior distribution of $\boldsymbol{\theta}_y$ is independent of the prior distribution of $\boldsymbol{\theta}_\omega$, that is, $p(\boldsymbol{\theta}) = p(\boldsymbol{\theta}_y)p(\boldsymbol{\theta}_\omega)$. Moreover, it follows from the reasoning given in Section 4.4.1 that the marginal conditional densities $\boldsymbol{\theta}_y$ and $\boldsymbol{\theta}_\omega$ given $(\mathbf{Y}, \boldsymbol{\Omega})$ are proportional to $p(\mathbf{Y}|\boldsymbol{\Omega}, \boldsymbol{\theta}_y)p(\boldsymbol{\theta}_y)$ and $p(\boldsymbol{\Omega}|\boldsymbol{\theta}_\omega)p(\boldsymbol{\theta}_\omega)$, respectively. Hence, these conditional densities can be treated separately as before.

Consider first the marginal conditional distribution of $\boldsymbol{\theta}_y$. Let $\boldsymbol{\Lambda}_y = (\mathbf{A}, \boldsymbol{\Lambda})$ with general elements λ_{ykj}, $j = 1, \cdots, r_1 + q$, $k = 1, \cdots, p$, and $\mathbf{u}_i = (\mathbf{x}_i^T, \boldsymbol{\omega}_i^T)^T$. It follows that $\mathbf{y}_i = \boldsymbol{\Lambda}_y\mathbf{u}_i + \boldsymbol{\epsilon}_i$. This simple transformation reformulates the model with fixed covariate $\mathbf{x}_i$ to the original factor analysis model. The positions of the fixed elements in $\boldsymbol{\Lambda}_y$ are identified via an index matrix $\mathbf{L}_y$ with the following elements:

$$l_{ykj} = \begin{cases} 0 & \text{if} \quad \lambda_{ykj} \quad \text{fixed;} \\ 1 & \text{if} \quad \lambda_{ykj} \quad \text{free;} \end{cases} \quad \text{for} \quad j = 1, \cdots, r_1 + q \quad \text{and} \quad k = 1, \cdots, p.$$

Let $\psi_{\epsilon k}$ be the kth diagonal element of $\boldsymbol{\Psi}_\epsilon$, and $\boldsymbol{\Lambda}_{yk}^T$ be the row vector that contains the unknown parameters in the kth row of $\boldsymbol{\Lambda}_y$. The following commonly used conjugate type prior distributions are used. For any $k \neq h$, we assume that the prior distribution of $(\psi_{\epsilon k}, \boldsymbol{\Lambda}_{yk})$ is independent of $(\psi_{\epsilon h}, \boldsymbol{\Lambda}_{yh})$, and

$$\psi_{\epsilon k}^{-1} \stackrel{D}{=} Gamma[\alpha_{0\epsilon k}, \beta_{0\epsilon k}] \text{ and } [\boldsymbol{\Lambda}_{yk}|\psi_{\epsilon k}] \stackrel{D}{=} N[\boldsymbol{\Lambda}_{0yk}, \psi_{\epsilon k}\mathbf{H}_{0yk}], \; k = 1, \cdots, p, \tag{A5.7}$$

where $\alpha_{0\epsilon k}, \beta_{0\epsilon k}, \boldsymbol{\Lambda}_{0yk}^T = (\mathbf{A}_{0k}^T, \boldsymbol{\Lambda}_{0k}^T)$ and the positive definite matrix $\mathbf{H}_{0yk}$ are hyperparameters whose values are assumed to be given from the prior information of previous studies or other sources.

Let $\mathbf{U} = (\mathbf{u}_1, \cdots, \mathbf{u}_n)$, and let $\mathbf{U}_k$ be the submatrix of $\mathbf{U}$ such that all the rows corresponding to $l_{ykj} = 0$ are deleted; let $\mathbf{Y}_k^T$ be the kth row of $\mathbf{Y}$ with general elements y_{ik} and $\mathbf{Y}_k^{*T} = (y_{1k}^*, \cdots, y_{nk}^*)$ with

$$y_{ik}^* = y_{ik} - \sum_{j=1}^{r_1+q} \lambda_{ykj} u_{ij}(1 - l_{ykj}),$$

where u_{ij} is jth element of $\mathbf{u}_i$. Then, for $k = 1, \cdots, p$, it can be shown (by similar reasoning as in Appendix 4.3) that

$$[\psi_{\epsilon k}^{-1}|\mathbf{Y}, \boldsymbol{\Omega}] \stackrel{D}{=} Gamma[n/2 + \alpha_{0\epsilon k}, \beta_{\epsilon k}], \; [\boldsymbol{\Lambda}_{yk}|\mathbf{Y}, \boldsymbol{\Omega}, \psi_{\epsilon k}^{-1}] \stackrel{D}{=} N[\mathbf{a}_{yk}, \psi_{\epsilon k}\mathbf{A}_{yk}], \tag{A5.8}$$

where $\mathbf{A}_{yk} = (\mathbf{H}_{0yk}^{-1} + \mathbf{U}_k\mathbf{U}_k^T)^{-1}$, $\mathbf{a}_{yk} = \mathbf{A}_{yk}(\mathbf{H}_{0yk}^{-1}\boldsymbol{\Lambda}_{0yk} + \mathbf{U}_k\mathbf{Y}_k^*)$ and

$$\beta_{\epsilon k} = \beta_{0\epsilon k} + \frac{1}{2}(\mathbf{Y}_k^{*T}\mathbf{Y}_k^* - \mathbf{a}_{yk}^T\mathbf{A}_{yk}^{-1}\mathbf{a}_{yk} + \boldsymbol{\Lambda}_{0yk}^T\mathbf{H}_{0yk}^{-1}\boldsymbol{\Lambda}_{0yk}).$$

Since $[\boldsymbol{\Lambda}_{yk}, \psi_{\epsilon k}^{-1}|\mathbf{Y}, \boldsymbol{\Omega}]$ is equal to $[\psi_{\epsilon k}^{-1}|\mathbf{Y}, \boldsymbol{\Omega}][\boldsymbol{\Lambda}_{yk}|\mathbf{Y}, \boldsymbol{\Omega}, \psi_{\epsilon k}^{-1}]$, it can be obtained via (A5.8). This gives the conditional distribution in relation to $\boldsymbol{\theta}_y$.

Now, consider the conditional distribution of $\boldsymbol{\theta}_\omega$ that is proportional to $p(\boldsymbol{\Omega}|\boldsymbol{\theta}_\omega)p(\boldsymbol{\theta}_\omega)$. Let $\boldsymbol{\Omega}_1 = (\boldsymbol{\eta}_1, \cdots, \boldsymbol{\eta}_n)$ and $\boldsymbol{\Omega}_2 = (\boldsymbol{\xi}_1, \cdots, \boldsymbol{\xi}_n)$. Since the distribution of $\boldsymbol{\xi}_i$ only involves $\boldsymbol{\Phi}$, $p(\boldsymbol{\Omega}_2|\boldsymbol{\theta}_\omega) = p(\boldsymbol{\Omega}_2|\boldsymbol{\Phi})$. Under the assumption that the prior distribution of $\boldsymbol{\Phi}$ is independent of the prior distributions of $\mathbf{B}, \boldsymbol{\Pi}, \boldsymbol{\Gamma}$ and $\boldsymbol{\Psi}_\delta$,we have

$$p(\boldsymbol{\Omega}|\boldsymbol{\theta}_\omega)p(\boldsymbol{\theta}_\omega) = [p(\boldsymbol{\Omega}_1|\boldsymbol{\Omega}_2, \mathbf{B}, \boldsymbol{\Pi}, \boldsymbol{\Gamma}, \boldsymbol{\Psi}_\delta)p(\mathbf{B}, \boldsymbol{\Pi}, \boldsymbol{\Gamma}, \boldsymbol{\Psi}_\delta)][p(\boldsymbol{\Omega}_2|\boldsymbol{\Phi})p(\boldsymbol{\Phi})].$$

Hence,the marginal conditional densities of $(\mathbf{B}, \mathbf{\Pi}, \mathbf{\Gamma}, \mathbf{\Psi}_\delta)$ and $\mathbf{\Phi}$ can be treated separately.

Consider a conjugate type prior distribution for $\mathbf{\Phi}$ with $\mathbf{\Phi}^{-1} \stackrel{D}{=} W_{r_2}[\mathbf{R}_0, \rho_0]$, with hyperparameters ρ_0 and $\mathbf{R}_0$. It can be shown by reasoning similar to that used in Section 4.4.1 that the conditional distribution of $\mathbf{\Phi}$ given $\mathbf{\Omega}_2$ is given by

$$[\mathbf{\Phi}|\mathbf{\Omega}_2] \stackrel{D}{=} IW_{q_2}[(\mathbf{\Omega}_2\mathbf{\Omega}_2^T + \mathbf{R}_0^{-1}), n + \rho_0]. \tag{A5.9}$$

Rewrite Equation (5.15) as $\boldsymbol{\eta}_i = \mathbf{\Lambda}_\omega \mathbf{v}_i + \boldsymbol{\delta}_i$, where $\mathbf{\Lambda}_\omega = (\mathbf{B}, \mathbf{\Pi}, \mathbf{\Gamma})$ with general elements $\lambda_{\omega kj}$ for $k = 1, \ldots, r_1$, and $\mathbf{v}_i = (\mathbf{z}_i^T, \boldsymbol{\eta}_i^T, \boldsymbol{\xi}_i^T)^T = (\mathbf{z}_i^T, \boldsymbol{\omega}_i^T)^T$ be a $(r_2 + q) \times 1$ vector. Let $\mathbf{V} = (\mathbf{v}_1, \cdots, \mathbf{v}_n)$, $\mathbf{L}_\omega$ be the index matrix with general elements $l_{\omega kj}$ that similarly defined as $\mathbf{L}_y$ to indicate the fixed known parameters in $\mathbf{\Lambda}_\omega$; let $\psi_{\delta k}$ be the kth diagonal element of $\mathbf{\Psi}_\delta$ and $\mathbf{\Lambda}_{\omega k}^T$ be the row vector that contains the unknown parameters in the kth row of $\mathbf{\Lambda}_\omega$. The prior distributions of $\mathbf{\Lambda}_{\omega k}$ and $\psi_{\delta k}^{-1}$ are similarly selected via the following conjugate type distributions:

$$\psi_{\delta k}^{-1} \stackrel{D}{=} Gamma[\alpha_{0\delta k}, \beta_{0\delta k}], \text{ and } [\mathbf{\Lambda}_{\omega k}|\psi_{\delta k}] \stackrel{D}{=} N[\mathbf{\Lambda}_{0\omega k}, \psi_{\delta k}\mathbf{H}_{0\omega k}], \quad k = 1, \cdots, q_1, \tag{A5.10}$$

where $\alpha_{0\delta k}, \beta_{0\delta k}, \mathbf{\Lambda}_{0\omega k}^T = (\mathbf{B}_{0k}^T, \mathbf{\Pi}_{0k}^T, \mathbf{\Gamma}_{0k}^T)$ and $\mathbf{H}_{0\omega k}$ are given hyperparameters. Moreover, it is assumed that for $h \neq k$, $(\psi_{\delta k}, \mathbf{\Lambda}_{\omega k})$ and $(\psi_{\delta h}, \mathbf{\Lambda}_{\omega h})$ are independent. Let $\mathbf{V}_k$ be the submatrix of $\mathbf{V}$ such that all the rows corresponding to $l_{\omega kj} = 0$ are deleted, and let $\mathbf{\Xi}_k^T = (\eta_{1k}^*, \cdots, \eta_{nk}^*)$, where

$$\eta_{ik}^* = \eta_{ik} - \sum_{j=1}^{r_1+q} \lambda_{\omega kj} v_{ij}(1 - l_{\omega kj}).$$

Then, following the same reasoning as before, it can be shown that:

$$\begin{aligned} &[\psi_{\delta k}^{-1}|\mathbf{\Omega}] \stackrel{D}{=} Gamma\ [n/2 + \alpha_{0\delta k}, \beta_{\delta k}] \text{ and} \\ &[\mathbf{\Lambda}_{\omega k}|\mathbf{\Omega}, \psi_{\delta k}^{-1}] \stackrel{D}{=} N[\mathbf{a}_{\omega k}, \psi_{\delta k}\mathbf{A}_{\omega k}], \end{aligned} \tag{A5.11}$$

where $\mathbf{A}_{\omega k} = (\mathbf{H}_{0\omega k}^{-1} + \mathbf{V}_k\mathbf{V}_k^{T})^{-1}$, $\mathbf{a}_{\omega k} = \mathbf{A}_{\omega k}(\mathbf{H}_{0\omega k}^{-1}\mathbf{\Lambda}_{0\omega k} + \mathbf{V}_k\mathbf{\Xi}_k)$ and

$$\beta_{\delta k} = \beta_{0\delta k} + \frac{1}{2}(\mathbf{\Xi}_k^T\mathbf{\Xi}_k - \mathbf{a}_{\omega k}^T\mathbf{A}_{\omega k}^{-1}\mathbf{a}_{\omega k} + \mathbf{\Lambda}_{0\omega k}^T\mathbf{H}_{0\omega k}^{-1}\mathbf{\Lambda}_{0\omega k}).$$

The conditional distribution $[\mathbf{B}, \mathbf{\Pi}, \mathbf{\Lambda}, \mathbf{\Psi}_\delta|\mathbf{\Omega}]$ can be obtained through Equations (A5.11).

APPENDIX 5.3: PP P-VALUES FOR MODEL ASSESSMENT

Based on the posterior predictive assessment as discussed in Rubin (1984), Gelman, Meng and Stern (1996) introduced a Bayesian counterpart of the classical p-value by defining a posterior predictive (PP) p-value for model-checking. To apply the approach for establishing a goodness-of-fit assessment of the posited model M_0 with parameter vector $\boldsymbol{\theta}$, observed data $\mathbf{Y}$ and latent data $\boldsymbol{\Omega}$, one needs to specify a discrepancy variable $D(\mathbf{Y}|\boldsymbol{\theta}, \boldsymbol{\Omega})$ for measuring the discrepancy between $\mathbf{Y}$ and the generated hypothetical replicate data $\mathbf{Y}^{\text{rep}}$. More specifically, the PP p-value is defined as

$$\begin{aligned} p_B(\mathbf{Y}) &= Pr\{D(\mathbf{Y}^{\text{rep}}|\boldsymbol{\theta}, \boldsymbol{\Omega}) \geq D(\mathbf{Y}|\boldsymbol{\theta}, \boldsymbol{\Omega})|\mathbf{Y}, M_0\}, \\ &= \int I\,[D(\mathbf{Y}^{\text{rep}}|\boldsymbol{\theta}, \boldsymbol{\Omega}) \geq D(\mathbf{Y}|\boldsymbol{\theta}, \boldsymbol{\Omega})]p(\mathbf{Y}^{\text{rep}}, \boldsymbol{\theta}, \boldsymbol{\Omega}|\mathbf{Y}, M_0)\text{d}\,\mathbf{Y}^{\text{rep}}\text{d}\boldsymbol{\theta}\text{d}\boldsymbol{\Omega}. \end{aligned}$$

where $I(\cdot)$ is an indicator function. The probability is taken over the following joint posterior distribution of $(\mathbf{Y}^{\text{rep}}, \boldsymbol{\theta}, \boldsymbol{\Omega})$ given $\mathbf{Y}$ and M_0:

$$p(\mathbf{Y}^{\text{rep}}, \boldsymbol{\theta}, \boldsymbol{\Omega}|\mathbf{Y}, M_0) = p(\mathbf{Y}^{\text{rep}}|\boldsymbol{\theta}, \boldsymbol{\Omega})p(\boldsymbol{\theta}, \boldsymbol{\Omega}|\mathbf{Y}).$$

In almost all the applications to SEMs considered in this book, we take the chi-square discrepancy variable such that $D(\mathbf{Y}^{\text{rep}}|\boldsymbol{\theta}, \boldsymbol{\Omega})$ has a chi-squared distribution with pn degrees of freedom. Thus, the PP p-value is equal to

$$\int p[\chi^2(pn) \geq D(\mathbf{Y}|\boldsymbol{\theta}, \boldsymbol{\Omega})]p(\boldsymbol{\theta}, \boldsymbol{\Omega}|\mathbf{Y})\text{d}\boldsymbol{\theta}\text{d}\boldsymbol{\Omega}.$$

A Rao–Blackwellized type estimate of this PP p-value is:

$$\hat{p}_B(\mathbf{Y}) = J^{-1}\sum_{j=1}^{J} Pr\{\chi^2(d^*) \geq D(\mathbf{Y}|\boldsymbol{\theta}^{(j)}, \boldsymbol{\Omega}^{(j)})\}, \tag{A5.12}$$

where $\{(\boldsymbol{\theta}^{(j)}, \boldsymbol{\Omega}^{(j)}); j = 1, \cdots, J\}$ are observations simulated during the estimation. The computational burden for obtaining this $\hat{p}_B(\mathbf{Y})$ is light.

REFERENCES

Akaike, H. (1973) Information theory and an extension of the maximum likelihood principle. In B. N. Petrox and F. Caski (eds), *Second International Symposium on Information Theory*, p. 267. Budapest, Hungary: Akademiai Kiado.

Bayarri, M. J. and Berger, J. O. (2000) P values for composite null models. *Journal of the American Statistical Association*, **95**, 1127–1142.

Bentler, P. M. (1992) *EQS: Structural Equation Program Manual*. Los Angeles, CA: BMDP Statistical Software.

Bentler, P. M. and Bonett, D. G. (1980) Significance tests and goodness of fit in the analysis of covariance structures. *Psychological Bulletin*, **88**, 588–606.

Berger, J. O. (1985) *Statistical Decision Theory and Bayesian Analysis*. New York: Springer-Verlag.

Berger, J. O. and Dalampady, M. (1987) Testing precise hypotheses. *Statistical Science*, **3**, 317–352.

Berger, J. O. and Pericchi, L. R. (1996) The intrinsic Bayes factor for model selection and prediction. *Journal of the American Statistical Association*, **91**, 109–122.

Berger, J. O. and Sellke, T. (1987) Testing a point null hypothesis: The irreconcilability of P values and evidence. *Journal of the American Statistical Association*, **82**, 112–122.

Bollen, K. A. (1989) A new incremental fit index for general structural equation models. *Sociological Methods and Research*, **17**, 303–316.

Carlin, B. P. and Louis, T. A. (1996) *Bayes and Empirical Bayes Methods for Data Analysis*. New York: Chapman and Hall.

Celeux, G., Forbes, F., Roberts, C. P. and Titterington, D. M. (2003) Deviance information criteria for missing data models. *Cahires du Ceremade, 0325*.

Chib, S. (1995) Marginal likelihood from the Gibbs output. *Journal of the American Statistical Association*, **90**, 1313–1321.

Chib, S. and Jeliazkov, I. (2001) Marginal likelihood from the Metropolis–Hastings outputs. *Journal of the American Statistical Association*, **96**, 270–281.

DiCiccio, T. J., Kass, R. E., Raftery, A. and Wasserman, L. (1997) Computing Bayes factors by combining simulation and asymptotic approximations. *Journal of the American Statistical Association*, **92**, 903–915.

Gelfand, A. E. and Dey, D. K. (1994) Bayesian model choice: Asymptotics and exact calculations. *Journal of the Royal Statistical Society, Series B*, **56**, 501–514.

Gelman, A. and Meng, X. L. (1998) Simulating normalizing constants: from importance sampling to bridge sampling to path sampling. *Statistical Science*, **13**, 163–185.

Gelman, A., Meng, X. L. and Stern, H. (1996) Posterior predictive assessment of model fitness via realized discrepancies. *Statistica Sinica*, **6**, 733–759.

Gelman, A., Carlin, J. B., Stern, H. and Rubin, D. B. (2003) *Bayesian Data Analysis*, (2nd edn). London: Chapman and Hall / CRC.

Jedidi, K., Jagpal, H. S. and DeSarbo, W. S. (1997) Finite-mixture structural equation models for response-based segmentation and unobserved heterogeneity. *Marketing Science*, **16**, 39–59.

Johnson, R. A. and Wichern, D. W. (1992) *Applied Multivariate Statistical Analysis* (3rd edn). New York: Prentice-Hall International, Inc.

Jöreskog, K. G. and Sörbom, D. (1996). *LISREL 8: Structural Equation Modeling with the SIMPLIS Command Language*. Hove and London: Scientific Software International.

Kass, R. E. and Raftery, A. E. (1995) Bayes factors. *Journal of the American Statistical Association*, **90**, 773–795.

Lee, S. Y. and Song, X. Y. (2001) Hypothesis testing and model comparison in two-level structural equation models. *Multivariate Behavioral Research*, **36**, 639–55.

Lee, S. Y. and Song, X. Y. (2003) Bayesian model selection for mixtures of structural equation models with an unknown number of components. *British Journal of Mathematical and Statistical Psychology*, **56**, 145–165.

Lee, S. Y. and Tang, N. S. (2006) Bayesian analysis of nonlinear structural equation models with nonignorable missing data. *Psychometrika*, in press.

Lempers, F. B. (1971) *Posterior Probabilities of Alternative Linear Models*. Rotterdam: University Press.

Meng, X. L. (1994) Posterior predictive *p*-values. *The Annals of Statistics*, **22**, 1142–1160.

Meng, X. L. and Wong, W. H. (1996) Simulating ratios of normalizing constants via simple identity, a theoretical exploration. *Statistica Sinica*. **6**, 831–860.

Ogata, Y. (1989) A Monte Carlo method for high dimensional integration. *Numerische Mathematik*, **55**, 137–157.

Ogata, Y. (1990) A Monte Carlo method for an objective Bayesian procedure. *Annals of the Institute of Statistical Mathematics*, **42**, 403–433.

O'Hagan, A. (1995) Fractional Bayes factor for model comparison. *Journal of the Royal Statistical Society, Series B*, **57**, 99–138.

Raftery, A. E. (1986) Choosing models for cross-classifications. *American Sociological Review*, **51**, 145–146.

Raftery, A. E. (1993) Bayesian model selection in structural equation models. In K. A. Bollen and J. S. Long (eds), *Testing Structural Equation Models*, pp. 163–180. Beverly Hills, CA: Sage.

Rubin, D. B. (1984) Bayesianly justifiable and relevant frequency calculations for the applied statistician. *The Annals of Statistics*, **12**, 1151–1172.

Sammel, M.D. and Ryan, L. M. (1996) Latent variable models with fixed effects. *Biometrics*, **52**, 650–663.

Schwarz, G. (1978) Estimating the dimension of a model. *The Annals of Statistics*, **6**, 461–464.

Spiegelhalter, D. J. and Smith, A. F. M. (1982) Bayes factor for linear and log-linear models with vague prior information. *Journal of the Royal Statistical Society, Series B*, **44**, 377–387.

Spiegelhalter, D. J., Best, N. G. and Lunn, D. (2003) *WinBUGS Version 1.4 User Manual*. Cambridge: MRC Biostatistics Unit, URL http://www.mrc-bsu.cam.ac.uk/bugs/.

Spiegelhalter, D. J., Best, N. G., Carlin, B. P. and van der Linde, A. (2002) Bayesian measure of model complexity and fit. *Journal of the Royal Statistical Society, Series B*, **64**, 583–639.

World Values Study Group (1994) World Values Survey, 1981–1984 and 1990–1993. ICPSR version. Ann Arbor, MI: Institute for Social Research (producer). Ann Arbon, MI: Inter-university Consortium for Political and Social Research (distributor).

6
Structural Equation Models with Continuous and Ordered Categorical Variables

6.1 INTRODUCTION

Due to the design of questionnaires and the nature of the problems in behavioral, educational, medical and social sciences, data are often coming from ordered categorical variables with observations in discrete form. Examples of such variables are attitude items, Likert items, rating scales and the like. Typical cases are when a subject is asked to report some attitude on scales like 'disapprove', 'no opinion', 'approve'; to report the effect of a drug on scale like 'getting worse', 'no change', 'getting better'; or to report the opinion of a policy on a scale such as 'strongly disagree', 'disagree', 'no opinion', 'agree', 'strongly agree'. Consider an ordered categorical variable with a five-point scale $\{1, 2, 3, 4, 5\}$ corresponding to the answer on the opinion of a policy. One common approach is to treat the assigned integers as continuous data from a normal distribution. This approach may not lead to serious problems if the histogram of the observations is symmetrical and with the highest frequency at the center. This is the situation where most subjects choose the category 'no opinion'. To claim multivariate normality of the observed variables, we need to have most subjects choosing the middle category, for example 'no opinion' or 'no change', in all the corresponding items. However, for an interesting item in the questionnaire, most subjects would be likely to select categories at both ends, for example, 'strongly agree (strongly disagree)' or 'agree (disagree)'. Hence, in practice, histograms corresponding to most variables are either skewed or bi-modal. Clearly, routinely treating ordered categorical variables as normal may lead to erroneous conclusions (see Olsson, 1979a,b; Lee, Poon and Bentler, 1990a,b).

A better approach for assessing these kinds of discrete data is to treat them as observations that are coming from a hidden continuous normal distribution with a threshold specification. Suppose for a given data set, the proportions of 1, 2, 3, 4 are 0.05, 0.05, 0.40 and 0.5, respectively. From the histogram given in Figure 6.1, we see that these discrete data are highly skewed to the right. The threshold approach for analyzing this highly skewed discreted variable is to treat the ordered categorical data as manifestations of an underlying normal variable y. The exact continuous measurements of y are not available, but are related to the observed ordered categorical variable z such as follows: for $k = 1, 2, 3, 4$

$$z = k \quad \text{if } \alpha_{k-1} < y \leq \alpha_k;$$

where $-\infty = \alpha_0 < \alpha_1 < \alpha_2 < \alpha_3 < \alpha_4 = \infty$, and α_1, α_2, and α_3 are thresholds. Then, the ordered categorical observations that give the histogram in Figure 6.1 can be captured by $N[0, 1]$ with appropriate thresholds (see Figure 6.2). As $\alpha_2 - \alpha_1$ can be different from $\alpha_3 - \alpha_2$, unequal-interval scales are allowed. Hence, this threshold approach allows flexible modeling. As it is related to a common normal distribution, it also provides easy interpretation of the parameters. It should be noted that the ad hoc integral values, here $k = 1, 2, 3, 4$, are solely used to represent the category; only their frequencies are important in the statistical analysis.

Analysis of SEMs with mixed continuous and ordered categorical data is not straightforward, because we need to compute the multiple integrals associated

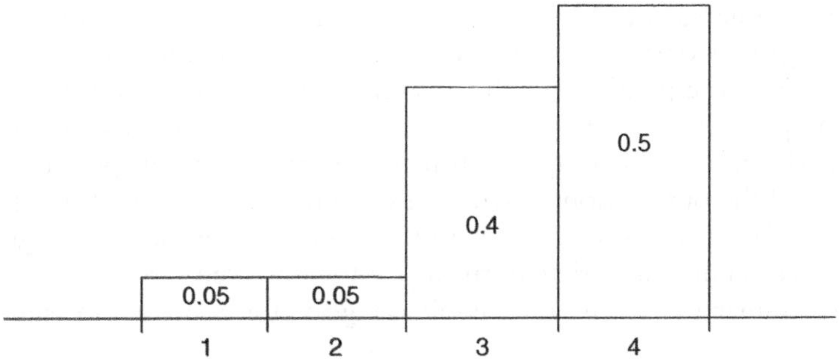

Figure 6.1 Histogram of a hypothetical ordinal categorical data set. This figure and Figure 6.4 are taken from Lee, Song, Skevington and Hao (2005).

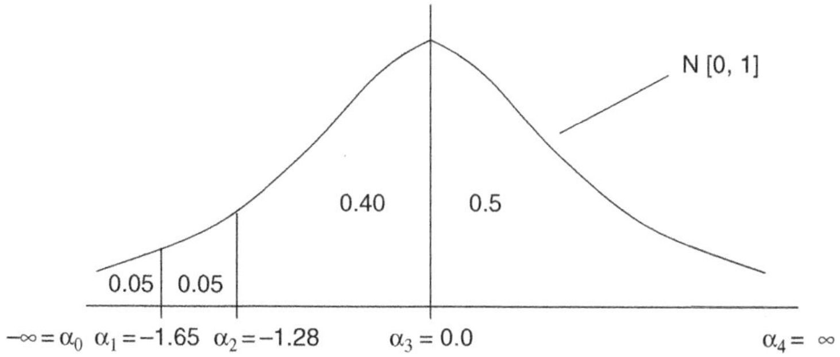

Figure 6.2 The underlying normal distribution with a threshold specification.

with the cell probabilities that are induced by the ordered categorical outcomes. Some multistage methods have been proposed to reduce the computational burden in evaluating these integrals. The basic procedure of these multistage methods is to estimate the polychoric and polyserial correlations, and the thresholds at the first stage, derive the asymptotic distribution of the estimates at the second stage, and analyze the SEM with a covariance structural analysis approach through a generalized least square (GLS) minimization at the final stage. Different methods employed at the first stage lead to the different procedures that are given in PRELIS and LISREL (Jöreskog and Sörbom, 1996), LISCOMP and Mplus (Muthen and Muthen, 2000) and Lee, Poon, and Bentler (1995). However, the multistage estimators are not statistically optimal and need to invert at each iteration of the GLS minimization a huge matrix whose dimension increases very rapidly with the number of manifest variables. Besides the multistage procedures, Reboussin and Liang (1998) proposed an estimating equation approach, and Shi and Lee (2000) developed a Monte Carlo EM type algorithm (Wei and Tanner, 1990) for ML analysis of a factor analysis model.

The main objective of this chapter is to introduce a Bayesian approach on the basis of the framework given in Chapters 4 and 5 for analyzing SEMs with mixed continuous and ordered categorical variables. The main idea in handling the ordered categorical variables in the Bayesian analysis is to treat the underlying latent continuous measurements as hypothetical missing data, and augment them with the observed data in the posterior analysis. Using this data augmentation strategy, the model that is based on the complete data set becomes one with continuous variables. In the posterior analysis, sequences of observations of the structural parameters, latent variables and thresholds, are simulated from the joint posterior distribution via a hybrid algorithm that combines the Gibbs sampler (Geman and Geman, 1984) and the MH algorithm (Metropolis *et al.*, 1953; Hastings, 1970). By means of the simulated observations, joint Bayesian estimates of the unknown thresholds, structural parameters and latent variables are produced together with their standard error estimates. In addition to these point estimates, we also consider Bayesian model selection via the Bayes factor.

This chapter is organized as follows. For completeness, SEMs with continuous and ordered categorical variables are described in Section 6.2. Identification of an ordered categorical variable is discussed. Bayesian estimation, and the goodness-of-fit test of a hypothesized model are addressed in Section 6.3. Detailed discussions of the posterior analysis are presented here. These include the derivation of the full conditional distributions that are required by the Gibbs sampler for simulating observations from the joint posterior distribution, and the implementation of the MH algorithm. The PP p-value for assessing goodness-of-fit is outlined. Section 6.4 considers Bayesian model comparison. A path sampling procedure for computing the Bayes factor is described. Two applications are presented in Sections 6.5 and 6.6 to illustrate the Bayesian methodologies. One is on the Bayesian selection of the number of factors in EFA, the other is on Bayesian analysis of quality of life (QOL) data. An application of the software WinBUGS (Spiegelhalter, Thomas, Best and Lunn, 2002) to obtain the Bayesian solution is outlined in Section 6.6.

6.2 THE BASIC MODEL

Consider the following measurement equation for a $p \times 1$ manifest random vector $\mathbf{v}_i$:

$$\mathbf{v}_i = \boldsymbol{\mu} + \boldsymbol{\Lambda}\boldsymbol{\omega}_i + \boldsymbol{\epsilon}_i \quad i = 1, \ldots, n, \tag{6.1}$$

where $\boldsymbol{\mu}(p \times 1)$ is the vector of intercepts, $\boldsymbol{\Lambda}(p \times q)$ is the factor loading matrix, $\boldsymbol{\omega}_i(q \times 1)$ is a latent random vector and $\boldsymbol{\epsilon}_i(p \times 1)$ is a random vector of error measurements with distribution $N[\mathbf{0}, \boldsymbol{\Psi}_\epsilon]$, $\boldsymbol{\Psi}_\epsilon$ is diagonal and $\boldsymbol{\epsilon}_i$ is independent with $\boldsymbol{\omega}_i$. We let $\boldsymbol{\eta}_i(q_1 \times 1)$ and $\boldsymbol{\xi}_i(q_2 \times 1)$ be latent subvectors of $\boldsymbol{\omega}_i$, and consider the following structural equation:

$$\boldsymbol{\eta}_i = \boldsymbol{\Pi}\boldsymbol{\eta}_i + \boldsymbol{\Gamma}\boldsymbol{\xi}_i + \boldsymbol{\delta}_i \tag{6.2}$$

where $\boldsymbol{\Pi}(q_1 \times q_1)$ and $\boldsymbol{\Gamma}(q_1 \times q_2)$ are matrices of regression coefficients, $\boldsymbol{\xi}_i$ and $\boldsymbol{\delta}_i$ are independently distributed as $N[\mathbf{0}, \boldsymbol{\Phi}]$ and $N[\mathbf{0}, \boldsymbol{\Psi}_\delta]$, where $\boldsymbol{\Psi}_\delta$ is a diagonal covariance matrix. It is assumed that $\boldsymbol{\Pi}_0 = |\mathbf{I}_{q_1} - \boldsymbol{\Pi}|$ is a nonzero constant that is independent with elements in $\boldsymbol{\Pi}$. Let $\boldsymbol{\Lambda}_\omega = (\boldsymbol{\Pi}, \boldsymbol{\Gamma})$, then Equation (6.2) can be written as $\boldsymbol{\eta}_i = \boldsymbol{\Lambda}_\omega\boldsymbol{\omega}_i + \boldsymbol{\delta}_i$.

Let $\mathbf{v} = \{\mathbf{x}, \mathbf{y}\}$, where $\mathbf{x} = \{x_1, \ldots, x_r\}$, is a subset of variables whose exact continuous measurements are observable, while $\mathbf{y} = \{y_1, \ldots, y_s\}$ is the remaining subset of variables such that $p \geq s = p - r \geq 0$ and the corresponding continuous measurements are unobservable. The information associated with $\mathbf{y}$ is given by an observable ordered categorical vector $\mathbf{z} = (z_1, \ldots, z_s)^T$. For any latent variable in $\boldsymbol{\eta}$ or $\boldsymbol{\xi}$, it may be related with manifest variables in either $\mathbf{x}$ or $\mathbf{y}$.

That is, any latent variable may have continuous and/or ordered categorical manifest variables as its indicators. The relationship between $\mathbf{y}$ and $\mathbf{z}$ is defined by a set of thresholds as follows:

$$\mathbf{z} = \begin{bmatrix} z_1 \\ \vdots \\ z_s \end{bmatrix} \quad \text{if} \quad \begin{matrix} \alpha_{1,z_1} < y_1 \leq \alpha_{1,z_1+1} \\ \vdots \\ \alpha_{s,z_s} < y_s \leq \alpha_{s,z_s+1} \end{matrix}, \tag{6.3}$$

where for $k = 1, \ldots, s$, z_k is an integral value in $\{0, 1, \ldots, b_k\}$, and $\alpha_{k,0} < \alpha_{k,1} < \cdots < \alpha_{k,b_k} < \alpha_{k,b_k+1}$. In general, we set $\alpha_{k,0} = -\infty, \alpha_{k,b_k+1} = \infty$. For the kth variable, there are $b_k + 1$ categories which are defined by the unknown thresholds $\alpha_{k,j}$. The integral values $\{0, 1, \ldots, b_k\}$ of z_k are just used for specifying the categories that contain the corresponding elements in y_k. For example, let z_k be any component of $\mathbf{z}$, '$z_k = 3$' provides only the information that y_k is in the interval $(\alpha_{k,3}, \alpha_{k,4}]$ that is defined by the threshold parameters $\alpha_{k,3} < \alpha_{k,4}$. Here, '3' is just used as a symbol and can be replaced by some other values. However, in order to indicate the 'ordered' nature of the categorical values and their thresholds, it is desirable to choose an ordered set of integers for $\mathbf{z}$. In the Bayesian analysis, these integral values are neither directly used in the posterior simulation nor in any actual computation.

Suppose that the model in relation to the subvector $\mathbf{y}_i = (y_{i1}, \ldots, y_{is})^T$ of $\mathbf{v}_i$ is given by:

$$\mathbf{y}_i = \boldsymbol{\mu}_y + \boldsymbol{\Lambda}_y \boldsymbol{\omega}_i + \boldsymbol{\epsilon}_{yi}, \tag{6.4}$$

where $\boldsymbol{\mu}_y(s \times 1)$ is a subvector of $\boldsymbol{\mu}$, $\boldsymbol{\Lambda}_y(s \times q)$ is a submatrix of $\boldsymbol{\Lambda}$, $\boldsymbol{\epsilon}_{yi}(s \times 1)$ is a subvector of $\boldsymbol{\epsilon}_i$ with diagonal covariance submatrix $\boldsymbol{\Psi}_{y\epsilon}$ of $\boldsymbol{\Psi}_\epsilon$. Let $\mathbf{z}_i = (z_{i1}, \ldots, z_{is})^T$ be the ordered categorical observation corresponding to $\mathbf{y}_i$, $i = 1, \ldots, n$. In this chapter, the SEM that is defined by Equations (6.1) and (6.2) is analyzed with the following mixed continuous and ordered categorical observations:

$$\begin{pmatrix} \mathbf{x}_1 \\ \mathbf{z}_1 \end{pmatrix}, \ldots, \begin{pmatrix} \mathbf{x}_n \\ \mathbf{z}_n \end{pmatrix}.$$

The model defined by Equations (6.1) and (6.2) is not identified without imposing appropriate identification conditions. There are two kinds of indeterminacies involved in this model. One is the common indeterminacy coming from the covariance structure of the model that can be solved by the common method of fixing appropriate elements in $\boldsymbol{\Lambda}$, $\boldsymbol{\Pi}$ and/or $\boldsymbol{\Gamma}$ at preassigned values. The other indeterminacy is induced by the ordered categorical variables.

To tackle the identification problem that is induced by an ordered categorical variable, we have to keep in mind the following two issues. First, as it is difficult to obtain a necessary and sufficient condition for identification, we are mainly interested in finding a reasonable and convenient way to solve the problem. Second, for an ordered categorical variable, the location and dispersion of its underlying continuous normal variable are unknown. As we have completely no idea about these latent values, it is desirable to take a unified scale to every ordered categorical variable, and the obtained statistical results should be interpreted in relative sense.

Now, let z_k be the ordered categorical variable that is defined with a set of thresholds and the underlying continuous variable y_k whose distribution is $N[\mu, \sigma^2]$. The indeterminacy is caused by the fact that the thresholds, μ and σ^2 are not identified. One method to solve the problem is to fix (μ, σ^2) at some fixed values. However, as these parameters are usually of primary interest, it is better not to impose restrictions on them. Hence, we propose to impose the identification conditions on the thresholds, which are the less interesting nuisance parameters. More specifically, we propose to fix the thresholds at both ends, $\alpha_{k,1}$, and α_{k,b_k}, at preassigned values. This method implicitly picks measures for the location and the dispersion of y_k. For instance, the range $\alpha_{k,b_k} - \alpha_{k,1}$ provides a standard for measuring the dispersion. This method can be applied to the multivariate case by imposing the above restrictions on the appropriate thresholds for every component in $\mathbf{z}$. If the model is scale invariant, the choice of the preassigned values for the fixed thresholds only changes the scale of the estimated covariance matrix (see Lee, Poon and Bentler, 1990a). For better interpretation of the statistical results, it is advantageous to assign the values of the fixed thresholds so that the scale of each variable is the same. One common method is to use the observed frequencies and the standard normal distribution, $N[0, 1]$. More specifically, for every k, we may fix $\alpha_{k,1} = \Phi^{*-1}(f^*_{k,1})$ and $\alpha_{k,b_k} = \Phi^{*-1}(f^*_{k,b_k})$, where $\Phi^*(\cdot)$ is the distribution function of $N[0, 1]$, $f^*_{k,1}$ and f^*_{k,b_k} are the frequency of the first category, and the cumulative frequencies of categories with $z_k < b_k$, respectively. For linear SEMs, these restrictions imply that the mean and the variance of the underlying continuous variable y_k are 0 and 1, respectively. They have been frequently used in Bayesian analyses of SEMs with ordered categorical variables, see for example Lee and Zhu (2000) and Song and Lee (2001). Although there are other possible methods, for convenience we will use the above method for solving the identification problem of the ordered categorical variables in this book.

6.3 BAYESIAN ESTIMATION AND GOODNESS-OF-FIT

We will utilize the useful strategy of data augmentation described in Chapter 4 in the Bayesian estimation for the current SEM with continuous and ordered categorical variables. Let $\mathbf{X} = (\mathbf{x}_1, \ldots, \mathbf{x}_n)$ and $\mathbf{Z} = (\mathbf{z}_1, \ldots, \mathbf{z}_n)$ be the observed continuous and ordered categorical data matrices, respectively; and let $\mathbf{Y} = (\mathbf{y}_1, \ldots, \mathbf{y}_n)$ and $\boldsymbol{\Omega} = (\boldsymbol{\omega}_1, \ldots, \boldsymbol{\omega}_n)$ be the matrices of latent continuous measurements and latent variables, respectively. The observed data $[\mathbf{X}, \mathbf{Z}]$ are augmented with the latent data $[\mathbf{Y}, \boldsymbol{\Omega}]$ in the posterior analysis. Joint Bayesian estimates of $\boldsymbol{\Omega}$, unknown thresholds in $\boldsymbol{\alpha} = (\boldsymbol{\alpha}_1, \ldots, \boldsymbol{\alpha}_s)$ and the structural parameter vector $\boldsymbol{\theta}$ that contains all unknown parameters in $\boldsymbol{\mu}, \boldsymbol{\Phi}, \boldsymbol{\Lambda}, \boldsymbol{\Lambda}_\omega, \boldsymbol{\Psi}_\delta$ and $\boldsymbol{\Psi}_\epsilon$, will be obtained.

In the Bayesian approach, we need to evaluate the posterior distribution $[\boldsymbol{\alpha}, \boldsymbol{\theta}, \boldsymbol{\Omega}|\mathbf{X}, \mathbf{Z}]$. This distribution is rather complicated. To capture its characteristics, we will try to draw a sufficiently large number of observations from it such that the empirical distribution of the generated observations is a close approximation to the true distribution. A good candidate for simulating observations from the posterior distribution is the Gibbs sampler (Geman and Geman, 1984), which iteratively simulates $\boldsymbol{\alpha}, \boldsymbol{\theta}$ and $\boldsymbol{\Omega}$ from the full conditional distributions. However, owing to the presence of the ordered categorical variables, these conditional distributions are rather complicated to derive and simulating observations from them is difficult. This motivates the further augmentation of the latent matrix $\mathbf{Y}$ in the posterior analysis, and the consideration of the joint posterior distribution $[\boldsymbol{\alpha}, \boldsymbol{\theta}, \boldsymbol{\Omega}, \mathbf{Y}|\mathbf{X}, \mathbf{Z}]$. To implement the Gibbs sampler for generating observations of this posterior distribution, we start with initial starting values $(\boldsymbol{\alpha}^{(0)}, \boldsymbol{\theta}^{(0)}, \boldsymbol{\Omega}^{(0)}, \mathbf{Y}^{(0)})$, then simulate $(\boldsymbol{\alpha}^{(1)}, \boldsymbol{\theta}^{(1)}, \boldsymbol{\Omega}^{(1)}, \mathbf{Y}^{(1)})$ and so on according to the following procedure. At the mth iteration with current values $\boldsymbol{\alpha}^{(m)}, \boldsymbol{\theta}^{(m)}, \boldsymbol{\Omega}^{(m)}, \mathbf{Y}^{(m)}$,

$$\text{Step(a)}: \quad \text{Generate } \boldsymbol{\Omega}^{(m+1)} \text{ from } p(\boldsymbol{\Omega}|\boldsymbol{\theta}^{(m)}, \boldsymbol{\alpha}^{(m)}, \mathbf{Y}^{(m)}, \mathbf{X}, \mathbf{Z});$$

$$\text{Step(b)}: \quad \text{Generate } \boldsymbol{\theta}^{(m+1)} \text{ from } p(\boldsymbol{\theta}|\boldsymbol{\Omega}^{(m+1)}, \boldsymbol{\alpha}^{(m)}, \mathbf{Y}^{(m)}, \mathbf{X}, \mathbf{Z});$$

$$\text{Step(c)}: \text{Generate } (\boldsymbol{\alpha}^{(m+1)}, \mathbf{Y}^{(m+1)}) \text{ from } p(\boldsymbol{\alpha}, \mathbf{Y}|\boldsymbol{\theta}^{(m+1)}, \boldsymbol{\Omega}^{(m+1)}, \mathbf{X}, \mathbf{Z}).$$

The cycle defined above generates $(\boldsymbol{\alpha}^{(m+1)}, \boldsymbol{\theta}^{(m+1)}, \boldsymbol{\Omega}^{(m+1)}, \mathbf{Y}^{(m+1)})$ after the mth iteration. As m approaches infinity, the joint distribution of $(\boldsymbol{\alpha}^{(m)}, \boldsymbol{\theta}^{(m)}, \boldsymbol{\Omega}^{(m)}, \mathbf{Y}^{(m)})$ can be shown to approach the joint posterior distribution $[\boldsymbol{\alpha}, \boldsymbol{\theta}, \boldsymbol{\Omega}, \mathbf{Y}|\mathbf{X}, \mathbf{Z}]$, see Geman and Geman (1984), and Geyer (1992). Convergence of the Gibbs sampler can be monitored by the EPSR values (Gelman, 1996), or plots of several simulated sequences of the individual parameters with different starting values. The sequences of the quantities simulated from the joint posterior distribution will be used to calculate the Bayesian estimates and other related statistics. Conditional distributions required by the Gibbs sampler are presented below.

6.3.1 Conditional Distributions

We first consider the conditional distribution in Step (a) of the Gibbs sampler. We note that as the underlying continuous measurements in $\mathbf{Y}$ are given, $\mathbf{Z}$ gives no additional information to this conditional distribution. As $\mathbf{v}_i$ are conditionally independent, and $\boldsymbol{\omega}_i$ are also mutually independent among themselves and independent with $\mathbf{Z}$, we have

$$p(\boldsymbol{\Omega}|\boldsymbol{\alpha}, \boldsymbol{\theta}, \mathbf{Y}, \mathbf{X}, \mathbf{Z}) = \prod_{i=1}^{n} p(\boldsymbol{\omega}_i|\mathbf{v}_i, \boldsymbol{\theta}).$$

There are two representations for $p(\boldsymbol{\omega}_i|\mathbf{v}_i, \boldsymbol{\theta})$. The first representation follows from a similar derivation to that in Chapter 5 that

$$[\boldsymbol{\omega}_i|\mathbf{v}_i, \boldsymbol{\theta}] \stackrel{D}{=} N[\boldsymbol{\Sigma}^{*-1}\boldsymbol{\Lambda}^T\boldsymbol{\Psi}_\epsilon^{-1}(\mathbf{v}_i - \boldsymbol{\mu}), \boldsymbol{\Sigma}^{*-1}], \tag{6.5}$$

in which $\boldsymbol{\Sigma}^* = (\boldsymbol{\Sigma}_\omega^{-1} + \boldsymbol{\Lambda}^T\boldsymbol{\Psi}_\epsilon^{-1}\boldsymbol{\Lambda})$, where

$$\boldsymbol{\Sigma}_\omega = \begin{bmatrix} \boldsymbol{\Pi}_0^{-1}(\boldsymbol{\Gamma}\boldsymbol{\Phi}\boldsymbol{\Gamma}^T + \boldsymbol{\Psi}_\delta)\boldsymbol{\Pi}_0^{-T} & \boldsymbol{\Pi}_0^{-1}\boldsymbol{\Gamma}\boldsymbol{\Phi} \\ \boldsymbol{\Phi}\boldsymbol{\Gamma}^T\boldsymbol{\Pi}_0^{-T} & \boldsymbol{\Phi} \end{bmatrix},$$

is the covariance matrix of $\boldsymbol{\omega}_i$. For the second representation, we note that

$$p(\boldsymbol{\omega}_i|\mathbf{v}_i, \boldsymbol{\theta}) \propto p(\mathbf{v}_i|\boldsymbol{\omega}_i, \boldsymbol{\theta})p(\boldsymbol{\eta}_i|\boldsymbol{\xi}_i, \boldsymbol{\theta})p(\boldsymbol{\xi}_i|\boldsymbol{\theta}).$$

Based on the definition of the model and assumptions,

$$p(\mathbf{v}_i|\boldsymbol{\omega}_i, \boldsymbol{\theta}) \propto \exp\left[-\frac{1}{2}(\mathbf{v}_i - \boldsymbol{\mu} - \boldsymbol{\Lambda}\boldsymbol{\omega}_i)^T\boldsymbol{\Psi}_\epsilon^{-1}(\mathbf{v}_i - \boldsymbol{\mu} - \boldsymbol{\Lambda}\boldsymbol{\omega}_i)\right],$$

$$p(\boldsymbol{\eta}_i|\boldsymbol{\xi}_i, \boldsymbol{\theta}) \propto \exp\left[-\frac{1}{2}(\boldsymbol{\eta}_i - \boldsymbol{\Lambda}_\omega\boldsymbol{\omega}_i)^T\boldsymbol{\Psi}_\delta^{-1}(\boldsymbol{\eta}_i - \boldsymbol{\Lambda}_\omega\boldsymbol{\omega}_i)\right],$$

$$p(\boldsymbol{\xi}_i|\boldsymbol{\theta}) \propto \exp\left(-\frac{1}{2}\boldsymbol{\xi}_i^T\boldsymbol{\Phi}^{-1}\boldsymbol{\xi}_i\right).$$

Consequently, $p(\boldsymbol{\omega}_i|\mathbf{v}_i, \boldsymbol{\theta})$ is proportional to

$$\begin{aligned} \exp\{-\frac{1}{2}[\boldsymbol{\xi}_i^T\boldsymbol{\Phi}^{-1}\boldsymbol{\xi}_i + (\mathbf{v}_i - \boldsymbol{\mu} - \boldsymbol{\Lambda}\boldsymbol{\omega}_i)^T\boldsymbol{\Psi}_\epsilon^{-1}(\mathbf{v}_i - \boldsymbol{\mu} - \boldsymbol{\Lambda}\boldsymbol{\omega}_i) \\ + (\boldsymbol{\eta}_i - \boldsymbol{\Lambda}_\omega\boldsymbol{\omega}_i)^T\boldsymbol{\Psi}_\delta^{-1}(\boldsymbol{\eta}_i - \boldsymbol{\Lambda}_\omega\boldsymbol{\omega}_i)]\}. \end{aligned} \tag{6.6}$$

Hence, $p(\boldsymbol{\Omega}|\boldsymbol{\theta}, \boldsymbol{\alpha}, \mathbf{Y}, \mathbf{X}, \mathbf{Z})$ is a product of $p(\boldsymbol{\omega}_i|\mathbf{v}_i, \boldsymbol{\theta})$, whose distribution can be obtained from either Equations (6.5) or (6.6). Based on the practical experience available so far, simulating observations on the basis of Equations (6.5) or (6.6) gives similar and acceptable results for statistical inference.

To derive the conditional distributions with respect to the structural parameters in Step (b), we note that as $\boldsymbol{\Omega}$ and $\mathbf{Y}$ are given, the model defined by Equations (6.1) and (6.2) reduces to the standard linear model with continuous data. Hence, these conditional distributions are independent of $\boldsymbol{\alpha}$ and $\mathbf{Z}$. Let $\boldsymbol{\theta}_v$ be the unknown parameters in $\boldsymbol{\mu}$, $\boldsymbol{\Lambda}$ and $\boldsymbol{\Psi}_\epsilon$ associated with Equation (6.1), and let $\boldsymbol{\theta}_\omega$ be the unknown parameters in $\boldsymbol{\Lambda}_\omega$, $\boldsymbol{\Phi}$ and $\boldsymbol{\Psi}_\delta$ associated with Equation (6.2). It is natural to take prior distributions such that $p(\boldsymbol{\theta}) = p(\boldsymbol{\theta}_v)p(\boldsymbol{\theta}_\omega)$.

We first consider the conditional distributions corresponding to $\boldsymbol{\theta}_v$: $p(\boldsymbol{\mu}|\boldsymbol{\Lambda}, \boldsymbol{\Psi}_\epsilon, \boldsymbol{\Omega}, \mathbf{V})$, $p(\boldsymbol{\Lambda}|\boldsymbol{\mu}, \boldsymbol{\Psi}_\epsilon, \boldsymbol{\Omega}, \mathbf{V})$ and $p(\boldsymbol{\Psi}_\epsilon|\boldsymbol{\Lambda}, \boldsymbol{\mu}, \boldsymbol{\Omega}, \mathbf{V})$. In the same way as before, the following commonly used conjugate type prior distributions are used:

$$\boldsymbol{\mu} \stackrel{D}{=} N[\boldsymbol{\mu}_0, \boldsymbol{\Sigma}_0], \quad \psi_{\epsilon k}^{-1} \stackrel{D}{=} Gamma[\alpha_{0\epsilon k}, \beta_{0\epsilon k}],$$

$$[\boldsymbol{\Lambda}_k|\psi_{\epsilon k}] \stackrel{D}{=} N[\boldsymbol{\Lambda}_{0k}, \psi_{\epsilon k}\mathbf{H}_{0vk}], \quad k = 1, \ldots, p,$$

where $\psi_{\epsilon k}$ is the kth diagonal element of $\boldsymbol{\Psi}_\epsilon$, $\boldsymbol{\Lambda}_k^T$ is an $1 \times r_k$ row vector that only contains the unknown parameters in the kth row of $\boldsymbol{\Lambda}$; $\alpha_{0\epsilon k}$, $\beta_{0\epsilon k}$, $\boldsymbol{\mu}_0$, $\boldsymbol{\Lambda}_{0k}$, $\mathbf{H}_{0vk}$ and $\boldsymbol{\Sigma}_0$ are hyperparameters whose values are assumed to be given. For $k \neq h$, it is assumed that $(\psi_{\epsilon k}, \boldsymbol{\Lambda}_k)$ and $(\psi_{\epsilon h}, \boldsymbol{\Lambda}_h)$ are independent.

To cope with the case with fixed known elements in $\boldsymbol{\Lambda}$, let $\mathbf{C} = (c_{kj})$ be the index matrix such that $c_{kj} = 0$ if λ_{kj} is known and $c_{kj} = 1$ if λ_{kj} is unknown, and $r_k = \sum_{j=1}^q c_{kj}$. Let $\boldsymbol{\Omega}_k(r_k \times n)$ be a submatrix of $\boldsymbol{\Omega}$ such that all the jth rows with $c_{kj} = 0$ are deleted, and let $\mathbf{V}_k^{*T} = (\mathrm{v}_{1k}^*, \ldots, \mathrm{v}_{nk}^*)$ with

$$\mathrm{v}_{ik}^* = \mathrm{v}_{ik} - \mu_k - \sum_{j=1}^q \lambda_{kj}\omega_{ij}(1 - c_{kj}),$$

where v_{ik} is the kth element of $\mathbf{v}_i$, and μ_k is the kth element of $\boldsymbol{\mu}$. Let $\mathbf{A}_k = (\mathbf{H}_{0vk}^{-1} + \boldsymbol{\Omega}_k\boldsymbol{\Omega}_k^T)^{-1}$, $\mathbf{a}_k = \mathbf{A}_k(\mathbf{H}_{0vk}^{-1}\boldsymbol{\Lambda}_{0k} + \boldsymbol{\Omega}_k\mathbf{V}_k^*)$ and $\beta_{\epsilon k} = \beta_{0\epsilon k} + 2^{-1}(\mathbf{V}_k^{*T}\mathbf{V}_k^* - \mathbf{a}_k^T\mathbf{A}_k^{-1}\mathbf{a}_k + \boldsymbol{\Lambda}_{0k}^T\mathbf{H}_{0vk}^{-1}\boldsymbol{\Lambda}_{0k})$. Then it can be shown (by similar reasoning to that in Appendix 4.3) that for $k = 1, \ldots, p$,

$$p(\psi_{\epsilon k}^{-1}|\boldsymbol{\mu}, \mathbf{V}, \boldsymbol{\Omega}) \stackrel{D}{=} Gamma[n/2 + \alpha_{0\epsilon k}, \beta_{\epsilon k}], \quad p(\boldsymbol{\Lambda}_k|\psi_{\epsilon k}^{-1}, \boldsymbol{\mu}, \mathbf{V}, \boldsymbol{\Omega}) \stackrel{D}{=} N[\mathbf{a}_k, \psi_{\epsilon k}\mathbf{A}_k],$$

$$p(\boldsymbol{\mu}|\boldsymbol{\Lambda}, \boldsymbol{\Psi}_\epsilon, \mathbf{V}, \boldsymbol{\Omega}) \stackrel{D}{=} N[(\boldsymbol{\Sigma}_0^{-1} + n\boldsymbol{\Psi}_\epsilon^{-1})^{-1}(n\boldsymbol{\Psi}_\epsilon^{-1}\overline{\mathbf{V}} + \boldsymbol{\Sigma}_0^{-1}\boldsymbol{\mu}_0), \ (\boldsymbol{\Sigma}_0^{-1} + n\boldsymbol{\Psi}_\epsilon^{-1})^{-1}], \tag{6.7}$$

where $\overline{\mathbf{V}} = \sum_{i=1}^n(\mathbf{v}_i - \boldsymbol{\Lambda}\boldsymbol{\omega}_i)/n$.

Now, consider the conditional distribution of $\boldsymbol{\theta}_\omega$. As the parameters in $\boldsymbol{\theta}_\omega$ are just involved in the structural equation, this conditional distribution is proportional to $p(\boldsymbol{\Omega}|\boldsymbol{\theta}_\omega)p(\boldsymbol{\theta}_\omega)$, which is independent of $\mathbf{V}$ and $\mathbf{Z}$. Let $\boldsymbol{\Omega}_1 = (\boldsymbol{\eta}_1, \ldots, \boldsymbol{\eta}_n)$ and $\boldsymbol{\Omega}_2 = (\boldsymbol{\xi}_1, \ldots, \boldsymbol{\xi}_n)$. Since the distribution of $\boldsymbol{\xi}_i$ only involves $\boldsymbol{\Phi}$, $p(\boldsymbol{\Omega}_2|\boldsymbol{\theta}_\omega) = p(\boldsymbol{\Omega}_2|\boldsymbol{\Phi})$. Moreover, it is natural to take prior distribution of $\boldsymbol{\Phi}$ such that it is independent with the prior distributions of $\boldsymbol{\Lambda}_\omega$ and $\boldsymbol{\Psi}_\delta$. It follows that

$$p(\boldsymbol{\Omega}|\boldsymbol{\theta}_\omega)p(\boldsymbol{\theta}_\omega) \propto [p(\boldsymbol{\Omega}_1|\boldsymbol{\Omega}_2, \boldsymbol{\Lambda}_\omega, \boldsymbol{\Psi}_\delta)p(\boldsymbol{\Lambda}_\omega, \boldsymbol{\Psi}_\delta)][p(\boldsymbol{\Omega}_2|\boldsymbol{\Phi})p(\boldsymbol{\Phi})].$$

Hence, the marginal conditional densities of $(\boldsymbol{\Lambda}_\omega, \boldsymbol{\Psi}_\delta)$ and $\boldsymbol{\Phi}$ can again be treated separately.

Consider a conjugate type prior distribution for $\boldsymbol{\Phi}$ with $\boldsymbol{\Phi}^{-1} \stackrel{D}{=} W_{q_2}(\mathbf{R}_0, \rho_0)$, where $W_{q_2}(\cdot, \cdot)$ denotes the Wishart distribution with q_2 degrees of freedom, ρ_0 and the positive definite matrix $\mathbf{R}_0$ are the given hyperparameters. It can be shown (see Chapter 4, Section 4.3.1) that

$$p(\boldsymbol{\Phi}|\boldsymbol{\Omega}_{(2)}) \stackrel{D}{=} IW_{q_2}[(\boldsymbol{\Omega}_{(2)}\boldsymbol{\Omega}_{(2)}^T + \mathbf{R}_0^{-1}), n+\rho_0] \tag{6.8}$$

where $IW[\cdot, \cdot,]$ denotes the inverted Wishart distribution.

In a similar way to before, the prior distributions of elements in $(\boldsymbol{\Psi}_\delta, \boldsymbol{\Lambda}_\omega)$ are taken as

$$\psi_{\delta k}^{-1} \stackrel{D}{=} Gamma[\alpha_{0\delta k}, \beta_{0\delta k}], \quad [\boldsymbol{\Lambda}_{\omega k}|\psi_{\delta k}] \stackrel{D}{=} N[\boldsymbol{\Lambda}_{0\omega k}, \psi_{\delta k}\mathbf{H}_{0\omega k}],$$

where $k = 1, \ldots, q_1$, $\boldsymbol{\Lambda}_{\omega k}^T$ is an $1 \times r_{\omega k}$ row vector that contains the unknown parameters in the kth row of $\boldsymbol{\Lambda}_\omega$; $\alpha_{0\delta k}, \beta_{0\delta k}, \boldsymbol{\Lambda}_{0\omega k}$ and $\mathbf{H}_{0\omega k}$ are given hyperparameters. For $h \neq k$, $(\psi_{\delta k}, \boldsymbol{\Lambda}_{\omega k})$ and $(\psi_{\delta h}, \boldsymbol{\Lambda}_{\omega h})$ are assumed to be independent.

Let $\mathbf{C}_\omega = (c_{\omega kj})$ be the index matrix associated with $\boldsymbol{\Lambda}_\omega$, here $k = 1, \ldots, q_1$ and $j = 1, \ldots, q$. Let $\boldsymbol{\Omega}_k^*$ be the submatrix of $\boldsymbol{\Omega}$ such that all the jth rows corresponding to $c_{\omega kj} = 0$ are deleted; and $\boldsymbol{\Omega}_{\eta k}^{*T} = (\eta_{1k}^*, \ldots, \eta_{nk}^*)$ with

$$\eta_{ik}^* = \eta_{ik} - \sum_{j=1}^{q} \lambda_{\omega kj}\omega_{ij}(1 - c_{\omega kj}),$$

where ω_{ij} is the jth element of $\boldsymbol{\omega}_i$. Then, it can be shown that

$$p(\psi_{\delta k}^{-1}|\boldsymbol{\Omega}) \stackrel{D}{=} Gamma[n/2+\alpha_{0\delta k}, \beta_{\delta k}], \; p(\boldsymbol{\Lambda}_{\omega k}|\boldsymbol{\Omega}, \psi_{\delta k}^{-1}) \stackrel{D}{=} N[\mathbf{a}_{\omega k}, \psi_{\delta k}\mathbf{A}_{\omega k}] \tag{6.9}$$

where $\mathbf{A}_{\omega k} = (\mathbf{H}_{0\omega k}^{-1} + \mathbf{\Omega}_k^* \mathbf{\Omega}_k^{*T})^{-1}$, $\mathbf{a}_{\omega k} = \mathbf{A}_{\omega k}(\mathbf{H}_{0\omega k}^{-1}\mathbf{\Lambda}_{0\omega k} + \mathbf{\Omega}_k^* \mathbf{\Omega}_{\eta k}^*)$, and $\beta_{\delta k} = \beta_{0\delta k} + 2^{-1}(\mathbf{\Omega}_{\eta k}^{*T}\mathbf{\Omega}_{\eta k}^* - \mathbf{a}_{\omega k}^T \mathbf{A}_{\omega k}^{-1}\mathbf{a}_{\omega k} + \mathbf{\Lambda}_{0\omega k}^T \mathbf{H}_{0\omega k}^{-1}\mathbf{\Lambda}_{0\omega k})$.

Finally, we consider the joint conditional distribution of $(\boldsymbol{\alpha}, \mathbf{Y})$ given $\boldsymbol{\theta}, \mathbf{\Omega}, \mathbf{X}$ and $\mathbf{Z}$. Due to the ordinal nature of the thresholds, and for less complicated derivation of the conditional distributions, it is natural to use the following non-informative prior distribution for the unknown thresholds in $\boldsymbol{\alpha}_k$:

$$p(\alpha_{k,2}, \ldots, \alpha_{k,b_k-1}) \propto c, \quad \text{for } \alpha_{k,2} < \cdots < \alpha_{k,b_k-1}, \quad k = 1, \ldots, s,$$

where c is a constant. Given $\mathbf{\Omega}$, and the fact that the covariance matrix $\mathbf{\Psi}_\epsilon$ is diagonal, the ordered categorical data $\mathbf{Z}$ and the thresholds corresponding to different rows are also conditionally independent. For $k = 1, \ldots, s$, let $\mathbf{Y}_k$ and $\mathbf{Z}_k$ be the kth rows of $\mathbf{Y}$ and $\mathbf{Z}$, respectively, it follows from Equation (6.4) and Cowles (1996) that

$$p(\boldsymbol{\alpha}_k, \mathbf{Y}_k | \mathbf{Z}_k, \boldsymbol{\theta}, \mathbf{\Omega}) = p(\boldsymbol{\alpha}_k | \mathbf{Z}_k, \boldsymbol{\theta}, \mathbf{\Omega}) p(\mathbf{Y}_k | \boldsymbol{\alpha}_k, \mathbf{Z}_k, \boldsymbol{\theta}, \mathbf{\Omega}), \tag{6.10}$$

with

$$p(\boldsymbol{\alpha}_k | \mathbf{Z}_k, \boldsymbol{\theta}, \mathbf{\Omega}) \propto \prod_{i=1}^{n} \left\{ \Phi^* \left[\psi_{yk}^{-1/2} (\alpha_{k,z_{ik}+1} - \mu_{yk} - \mathbf{\Lambda}_{yk}^T \boldsymbol{\omega}_i) \right] - \Phi^* \left[\psi_{yk}^{-1/2} (\alpha_{k,z_{ik}} - \mu_{yk} - \mathbf{\Lambda}_{yk}^T \boldsymbol{\omega}_i) \right] \right\}, \tag{6.11}$$

and $p(\mathbf{Y}_k | \boldsymbol{\alpha}_k, \mathbf{Z}_k, \boldsymbol{\theta}, \mathbf{\Omega})$ is a product of $p(y_{ik} | \boldsymbol{\alpha}_k, \mathbf{Z}_k, \boldsymbol{\theta}, \mathbf{\Omega})$, where

$$p(y_{ik} | \boldsymbol{\alpha}_k, \mathbf{Z}_k, \boldsymbol{\theta}, \mathbf{\Omega}) \stackrel{D}{=} N(\mu_{yk} + \mathbf{\Lambda}_{yk}^T \boldsymbol{\omega}_i, \psi_{yk}) I_{(\alpha_{k,z_{ik}}, \alpha_{k,z_{ik}+1}]}(y_{ik}), \tag{6.12}$$

in which ψ_{yk} is the kth diagonal element of $\mathbf{\Psi}_{y\epsilon}$, μ_{yk} is the kth element of $\boldsymbol{\mu}_y$, $\mathbf{\Lambda}_{yk}^T$ is the kth row of $\mathbf{\Lambda}_y$, $I_A(y)$ is an indicator function which takes 1 if $y \in A$ and 0 otherwise, and $\Phi^*(\cdot)$ denotes the standard normal cumulative distribution function. As a result,

$$p(\boldsymbol{\alpha}_k, \mathbf{Y}_k | \mathbf{Z}_k, \boldsymbol{\theta}, \mathbf{\Omega}) \propto \prod_{i=1}^{n} \phi \left[\psi_{yk}^{-1/2} (y_{ik} - \mu_{yk} - \mathbf{\Lambda}_{yk}^T \boldsymbol{\omega}_i) \right] I_{(\alpha_{k,z_{ik}}, \alpha_{k,z_{ik}+1}]}(y_{ik}), \tag{6.13}$$

where $\phi(\cdot)$ is the standard normal density. It should be noted that the integral values of z_{ik} are only used in specifying the thresholds in deciding the interval $I_{(\alpha_{k,z_{ik}}, \alpha_{k,z_{ik}+1}]}(y_{ik})$ in Equation (6.13). As long as the integral values are ordered, the intervals can be precisely defined and the exact integral values are not important.

6.3.2 Implementation

The efficiency of the Gibbs sampler algorithm heavily depends on how easily one can sample observations from the conditional distributions. It can be seen that conditional distributions associated with Equations (6.5), (6.7)–(6.9) are the familiar normal, Gamma and inverted Wishart distributions. Drawing observations from these distributions is straightforward and fast. From Equations (6.6), (6.11), and (6.12), we see that it is not easy to sample from $p(\boldsymbol{\omega}_i|\mathbf{v}_i, \boldsymbol{\theta})$ and $p(\boldsymbol{\alpha}_k, \mathbf{Y}_k|\mathbf{Z}_k, \boldsymbol{\theta}, \boldsymbol{\Omega})$, which are nonstandard and complex. As the number of latent continuous measurements $\mathbf{Y}$ is large, simulating observations in Step (c) of the Gibbs sampler plays a more important role in the algorithm.

The Metropolis–Hastings (MH) algorithm (Metropolis *et al.*, 1953; Hastings, 1970) is a well-known MCMC method that has been widely used to simulate observations from a target density via the help of a proposal distribution from which it is easy to sample. Inspired by the work of Zhu and Lee (1998), this algorithm is applied here to generate observations from our target densities $p(\boldsymbol{\omega}_i|\mathbf{v}_i, \boldsymbol{\theta})$ in Equation (6.6) and $p(\boldsymbol{\alpha}_k, \mathbf{Y}_k|\mathbf{Z}_k, \boldsymbol{\theta}, \boldsymbol{\Omega})$ in Equations (6.11) and (6.12).

For $p(\boldsymbol{\omega}_i|\mathbf{v}_i, \boldsymbol{\theta})$, we choose $N[\mathbf{0}, \sigma^2\boldsymbol{\Sigma}^{*-1}]$ as the proposal distribution, where $\boldsymbol{\Sigma}^* = \boldsymbol{\Sigma}_\omega^{-1} + \boldsymbol{\Lambda}^T\boldsymbol{\Psi}_\epsilon^{-1}\boldsymbol{\Lambda}$, with

$$\boldsymbol{\Sigma}_\omega^{-1} = \begin{bmatrix} \boldsymbol{\Pi}_0^T\boldsymbol{\Psi}_\delta^{-1}\boldsymbol{\Pi}_0, & -\boldsymbol{\Pi}_0^T\boldsymbol{\Psi}_\delta^{-1}\boldsymbol{\Gamma} \\ -\boldsymbol{\Gamma}^T\boldsymbol{\Psi}_\delta^{-1}\boldsymbol{\Pi}_0, & \boldsymbol{\Phi}^{-1} + \boldsymbol{\Gamma}^T\boldsymbol{\Psi}_\delta^{-1}\boldsymbol{\Gamma} \end{bmatrix},$$

where $\boldsymbol{\Pi}_0 = \mathbf{I}_{q_1} - \boldsymbol{\Pi}$. Let $p(\cdot|\mathbf{0}, \sigma^2, \boldsymbol{\Sigma}^{*-1})$ be the proposal density corresponding to $N[\mathbf{0}, \sigma^2\boldsymbol{\Sigma}^{*-1}]$. The MH algorithm is implemented as follows. At the lth MH iteration with a current value $\boldsymbol{\omega}_i^{(l)}$, a new candidate $\boldsymbol{\omega}_i$ is generated from $p(\cdot|\boldsymbol{\omega}_i^{(l)}, \sigma^2, \boldsymbol{\Sigma}^{*-1})$, and accepting this new candidate with the probability

$$\min\left[1, \frac{p(\boldsymbol{\omega}_i|\mathbf{v}_i, \boldsymbol{\theta})}{p(\boldsymbol{\omega}_i^{(l)}|\mathbf{v}_i, \boldsymbol{\theta})}\right],$$

where $p(\boldsymbol{\omega}_i|\mathbf{v}_i, \boldsymbol{\theta})$ is given by Equation (6.6). The variance σ^2 can be chosen such that the average acceptance rate is approximately 0.25 or more, see Gelman, Carlin, Stern and Rubin (1995). Note that $p(\boldsymbol{\omega}_i|\mathbf{v}_i, \boldsymbol{\theta})$ could be simulated from Equation (6.5) without using the MH algorithm.

In constructing a suitable joint proposal density for $\boldsymbol{\alpha}_k$, and $\mathbf{Y}_k$ in the MH algorithm for generating observations from the target distribution $p(\boldsymbol{\alpha}_k, \mathbf{Y}_k|\mathbf{Z}_k, \boldsymbol{\theta}, \boldsymbol{\Omega})$ in Equation (6.13), we follow Cowles (1996) to consider the same factorization of the target distribution as in Equation (6.10):

$$p(\boldsymbol{\alpha}_k, \mathbf{Y}_k|\mathbf{Z}_k, \boldsymbol{\theta}, \boldsymbol{\Omega}) = p(\boldsymbol{\alpha}_k|\mathbf{z}_k, \boldsymbol{\theta}, \boldsymbol{\Omega})p(\mathbf{Y}_k|\boldsymbol{\alpha}_k, \mathbf{Z}_k, \boldsymbol{\theta}, \boldsymbol{\Omega}).$$

However, in the proposal distribution, $\boldsymbol{\alpha}_k$ is not simulated from Equation (6.11). Rather, at the lth MH iteration, we generate a vector of thresholds $(\alpha_{k,2}, \ldots, \alpha_{k,b_k-1})$ from the following univariate truncated normal distribution:

$$\alpha_{k,z} \sim N(\alpha_{k,z}^{(l)}, \sigma_{\alpha_k}^2) I_{(\alpha_{k,z-1}, \alpha_{k,z+1}^{(l)}]}(\alpha_{k,z}) \quad \text{for } z = 2, \ldots, b_k - 1, \tag{6.14}$$

where $\alpha_{k,z}^{(l)}$ is the current value of $\alpha_{k,z}$ at the lth iteration of the Gibbs sampler, and $\sigma_{\alpha_k}^2$ is an appropriate preassigned constant. Random observations from the above univariate truncated normal are simulated via the algorithm of Roberts (1995). It follows from the MH algorithm that the acceptance probability for $(\boldsymbol{\alpha}_k, \mathbf{Y}_k)$ as a new observation is $\min(1, R_k)$, where

$$R_k = \frac{p(\boldsymbol{\alpha}_k, \mathbf{Y}_k | \boldsymbol{\theta}, \boldsymbol{\Omega}, \mathbf{Z}_k) p(\boldsymbol{\alpha}_k^{(l)}, \mathbf{Y}_k^{(l)} | \boldsymbol{\alpha}_k, \mathbf{Y}_k, \boldsymbol{\theta}, \boldsymbol{\Omega}, \mathbf{Z}_k)}{p(\boldsymbol{\alpha}_k^{(l)}, \mathbf{Y}_k^{(l)} | \boldsymbol{\theta}, \boldsymbol{\Omega}, \mathbf{Z}_k) p(\boldsymbol{\alpha}_k, \mathbf{Y}_k | \boldsymbol{\alpha}_k^{(l)}, \mathbf{Y}_k^{(l)}, \boldsymbol{\theta}, \boldsymbol{\Omega}, \mathbf{Z}_k)}.$$

It can be shown from Equations (6.13) and (6.14) that

$$R_k = \prod_{z=2}^{b_k-1} \frac{\Phi^*\left[(\alpha_{k,z+1}^{(l)} - \alpha_{k,z}^{(l)})/\sigma_{\alpha_k}\right] - \Phi^*\left[(\alpha_{k,z-1} - \alpha_{k,z}^{(l)})/\sigma_{\alpha_k}\right]}{\Phi^*\left[(\alpha_{k,z+1} - \alpha_{k,z})/\sigma_{\alpha_k}\right] - \Phi^*\left[(\alpha_{k,z-1}^{(l)} - \alpha_{k,z})/\sigma_{\alpha_k}\right]} \times$$
$$\prod_{i=1}^{n} \frac{\Phi^*\left[\psi_{yk}^{-1/2}(\alpha_{k,z_{ik}+1} - \mu_{yk} - \boldsymbol{\Lambda}_{yk}^T \boldsymbol{\omega}_i)\right] - \Phi^*\left[\psi_{yk}^{-1/2}(\alpha_{k,z_{ik}} - \mu_{yk} - \boldsymbol{\Lambda}_{yk}^T \boldsymbol{\omega}_i)\right]}{\Phi^*\left[\psi_{yk}^{-1/2}(\alpha_{k,z_{ik}+1}^{(l)} - \mu_{yk} - \boldsymbol{\Lambda}_{yk}^T \boldsymbol{\omega}_i)\right] - \Phi^*\left[\psi_{yk}^{-1/2}(\alpha_{k,z_{ik}}^{(l)} - \mu_{yk} - \boldsymbol{\Lambda}_{yk}^T \boldsymbol{\omega}_i)\right]}. \tag{6.15}$$

As R_k only depends on the old and new values of $\boldsymbol{\alpha}_k$ and not on the $\mathbf{Y}_k$, so it is not necessary to generate a new $\mathbf{Y}_k$ in any iteration in which the new value of $\boldsymbol{\alpha}_k$ is not accepted (see Cowles, 1996). For an accepted $\boldsymbol{\alpha}_k$, a new $\mathbf{Y}_k$ is simulated from Equation (6.12).

6.3.3 Bayesian Estimates

It has been shown (Geman and Geman, 1984; Geyer, 1992) that under mild conditions and for sufficiently large m, the joint distribution of $(\boldsymbol{\alpha}^{(m)}, \boldsymbol{\theta}^{(m)}, \mathbf{Y}^{(m)}, \boldsymbol{\Omega}^{(m)})$ converges at an exponential rate to the desired posterior distribution $[\boldsymbol{\alpha}, \boldsymbol{\theta}, \boldsymbol{\Omega}, \mathbf{Y}|\mathbf{X}, \mathbf{Z}]$. Hence, $[\boldsymbol{\alpha}, \boldsymbol{\theta}, \boldsymbol{\Omega}, \mathbf{Y}|\mathbf{X}, \mathbf{Z}]$ can be approximated by the empirical distribution of a sufficiently large number of simulated observations collected after convergence of the algorithm. The convergence of the algorithm is monitored by the 'estimated potential scale reduction (EPSR)' values suggested by Gelman and Rubin (1992), or by plots of the generated parameters values from different starting points.

For brevity, let $\{(\boldsymbol{\alpha}^{(j)}, \boldsymbol{\theta}^{(j)}, \boldsymbol{\Omega}^{(j)}, \mathbf{Y}^{(j)}), j = 1, \ldots, J\}$ be the observations of $(\boldsymbol{\alpha}, \boldsymbol{\theta}, \boldsymbol{\Omega}, \mathbf{Y})$ generated from $[\boldsymbol{\alpha}, \boldsymbol{\theta}, \boldsymbol{\Omega}, \mathbf{Y}|\mathbf{X}, \mathbf{Z}]$ by the proposed hybrid algorithm. The Bayesian estimates of $\boldsymbol{\alpha}$ and $\boldsymbol{\theta}$, and a Bayesian estimate of $\boldsymbol{\Omega}$ can be obtained easily via the corresponding sample means of the generated observations as follows:

$$\hat{\boldsymbol{\alpha}} = J^{-1}\sum_{j=1}^{J}\boldsymbol{\alpha}^{(j)}, \quad \hat{\boldsymbol{\theta}} = J^{-1}\sum_{j=1}^{J}\boldsymbol{\theta}^{(j)}, \quad \hat{\boldsymbol{\Omega}} = J^{-1}\sum_{j=1}^{J}\boldsymbol{\Omega}^{(j)}. \tag{6.16}$$

Clearly, these Bayesian estimates are consistent estimates of the corresponding posterior means, see Geyer (1992). Estimates of the latent variables scores can be obtained directly from $\hat{\boldsymbol{\omega}}_i$ in $\hat{\boldsymbol{\Omega}}$. It is rather difficult to derive analytic forms for the covariance matrices $\text{Var}(\boldsymbol{\alpha}|\mathbf{X}, \mathbf{Z})$, $\text{Var}(\boldsymbol{\theta}|\mathbf{X}, \mathbf{Z})$ and $\text{Var}(\boldsymbol{\omega}_i|\mathbf{X}, \mathbf{Z})$. However, estimates of these covariance matrices can be obtained as the corresponding sample covariance matrices based on the simulated observations. For example, consistent estimates of $\text{Var}(\boldsymbol{\theta}|\mathbf{X}, \mathbf{Z})$ and $\text{Var}(\boldsymbol{\omega}_i|\mathbf{X}, \mathbf{Z})$ can be obtained as follows:

$$\widehat{\text{Var}(\boldsymbol{\theta}|\mathbf{X}, \mathbf{Z})} = (J-1)^{-1}\sum_{j=1}^{J}(\boldsymbol{\theta}^{(j)} - \hat{\boldsymbol{\theta}})(\boldsymbol{\theta}^{(j)} - \hat{\boldsymbol{\theta}})^T,$$

$$\widehat{\text{Var}(\boldsymbol{\omega}_i|\mathbf{X}, \mathbf{Z})} = (J-1)^{-1}\sum_{j=1}^{J}(\boldsymbol{\omega}_i^{(j)} - \hat{\boldsymbol{\omega}}_i)(\boldsymbol{\omega}_i^{(j)} - \hat{\boldsymbol{\omega}}_i)^T.$$

Finally, estimated residual can be obtained through the estimates of the latent variables and the latent continuous measurements of $\mathbf{y}$, and the parameter estimates. For example, it follows from Equation (6.4) that

$$\hat{\boldsymbol{\epsilon}}_{yi} = \hat{\mathbf{y}}_i - \hat{\boldsymbol{\mu}}_y - \hat{\boldsymbol{\Lambda}}_y\hat{\boldsymbol{\omega}}_i.$$

6.3.4 Goodness-of-fit of the Model

The goodness-of-fit of a posited model can be tested via the posterior predictive (PP) p-value (Gelman, Meng and Stern, 1996) as introduced in Chapter 5. Let H_0 be the null hypothesis that the proposed model defined in Equations (6.1) and (6.2) is plausible. Let $\mathbf{V} = (\mathbf{v}_1, \ldots, \mathbf{v}_n)$ and $\mathbf{V}^{\text{rep}}$ denotes a replication of $\mathbf{V}$, the PP p-value is defined by

$$p_B = Pr\,[D(\mathbf{V}^{\text{rep}}|\boldsymbol{\theta}, \boldsymbol{\Omega}, \mathbf{Y}) \geq D(\mathbf{V}|\boldsymbol{\theta}, \boldsymbol{\Omega}, \mathbf{Y})|\mathbf{X}, \mathbf{Z}, H_0],$$

where $D(\cdot|\cdot)$ is a discrepancy variable. The probability is taken over the joint posterior distribution of $(\mathbf{V}^{\text{rep}}, \boldsymbol{\theta}, \boldsymbol{\Omega}, \mathbf{Y})$ given $H_0, \mathbf{X}$ and $\mathbf{Z}$, where

$$p(\mathbf{V}^{\text{rep}}, \boldsymbol{\theta}, \boldsymbol{\Omega}, \mathbf{Y}|\mathbf{X}, \mathbf{Z}, H_0) = p(\mathbf{V}^{\text{rep}}|\boldsymbol{\theta}, \boldsymbol{\Omega}, \mathbf{Y})p(\boldsymbol{\theta}, \boldsymbol{\Omega}, \mathbf{Y}|\mathbf{X}, \mathbf{Z}).$$

For our model, we choose the following χ^2 discrepancy variable

$$D(\mathbf{V}^{\text{rep}}|\boldsymbol{\theta}, \boldsymbol{\Omega}, \mathbf{Y}) = \sum_{i=1}^{n}(\mathbf{V}_i^{\text{rep}} - \boldsymbol{\mu} - \boldsymbol{\Lambda}\boldsymbol{\omega}_i)^T \boldsymbol{\Psi}_\epsilon^{-1}(\mathbf{V}_i^{\text{rep}} - \boldsymbol{\mu} - \boldsymbol{\Lambda}\boldsymbol{\omega}_i),$$

which is distributed as $\chi^2(pn)$, a chi-square distribution with pn degrees of freedom. Here, implicitly, the partition $(\boldsymbol{\eta}_i^T, \boldsymbol{\xi}_i^T)^T$ of $\boldsymbol{\omega}_i$ is required to satisfy the model as defined in Equation (6.2). The PP p-value based on this discrepancy variable is given by

$$p_B(\mathbf{X}, \mathbf{Z}) = \int Pr\,[(\chi^2(pn) \geq D(\mathbf{V}|\boldsymbol{\Omega}, \boldsymbol{\theta}, \mathbf{Y})]p(\boldsymbol{\theta}, \boldsymbol{\Omega}, \mathbf{Y}|\mathbf{X}, \mathbf{Z})\, \mathrm{d}\boldsymbol{\theta}\, \mathrm{d}\boldsymbol{\Omega}\, \mathrm{d}\mathbf{Y}.$$

A Rao–Blackwellized type estimate of $p_B(\mathbf{X}, \mathbf{Z})$ is equal to

$$\hat{p}_B(\mathbf{X}, \mathbf{Z}) = J^{-1}\sum_{j=1}^{J} Pr\,[\chi^2(pn) \geq D(\mathbf{X}, \mathbf{Y}^{(j)}|\boldsymbol{\theta}^{(j)}, \boldsymbol{\Omega}^{(j)})],$$

where $\chi^2(pn)$ is a chi-square distribution with pn degrees of freedom. The computation of $\hat{p}_B(\mathbf{X}, \mathbf{Z})$ is straightforward, since $D(\mathbf{X}, \mathbf{Y}^{(j)}|\boldsymbol{\theta}^{(j)}, \mathbf{Z}^{(j)})$ can be calculated in each iteration and the tail-area probability of the χ^2 distribution can be obtained in any standard statistical software. H_0 is not rejected if $\hat{p}_B(\mathbf{X}, \mathbf{Z})$ is not far away from 0.5.

6.3.5 An Illustrative Example

A small portion of the ICPSR data set collected in the project World Values Survey 1981–1984 and 1990–1993 (World Value Study Group, 1994) is analyzed for illustration. The data obtained from Canada are used to illustrate our proposed method. Eight variables in the original data set (with corresponding question numbers 180, 96, 116, 117, 62, 179, 252 and 254, see Appendix 1.1) that are related with respondents' job, religious belief and home-life are taken as manifest variables in $(v_{(1)}, \ldots, v_{(8)})^T$. After deleting the cases with missing entries, the sample size is 470. Among them, $(v_{(1)}, v_{(2)})$ are related to life, $(v_{(3)}, v_{(4)})$ are related to job satisfaction, $(v_{(5)}, v_{(6)})$ are related to religious belief and $(v_{(7)}, v_{(8)})$ are related to job attitude. Measurements associated with $v_{(5)}$ and $v_{(6)}$ are based on a five-point scale, while all others are measured by a 10-point scale. For convenience, variables $v_{(5)}$ and $v_{(6)}$ are treated as ordered categorical and the others are treated as continuous.

As an illustration, consider the structural equation model with the following measurement equation:

$$\mathbf{v} = \boldsymbol{\mu} + \boldsymbol{\Lambda}\boldsymbol{\omega} + \boldsymbol{\epsilon}, \tag{6.17}$$

$$\boldsymbol{\Lambda}^T = \begin{bmatrix} 1 & \lambda_{21} & 0 & 0 & 0 & 0 & 0 & 0 \\ 0 & 0 & 1 & \lambda_{42} & 0 & 0 & 0 & 0 \\ 0 & 0 & 0 & 0 & 1 & \lambda_{63} & 0 & 0 \\ 0 & 0 & 0 & 0 & 0 & 0 & 1 & \lambda_{84} \end{bmatrix},$$

where 1's and 0's are fixed parameters, and $\boldsymbol{\omega} = (\eta_1, \eta_2, \xi_1, \xi_2)^T$. In this specification, the endogenous latent variables η_1 and η_2 in $\boldsymbol{\omega}$ can be interpreted as factors about life and job satisfaction, and the exogenous latent variables ξ_1 and ξ_2 can be interpreted as factors about religious belief and job attitude, respectively. The structural equation for the latent variables is given by

$$\begin{pmatrix} \eta_1 \\ \eta_2 \end{pmatrix} = \begin{pmatrix} 0 & 0 \\ \pi_{21} & 0 \end{pmatrix} \begin{pmatrix} \eta_1 \\ \eta_2 \end{pmatrix} + \begin{pmatrix} \gamma_{11} & 0 \\ 0 & \gamma_{22} \end{pmatrix} \begin{pmatrix} \xi_1 \\ \xi_2 \end{pmatrix} + \begin{pmatrix} \delta_1 \\ \delta_2 \end{pmatrix},$$

here the 0's are also fixed parameters. There are four thresholds for each ordered categorical variable that is measured with a five-point scale. After fixing the smallest and the largest thresholds for identification, each ordered categorical variable associates with two unknown thresholds. Hence there are a total of four unknown thresholds. Including the other structural parameters, there are a total of 32 unknown parameters in this model.

The MCMC method described in previous subsections is used to obtain the Bayesian estimates of the unknown parameters. Hyperparameter values in the conjugate prior distributions are set equal to the estimates that are obtained from an initial run with non-informative priors. The Gibbs sampler in the actual estimation converged fairly quickly within 500 iterations. To give some idea about the convergence of the MCMC method, plots of sequences of some parameters' values with different starting values are displayed in Figure 6.3. The PP p-value is equal to 0.563, indicating that the proposed model fits the sample data. The Bayesian estimates and their associated standard errors estimates are reported in Table 6.1. As we fix the smallest and the largest thresholds of the ordered categorical variables associated with $v_{(5)}$ and $v_{(6)}$ according to the commutative distribution of $N[0, 1]$, the estimates of μ_5 and μ_6 are close to zero and the estimates of $\psi_{\epsilon5}$ and $\psi_{\epsilon6}$ are less than the other estimates of elements in $\boldsymbol{\Psi}_{\epsilon}$.

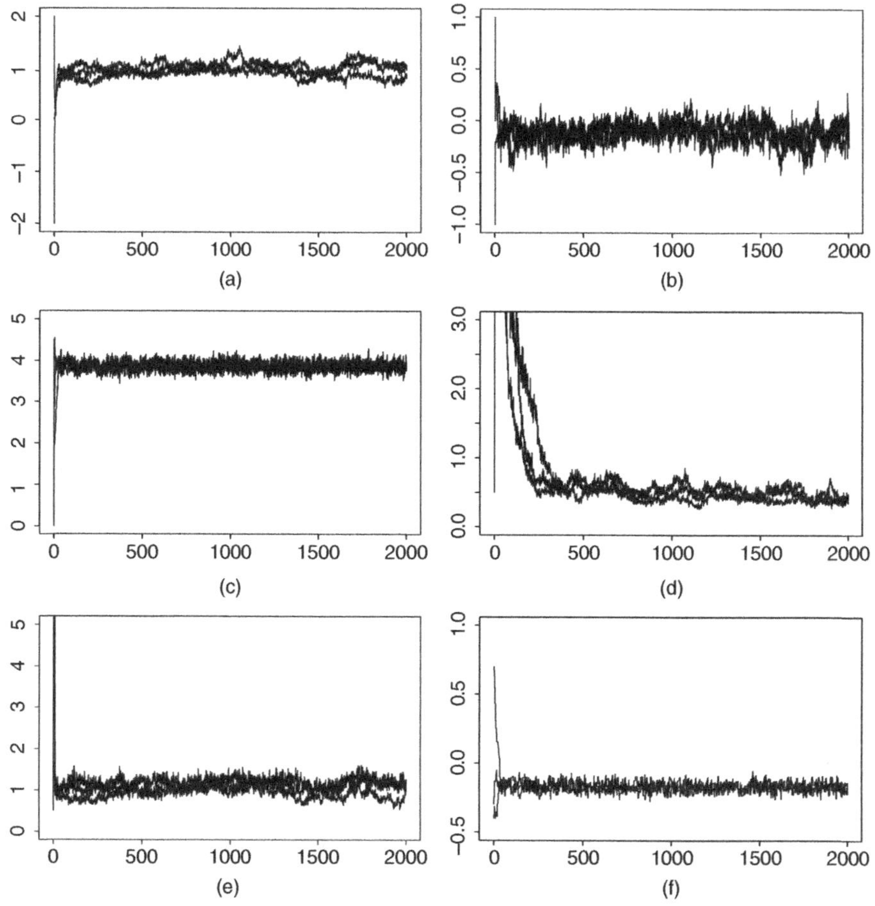

Figure 6.3 (a), (b), (c), (d), (e) and (f) are plots of parallel sequences corresponding to different starting values of λ_{21}, π_{21}, μ_3, ϕ_{11}, $\psi_{\epsilon 1}$ and α_{12} against iterations, in the ICPSR example.

6.4 BAYESIAN MODEL COMPARISON

The Bayes factor introduced in Chapter 5 is applied to compare different competing models that are defined in the context of Equations (6.1) and (6.2). Let M_0 and M_1 be two competing models of interest. Given the observed data $(\mathbf{X}, \mathbf{Z})$, the Bayes factor for comparing M_1 with M_0 is

$$B_{10} = \frac{p(\mathbf{X}, \mathbf{Z}|M_1)}{p(\mathbf{X}, \mathbf{Z}|M_0)}.$$

Table 6.1 Bayesian estimates and standard error estimates.

Parameters	EST	SE	Parameters	EST	SE
μ_1	8.414	0.079	π_{21}	−0.112	0.094
μ_2	7.999	0.076	γ_{11}	0.136	0.132
μ_3	3.871	0.113	γ_{22}	−0.539	0.141
μ_4	3.204	0.096	ϕ_{11}	0.453	0.077
μ_5	0.007	0.053	ϕ_{12}	0.086	0.052
μ_6	−0.014	0.055	ϕ_{22}	0.890	0.147
μ_7	8.049	0.071	$\psi_{\delta 1}$	1.309	0.171
μ_8	7.446	0.091	$\psi_{\delta 2}$	1.925	0.511
λ_{21}	1.006	0.105	$\psi_{\epsilon 1}$	1.112	0.144
λ_{42}	1.015	0.269	$\psi_{\epsilon 2}$	0.853	0.149
λ_{63}	−1.004	0.137	$\psi_{\epsilon 3}$	4.087	0.668
λ_{84}	1.584	0.230	$\psi_{\epsilon 4}$	2.714	0.665
α_{12}	−0.166	0.032	$\psi_{\epsilon 5}$	0.547	0.072
α_{13}	0.332	0.041	$\psi_{\epsilon 6}$	0.558	0.079
α_{22}	0.479	0.042	$\psi_{\epsilon 7}$	1.669	0.174
α_{23}	0.733	0.035	$\psi_{\epsilon 8}$	2.289	0.351

A path sampling procedure is used here for computing logarithm B_{10}. Similar to the Bayesian estimation, the observed data $(\mathbf{X}, \mathbf{Z})$ are augmented with the latent data $(\mathbf{Y}, \boldsymbol{\Omega})$ in the computation. Suppose M_t is a linked model that links M_1 and M_0. Let $p(\mathbf{X}, \mathbf{Z}, \mathbf{Y}, \boldsymbol{\Omega}|\boldsymbol{\theta}, \boldsymbol{\alpha}, t)$ be the complete-data likelihood function under the linked model M_t, and

$$U(\mathbf{X}, \mathbf{Z}, \boldsymbol{\alpha}, \boldsymbol{\theta}, \mathbf{Y}, \boldsymbol{\Omega}, t) = \frac{\mathrm{d}}{\mathrm{d}t} \log [p(\mathbf{X}, \mathbf{Z}, \mathbf{Y}, \boldsymbol{\Omega}|\boldsymbol{\theta}, \boldsymbol{\alpha}, t)]. \tag{6.18}$$

Moreover, let $\{t_{(s)}, s = 0, \ldots, S+1\}$ be grids in [0,1] such that $t_{(0)} = 0 < t_{(1)} < \cdots < t_{(S)} = t_{(S+1)} = 1$. The logarithm Bayes factor is computed as

$$\log B_{10} = \frac{1}{2} \sum_{s=0}^{S} (t_{(s+1)} - t_{(s)})(\overline{U}_{(s+1)} + \overline{U}_{(s)}), \tag{6.19}$$

where

$$\overline{U}_{(s)} = J^{-1} \sum_{j=1}^{J} U(\mathbf{X}, \mathbf{Z}, \boldsymbol{\alpha}^{(j)}, \boldsymbol{\theta}^{(j)}, \mathbf{Y}^{(j)}, \boldsymbol{\Omega}^{(j)}, t_{(s)}), \tag{6.20}$$

in which $\{(\boldsymbol{\alpha}^{(j)}, \boldsymbol{\theta}^{(j)}, \mathbf{Y}^{(j)}, \boldsymbol{\Omega}^{(j)}); j = 1, \ldots, J\}$ are simulated observations from the posterior distribution $p(\boldsymbol{\alpha}, \boldsymbol{\theta}, \mathbf{Y}, \boldsymbol{\Omega}|\mathbf{X}, \mathbf{Z}, t_{(s)})$. These observations are drawn via the MCMC methods in the estimation described in Section 6.3.

The prior distributions of the thresholds and the structural parameters are only involved in the posterior simulation in generating the $(\boldsymbol{\alpha}^{(j)}, \boldsymbol{\theta}^{(j)}, \mathbf{Y}^{(j)}, \boldsymbol{\Omega}^{(j)})$, but not in the direct computation of $\log B_{10}$. As the thresholds are nuisance parameters that are not involved in the competing models, the choice of non-informative priors is acceptable (see discussions at the end of Chapter 5). Informative prior distributions should be used for the structural parameters that are involved in the competing models. The conjugate prior families used in the estimation are convenient and good choices.

The choice of the linked model M_t depends on the competing models M_0 and M_1. We have to approach this on a problem-by-problem basis. Usually, defining a suitable M_t is straightforward. To give a more specific illustration, consider the following models with different measurement and structural equations:

$$\begin{aligned} M_0: &\quad \mathbf{v} = \boldsymbol{\mu} + \boldsymbol{\Lambda}_0 \boldsymbol{\omega} + \boldsymbol{\epsilon}, \quad \text{and} \quad \boldsymbol{\eta} = \boldsymbol{\Pi}_0 \boldsymbol{\eta} + \boldsymbol{\Gamma}_0 \boldsymbol{\xi} + \boldsymbol{\delta}; \\ M_1: &\quad \mathbf{v} = \boldsymbol{\mu} + \boldsymbol{\Lambda}_1 \boldsymbol{\omega} + \boldsymbol{\epsilon}, \quad \text{and} \quad \boldsymbol{\eta} = \boldsymbol{\Pi}_1 \boldsymbol{\eta} + \boldsymbol{\Gamma}_1 \boldsymbol{\xi} + \boldsymbol{\delta}. \end{aligned} \tag{6.21}$$

These two models can be linked up by the following M_t with $t \in [0, 1]$:

$$\begin{aligned} M_t: &\quad \mathbf{v} = \boldsymbol{\mu} + [(1-t)\boldsymbol{\Lambda}_0 + t\boldsymbol{\Lambda}_1]\boldsymbol{\omega} + \boldsymbol{\epsilon}, \quad \text{and} \\ &\quad \boldsymbol{\eta} = [(1-t)\boldsymbol{\Pi}_0 + t\boldsymbol{\Pi}_1]\boldsymbol{\eta} + [(1-t)\boldsymbol{\Gamma}_0 + t\boldsymbol{\Gamma}_1]\boldsymbol{\xi} + \boldsymbol{\delta}. \end{aligned} \tag{6.22}$$

Clearly, when $t = 0$, $M_t = M_0$; when $t = 1$, $M_t = M_1$.

Based on the model defined in Equations (6.1) and (6.2), and the nature of the ordered categorical data, the complete-data likelihood function is equal to

$$\begin{aligned} &p(\mathbf{X}, \mathbf{Z}, \mathbf{Y}, \boldsymbol{\Omega}|\boldsymbol{\theta}, \boldsymbol{\alpha}, t) = p(\mathbf{Z}|\mathbf{X}, \mathbf{Y}, \boldsymbol{\Omega}, \boldsymbol{\theta}, \boldsymbol{\alpha}, t) p(\mathbf{X}, \mathbf{Y}|\boldsymbol{\Omega}, \boldsymbol{\theta}, \boldsymbol{\alpha}, t) p(\boldsymbol{\Omega}|\boldsymbol{\theta}, t) \\ &= (2\pi)^{-np/2} |\boldsymbol{\Psi}_\epsilon|^{-n/2} \exp\left[-\frac{1}{2} \sum_{i=1}^n (\mathbf{v}_i - \boldsymbol{\mu} - \boldsymbol{\Lambda}_t \boldsymbol{\omega}_i)^T \boldsymbol{\Psi}_\epsilon^{-1} (\mathbf{v}_i - \boldsymbol{\mu} - \boldsymbol{\Lambda}_t \boldsymbol{\omega}_i) \right] I_{Ai}(\mathbf{y}_i) \\ &\quad \times (2\pi)^{-nq_2/2} |\boldsymbol{\Phi}|^{-n/2} \exp\left(-\frac{1}{2} \sum_{i=1}^n \boldsymbol{\xi}_i^T \boldsymbol{\Phi}^{-1} \boldsymbol{\xi}_i \right) \\ &\quad \times (2\pi)^{-nq_1/2} |(\mathbf{I}_{q_1} - \boldsymbol{\Pi}_t)|^n |\boldsymbol{\Psi}_\delta|^{-n/2} \\ &\quad \times \exp\left[-\frac{1}{2} \sum_{i=1}^n (\boldsymbol{\eta}_i - \boldsymbol{\Pi}_t \boldsymbol{\eta}_i - \boldsymbol{\Gamma}_t \boldsymbol{\xi}_i)^T \boldsymbol{\Psi}_\delta^{-1} (\boldsymbol{\eta}_i - \boldsymbol{\Pi}_t \boldsymbol{\eta}_i - \boldsymbol{\Gamma}_t \boldsymbol{\xi}_i) \right], \end{aligned}$$

where $\boldsymbol{\Lambda}_t = (1-t)\boldsymbol{\Lambda}_0 + t\boldsymbol{\Lambda}_1$, $\boldsymbol{\Pi}_t = (1-t)\boldsymbol{\Pi}_0 + t\boldsymbol{\Pi}_1$, $\boldsymbol{\Gamma}_t = (1-t)\boldsymbol{\Gamma}_0 + t\boldsymbol{\Gamma}_1$, $\mathrm{I}_A(\mathbf{y})$ is an indicator function which takes the value 1 if $\mathbf{y} \in A$ and zero otherwise, and

$$A_i = (\alpha_{1,z_{i1}}, \alpha_{1,z_{i1}+1}] \times \cdots \times (\alpha_{s,z_{is}}, \alpha_{s,z_{is}+1}].$$

Note that for every $\mathbf{v}_i$, there exists one and only one A_i such that $\mathbf{y}_i$ is in A_i and $I_{Ai}(\mathbf{y}_i) = 1$, hence the corresponding value of the density function is nonzero. The logarithm of complete-data likelihood corresponding to M_t is equal to

$$\log[p(\mathbf{X}, \mathbf{Z}, \mathbf{Y}, \boldsymbol{\Omega}|\boldsymbol{\theta}, \boldsymbol{\alpha}, t)] = \log[p(\mathbf{X}, \mathbf{Z}, \mathbf{Y}|\boldsymbol{\Omega}, \boldsymbol{\theta}, \boldsymbol{\alpha}, t)] + \log[p(\boldsymbol{\Omega}|\boldsymbol{\theta}, t)]$$

$$= -\frac{1}{2}\Big[(p+q)n\log(2\pi) + n\log|\boldsymbol{\Psi}_\epsilon| + n\log|\boldsymbol{\Psi}_\delta| - 2n\log|\mathbf{I}_{q1} - \boldsymbol{\Pi}_t| + n\log|\boldsymbol{\Phi}|$$

$$+ \sum_{i=1}^{n}(\mathbf{v}_i - \boldsymbol{\mu} - \boldsymbol{\Lambda}_t\boldsymbol{\omega}_i)^T\boldsymbol{\Psi}_\epsilon^{-1}(\mathbf{v}_i - \boldsymbol{\mu} - \boldsymbol{\Lambda}_t\boldsymbol{\omega}_i)$$

$$+ \sum_{i=1}^{n}(\boldsymbol{\eta}_i - \boldsymbol{\Pi}_t\boldsymbol{\eta}_i - \boldsymbol{\Gamma}_t\boldsymbol{\xi}_i)^T\boldsymbol{\Psi}_\delta^{-1}(\boldsymbol{\eta}_i - \boldsymbol{\Pi}_t\boldsymbol{\eta}_i - \boldsymbol{\Gamma}_t\boldsymbol{\xi}_i)$$

$$+ \sum_{i=1}^{n}\boldsymbol{\xi}_i^T\boldsymbol{\Phi}^{-1}\boldsymbol{\xi}_i\Big].$$

Note that as $I_{Ai}(\mathbf{y}_i) = 1$, it does not appear in the log-likelihood. By differentiation with respect to t, we have

$$U(\mathbf{X}, \mathbf{Z}, \boldsymbol{\alpha}, \boldsymbol{\theta}, \mathbf{Y}, \boldsymbol{\Omega}, t) = \sum_{i=1}^{n}\Big[(\mathbf{v}_i - \boldsymbol{\mu} - \boldsymbol{\Lambda}_t\boldsymbol{\omega}_i)^T\boldsymbol{\Psi}_\epsilon^{-1}\boldsymbol{\Lambda}_{t0}\boldsymbol{\omega}_i$$

$$+ (\boldsymbol{\eta}_i - \boldsymbol{\Pi}_t\boldsymbol{\eta}_i - \boldsymbol{\Gamma}_t\boldsymbol{\xi}_i)^T\boldsymbol{\Psi}_\delta^{-1}(\boldsymbol{\Pi}_{t0}\boldsymbol{\eta}_i + \boldsymbol{\Gamma}_{t0}\boldsymbol{\xi}_i)\Big],$$

where $\boldsymbol{\Lambda}_{t0} = \boldsymbol{\Lambda}_1 - \boldsymbol{\Lambda}_0$, $\boldsymbol{\Pi}_{t0} = \boldsymbol{\Pi}_1 - \boldsymbol{\Pi}_0$ and $\boldsymbol{\Gamma}_{t0} = \boldsymbol{\Gamma}_1 - \boldsymbol{\Gamma}_0$. Observations from the posterior distribution $p(\boldsymbol{\alpha}, \boldsymbol{\theta}, \mathbf{Y}, \boldsymbol{\Omega}|\mathbf{X}, \mathbf{Z}, t)$ are then simulated by the MCMC method as described in Section 6.3. The Bayes factor is estimated by using Equations (6.19) and (6.20).

We use the real example given in Section 6.3.5 to illustrate path sampling in computing the Bayes factor for model comparison. Suppose we are interested in the following non-nested models whose measurement equation is the same as given in Equation (6.17) but with the following different structural equations:

$$M_0: \quad \begin{pmatrix}\eta_1\\ \eta_2\end{pmatrix} = \begin{pmatrix}\gamma_{11} & \gamma_{12}\\ \gamma_{21} & \gamma_{22}\end{pmatrix}\begin{pmatrix}\xi_1\\ \xi_2\end{pmatrix} + \begin{pmatrix}\delta_1\\ \delta_2\end{pmatrix},$$

$$M_1: \quad \begin{pmatrix}\eta_1\\ \eta_2\end{pmatrix} = \begin{pmatrix}0 & 0\\ \pi_{21} & 0\end{pmatrix}\begin{pmatrix}\eta_1\\ \eta_2\end{pmatrix} + \begin{pmatrix}\gamma_{11} & 0\\ 0 & \gamma_{22}\end{pmatrix}\begin{pmatrix}\xi_1\\ \xi_2\end{pmatrix} + \begin{pmatrix}\delta_1\\ \delta_2\end{pmatrix},$$

$$M_2: \quad \begin{pmatrix}\eta_1\\ \eta_2\end{pmatrix} = \begin{pmatrix}0 & 0\\ \pi_{21} & 0\end{pmatrix}\begin{pmatrix}\eta_1\\ \eta_2\end{pmatrix} + \begin{pmatrix}0 & \gamma_{12}\\ \gamma_{21} & 0\end{pmatrix}\begin{pmatrix}\xi_1\\ \xi_2\end{pmatrix} + \begin{pmatrix}\delta_1\\ \delta_2\end{pmatrix}.$$

These competing models are special cases of the models given in Equation (6.21), hence the linked models can be easily obtained as special cases of

M_t given in Equation (6.22). In the application of path sampling, the number of grids in [0, 1] is taken to be 20. Again we use the same prior inputs as before and take 500 burn-in iterations in the MCMC method in simulating the observations. After convergence, 2000 observations are collected for each $t_{(s)}$ for computing $\overline{U}_{(s)}$. We find that $\widehat{\log B_{01}} = -0.206$ and $\widehat{\log B_{12}} = 6.588$. According to the criterion given in Table 5.1, these results suggest that M_1 is better than M_0 and is significantly better than M_2. Hence, M_1 is selected. Bayesian estimates of the parameters in M_1 can be obtained in the computation of $\widehat{\log B_{12}}$. The data have been reanalyzed using different prior inputs. We obtain close values of the logarithm of Bayes factors and the same conclusion as above.

6.5 APPLICATION 1: BAYESIAN SELECTION OF THE NUMBER OF FACTORS IN EFA

Factor analysis (Lawley and Maxwell, 1971) is the most basic structural equation model and it is still very useful in behavioral, social and psychological research. One fundamental problem associated with factor analysis is to select the appropriate number of factors for an observed data set. For continuous data, this problem has been addressed by Akaike (1987) via the Akaike's information criterion (AIC) that is analogous to the final prediction error (see Akaike, 1970). Another related criterion is the BIC which is a rough approximation of the Bayes factor. For factor analysis models with ordered categorical variables, AIC and BIC depend on the observed-data likelihood function that involves rather complicated multiple integrals. The computation of these criteria is not straightforward. The main purpose of this section is to illustrate the application of path sampling for computing the Bayes factor in solving this fundamental problem.

For completeness, we present the following exploratory factor analysis model with a $p \times 1$ manifest random vector $\mathbf{v}_i$:

$$\mathbf{v}_i = \boldsymbol{\Lambda}\boldsymbol{\omega}_i + \boldsymbol{\epsilon}_i, \quad i = 1, \ldots, n \tag{6.23}$$

where $\boldsymbol{\Lambda}(p \times q)$ is the factor loading matrix, $\boldsymbol{\omega}_i(q \times 1)$ is a latent random vector with distribution $N[\mathbf{0}, \mathbf{I}]$, and $\boldsymbol{\epsilon}_i(p \times 1)$ is a random vector of error measurements with distribution $N[\mathbf{0}, \boldsymbol{\Psi}_\epsilon]$, where $\boldsymbol{\Psi}_\epsilon$ is diagonal and $\boldsymbol{\epsilon}_i$ is independent of $\boldsymbol{\omega}_i$. Let $\mathbf{v} = \{\mathbf{x}, \mathbf{y}\}$, where $\mathbf{x}$ is an $r \times 1$ vector of variables whose exact continuous measurements are observable, whereas $\mathbf{y}$ is an $s \times 1$ vector of the remaining variables such that the corresponding continuous measurements are unobservable, and their information is given by an observable ordered categorical vector $\mathbf{z}$ as described in Section 6.2. Let $\mathbf{X} = (\mathbf{x}_1, \ldots, \mathbf{x}_n)$ and $\mathbf{Z} = (\mathbf{z}_1, \ldots, \mathbf{z}_n)$ be the observed data matrices. Let $\boldsymbol{\Omega} = (\boldsymbol{\omega}_1, \ldots, \boldsymbol{\omega}_n)$ be the $q \times n$ matrix of latent factors and let $\mathbf{Y} = (\mathbf{y}_1, \ldots, \mathbf{y}_n)$ be the latent continuous measurements underlying $\mathbf{Z}$. Let $\boldsymbol{\alpha}$ be the vector that contains all the unknown thresholds and $\boldsymbol{\theta}$ be the vector that contains all the unknown parameters in $\boldsymbol{\Lambda}$ and $\boldsymbol{\Psi}_\epsilon$.

Consider the application of the path sampling to select one of the following two competing models M_0 and M_1:

$$M_0: \quad \mathbf{v} = \boldsymbol{\Lambda}_{10}\omega_1 + \cdots + \boldsymbol{\Lambda}_{q0}\omega_q + \boldsymbol{\epsilon};$$

$$M_1: \quad \mathbf{v} = \boldsymbol{\Lambda}_{11}\omega_1 + \cdots + \boldsymbol{\Lambda}_{r1}\omega_r + \boldsymbol{\epsilon};$$

where $q < r$, and $\boldsymbol{\Lambda}_{h0}$ and $\boldsymbol{\Lambda}_{h1}$ are the columns of the loading matrices under M_0 and M_1 respectively. Clearly, M_0 corresponds to a model with q factors whilst M_1 corresponds to a model with r factors. These two competing models are linked up by t in $[0, 1]$ as below:

$$M_t: \quad \mathbf{v} = [(1-t)\boldsymbol{\Lambda}_{10} + t\boldsymbol{\Lambda}_{11}]\omega_1 + \cdots + [(1-t)\boldsymbol{\Lambda}_{q0} + t\boldsymbol{\Lambda}_{q1}]\omega_q + t\boldsymbol{\Lambda}_{(q+1)1}\omega_{q+1}$$
$$+ \cdots + t\boldsymbol{\Lambda}_{r1}\omega_r + \boldsymbol{\epsilon}.$$

Clearly M_t reduces to M_0 if $t = 0$, and it reduces to M_1 if $t = 1$. Let $\boldsymbol{\Lambda}_t = [(1-t)\boldsymbol{\Lambda}_{10} + t\boldsymbol{\Lambda}_{11}, \ldots, (1-t)\boldsymbol{\Lambda}_{q0} + t\boldsymbol{\Lambda}_{q1}, t\boldsymbol{\Lambda}_{(q+1)1}, \ldots, t\boldsymbol{\Lambda}_{r1}]$, the corresponding logarithm complete-data likelihood function is given by

$$\log p(\mathbf{Y}, \boldsymbol{\Omega}, \mathbf{X}, \mathbf{Z}|\boldsymbol{\alpha}, \boldsymbol{\theta}, t) = -\frac{1}{2}[pn\log(2\pi) + n\log|\boldsymbol{\Psi}_\epsilon| +$$
$$\sum_{i=1}^n \boldsymbol{\omega}_i^T\boldsymbol{\omega}_i + \sum_{i=1}^n (\mathbf{v}_i - \boldsymbol{\Lambda}_t\boldsymbol{\omega}_i)^T\boldsymbol{\Psi}_\epsilon^{-1}(\mathbf{v}_i - \boldsymbol{\Lambda}_t\boldsymbol{\omega}_i)],$$

where $\mathbf{v}_i = \{\mathbf{x}_i, \mathbf{y}_i\}$ and $\mathbf{y}_i$ is in one and only one $A_i = (\alpha_{1,z_{i1}}, \alpha_{1,z_{i1}+1}] \times \cdots \times (\alpha_{s,z_{is}}, \alpha_{s,z_{is}+1}]$. By differentiation with respect to t, we have

$$U(\boldsymbol{\theta}, \boldsymbol{\alpha}, \mathbf{Y}, \boldsymbol{\Omega}, \mathbf{X}, \mathbf{Z}, t) = \sum_{i=1}^n (\mathbf{v}_i - \boldsymbol{\Lambda}_t\boldsymbol{\omega}_i)^T\boldsymbol{\Psi}_\epsilon^{-1}\boldsymbol{\Lambda}_{t0}\boldsymbol{\omega}_i, \tag{6.24}$$

where $\boldsymbol{\Lambda}_{t0} = (\boldsymbol{\Lambda}_{11} - \boldsymbol{\Lambda}_{10}, \ldots, \boldsymbol{\Lambda}_{q1} - \boldsymbol{\Lambda}_{q0}, \boldsymbol{\Lambda}_{(q+1)1}, \ldots, \boldsymbol{\Lambda}_{r1})$. The logarithm of Bayes factor for selecting M_0 or M_1 can be estimated by using expressions similar to those in Equations (6.19) and (6.20). The main computation is on simulating a sample $\{(\boldsymbol{\theta}^{(j)}, \mathbf{Y}^{(j)}, \boldsymbol{\Omega}^{(j)}), j = 1, \ldots, J\}$ from the posterior distribution $p(\boldsymbol{\alpha}, \boldsymbol{\theta}, \mathbf{Y}, \boldsymbol{\Omega}|\mathbf{X}, \mathbf{Z}, t)$. The task can be done by the Gibbs sampler which iteratively generates observations from conditional distributions $p(\boldsymbol{\theta}|\boldsymbol{\alpha}, \mathbf{Y}, \boldsymbol{\Omega}, \mathbf{X}, \mathbf{Z})$, $p(\boldsymbol{\alpha}, \mathbf{Y}|\boldsymbol{\theta}, \boldsymbol{\Omega}, \mathbf{X}, \mathbf{Z})$ and $p(\boldsymbol{\Omega}|\boldsymbol{\theta}, \boldsymbol{\alpha}, \mathbf{Y}, \mathbf{X}, \mathbf{Z})$ as before. In the derivation of these conditional distributions, the following conjugate prior distributions for the factor loadings and unique variance are used: for $k = 1, \ldots, p$

$$\psi_{\epsilon k}^{-1} \overset{D}{=} Gamma(\alpha_{0\epsilon k}, \beta_{0\epsilon k}); \quad [\boldsymbol{\Lambda}_k|\psi_{\epsilon k}] \overset{D}{=} N[\boldsymbol{\Lambda}_{0k}, \psi_{\epsilon k}\mathbf{H}_{0vk}],$$

where $\boldsymbol{\Lambda}_k^T$ is the kth row of $\boldsymbol{\Lambda}$, $\psi_{\epsilon k}$ is the kth diagonal element of $\boldsymbol{\Psi}_\epsilon$, and $\boldsymbol{\Lambda}_{0k}, \mathbf{H}_{0vk}, \alpha_{0\epsilon k}$ and $\beta_{0\epsilon k}$ are given hyperparameters. Non-informative prior is used for the nuisance threshold parameters.

6.5.1 A Simulation Study

Results obtained from a simulation study are presented here to illustrate the path sampling approach for computing the Bayes factor in selecting the number of factors in the model. The true model is a two-factor model with the following population values:

$$\boldsymbol{\Lambda}^T = \begin{bmatrix} 0.8 & 0.0 & 0.8 & 0.8 & 0.0 & 0.0 & 0.8 & 0.6 \\ 0.0 & 0.8 & 0.0 & 0.0 & 0.8 & 0.8 & 0.0 & 0.6 \end{bmatrix},$$
$$\boldsymbol{\Psi}_\epsilon = \text{diag}\{0.36, 0.36, 0.36, 0.36, 0.36, 0.36, 0.36, 0.28\}.$$

A data set $\{\mathbf{v}_i, i = 1, \ldots, n\}$ is generated according to Equation (6.23). The continuous measurements $\mathbf{v}_{(7)}$ and $\mathbf{v}_{(8)}$ are transformed to ordered categorical observations via the following thresholds: $\boldsymbol{\alpha}_1 = \boldsymbol{\alpha}_2 = (-1.0, -0.2, 0.2, 1.0)$. Here, the first six variables are continuous and the last two are ordered categorical.

To identify the ordered categorical variables, the first and the last elements of $\boldsymbol{\alpha}_1$ and $\boldsymbol{\alpha}_2$ are fixed at -1.0 and 1.0, respectively. To eliminate the indeterminacy of rotation in a model with q factors, we fix $\lambda_{ab} = 0.0$ with $a, b = 1, \ldots, q$ and $a < b$. This condition is sufficient to identify the covariance structure of the model but not enough to identify $\boldsymbol{\Lambda}$ completely because each column of $\boldsymbol{\Lambda}$ can have a change of sign. Thus, we require to fix more elements in $\boldsymbol{\Lambda}$ in order to remove this indeterminacy. We select these elements by the information obtained from an initial estimation on the basis of non-informative prior distributions and λ_{ab} with fixed at 0.0 for $a, b = 1, \ldots, q$ and $a < b$. Then, for each column, we pick the entry associated with the largest estimate and fix it at some value. Using this method in the simulation study, we fix $\lambda_{12} = 0.0$, $\lambda_{31} = 0.8$ and $\lambda_{52} = 0.8$ in assessing the two-factor model; and additionally fix $\lambda_{13} = 0.0$, $\lambda_{23} = 0.0$ and $\lambda_{43} = 0.8$ in assessing the three-factor model.

Using a simulated sample of size $n = 300$, the path sampling procedure is implemented using 20 grids in [0,1], and the following hyperparameter values in the prior distributions: $\mathbf{\Lambda}_{0k}$ fixed at the true values, $\alpha_{0\epsilon k} = 10$ and $\beta_{0\epsilon k} = 8$, for all $k = 1, \ldots, 8$. We observe that the hybrid algorithm for simulating observations in the calculation of $\bar{U}_{(s)}$ converges quickly within 200 iterations. At each $t_{(s)}$, a total of $J = 2000$ observations are collected after a 'burn-in' phase of 200 iterations. The logarithm Bayes factors are computed via Equations (6.24) and (6.20). We find that $\widehat{\log B_{21}} = 195.23$ and $\widehat{\log B_{32}} = -31.45$. These results indicate that a two-factor model is better than either a one-factor model or a three-factor model. Hence, the two-factor model is selected. This agrees with the true situation.

In providing some insight into the sensitivity of the Bayes factor to the prior inputs, we perturb the hyperparameters alternatively as follows: (I) For all $k = 1, \ldots, p$, fix $\alpha_{0\epsilon k} = 10$, and $\beta_{0\epsilon k} = 8$; hyperparameters in $\mathbf{\Lambda}_{0k}$ are selected to be (i) half of the true values and (ii) twice the true values. (II) For all $k = 1, \ldots, p$, fix $\mathbf{\Lambda}_{0k}$ at the true values, $\alpha_{0\epsilon k}$ and $\beta_{0\epsilon k}$ are selected as follows: (i) $\alpha_{0\epsilon k} = 8$, $\beta_{0\epsilon k} = 6$, and (ii) $\alpha_{0\epsilon k} = 12$, $\beta_{0\epsilon k} = 10$. The averages, minimum and maximum of the estimated $\log B_{21}$ on the basis of 50 replications under (I, i), (I, ii), (II, i) and (II, ii) are equal to $\{189.05, 136.76, 240.58\}$, $\{188.04, 137.74, 241.08\}$, $\{195.08, 141.90, 248.22\}$ and $\{184.20, 135.75, 236.43\}$ respectively. The values corresponding to the estimated $\log B_{32}$ are $\{-33.93, -43.35, -24.68\}$, $\{-34.74, -44.40, -25.34\}$, $\{-32.06, -41.20, -22.76\}$ and $\{-35.75, -45.35, -26.54\}$, respectively. Hence, the Bayes factors estimated through the proposed method give the same statistical conclusion under these prior inputs and $n = 300$. As expected, the Bayesian estimates of the unknown parameters under different prior inputs are close to each other. To provide some idea about the accuracy of the estimates, the mean of the Bayesian estimates and the root mean squares (RMS) of the estimates and the true values of the selected two-factor model with prior inputs $\alpha_{0\epsilon k} = 10$, $\beta_{0\epsilon k} = 8$ and $\mathbf{\Lambda}_{0k}$ is equal to twice the true values are reported in Table 6.2.

Using a SUN Enterprise 4500 Server, the computational time for calculating $\log B_{21}$ and $\log B_{32}$ in each replication is roughly 6 and 8 minutes, respectively.

6.5.2 A Real Example

To provide a real example for illustrating the proposed method, a small portion of the ICPSR data set collected in the project World Values Survey 1981–1984 and 1990–1993 (World Values Study Group, 1994) is analyzed. As an illustration, only the data obtained from USA were used. Nine variables in the original data set (variables 132, 116, 117, 129, 115, 241, 336, 337 and 339) were taken

Table 6.2 Mean and RMS of Bayesian estimates corresponding to a two-factor model: simulation study.

Par	Mean	RMS	Par	Mean	RMS
$\lambda_{11} = 0.8$	0.776	0.047	$\psi_{\epsilon 1} = 0.36$	0.404	0.054
$\lambda_{21} = 0.0$	−0.011	0.063	$\psi_{\epsilon 2} = 0.36$	0.412	0.062
$\lambda_{22} = 0.8$	0.792	0.047	$\psi_{\epsilon 3} = 0.36$	0.390	0.042
$\lambda_{32} = 0.0$	−0.004	0.041	$\psi_{\epsilon 4} = 0.36$	0.409	0.058
$\lambda_{41} = 0.8$	0.783	0.051	$\psi_{\epsilon 5} = 0.36$	0.396	0.049
$\lambda_{42} = 0.0$	0.002	0.048	$\psi_{\epsilon 6} = 0.36$	0.409	0.057
$\lambda_{51} = 0.0$	−0.004	0.070	$\psi_{\epsilon 7} = 0.36$	0.443	0.092
$\lambda_{61} = 0.0$	−0.011	0.075	$\psi_{\epsilon 8} = 0.28$	0.381	0.108
$\lambda_{62} = 0.8$	0.788	0.045	$\alpha_{7,2} = -0.2$	−0.207	0.054
$\lambda_{71} = 0.8$	0.809	0.059	$\alpha_{7,3} = 0.2$	0.221	0.058
$\lambda_{72} = 0.0$	−0.040	0.062	$\alpha_{8,2} = -0.2$	−0.193	0.058
$\lambda_{81} = 0.6$	0.605	0.057	$\alpha_{8,3} = 0.2$	0.214	0.048
$\lambda_{82} = 0.6$	0.609	0.066			

This table and Table 6.3 are taken from Lee and Song (2002).

as manifest variables in $\mathbf{v} = (v_{(1)}, \ldots, v_{(9)})^T$. Questions corresponding to these variables are presented in Appendix 1.1. Variables $v_{(1)}, v_{(2)}, v_{(3)}$ were measured via a 10-point scale and hence were treated as continuous for convenience; variables $v_{(4)}$ and $v_{(5)}$ are measured via a three-point scale, variable $v_{(6)}$ via a four-point scale and variables $v_{(7)}, v_{(8)}$ and $v_{(9)}$ via a five-point scale; these variables are treated as ordered categorical. For brevity, observations with missing entries are deleted and the remaining sample size is 1048.

The data set is first analyzed with a three-factor model using non-informative prior distributions, in which $\lambda_{12}, \lambda_{13}$ and λ_{23} are fixed at 0.0. On the basis of the results in this initial analysis, we obtain hyperparameter values, and fixed $\lambda_{21}, \lambda_{72}$ and λ_{93} at 1.0 to identify the model completely. After fixing these parameters, there are a total of 37 unknown parameters which involve the unknown loadings, unique variances and thresholds $\alpha_{6,2}, \alpha_{7,2}, \alpha_{7,3}, \alpha_{8,2}, \alpha_{8,3}, \alpha_{9,2}$ and $\alpha_{9,3}$. The path sampling procedure is applied to estimate the Bayes factors for selecting the appropriate number of factors, using 20 grids in [0, 1]. We observe that the Gibbs sampler again converges very quickly within 200 iterations. Again, a total of $J = 2000$ observations are collected after a 'burn-in' phase of 200 iterations. We find that $\widehat{\log B_{21}} = 11.66, \widehat{\log B_{32}} = 19.70$ and $\widehat{\log B_{43}} = -11.29$. Hence, the model with three factors is selected. The corresponding PP p-value of this model is 0.558, which indicates that the three factor model fits the observed data. The Bayesian estimates and their standard error estimates associated with this model are reported in Table 6.3. To achieve better interpretation, the estimated factor loading matrix can be rotated.

Table 6.3 Bayesian estimates of parameters and their standard error estimates for a three-factor model: ICPSR data set.

Par	EST	SE	Par	EST	SE	Par	EST	SE
λ_{11}	0.859	0.075	λ_{62}	0.086	0.043	$\psi_{\epsilon 4}$	0.976	0.088
λ_{22}	−0.304	0.130	λ_{63}	−0.271	0.043	$\psi_{\epsilon 5}$	0.616	0.061
λ_{31}	1.507	0.099	λ_{71}	0.269	0.090	$\psi_{\epsilon 6}$	0.915	0.057
λ_{32}	−0.443	0.144	λ_{73}	0.098	0.058	$\psi_{\epsilon 7}$	0.366	0.050
λ_{33}	0.344	0.087	λ_{81}	0.077	0.049	$\psi_{\epsilon 8}$	0.858	0.051
λ_{41}	0.122	0.051	λ_{82}	0.382	0.044	$\psi_{\epsilon 9}$	0.535	0.076
λ_{42}	0.016	0.056	λ_{83}	0.055	0.046	$\alpha_{6,2}$	0.328	0.034
λ_{43}	0.230	0.060	λ_{91}	−0.104	0.052	$\alpha_{7,2}$	1.116	0.052
λ_{51}	−0.487	0.059	λ_{92}	0.036	0.073	$\alpha_{7,3}$	1.657	0.062
λ_{52}	0.316	0.078	$\psi_{\epsilon 1}$	4.190	0.195	$\alpha_{8,2}$	−0.107	0.034
λ_{53}	−0.111	0.059	$\psi_{\epsilon 2}$	1.798	0.115	$\alpha_{8,3}$	0.472	0.036
λ_{61}	−0.048	0.039	$\psi_{\epsilon 3}$	2.604	0.224	$\alpha_{9,2}$	−0.782	0.046
						$\alpha_{9,3}$	0.162	0.046

6.6 APPLICATION 2: BAYESIAN ANALYSIS OF QUALITY OF LIFE DATA

There is increasing recognition that measures of quality of life (QOL) and/or health-related QOL have great value for clinical work and the planning and evaluation of health care as well as for medical research. It has been generally accepted that QOL is a multidimensional concept (Staquet, Hayes and Fayes, 1998) that is best evaluated by a number of different latent constructs such as physical function, health status, mental status and social relationships. As these latent constructs often cannot be measured objectively and directly, they are treated as latent variables in QOL analysis. The most popular method that is used to assess a latent construct is by a survey which incorporates a number of related items that are intended to reflect the underlying latent construct of interest.

EFA has been used as a method for exploring the structure of a new QOL instrument (The WHOQOL Group, 1998; Fayer and Machin, 1998), while CFA has been used to confirm the factor structure of the instrument. SEMs that are based on continuous observations with a normal distribution have also been applied to QOL analyses (Power, Bullingen and Hazper, 1999).

Items in a QOL instrument are usually measured on an ordered categorical scale, typically with three- to five-points. The discrete ordinal nature of the items also attracts much attention in QOL analysis (Fayer and Machin, 1998; Fayer and Hand, 1997). It has been pointed out that non-rigorous treatments of the ordinal items as continuous can be subjected to criticism (Glonek and McCullagh, 1995), and models such as the item response model and ordinal regression that take into account the ordinal nature are more appropriate (Olschewski and Schumacker, 1990). The aim of this section is to apply the Bayesian methods for analyzing a common QOL instrument with ordered categorical items.

6.6.1 A Synthetic Illustrative Example

This instrument WHOQOL-100 (Power, Bullingen and Hazper, 1999) was established to evaluate four latent constructs. The first seven items (Q3 to Q9) are intended to address physical health, the next six items (Q10 to Q15) are intended to address psychological health, the three items (Q16, Q17, Q18) that follow are for social relationships, and the last eight items (Q19 to Q26) are intended to address environment. In addition to the 24 ordered categorical items, the instrument also includes two ordered categorical items for the overall QOL (Q1) and general health (Q2), giving a total of 26 items. All of the items are measured with a five-point scale (1 = 'not at all/very dissatisfied'; 2 = 'a little/dissatisfied'; 3 = 'moderate/neither'; 4 = 'very much/satisfied'; 5 = 'extremely/very satisfied'). The sample size of the whole data set is extremely large. To illustrate the Bayesian methods, we only analyze a synthetic data set with sample size $n = 338$. The frequencies of all the ordered categorical items are presented in Table 6.4. As can be seen from the table, many items are skewed to the right. Treating these ordered categorical data as coming from normal is not correct. Hence, our Bayesian approach that takes into account the discrete nature of the data is applied to analyze this ordered categorical data set.

To illustrate the path sampling procedure, we compare an SEM with four exogenous latent variables with another SEM with three exogenous latent variables, see Lee, Song, Skevington and Hao (2005). Let M_1 be the SEM whose measurement equation is defined by

$$y_i = \mathbf{\Lambda}_1 \boldsymbol{\omega}_{1i} + \boldsymbol{\epsilon}_i, \tag{6.25}$$

where $\boldsymbol{\omega}_{1i} = (\eta_i, \xi_{i1}, \xi_{i2}, \xi_{i3}, \xi_{i4})^T$, $\boldsymbol{\epsilon}_i$ is distributed according to $N[0, \boldsymbol{\Psi}_{\epsilon 1}]$, and

$$\mathbf{\Lambda}_1^T = \begin{bmatrix} 1 & \lambda_{21} & 0 & 0 & \cdots & 0 & 0 & 0 & \cdots & 0 & 0 & 0 & 0 & 0 & 0 & \cdots & 0 \\ 0 & 0 & 1 & \lambda_{42} & \cdots & \lambda_{92} & 0 & 0 & \cdots & 0 & 0 & 0 & 0 & 0 & 0 & \cdots & 0 \\ 0 & 0 & 0 & 0 & \cdots & 0 & 1 & \lambda_{11,3} & \cdots & \lambda_{15,3} & 0 & 0 & 0 & 0 & 0 & \cdots & 0 \\ 0 & 0 & 0 & 0 & \cdots & 0 & 0 & 0 & \cdots & 0 & 1 & \lambda_{17,4} & \lambda_{18,4} & 0 & 0 & \cdots & 0 \\ 0 & 0 & 0 & 0 & \cdots & 0 & 0 & 0 & \cdots & 0 & 0 & 0 & 0 & 1 & \lambda_{20,5} & \cdots & \lambda_{26,5} \end{bmatrix}.$$

The structural equation of M_1 is defined by

$$\eta = \gamma_1 \xi_1 + \gamma_2 \xi_2 + \gamma_3 \xi_3 + \gamma_4 \xi_4 + \delta, \tag{6.26}$$

where the distributions of $(\xi_1, \xi_2, \xi_3, \xi_4)^T$ and δ are independently distributed as $N[\mathbf{0}, \mathbf{\Phi}_1]$, and $N[0, \sigma_{1\delta}^2]$, respectively. Let M_2 be the SEM whose measurement is defined by

$$y_i = \mathbf{\Lambda}_2 \boldsymbol{\omega}_{2i} + \boldsymbol{\epsilon}_i, \tag{6.27}$$

Table 6.4 Frequencies of the questions in the WHOQOL data set.

	1	2	3	4	5
Q1 Overall QOL	2	34	75	160	67
Q2 Overall health	25	89	71	117	36
Q3 Pain and discomfort	16	49	78	111	84
Q4 Medical treatment dependence	17	48	65	83	125
Q5 Energy and fatigue	16	53	107	86	76
Q6 Mobility	13	33	62	95	135
Q7 Sleep and rest	23	62	73	116	64
Q8 Daily activities	9	55	63	158	53
Q9 Work capacity	19	71	79	116	53
Q10 Positive feeling	8	22	93	165	50
Q11 Spirituality/personal beliefs	8	29	99	137	65
Q12 Memory and concentration	4	22	148	133	31
Q13 Bodily image and appearance	3	30	106	112	87
Q14 Self-esteem	7	38	104	148	41
Q15 Negative feeling	4	35	89	171	39
Q16 Personal relationship	5	16	59	165	93
Q17 Sexual activity	25	48	112	100	53
Q18 Social support	7	6	73	164	88
Q19 Physical safety and security	4	20	147	129	38
Q20 Physical environment	7	20	142	126	43
Q21 Financial resources	15	34	140	87	62
Q22 Daily life information	4	22	102	154	56
Q23 Participation in leisure activity	15	76	102	108	37
Q24 Living condition	4	12	35	173	114
Q25 Health accessibility and quality	4	20	59	205	50
Q26 Transportation	5	16	43	188	86
Total	269	960	2326	3507	1726

where $\boldsymbol{\omega}_{2i} = (\eta_i, \xi_{i1}, \xi_{i2}, \xi_{i3})^T$. $\boldsymbol{\epsilon}_i$ is distributed according to $N[\mathbf{0}, \boldsymbol{\Psi}_{\epsilon 2}]$, and

$$\boldsymbol{\Lambda}_2^T = \begin{bmatrix} 1 & \lambda_{21} & 0 & 0 & \cdots & 0 & 0 & 0 & \cdots & 0 & 0 & 0 & \cdots & 0 \\ 0 & 0 & 1 & \lambda_{42} & \cdots & \lambda_{92} & 0 & 0 & \cdots & 0 & 0 & 0 & \cdots & 0 \\ 0 & 0 & 0 & 0 & \cdots & 0 & 1 & \lambda_{11,3} & \cdots & \lambda_{15,3} & 0 & 0 & \cdots & 0 \\ 0 & 0 & 0 & 0 & \cdots & 0 & 0 & 0 & \cdots & 0 & 1 & \lambda_{17,4} & \cdots & \lambda_{26,4} \end{bmatrix}.$$

The structural equation of M_2 is defined by

$$\eta = \gamma_1\xi_1 + \gamma_2\xi_2 + \gamma_3\xi_3 + \delta, \tag{6.28}$$

where the distributions of $(\xi_1, \xi_2, \xi_3)^T$ and δ are independently distributed as $N[\mathbf{0}, \boldsymbol{\Phi}_2]$, and $N[0, \sigma_{2\delta}^2]$, respectively. Bayesian analyses are conducted using the conjugate prior distributions. The hyperparameter values corresponding to the prior distributions of the unknown loadings in $\mathbf{\Lambda}_1$ or $\mathbf{\Lambda}_2$ are all taken to be 0.8; those corresponding to $\{\gamma_1, \gamma_2, \gamma_3, \gamma_4\}$ are $\{0.6, 0.6, 0.4, 0.4\}$; those corresponding to $\mathbf{\Phi}_1$ and $\mathbf{\Phi}_2$ are $\rho_0 = 30$, $\boldsymbol{R}_0^{-1} = 8\mathbf{I}_4$ and $\boldsymbol{R}_0^{-1} = 8\mathbf{I}_3$, respectively; $\mathbf{H}_{0vk} = 0.25\mathbf{I}_{26}$, $\mathbf{H}_{0\omega k} = 0.25\mathbf{I}_4$, or $\mathbf{H}_{0\omega k} = 0.25\mathbf{I}_3$, $\alpha_{0\epsilon k} = \alpha_{0\delta k} = 10$, and $\beta_{0\epsilon k} = \beta_{0\delta k} = 8$. In the path sampling procedure in computing the Bayes factor, we take $S = 10$ and $J = 2000$ after a 'burn-in' phase of 1000 iterations.

We first compare M_1 with the following simple model M_0:

$$M_0 : \mathbf{y}_i = \boldsymbol{\epsilon}_i$$

where $\boldsymbol{\epsilon}_i \stackrel{D}{=} N[\mathbf{0}, \mathbf{\Psi}_\epsilon]$ and $\mathbf{\Psi}_\epsilon$ is a diagonal matrix. The measurement equation of the linked model is defined by $M_t : \mathbf{y}_i = t\mathbf{\Lambda}_1\boldsymbol{\omega}_i + \boldsymbol{\epsilon}_i$. We obtain $\widehat{\log B_{10}} = 81.05$. Clearly, M_1 is better than M_0. Similarly, M_2 and M_0 can be compared via the path sampling procedure. We find that $\widehat{\log B_{20}} = 57.65$, which suggests that M_2 is better than M_0. From the above results, we can obtain an estimate of $\log B_{12}$, which is equal to 23.40. Hence, M_1, the SEM with one endogenous and four exogenous latent variables is selected. Bayesian estimates of the unknown structural parameters in M_1 are presented in Figure 6.4. The less interesting threshold estimates are not presented. All the factor loading estimates, except $\hat{\lambda}_{17,4}$ that associates with the indicator 'sexual activity', are high. This indicates a strong association between each of the latent variables and their corresponding indicators. From the meaning of the items, η, ξ_1, ξ_2, ξ_3 and ξ_4 can be interpreted as the overall QOL, physical health, psychological health, social relationship and environment, respectively. The estimates of correlations among the exogenous latent variables are equal to $\{0.680, 0.432, 0.632, 0.685, 0.736, 0.698\}$. As expected, these correlations are high. The estimated structural equation that addresses the relations of QOL with the latent constructs about physical and psychological health, social relationship and environment is

$$\eta = 0.724\xi_1 + 0.319\xi_2 + 0.169\xi_3 - 0.041\xi_4.$$

Hence, physical health has the most important effect on QOL, followed in turn by psychological health and social relationship, while the effect of environment is not important.

6.6.2 Application of WinBUGS

For SEMs with ordered categorical variables, the software WinBUGS (Spiegelhalter, Thomas, Best and Lunn, 2003) can produce Bayesian estimates of the structural parameters and latent variable estimates in the model, as well

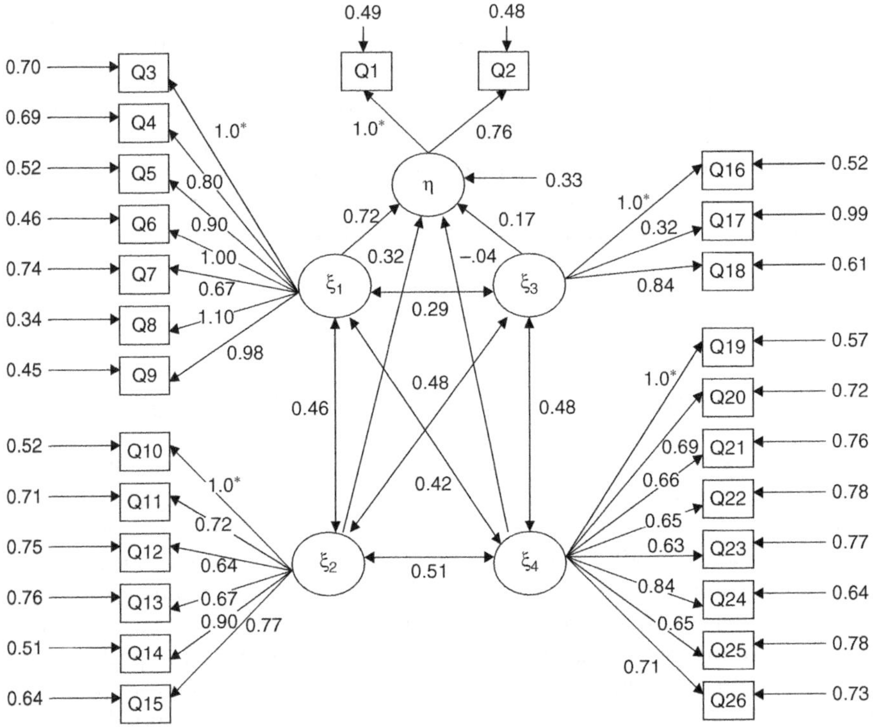

Figure 6.4 Path diagram and Bayesian estimates of parameters in the analysis of the QOL data. Note that Bayesian estimates of ϕ_{11}, ϕ_{22}, ϕ_{33} and ϕ_{44} are 0.648, 0.706, 0.694 and 0.680, respectively.

as their standard error estimates, by means of a sufficiently large number of observations simulated by MCMC methods. In our Bayesian treatment of the ordered categorical variables, we fix the thresholds at both ends in order to solve the identification problem, then the other unknown thresholds are simultaneously estimated with the structural parameters $\boldsymbol{\mu}$, $\boldsymbol{\Lambda}$, $\boldsymbol{\Psi}_\epsilon$, $\boldsymbol{\Phi}$, $\boldsymbol{\Lambda}_\omega$ and $\boldsymbol{\Psi}_\delta$ in the model. However, according to our understanding of WinBUGS, it is not straightforward to apply this software to estimate the unknown thresholds and structural parameters simultaneously. Hence, in applying WinBUGS, we first estimate all the thresholds through the method as described in Section 6.2, using the observed frequencies and the distribution function of $N[0, 1]$. Then, the thresholds are fixed in the WinBUGS program in producing the Bayesian solutions. Note that this procedure may underestimate the standard error estimates. Hence, hypothesis testing should be conducted through DIC, rather than the z-score that depends on the standard error estimate.

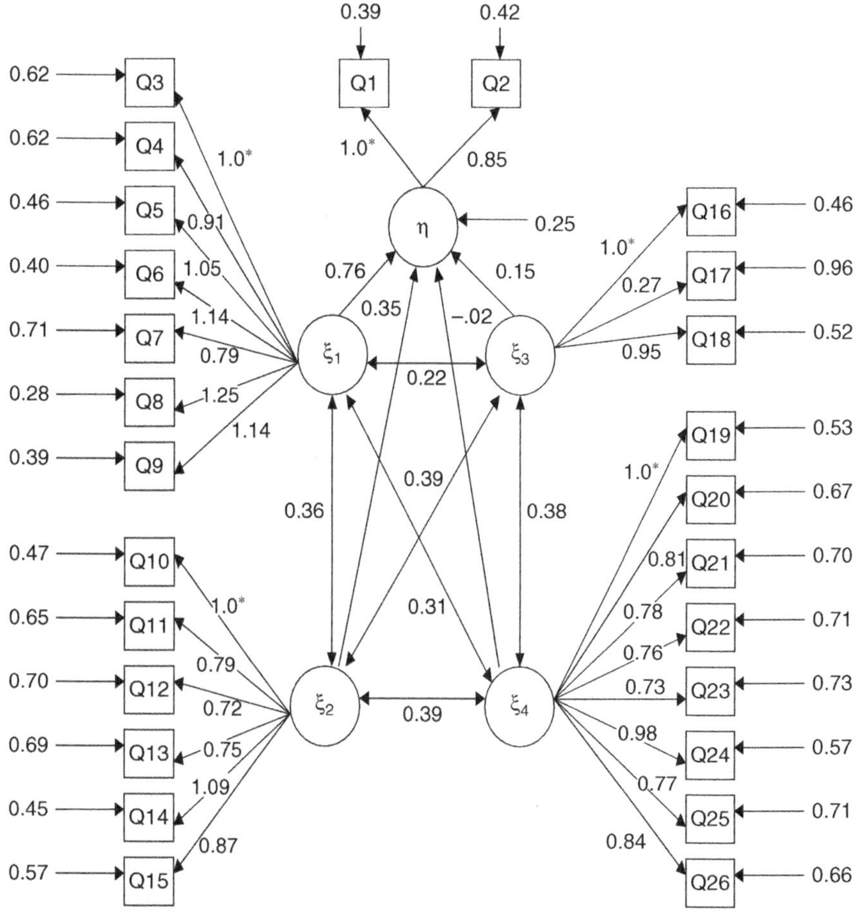

Figure 6.5 Path diagram and Bayesian estimates of parameters in the analysis of the QOL data obtained via WinBUGS. Note that Bayesian estimates of ϕ_{11}, ϕ_{22}, ϕ_{33} and ϕ_{44} are 0.493, 0.584, 0.600 and 0.530, respectively.

We apply WinBUGS to the synthetic QOL data set as discussed in the previous subsection. Bayesian solutions are obtained under the selected model M_1 with the same conjugate prior distributions and hyperparameter inputs. Note that it is desirable to give initial values for $\boldsymbol{\theta}$ and $\boldsymbol{\Omega}$ in order to save computer time. The DIC value corresponding to this model is 19 537.9. Bayesian estimates of the unknown structural parameters are presented in Figure 6.5. We observe that the estimates respectively presented in Figures 6.4 and 6.5 are not as close as expected. The possible reason may be that the treatments of the thresholds by the two programs are different. The completely

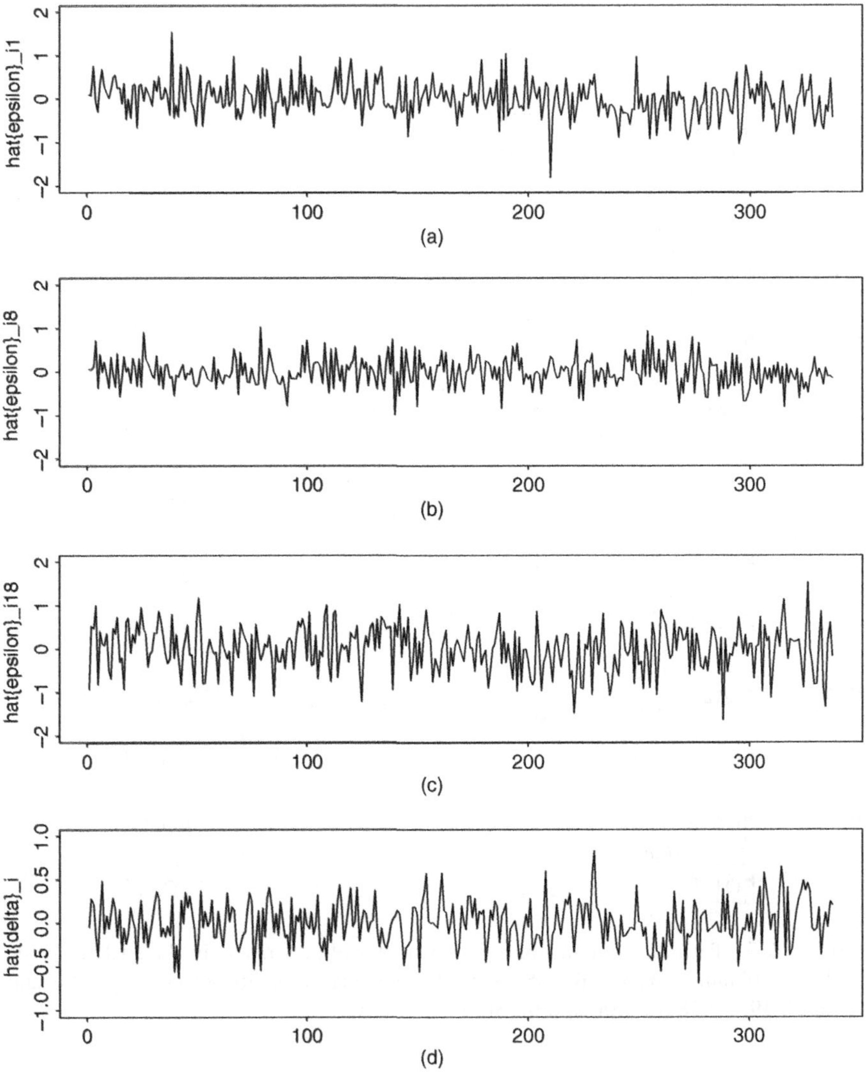

Figure 6.6 Estimated residual plots, (a) $\hat{\epsilon}_{i1}$, (b) $\hat{\epsilon}_{i8}$, (c) $\hat{\epsilon}_{i18}$ and (d) $\hat{\delta}_i$.

standardized solutions corresponding to these two sets of estimates are closer, for example, the estimates of the correlations among the exogenous latent variables are equal to $\{0.664, 0.406, 0.610, 0.661, 0.705, 0.679\}$. Some estimated residual plots, $\hat{\epsilon}_{i1}$, $\hat{\epsilon}_{i8}$, $\hat{\epsilon}_{i18}$ and $\hat{\delta}_i$ versus the case number are displayed in Figure 6.6, while plots of estimated residual $\hat{\epsilon}_{i1}$ and $\hat{\delta}_i$ versus $\hat{\xi}_{i1}$, $\hat{\xi}_{i2}$, $\hat{\xi}_{i3}$, $\hat{\xi}_{i4}$ and $\hat{\eta}_i$ are presented in Figures 6.7 and 6.8 respectively. Other estimated

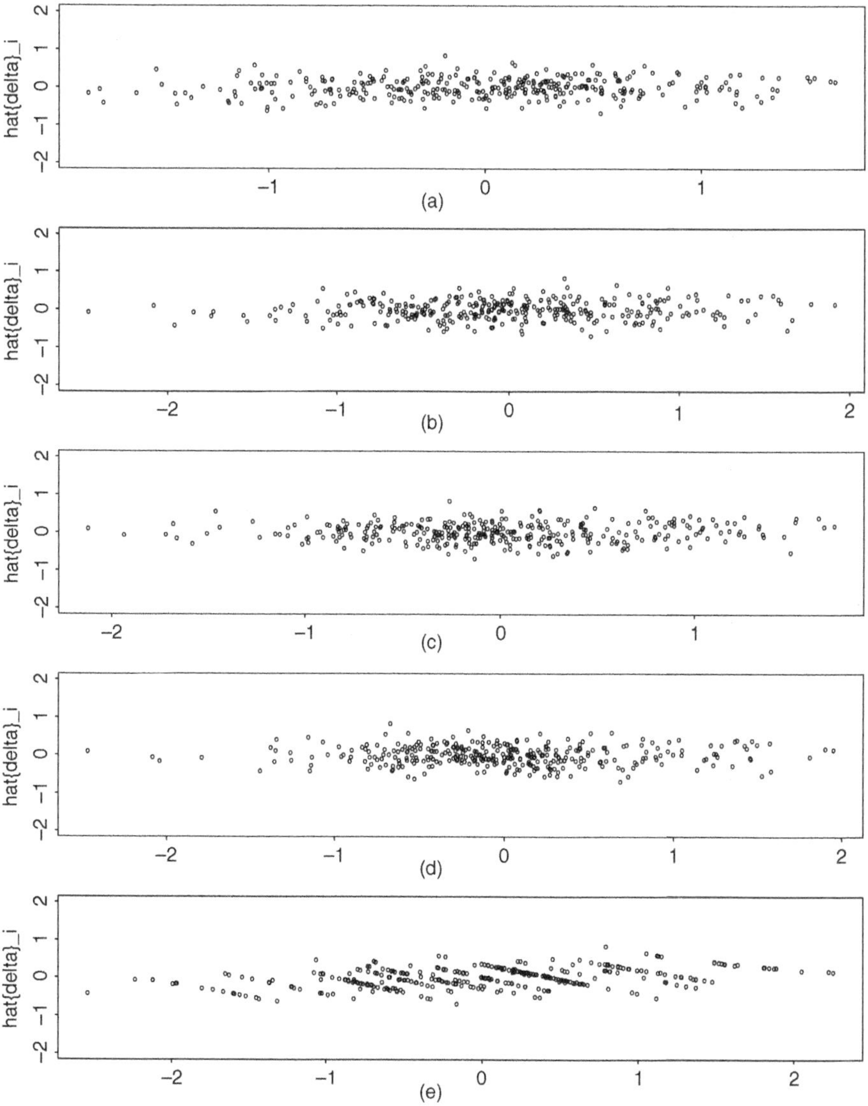

Figure 6.7 Plots of estimated residuals $\hat{\epsilon}_{i1}$ versus (a) $\hat{\xi}_{i1}$, (b) $\hat{\xi}_{i2}$, (c) $\hat{\xi}_{i3}$, (d) $\hat{\xi}_{i4}$ and (e) $\hat{\eta}_i$.

residual plots are similar. These estimated residual plots lie within two parallel horizontal lines that are centered at zero, and no linear or quadratic trends are detected. This indicates that the proposed measurement and structural equations are adequate. The WinBUGS codes and the data are given in the following website: http://www.wiley.com/go/lee_structural.

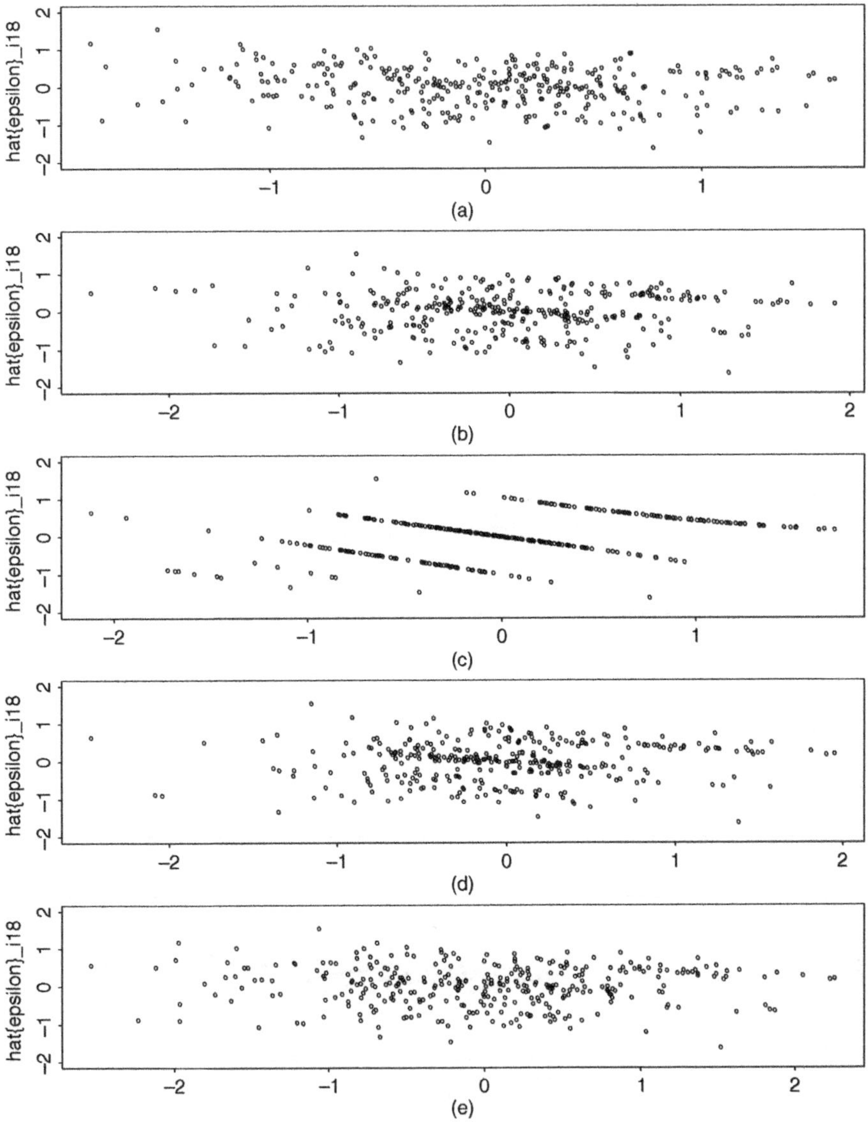

Figure 6.8 Plots of estimated residuals $\hat{\delta}_i$ versus (a) $\hat{\xi}_{i1}$, (b) $\hat{\xi}_{i2}$, (c) $\hat{\xi}_{i3}$, (d) $\hat{\xi}_{i4}$ and (e) $\hat{\eta}_i$.

REFERENCES

Akaike, H (1970). Statistical predictor identification. *Annals of the Institute of Statistical Mathematics*, **22**, 203–217.

Akaike, H. (1987) Factor analysis and AIC. *Psychometrika*, **52**, 317–332.

Cowles, M. K. (1996) Accelerating Monte Carlo Markov chain convergence for cumulative-link generalized linear models. *Statistics and Computing*, **6**, 101–111.

Fayer, P. M. and Hand, D. J. (1997) Factor analysis, causal indicators and quality of life. *Quality of Life Research*, **8**, 139–150.

Fayer, P. M. and Machin, D. (1998) Factor analysis. In M. J. Staquet, R. D. Hayes and P. M. Fayer (eds), *Quality of Life Assessment in Clinical Trials*. New York: Oxford University Press.

Gelman, A. (1996) Inference and monitoring convergence. In W. R. Gilks, S. Richardson and D. J. Spiegelhalter (eds), *Markov Chain Monte Carlo in Practice* pp. 131–144. London: Chapman and Hall.

Gelman, A. and Rubin, D. B. (1992) Inference from iterative simulation using multiple sequences. *Statistical Science*, **7**, 457–472.

Gelman, A., Meng, S. L. and Stern, H. (1996) Posterior predictive assessment of model fitness via realized discrepancies. *Statistica Sinica*, **6**, 733–759.

Gelman, A., Carlin, J. B., Stern, H. S. and Rubin, D. B. (1995) *Bayesian Data Analysis*. London: Chapman and Hall.

Geman, S. and Geman, D. (1984) Stochastic relaxation, Gibbs distribution and the Bayesian restoration of images. *IEEE Transactions on Pattern Analysis and Machine Intelligence*, **6**, 721–741.

Geyer, C. J. (1992) Practical Markov chain Monte Carlo. *Statistical Science*, **7**, 473–511.

Glonek, G. F. V. and McCullagh, P. (1995) Multivariate logistic models. *Journal of the Royal Statistical Society, Series B*, **57**, 533–546.

Hastings, W. K. (1970) Monte Carlo sampling methods using Markov chains and their application. *Biometrika*, **57**, 97–109.

Jöreskog, K. G. and Sörbom, D. (1996) *LISREL 8: Structural Equation Modeling with the SIMPLIS Command Language*. Hove and London: Scientific Software International.

Lawley, D. N. and Maxwell, A. E. (1971) *Factor Analysis as a Statistical Method* (2nd edn). New York: Elsevier.

Lee, S. Y. and Song, X. Y. (2002) Bayesian selection on the number of factors in a factor analysis model. *Behaviormetrika*, **29**, 23–39.

Lee, S. Y. and Zhu, H. T. (2000) Statistical analysis of nonlinear structural equation models with continuous and polytomous data. *British Journal of Mathematical and Statistical Psychology*, **53**, 209–232.

Lee, S. Y., Poon, W. Y. and Bentler, P. M. (1990a) Full maximum likelihood analysis of structural equation models with polytomous variables. *Statistics and Probability Letters*, **9**, 91–97.

Lee, S. Y., Poon, W. Y. and Bentler, P. M. (1990b). A three-stage estimation procedure for structural equation models with polytomous variables. *Psychometrika*, **55**, 45–51.

Lee, S. Y., Poon, W. Y. and Bentler, P. M. (1995). A two-stage estimation of structural equation models with continuous and polytomous variables. *British Journal of Mathematical and Statistical Psychology*, **48**, 339–358.

Lee, S. Y., Song, X. Y., Skevington, S. and Hao, Y. T. (2005) Application of structural equation models to quality of life. *Structural Equation Modeling – A Multidisciplinary Journal*, **12**, 435–453.

Metropolis, N. *et al.* (1953) Equations of state calculations by fast computing machine. *Journal of Chemical Physics*, **21**, 1087–1091.

Muthen, L. and Muthen, B. (2000) *Mplus User's Guide*. Los Angeles, CA: Muthen and Muthen.

Olschewski, M. and Schumacker, M. (1990) Statistical analysis of quality of life data in cancer clinical trials. *Statistics in Medicine*, **9**, 749–763.

Olsson, U. (1979a) Maximum likelihood estimation of the polychoric correlation coefficient. *Psychometrika*, **44**, 443–460.

Olsson, U. (1979b) On the robustness of factor analysis against crude classification of the observations. *Multivariate Behavioral Research*, **14**, 485–500.

Power, M., Bullingen, M. and Hazper, A. (1999) The World Health Organization WHOQOL-100: Tests of the universality of quality of life in 15 different cultural groups worldwide. *Health Psychology*, **18**, 495–505.

Reboussin, B. A. and Liang, K. Y. (1998) An estimating equation approach for the LISCOMP model. *Psychometrika*, **63**, 165–182.

Roberts, C. P. (1995) Simulation of truncated normal variables. *Statistics and Computing*, **5**, 121–125.

Shi, J. Q. and Lee, S. Y. (2000) Latent variable models with mixed continuous and polytomous data. *Journal of the Royal Statistical Society, Series B*, **62**, 77–87.

Song, X. Y. and Lee, S. Y. (2001) Bayesian estimation and test for factor analysis model with continuous and polytomous data in several populations. *British Journal of Mathematical and Statistical Psychology*, **54**, 237–263.

Song, X. Y. and Lee, S. Y. (2002) Analysis of structural equation model with ignorable missing continuous and polytomous data. *Psychometrika*, **67**, 261–288.

Spiegelhalter, D. J., Thomas, A., Best, N. G. and Lunn, D. (2003) *WinBUGS User Manual. Version 1.4.* Cambridge, England: MRC Biostatistics Unit.

Staquet, M. J., Hayes, R. D. and Fayes, P. M. (1998) *Quality of Life Assessment in Clinical Trials.* New York: Oxford University Press.

The WHOQOL Group (1998) Development of the World Health Organization WHOQOL-BREF quality of life assessment. *Psychological Medicine*, **28**, 551–558.

Wei, G. C. G. and Tanner, M. A. (1990) A Monte Carlo implementation of the EM algorithm and the poor man's data augmentation algorithm. *Journal of the American Statistical Association*, **85**, 699–704.

World Value Study Group (1994) World Values Survey, 1981–1984 and 1990–1993. ICPSR version. Ann Arbor, MI: Institute for Social Research (producer). Ann Arbon, MI: Inter-university Consortium for Political and Social Research (distributor).

Zhu, H. T. and Lee, S. Y. (1998) Statistical analysis of nonlinear factor analysis models. *British Journal of Mathematical and Statistical Psychology*, **52**, 225–242.

7 Structural Equation Models with Dichotomous Variables

7.1 INTRODUCTION

A Bayesian approach for analyzing SEMs with ordered categorical variables that are defined by more than two categories discussed in Chapter 6. Another kind of discrete variable that is frequently encountered in behavioral, educational, medical and social research is the dichotomous variable which only involves two categories. In this chapter, we will focus on dichotomous variables that are ordered binary and defined with one threshold. Dichotomous variables arise when respondents are asked to select answers from 'Yes or No' about the presence of a symptoms, 'Feeling better or Feeling worse' about the effect of a drug, 'Agree or Disagree' about a policy, 'Satisfactory or Unsatisfactory' about a public service, etc.. The usual numerical values assigned to these variables are the ad hoc numbers with an ordering such as '0' and '1', or '1' and '2'. For example, if 'Unsatisfactory' and 'Satisfactory' are respectively coded by 'a' and 'b', then we have a natural ordering that $a < b$. In analyzing dichotomous data, the basic assumption in SEM that the data come from a continuous normal distribution is clearly violated, and rigorous analysis that takes into account the dichotomous nature is necessary. As we will see, the analysis of SEMs with dichotomous variables is similar to, but not exactly the same as, the analysis with ordered categorical variables.

Structural Equation Modeling: A Bayesian Approach S-Y. Lee

In many substantive research, particularly in education, it is important to explore and determine a small number of intrinsic latent factors under a number of test items. Item factor analysis is an important model that has been proposed for explaining the underlying factor structures. An important direction of analysis is the ML full information item factor (FIIF) approach of Bock and Aitkin (1981). Their strategy in solving the computational difficulties induced by the dichotomous variables was to treat the latent factors as missing data, and then to apply the EM algorithm (Dempster, Laird and Rubin, 1977), with the E-step completed by numerical integration via fixed-point Gauss–Hermite quadrature. More recently, Meng and Schilling (1996) improved their EM algorithm by using a Monte Carlo EM (MCEM) algorithm (Wei and Tanner, 1990) for achieving the ML estimates. A Bayesian method for analyzing the more general two-level factor analysis model with dichotomous data has been developed by Ansari and Jedidi (2000), by using some MCMC methods such as the Gibbs sampler (Geman and Geman, 1984) and the MH algorithm (Metropolis *et al.*, 1953; Hastings, 1970). More recently, Lee and Song (2003) developed a Bayesian approach for a general SEM with dichotomous variables that is compose of a measurement equation and a structural equation, based on the Gibbs sampler. In this type of analysis, the emphasis is on the relationships between the manifest and latent variables, or the related covariance structure, rather than the mean vector.

Another direction of analysis is motivated from the fact that correlated dichotomous data frequently arise in many medical and biological studies, ranging from measurements of random cross section subjects to repeated measurements in longitudinal studies. The multivariate probit (MP) model is a popular method in biostatistics for analyzing this kind of data. This model is described in terms of a correlated multivariate normal distribution of the underlying latent variables that are manifested as discrete variables through a threshold specification, and hence allows the flexible modeling of the correlation structure and easy interpretation of the parameters. The emphasis of this approach is focused on the mean structure, and the main difficulty in the analysis is in evaluating the multivariate normal orthant probabilities induced by the dichotomous variables. One approach uses models with less restrictive covariance structures to reduce the computational burden of evaluating the probabilities. For example, Bock and Gibbons (1996), and Gibbons and Wilcox-Gök (1998) considered the exploratory factor analysis model for the covariance structure, fixed covariates for accounting the mean structure, and applied the fixed point Gauss–Hermite quadrature to approximate the orthant probabilities. Their model can be regarded as a generalization of the FIIF model that allows fixed covariates. Chib and Greenberg (1998) developed a Bayesian approach and an ML approach for an MP model with a generic residual covariance structure, and applied the method to various real data sets, including the canonical 4 year data set from the Six Cities Study of health effects. Both their Bayesian and ML approaches require the simulation of observations from a multivariate truncated normal distribution with an arbitrary covariance matrix. Even with the efficient methods of statistical computing, the underlying computational effort is rather heavy. We will show in Section 7.3 that this computational burden can be reduced by adopting an SEM approach.

This chapter is organized as follows. In Section 7.2, we discuss the Bayesian approach proposed in Lee and Song (2003) for analyzing SEMs with dichotomous variables. Bayesian estimation and model comparison will be discussed and a real example will be given. An SEM with fixed covariates for analyzing the MP model will be discussed in Section 7.3 and an illustrative example is provided (WinBUGS codes for analyzing an example are given in a web site). A discussion is given in Section 7.4.

7.2 BAYESIAN ANALYSIS

Again, we consider a commonly used SEM that is compose of a measurement equation and a structural equation. The measurement equation is defined by the following confirmatory factor analysis model:

$$\mathbf{y}_i = \boldsymbol{\mu} + \boldsymbol{\Lambda}\boldsymbol{\omega}_i + \boldsymbol{\epsilon}_i, \quad i = 1, \ldots, n, \tag{7.1}$$

where $\mathbf{y}_i$ is a $p \times 1$ random vector of manifest variables, $\boldsymbol{\mu}$ is a $p \times 1$ mean vector, $\boldsymbol{\omega}_i$ is a $q \times 1$ random vector of latent variables, $\boldsymbol{\epsilon}_i$ is a $p \times 1$ random vector of residuals, and $\boldsymbol{\Lambda}$ is a $p \times q$ unknown factor loading matrix. It is assumed that for $i = 1, \ldots, n$, $\boldsymbol{\omega}_i$ is independently distributed as $N[\mathbf{0}, \boldsymbol{\Phi}]$, $\boldsymbol{\epsilon}_i$ is independently distributed as $N[\mathbf{0}, \boldsymbol{\Psi}_\epsilon]$, where $\boldsymbol{\Psi}_\epsilon$ is a diagonal matrix, and $\boldsymbol{\omega}_i$ and $\boldsymbol{\epsilon}_i$ are uncorrelated. Let the latent vector $\boldsymbol{\omega}_i$ be partitioned into $(\boldsymbol{\eta}_i^T, \boldsymbol{\xi}_i^T)^T$, where $\boldsymbol{\eta}_i$ and $\boldsymbol{\xi}_i$ are $q_1 \times 1$, and $q_2 \times 1$ vectors of latent variables respectively. Moreover, suppose that these latent vectors satisfy the following structural equation:

$$\boldsymbol{\eta}_i = \boldsymbol{\Pi}\boldsymbol{\eta}_i + \boldsymbol{\Gamma}\boldsymbol{\xi}_i + \boldsymbol{\delta}_i, \tag{7.2}$$

where $\boldsymbol{\Pi}$ and $\boldsymbol{\Gamma}$ are $q_1 \times q_1$ and $q_1 \times q_2$ matrices of unknown parameters, and $\boldsymbol{\delta}_i$ is distributed as $N[\mathbf{0}, \boldsymbol{\Psi}_\delta]$, where $\boldsymbol{\Psi}_\delta$ is a diagonal matrix, and $\boldsymbol{\xi}_i$ and $\boldsymbol{\delta}_i$ are independent. It is assumed that $\boldsymbol{\Pi}_0 = \mathbf{I} - \boldsymbol{\Pi}$ is nonsingular, and its determinant is a nonzero constant independent with elements of $\boldsymbol{\Pi}$. For convenience, let $\boldsymbol{\Lambda}_\omega = (\boldsymbol{\Pi}, \boldsymbol{\Gamma})$, then the structural Equation (7.2) can be written as:

$$\boldsymbol{\eta}_i = \boldsymbol{\Lambda}_\omega\boldsymbol{\omega}_i + \boldsymbol{\delta}_i. \tag{7.3}$$

Now suppose that the exact measurement of $\mathbf{y}_i = (y_{i1}, \ldots, y_{ip})^T$ is not available and its information is given by an observed dichotomous vector $\mathbf{z}_i = (z_{i1}, \ldots, z_{ip})^T$ such that, for $k = 1, \ldots, p$,

$$z_{ik} = \begin{cases} 1 & \text{if } y_{ik} > 0, \\ 0 & \text{otherwise.} \end{cases} \tag{7.4}$$

The available observed data set is $\{\mathbf{z}_i; i = 1, \ldots, n\}$. As the density function of $\mathbf{z}_i$ involves a multidimensional integral, analysis of SEMs with dichotomous variables is not straightforward.

Consider the relationship between the factor analysis model defined by Equation (7.1) with the dichotomous variables in $\mathbf{z}$. Let $\mathbf{\Lambda}_k^T$, μ_k and $\psi_{\epsilon k}$ be the kth row of $\mathbf{\Lambda}$, the kth element of $\boldsymbol{\mu}$, and the kth diagonal element of $\mathbf{\Psi}_\epsilon$, respectively. It follows from Equation (7.4) that

$$\begin{aligned} \Pr(z_{ik} = 1 | \boldsymbol{\omega}_i, \mu_k, \mathbf{\Lambda}_k, \psi_{\epsilon k}) &= \Pr(y_{ik} > 0 | \boldsymbol{\omega}_i, \mu_k, \mathbf{\Lambda}_k, \psi_{\epsilon k}) \\ &= \mathbf{\Phi}^*[(\mathbf{\Lambda}_k^T / \psi_{\epsilon k}^{1/2})\boldsymbol{\omega}_i + \mu_k / \psi_{\epsilon k}^{1/2}], \end{aligned} \tag{7.5}$$

where $\mathbf{\Phi}^*$ is the distribution function of $N[0, 1]$. Note that $\mu_k, \mathbf{\Lambda}_k$ and $\psi_{\epsilon k}$ are not estimable, because $C\mathbf{\Lambda}_k/(C\psi_{\epsilon k}^{1/2}) = \mathbf{\Lambda}_k/\psi_{\epsilon k}^{1/2}$ and $C\mu_k/(C\psi_{\epsilon k}^{1/2}) = \mu_k/\psi_{\epsilon k}^{1/2}$ for any positive constant C. There are many ways to solve this identification problem. Meng and Schilling (1996) suggested fixing $\psi_{\epsilon k}$ implicitly and only estimating $\mathbf{\Lambda}_k$. Following this basic idea, we fix $\psi_{\epsilon k} = 1.0$. Note that the value 1.0 is chosen for convenience and any other value would give an equivalent solution up to a change of scale. Again the measurement and structural equations are not identified. We follow the common method in SEMs to solve this problem by fixing the approximate elements of $\mathbf{\Lambda}$ and $\mathbf{\Lambda}_\omega$ at preassigned values.

Now, we consider the interpretation of the factor analysis model in the measurement equation by looking at the relation between $\boldsymbol{\mu}$ and the observed dichotomous vector $\mathbf{z}$. Let $y_{ik}^* = \mathbf{\Lambda}_k^T \boldsymbol{\omega}_i + \epsilon_{ik}$, that is $y_{ik}^* + \mu_k = y_{ik}$. Because $y_{ik} > 0$ if and only if $y_{ik}^* \geq -\mu_k$, we have $z_{ik} = 1$ if and only if $y_{ik}^* > -\mu_k$. Consequently, $-\mu_k$ is the threshold corresponding to the factor analysis model associated with y_{ik}^*. A special case of the current model with $\boldsymbol{\omega}_i$ distributed as $N[\mathbf{0}, \mathbf{I}]$ is the FIIF model (see Bock and Aitkin, 1981) that has wide applications in education. Under this special case, $-\mu_k$ can be interpreted as the kth item difficulty, and the elements in $\mathbf{\Lambda}_k^T$ are the item slopes of the latent factors. Using the above natural formulation of the model and identification conditions, we can estimate the thresholds via $-\mu_k$. Moreover, as $\psi_{\epsilon k}$ are not the parameters of main interest, fixing them at 1.0 does not have a great influence on the interpretation of the structural equation model. However, we emphasize that this is not the unique way to formulate the model and/or to impose its identification conditions.

Note that as there are at least two thresholds associated with an ordered categorical variable, the relation between the thresholds and $-\mu_k$ is not as clear. Also, the identification conditions are slightly different. Thus, methods for analyzing these two types of discrete variables are not exactly the same.

Let $\mathbf{Z} = (\mathbf{z}_1, \ldots, \mathbf{z}_n)$ be the observed data set of the dichotomous variables, and our objective is to develop a Bayesian procedure to estimate the unknown parameter vector $\boldsymbol{\theta}$, which contains parameters in $\boldsymbol{\mu}$, $\boldsymbol{\Lambda}$, $\boldsymbol{\Lambda}_\omega$, $\boldsymbol{\Phi}$ and $\boldsymbol{\Psi}_\delta$, based on $\mathbf{Z}$. As usual, this is done by analyzing the following log posterior density of $\boldsymbol{\theta}$ given $\mathbf{Z}$,

$$\log p(\boldsymbol{\theta}|\mathbf{Z}) \propto \log p(\mathbf{Z}|\boldsymbol{\theta}) + \log p(\boldsymbol{\theta}), \tag{7.6}$$

where $p(\mathbf{Z}|\boldsymbol{\theta})$ is the likelihood function, and $p(\boldsymbol{\theta})$ is the prior density of $\boldsymbol{\theta}$. Due to the complexity of the model and the nature of the data, $p(\mathbf{Z}|\boldsymbol{\theta})$ involves complicated multidimensional integrals and the posterior distribution is very complicated. It is both difficult and tedious to compute the posterior mean directly from $p(\boldsymbol{\theta}|\mathbf{Z})$. Hence, the technique of data augmentation (Tanner and Wong, 1987) is again employed in the posterior analysis. Let $\boldsymbol{\Omega} = (\boldsymbol{\omega}_1, \ldots, \boldsymbol{\omega}_n)$ be the matrix of latent variables of the model, and $\mathbf{Y} = (\mathbf{y}_1, \ldots, \mathbf{y}_n)$ be the matrix of latent continuous measurements underlying the matrix of observed dichotomous data $\mathbf{Z}$. In the analysis, the observed data $\mathbf{Z}$ is augmented with $\boldsymbol{\Omega}$ and $\mathbf{Y}$, which can be considered as hypothetical missing data, to form a complete data set $(\mathbf{Z}, \boldsymbol{\Omega}, \mathbf{Y})$. Again, a large sample of observations will be sampled from $p(\boldsymbol{\theta}, \boldsymbol{\Omega}, \mathbf{Y}|\mathbf{Z})$ by the Gibbs sampler (Geman and Geman, 1984). The statistics of interest, for example the Bayesian estimates, will be obtained from the generated sample via standard data analysis methods.

To implement the Gibbs sampler, we start with initial values $(\boldsymbol{\theta}^{(0)}, \boldsymbol{\Omega}^{(0)}, \mathbf{Y}^{(0)})$, simulate $(\boldsymbol{\theta}^{(1)}, \boldsymbol{\Omega}^{(1)}, \mathbf{Y}^{(1)}), \ldots,$ and continue as follows. At the mth iteration with current values $\boldsymbol{\theta}^{(m)}$, $\boldsymbol{\Omega}^{(m)}$ and $\mathbf{Y}^{(m)}$:

$$\begin{aligned}
&\text{Step (a)}: \text{ Generate } \boldsymbol{\Omega}^{(m+1)} \text{ from } p(\boldsymbol{\Omega}|\boldsymbol{\theta}^{(m)}, \mathbf{Y}^{(m)}, \mathbf{Z});\\
&\text{Step (b)}: \text{ Generate } \boldsymbol{\theta}^{(m+1)} \text{ from } p(\boldsymbol{\theta}|\boldsymbol{\Omega}^{(m+1)}, \mathbf{Y}^{(m)}, \mathbf{Z});\\
&\text{Step (c)}: \text{ Generate } \mathbf{Y}^{(m+1)} \text{ from } p(\mathbf{Y}|\boldsymbol{\theta}^{(m+1)}, \boldsymbol{\Omega}^{(m+1)}, \mathbf{Z}).
\end{aligned} \tag{7.7}$$

Under mild conditions and after a sufficiently large number of iterations, the joint distribution of $(\boldsymbol{\theta}^{(m)}, \boldsymbol{\Omega}^{(m)}, \mathbf{Y}^{(m)})$ converges at an exponential rate to the desired posterior distribution $[\boldsymbol{\theta}, \boldsymbol{\Omega}, \mathbf{Y}|\mathbf{Z}]$. Convergence of the Gibbs sampler can be revealed by plots of the simulated sequences of the individual parameters, and/or by the EPSR values (Gelman, 1996) corresponding to the parameters that are calculated sequentially as the runs proceed.

In deriving the conditional distribution $p(\boldsymbol{\Omega}|\boldsymbol{\theta}, \mathbf{Y}, \mathbf{Z})$ in Step (a), we first note that given $\mathbf{Y}$, the underlying model becomes one with continuous data and is relatively easy to handle. Moreover, as $\boldsymbol{\omega}_i, i = 1, \ldots, n$, are mutually independent,

$$p(\boldsymbol{\Omega}|\boldsymbol{\theta}, \mathbf{Y}, \mathbf{Z}) = p(\boldsymbol{\Omega}|\boldsymbol{\theta}, \mathbf{Y}) = \prod_{i=1}^{n} p(\boldsymbol{\omega}_i|\boldsymbol{\theta}, \mathbf{y}_i). \tag{7.8}$$

It follows from derivations similar to these in Section 6.3.1 that one representation of the conditional distribution $[\boldsymbol{\omega}_i|\boldsymbol{\theta}, \mathbf{y}_i]$ is given by:

$$[\boldsymbol{\omega}_i|\boldsymbol{\theta}, \mathbf{y}_i] \stackrel{D}{=} N[\boldsymbol{\Sigma}^* \boldsymbol{\Lambda}^T \boldsymbol{\Psi}_\epsilon^{-1}(\mathbf{y}_i - \boldsymbol{\mu}), \boldsymbol{\Sigma}^*], \quad i = 1, \ldots, n, \tag{7.9}$$

where $\boldsymbol{\Sigma}^* = (\boldsymbol{\Sigma}_\omega^{-1} + \boldsymbol{\Lambda}^T \boldsymbol{\Psi}_\epsilon^{-1} \boldsymbol{\Lambda})^{-1}$, $\boldsymbol{\Sigma}_\omega$ is the covariance matrix of $\boldsymbol{\omega}_i$ (see Equation (6.5)), and $\boldsymbol{\Psi}_\epsilon$ is fixed as the identity matrix for identifying the model. The conditional distribution of $\boldsymbol{\Omega}$ given $\boldsymbol{\theta}$, $\mathbf{Y}$ and $\mathbf{Z}$ can be obtained via Equations (7.8) and (7.9).

Consider the conditional distribution of $\boldsymbol{\theta}$ given $\boldsymbol{\Omega}, \mathbf{Y}$ and $\mathbf{Z}$ in Step (b). It involves components that correspond to $\boldsymbol{\mu}, \boldsymbol{\Lambda}, \boldsymbol{\Lambda}_\omega, \boldsymbol{\Phi}$ and $\boldsymbol{\Psi}_\delta$. Hence, it is divided into the following substeps which generate $\boldsymbol{\mu}^{(m+1)}$ from $p(\boldsymbol{\mu}|\boldsymbol{\Lambda}^{(m)}, \boldsymbol{\Lambda}_\omega^{(m)}, \boldsymbol{\Phi}^{(m)}, \boldsymbol{\Psi}_\delta^{(m)}, \boldsymbol{\Omega}^{(m)}, \mathbf{Y}^{(m)}, \mathbf{Z})$, $\boldsymbol{\Lambda}^{(m+1)}$ from $p(\boldsymbol{\Lambda}|\boldsymbol{\mu}^{(m+1)}, \boldsymbol{\Lambda}_\omega^{(m)}, \boldsymbol{\Phi}^{(m)}, \boldsymbol{\Psi}_\delta^{(m)}, \boldsymbol{\Omega}^{(m)}, \mathbf{Y}^{(m)}, \mathbf{Z})$, $\boldsymbol{\Lambda}_\omega^{(m+1)}$ from $p(\boldsymbol{\Lambda}_\omega|\boldsymbol{\mu}^{(m+1)}, \boldsymbol{\Lambda}^{(m+1)}, \boldsymbol{\Phi}^{(m)}, \boldsymbol{\Psi}_\delta^{(m)}, \boldsymbol{\Omega}^{(m)}, \mathbf{Y}^{(m)}, \mathbf{Z})$, $\boldsymbol{\Phi}^{(m+1)}$ from $p(\boldsymbol{\Phi}|\boldsymbol{\mu}^{(m+1)}, \boldsymbol{\Lambda}^{(m+1)}, \boldsymbol{\Lambda}_\omega^{(m+1)}, \boldsymbol{\Psi}_\delta^{(m)}, \boldsymbol{\Omega}^{(m)}, \mathbf{Y}^{(m)}, \mathbf{Z})$, and $\boldsymbol{\Psi}_\delta^{(m+1)}$ from $p(\boldsymbol{\Psi}_\delta|\boldsymbol{\mu}^{(m+1)}, \boldsymbol{\Lambda}^{(m+1)}, \boldsymbol{\Lambda}_\omega^{(m+1)}, \boldsymbol{\Phi}^{(m+1)}, \boldsymbol{\Omega}^{(m)}, \mathbf{Y}^{(m)}, \mathbf{Z})$. We first note that the SEM defined in Equations (7.1) and (7.2) is equal to that given in Equations (6.1) and (6.2). Moreover, as $\mathbf{Y}$ and $\boldsymbol{\Omega}$ are given, we essentially obtain the same information about the underlying continuous measurements and the latent variables as in the conditional distribution of $\boldsymbol{\theta}$ involved in Step (b) of the Gibbs sampler described in Section 6.3. Hence, the model defined in Equations (7.1) and (7.2) with given $\mathbf{Y}$ and $\boldsymbol{\Omega}$ reduces to a similar linear model as before. Consequently, the conditional distributions involved in Step (b) can be obtained from Equations (6.7), (6.8) and (6.9), under the following conjugate prior distributions that are similar to those given in Section 6.3.1:

$$\begin{aligned} &\boldsymbol{\mu} \stackrel{D}{=} N(\boldsymbol{\mu}_0, \Sigma_0), \quad \boldsymbol{\Lambda}_k \stackrel{D}{=} N[\boldsymbol{\Lambda}_{0k}, \mathbf{H}_{0yk}], \quad \psi_{\delta k}^{-1} \stackrel{D}{=} Gamma[\alpha_{0\delta k}, \beta_{0\delta k}], \\ &[\boldsymbol{\Lambda}_{\omega k}|\psi_{\delta k}] \stackrel{D}{=} N[\boldsymbol{\Lambda}_{0\omega k}, \psi_{\delta k}\mathbf{H}_{0\omega k}], \quad \boldsymbol{\Phi}^{-1} \stackrel{D}{=} W_q[\mathbf{R}_0, \rho_0], \end{aligned} \tag{7.10}$$

where $\psi_{\delta k}$ is the kth diagonal element of $\boldsymbol{\Psi}_\delta$, $\boldsymbol{\Lambda}_k^T$ and $\boldsymbol{\Lambda}_{\omega k}^T$ are the kth rows of $\boldsymbol{\Lambda}$ and $\boldsymbol{\Lambda}_\omega$, respectively; and $\boldsymbol{\Lambda}_{0k}$, $\boldsymbol{\Lambda}_{0\omega k}$ $\alpha_{0\delta k}$, $\beta_{0\delta k}$, $\boldsymbol{\mu}_0$ and ρ_0, and positive definite matrices $\mathbf{H}_{0yk}$, $\mathbf{H}_{0\omega k}$ and $\mathbf{R}_0$ are hyperparameters whose values are assumed to be given by the prior information.

We now consider $p(\mathbf{Y}|\boldsymbol{\theta}, \boldsymbol{\Omega}, \mathbf{Z})$. As the $\mathbf{y}_i$ are mutually independent, it follows that

$$p(\mathbf{Y}|\boldsymbol{\theta}, \boldsymbol{\Omega}, \mathbf{Z}) = \prod_{i=1}^{n} p(\mathbf{y}_i|\boldsymbol{\theta}, \boldsymbol{\omega}_i, \mathbf{z}_i).$$

Moreover, it follows from the definition of the model and Equation (7.4) that

$$p(y_{ik}|\boldsymbol{\theta}, \boldsymbol{\omega}_i, \mathbf{z}_i) \stackrel{D}{=} \begin{cases} N[\mu_k + \boldsymbol{\Lambda}_k^T \boldsymbol{\omega}_i, 1] I_{(-\infty,0)}(y_{ik}), & \text{if } z_{ik} = 0 \\ N[\mu_k + \boldsymbol{\Lambda}_k^T \boldsymbol{\omega}_i, 1] I_{(0,\infty)}(y_{ik}), & \text{if } z_{ik} = 1, \end{cases} \tag{7.11}$$

where μ_k is the kth element of $\boldsymbol{\mu}$, and $I_A(y)$ is an indicator function that takes the value 1 if y is in A, and 0 otherwise.

As the conditional distributions that are involved in Steps (a), (b) and (c) are familiar, drawing observations from them is straightforward and fast. Consequently, the programming and computational burden involved in the Gibbs sampler is not heavy.

Statistical inference of the model can be obtained on the basis of the simulated sample of observations from $p(\boldsymbol{\theta}, \boldsymbol{\Omega}, \mathbf{Y}|\mathbf{Z})$, as before. For example, the Bayesian estimates of elements in $\boldsymbol{\theta}$ as well as their numerical standard error estimates can be obtained from the sample mean and the sample covariance matrix, respectively. Note that for dichotomous data analysis, a vector of dichotomous observation $\mathbf{z}_i$ rather than $\mathbf{y}_i$ is observed; hence a lot of information of $\mathbf{y}_i$ is lost. As a result, it requires a rather large sample size to achieve accurate estimates. The PP p-value for assessing the goodness-of-fit of a posited model, and the Bayes factor for model comparison can be obtained via similar developments as given in Chapter 6. Theoretically, for each $\mathbf{y}_i$, we can estimate its associative vector of latent variables $\boldsymbol{\omega}_i$. Since the only data information available in the estimation is $\mathbf{y}_i$, $\hat{\boldsymbol{\omega}}_i$ is not expected to be an accurate estimate of the true $\boldsymbol{\omega}_i$. This is particularly true in dealing with dichotomous data. As the estimate of the residual $\hat{\boldsymbol{\epsilon}}_i$ depends on an estimate of $\mathbf{y}_i$, its accuracy is further affected.

7.2.1 Illustrative Example 1

It has been pointed out that patient adherence to prescribed medication is crucial to the success of medical treatment, and that nonadherence can lead to a misjudgment of the effectiveness of the medication, see Czajkowski, Chesney, and Smith (1998), and Rand and Weeks (1998). In the promotion of adherence, it is desirable to establish a statistical model to study the relationships between nonadherence and its core factors such as a patient's knowledge, attitudes and beliefs concerning medication, the health condition of the patients, a physician's interaction and communication, and so on.

To enrich existing knowledge about patient nonadherence, the Department of Medicine and Therapeutics, Community and Family Medicine, and Pharmacy at the Chinese University of Hong Kong conducted a survey of ethnic Chinese patients diagnosed as suffering from hypertension (Chan, 2002). The main objectives were to measure and examine relationships among latent variables such as physician advice and concern, patient knowledge and belief, social cognition and social influence, and the subsequent study reported nonadherence with reference to a structural equation model. Inspired by the fact that this study involved many dichotomous variables, we apply the methodology developed here to assess the effects of some latent variables on patients' nonadherence to medication.

For the sake of illustration, suppose that we are interested in studying how 'nonadherence' of patients is affected by their 'knowledge of medication' and 'health condition' by analyzing the related portion of the whole data set. Six dichotomous manifest variables are selected as indicators for the first two latent variables mentioned above. Three ordered categorical manifest variables measured by a five-point scale are used as indicators for the last latent variable about 'health condition'. As these ordered categorical variables are heavily skewed (see frequencies of the last three variables in Appendix 7.1), and in order to provide an illustration of the proposed method for dichotomous variables, we transform them to dichotomous by grouping the first four categories on the left together. The information lost due to this grouping should not be substantial. Translations of the corresponding questions from Chinese into English are listed in Appendix 7.1, together with their frequencies. For brevity, we deleted a small number of observations with missing entries, and the remaining sample size is 837.

The resulting data set of dichotomous observations is analyzed using a model as defined in Equations (7.1) and (7.2) (see also Lee and Song, 2003). Although other structures for the loading matrix can be considered, we choose the structure that gives nonoverlapping latent factors for clear interpretation. Hence the following specification of the loading matrix $\mathbf{\Lambda}$ is used:

$$\mathbf{\Lambda}^T = \begin{bmatrix} 1 & \lambda_{21} & \lambda_{31} & 0 & 0 & 0 & 0 & 0 & 0 \\ 0 & 0 & 0 & 1 & \lambda_{52} & \lambda_{62} & 0 & 0 & 0 \\ 0 & 0 & 0 & 0 & 0 & 0 & 1 & \lambda_{83} & \lambda_{93} \end{bmatrix},$$

where λ_{ij}'s are the unknown factor loading parameters, but 1's and 0's are fixed in the estimation for achieving an identified model. From the meanings of the questions, see Appendix 7.1, it is clear that this structure gives three nonoverlapping factors (latent variables), which can be interpreted as the 'nonadherence, η', 'knowledge of medication, ξ_1' and 'health condition, ξ_2' of the patients. As our main interest is to study the linear effects of 'knowledge of medication' and 'health condition' on 'nonadherence', the structural equation is chosen to be a linear regression model which regresses η on ξ_1 and ξ_2 as follows:

$$\eta = \gamma_1 \xi_1 + \gamma_2 \xi_2 + \delta, \tag{7.12}$$

where γ_1 and γ_2 are unknown parameters. Other unknown parameters include the variances and covariance of ξ_1 and ξ_2: ϕ_{11}, ϕ_{22} and ϕ_{12}, and the variance of δ, ψ_δ. There are a total of 12 unknown parameters in this model.

The proposed Gibbs sampler algorithm is used to obtain Bayesian estimates of the parameters. Estimates of the latent variables can also be obtained as by-products. As an illustration for analyzing data sets with no good prior information, we use the following data-dependent prior inputs. An auxiliary Bayesian estimation with non-informative prior for obtaining prior inputs of the hyperparameters values for $\mathbf{\Lambda}_{0k}$, $\mathbf{\Lambda}_{0\omega k}$ and $\mathbf{R}_0$ in the conjugate prior distributions is first conducted. Consequently, the values of hyperparameters are given by $\alpha_{0\delta k} = 8$, $\beta_{0\delta k} = 10$, $\rho_0 = 8$, $\mathbf{H}_{0yk}$ and $\mathbf{H}_{0\omega k}$ are identity matrices, $\mathbf{\Lambda}_{0k} = \tilde{\mathbf{\Lambda}}_k$, $\mathbf{\Lambda}_{0\omega k} = \tilde{\mathbf{\Lambda}}_{\omega k}$ and $\mathbf{R}_0 = \mathbf{I}$, where $\tilde{\mathbf{\Lambda}}_k$ and $\tilde{\mathbf{\Lambda}}_{\omega k}$ are estimates obtained from the auxiliary estimation. With some arbitrary starting values, we observe that the Gibbs sampler in the actual estimation converges in about 5000 iterations. To

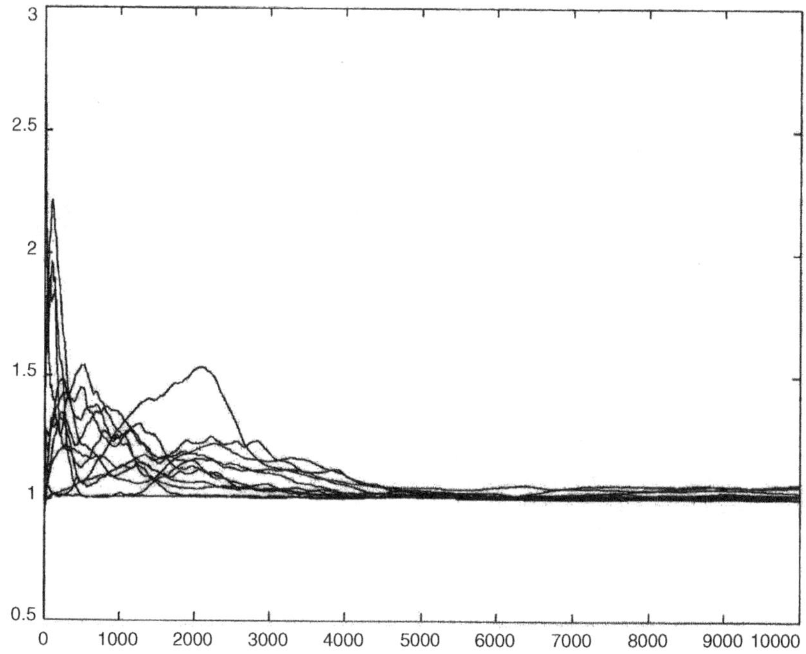

Figure 7.1 EPSR values against the numbers of iterations. This figure and Figure 7.2 are taken from Lee and Song (2003).

reveal the convergence, plots of the EPSR values and some parameters' values against the iteration numbers are presented in Figures 7.1 and 7.2 respectively. Comparing the numbers of iterations required in analyzing dichotomous and ordered categorical data, we find that dichotomous data require more iterations. This is probably due to the fact that dichotomous data carry less information than ordered categorical data. To obtain the Bayesian estimates, 5000 observations are collected after discarding the first 5000 burn-in iterations. Results are presented in Table 7.1. Based on these estimates, $\hat{\boldsymbol{\Sigma}}$ and $\hat{\boldsymbol{\Phi}}$, the estimated covariance matrices of the manifest random vector $\mathbf{y}$ and the latent random vector $\boldsymbol{\omega}$, can be computed. Sometimes, it is also desirable to transform these estimates to a completely standardized (CS) solution such that both manifest and latent variables are standardized, and $\hat{\boldsymbol{\Sigma}}$ and $\hat{\boldsymbol{\Phi}}$ are correlation matrices

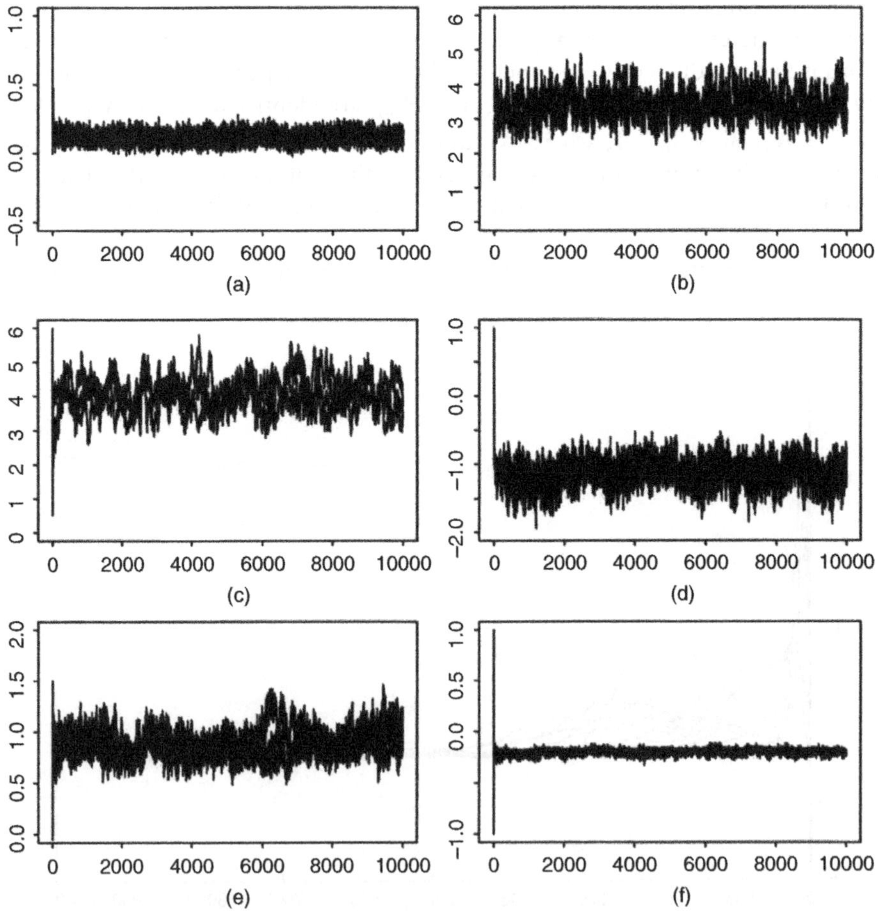

Figure 7.2 (a), (b), (c), (d), (e) and (f) are plots of parallel sequences corresponding to different starting values of λ_{31}, λ_{52}, λ_{83}, γ_1, γ_2 and ϕ_{12} against iterations.

Table 7.1 Bayesian estimates and its completely standardized solution.

Parameter	Bayesian estimates	CS solution
λ_{11}	1.0*	0.97
λ_{21}	3.66	0.97
λ_{31}	0.12	1.31
λ_{42}	1.0*	0.40
λ_{52}	4.32	0.88
λ_{62}	3.67	0.85
λ_{73}	1.0*	0.57
λ_{83}	4.32	0.95
λ_{93}	4.14	0.95
ϕ_{11}	0.19	1.0
ϕ_{21}	−0.18	−0.59
ϕ_{22}	0.49	1.0
γ_1	−1.21	−0.38
γ_2	0.90	0.45
ψ_δ	0.89	0.46

Note: λ_{11}, λ_{42} and λ_{73} are fixed in the Bayesian estimation. This table is taken from Lee and Song (2003).

(see Jöreskog and Sörbom, 1996). For completeness, the CS solution obtained from our Bayesian estimates is also presented in Table 7.1. The PP p-value for goodness-of-fit testing of the posited model is 0.414. On the basis of the result given in Gelman, Meng and Stern (1996) which suggested that a PP p-value close to 0.5 indicates good fit of the posited model, the PP p-value obtained indicates that the proposed model fits the data.

The most important interpretations of the results are as follows. (i) From the estimates of γ_1 and γ_2 in the structural equation, we see that ξ_1 and ξ_2 have significant effects on η. Moreover, it is clear from the estimated structural equation: $\eta = -1.21\xi_1 + 0.90\xi_2$, that better 'knowledge of medication, ξ_1' has a negative effect on 'nonadherence', whilst weaker or worse 'health condition, ξ_2' has a positive effect on 'nonadherence'. Hence, it is desirable to better educate patients about their illnesses and encourage them to pay more attention to their health. (ii) From the completely standardized solution, we see that the correlation estimate between ξ_1 and ξ_2 is −0.59. Hence, we arrive at the expected conclusion that better 'knowledge of medication, ξ_1' and weaker 'health condition, ξ_2' are negatively correlated.

This data set has been reanalyzed by using WinBUGS. We find that the WinBUGS program converged after 5000 iterations. The Bayesian estimates that are obtained by using 5000 observations after convergence are reported in Table 7.2. Under the expected MCMC sampling errors, we observe that the Bayesian estimates obtained by WinBUGS are reasonable close to those that are reported in Table 7.1. The DIC value corresponding to this model is

Table 7.2 Bayesian estimates and their standard error estimates obtained via WinBUGS.

Parameter	Estimate	SE	Parameter	Estimate	SE
λ_{11}	1.0*	–	ϕ_{11}	0.13	0.02
λ_{21}	3.36	0.38	ϕ_{21}	−0.14	0.02
λ_{31}	0.12	0.04	ϕ_{22}	0.42	0.06
λ_{42}	1.0*	–			
λ_{52}	4.69	0.37	γ_1	−1.51	0.26
λ_{62}	4.14	0.36	γ_2	1.01	0.15
λ_{73}	1.0*	–			
λ_{83}	4.90	0.38	ψ_δ	0.80	0.14
λ_{93}	4.26	0.36			

Note: λ_{11}, λ_{42} and λ_{73} are fixed in the Bayesian estimation.

7403.5. The codes and the data set are presented in the following website: http://www.wiley.com/go/lee_structural.

7.3 ANALYSIS OF A MULTIVARIATE PROBIT CONFIRMATORY FACTOR ANALYSIS MODEL

As the main objective of a multivariate probit (MP) model is on the mean structure, we consider a model such that each subject has vectors of fixed covariates that may come from any mixture of discrete and continuous variables. Specifically, we assume that each subject produces p distinct quantal responses or is classified with respect to p dichotomous categories. Let $\mathbf{z}_i = (z_{i1}, \ldots, z_{ip})^T$ denote the collection of observed dichotomous responses in p variables on the ith subject, $i = 1, \ldots, n$, $\mathbf{x}_{ik}$ be an $r_k \times 1$ vector of covariates, $R = r_1 + \cdots + r_p$, and let

$$\mathbf{X}_i = \begin{bmatrix} \mathbf{x}_{i1}^T & 0 & \cdots & 0 \\ 0 & \mathbf{x}_{i2}^T & \cdots & 0 \\ \vdots & \vdots & \vdots & \vdots \\ 0 & 0 & \cdots & \mathbf{x}_{ip}^T \end{bmatrix}$$

be a $p \times R$ matrix. Consider the following MP model that was formulated by Chib and Greenberg (1998). Let $\mathbf{y}_i = (y_{i1}, \ldots, y_{ip})^T$ denote a p-variate normal vector of 'response strengths' such that

$$\mathbf{y}_i = \mathbf{X}_i \mathbf{B} + \boldsymbol{\epsilon}_i, \quad i = 1, \ldots, n, \tag{7.13}$$

where $\mathbf{B}^T = (\mathbf{b}_1^T, \ldots, \mathbf{b}_p^T)$, $\mathbf{b}_k$ is an $r_k \times 1$ unknown parameter vector, $\boldsymbol{\epsilon}_i$ is a $p \times 1$ vector of residuals that is distributed as $N[\mathbf{0}, \boldsymbol{\Sigma}]$, and

$$z_{ik} = \begin{cases} 1 & \text{if } y_{ik} = \mathbf{x}_{ik}^T \mathbf{b}_k + \epsilon_{ik} = \mathbf{b}_k^T \mathbf{x}_{ik} + \epsilon_{ik} > 0, \quad k = 1, \ldots, p \\ 0 & \text{if otherwise.} \end{cases} \tag{7.14}$$

In this model, the exact measurement of 'response strengths' $\mathbf{y}_i$ is not observed, and its information is given by an observed dichotomous vector $\mathbf{z}_i = (z_{i1}, \ldots, z_{ip})^T$ with z_{ik} defined by Equation (7.14). Here, $\mathbf{B}$ is an $R \times 1$ vector of regression coefficients of $\mathbf{y}_i$ on $\mathbf{X}_i$. If we take the first component of $\mathbf{x}_{ik}$ to be 1.0 for all i and k, then the first component of $\mathbf{b}_k$ can be interpreted as the mean (or intercept) of $\mathbf{y}_{ik}$. This MP model has quite wide applications. Chib and Greenberg (1998) utilized some MCMC methods in developing a Bayesian approach and an ML approach for analyzing this model. As the covariance matrix, $\boldsymbol{\Sigma}$, of $\boldsymbol{\epsilon}_i$ is a general positive definite matrix, observations from a multivariate truncated normal distribution are simulated in their method. The computational burden for this task is fairly heavy. We will show below that this burden can be reduced by incorporating a confirmatory factor analysis in the MP model.

Consider the following extension of the MP model with a confirmatory factor analysis model for the underlying 'response strengths' $\mathbf{y}_i$ (see Song and Lee, 2005):

$$\mathbf{y}_i = \mathbf{X}_i \mathbf{B} + \boldsymbol{\Lambda} \boldsymbol{\omega}_i + \boldsymbol{\zeta}_i, \quad i = 1, \ldots, n, \tag{7.15}$$

where $\boldsymbol{\Lambda}$ is a $p \times q$ loading matrix of parameters which may be unknown or known, $\boldsymbol{\omega}_i$ is a $q \times 1$ vector of latent factors, and $\boldsymbol{\zeta}_i$ is a $p \times 1$ vector of residuals. For $i = 1, \ldots, n$, we assume that $\boldsymbol{\omega}_i$ is independently distributed as $N[\mathbf{0}, \boldsymbol{\Phi}]$, $\boldsymbol{\zeta}_i$ is independently distributed as $N[\mathbf{0}, \boldsymbol{\Psi}_\zeta]$, where $\boldsymbol{\Phi}$ is an arbitrary covariance or correlation matrix, $\boldsymbol{\Psi}_\zeta$ is a diagonal covariance matrix with diagonal elements $\psi_{\zeta k}$, and $\boldsymbol{\omega}_i$ and $\boldsymbol{\zeta}_i$ are uncorrelated. For brevity, we call this the MPCFA model. To identify the model, we fix $\boldsymbol{\Psi}_\zeta$ at a known diagonal matrix and appropriate elements in $\boldsymbol{\Lambda}$ at preassigned values.

The MP model (Chib and Greenberg, 1998), as defined by Equations (7.13) and (7.14) with an arbitrary covariance matrix $\boldsymbol{\Sigma}$, can be analyzed with the MPCFA model by setting $\boldsymbol{\Lambda} = \mathbf{I}_p$, where $\mathbf{I}_p$ is a $p \times p$ identity matrix, $\boldsymbol{\epsilon}_i = \boldsymbol{\omega}_i + \boldsymbol{\zeta}_i$ and $\boldsymbol{\Sigma} = \boldsymbol{\Phi} + \boldsymbol{\Psi}_\zeta$. In practice, we can set $\boldsymbol{\Psi}_\zeta = c\mathbf{I}_p$, where c is a small value, and use $\boldsymbol{\Phi}$ to capture covariances in $\boldsymbol{\Sigma}$.

If all $\mathbf{x}_{i1}, \ldots, \mathbf{x}_{ip}$ equal to an $r^* \times 1$ vector $\mathbf{x}_i$, then $\mathbf{X}_i\mathbf{B}$ in Equation (7.15) can be written as $\mathbf{B}^*\mathbf{x}_i$, where $\mathbf{B}^*$ is a $p \times r^*$ matrix with rows equal to $\mathbf{b}_1^T, \ldots, \mathbf{b}_p^T$. The form of Equation (7.15) then reduces to the model given by Gibbons and Wilcox-Gök (1998). Hence, the covariates that are considered in the proposed

MPCFA model are more general. Another generalization is that the covariance matrix of the latent factors in the MPCFA model is equal to a general covariance or correlation matrix $\mathbf{\Phi}$ rather than an identity matrix as in the Gibbons and Wilcox-Gök (1998) model. This extension requires very little extra computing effort. On the other hand, because $\mathbf{\Phi}$ is an arbitrary positive definite matrix, it is useful for capturing $\mathbf{\Sigma}$ as $\mathbf{\Phi}+\mathbf{\Psi}_\zeta$ in analyzing models with arbitrary covariance structures. Another special case of the MPCFA model is given by

$$\mathbf{y}_i = \mathbf{\Lambda}\boldsymbol{\omega}_i + \boldsymbol{\zeta}_i, \quad i = 1, \ldots, n \tag{7.16}$$

without any covariates. This model can be viewed as a confirmatory factor analysis model for dichotomous variables, or the FIIF model with correlated factors.

Let $\mathbf{Z} = (\mathbf{z}_1, \ldots, \mathbf{z}_n)$ be the observed data matrix of the dichotomous outcomes, $\mathbf{Y} = (\mathbf{y}_1, \ldots, \mathbf{y}_n)$ be the matrix of latent continuous measurements underlying $\mathbf{Z}$, $\mathbf{\Omega} = (\boldsymbol{\omega}_1, \ldots, \boldsymbol{\omega}_n)$ be the matrix of latent variables, $\mathbf{X} = (\mathbf{X}_1, \ldots, \mathbf{X}_n)$ be the observed values of the fixed covariates, and $\boldsymbol{\theta}$ be the parameter vector that contains unknown parameters in $\mathbf{B}$, $\mathbf{\Lambda}$ and $\mathbf{\Phi}$. The Bayesian estimate of $\boldsymbol{\theta}$, and the latent vectors in $\mathbf{\Omega}$ can be obtained via the following Gibbs sampler: At the mth iteration with current values $\boldsymbol{\theta}^{(m)}$, $\mathbf{\Omega}^{(m)}$ and $\mathbf{Y}^{(m)}$,

$$\begin{aligned}
&\text{Step (a): Generate } \mathbf{\Omega}^{(m+1)} \text{ from } p(\mathbf{\Omega}|\boldsymbol{\theta}^{(m)}, \mathbf{Y}^{(m)}, \mathbf{Z});\\
&\text{Step (b): Generate } \boldsymbol{\theta}^{(m+1)} \text{ from } p(\boldsymbol{\theta}|\mathbf{\Omega}^{(m+1)}, \mathbf{Y}^{(m)}, \mathbf{Z});\\
&\text{Step (c): Generate } \mathbf{Y}^{(m+1)} \text{ from } p(\mathbf{Y}|\boldsymbol{\theta}^{(m+1)}, \mathbf{\Omega}^{(m+1)}, \mathbf{Z}).
\end{aligned}$$

The basic structure of this Gibbs sampler is similar to that given in (7.7). For completeness, the conditional distributions are given as follows.

To derive the conditional distribution in Step (a), we let $\mathbf{\Sigma}^* = (\mathbf{\Phi}^{-1} + \mathbf{\Lambda}^T\mathbf{\Psi}_\zeta^{-1}\mathbf{\Lambda})^{-1}$. We have

$$p(\mathbf{\Omega}|\boldsymbol{\theta}, \mathbf{Y}, \mathbf{Z}) = p(\mathbf{\Omega}|\boldsymbol{\theta}, \mathbf{Y}) = \prod_{i=1}^{n} p(\boldsymbol{\omega}_i|\boldsymbol{\theta}, \mathbf{y}_i)$$

where $[\boldsymbol{\omega}_i|\boldsymbol{\theta}, \mathbf{y}_i] \stackrel{D}{=} N[\mathbf{\Sigma}^*\mathbf{\Lambda}^T\mathbf{\Psi}_\zeta^{-1}(\mathbf{y}_i - \mathbf{X}_i\mathbf{B}), \mathbf{\Sigma}^*]$. To obtain the conditional distribution in Step (c), it follows from the definition of the model in Equation (7.15) that

$$\begin{aligned}
p(\mathbf{Y}|\boldsymbol{\theta}, \mathbf{\Omega}, \mathbf{Z}) &= \prod_{i=1}^{n} p(\mathbf{y}_i|\boldsymbol{\theta}, \boldsymbol{\omega}_i, \mathbf{z}_i)\\
&= \prod_{i=1}^{n}\prod_{k=1}^{p} p(y_{ik}|\boldsymbol{\theta}, \boldsymbol{\omega}_i, z_{ik}),
\end{aligned}$$

where

$$[y_{ik}|\boldsymbol{\theta}, \boldsymbol{\omega}_i, z_{ik}] \stackrel{D}{=} \begin{cases} N[\mathbf{x}_{ik}^T\mathbf{b}_k + \boldsymbol{\Lambda}_k^T\boldsymbol{\omega}_i, \psi_{\zeta k}]I_{(-\infty,0)}(y_{ik}) & \text{if} \quad z_{ik}=0, \\ N[\mathbf{x}_{ik}^T\mathbf{b}_k + \boldsymbol{\Lambda}_k^T\boldsymbol{\omega}_i, \psi_{\zeta k}]I_{(0,\infty)}(y_{ik}) & \text{if} \quad z_{ik}=1. \end{cases}$$

The following conjugate prior distributions that are similar to those given in Equation (7.10) are used in obtaining the conditional distributions involved in Step (b). For $k=1,\ldots,p$

$$\mathbf{b}_k \stackrel{D}{=} N[\mathbf{b}_{0k}, \mathbf{H}_{0bk}], \ \boldsymbol{\Lambda}_k \stackrel{D}{=} N[\boldsymbol{\Lambda}_{0k}, \mathbf{H}_{0yk}] \ \text{and} \ \boldsymbol{\Phi}^{-1} \stackrel{D}{=} W_q[\mathbf{R}_0, \rho_0], \qquad (7.17)$$

where $\boldsymbol{\Lambda}_k^T$ is the kth row of $\boldsymbol{\Lambda}$, and $\mathbf{b}_{0k}, \mathbf{H}_{0bk}, \boldsymbol{\Lambda}_{0k}, \mathbf{H}_{0yk}, \mathbf{R}_0$ and ρ_0 are given hyperparameters. The conditional distribution corresponding to $\mathbf{b}_k$ is given by

$$p(\mathbf{b}_k|\boldsymbol{\Omega}, \mathbf{Y}, \mathbf{Z}) \stackrel{D}{=} N[\mathbf{a}_k, \mathbf{A}_k], \qquad (7.18)$$

where $\mathbf{A}_k = (\mathbf{H}_{0bk}^{-1} + \mathbf{X}_k^*\mathbf{X}_k^{*T})^{-1}, \mathbf{a}_k = \mathbf{A}_k(\mathbf{H}_{0bk}^{-1}\mathbf{b}_{0k} + \mathbf{X}_k^*\mathbf{Y}_k^{*T})$ with $\mathbf{X}_k^* = (\mathbf{x}_{1k}, \ldots, \mathbf{x}_{nk})$ and $\mathbf{Y}_k^* = \{(\mathbf{y}_{ik} - \boldsymbol{\Lambda}_k^T\boldsymbol{\omega}_i), i=1,\ldots,n\}$. The conditional distributions corresponding to $\boldsymbol{\Lambda}_k$ and $\boldsymbol{\Phi}$ can be obtained from the results given in Section 6.3.1 (see also Section 7.2).

7.3.1 Illustrative Example 2

The same data set used in the illustrative example in Section 7.2.1 is reanalyzed. To demonstrate the MPCFA model, we include two fixed covariates, one about patients' education, x_{ik1} (coded by 0, 1, 2, 3), and the other about the presence of side effects of drugs, x_{ik2} (coded by 0 and 1) in the analysis. The loading matrix $\boldsymbol{\Lambda}$ for the three latent factors 'nonadherence, $\eta = \omega_1$', 'knowledge of medication $\xi_1 = \omega_2$' and 'health condition, $\xi_2 = \omega_3$' is taken to be the same nonoverlapping structure as given in Section 7.2.1. To identify the model, $\boldsymbol{\Psi}_\zeta$ is fixed to be an identity matrix.

To illustrate the use of the Bayes factor for model comparison, we consider the following models with or without fixed covariates:

$$M_0: \ y_{ik} = b_{k1}x_{ik1} + b_{k2}x_{ik2} + \boldsymbol{\Lambda}_k^T\boldsymbol{\omega}_i + \zeta_{ik}$$
$$M_1: \ y_{ik} = b_{k1}x_{ik1} + \boldsymbol{\Lambda}_k^T\boldsymbol{\omega}_i + \zeta_{ik}$$
$$M_2: \ y_{ik} = b_{k2}x_{ik2} + \boldsymbol{\Lambda}_k^T\boldsymbol{\omega}_i + \zeta_{ik}$$
$$M_3: \ y_{ik} = \boldsymbol{\Lambda}_k\boldsymbol{\omega}_i + \zeta_{ik}$$

where $\boldsymbol{b}_k^T = (b_{k1}, b_{k2})$. In the same way as before, the hyperparameters $\mathbf{b}_{0k}$ and $\boldsymbol{\Lambda}_{0k}$ are obtained on the basis of an auxiliary estimation with noninformative

Table 7.3 Bayesian estimates and their standard error estimates obtained via WinBUGS for the MPCFA model.

Parameters	Estimate	SE	Parameters	Estimate	SE
b_{11}	−0.658	0.061	λ_{21}	2.485	0.628
b_{12}	−1.658	0.314	λ_{31}	0.298	0.081
b_{13}	−0.024	0.034	λ_{52}	4.157	0.637
b_{14}	−0.124	0.034	λ_{62}	4.693	0.676
b_{15}	0.838	0.089	λ_{83}	4.145	0.634
b_{16}	0.718	0.090	λ_{93}	4.264	0.678
b_{17}	−0.090	0.039			
b_{18}	−1.430	0.204	ϕ_{11}	0.778	0.216
b_{19}	−1.469	0.224	ϕ_{21}	−0.101	0.024
b_{21}	0.074	0.147	ϕ_{31}	0.301	0.057
b_{22}	0.170	0.269	ϕ_{22}	0.094	0.021
b_{23}	0.103	0.112	ϕ_{32}	−0.087	0.017
b_{24}	0.328	0.115	ϕ_{33}	0.450	0.088
b_{25}	0.285	0.190			
b_{26}	−0.054	0.194			
b_{27}	0.261	0.133			
b_{28}	0.633	0.317			
b_{29}	0.646	0.320			

priors, $\mathbf{H}_{0bk}$ and $\mathbf{H}_{0yk}$ are taken to be the identity matrices, $\mathbf{R}_0 = 0.5\mathbf{I}$ and $\rho_0 = 8$. In the path sampling procedure, we take 20 grids in [0, 1], 1000 burn-in iterations, and further collect 1000 observations at each grid for computing the Bayes factor. The estimated logarithm Bayes factors are $\log\widehat{B_{01}} = 4.87$, $\log\widehat{B_{02}} = 105.56$ and $\log\widehat{B_{03}} = 106.47$. Based on the selection criterion of the Bayes factor, M_0 is selected. Hence, the model with two fixed covariates about education and side effects is the best model among the competitive models.

The Bayesian estimates of the parameters in M_0 obtained through WinBUGS are presented in Table 7.3, together with the standard error estimates. Interpretation of these estimates is obvious. The DIC value corresponding to M_0 is 7200.51. The WinBUGS codes and data are given in the website: http://www.wiley.com/go/lee_structural.

7.4 DISCUSSION

Based on the definition of a dichotomous observation, see Equation (7.4), we understand that the exact measurement of the underlying continuous variable y in the whole range $(-\infty, \infty)$ is not available. The only information available is whether y is less than zero (in that case $z = 0$), or y is larger than zero (in that case $z = 1$). As the information carried by a dichotomous observation is very rough, we need to pay special attention in analyzing SEMs or MPCFA models with dichotomous variables. We find from our experience that: (i) It requires comparatively large sample sizes to achieve accurate estimates. Recall that in Section 4.3.3, we reported results of simulation studies to show that for continuous data, the sufficient sample sizes to achieve accurate Bayesian estimates in analyzing a model with a unknown parameters is $5a$. Roughly speaking, to achieve reasonably accurate results in analyzing dichotomous data, the available sample sizes should be larger than $30a$. (ii) As the information given by the data is rough, the prior information plays a more important role in the Bayesian analysis. (iii) It requires more iterations for the MCMC algorithm to converge. It is important to monitor convergence with more care. (iv) The standard error estimates usually overestimate the true standard derivations. Thus, these estimates should be interpreted with great caution, and they should not be used to obtain the z-score for hypothesis testing. We emphasize that hypothesis testing should be approached by means of model comparison through Bayes factor or DIC, particularly in analyzing models with dichotomous variables.

In this chapter, we focused on the dichotomous variables that are basically ordered binary variables. These kind of variables are defined by an underlying normal distribution with a threshold at zero, and their coded values have a natural ordering. It is important to note that an ordered binary variable is different from an unordered binary variable which has a binomial distribution. Unordered binary variables come from items which give unordered responses. For example, they arise when subjects are asked to select one of the two different products $\{A, B\}$, etc.. As pointed out by Lee, Song and Cai (2006), treating unordered binary variables as ordered binary variables will lead to biased results. In applying the methodologies presented in this chapter, it is important to make sure that the dichotomous variables are ordered binary. Analysis of unordered binary variables will be discussed in Chapter 13, under the framework of the exponential family of distributions.

APPENDIX 7.1: QUESTIONS ASSOCIATED WITH THE MANIFEST VARIABLES

Frequencies of (Yes '1'/No '0') are in parentheses.

y_1: Did you have any surplus in the previous prescribed drugs? (175/662)
y_2: Did you stop/reduce/increase the dosage? (69/768)
y_3: Did you forget to take medications? (391/446)
y_4: Do you feel you have hypertension? (363/474)
y_5: Do you know the reasons for taking drugs? (650/187)
y_6: Do you know the reasons for taking drugs long term? (605/232)
y_7: In the past 2 weeks, did you have emotional problems such as feeling upset, bad tempered, etc.? (387/450)
y_8: In the past 2 weeks, did your health cause any difficulties in daily activities? (181/656)

y_9: In the past 2 weeks, did your health cause any difficulties in social activities? (177/660)

This Dataset was supplied by Professor Juliana C.N. Chan and Dr Grace Chan of the Chinese University of Hong Kong, and was taken from Lee and Song (2003).

REFERENCES

Ansari, A. and Jedidi, K. (2000) Bayesian factor analysis for multilevel binary observations. *Psychometrika*, **65**, 475–498.

Bock, R. D. and Aitkin, M. (1981) Marginal maximum likelihood estimation of item parameters: Application of an EM algorithm. sl Psychometrika, **46**, 443–461.

Bock, R. D. and Gibbons, R. D. (1996) High dimensional multivariate probit analysis. *Biometrics*, **52**, 1183–1194.

Chan, G. M. C. (2002) *The Effects of Treatment Compliance on Clinical Outcomes, in Patients with Chronic Diseases.* Ph.D. Thesis, Department of Medicine and Therapeutics, the Chinese University of Hong Kong.

Chib, S. and Greenberg, E. (1998) Analysis of multivariate probit models. *Biometrika*, **85**, 347–361.

Czajkowski, S. M., Chesney, M. A. and Smith, A. W. (1998) Adherence and the placebo effect. In Shumaker, S. A., Schran, E. B., Ockene, J. K. and McBee, W. L. (eds), *The Handbook of Health Behavior Change* (2nd edn), pp. 513–534. New York: Springer.

Dempster, A. P., Laird, N. M. and Rubin, D. B. (1977) Maximum likelihood from incomplete data via the EM algorithm (with discussion), *Journal of Royal Statistical Society, Series B*, **39**, 1–38.

Gelman, A. (1996) Inference and monitoring convergence. In W. R. Gilks, S. Richardson and D. J. Spiegelhalter (eds), *Markov Chain Monte Carlo in Practice*, pp. 131–144. London: Chapman and Hall.

Gelman, A., Meng, X. L. and Stern, H. (1996) Posterior predictive assessment of model fitness via realized discrepancies. *Statistica Sinica*, **6**, 733–759.

Geman, S. and Geman, D. (1984) Stochastic relaxation, Gibbs distribution and the Bayesian restoration of images. *IEEE Transactions on Pattern Analysis and Machine Intelligence*, **6**, 721–741.

Gibbons, R. D. and Wilcox-Gök, V. (1998) Health service utilization and insurance coverage: a multivariate probit model. *Journal of the American Statistical Association*, **93**, 63–72.

Hastings, W. K. (1970) Monte Carlo sampling methods using Markov chains and their application. *Biometrika*, **57**, 97–109.

Jöreskog, K. G. and Sörbom, D. (1996) *LISREL 8: Structural Equation Modeling with the SIMPLIS Command Language*. Hove and London: Scientific Software International.

Lee, S. Y. and Song X. Y. (2003) Bayesian analysis of structural equation models with dichotomous variables. *Statistics in Medicine*, **22**, 3073–3088.

Lee, S. Y., Song, X. Y. and Cai, J. H. (2006) A Bayesian approach for nonlinear structural equation models with ordered and unordered binary variables using WinBUGS. Submitted.

Meng, X. L. and Schilling, S. (1996) Fitting full-information item factor models and an empirical investigation of bridge sampling. *Journal of the American Statistical Association*, **91**, 1254–1267.

Metropolis, N. *et al.* (1953) Equations of state calculations by fast computing machine. *Journal of Chemical Physics*, **21**, 1087–1091.

Rand, C. S. and Weeks, K. (1998) Measuring adherence with medication regiments in clinical care and research, In Shumaker, S. A., Schron, E. B., Ockene, J. K. and McBee, W. L. (eds), *The Handbook of Health Behavior Change* (2nd edn), pp. 71–103. New York: Springer-Vertag.

Song, X. Y. and Lee, S. Y. (2005) A multivariate probit latent variable model for analyzing dichotomous responses. *Statistica Sinica*, **15**, 645–664.

Tanner, M. A. and Wong, W. H. (1987) The calculation of posterior distributions by data augmentation(with discussion). *Journal of the American statistical Association*, **82**, 528–550.

Wei, G. C. G. and Tanner, M. A. (1990) A Monte Carlo implementation of the EM algorithm and the poor man's data augmentation algorithm. *Journal of the American Statistical Association*, **85**, 699–704.

8 Nonlinear Structural Equation Models

8.1 INTRODUCTION

Structural equation models considered in previous chapters, for example the factor analysis model and the LISREL type models, are models in which the latent variables are related by linear functions. Recently, it has been recognized that nonlinear relations among the latent variables are important for establishing more meaningful and correct models for some complex situations. For example, see Busemeyer and Jones (1983), Jonsson (1998), Ping (1996), Kenny and Judd (1984), Bagozzi, Baumgartner and Yi (1992), Schumacker and Marcoulides (1998), and references therein on the importance of quadratic and interaction effects of latent variables in various applied research.

Due to the complex distributions associated with the nonlinear latent variables, methods for analyzing nonlinear structural equation models are more difficult. Nonlinear factor analysis models with polynomial relationships were first explored by McDonald (1962), then followed by Etezadi-Amoli and McDonald (1983) and Mooijaart and Bentler (1986). More recently, methods that used the LISREL (Jöreskog and Sörbom, 1996) program have been proposed to analyze some nonlinear structural equation models with interaction terms of latent variables, see, for example, Kenny and Judd (1984), Ping (1996), Jaccard and Wan (1995), Jöreskog and Yang (1996) and Marsh, Wen and Hau (2004) among others. The basic approach of these contributions is to include artificial nonlinear manifest variables in the analysis to account for nonlinear relationships among variables. However, this approach has some remaining issues that need to be addressed. For example, there is no sound

Structural Equation Modeling: A Bayesian Approach S-Y. Lee

theoretical basis for forming the products of the indicators, the conclusions that are obtained using different indicators can be different, and the problems on the robustness of the ML to violations of multivariate normality still exist (see Lee, Song and Poon, 2004; Marsh, Wen and Han, 2004). The asymptotically distribution-free (ADF) theory (Browne, 1984; Bentler, 1983, 1992) is an alternative for rigorous treatment for the non-normality induced by the nonlinear latent variables. However, it is well known that (see, for example, Hu, Bentler and Kano, 1992; Bentler and Dudgeon, 1996) ADF theory requires very large sample sizes to attain its large sample properties. Other methodological developments include the moment-based approaches (Wall and Amemiya, 2000, 2003), the latent moderated structural equations (LMS) approach that utilizes the mixture distributions (Klein and Moosbrugger, 2000), and the exact ML approach (Lee and Zhu, 2002). Moreover, the Bayesian approach has also been developed for nonlinear SEMs with general forms, see Zhu and Lee (1999), Lee and Song (2003). Lee, Song and Poon (2004) conducted simulation studies to compare the Bayesian approach with the product indicator approach. Their simulation results indicate that the Bayesian approach is better. Recently, Lee and Zhu (2000) and Lee and Song (2004) developed Bayesian methods for analyzing nonlinear SEMs with mixed continuous and ordered categorical variables.

Model comparison (or hypothesis testing) is a very important component in analyzing SEMs, especially nonlinear SEMs. A fundamental problem is to decide whether a nonlinear SEM is better than a linear one in fitting a particular data set. It is also important to compare various non-nested nonlinear models and to select the better model. As the covariance matrix and distribution of the nonlinear latent variables are rather complex, the covariance structure and the distribution of the manifest variables in the nonlinear model are complicated. The traditional likelihood ratio test that requires the normality assumption of the observed random vector cannot be used. Moreover, this traditional statistic cannot be applied to compare non-nested models or hypotheses. Here, we again use the Bayes factor to address the important issue of model comparison and propose path sampling to compute this statistic.

The main objective of this chapter is to describe Bayesian estimation and model comparison (and/or hypothesis testing) procedures for analyzing nonlinear SEMs. The basic model is defined by a linear measurement equation and a nonlinear structural equation that involves differentiable functions of the exogenous latent variables. Under this framework, nonlinear causal relationships among the latent variables in the model can be analyzed. Nonlinear SEMs with continuous data will be discussed in Section 8.2. Bayesian estimates are obtained through a hybrid algorithm which combines the Gibbs sampler (Geman and Geman, 1984) and the MH algorithm (Metropolis *et al.*, 1953; Hastings, 1970). Joint Bayesian estimates of the random variables and the unknown parameters are obtained together with their standard error estimates. We also report simulation results in relation to the comparisons of a

Bayesian approach with a product indicator approach (Jaccard and Wan, 1995) as well as the LMS (Klein and Moosbrugger, 2000) approach. Bayesian analysis of nonlinear SEMs with mixed type of continuous and ordered categorical data is discussed in Section 8.3. Then, in Section 8.4, we consider a general SEM which involves nonlinear terms among covariates and exogenous latent variables. A detailed real example is presented. In addition, we discuss the application of WinBUGS (Spiegelhalter, Thomas, Best and Lunn, 2003) in analyzing an artificial example. A procedure based on the path sampling for computing the Bayes factor for model comparison is described in Section 8.5.

8.2 BAYESIAN ANALYSIS OF A NONLINEAR SEM

8.2.1 The Model

Consider the following nonlinear structural equation model (NSEM) with a $p \times 1$ manifest random vector $\mathbf{y} = (y_1, \ldots, y_p)^T$:

$$\mathbf{y} = \boldsymbol{\mu} + \boldsymbol{\Lambda}\boldsymbol{\omega} + \boldsymbol{\epsilon}, \tag{8.1}$$

where $\boldsymbol{\mu}$ is a vector of intercepts, $\boldsymbol{\Lambda}$ is a $p \times q$ matrix of factor loadings, $\boldsymbol{\omega} = (\omega_1, \ldots, \omega_q)^T$ is a random vector of latent factors with $q < p$, $\boldsymbol{\epsilon}$ is a $p \times 1$ random vector of error measurements with distribution $N[\mathbf{0}, \boldsymbol{\Psi}_\epsilon]$, where $\boldsymbol{\Psi}_\epsilon$ is diagonal and $\boldsymbol{\epsilon}$ is independent of $\boldsymbol{\omega}$. To handle more complex situations, we partition $\boldsymbol{\omega}$ as $(\boldsymbol{\eta}^T, \boldsymbol{\xi}^T)^T$ and further model this latent vector by the following nonlinear structural model:

$$\boldsymbol{\eta} = \boldsymbol{\Pi}\boldsymbol{\eta} + \boldsymbol{\Gamma}\mathbf{H}(\boldsymbol{\xi}) + \boldsymbol{\delta}, \tag{8.2}$$

where $\boldsymbol{\eta}$ and $\boldsymbol{\xi}$ are $q_1 \times 1$ and $q_2 \times 1$ latent subvectors of $\boldsymbol{\omega}$ respectively; $\mathbf{H}(\boldsymbol{\xi}) = (h_1(\boldsymbol{\xi}), \ldots, h_t(\boldsymbol{\xi}))^T$ is a $t \times 1$ nonzero vector-valued function with nonzero, known, and linearly independent differential functions $h_1, \ldots, h_t$, and $t \geq q_2$; $\boldsymbol{\Pi}(q_1 \times q_1)$ and $\boldsymbol{\Gamma}(q_1 \times t)$ are matrices of regression coefficients of $\boldsymbol{\eta}$ on $\boldsymbol{\eta}$ and $H(\boldsymbol{\xi})$, respectively. It is assumed that $\boldsymbol{\xi}$ and $\boldsymbol{\delta}$ are independently distributed as $N[\mathbf{0}, \boldsymbol{\Phi}]$ and $N[\mathbf{0}, \boldsymbol{\Psi}_\delta]$ respectively, where $\boldsymbol{\Psi}_\delta$ is a diagonal covariance matrix. If some $h_j(\boldsymbol{\xi})$ are nonlinear, the distribution of the manifest random vector $\mathbf{y}$ is non-normal. Let $\boldsymbol{\Pi}_0 = \mathbf{I}_{q_1} - \boldsymbol{\Pi}$, we assume that $|\boldsymbol{\Pi}_0|$ is a nonzero constant independent of $\boldsymbol{\Pi}$. The structural model in Equation (8.2) is linear in parameter matrices $\boldsymbol{\Pi}$ and $\boldsymbol{\Gamma}$, but may be nonlinear in the latent variables in $\boldsymbol{\xi}$. This allows the assessment of the nonlinear causal effects of latent variables in $\boldsymbol{\xi}$ on latent variables in $\boldsymbol{\eta}$. Let $\boldsymbol{\Lambda}_\omega = (\boldsymbol{\Pi}, \boldsymbol{\Gamma})$ and $\mathbf{G}(\boldsymbol{\omega}) = (\boldsymbol{\eta}^T, \mathbf{H}(\boldsymbol{\xi})^T)^T$, then Equation (8.2) can be written as

$$\boldsymbol{\eta} = \boldsymbol{\Lambda}_\omega \mathbf{G}(\boldsymbol{\omega}) + \boldsymbol{\delta}. \tag{8.3}$$

The components of $\mathbf{H}(\boldsymbol{\xi})$ are any differential functions, hence they are general enough to deal with the common polynomial relationships among the latent variables. However, the choice of $\mathbf{H}(\boldsymbol{\xi})$ is not completely arbitrary. For example, the following obvious cases are not allowed: $\mathbf{H}_1(\boldsymbol{\xi}) = (\xi_1, \xi_2, \xi_1^2, \xi_1^2)$ and $\mathbf{H}_2(\boldsymbol{\xi}) = (\xi_1, \xi_2, \xi_1\xi_2, 0)$. Clearly, $\mathbf{H}_1(\boldsymbol{\xi})$ and $\mathbf{H}_2(\boldsymbol{\xi})$ should be modified as (ξ_1, ξ_2, ξ_1^2) and $(\xi_1, \xi_2, \xi_1\xi_2)$, respectively. An example of an identified structural equation is:

$$\begin{pmatrix} \eta_1 \\ \eta_2 \end{pmatrix} = \begin{pmatrix} 0 & \pi \\ 0 & 0 \end{pmatrix} \begin{pmatrix} \eta_1 \\ \eta_2 \end{pmatrix} + \begin{pmatrix} \gamma_{11} & \gamma_{12} & 0 & 0 & 0 \\ \gamma_{21} & \gamma_{22} & \gamma_{23} & \gamma_{24} & \gamma_{25} \end{pmatrix} \begin{pmatrix} \xi_1 \\ \xi_2 \\ \xi_1^2 \\ \xi_1\xi_2 \\ \xi_2^2 \end{pmatrix} + \begin{pmatrix} \delta_1 \\ \delta_2 \end{pmatrix}.$$

The proposed nonlinear model is over parameterized without appropriate identification conditions. For instance, an equivalent form of model Equation (8.1) is $\mathbf{y} = \boldsymbol{\mu} + \boldsymbol{\Lambda}^*\boldsymbol{\omega}^* + \boldsymbol{\epsilon}$, where $\boldsymbol{\Lambda}^* = \boldsymbol{\Lambda}\mathbf{R}$ and $\boldsymbol{\omega}^* = \mathbf{R}^{-1}\boldsymbol{\omega}$, for any nonsingular matrix $\mathbf{R}$. One common method to solve the identification problem in structural equation modeling is to fix appropriate elements in $\boldsymbol{\Lambda}$ at some known values so that the only possible choice of $\mathbf{R}$ is the identity matrix. Similarly, appropriate elements in $\boldsymbol{\Lambda}_\omega$ may also be fixed at known values if necessary.

In the context of nonlinear SEMs, more care is needed to interpret the mean vector of $\mathbf{y}$. Let $\boldsymbol{\Lambda}_k^T$ be the kth row of $\boldsymbol{\Lambda}$. For $k = 1, \ldots, p$, it follows from Equation (8.1) that $E(y_k) = \mu_k + \boldsymbol{\Lambda}_k^T E(\boldsymbol{\omega})$. Although $E(\boldsymbol{\xi}) = \mathbf{0}$, it follows from Equation (8.2) that $E(\boldsymbol{\eta}) \neq \mathbf{0}$ if $\mathbf{H}(\boldsymbol{\xi})$ is a nonlinear function of $\boldsymbol{\xi}$. Hence $E(\boldsymbol{\omega}) \neq \mathbf{0}$ and $E(y_k) \neq \mu_k$. Let $\boldsymbol{\Lambda}_k^T = (\boldsymbol{\Lambda}_{k\eta}^T, \boldsymbol{\Lambda}_{k\xi}^T)$ be a partition of $\boldsymbol{\Lambda}_k^T$ that corresponds to the partition of $\boldsymbol{\omega} = (\boldsymbol{\eta}^T, \boldsymbol{\xi}^T)^T$. Because $E(\boldsymbol{\xi}) = \mathbf{0}$ and $\boldsymbol{\eta} = (\mathbf{I} - \boldsymbol{\Pi})^{-1}[\boldsymbol{\Gamma}\mathbf{H}(\boldsymbol{\xi})]$, it follows from Equation (8.2) that

$$E(y_k) = \mu_k + \boldsymbol{\Lambda}_{k\eta}^T E(\boldsymbol{\eta}) + \boldsymbol{\Lambda}_{k\xi}^T E(\boldsymbol{\xi}) = \mu_k + \boldsymbol{\Lambda}_{k\eta}^T[(\mathbf{I} - \mathbf{B})^{-1}\boldsymbol{\Gamma}]E(\mathbf{H}(\boldsymbol{\xi})). \quad (8.4)$$

As $\mathbf{H}(\boldsymbol{\xi})$ is usually not very complicated in most practical applications, $E(\mathbf{H}(\boldsymbol{\xi}))$ is not very complex and the computation of $E(y_k)$ is not difficult. See the illustrative example presented in subsequent sections.

For a linear SEM, the saturated model is defined by $\mathbf{y} = \boldsymbol{\mu} + \boldsymbol{\epsilon}^*$, where $\boldsymbol{\epsilon}^*$ is distributed as $N[\mathbf{0}, \boldsymbol{\Sigma}]$, in which $\boldsymbol{\Sigma}$ is a generic covariance matrix without any structure. The above model cannot be regarded as a saturated model for assessing the goodness-of-fit of the posited model in the context of nonlinear SEMs, because in the presence of the nonlinear terms, the distribution of $\boldsymbol{\epsilon}^*$ is not normal. Hence, the assessment about the significance of the nonlinear terms should be approached with care. This issue will be discussed further in Section 8.5.

8.2.2 The Gibbs Sampler For Posterior Simulation

Let $\mathbf{Y} = (\mathbf{y}_1, \ldots, \mathbf{y}_n)$ be the observed data matrix and $\boldsymbol{\theta}$ be the parameter vector that contains the unknown parameters in $\boldsymbol{\mu}$, $\boldsymbol{\Lambda}$, $\boldsymbol{\Lambda}_\omega$, $\boldsymbol{\Phi}$, $\boldsymbol{\Psi}_\epsilon$ and $\boldsymbol{\Psi}_\delta$. Due to the nonlinearity of $\mathbf{H}(\boldsymbol{\xi})$, the classical GLS and ML approaches would encounter great difficulties in estimating $\boldsymbol{\theta}$. The main objective here is to obtain a Bayesian estimate of $\boldsymbol{\theta}$ with the given observed data $\mathbf{Y}$. To alleviate the difficulty induced by $\mathbf{H}(\boldsymbol{\xi})$, $\mathbf{Y}$ is augmented with the matrix of latent variables, $\boldsymbol{\Omega} = (\boldsymbol{\omega}_1, \ldots, \boldsymbol{\omega}_n)$ in the posterior analysis. Then, the following Gibbs sampler (Geman and Geman, 1984) is used to generate a sequence of observations from the posterior distribution $[\boldsymbol{\theta}, \boldsymbol{\Omega}|\mathbf{Y}]$: At the jth iteration with current values $\boldsymbol{\theta}^{(j)}$ and $\boldsymbol{\Omega}^{(j)}$, it simulates in turn,

Step (a): $\boldsymbol{\theta}^{(j+1)}$ from $p(\boldsymbol{\theta}|\boldsymbol{\Omega}^{(j)}, \mathbf{Y})$, and

Step (b): $\boldsymbol{\Omega}^{(j+1)}$ from $p(\boldsymbol{\Omega}|\boldsymbol{\theta}^{(j+1)}, \mathbf{Y})$.

In Step (a), $\boldsymbol{\theta}$ is composed of $\boldsymbol{\mu}$, $\boldsymbol{\Lambda}$, $\boldsymbol{\Lambda}_\omega$, $\boldsymbol{\Phi}$, $\boldsymbol{\Psi}_\epsilon$ and $\boldsymbol{\Psi}_\delta$. As $\boldsymbol{\Omega}$ is given, the nonlinear SEM defined by Equations (8.1) and (8.2) reduces to a simultaneous regression model and it is not too difficult to derive the full conditional distributions.

8.2.3 Full Conditional Distributions

Let $\boldsymbol{\theta}_y$ be the unknown parameters in $\boldsymbol{\mu}$, $\boldsymbol{\Lambda}$ and $\boldsymbol{\Psi}_\epsilon$ that are associated with the measurement equation; and let $\boldsymbol{\theta}_\omega$ be the unknown parameters in $\boldsymbol{\Lambda}_\omega$, $\boldsymbol{\Phi}$ and $\boldsymbol{\Psi}_\delta$ that are associated with the structural equation. For simplicity, we assume no fixed parameters. Slight modification as given in Section 5.3.1 can be incorporated for situations with fixed parameters. It is natural to assume that the prior distribution of $\boldsymbol{\theta}_y$ is independent of the prior distribution of $\boldsymbol{\theta}_\omega$, i.e. $p(\boldsymbol{\theta}) = p(\boldsymbol{\theta}_y)p(\boldsymbol{\theta}_\omega)$. Moreover, as $p(\mathbf{Y}, \boldsymbol{\Omega}|\boldsymbol{\theta}) = p(\mathbf{Y}|\boldsymbol{\Omega}, \boldsymbol{\theta}_y)p(\boldsymbol{\Omega}|\boldsymbol{\theta}_\omega)$, we have

$$\begin{aligned} p(\boldsymbol{\theta}_y, \boldsymbol{\theta}_\omega|\mathbf{Y}, \boldsymbol{\Omega}) &\propto p(\mathbf{Y}, \boldsymbol{\Omega}|\boldsymbol{\theta}_y, \boldsymbol{\theta}_\omega)p(\boldsymbol{\theta}_y)p(\boldsymbol{\theta}_\omega) \\ &= [p(\mathbf{Y}|\boldsymbol{\Omega}, \boldsymbol{\theta}_y)p(\boldsymbol{\theta}_y)][p(\boldsymbol{\Omega}|\boldsymbol{\theta}_\omega)p(\boldsymbol{\theta}_\omega)]. \end{aligned}$$

As a result, the conditional densities of $\boldsymbol{\theta}_y$ and $\boldsymbol{\theta}_\omega$ can be treated separately. According to the reasoning given in previous chapters, the following commonly used conjugate type prior distributions are used. For $k = 1, \ldots, p$

$$\begin{aligned} &\boldsymbol{\mu} \stackrel{D}{=} N[\boldsymbol{\mu}_0, \Sigma_0], \quad \psi_{\epsilon k}^{-1} \stackrel{D}{=} Gamma\,[\alpha_{0\epsilon k}, \beta_{0\epsilon k}], \\ &[\boldsymbol{\Lambda}_k|\psi_{\epsilon k}] \stackrel{D}{=} N[\boldsymbol{\Lambda}_{0k}, \psi_{\epsilon k}\mathbf{H}_{0yk}], \end{aligned} \qquad (8.5)$$

where $\mathbf{\Lambda}_k^T$ is the kth row of $\mathbf{\Lambda}$; $\alpha_{0\epsilon k}, \beta_{0\epsilon k}, \boldsymbol{\mu}_0, \mathbf{\Lambda}_{0k}, \mathbf{H}_{0yk}$ and Σ_0 are hyperparameters whose values are assumed to be given. For $k \neq h$, it is assumed that $(\psi_{\epsilon k}, \mathbf{\Lambda}_k)$ and $(\psi_{\epsilon h}, \mathbf{\Lambda}_h)$ are independent. The conditional distribution of $\mathbf{\Lambda}$ is given for the case where all its elements are unknown parameters. The conditional distribution with fixed elements can be obtained with slight modification as before.

Let $\mathbf{A}_k = (\mathbf{H}_{0yk}^{-1} + \mathbf{\Omega}\mathbf{\Omega}^T)^{-1}, \mathbf{Y}_k^* = (y_{1k}^*, \ldots, y_{nk}^*)^T$ with $y_{ik}^* = y_{ik} - \mu_k$, $\mathbf{a}_k = \mathbf{A}_k(\mathbf{H}_{0yk}^{-1}\mathbf{\Lambda}_{0k} + \mathbf{\Omega}\mathbf{Y}_k^*)$ and $\beta_{\epsilon k} = \beta_{0\epsilon k} + 2^{-1}(\mathbf{Y}_k^{*T}\mathbf{Y}_k^* - \mathbf{a}_k^T\mathbf{A}_k^{-1}\mathbf{a}_k + \mathbf{\Lambda}_{0k}^T\mathbf{H}_{0yk}^{-1}\mathbf{\Lambda}_{0k})$. Then, it can be shown by similar reasoning to that in Section 4.3.1 that for $k = 1, \ldots, p$,

$$p(\psi_{\epsilon k}^{-1}|\mathbf{Y}, \mathbf{\Omega}, \boldsymbol{\mu}) \stackrel{D}{=} Gamma[n/2 + \alpha_{0\epsilon k}, \beta_{\epsilon k}],\ p(\mathbf{\Lambda}_k|\mathbf{Y}, \mathbf{\Omega}, \psi_{\epsilon k}^{-1}, \boldsymbol{\mu}) \stackrel{D}{=} N[\mathbf{a}_k, \psi_{\epsilon k}\mathbf{A}_k], \tag{8.6}$$

$$p(\boldsymbol{\mu}|\mathbf{Y}, \mathbf{\Omega}, \mathbf{\Lambda}, \mathbf{\Psi}_\epsilon) \stackrel{D}{=} N[(\mathbf{\Sigma}_0^{-1} + n\mathbf{\Psi}_\epsilon^{-1})^{-1}(n\mathbf{\Psi}_\epsilon^{-1}\bar{\mathbf{Y}} + \mathbf{\Sigma}_0^{-1}\boldsymbol{\mu}_0), (\mathbf{\Sigma}_0^{-1} + n\mathbf{\Psi}_\epsilon^{-1})^{-1}], \tag{8.7}$$

where $\bar{\mathbf{Y}} = \Sigma_{i=1}^n(\mathbf{y}_i - \mathbf{\Lambda}\boldsymbol{\omega}_i)/n$.

Now, consider the conditional distribution of $\boldsymbol{\theta}_\omega$ that is proportional to $p(\mathbf{\Omega}|\boldsymbol{\theta}_\omega)p(\boldsymbol{\theta}_\omega)$. Let $\mathbf{\Omega}_1 = (\boldsymbol{\eta}_1, \ldots, \boldsymbol{\eta}_n), \mathbf{\Omega}_2 = (\boldsymbol{\xi}_1, \cdots, \boldsymbol{\xi}_n)$ and $\mathbf{G} = [\mathbf{G}(\boldsymbol{\omega}_1), \ldots, \mathbf{G}(\boldsymbol{\omega}_n)]$. Since the distribution of $\boldsymbol{\xi}_i$ only involves $\mathbf{\Phi}$, $p(\mathbf{\Omega}_2|\boldsymbol{\theta}_\omega) = p(\mathbf{\Omega}_2|\mathbf{\Phi})$. Moreover, it is assumed that the prior distribution of $\mathbf{\Phi}$ is independent of the prior distribution of $\mathbf{\Lambda}_\omega$ and $\mathbf{\Psi}_\delta$. It follows that

$$p(\mathbf{\Omega}|\boldsymbol{\theta}_\omega)p(\boldsymbol{\theta}_\omega) \propto [p(\mathbf{\Omega}_1|\mathbf{\Omega}_2, \mathbf{\Lambda}_\omega, \mathbf{\Psi}_\delta)p(\mathbf{\Lambda}_\omega, \mathbf{\Psi}_\delta)][p(\mathbf{\Omega}_2|\mathbf{\Phi})p(\mathbf{\Phi})].$$

Hence, the marginal conditional densities of $(\mathbf{\Lambda}_\omega, \mathbf{\Psi}_\delta)$ and $\mathbf{\Phi}$ can be treated separately, as before.

Consider a similar conjugate type prior distribution for $\mathbf{\Phi}$ with $\mathbf{\Phi}^{-1} \stackrel{D}{=} W_{q_2}[\mathbf{R}_0, \rho_0]$, where ρ_0 and the positive definite matrix $\mathbf{R}_0$ are the given hyperparameters. It can be shown by reasoning similar to that in Section 4.3.1 that

$$p(\mathbf{\Phi}|\mathbf{\Omega}_2) \stackrel{D}{=} IW_{q_2}[(\mathbf{\Omega}_2\mathbf{\Omega}_2^T + \mathbf{R}_0^{-1}), n + \rho_0]. \tag{8.8}$$

Similarly as before, we use conjugate prior distributions for $\boldsymbol{\theta}_\omega$, so that for $k = 1, \cdots, q_1$,

$$\psi_{\delta k}^{-1} \stackrel{D}{=} Gamma[\alpha_{0\delta k}, \beta_{0\delta k}], \quad [\mathbf{\Lambda}_{\omega k}|\psi_{\delta k}] \stackrel{D}{=} N[\mathbf{\Lambda}_{0\omega k}, \psi_{\delta k}\mathbf{H}_{0\omega k}], \tag{8.9}$$

where $\mathbf{\Lambda}_{\omega k}^T$ is the kth row of $\mathbf{\Lambda}_\omega$; $\alpha_{0\delta k}, \beta_{0\delta k}, \mathbf{\Lambda}_{0\omega k}$ and $\mathbf{H}_{0\omega k}$ are all given hyper parameters. For $h \neq k$, $(\psi_{\delta k}, \mathbf{\Lambda}_{\omega k})$ and $(\psi_{\delta h}, \mathbf{\Lambda}_{\omega h})$ are assumed to be independent. Similar to $\mathbf{\Lambda}_k$, we assume no fixed elements in $\mathbf{\Lambda}_{\omega k}$.

Let $\mathbf{E}_k^* = (\eta_{1k}, \ldots, \eta_{nk})^T$ where $k = 1, \ldots, q_1$, π_{kj} and γ_{kl} are elements in $\mathbf{\Pi}$ and $\mathbf{\Gamma}$, respectively. Then, it can be shown that for $k = 1, \ldots, q_1$,

$$p(\psi_{\delta k}^{-1}|\mathbf{\Omega}) \stackrel{D}{=} Gamma[n/2 + \alpha_{0\delta k}, \beta_{\delta k}], \quad p(\mathbf{\Lambda}_{\omega k}|\mathbf{\Omega}, \psi_{\delta k}^{-1}) \stackrel{D}{=} N[\mathbf{a}_{\omega k}, \psi_{\delta k}\mathbf{A}_{\omega k}] \tag{8.10}$$

where $\mathbf{A}_{\omega k} = (\mathbf{H}_{0\omega k}^{-1} + \mathbf{G}\mathbf{G}^T)^{-1}$, $\mathbf{a}_{\omega k} = \mathbf{A}_{\omega k}(\mathbf{H}_{0\omega k}^{-1}\mathbf{\Lambda}_{0\omega k} + \mathbf{G}\mathbf{E}_k^*)$, and $\beta_{\delta k} = \beta_{0\delta k} + 2^{-1}(\mathbf{E}_k^{*T}\mathbf{E}_k^* - \mathbf{a}_{\omega k}^T\mathbf{A}_{\omega k}^{-1}\mathbf{a}_{\omega k} + \mathbf{\Lambda}_{0\omega k}^T\mathbf{H}_{0\omega k}^{-1}\mathbf{\Lambda}_{0\omega k})$.

Hence, the conditional distributions associated with Step (a) in Equation (8.3) are the familiar gamma, normal and inverted Wishart distributions. Simulating observations from these distributions is straightforward and fast.

Now, consider the conditional distributions associated with Step (b) in Equation (8.3). It can be shown on the basis of the definition and assumptions that

$$p(\mathbf{\Omega}|\boldsymbol{\theta}, \mathbf{Y}) = \mathbf{\Pi}_{i=1}^n p(\boldsymbol{\omega}_i|\mathbf{y}_i, \boldsymbol{\theta}) \propto \mathbf{\Pi}_{i=1}^n p(\mathbf{y}_i|\boldsymbol{\omega}_i, \boldsymbol{\theta})p(\boldsymbol{\eta}_i|\boldsymbol{\xi}_i, \boldsymbol{\theta})p(\boldsymbol{\xi}_i|\boldsymbol{\theta}).$$

As $\boldsymbol{\omega}_i$ are mutually independent, and $\mathbf{y}_i$ are also mutually independent, $p(\boldsymbol{\omega}_i|\mathbf{y}_i, \boldsymbol{\theta})$ is proportional to

$$\exp\left\{-\frac{1}{2}\boldsymbol{\xi}_i^T\mathbf{\Phi}^{-1}\boldsymbol{\xi}_i - \frac{1}{2}(\mathbf{y}_i - \boldsymbol{\mu} - \mathbf{\Lambda}\boldsymbol{\omega}_i)^T\mathbf{\Psi}_\epsilon^{-1}(\mathbf{y}_i - \boldsymbol{\mu} - \mathbf{\Lambda}\boldsymbol{\omega}_i) - \frac{1}{2}[\boldsymbol{\eta}_i - \mathbf{\Lambda}_\omega\mathbf{G}(\boldsymbol{\omega}_i)]^T\mathbf{\Psi}_\delta^{-1}[\boldsymbol{\eta}_i - \mathbf{\Lambda}_\omega\mathbf{G}(\boldsymbol{\omega}_i)]\right\}. \tag{8.11}$$

Hence, the conditional distribution required in Step (b) is achieved. This distribution is nonstandard and complex. The MH algorithm is used to generate observations from the target density $p(\boldsymbol{\omega}_i|\mathbf{y}_i, \boldsymbol{\theta})$ as given in (8.11). In this algorithm, we choose $N[\mathbf{0}, \sigma^2\mathbf{\Sigma}_\omega]$ as the proposal distribution, where $\mathbf{\Sigma}_\omega^{-1} = \mathbf{\Sigma}_\delta^{-1} + \mathbf{\Lambda}^T\mathbf{\Psi}_\epsilon^{-1}\mathbf{\Lambda}$ and $\mathbf{\Sigma}_\delta^{-1}$ is given by

$$\mathbf{\Sigma}_\delta^{-1} = \begin{bmatrix} \mathbf{\Pi}_0^T\mathbf{\Psi}_\delta^{-1}\mathbf{\Pi}_0, & -\mathbf{\Pi}_0^T\mathbf{\Psi}_\delta^{-1}\mathbf{\Gamma}\mathbf{\Delta} \\ -\mathbf{\Delta}^T\mathbf{\Gamma}^T\mathbf{\Psi}_\delta^{-1}\mathbf{\Pi}_0, & \mathbf{\Phi}^{-1} + \mathbf{\Delta}^T\mathbf{\Gamma}^T\mathbf{\Psi}_\delta^{-1}\mathbf{\Gamma}\mathbf{\Delta} \end{bmatrix}, \tag{8.12}$$

where $\mathbf{\Delta} = \partial\mathbf{H}(\boldsymbol{\xi}_i)/\partial\boldsymbol{\xi}_i^T|_{\boldsymbol{\xi}_i=0}$. Let $p(\cdot|\boldsymbol{\omega}, \sigma^2\mathbf{\Sigma}_\omega)$ be the proposal density corresponding to $N[\boldsymbol{\omega}, \sigma^2\mathbf{\Sigma}_\omega]$, the MH algorithm for our problem is implemented as follows. At the rth iteration with a current value $\boldsymbol{\omega}_i^{(r)}$, a new candidate $\boldsymbol{\omega}_i$ is generated from $p(\cdot|\boldsymbol{\omega}_i^{(r)}, \sigma^2\mathbf{\Sigma}_\omega)$, and accepting this new candidate with the probability

$$\min\left[1, \frac{p(\boldsymbol{\omega}_i|\mathbf{y}_i, \boldsymbol{\theta})}{p(\boldsymbol{\omega}_i^{(r)}|\mathbf{y}_i, \boldsymbol{\theta})}\right]. \tag{8.13}$$

The variance σ^2 is chosen such that the acceptance rate is approximately 0.25 or more, see Gelman, Roberts and Gilks (1995).

It should be noted that the components of the conditional distribution $[\boldsymbol{\theta}|\boldsymbol{\Omega}, \mathbf{Y}]$ involved in the Gibbs sampler in analyzing the nonlinear SEMs are very similar to those in analyzing the linear SEMs, whereas the conditional distribution $[\boldsymbol{\Omega}|\boldsymbol{\theta}, \mathbf{Y}]$ is slightly more complicated (see Equations (8.11), (4.25) and (4.39)). Hence, the generalization of the Bayesian methods for the linear SEMs to nonlinear SEMs is not very difficult. Indeed, after data augmentation and given the latent variables, both the linear and nonlinear SEMs are just regression models.

8.2.4 Bayesian Estimates

For brevity, let $\{(\boldsymbol{\theta}^{(t)}, \boldsymbol{\Omega}^{(t)}), t = 1, \ldots, T^*\}$ be the random observations of $(\boldsymbol{\theta}, \boldsymbol{\Omega})$ generated by the Gibbs sampler and the MH algorithm from $[\boldsymbol{\theta}, \boldsymbol{\Omega}|\mathbf{Y}]$. As before, Bayesian estimates of $\boldsymbol{\theta}$ and $\boldsymbol{\omega}_i$ can easily be obtained via the corresponding sample means of the generated observations as given in Equations (4.17) and (4.19) of Chapter 4. Clearly, these Bayesian estimates are consistent estimates of the corresponding posterior means (see Geyer (1992)). It is rather difficult to derive analytic forms for the covariance matrices $\text{Var}(\boldsymbol{\theta}|\mathbf{Y})$ and $\text{Var}(\boldsymbol{\omega}_i|\mathbf{Y})$. However, estimates of these covariance matrices can be obtained via the corresponding sample covariance matrices based on the simulated observations (see Equations (4.18) and (4.20) of Chapter 4). Moreover, other statistical inferences can be conducted on the basis of the generated observations, see Besag, Green, Higdon and Mengersen (1995) and Gilks, Richardson and Spiegehalter, (1996). For instance, estimated residuals can be obtained by means of the parameter estimates and latent variable estimates (see Section 8.4.3).

8.2.5 Illustrative Example 1

To illustrate the Bayesian approach with a real example, a small portion of the Inter-university Consortium for Political and Social Research (ICPSR) data set collected in the project World Values Survey 1981–1984 and 1990–1993 (World Values Study Group, 1994) is analyzed. In our illustration, only the data obtained from Canada were used. Six variables in the full data set (variables 180, 96, 62, 176, 116 and 117) that are related with the respondents' job, religious belief and homelife were taken as manifest variables in $\mathbf{Y}$. Details of these variables are given in Appendix 1.1. Variable 62 was measured by a five-point scale, while all others were measured by a 10-point scale. As the purpose of this example is for illustration, they were all treated as continuous for brevity. Moreover, cases with missing data were not used, and the remaining sample size was 451.

Parameters in $\boldsymbol{\mu}$ and $\boldsymbol{\Lambda}$ of the measurement Equation (8.1) were specified as follows:

$$\boldsymbol{\mu} = (\mu_1, \ldots, \mu_6)^T, \text{ and } \boldsymbol{\Lambda}^T = \begin{bmatrix} 1.0 & \lambda_{21} & 0 & 0 & 0 & 0 \\ 0 & 0 & 1.0 & \lambda_{42} & 0 & 0 \\ 0 & 0 & 0 & 0 & 1.0 & \lambda_{63} \end{bmatrix}, \quad (8.14)$$

where 1's and 0's in $\boldsymbol{\Lambda}$ were treated as fixed parameters to identify the model. In this model, there were three latent factors: η, ξ_1 and ξ_2, which can be roughly interpreted as 'life', 'religious belief' and 'job satisfaction' respectively. These latent variables were related by the following nonlinear structural equation:

$$\eta = \gamma_1\xi_1 + \gamma_2\xi_2 + \gamma_3\xi_1^2 + \gamma_4\xi_1\xi_2 + \gamma_5\xi_2^2 + \delta \quad (8.15)$$

To obtain some prior inputs for the hyperparameters, we first conducted an auxiliary Bayesian estimation with noninformative prior distributions. Then, we carried out an actual estimation with the following hyperparameter values for obtaining the conditional distribution $p(\boldsymbol{\theta}|\mathbf{Y}, \boldsymbol{\Omega}) : \alpha_{0\epsilon k} = \alpha_{0\delta} = 10, \beta_{0\epsilon k} = \beta_{0\delta} = 10, \rho_0 = 8, \mathbf{H}_{0yk} = \mathbf{I}, \mathbf{H}_{0\omega k} = \mathbf{I}, \boldsymbol{\Lambda}_{0k} = \widetilde{\boldsymbol{\Lambda}}_k, \boldsymbol{\Lambda}_{0\omega k} = \widetilde{\boldsymbol{\Lambda}}_{\omega k}$ and $\mathbf{R}_0 = (\rho_0 - q_2 - 1)\widetilde{\boldsymbol{\Phi}}$, where $\widetilde{\boldsymbol{\Lambda}}_k$, $\widetilde{\boldsymbol{\Lambda}}_{\omega k}$ and $\widetilde{\boldsymbol{\Phi}}$ are estimates obtained from the auxiliary estimation. The σ^2 in the proposed distribution of the MH algorithm was equal to 1.0, giving an approximately average acceptance rate of 0.46. The algorithm converged in about 2000 iterations. We took a burn-in phase of 2000 iterations and collected $J = 2000$ observations after convergence for obtaining the Bayesian estimates and their standard errors estimates. Results are reported in Table 8.1. It can be seen from $\hat{\gamma}_j$ that the linear effects and the quadratic effect of ξ_1 are significant, whilst the interaction effect and the quadratic effect that involve ξ_2 are not significant.

8.2.6 Comparison with a Product Indicator Method

In the past few years, a number of methods that used the key idea of Kenny and Judd (1984) incorporating appropriate products of indicators in the model have been proposed. Results were obtained via the LISREL program, based on the ML option with nonlinear constraints. See, for example, Jaccard and Wan (1995), Jöreskog and Yang (1996), Ping (1996), among others. Li *et al.* (1998) provided a comparison of these methods and concluded that no substantial discrepancy was found in parameters estimates across these methods. Owing to this conclusion, the results obtained by the product indicator (PI) approach as given in Jaccard and Wan (1995) are compared here with the Bayesian estimates via a simulation study as reported in Lee, Song and Poon (2004).

Table 8.1 Bayesian estimates (EST) and standard error estimates (SE) of illustrative example 1.

Parameter	EST	SE	Parameter	EST	SE
μ_1	8.296	0.112	λ_{21}	0.971	0.084
μ_2	7.867	0.102	λ_{42}	2.400	0.194
μ_3	2.758	0.073	λ_{63}	0.822	0.101
μ_4	6.951	0.121	γ_1	0.183	0.081
μ_5	8.027	0.076	γ_2	0.476	0.071
μ_6	7.389	0.099	γ_3	0.164	0.082
$\psi_{\epsilon 1}$	1.147	0.137	γ_4	0.033	0.080
$\psi_{\epsilon 2}$	0.872	0.114	γ_5	−0.025	0.041
$\psi_{\epsilon 3}$	1.476	0.124	ϕ_{11}	0.882	0.114
$\psi_{\epsilon 4}$	2.242	0.420	ϕ_{21}	0.270	0.087
$\psi_{\epsilon 5}$	0.897	0.192	ϕ_{22}	1.829	0.259
$\psi_{\epsilon 6}$	3.257	0.255	ψ_δ	0.870	0.124

The simulation study was conducted on the basis of a nonlinear LISREL type model. According to slightly different notation in Jaccard and Wan (1995), let $\mathbf{y} = (y_1, y_2, x_1, x_2, x_3, x_4)^T$. The measurement equations are given by

$$\begin{pmatrix} y_1 \\ y_2 \end{pmatrix} = \begin{pmatrix} 1 \\ \lambda_y \end{pmatrix} \eta + \begin{pmatrix} \varepsilon_1 \\ \varepsilon_2 \end{pmatrix}, \tag{8.16}$$

$$\begin{pmatrix} x_1 \\ x_2 \\ x_3 \\ x_4 \end{pmatrix} = \begin{pmatrix} 1 & 0 \\ \lambda_{21} & 0 \\ 0 & 1 \\ 0 & \lambda_{42} \end{pmatrix} \begin{pmatrix} \xi_1 \\ \xi_2 \end{pmatrix} + \begin{pmatrix} \zeta_1 \\ \zeta_2 \\ \zeta_3 \\ \zeta_4 \end{pmatrix}, \tag{8.17}$$

where the y's and x's are manifest variables; η, ξ_1 and ξ_2 are latent variables, ε's and ζ's are error measurements; λ_y, λ_{21} and λ_{42} are unknown parameters, 1's and 0's are fixed parameters. It is assumed that $\boldsymbol{\xi} = (\xi_1, \xi_2)^T$ is distributed as $N[\mathbf{0}, \boldsymbol{\Phi}]$, where the covariance matrix $\boldsymbol{\Phi}$ contains unknown parameters ϕ_{11}, ϕ_{21} and ϕ_{22}; $\boldsymbol{\varepsilon} = (\varepsilon_1, \varepsilon_2)^T$ is distributed as $N[\mathbf{0}, \boldsymbol{\Psi}_\varepsilon]$, where $\boldsymbol{\Psi}_\varepsilon$ is a diagonal covariance matrix with diagonal elements $\psi_{\varepsilon 1}$ and $\psi_{\varepsilon 2}$; $\boldsymbol{\zeta} = (\zeta_1, \zeta_2, \zeta_3, \zeta_4)^T$ is distributed as $N[\mathbf{0}, \boldsymbol{\Psi}_\zeta]$, where $\boldsymbol{\Psi}_\zeta$ is a diagonal covariance matrix with diagonal elements $\psi_{\zeta k}$, $k = 1, \ldots, 4$. The latent vectors $\boldsymbol{\varepsilon}$, $\boldsymbol{\xi}$ and $\boldsymbol{\zeta}$ are uncorrelated to each other as usual.

The following structural equation is considered:

$$\eta = \gamma_1 \xi_1 + \gamma_2 \xi_2 + \gamma_3 \xi_1 \xi_2 + \gamma_4 \xi_1^2 + \gamma_5 \xi_2^2 + \delta, \tag{8.18}$$

where γ's are unknown regression coefficients that describe the relations between the endogenous latent variable η with exogenous variables ξ's, δ is distributed as $N[0, \psi_\delta]$ and is uncorrelated with the $\xi's$. In this structural equation, both interaction and quadratic effects are involved. It is linear in the regression coefficients, but nonlinear in the latent variables. The estimates of the unknown parameters obtained from the Bayesian and PI approaches are compared.

Consider the application of the PI approach for obtaining the estimates. It is assumed that the data are in mean deviation form. This approach requires four product indicators for $\xi_1\xi_2$, three product indicators for ξ_1^2 and three product indicators for ξ_2^2. It can be shown via similar derivations to those in Jaccard and Wan (1995) or Jöreskog and Yang (1996) that the equation relating the exogenous latent variables and their indicators is given by:

$$\begin{pmatrix} x_1 \\ x_2 \\ x_3 \\ x_4 \\ x_1x_3 \\ x_1x_4 \\ x_2x_3 \\ x_2x_4 \\ x_1^2 \\ x_1x_2 \\ x_2^2 \\ x_3^2 \\ x_3x_4 \\ x_4^2 \end{pmatrix} = \begin{pmatrix} 1 & 0 & 0 & 0 & 0 \\ \lambda_{21} & 0 & 0 & 0 & 0 \\ 0 & 1 & 0 & 0 & 0 \\ 0 & \lambda_{42} & 0 & 0 & 0 \\ 0 & 0 & 1 & 0 & 0 \\ 0 & 0 & \lambda_{42} & 0 & 0 \\ 0 & 0 & \lambda_{21} & 0 & 0 \\ 0 & 0 & \lambda_{21}\lambda_{42} & 0 & 0 \\ 0 & 0 & 0 & 1 & 0 \\ 0 & 0 & 0 & \lambda_{21} & 0 \\ 0 & 0 & 0 & \lambda_{21}^2 & 0 \\ 0 & 0 & 0 & 0 & 1 \\ 0 & 0 & 0 & 0 & \lambda_{42} \\ 0 & 0 & 0 & 0 & \lambda_{42}^2 \end{pmatrix} \begin{pmatrix} \xi_1 \\ \xi_2 \\ \xi_1\xi_2 \\ \xi_1^2 \\ \xi_2^2 \end{pmatrix} + \begin{pmatrix} \zeta_1 \\ \zeta_2 \\ \zeta_3 \\ \zeta_4 \\ \zeta_5 \\ \zeta_6 \\ \zeta_7 \\ \zeta_8 \\ \zeta_9 \\ \zeta_{10} \\ \zeta_{11} \\ \zeta_{12} \\ \zeta_{13} \\ \zeta_{14} \end{pmatrix}.$$

The covariance matrix of $(\xi_1, \xi_2, \xi_1\xi_2, \xi_1^2, \xi_2^2)^T$ and the variances of $\zeta_5, \ldots, \zeta_{14}$ can be derived by straightforward calculations of variances and covariances (see Lee, Song and Poon (2004)). Moreover, following Jaccard and Wan (1995), the following pairs of measurement errors associated with the products of indicators must be permitted to be correlated. More specifically, $\text{cov}(\zeta_5, \zeta_6)$, $\text{cov}(\zeta_5, \zeta_7)$, $\text{cov}(\zeta_6, \zeta_8)$, $\text{cov}(\zeta_7, \zeta_8)$, $\text{cov}(\zeta_9, \zeta_{10})$, $\text{cov}(\zeta_{10}, \zeta_{11})$, $\text{cov}(\zeta_{12}, \zeta_{13})$ and $\text{cov}(\zeta_{13}, \zeta_{14})$ are not equal to zero. All these nonlinear constraints and specifications are required to cooperate in the LISREL program; see Lee, Song and Poon (2004) for the program setup.

In the simulation study, two sets of true population values of the unknown parameters are considered:

(I): $\lambda_y = 0.6$, $\lambda_{21} = 0.5$, $\lambda_{42} = 0.6$, $\psi_{\varepsilon1} = \psi_{\varepsilon2} = 0.5$, $\psi_{\zeta1} = \psi_{\zeta2} = \psi_{\zeta3} = \psi_{\zeta4} = 0.5$, $\phi_{11} = \phi_{22} = 1.0$, $\phi_{21} = 0.6$, $\gamma_1 = 0.4$, $\gamma_2 = 0.6$, $\gamma_3 = \gamma_4 = \gamma_5 = 0.5$.
(II): The unknown parameters are the same as the first set, except $\phi_{21} = 0.2$, $\gamma_3 = \gamma_4 = \gamma_5 = 0.15$.

The first set of true values is corresponding to a situation with larger latent factors correlation and more significant interaction and/or quadratic effects.

On the basis of these true values, random vectors $\{(\varepsilon_{i1}, \varepsilon_{i2})^T, (\xi_{i1}, \xi_{i2})^T, (\zeta_{i1}, \zeta_{i2}, \zeta_{i3}, \zeta_{i4})^T, \delta_i : i = 1, \ldots, n\}$, with $n = 150$ and 300, are simulated from the corresponding normal distributions. Then $\eta_i, i = 1, \ldots, n$ are obtained from (ξ_{i1}, ξ_{i2}) and δ_i via Equation (8.18). Finally, for $i = 1, \ldots, n, \mathbf{y}_i$ is obtained from η_i and $(\varepsilon_{i1}, \varepsilon_{i2})$ via Equation (8.16), and $\mathbf{x}_i$ is obtained from (ξ_{i1}, ξ_{i2}) and $(\zeta_{i1}, \zeta_{i2}, \zeta_{i3}, \zeta_{i4})$ via Equation (8.17). Although the population means of $\mathbf{y}$ and $\mathbf{x}$ are zero, the data are transformed to mean deviation form in order to satisfy the assumption of the PI approach. The same data are analyzed by the PI approach with the LISREL program and the Bayesian approach.

In the simulation study, the hyperparameters in the conjugate prior distribution are specified as follows. We first conduct a Bayesian estimation on a simulated data set with non-information prior distribution which involves ad hoc prior hyperparameters, then we choose the hyperparameter values from the solution of this preliminary estimation. A few test runs have also been taken to decide the required number of 'burn-in' iterations. We observe that the Gibbs sampler with the MH algorithm converges at about 450 iterations. To be conservative, we take 500 'burn-in' iterations. Then, a total of 2000 additional iterations are taken to collect the sample $\{\boldsymbol{\theta}^{(t)} : t = 1, \ldots, T^*\}$ for computing the Bayesian estimates and standard error estimates.

The empirical performances of the approaches are evaluated on the basis of $\{\hat{\boldsymbol{\theta}}_r : r = 1, \ldots, 100\}$ obtained in 100 replications. The means (MEAN) of the estimates and the following root mean squares (RMS) between the true values and the corresponding estimates are computed:

$$\text{RMS of } \hat{\theta}(k) = \left\{ 100^{-1} \sum_{r=1}^{100} [\hat{\theta}_r(k) - \theta_0(k)]^2 \right\}^{1/2},$$

where $\hat{\theta}(k)$ and $\theta_0(k)$ are the kth element of the parameter vector and its true value, respectively. To evaluate the accuracy of the standard errors estimates, let $\text{SD}(\theta(k))$ be the sample standard deviation obtained from $\{\hat{\theta}_r(k); r = 1, \ldots, 100\}$, and $\text{SE}(\theta(k))$ be the mean of standard errors estimates of $\hat{\theta}(k)$ obtained from the various approaches. If the standard errors estimates obtained by the method are accurate, $\text{SE}(\theta(k))$ should be close to $\text{SD}(\theta(k))$ and $\text{SE}(\theta(k))/\text{SD}(\theta(k))$ should be close to 1.0. Hence the ratio $\text{SE}(\theta(k))/\text{SD}(\theta(k))$ is used to evaluate the accuracy of standard error estimates (see Lee, Poon and Bentler, 1995).

Results obtained from the PI approach on the basis of the model with structural Equation (8.18) under true values $\{\gamma_3 = \gamma_4 = \gamma_5 = 0.15 \text{ and } \phi_{21} = 0.2\}$ and $\{\gamma_3 = \gamma_4 = \gamma_5 = 0.5 \text{ and } \phi_{21} = 0.6\}$ are presented in Tables 8.2 and 8.3 respectively. From the RMS columns, it is obvious that the estimates produced

Table 8.2 PI Estimates in model with true quadratic and interaction coefficients $\gamma_3 = \gamma_4 = \gamma_5 = 0.15$ and $\phi_{21} = 0.2$.

Parameter	True value	$n = 150$			$n = 300$		
		MEAN	RMS	SE/SD	MEAN	RMS	SE/SD
λ_y	0.6	0.620	0.111	0.820	0.587	0.070	0.921
λ_{21}	0.5	0.550	0.544	0.179	0.507	0.158	0.409
λ_{42}	0.6	0.644	0.253	0.365	0.619	0.183	0.323
γ_1	0.4	0.385	0.158	0.747	0.388	0.128	0.646
γ_2	0.6	0.586	0.144	0.847	0.589	0.126	0.670
γ_3	0.15	0.103	0.251	0.653	0.119	0.122	0.852
γ_4	0.15	0.147	0.162	0.726	0.162	0.096	0.813
γ_5	0.15	0.169	0.176	0.663	0.140	0.081	0.847
ϕ_{11}	1.0	1.156	0.868	0.451	1.076	0.843	0.388
ϕ_{21}	0.2	0.251	0.176	0.367	0.229	0.100	0.384
ϕ_{22}	1.0	1.017	0.587	0.378	1.037	0.401	0.395
$\psi_{\varepsilon 1}$	0.5	0.492	0.655	0.898	0.490	0.613	0.984
$\psi_{\varepsilon 2}$	0.5	0.487	0.248	0.943	0.503	0.197	0.939
$\psi_{\zeta 1}$	0.5	0.327	0.165	0.461	0.422	0.113	0.402
$\psi_{\zeta 2}$	0.5	0.468	0.081	0.208	0.490	0.057	0.487
$\psi_{\zeta 3}$	0.5	0.462	0.561	0.405	0.462	0.371	0.382
$\psi_{\zeta 4}$	0.5	0.466	0.286	0.426	0.488	0.080	0.374
ψ_δ	0.36	0.372	0.196	0.789	0.391	0.174	0.728

Note: Tables 8.2 to 8.5 are Taken from Lee, Song and Poon (2004).

from the PI approach are inaccurate. Also, from the SE/SD columns, it is obvious that the PI approach seriously underestimates the standard errors. For this model, parameter estimates and standard error estimates are also obtained via the Bayesian approach. To give some idea about the empirical performances, Bayesian estimates with true values $\{\gamma_3 = \gamma_4 = \gamma_5 = 0.15$ and $\phi_{21} = 0.2\}$ and $\{\gamma_3 = \gamma_4 = \gamma_5 = 0.5$ and $\phi_{21} = 0.6\}$ are reported in Tables 8.4 and 8.5, respectively. It can be observed from these tables that the parameter estimates and the standard error estimates obtained from the Bayesian approach are accurate, and clearly better than those that are obtained from the PI approach.

8.2.7 Comparison with the LMS Approach via the Kenny–Judd Model

The objective of this simulation study is to investigate the empirical performances of the Bayesian approach in analyzing the basic Kenny–Judd model (Kenny and Judd, 1984), and to compare the efficiency of the Bayesian method with the results that were obtained by the LMS approach (Klein and Moosbrugger, 2000), and

Table 8.3 PI estimates in model with quadratic and interaction effects coefficients $\gamma_3 = \gamma_4 = \gamma_5 = 0.5$ and $\phi_{21} = 0.6$.

Parameter	True value	$n = 150$			$n = 300$		
		MEAN	RMS	SE/SD	MEAN	RMS	SE/SD
λ_y	0.6	0.608	0.044	1.003	0.599	0.037	0.836
λ_{21}	0.5	0.460	0.126	0.314	0.470	0.089	0.351
λ_{42}	0.6	0.580	0.134	0.350	0.575	0.098	0.334
γ_1	0.4	0.401	0.261	0.694	0.372	0.194	0.684
γ_2	0.6	0.580	0.254	0.746	0.590	0.179	0.714
γ_3	0.5	0.019	1.051	0.683	0.107	0.661	0.753
γ_4	0.5	0.557	0.557	0.649	0.526	0.367	0.617
γ_5	0.5	0.653	0.577	0.646	0.561	0.308	0.745
ϕ_{11}	1.0	1.145	0.706	0.398	1.111	0.599	0.394
ϕ_{21}	0.6	0.666	0.560	0.438	0.652	0.305	0.507
ϕ_{22}	1.0	1.081	0.316	0.359	1.059	0.233	0.381
$\psi_{\varepsilon 1}$	0.5	0.507	0.639	0.966	0.490	0.591	0.950
$\psi_{\varepsilon 2}$	0.5	0.486	0.566	0.934	0.510	0.548	1.050
$\psi_{\zeta 1}$	0.5	0.314	0.194	0.390	0.353	0.149	0.374
$\psi_{\zeta 2}$	0.5	0.508	0.093	0.355	0.504	0.062	0.370
$\psi_{\zeta 3}$	0.5	0.374	0.291	0.427	0.378	0.227	0.386
$\psi_{\zeta 4}$	0.5	0.488	0.100	0.434	0.503	0.074	0.417
ψ_{δ}	0.36	0.699	0.335	0.746	0.872	0.305	0.612

Table 8.4 Bayesian estimates in model with true quadratic and interaction coefficients $\gamma_3 = \gamma_4 = \gamma_5 = 0.15$ and $\phi_{21} = 0.2$.

Parameter	True value	$n = 150$			$n = 300$		
		MEAN	RMS	SE/SD	MEAN	RMS	SE/SD
λ_y	0.6	0.613	0.093	0.871	0.593	0.059	0.963
λ_{21}	0.5	0.582	0.146	1.102	0.557	0.104	1.055
λ_{42}	0.6	0.668	0.125	1.131	0.629	0.089	0.941
γ_1	0.4	0.447	0.121	1.241	0.430	0.102	0.921
γ_2	0.6	0.639	0.114	1.278	0.616	0.087	1.042
γ_3	0.15	0.230	0.179	1.086	0.192	0.113	0.956
γ_4	0.15	0.177	0.108	1.132	0.186	0.079	1.028
γ_5	0.15	0.191	0.115	0.976	0.149	0.061	0.968
ϕ_{11}	1.0	0.874	0.254	1.035	0.905	0.194	0.992
ϕ_{21}	0.2	0.187	0.103	1.057	0.189	0.078	1.008

ϕ_{22}	1.0	0.896	0.237	0.950	0.978	0.152	1.019
$\psi_{\varepsilon 1}$	0.5	0.511	0.143	0.969	0.518	0.097	1.028
$\psi_{\varepsilon 2}$	0.5	0.506	0.079	0.994	0.511	0.054	1.003
$\psi_{\zeta 1}$	0.5	0.650	0.250	0.951	0.610	0.174	1.018
$\psi_{\zeta 2}$	0.5	0.498	0.078	1.062	0.486	0.055	1.090
$\psi_{\zeta 3}$	0.5	0.617	0.191	1.010	0.540	0.135	0.882
$\psi_{\zeta 4}$	0.5	0.494	0.073	1.135	0.501	0.059	0.998
ψ_{δ}	0.36	0.345	0.119	1.053	0.333	0.102	0.947

Table 8.5 Bayesian estimates in model with true quadratic and interaction coefficients $\gamma_3 = \gamma_4 = \gamma_5 = 0.5$ and $\phi_{21} = 0.6$.

Parameter	True value	$n = 150$			$n = 300$		
		MEAN	RMS	SE/SD	MEAN	RMS	SE/SD
λ_y	0.6	0.608	0.037	0.883	0.604	0.027	1.004
λ_{21}	0.5	0.510	0.082	1.047	0.523	0.063	1.047
λ_{42}	0.6	0.615	0.081	1.102	0.615	0.063	1.004
γ_1	0.4	0.375	0.185	1.113	0.367	0.137	1.127
γ_2	0.6	0.597	0.182	1.127	0.628	0.128	1.207
γ_3	0.5	0.519	0.238	1.275	0.475	0.184	1.289
γ_4	0.5	0.514	0.176	1.236	0.565	0.168	1.099
γ_5	0.5	0.582	0.197	1.269	0.539	0.129	1.326
ϕ_{11}	1.0	0.974	0.194	0.989	0.915	0.164	0.941
ϕ_{21}	0.6	0.617	0.126	1.000	0.581	0.091	0.968
ϕ_{22}	1.0	0.955	0.204	0.912	0.933	0.147	0.972
$\psi_{\varepsilon 1}$	0.5	0.539	0.179	0.856	0.521	0.085	1.312
$\psi_{\varepsilon 2}$	0.5	0.501	0.087	0.909	0.506	0.052	1.067
$\psi_{\zeta 1}$	0.5	0.562	0.157	0.886	0.560	0.111	0.929
$\psi_{\zeta 2}$	0.5	0.524	0.076	0.983	0.504	0.050	0.946
$\psi_{\zeta 3}$	0.5	0.560	0.126	1.027	0.529	0.085	0.961
$\psi_{\zeta 4}$	0.5	0.515	0.065	1.144	0.509	0.052	0.964
ψ_{δ}	0.36	0.353	0.129	1.085	0.353	0.059	1.327

published in Schermelleh-Engel, Klein and Moosbrugger (1998). Using our notation that is slightly different from that in Schermelleh-Engel, Klein and Moosbrugger (1998), the Kenny–Judd model is defined by the following structural equation and measurement models:

$$
\begin{aligned}
y_1 &= \xi_1 + \epsilon_1, \ y_2 = \lambda_{21}\xi_1 + \epsilon_2, \\
y_3 &= \xi_2 + \epsilon_3, \ y_4 = \lambda_{42}\xi_2 + \epsilon_4, \qquad (8.19) \\
y^* &= \eta = \mu + \gamma_1\xi_1 + \gamma_2\xi_2 + \gamma_3\xi_1\xi_2 + \delta
\end{aligned}
$$

Based on the Kenny–Judd model, Schermelleh-Engel, Klein and Moosbrugger (1998) conducted a simulation study to examine the performances of the LMS method. In their simulation study, three models that correspond to the following true values of the latent interaction effect were considered: $\gamma_3 = 0.3$ (Model A), $\gamma_3 = 0.7$ (Model B) and $\gamma_3 = 1.5$ (Model C). The true values of the other parameters in Models A, B and C were given by

$$\gamma_1 = 0.2,\ \gamma_2 = 0.4,\ \mu = 1.0,\ \psi_\delta = 0.2,\ \phi_{11} = 0.49,\ \phi_{21} = 0.2352,$$
$$\phi_{22} = 0.64, \lambda_{21} = 0.6,\ \lambda_{42} = 0.7,\ \psi_{\epsilon 1} = 0.51,\ \psi_{\epsilon 2} = 0.64,$$
$$\psi_{\epsilon 3} = 0.36, \psi_{\epsilon 4} = 0.51,$$

where ψ_δ is the variance of δ, ϕ_{ij} are the variances and covariance of (ξ_1, ξ_2) and $\psi_{\epsilon i}$ are the variances of ϵ_i. (Note that in the notations of Schermelleh-Engel, Klein and Moosbrugger (1998), $\psi_{11} = \psi_\delta$ and $\theta_{ii} = \psi_{\epsilon i}$.) The sample sizes were varied by 200, 400 and 800. For each case in the 3×3 design, they used 500 replications (except for nonconvergent replications) to obtain the results. The means (MEAN) and standard deviation (MC-SD, same as our SD($\theta(k)$)) of the parameter estimates, and the means of the estimated standard error (EST-SD, same as our SE($\theta(k)$)) were computed. The simulated results were reported in Appendix C of Schermelleh-Engel, Klein and Moosbrugger (1998).

To study the impact of the prior inputs of the hyperparameters (see Equations (8.8) and (8.9)) in the Bayesian method, we consider accurate prior inputs (Type I) and inaccurate prior inputs (Type II). The accurate prior inputs are given by the following: (i) in the prior distribution of λ's, the means of the normal distributions are taken to be the true values and the covariance matrices are taken to be the identity matrix; (ii) in the prior distribution of μ, γ_1, γ_2 and γ_3, the elements in the mean vector of the normal distribution are taken to be the true values and the covariance matrix is taken to be the identity matrix; (iii) in the prior distribution of $\mathbf{\Phi}$, ρ_0 and $\mathbf{R}_0$ in the Wishart distribution are respectively taken to be 4 and $\mathbf{\Phi}_0^{-1}$, where $\mathbf{\Phi}_0$ is the matrix with true values of ϕ_{11}, ϕ_{21} and ϕ_{22}; (iv) in the prior distributions of ψ_δ and $\psi_{\epsilon i}$, the hyperparameters in the inverted Gamma distribution are taken to be $\alpha_{0\epsilon k} = 5$, $\beta_{0\epsilon k} = 2$ and $\alpha_{0\delta k} = 10$, $\beta_{0\delta k} = 2$. The inaccurate prior inputs are given by the following ad hoc values. In (i) and (ii), the means of prior normal distributions are all zero, and the covariance matrices are equal to twice the identity matrices; in (iii), $\rho_0 = 10$, and $\mathbf{R}_0$ is the identity matrix; and in (iv), $\alpha_{0\epsilon k} = 8$, $\beta_{0\epsilon k} = 5$ and $\alpha_{0\delta k} = 20$, $\beta_{0\delta k} = 5$. The other settings of the simulation study, for example the sample sizes, are exactly equal to those in Schermelleh-Engel, Klein and Moosbrugger (1998). The data sets for the simulation study were generated by using the true values and the model defined by Equation (8.19). Although these data sets are not exactly the same as those used in Schermelleh-Engel, Klein and Moosbrugger (1998), they are simulated from the same distribution. Bayesian solutions were obtained on the bases of 3000 observations after 3000 burn-in iterations.

Table 8.6 Results of Monte Carlo studies with $\gamma_3 = 0.3$ (Model A): mean and standard deviation of parameter estimates and estimated standard error.

		$N = 200$			$N = 400$			$N = 800$		
Par.	True value	Mean	MC-SD	EST-SD	Mean	MC-SD	EST-SD	Mean	MC-SD	EST-SD
Prior input Type I										
μ	1.00	1.002	.040	.039	1.001	.027	.028	1.001	.019	.019
γ_1	0.20	.196	.049	.050	.199	.036	.035	.198	.025	.025
γ_2	0.40	.396	.045	.044	.399	.030	.031	.400	.022	.022
γ_3	0.30	.302	.048	.047	.302	.034	.033	.299	.022	.023
ψ_δ	0.20	.198	.018	.019	.200	.013	.014	.200	.010	.010
λ_{21}	0.60	.590	.086	.080	.595	.056	.057	.597	.044	.040
λ_{42}	0.70	.696	.062	.063	.697	.044	.045	.698	.032	.032
ϕ_{11}	0.49	.498	.050	.050	.494	.034	.035	.492	.024	.025
ϕ_{21}	0.2352	.236	.041	.043	.235	.031	.031	.237	.022	.022
ϕ_{22}	0.64	.644	.063	.065	.646	.047	.046	.643	.031	.032
$\psi_{\epsilon 1}$	0.51	.513	.050	.051	.512	.034	.036	.512	.027	.025
$\psi_{\epsilon 2}$	0.64	.633	.062	.062	.636	.043	.045	.639	.032	.032
$\psi_{\epsilon 3}$	0.36	.364	.033	.036	.365	.025	.026	.362	.017	.018
$\psi_{\epsilon 4}$	0.51	.511	.051	.050	.513	.037	.036	.512	.025	.026
Prior input Type II										
μ	1.00	.995	.039	.040	.999	.029	.028	1.000	.018	.020
γ_1	0.20	.197	.049	.052	.201	.035	.036	.200	.023	.025
γ_2	0.40	.396	.043	.045	.397	.031	.031	.398	.022	.022
γ_3	0.30	.301	.048	.049	.299	.034	.033	.300	.023	.023
ψ_δ	0.20	.212	.017	.020	.207	.012	.014	.204	.009	.010
λ_{21}	0.60	.599	.076	.081	.596	.057	.057	.600	.041	.041
λ_{42}	0.70	.698	.063	.064	.697	.044	.045	.698	.031	.032
ϕ_{11}	0.49	.482	.046	.048	.490	.035	.034	.490	.025	.024
ϕ_{21}	0.2352	.231	.042	.042	.233	.030	.030	.234	.021	.021
ϕ_{22}	0.64	.625	.061	.062	.636	.044	.045	.638	.032	.032
$\psi_{\epsilon 1}$	0.51	.526	.050	.051	.519	.034	.036	.515	.026	.026
$\psi_{\epsilon 2}$	0.64	.647	.061	.063	.646	.044	.045	.644	.032	.032
$\psi_{\epsilon 3}$	0.36	.386	.033	.037	.373	.024	.026	.366	.017	.018
$\psi_{\epsilon 4}$	0.51	.527	.050	.051	.519	.034	.036	.514	.027	.026

The simulation results obtained by the Bayesian approach under Models A, B and C are reported in Tables 8.6, 8.7, and 8.8, respectively. Our focus is on the comparison of the results in these tables with those reported in Table 10.C1, Table 10.C2, and Table 10.C3 in Schermelleh-Engel, Klein and Moosbrugger (1998), in terms of the bias of parameter estimates and standard error estimates (with more emphasis on the interaction effect γ_3), and the efficiency of parameter estimates.

Table 8.7 Results of Monte Carlo studies with $\gamma_3 = 0.7$ (Model B): mean and standard deviation of parameter estimates and estimated standard error.

		$N = 200$			$N = 400$			$N = 800$		
Par.	True value	Mean	MC-SD	EST-SD	Mean	MC-SD	EST-SD	Mean	MC-SD	EST-SD
Prior input Type I										
μ	1.00	1.001	.040	.039	1.002	.028	.028	1.003	.019	.019
γ_1	0.20	.200	.050	.050	.199	.035	.035	.199	.025	.025
γ_2	0.40	.399	.042	.044	.400	.030	.031	.400	.023	.022
γ_3	0.70	.695	.048	.047	.699	.033	.033	.698	.023	.023
ψ_δ	0.20	.197	.018	.019	.200	.014	.014	.200	.010	.010
λ_{21}	0.60	.594	.081	.081	.596	.056	.057	.600	.039	.040
λ_{42}	0.70	.697	.065	.063	.693	.043	.045	.702	.033	.032
ϕ_{11}	0.49	.497	.051	.050	.493	.034	.035	.493	.025	.025
ϕ_{21}	0.2352	.237	.043	.044	.236	.030	.031	.237	.021	.022
ϕ_{22}	0.64	.648	.066	.065	.647	.045	.046	.645	.031	.032
$\psi_{\epsilon 1}$	0.51	.507	.049	.050	.511	.037	.036	.511	.027	.025
$\psi_{\epsilon 2}$	0.64	.637	.060	.063	.639	.046	.045	.640	.032	.032
$\psi_{\epsilon 3}$	0.36	.363	.034	.036	.364	.025	.026	.363	.018	.018
$\psi_{\epsilon 4}$	0.51	.514	.049	.051	.511	.035	.036	.512	.026	.026
Prior input Type II										
μ	1.00	.997	.041	.040	1.000	.027	.028	1.001	.021	.020
γ_1	0.20	.201	.052	.052	.200	.036	.036	.198	.024	.025
γ_2	0.40	.395	.044	.045	.400	.032	.031	.400	.023	.022
γ_3	0.70	.699	.049	.049	.698	.032	.033	.701	.024	.023
ψ_δ	0.20	.214	.015	.020	.207	.013	.014	.205	.010	.010
λ_{21}	0.60	.596	.083	.081	.594	.055	.057	.597	.040	.040
λ_{42}	0.70	.698	.063	.064	.695	.044	.045	.699	.033	.032
ϕ_{11}	0.49	.481	.046	.047	.487	.033	.034	.490	.024	.024
ϕ_{21}	0.2352	.227	.041	.042	.231	.030	.030	.235	.021	.021
ϕ_{22}	0.64	.626	.060	.062	.632	.045	.045	.638	.032	.032
$\psi_{\epsilon 1}$	0.51	.526	.047	.051	.518	.033	.036	.517	.025	.026
$\psi_{\epsilon 2}$	0.64	.647	.062	.063	.645	.044	.045	.644	.033	.032
$\psi_{\epsilon 3}$	0.36	.385	.035	.037	.373	.025	.026	.368	.017	.018
$\psi_{\epsilon 4}$	0.51	.521	.046	.051	.518	.036	.036	.516	.025	.026

Bias of parameters and standard error estimates. The means of the parameter estimates as reported in the columns under 'Mean' in Tables 8.6, 8.7 and 8.8 indicate that there is no sign of substantial bias in the Bayesian estimates. Schermelleh-Engel, Klein and Moosbrugger (1998) defined the following relative bias of the estimated standard errors:

$$\text{Rel-Bias}(\hat{\theta}) = [\{[\text{EST-SD}(\hat{\theta})]/[\text{MC-SD}(\hat{\theta})]\} - 1.0] \times 100\%,$$

Table 8.8 Results of Monte Carlo studies with $\gamma_3 = 1.5$ (Model C): mean and standard deviation of parameter estimates and estimated standard error.

		$N = 200$			$N = 400$			$N = 800$		
Par.	True value	Mean	MC-SD	EST-SD	Mean	MC-SD	EST-SD	Mean	MC-SD	EST-SD
Prior input Type I										
μ	1.00	1.006	.038	.039	1.005	.027	.028	1.003	.021	.020
γ_1	0.20	.204	.050	.051	.201	.036	.036	.202	.025	.025
γ_2	0.40	.398	.044	.044	.399	.032	.031	.398	.023	.022
γ_3	1.50	1.488	.048	.048	1.497	.031	.033	1.497	.023	.023
ψ_δ	0.20	.203	.017	.020	.202	.013	.014	.201	.010	.010
λ_{21}	0.60	.597	.078	.080	.600	.059	.057	.602	.041	.040
λ_{42}	0.70	.695	.061	.063	.697	.044	.044	.700	.030	.031
ϕ_{11}	0.49	.499	.047	.050	.494	.034	.035	.492	.025	.025
ϕ_{21}	0.2352	.239	.042	.044	.235	.029	.031	.238	.021	.022
ϕ_{22}	0.64	.645	.063	.065	.648	.048	.046	.646	.034	.032
$\psi_{\epsilon 1}$	0.51	.507	.050	.050	.502	.033	.035	.510	.024	.025
$\psi_{\epsilon 2}$	0.64	.633	.059	.062	.633	.045	.045	.640	.031	.032
$\psi_{\epsilon 3}$	0.36	.363	.034	.036	.360	.025	.025	.363	.017	.018
$\psi_{\epsilon 4}$	0.51	.508	.049	.050	.507	.033	.036	.509	.026	.025
Prior input Type II										
μ	1.00	1.001	.038	.041	1.002	.027	.028	1.001	.019	.020
γ_1	0.20	.205	.050	.053	.201	.036	.036	.200	.025	.025
γ_2	0.40	.396	.044	.046	.398	.032	.032	.399	.022	.022
γ_3	1.50	1.489	.048	.050	1.497	.031	.034	1.498	.023	.023
ψ_δ	0.20	.221	.016	.020	.212	.012	.014	.207	.010	.010
λ_{21}	0.60	.597	.078	.081	.600	.059	.057	.598	.041	.040
λ_{42}	0.70	.695	.061	.064	.697	.044	.045	.697	.031	.032
ϕ_{11}	0.49	.484	.046	.048	.487	.034	.034	.491	.024	.025
ϕ_{21}	0.2352	.232	.040	.042	.234	.029	.030	.235	.021	.021
ϕ_{22}	0.64	.626	.062	.062	.638	.047	.045	.638	.033	.032
$\psi_{\epsilon 1}$	0.51	.521	.048	.050	.509	.032	.035	.510	.025	.025
$\psi_{\epsilon 2}$	0.64	.643	.058	.062	.638	.045	.045	.638	.032	.032
$\psi_{\epsilon 3}$	0.36	.381	.033	.037	.369	.025	.026	.364	.018	.018
$\psi_{\epsilon 4}$	0.51	.522	.047	.051	.514	.032	.036	.514	.026	.026

for comparing the standard error estimate with the 'true' standard error estimate (MC-SD($\hat{\theta}$)). From Tables 8.6, 8.7 and 8.8, we observe that the relative bias of the estimated standard error of the other estimates is low for all models, sample sizes and prior inputs Type I and Type II. Hence, the estimated standard errors provided by the Bayesian method are comparable to those in Schermelleh-Engel, Klein and Moosbrugger (1998) and can be effectively used for confidence intervals and hypothesis testing.

Table 8.9 Relative efficiency of Bayesian estimates $\hat{\gamma}_3$ and $\hat{\psi}_\delta$ in relation to LMS in Simulation Study I.

Model	Prior	$N = 200$		$N = 400$		$N = 800$	
		$\hat{\gamma}_3$	$\hat{\psi}_\delta$	$\hat{\gamma}_3$	$\hat{\psi}_\delta$	$\hat{\gamma}_3$	$\hat{\psi}_\delta$
A	Type I	23.5%	41.3%	22.9%	42.3%	20.2%	51.0%
	Type II	23.5%	36.9%	22.9%	36.0%	22.0%	41.3%
B	Type I	13.2%	29.8%	12.3%	31.4%	12.5%	39.1%
	Type II	13.8%	20.7%	11.6%	27.0%	13.6%	39.1%
C	Type I	4.1%	10.3%	3.7%	11.7%	4.4%	13.7%
	Type II	4.1%	9.1%	3.7%	10.0%	4.4%	13.7%

Efficiency of parameter estimates. Schermelleh-Engel, Klein and Moosbrugger (1998) compared the efficiency of different methods by means of the relative efficiency. Based on their definition, the relative efficiency of the Bayesian method versus LMS is calculated by

$$\text{Rel-Eff}(\hat{\theta}) = [\text{MC-SD}(\hat{\theta})_{\text{BAY}}/\text{MC-SD}(\hat{\theta})_{\text{LMS}}] \times 100\%.$$

The relative efficiency values of the Bayesian methods versus LMS for the interesting estimators $\hat{\gamma}_3$ and $\hat{\psi}_\delta$ are reported in Table 8.9. We observe that under prior inputs Type I and II, the relative efficiency percentage of parameter estimates $\hat{\gamma}_3$ and $\hat{\psi}_\delta$ for the Bayesian method versus LMS is significantly smaller than 100%. Comparing the other MC-SD($\hat{\boldsymbol{\theta}}$) values in Tables 8.6 to 8.9 with those values in Tables 10.C1, 10.C2 and 10.C3 in Schermelleh-Engel, Klein and Moosbrugger (1998), the above finding is also true for the other parameter estimates. These results indicate that the Bayesian estimates are significantly more efficient than the LMS estimates.

From our discussion on the Bayesian estimation presented in Chapter 4, Section 4.2.1, particularly Equation (4.1), we know that the asymptotic standard derivations of the Bayesian estimate and the ML estimate should be close. Here, we recall that the 'true' ML estimator is the one that maximizes the exact likelihood function (or equivalently its logarithm), and this 'true' ML estimator has the optimal (asymptotic) properties. However, the LMS estimator is not the 'true' ML estimator for the Kenny–Judd model. Recall that the LMS procedure at least involves the following rather complicated approximations. In Klein and Moosbrugger's notation, the density for the realization $(\mathbf{x} = \text{x}, \mathbf{y} = \text{y})$ is expressed as a high-dimensional integral, see their Equation (15). This high-dimensional integral was approximated by the Hermite–Gaussian quadrature, and then led to a finite mixture density of $M(= 24)$ components. Hence, the exact likelihood function is not expressed in closed form. The LMS procedure then applied the EM algorithm to find the solution of the 'approximated' likelihood function. As a result, the LMS method only gives an approximation of the 'true' ML estimate, which may not be able to achieve the optimal asymptotic properties. Practically, the large variablity of the LMS estimate may be due to the various complicated approximations.

8.3 BAYESIAN ESTIMATION OF NONLINEAR SEMs WITH MIXED CONTINUOUS AND ORDERED CATEGORICAL VARIABLES

In this section, NSEMs with mixed continuous and ordered categorical variables are analyzed. Using similar notations to those in Chapter 6, the $p \times 1$ manifest random vector is denoted by $\mathbf{v}$, which contains a subvector of variables $\mathbf{x} = (x_1, \ldots, x_r)^T$ whose exact continuous measurements are observable, and a subvector of variables $\mathbf{y} = (y_1, \ldots, y_s)^T$ whose continuous measurements are unobservable, but the underlying information is given by an observable ordered categorical vector $\mathbf{z} = (z_1, \ldots, z_s)^T$. The relationship between $\mathbf{y}$ and $\mathbf{z}$ is similarly given by Equation (6.3) by a set of thresholds $\boldsymbol{\alpha}_k = \{\alpha_{k,1}, \ldots, \alpha_{k,b_k}\}$ defined as before. Without loss of generality, suppose $\mathbf{v} = (\mathbf{x}^T, \mathbf{y}^T)^T$ and this manifest random vector satisfies the following measurement equation as in Equation (8.1):

$$\mathbf{v}_i = \boldsymbol{\mu} + \boldsymbol{\Lambda}\boldsymbol{\omega}_i + \boldsymbol{\epsilon}_i, \quad i = 1, \ldots, n \tag{8.20}$$

where $\boldsymbol{\Lambda}, \boldsymbol{\omega}_i$ and $\boldsymbol{\epsilon}_i$ are exactly defined as in Section 8.2.1. The nonlinear structural equation is defined either by Equation (8.2) or (8.3). Again, the same identification conditions are imposed and the variables in $\mathbf{x}$ and $\mathbf{y}$ can be indicators for $\boldsymbol{\eta}$ and/or $\boldsymbol{\xi}$. Similarly, as before, suppose that the model relating to the subvector $\mathbf{y}_i = (y_{i1}, \ldots, y_{is})^T$ of $\mathbf{v}_i$ is given by:

$$\mathbf{y}_i = \boldsymbol{\mu}_y + \boldsymbol{\Lambda}_y\boldsymbol{\omega}_i + \boldsymbol{\epsilon}_{yi},$$

where $\boldsymbol{\mu}_y(s \times 1)$ is a subvector of $\boldsymbol{\mu}$, $\boldsymbol{\Lambda}_y(s \times q)$ is a submatrix of $\boldsymbol{\Lambda}$, and $\boldsymbol{\epsilon}_{yi}(s \times 1)$ is a subvector of $\boldsymbol{\epsilon}_i$ with diagonal covariance matrix $\boldsymbol{\Psi}_{y\epsilon}$. Let $\mathbf{z}_i = (z_{i1}, \ldots, z_{is})^T$ be the ordered categorical observation corresponding to $\mathbf{y}_i$, $i = 1, \ldots, n$. In this section, the NSEM defined above is estimated with the data $\{(\mathbf{x}_i, \mathbf{z}_i), i = 1, \ldots, n\}$ of mixed continuous and ordered categorical observations.

8.3.1 Posterior Analysis

Let $\mathbf{X} = (\mathbf{x}_1, \ldots, \mathbf{x}_n)$ and $\mathbf{Z} = (\mathbf{z}_1, \ldots, \mathbf{z}_n)$ be the observed continuous and ordered categorical data matrices respectively; let $\mathbf{Y} = (\mathbf{y}_1, \ldots, \mathbf{y}_n)$ and

$\boldsymbol{\Omega} = (\boldsymbol{\omega}_1, \ldots, \boldsymbol{\omega}_n)$ be the latent continuous data and latent variables respectively. Inspired again by the technique of data augmentation, the observed data $[\mathbf{X}, \mathbf{Z}]$ are augmented with the latent data $[\mathbf{Y}, \boldsymbol{\Omega}]$ in the posterior analysis. As we will see later, the difficulty induced by the ordered categorical data can be solved. Joint Bayesian estimates of $\boldsymbol{\Omega}$, the thresholds in $\boldsymbol{\alpha} = (\boldsymbol{\alpha}_1, \ldots, \boldsymbol{\alpha}_s)$, and the structural parameter vector $\boldsymbol{\theta}$ that contains all unknown parameters in $\boldsymbol{\mu}, \boldsymbol{\Phi}, \boldsymbol{\Lambda}, \boldsymbol{\Lambda}_\omega, \boldsymbol{\Psi}_\delta$ and $\boldsymbol{\Psi}_\epsilon$, will be obtained via the following Gibbs sampler. At the jth iteration with current values $\boldsymbol{\alpha}^{(j)}, \boldsymbol{\theta}^{(j)}, \boldsymbol{\Omega}^{(j)}$ and $\mathbf{Y}^{(j)}$:

$$\begin{aligned} &\text{Generate } \boldsymbol{\Omega}^{(j+1)} \text{ from } p(\boldsymbol{\Omega}|\boldsymbol{\theta}^{(j)}, \boldsymbol{\alpha}^{(j)}, \mathbf{Y}^{(j)}, \mathbf{X}, \mathbf{Z}); \\ &\text{Generate } \boldsymbol{\theta}^{(j+1)} \text{ from } p(\boldsymbol{\theta}|\boldsymbol{\Omega}^{(j+1)}, \boldsymbol{\alpha}^{(j)}, \mathbf{Y}^{(j)}, \mathbf{X}, \mathbf{Z}); \\ &\text{Generate } (\boldsymbol{\alpha}^{(j+1)}, \mathbf{Y}^{(j+1)}) \text{ from } p(\boldsymbol{\alpha}, \mathbf{Y}|\boldsymbol{\theta}^{(j+1)}, \boldsymbol{\Omega}^{(j+1)}, \mathbf{X}, \mathbf{Z}). \end{aligned} \tag{8.21}$$

To derive the conditional distributions involved in Equation (8.21), we need to impose some natural assumptions on the prior distributions of $\boldsymbol{\theta}$. Let $\mathbf{V} = (\mathbf{v}_1, \ldots, \mathbf{v}_n)$ with $\mathbf{v}_i = (\mathbf{x}_i, \mathbf{y}_i)$; thus $\mathbf{V} = (\mathbf{X}, \mathbf{Y})$. Moreover, let $\boldsymbol{\theta}_\epsilon$ be the unknown parameters in $\boldsymbol{\mu}$, $\boldsymbol{\Lambda}$ and $\boldsymbol{\Psi}_\epsilon$ that are associated with Equation (8.20), and let $\boldsymbol{\theta}_\omega$ be the unknown parameters in $\boldsymbol{\Lambda}_\omega$, $\boldsymbol{\Phi}$ and $\boldsymbol{\Psi}_\delta$ that are associated with Equation (8.2). We assume that the prior distribution of $\boldsymbol{\theta}_\epsilon$ is independent of the prior distribution of $\boldsymbol{\theta}_\omega$. Hence $p(\boldsymbol{\theta}) = p(\boldsymbol{\theta}_\epsilon)p(\boldsymbol{\theta}_\omega)$. Conditional on the given $\mathbf{X}$ and $\mathbf{Y}$, $\boldsymbol{\Omega}$ and $\boldsymbol{\alpha}$ are not relevant. Hence,

$$p(\boldsymbol{\Omega}|\boldsymbol{\theta}, \boldsymbol{\alpha}, \mathbf{Y}, \mathbf{X}, \mathbf{Z}) = p(\boldsymbol{\Omega}|\boldsymbol{\theta}, \boldsymbol{\alpha}, \mathbf{X}, \mathbf{Y}) = p(\boldsymbol{\Omega}|\boldsymbol{\theta}, \mathbf{V}). \tag{8.22}$$

For the conditional distribution corresponding to $\boldsymbol{\theta}$,

$$p(\boldsymbol{\theta}|\boldsymbol{\Omega}, \boldsymbol{\alpha}, \mathbf{Y}, \mathbf{X}, \mathbf{Z}) = p(\boldsymbol{\theta}_\epsilon, \boldsymbol{\theta}_\omega|\boldsymbol{\Omega}, \mathbf{V}) = p(\boldsymbol{\theta}_\epsilon|\boldsymbol{\Omega}, \mathbf{V})\, p(\boldsymbol{\theta}_\omega|\boldsymbol{\Omega}, \mathbf{V}). \tag{8.23}$$

Moreover, when $\boldsymbol{\Omega}$ is given, $\mathbf{X}$ is conditionally independent of $\mathbf{Y}$, hence

$$p(\boldsymbol{\alpha}, \mathbf{Y}|\boldsymbol{\theta}, \boldsymbol{\Omega}, \mathbf{X}, \mathbf{Z}) = p(\boldsymbol{\alpha}, \mathbf{Y}|\boldsymbol{\theta}, \boldsymbol{\Omega}, \mathbf{Z}). \tag{8.24}$$

Note that given $\mathbf{V}$, the model with the ordered categorical data is the same as the model with the continuous data. Hence, using the same conjugate prior distributions, the conditional distribution in Equation (8.22) is exactly the same as that given in Equation (8.11), and the conditional distributions involved in Equation (8.23) are exactly the same as those given in Equations (8.6), (8.7), (8.8), and (8.10). Moreover, as $(\boldsymbol{\alpha}, \mathbf{Y})$ is associated with the measurement Equation (8.20) which is exactly the same as that given in Equation (6.1),

$p(\boldsymbol{\alpha}, \mathbf{Y}|\boldsymbol{\theta}, \boldsymbol{\Omega}, \mathbf{Z})$ can be derived by the same reasoning as before. Hence, under the same noninformative prior distribution for the unknown threshold parameters as given before, that is, for $k = 1, \ldots, s$,

$$p(\alpha_{k,2}, \ldots, \alpha_{k,b_{k-1}}) \propto c, \text{ for } \alpha_{k,2} < \cdots < \alpha_{k,b_{k-1}},$$

where c is a constant; the conditional distribution $p(\boldsymbol{\alpha}, \mathbf{Y}|\boldsymbol{\theta}, \boldsymbol{\Omega}, \mathbf{Z})$ is given by Equation (6.13). For completeness, it is reproduced here. For $k = 1, \ldots, s$, let $\mathbf{Y}_k$ and $\mathbf{Z}_k$ be the kth rows of $\mathbf{Y}$ and $\mathbf{Z}$, respectively, it follows that

$$p(\boldsymbol{\alpha}_k, \mathbf{Y}_k|\mathbf{Z}_k, \boldsymbol{\theta}, \boldsymbol{\Omega}) = p(\boldsymbol{\alpha}_k|\mathbf{Z}_k, \boldsymbol{\theta}, \boldsymbol{\Omega})p(\mathbf{Y}_k|\boldsymbol{\alpha}_k, \mathbf{Z}_k, \boldsymbol{\theta}, \boldsymbol{\Omega}),$$

with

$$p(\boldsymbol{\alpha}_k|\mathbf{Z}_k, \boldsymbol{\theta}, \boldsymbol{\Omega}) \propto \prod_{i=1}^{n} \left\{ \Phi^* \left[\psi_{yk}^{-1/2}(\alpha_{k,z_{ik}+1} - \mu_{yk} - \boldsymbol{\Lambda}_{yk}^T \boldsymbol{\omega}_i) \right] - \Phi^* \left[\psi_{yk}^{-1/2}(\alpha_{k,z_{ik}} - \mu_{yk} - \boldsymbol{\Lambda}_{yk}^T \boldsymbol{\omega}_i) \right] \right\},$$

and $p(\mathbf{Y}_k|\boldsymbol{\alpha}_k, \mathbf{Z}_k, \boldsymbol{\theta}, \boldsymbol{\Omega})$ is a product of $p(y_{ik}|\boldsymbol{\alpha}_k, \mathbf{Z}_k, \boldsymbol{\theta}, \boldsymbol{\Omega})$, where

$$p(y_{ik}|\boldsymbol{\alpha}_k, \mathbf{Z}_k, \boldsymbol{\theta}, \boldsymbol{\Omega}) \stackrel{D}{=} N(\mu_{yk} + \boldsymbol{\Lambda}_{yk}^T \boldsymbol{\omega}_i, \psi_{yk}) I_{(\alpha_{k,z_{ik}}, \alpha_{k,z_{ik+1}})}(y_{ik}),$$

in which ψ_{yk} is the kth diagonal element of $\boldsymbol{\Psi}_{y\epsilon}$, μ_{yk} is the kth element of $\boldsymbol{\mu}_y$, $\boldsymbol{\Lambda}_{yk}^T$ is the kth row of $\boldsymbol{\Lambda}_y$, $I_A(y)$ is an index function which takes 1 if $y \in A$ and 0 otherwise, and $\Phi^*(\cdot)$ denotes the standard normal cumulative distribution function. As a result,

$$p(\boldsymbol{\alpha}_k, \mathbf{Y}_k|\mathbf{Z}_k, \boldsymbol{\theta}, \boldsymbol{\Omega}) \propto \prod_{i=1}^{n} \phi[\psi_{yk}^{-1/2}(y_{ik} - \mu_{yk} - \boldsymbol{\Lambda}_{yk}^T \boldsymbol{\omega}_i)] I_{(\alpha_{k,z_{ik}}, \alpha_{k,z_{ik+1}})}(y_{ik}),$$

where $\phi(\cdot)$ is the standard normal density. Note that the above result is almost identical to Equation (6.11), except that the underlying model is different.

In a similar way to before, because the conditional distributions involved in $p(\boldsymbol{\theta}|\boldsymbol{\Omega}, \boldsymbol{\alpha}, \mathbf{Y}, \mathbf{X}, \mathbf{Z})$ are the standard Gamma, normal and inverted Wishart distributions, simulating observations from them is straightforward and fast. The MH algorithm as described in Sections 8.2.3 and 6.3.2 can be applied in exactly the same manner for simulating observations from $p(\boldsymbol{\Omega}|\boldsymbol{\theta}, \boldsymbol{\alpha}, \mathbf{Y}, \mathbf{X}, \mathbf{Z})$ and $p(\boldsymbol{\alpha}, \mathbf{Y}|\boldsymbol{\theta}, \boldsymbol{\Omega}, \mathbf{X}, \mathbf{Z})$, respectively. Bayesian estimates and other statistics for $\boldsymbol{\Omega}$

and $\boldsymbol{\theta}$ can be obtained via a sufficiently large number of simulated observations collected from the hybrid algorithm after convergence (see Equations (8.14) and (8.15)).

We mentioned at the end of Section 8.2.3 that the generalization of the Bayesian approach from a standard linear SEM to an NSEM is not difficult. From the discussion in this section, we understand that the Bayesian approach for analyzing an NSEM with continuous variables can be generalized to handle NSEMs with mixed continuous and ordered categorical variables without much difficulty. It is conceptually simple, and just involves one additional component in simulating $p(\boldsymbol{\alpha}_k, \mathbf{Y}_k|\mathbf{Z}_k, \boldsymbol{\theta}, \boldsymbol{\Omega})$ in the Gibbs sampler. In contrast, it is rather difficult to apply either the covariance structure analysis approach, the product indicator approach, or the multistage approach to analyze nonlinear SEMs with ordered categorical variables.

8.3.2 Illustrative Example 2

A small portion of the ICPSR data set collected in the project World Values Survey 1981–1984 and 1990–1993 (World Values Study Group, 1994) is analyzed in this example. In this illustration, only the data obtained from the UK were used. Six variables in the original data set (variables 180, 96, 62, 179, 116 and 117; see Appendix 1.1) that related to the respondents' job, religious belief and homelife were taken as manifest variables in $\mathbf{v} = (v_1, \cdots, v_6)^T$. After deleting cases with missing data, the sample size was 197. Among them, (v_1, v_2) were related to life, (v_3, v_4) were related to religions belief and (v_5, v_6) were related to job satisfaction. Variables (v_3, v_4) were ordered categorical with five categories. Other variables were measured on a 10-point scale and for convenience they were treated as continuous variables.

A nonlinear structural equation model with the latent factors η and $\boldsymbol{\xi} = (\xi_1, \xi_2)^T$ was proposed with the following specifications: $\boldsymbol{\Pi} = \mathbf{0}, \mathbf{H}(\boldsymbol{\xi}) = (\xi_1, \xi_2, \xi_1\xi_2)^T$, and

$$\boldsymbol{\Gamma}^T = \begin{bmatrix} \gamma_1 \\ \gamma_2 \\ \gamma_3 \end{bmatrix}, \quad \boldsymbol{\Lambda}^T = \begin{bmatrix} 1.0 & \lambda_{21} & 0 & 0 & 0 & 0 \\ 0 & 0 & 1.0 & \lambda_{42} & 0 & 0 \\ 0 & 0 & 0 & 0 & 1.0 & \lambda_{63} \end{bmatrix},$$

where the 1's and 0's in $\boldsymbol{\Lambda}$ were treated as fixed parameters. Here, the nonlinear structural equation is $\eta = \gamma_1\xi_1 + \gamma_2\xi_2 + \gamma_3\xi_1\xi_2 + \delta$. From the structure of $\boldsymbol{\Lambda}$, latent factors η, ξ_1 and ξ_2 can be roughly interpreted as 'life', 'religious belief' and 'job satisfaction', respectively, while $\xi_1\xi_2$ represents the interaction of 'religious belief' and 'job satisfaction'. To identify the ordered categorical variables, we set $\alpha_{k,1} = \Phi^{*-1}(n_{k,1}/n)$ and $\alpha_{k,4} = \Phi^{*-1}(\sum_{i=1}^{4} n_{k,i}/n)$, where $n_{k,i}$ denotes the number of ordered categorical observations that are equal to i, and Φ^* denotes the distribution function of the standard normal distribution. In this nonlinear model, there were a total of 26 unknown parameters which are the unknown elements in $\boldsymbol{\mu}$ and $\boldsymbol{\Lambda}$, $\alpha_{k,i}(k=3,4,i=2,3)$, $\boldsymbol{\Lambda}_\omega$, variances and covariance in $\boldsymbol{\Phi}$ and the diagonal elements in $\boldsymbol{\Psi}_\varepsilon$ and $\boldsymbol{\Psi}_\delta$. Bayesian estimates of these structural parameters, and estimates of η_i, ξ_{i1} and ξ_{i2} were obtained via the hybrid algorithm that combines the Gibbs sampler and the MH algorithm. The variances of the proposal distributions in the MH algorithm were chosen such that the approximately average acceptance rate is 0.35. The following hyperparameters were selected: $\alpha_{0\varepsilon k}=\alpha_{0\delta}=10$, $\beta_{0\varepsilon k}=\beta_{0\delta}=8$, $\mathbf{H}_{0yk}$ and $\mathbf{H}_{0\omega k}$ are diagonal matrices with diagonal elements 0.25, $\rho_0=20$, $\boldsymbol{\Sigma}_0=\mathbf{I}_6$, $\mathbf{R}_0^{-1}=2\tilde{\boldsymbol{\Phi}}$, $\boldsymbol{\mu}_0=\tilde{\boldsymbol{\mu}}$ and $\boldsymbol{\Lambda}_{0k}=\tilde{\boldsymbol{\Lambda}}_{0k}$, where $\tilde{\boldsymbol{\mu}}$, $\tilde{\boldsymbol{\Phi}}$ and $\tilde{\boldsymbol{\Lambda}}_{0k}$ were the Bayesian estimates obtained using noninformative prior distributions. Based on different starting values of the parameters, three parallel sequences of observations were generated. We found that the hybrid algorithm converged after about 2000 iterations. After convergence, a total of 2000 observations were collected to obtain the results.

Bayesian estimates of the structural parameters and their standard error estimates are reported in Table 8.10. The estimated nonlinear structural equation is

$$\eta_i = 0.723\xi_{i1} + 0.710\xi_{i2} - 0.494\xi_{i1}\xi_{i2}.$$

From the standard error estimates of $\hat{\gamma}_1$, $\hat{\gamma}_2$ and $\hat{\gamma}_3$, we see that all these coefficients are significantly different from zero. Hence, 'religious belief' and 'job satisfaction' have significant positive linear effects to 'life'. This finding is reasonable. Moreover, 'religious belief' and 'job satisfaction' also have a

Table 8.10 The Bayesian estimates (EST) and their standard error estimates (SE) of illustrative example 2.

Parameter	EST	SE	Parameter	EST	SE
$\alpha_{3,2}$	0.305	0.057	μ_1	8.414	0.121
$\alpha_{3,3}$	0.608	0.060	μ_2	7.824	0.123
$\alpha_{4,2}$	−0.140	0.054	μ_3	−0.001	0.081
$\alpha_{4,3}$	0.075	0.058	μ_4	0.030	0.083
γ_1	0.723	0.183	μ_5	7.573	0.142
γ_2	0.710	0.142	μ_6	7.430	0.167
γ_3	−0.494	0.150			
			$\psi_{\epsilon 1}$	0.697	0.153
λ_{21}	0.869	0.096	$\psi_{\epsilon 2}$	1.287	0.191
λ_{42}	1.050	0.148	$\psi_{\epsilon 3}$	0.447	0.085
λ_{63}	1.151	0.270	$\psi_{\epsilon 4}$	0.528	0.092
ϕ_{11}	0.519	0.122	$\psi_{\epsilon 5}$	2.439	0.507
ϕ_{21}	−0.110	0.108	$\psi_{\epsilon 6}$	3.356	0.543
ϕ_{22}	1.792	0.587	ψ_δ	0.828	0.168

This table is taken from Lee and Zhu (2000).

significant negative interaction effect on 'life'. This shows that 'life' cannot be adequately accounted by the linear effects of 'religious belief' and 'job satisfaction', and an interaction effect of these two independent latent variables has to be added. Depending on the signs of ξ_{1i} and ξ_{2i}, the interaction term has different impacts on η_i. For example, if both 'religious belief' and 'job satisfaction' are positive, then their linear effects would have too strong a positive effect on 'life' and need to be adjusted by a negative interaction term. If both of them are negative, then their linear effect would have too weak a negative effect on 'life' and need to be further adjusted by a negative interaction term. Finally, if one of them is negative and one of them is positive, then their interaction would add a positive effect to 'life'. Now, we consider the intercepts and the means of the manifest variables. As an example, we consider the mean of the second manifest variable, say $E(y_2)$. Since $E(\xi_1) = E(\xi_2) = 0$, $\hat{\mu}_2 = 7.824$, $\hat{\lambda}_{21} = 0.869$, $\hat{\gamma}_3 = -0.494$ and $\hat{E}(\xi_1\xi_2) = \hat{\phi}_{21} = -0.110$, it follows from Equation (8.4) that

$$\widehat{E(y_2)} = \hat{\mu}_2 + \hat{\lambda}_{21}\hat{\gamma}_3\hat{E}(\xi_1\xi_2) = 7.824 + 0.869(-0.494)(-0.110) = 7.871.$$

8.4 BAYESIAN ESTIMATION OF SEMs WITH NONLINEAR COVARIATES AND LATENT VARIABLES

As we discussed in Section 5.4, a useful extension of the basic SEM is the accommodation of covariates in the model for establishing better relationships, or for prediction. In the measurement equation that is used for exploring and/or identifying the latent variables, the inclusion of covariates is helpful in providing a better relationship between the manifest and latent variables. As the main purpose of the structural equation is to achieve accurate prediction or relation among latent variables, the accommodation of covariates is even more important. A covariate can be an explanatory variable such as age, gender, social status, or other important variable that is defined by a single manifest variable for relating or predicting the endogenous latent variables. It may come from discrete and continuous distributions.

From the above discussion, we see that it is useful to incorporate linear and nonlinear terms among exogenous latent variables and covariates in the structural equation; especially for situations where some covariates that represent important predictors of the endogenous latent variables. In this section, we formulate an NSEM which includes a measurement equation that is defined with linear covariates and linear latent variables, and a structural equation that is defined by a general vector-valued function that accommodates general nonlinear terms of both the covariates and exogenous latent variables.

8.4.1 A model with Ordered Categorical Variables and Nonlinear Covariates and Latent Variables

The discussion in this section is based on the NSEM proposed in Song and Lee (2006). In this model, the measurement Equation (8.1) is generalized as follows:

$$\mathbf{y}_i = \mathbf{A}\mathbf{c}_i + \boldsymbol{\Lambda}\boldsymbol{\omega}_i + \boldsymbol{\epsilon}_i, \quad i = 1, \ldots, n, \tag{8.25}$$

where $\mathbf{c}_i(m_1 \times 1)$ is a vector of covariates and $\mathbf{A}(p \times m_1)$ is an unknown parameter matrix. Elements of $\mathbf{c}_i$ could be constants or values that come from continuous or discrete distributions. To assess the possible important causal effects of a vector of covariates $\mathbf{x}_i(m_2 \times 1)$ to $\boldsymbol{\eta}_i$, the structural Equation (8.2) is generalized as follows:

$$\boldsymbol{\eta}_i = \boldsymbol{\Pi}\boldsymbol{\eta}_i + \boldsymbol{\Gamma}\mathbf{F}(\mathbf{x}_i, \boldsymbol{\xi}_i) + \boldsymbol{\delta}_i \quad i = 1, \ldots, n, \tag{8.26}$$

where $\mathbf{F}(\mathbf{x}_i, \boldsymbol{\xi}_i) = (f_1(\mathbf{x}_i, \boldsymbol{\xi}_i), \ldots, f_r(\mathbf{x}_i, \boldsymbol{\xi}_i))^T$ is a nonzero vector-valued function with nonzero, known and differentiable functions $f_1, \ldots, f_r$, and $\boldsymbol{\Gamma}(q_1 \times r)$ is a matrix of unknown parameters. To cope with some situations in substantive research, we allow that components in $\mathbf{y}_i$ are either continuous or ordered categorical. Identification of this model can be achieved via the methods given in Sections 6.2 and 8.3.1.

Clearly, the structural equation with the nonlinear term $\boldsymbol{\Gamma}\mathbf{F}(\mathbf{x}_i, \boldsymbol{\xi}_i)$ is rather general. A more concrete example of the general structural equation defined in Equation (8.26) with $\boldsymbol{\eta}_i = (\eta_i)$, $\boldsymbol{\xi}_i = (\xi_{i1}, \xi_{i2})^T$ and $\mathbf{x}_i = (x_{i1}, x_{i2})^T$ is:

$$\begin{aligned}\eta_i =& \gamma_1 x_{i1} + \gamma_2 x_{i2} + \gamma_3 x_{i1} x_{i2} + \alpha_1 \xi_{i1} + \alpha_2 \xi_{i2} + \alpha_3 \xi_{i1}\xi_{i2} + \beta_1 x_{i1}\xi_{i1} + \beta_2 x_{i1}\xi_{i2} \\ &+ \beta_3 x_{i2}\xi_{i1} + \beta_4 x_{i2}\xi_{i2} + \beta_5 x_{i1}\xi_{i1}\xi_{i2} + \beta_6 x_{i2}\xi_{i1}\xi_{i2} + \beta_7 x_{i1}\xi_{i1}^2 + \beta_8 x_{i2}\xi_{i2}^2 \\ &+ \beta_9 x_{i1}x_{i2}\xi_{i1}\xi_{i2} + \delta_i.\end{aligned}$$

Here, $\boldsymbol{\Gamma} = (\gamma_1, \gamma_2, \gamma_3, \alpha_1, \alpha_2, \alpha_3, \beta_1, \beta_2, \beta_3, \beta_4, \beta_5, \beta_6, \beta_7, \beta_8, \beta_9)^T$ and $\mathbf{F}(\mathbf{x}_i, \boldsymbol{\xi}_i) = (x_{i1}, x_{i2}, x_{i1}x_{i2}, \xi_{i1}, \xi_{i2}, \xi_{i1}\xi_{i2}, x_{i1}\xi_{i1}, x_{i1}\xi_{i2}, x_{i2}\xi_{i1}, x_{i2}\xi_{i2}, x_{i1}\xi_{i1}\xi_{i2}, x_{i2}\xi_{i1}\xi_{i2}, x_{i1}\xi_{i1}^2, x_{i2}\xi_{i2}^2, x_{i1}x_{i2}\xi_{i1}\xi_{i2})^T$. More complex product terms of elements in $\mathbf{x}_i$ and $\boldsymbol{\xi}_i$ can be assessed via appropriately defined structural equations. As the covariates $\mathbf{x}_i$ may come from any arbitrary distributions that give continuous data, ordered or unordered categorical data, this NSEM can handle a wide range of situations. Again, more care is needed to interpret the mean vector of $\mathbf{y}_i$ under the context of nonlinear SEM. Let $\mathbf{A}_k^T$ and $\boldsymbol{\Lambda}_k^T$ be the kth rows of $\mathbf{A}$ and $\boldsymbol{\Lambda}$, respectively. For $k = 1, \cdots, p$, it follows from Equation (8.25) that $E(y_{ik}) = \mathbf{A}_k^T\mathbf{c}_i + \boldsymbol{\Lambda}_k^T E(\boldsymbol{\omega}_i)$. Although $E(\boldsymbol{\xi}_i) = \mathbf{0}$, it follows from Equation (8.26) that $E(\boldsymbol{\eta}_i) \neq \mathbf{0}$ if $\mathbf{F}(\mathbf{x}_i, \boldsymbol{\xi}_i)$ is a nonlinear function of $\boldsymbol{\xi}_i$. This implies that $E(\boldsymbol{\omega}_i) \neq \mathbf{0}$ and $E(y_{ik}) \neq \mathbf{A}_k^T \mathbf{c}_i$. Let $\boldsymbol{\Lambda}_k^T = (\boldsymbol{\Lambda}_{k\eta}^T, \boldsymbol{\Lambda}_{k\xi}^T)$ be a partition of $\boldsymbol{\Lambda}_k^T$ that corresponds to the partition of $\boldsymbol{\omega}_i = (\boldsymbol{\eta}_i^T, \boldsymbol{\xi}_i^T)^T$. Because $E(\boldsymbol{\xi}_i) = \mathbf{0}$ and $\boldsymbol{\eta}_i = (\mathbf{I} - \boldsymbol{\Pi})^{-1}[\boldsymbol{\Gamma}\mathbf{F}(\mathbf{x}_i, \boldsymbol{\xi}_i) + \boldsymbol{\delta}_i]$, it follows from Equation (8.26) that

$$E(y_{ik}) = \mathbf{A}_k^T \mathbf{c}_i + \boldsymbol{\Lambda}_{k\eta}^T E(\boldsymbol{\eta}_i) + \boldsymbol{\Lambda}_{k\xi}^T E(\boldsymbol{\xi}_i) = \mathbf{A}_k^T \mathbf{c}_i + \boldsymbol{\Lambda}_{k\eta}^T [(\mathbf{I} - \boldsymbol{\Pi})^{-1}\boldsymbol{\Gamma}] E[\mathbf{F}(\mathbf{x}_i, \boldsymbol{\xi}_i)].$$

As $\mathbf{F}(\mathbf{x}_i, \boldsymbol{\xi}_i)$ is usually not very complicated in most practical applications, $E[\mathbf{F}(\mathbf{x}_i, \boldsymbol{\xi}_i)]$ is not very complex, and the computation of $E(y_{ik})$ is not difficult.

Treating a dichotomous variable as an ordered categorical variable with two categories and a fixed threshold at zero, Bayesian estimates of the parameters in the above NSEM can be obtained by a modified Gibbs sampler algorithm as described in Section 7.3, with additional modifications to take care with the covariates. Because the full conditional distributions have been presented in Song and Lee (2006), or can be similarly derived as in previous sections, they are not reported here.

8.4.2 Illustrative Example 3

A portion of the data that were obtained from the UK in the project World Values Survey 1981–1984 and 1990–1993 was analyzed in this illustrative example, see Song and Lee (2006). Seven variables related to respondents' homelife, job satisfaction, and their attitude on job benefit and working environment were taken as manifest variables. As covariates are less important in the measurement equation, only the intercept $\mathbf{A} = (a_1, \cdots, a_7)^T$ with all $c_i = 1.0$ is considered in defining Equation (8.25). For the structural equation, a binary variable x that measures whether respondents can get comfort and strength from religion is incorporated as a covariate. Details of these variables (variables 180, 96, 116, 117, 99, 100 and 101) are given in Appendix 1.1. For brevity, data points with missing entries were deleted, and the remaining sample size is 816. From the meaning of the corresponding questions, the first two manifest variables are indicators for the latent variable 'home life', the next two manifest variables are indicators for the latent variable 'job satisfaction', and the last three manifest variables are indicators for the latent variable ' job benefit–environment attitude'. The manifest variables for 'homelife' and 'job satisfaction' were measured by a 10-point scale and they are treated as continuous. The remaining three manifest variables were dichotomous. For clear interpretation, we related these seven manifest variables with three latent variables via a $\boldsymbol{\Lambda}$ with the following non-overlapping structure:

$$\boldsymbol{\Lambda}^T = \begin{bmatrix} 1 & \lambda_{21} & 0 & 0 & 0 & 0 & 0 \\ 0 & 0 & 1 & \lambda_{42} & 0 & 0 & 0 \\ 0 & 0 & 0 & 0 & 1 & \lambda_{63} & \lambda_{73} \end{bmatrix},$$

where 1's and 0's are fixed to identify the model. Parameters in $\mathbf{\Phi}$ are $(\phi_{11}, \phi_{21}, \phi_{22})$. The covariance matrix $\mathbf{\Psi}_\epsilon$ is taken to be a diagonal matrix, with diagonal elements $(\psi_{\epsilon 1}, \psi_{\epsilon 2}, \psi_{\epsilon 3}, \psi_{\epsilon 4}, 1, 1, 1)$, in which $\psi_{\epsilon 1}, \psi_{\epsilon 2}, \psi_{\epsilon 3}$ and $\psi_{\epsilon 4}$ are unknown parameters, and the 1's are fixed to identify the dichotomous variables. As an illustration, we considered 'homelife' as the endogenous latent variable η, which is related with exogenous latent variables 'job satisfaction, ξ_1', 'job benefit–environment attitude, ξ_2', and the covariate x is about religion (variable 177). The following nonlinear structural equation is used to assess the interaction effects from the covariate x and the exogenous latent variables:

$$\eta_i = \gamma x_i + \alpha_1 \xi_{i1} + \alpha_2 \xi_{i2} + \beta_1 x_i \xi_{i1} + \beta_2 x_i \xi_{i2} + \beta_3 \xi_{i1} \xi_{i2} + \delta_i. \tag{8.27}$$

In this model, there are 25 unknown parameters. The path diagram describing the proposed NSEM is presented in Figure 8.1.

The Bayesian estimates of the parameters were obtained with conjugate prior distributions. The hyperparameter values were produced by data-dependent prior inputs as in Section 8.3.2. The convergence of the Gibbs sampler was monitored by the EPSR values (see Gelman, 1996). To give some idea about the convergence, plots of these values against the iteration number are displayed in Figure 8.2. We observe that the sequence of observations converged after about 12 000 iterations. To obtain the Bayesian estimates and the numerical standard errors, an additional $T = 8000$ observations were collected after convergence.

The Bayesian estimates, and the standard error estimates are presented in Table 8.11. Some interesting interpretations are given below.

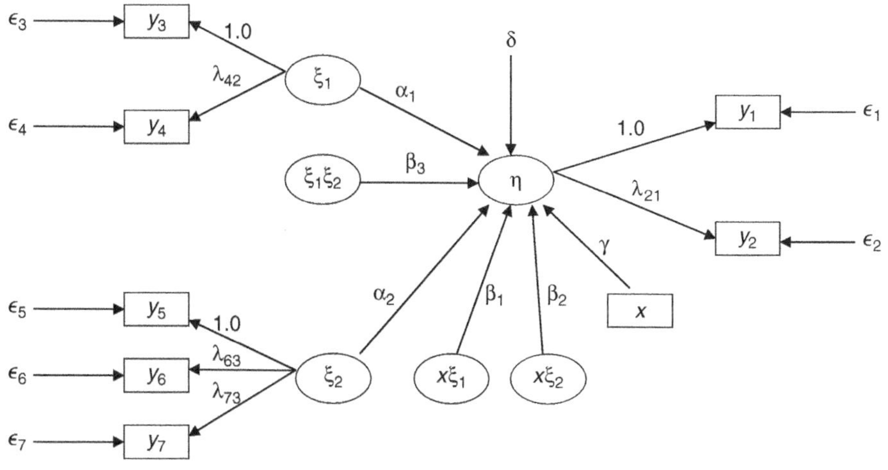

Figure 8.1 Path diagram of the SEM with interactions of covariates and latent variables. This figure and Figure 8.2 are taken from Song and Lee (2006).

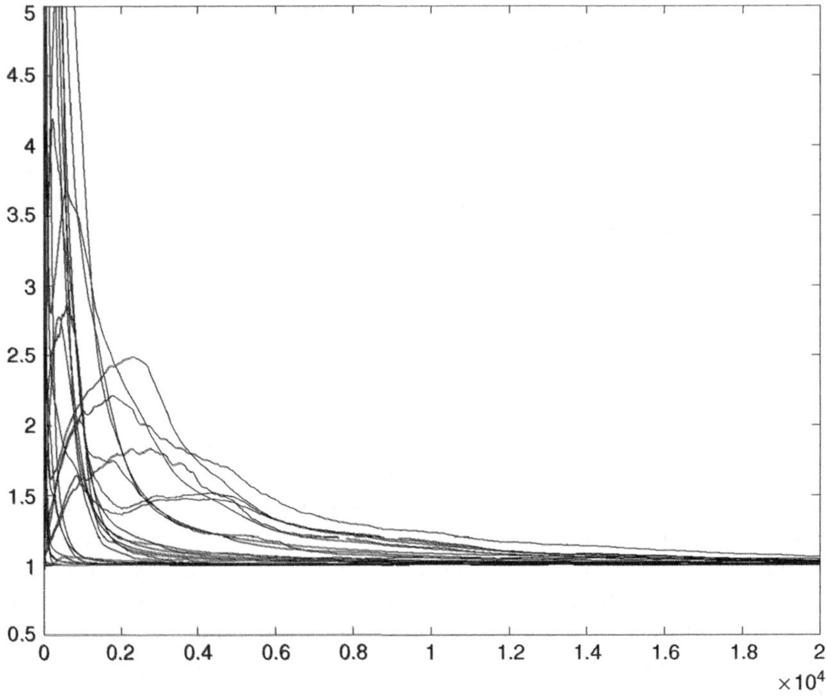

Figure 8.2 EPSR values against the number of iterations in the analysis of ICPSR data.

Table 8.11 Bayesian estimates (EST) and the standard error estimates (SE) of illustrative example 3.

Par	EST	SE	Par	EST	SE
a_1	8.120	0.06	γ	−0.197	0.10
a_2	7.457	0.06	α_1	0.621	0.09
a_3	7.425	0.07	α_2	−0.991	0.27
a_4	6.972	0.09	β_1	0.064	0.08
a_5	−0.755	0.06	β_2	−0.228	0.25
a_6	−0.462	0.05	β_3	0.603	0.15
a_7	0.989	0.07			
λ_{21}	0.908	0.08	ϕ_{11}	2.489	0.06
λ_{42}	0.768	0.09	ϕ_{12}	0.186	0.06
λ_{63}	0.644	0.17	ϕ_{22}	0.355	0.07
λ_{73}	0.894	0.22			
$\psi_{\epsilon 1}$	1.120	0.18	ψ_δ	0.718	0.19
$\psi_{\epsilon 2}$	1.391	0.15			
$\psi_{\epsilon 3}$	1.625	0.26			
$\psi_{\epsilon 4}$	4.869	0.29			

This table is taken from Song and Lee (2006).

(i) All of the non-fixed factor loading estimates are quite large. From the standard error estimates, it can be seen that these factor loadings are different from zero. These results indicate a strong association between each of the latent variables and their respective indicators. As expected, the standard errors of the loading estimates that correspond to the dichotomous manifest variables are relatively large.

(ii) The estimated nonlinear structural equation is given by

$$\eta_i = -0.197x_i + 0.621\xi_{i1} - 0.991\xi_{i2} + 0.064x_i\xi_{i1} - 0.228x_i\xi_{i2} + 0.603\xi_{i1}\xi_{i2}. \quad (8.28)$$

From the standard error estimates, we observe that all the linear effects of the covariate and the exogenous latent variables, and the interaction effects corresponding to $\xi_{i1}\xi_{i2}$ and $x_i\xi_{i2}$ are quite different from zero, whilst the interaction effect corresponding to $x_i\xi_{i1}$ is small. Hence, the interaction causal effect of religion (x_i) and 'job satisfaction, ξ_{i1}' is not significant when given the other linear and interaction effects in the structural equation. As ξ_{i2} is related to the dichotomous variables, the standard error estimate (= 0.250) corresponding to $\hat{\beta}_2(= -0.228)$ is quite large. However, as -0.228 is quite different from zero in magnitude, we tend to believe that the corresponding interaction effect is substantial.

(iii) Before giving more detailed interpretations of the nonlinear interaction terms, we note from the scales of the indicators in relation to 'life, η' and 'job satisfaction, ξ_1' that a comparatively larger (positive) value of η or ξ_1 implies that the individual has better 'life' or 'job satisfaction', respectively. From the scale of the indicators in relation to 'job benefit–environment attitude, ξ_2', a comparatively smaller (negative) value of ξ_2 implies that the individual is more concerned about benefits and working environment. With the above understanding of the exogenous latent variables, it follows from $\hat{\alpha}_1 = 0.621$ and $\hat{\alpha}_2 = -0.991$ that job satisfaction and more concern about benefits and working environment have a positive impact on 'life, η'. From $\hat{\beta}_3 = 0.603$, 'job satisfaction, ξ_1', and 'job benefit–environment attitude, ξ_2' have an interaction effect on 'life, η'. The basic interpretation is that the 'additive' effect of the linear latent variables of ξ_1 and ξ_2 in the structural equation is inadequate to account for their relationships with the latent variable 'life, η', and an interaction term of ξ_1 and ξ_2 has to be added. Depending on various situations, this interaction term has different effects. For example, it is interesting to observe that for employees with good job satisfaction (positive ξ_{i1}), and more concern about benefits and working environment (negative ξ_{i2}) would have an additive positive effect on 'life, η' which is too strong; hence a negative adjustment via an interaction effect (0.603 times a positive ξ_{i1} and a negative ξ_{i2} would be negative) is necessary. For the covariate about religion, from the coding of the corresponding question, we note that an individual with a small positive (or negative) value of x_i has more comfort and strength from religion, and vice versa. From $\hat{\gamma}(=-0.197)$, we observe the expected finding that individuals who can gain comfort and strength from religion tend to have a better life. The interpretation about the estimate of the interaction term ($\hat{\beta}_2=-0.228$) of x_i and ξ_{i2} is similar to the interpretation of $\hat{\beta}_3$.

(iv) From $\hat{\phi}_{11}$, $\hat{\phi}_{12}$ and $\hat{\phi}_{22}$, we note that the estimate of the correlation between 'job satisfaction, ξ_1' and 'job benefit–environment attitude, ξ_2' is 0.198. It indicates that job satisfaction and less concern about the benefit and environment are positively correlated.

(v) Now, we consider the interpretation of the intercepts and the means of the manifest variables. As an example, we consider $\widehat{E(y_{i2})}$. Since $E(\xi_1)=E(\xi_2)=0$, $\hat{a}_2=7.457$, $\hat{\lambda}_{21}=0.768$, $\hat{\gamma}=-0.197$, $\hat{\beta}_3=0.603$ and $\hat{E}(\xi_1\xi_2)=\hat{\phi}_{12}=0.186$, we have

$$\widehat{E(y_{i2})}=\hat{a}_2+\hat{\lambda}_{21}[\hat{\gamma}x_i+\hat{\beta}_3\hat{E}(\xi_{i1}\xi_{i2})]=7.543-0.151x_i. \tag{8.29}$$

This provides an estimate of the mean value of the second manifest variable of the individual i, which depends on the value of the covariate x_i. Based on the meaning of y_2 and x, the interpretation of Equation (8.29) is that the mean life satisfaction is increased by perceived comfort and strength from religion.

8.4.3 Analysis of an Artificial Example using WinBUGS

The software WinBUGS (Spiegelhalter, Thomas, Best and Lunn, 2003) can produce Bayesian estimates of the parameters in an NSEM with covariates for continuous, ordered categorical and/or dichotomous data. As we mentioned in Section 6.6.2, for models with ordered categorical data, it is rather difficult to simultaneously estimate the unknown thresholds and the unknown structural parameters; and a convenient way is to fix the thresholds at some preassigned values. For simplicity, we illustrate the use of WinBUGS in analyzing an artificial example that is based on the following NSEM with a linear covariate (see Lee, Song and Tang, 2006) Let $y_{ik}\stackrel{D}{=}N[\mu_{ik},\psi_{ik}]$, where

$$\begin{aligned}\mu_{i1}&=\alpha_1+\eta_i,\ \mu_{ik}=\alpha_k+\lambda_{k1}\eta_i,\ k=2,3,\\ \mu_{i4}&=\alpha_4+\xi_{i1},\ \mu_{ik}=\alpha_k+\lambda_{k2}\xi_{i1},\ k=5,6,7,\ \text{and}\\ \mu_{i8}&=\alpha_8+\xi_{i2},\ \mu_{ik}=\alpha_k+\lambda_{k3}\xi_{i2},\ k=9,10,\end{aligned} \tag{8.30}$$

and the structural equation is reformulated by defining the conditional distribution of η_i given ξ_{i1} and ξ_{i2} as $N[\nu_i, \psi_\delta]$, where

$$\nu_i = \beta_1 x_i + \gamma_1 \xi_{i1} + \gamma_2 \xi_{i2} + \gamma_3 \xi_{i1}\xi_{i2} + \gamma_4 \xi_{i1}^2 + \gamma_5 \xi_{i2}^2. \tag{8.31}$$

The true population values of the free parameters in the model were taken to be:

$$\begin{aligned}
&\alpha_1 = \cdots = \alpha_{10} = 0.0,\ \lambda_{21} = \lambda_{52} = \lambda_{93} = 0.9, \lambda_{31} = \lambda_{62} = \lambda_{10,3} = 0.7,\\
&\lambda_{72} = 0.5, \psi_{\epsilon 1} = \psi_{\epsilon 2} = \psi_{\epsilon 3} = 0.3, \psi_{\epsilon 4} = \cdots = \psi_{\epsilon 7} = 0.5,\\
&\quad \psi_{\epsilon 8} = \psi_{\epsilon 9} = \psi_{\epsilon 10} = 0.4,\\
&\beta_1 = 0.5,\ \gamma_1 = \gamma_2 = 0.4,\ \gamma_3 = 0.3,\ \gamma_4 = 0.2,\ \gamma_5 = 0.5,\ \text{and}\\
&\phi_{11} = \phi_{22} = 1.0,\ \phi_{12} = 0.3,\ \text{and}\ \psi_\delta = 0.36.
\end{aligned} \tag{8.32}$$

Based on the model formulation and these true parameter values, a random sample of continuous observations $\{\mathbf{y}_i, i = 1, \ldots, 500\}$ was simulated, which gave the observed data set $\mathbf{Y}$. The following hyperparameter values were taken for the conjugate prior distributions:

$$\begin{aligned}
&\mathbf{a}_0 = (0.0, \ldots, 0.0)^T,\ \boldsymbol{\Sigma}_0 = \mathbf{I}_{10},\ \alpha_{0\epsilon k} = \alpha_{0\delta} = 9,\ \beta_{0\epsilon k} = \beta_{0\delta} = 4,\\
&\text{elements in } \boldsymbol{\Lambda}_{0k} \text{ and } \boldsymbol{\Gamma}_0 \text{ are taken to be the true values,}\\
&\mathbf{H}_{0yk} = \mathbf{I}_{10},\ \mathbf{H}_{0\omega k} = \mathbf{I}_5,\ \rho_0 = 4,\ \mathbf{R}_0 = \boldsymbol{\Phi}_0^{-1},
\end{aligned} \tag{8.33}$$

where $\boldsymbol{\Phi}_0$ is the matrix with true values of ϕ_{11}, ϕ_{22} and ϕ_{12}. These hyperparameter values represent accurate prior inputs. We observe that the WinBUGS program converged in less than 4000 iterations. The plots of some of the simulated sequences of observations for monitoring convergence are presented in Figure 8.3. Bayesian estimates of the parameters and their standard error estimates as obtained from 6000 iterations after the 4000 'burn-in' iterations are presented in Table 8.12. We observe that the Bayesian estimates are close to the true values, and that the standard error estimates are reasonable. WinBUGS also produced estimates of the latent variables $\{\hat{\boldsymbol{\omega}}_i = (\hat{\eta}_i, \hat{\xi}_{i1}, \hat{\xi}_{i2})^{\mathrm{T}}; i = 1, \ldots, n\}$. Histograms that correspond to the sets of latent variable estimates $\hat{\xi}_{i1}$ and $\hat{\xi}_{i2}$ are presented in Figure 8.4. The QQ plots are presented in Figure 8.5. We observe from these histograms that the corresponding empirical distributions are close to normal. The elements in the sample covariance matrix of $\{\boldsymbol{\xi}_i; i = 1, \ldots, n\}$ are $s_{11} = 0.902, s_{12} = 0.311$ and $s_{22} = 0.910$, and thus this sample covariance matrix is close to the true covariance matrix of $\boldsymbol{\xi}_i$, see Equation (8.32).

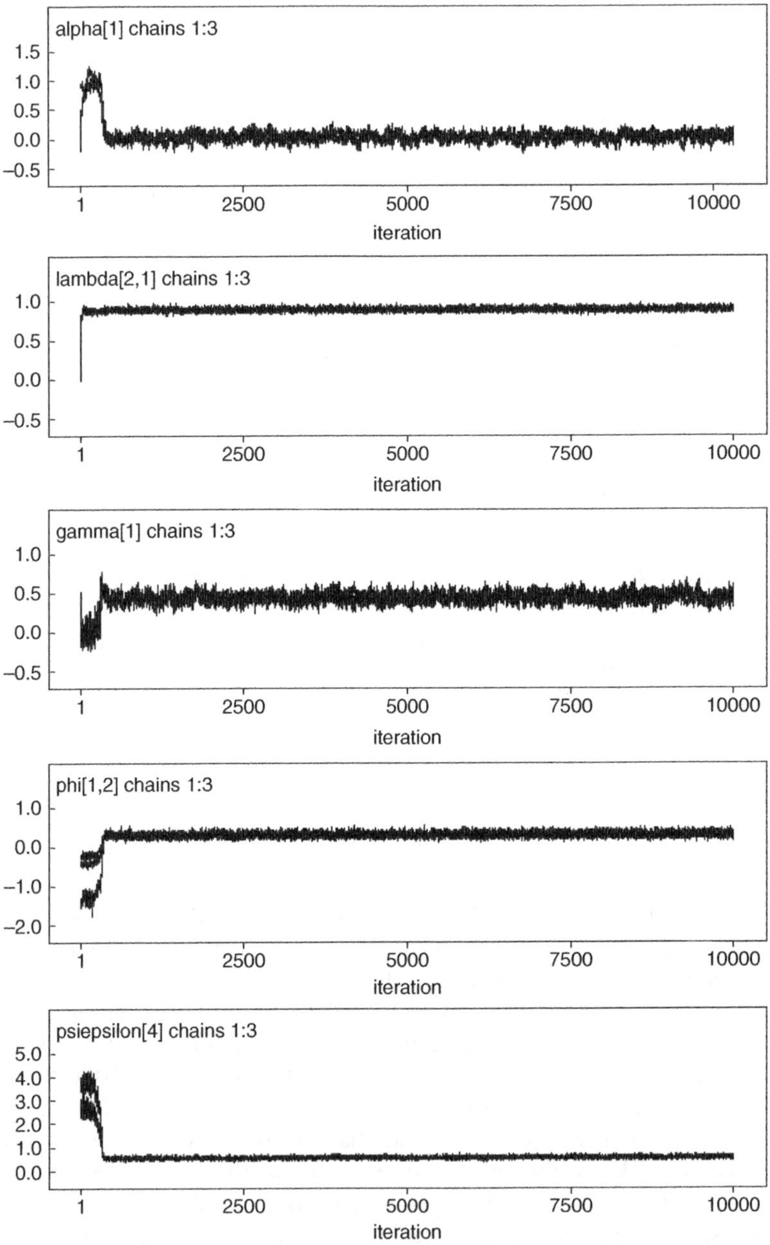

Figure 8.3 (a), (b), (c), (d) and (e) Three chains of observations corresponding to $\alpha_1, \lambda_{21}, \gamma_1, \phi_{12}$ and $\psi_{\epsilon 4}$, generated by different initial values.

Table 8.12 Bayesian estimates of the artificial example obtained from WinBUGS.

Par	True value	EST	SE	Par	True value	EST	SE
α_1	0.0	0.022	0.069	$\psi_{\epsilon 1}$	0.3	0.324	0.032
α_2	0.0	0.065	0.062	$\psi_{\epsilon 2}$	0.3	0.285	0.027
α_3	0.0	0.040	0.052	$\psi_{\epsilon 3}$	0.3	0.284	0.022
α_4	0.0	0.003	0.058	$\psi_{\epsilon 4}$	0.5	0.558	0.050
α_5	0.0	0.036	0.056	$\psi_{\epsilon 5}$	0.5	0.480	0.045
α_6	0.0	0.002	0.047	$\psi_{\epsilon 6}$	0.5	0.554	0.041
α_7	0.0	0.004	0.042	$\psi_{\epsilon 7}$	0.5	0.509	0.035
α_8	0.0	0.092	0.053	$\psi_{\epsilon 8}$	0.4	0.382	0.035
α_9	0.0	0.032	0.050	$\psi_{\epsilon 9}$	0.4	0.430	0.035
α_{10}	0.0	−0.000	0.044	$\psi_{\epsilon 10}$	0.4	0.371	0.029
λ_{21}	0.9	0.889	0.022	β_1	0.5	0.525	0.075
λ_{31}	0.7	0.700	0.019	γ_1	0.4	0.438	0.059
λ_{52}	0.9	0.987	0.053	γ_2	0.4	0.461	0.034
λ_{62}	0.7	0.711	0.046	γ_3	0.3	0.304	0.045
λ_{72}	0.5	0.556	0.040	γ_4	0.2	0.184	0.060
λ_{93}	0.9	0.900	0.042	γ_5	0.5	0.580	0.050
$\lambda_{10,3}$	0.7	0.766	0.038	ϕ_{11}	1.0	1.045	0.120
				ϕ_{12}	0.3	0.302	0.057
				ϕ_{22}	1.0	1.023	0.089
				ψ_δ	0.36	0.376	0.045

Inspired by the data analysis techniques in regression, we obtain the following estimated residuals by means of the $\hat{\boldsymbol{\theta}}$ and $\hat{\boldsymbol{\omega}}_i = (\hat{\eta}_i, \hat{\xi}_{i1}, \hat{\xi}_{i2})^T$ for $i = 1, \ldots, n$

$$\hat{\epsilon}_{i1} = y_{i1} - \hat{\alpha}_1 - \hat{\eta}_i, \quad \hat{\epsilon}_{ik} = y_{ik} - \hat{\alpha}_k - \hat{\lambda}_{k1}\hat{\eta}_i, \quad k = 2, 3$$

$$\hat{\epsilon}_{i4} = y_{i4} - \hat{\alpha}_4 - \hat{\xi}_{i1}, \quad \hat{\epsilon}_{ik} = y_{ik} - \hat{\alpha}_k - \hat{\lambda}_{k2}\hat{\xi}_{i1}, \quad k = 5, 6, 7$$

$$\hat{\epsilon}_{i8} = y_{i8} - \hat{\alpha}_8 - \hat{\xi}_{i2}, \quad \hat{\epsilon}_{ik} = y_{ik} - \hat{\alpha}_k - \hat{\lambda}_{k3}\hat{\xi}_{i2}, \quad k = 9, 10$$

$$\hat{\delta}_i = \hat{\eta}_i - \hat{\beta}_1 x_i - \hat{\gamma}_1\hat{\xi}_{i1} - \hat{\gamma}_2\hat{\xi}_{i2} - \hat{\gamma}_3\hat{\xi}_{i1}\hat{\xi}_{i2} - \hat{\gamma}_4\hat{\xi}_{i1}^2 - \hat{\gamma}_5\hat{\xi}_{i2}^2.$$

Some estimated residual plots, $\hat{\epsilon}_{i2}$, $\hat{\epsilon}_{i3}$, $\hat{\epsilon}_{i8}$ and $\hat{\delta}_i$, against the case number are displayed in Figure 8.6. And the plots of estimated residual $\hat{\delta}_i$ versus $\hat{\xi}_{i1}$ and $\hat{\xi}_{i2}$; and $\hat{\epsilon}_{i2}$ versus $\hat{\xi}_{i1}$, $\hat{\xi}_{i2}$ and $\hat{\eta}_i$ are presented in Figures 8.7 and 8.8, respectively. The other residual plots are similar. These plots lie within two parallel horizontal lines that are centered at zero, and no linear or quadratic trends are detected. This roughly indicates that the proposed measurement equation and structural equation are adequate. The WinBUGS codes and data are given in the website: http://www.wiley.com/go/lee_structural.

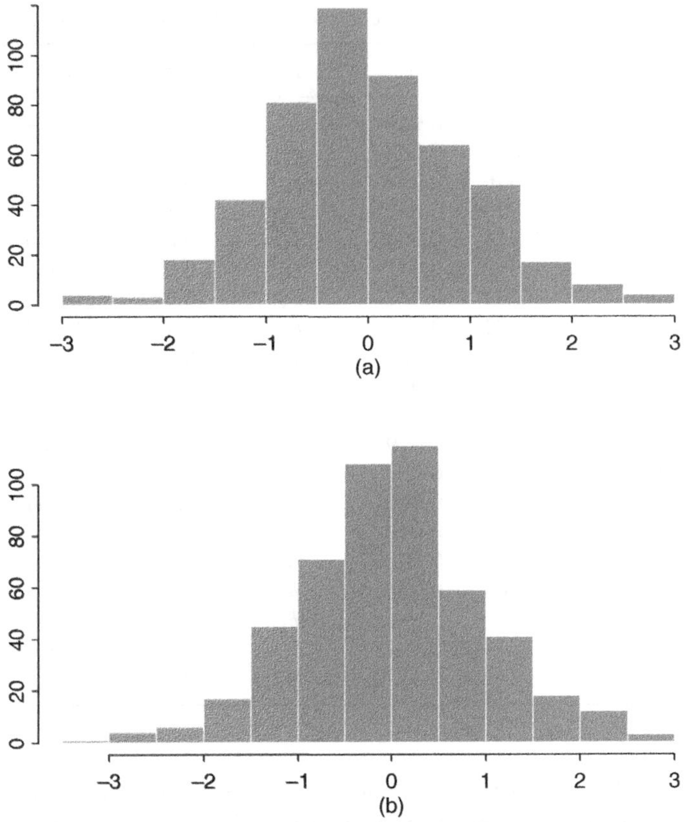

Figure 8.4 Histograms of the latent variables (a) $\hat{\xi}_{i1}$ and (b) $\hat{\xi}_{i2}$.

8.5 BAYESIAN MODEL COMPARISON

In this section, we address the important issue about model comparison. For NSEMs, the nonlinear terms of latent variables in the structural equation are of particular interest. To assess the significance of the unknown coefficient associated with a nonlinear term, the classical non-Bayesian approach uses the z-score that is obtained by the estimate divided by its standard error estimate. Due to the disadvantages of this approach as given in Chapter 5, Section 5.1, we recommend that this issue should be addressed through Bayesian model comparison using the Bayes factor. Basically, in assessing the significance of one or more than one nonlinear terms, we compare a nonlinear SEM with the nonlinear terms with an SEM without the nonlinear terms. The next section describes the path sampling procedure for computing the Bayes factor.

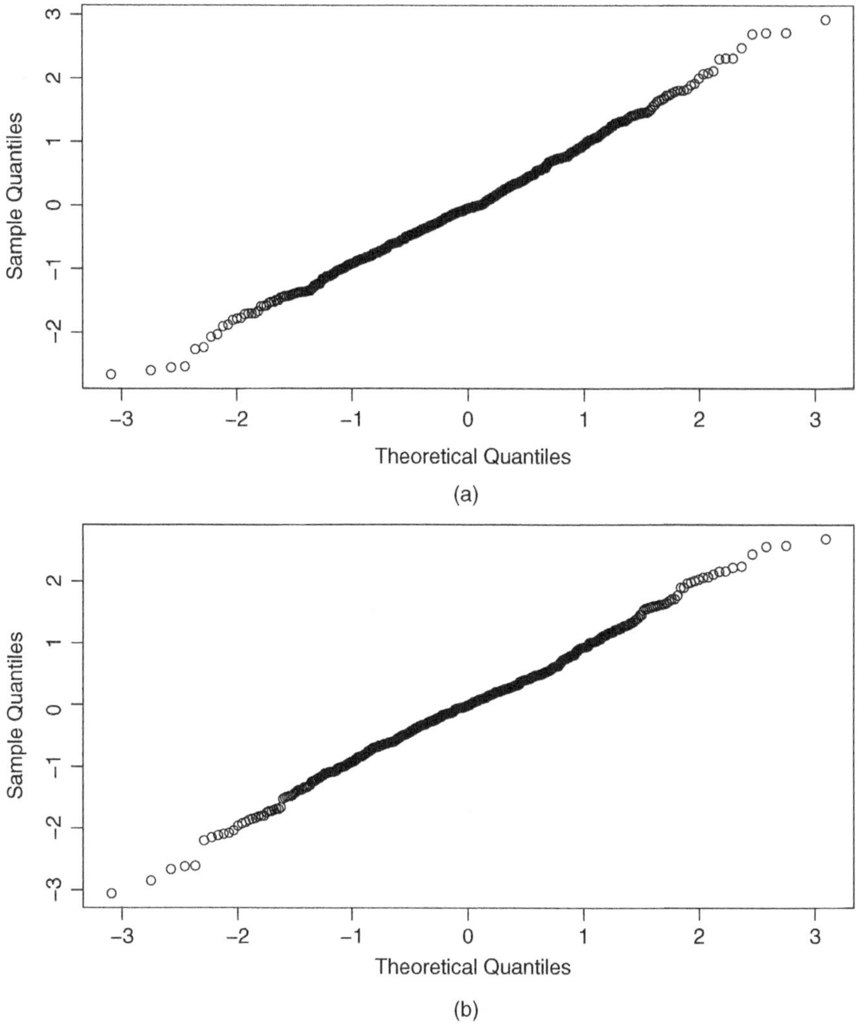

Figure 8.5 QQ plots of the latent variables (a) $\hat{\xi}_{i1}$ and (b) $\hat{\xi}_{i2}$

8.5.1 Path Sampling for Computing the Bayes Factor

We first consider the situation with continuous data. Let M_0 and M_1 be two competing models under the framework of an NSEM defined in Equations (8.1) and (8.2); and let B_{10} be the Bayes factor for comparing M_0 and M_1 as defined

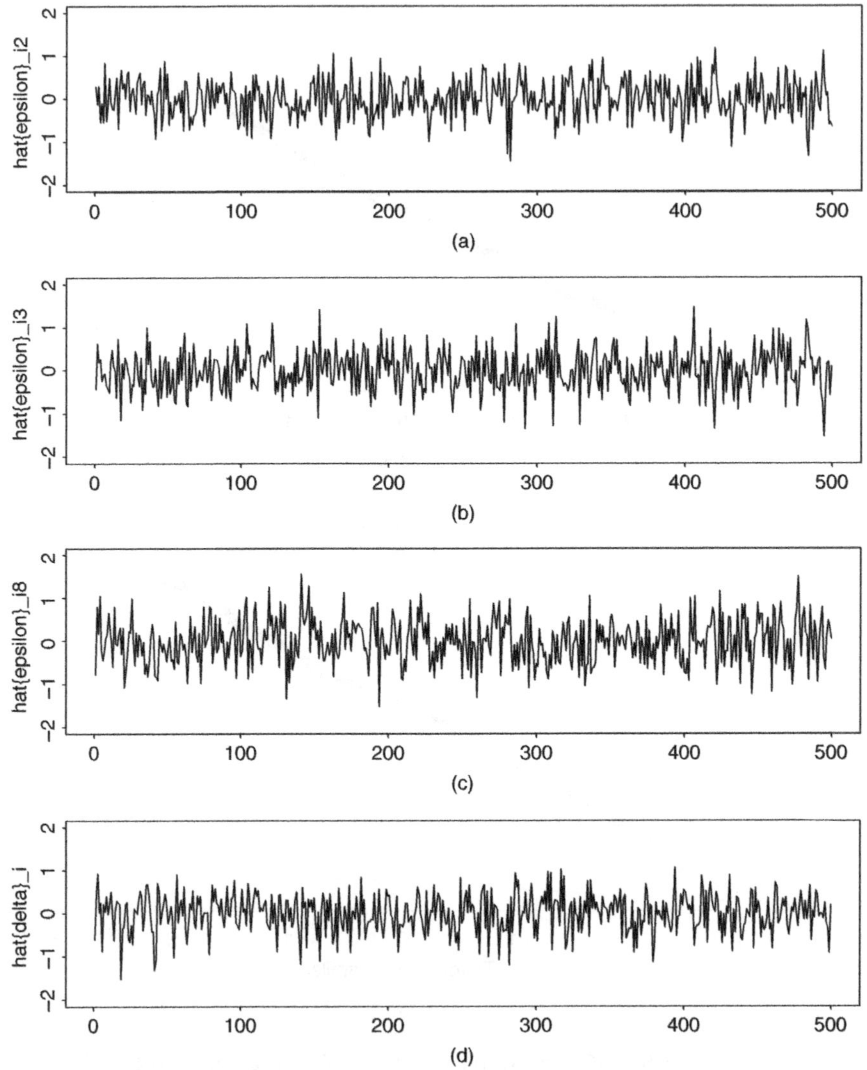

Figure 8.6 Estimated residual plots (a) $\hat{\epsilon}_{i2}$, (b) $\hat{\epsilon}_{i3}$, (c) $\hat{\epsilon}_{i8}$ and (d) $\hat{\delta}_i$.

in Chapter 5. Let $\mathbf{Y} = (\mathbf{y}_1, \ldots, \mathbf{y}_n)$ be the observed data matrix and $\mathbf{\Omega} = (\boldsymbol{\omega}_1, \ldots, \boldsymbol{\omega}_n)$ be the matrix of latent variables. Let $\log p(\mathbf{Y}, \mathbf{\Omega}|\boldsymbol{\theta}, t)$ be the log likelihood function with a continuous parameter t in $[0, 1]$ and

$$U(\mathbf{Y}, \mathbf{\Omega}, \boldsymbol{\theta}, t) = \frac{\mathrm{d}}{\mathrm{dt}} \log p(\mathbf{Y}, \mathbf{\Omega}|\boldsymbol{\theta}, t). \tag{8.34}$$

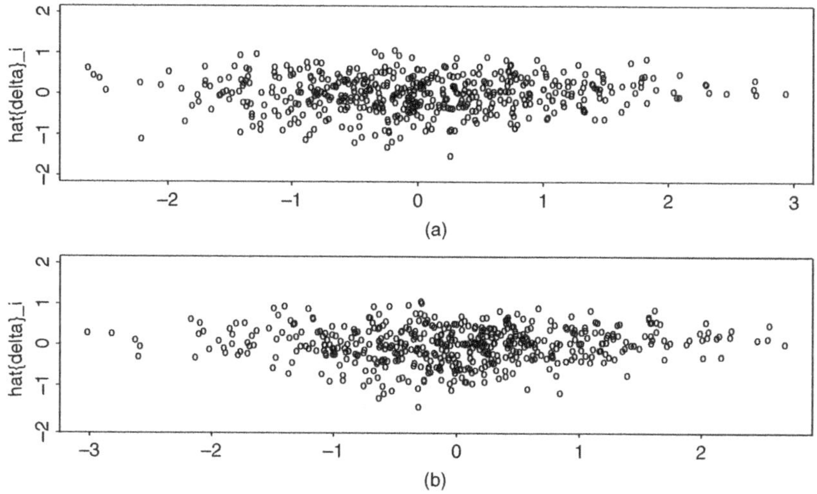

Figure 8.7 Plots of estimated residuals $\hat{\delta}_i$ versus (a) $\hat{\xi}_{i1}$ and (b) $\hat{\xi}_{i2}$.

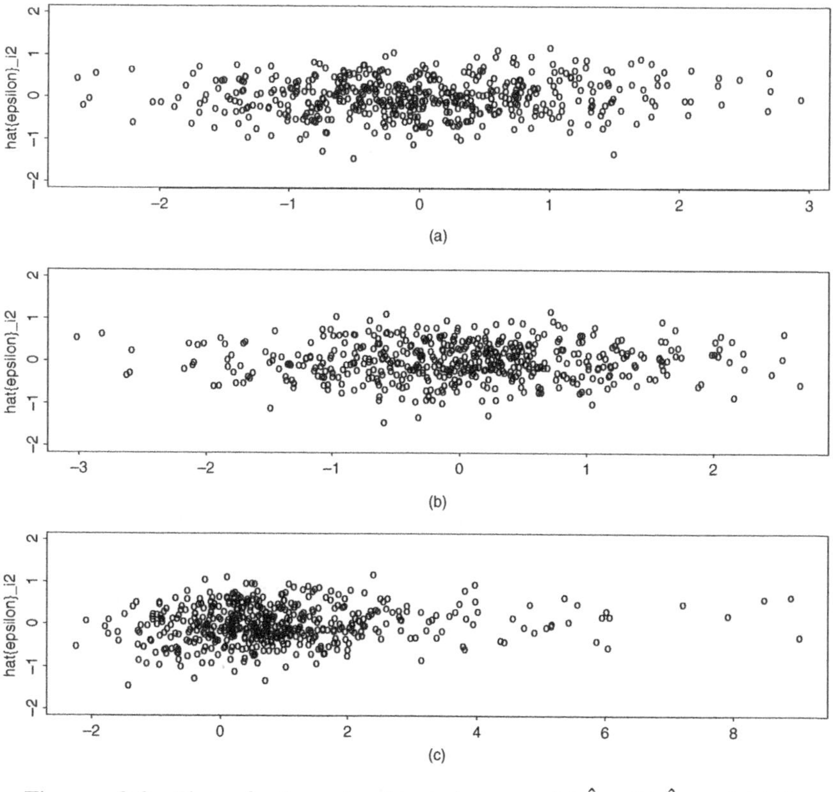

Figure 8.8 Plots of estimated residuals $\hat{\epsilon}_{i2}$ versus (a) $\hat{\xi}_{i1}$, (b) $\hat{\xi}_{i2}$ and (c) $\hat{\eta}_i$.

Further, let $t_{(0)} = 0 < t_{(1)} < \cdots < t_{(S)} < t_{(S+1)} = 1$ be fixed grids in [0, 1]. It follows from the reasoning in Section 5.3 that an estimate of logarithm B_{10} obtained via path sampling is given by

$$\widehat{\log B_{10}} = \frac{1}{2} \sum_{s=0}^{S} (t_{(s+1)} - t_{(s)})(\overline{U}_{(s+1)} + \overline{U}_{(s)}), \tag{8.35}$$

where $\overline{U}_{(s)}$ is the average of the values of $U(\mathbf{Y}, \mathbf{\Omega}, \boldsymbol{\theta}, t)$ based on all simulated draws for which $t = t_{(s)}$, that is,

$$\overline{U}_{(s)} = J^{-1} \sum_{j=1}^{J} U(\mathbf{Y}, \mathbf{\Omega}^{(j)}, \boldsymbol{\theta}^{(j)}, t_{(s)}), \tag{8.36}$$

in which $\{(\mathbf{\Omega}^{(j)}, \boldsymbol{\theta}^{(j)}), j = 1, \ldots, J\}$ are simulated draws from $p(\mathbf{\Omega}, \boldsymbol{\theta}|\mathbf{Y}, t_{(s)})$ via the algorithm developed in the estimation of the model.

To give a more specific illustration in applying the path sampling procedure to NSEMs, consider the following models M_1 and M_0 which satisfy the same measurement Equation (8.1), but with the following different nonlinear structural equations for the latent variables:

$$M_1 : \boldsymbol{\eta} = \mathbf{\Pi}\boldsymbol{\eta} + \mathbf{\Gamma}_1\mathbf{H}_1(\boldsymbol{\xi}) + \boldsymbol{\delta},$$
$$M_0 : \boldsymbol{\eta} = \mathbf{\Pi}\boldsymbol{\eta} + \mathbf{\Gamma}_0\mathbf{H}_0(\boldsymbol{\xi}) + \boldsymbol{\delta},$$

where $\mathbf{H}_1$ and $\mathbf{H}_0$ may involve different numbers of distinct nonlinear functions of $\boldsymbol{\xi}$, and $\mathbf{\Gamma}_1$ and $\mathbf{\Gamma}_0$ are the corresponding unknown coefficient matrices. In this illustration, differences between M_1 and M_0 are on the more interesting nonlinear relationships among the latent variables. The competing models M_1 and M_0 are linked up by a model M_t which is defined by the same measurement equation, and the following structural equation with t in [0, 1]:

$$M_t : \boldsymbol{\eta} = \mathbf{\Pi}\boldsymbol{\eta} + (1-t)\mathbf{\Gamma}_0\mathbf{H}_0(\boldsymbol{\xi}) + t\mathbf{\Gamma}_1\mathbf{H}_1(\boldsymbol{\xi}) + \boldsymbol{\delta} = \mathbf{\Lambda}_{tw}\mathbf{G}(\boldsymbol{\omega}) + \boldsymbol{\delta},$$

where $\mathbf{\Lambda}_{tw} = (\mathbf{\Pi}, (1-t)\mathbf{\Gamma}_0, t\mathbf{\Gamma}_1)$ and $\mathbf{G}(\boldsymbol{\omega}) = (\boldsymbol{\eta}^T, \mathbf{H}_0(\boldsymbol{\xi})^T, \mathbf{H}_1(\boldsymbol{\xi})^T)^T$. It can be shown that

$$\log\{p(\mathbf{Y}, \mathbf{\Omega}|\boldsymbol{\theta}, t)\} = -\frac{1}{2}\{(p+q)n\log(2\pi) + n\log|\mathbf{\Psi}_\epsilon| + n\log|\mathbf{\Psi}_\delta| + n\log|\mathbf{\Phi}|$$
$$-2n\log|\mathbf{\Pi}_0| + \sum_{i=1}^{n} \boldsymbol{\xi}_i^T\mathbf{\Phi}^{-1}\boldsymbol{\xi}_i$$

$$+\sum_{i=1}^{n}(\mathbf{y}_i-\boldsymbol{\mu}-\boldsymbol{\Lambda}\boldsymbol{\omega}_i)^T\boldsymbol{\Psi}_\epsilon^{-1}(\mathbf{y}_i-\boldsymbol{\mu}-\boldsymbol{\Lambda}\boldsymbol{\omega}_i)$$

$$+\sum_{i=1}^{n}[\boldsymbol{\eta}_i-\boldsymbol{\Lambda}_{t\omega}\mathbf{G}(\boldsymbol{\omega}_i)]^T\boldsymbol{\Psi}_\delta^{-1}[\boldsymbol{\eta}_i-\boldsymbol{\Lambda}_{t\omega}\mathbf{G}(\boldsymbol{\omega}_i)]\}.$$

Here, $\boldsymbol{\theta}$ is the parameter vector in the linked model. It contains all the common and distinct unknown parameters in M_0 and M_1; that is, unknown parameters in $\boldsymbol{\Lambda}$, $\boldsymbol{\Psi}_\epsilon$, $\boldsymbol{\Pi}$, $\boldsymbol{\Gamma}_0$, $\boldsymbol{\Gamma}_1$, $\boldsymbol{\Phi}$ and $\boldsymbol{\Psi}_\delta$. By differentiation with respect to t, we have

$$U(\mathbf{Y},\boldsymbol{\Omega},\boldsymbol{\theta},t)=\sum_{i=1}^{n}[\boldsymbol{\eta}_i-\boldsymbol{\Lambda}_{t\omega}\mathbf{G}(\boldsymbol{\omega}_i)]^T\boldsymbol{\Psi}_\delta^{-1}\boldsymbol{\Lambda}_0\mathbf{G}(\boldsymbol{\omega}_i),$$

where $\boldsymbol{\Lambda}_0=(\mathbf{0},-\boldsymbol{\Gamma}_0,\boldsymbol{\Gamma}_1)$.

Applications of this procedure to other special cases of the general model are similar. The main computation involved in the applications is on generating observations $(\boldsymbol{\Omega}^{(j)},\boldsymbol{\theta}^{(j)})$ from $p(\boldsymbol{\Omega},\boldsymbol{\theta}|\mathbf{Y},t_{(s)})$ for evaluating $\overline{U}_{(s)}$, see Equations (8.35) and (8.36).

Now consider the situation with mixed continuous and ordered categorical data. The basic NSEM is the same as above. However, to cope with the ordered categorical variables the notation for the measurement equation is slightly different, see Equation (8.20). Note that the nonlinear structural equation is again given by Equation (8.2) or (8.3). Let M_0 and M_1 be two competing models with observed data matrices $\mathbf{X}=(\mathbf{x}_1,\ldots,\mathbf{x}_n)$ of continuous measurements and $\mathbf{Z}=(\mathbf{z}_1,\ldots,\mathbf{z}_n)$ of ordered categorical measurements. Let $\boldsymbol{\Omega}$ be the matrix of latent variables, and let $\mathbf{Y}=(\mathbf{y}_1,\ldots,\mathbf{y}_n)$ be the latent continuous data matrix that underlies the ordered categorical data matrix $\mathbf{Z}$. The observed data $(\mathbf{X},\mathbf{Z})$ are augmented with the latent data matrix $\mathbf{Y}$ and the latent variables matrix $\boldsymbol{\Omega}$ in the path sampling procedure for computing the Bayes factor.

Let $\boldsymbol{\alpha}$ be the vector which contains all the unknown thresholds, and

$$U(\mathbf{X},\mathbf{Z},\boldsymbol{\theta},\boldsymbol{\Omega},\boldsymbol{\alpha},\mathbf{Y},t)=\frac{\mathrm{d}}{\mathrm{d}t}\log\ p(\mathbf{X},\mathbf{Z},\boldsymbol{\Omega},\mathbf{Y}|\boldsymbol{\theta},\boldsymbol{\alpha},t). \tag{8.37}$$

The $\log B_{10}$ can be estimated by Equation (8.35) with

$$\overline{U}_{(s)}=J^{-1}\sum_{j=1}^{J}U(\mathbf{X},\mathbf{Z},\boldsymbol{\theta}^{(j)},\boldsymbol{\Omega}^{(j)},\boldsymbol{\alpha}^{(j)},\mathbf{Y}^{(j)},t_{(s)}), \tag{8.38}$$

where $\{(\boldsymbol{\theta}^{(j)},\boldsymbol{\Omega}^{(j)},\boldsymbol{\alpha}^{(j)},\mathbf{Y}^{(j)}),j=1,\ldots,J\}$ are simulated observations drawn from $p(\boldsymbol{\theta},\boldsymbol{\Omega},\boldsymbol{\alpha},\mathbf{Y}|\mathbf{X},\mathbf{Z},t_{(s)})$ via the hybrid algorithm presented in Section 8.3 for the estimation.

8.5.2 Illustrative Example 4

The example given in Section 8.2.5 on the basis of an NSEM with continuous variables is used to illustrate the path sampling procedure in computing the logarithm Bayes factor for model comparison. Let M_0 be the model with the measurement equation and the nonlinear structural equation given in Equations (8.14) and (8.15), respectively. The competing models are defined by the same measurement equation and the following different structural equations:

$$M_0 : \eta = \gamma_1\xi_1 + \gamma_2\xi_2 + \gamma_3\xi_1^2 + \gamma_4\xi_1\xi_2 + \gamma_5\xi_2^2 + \delta, \tag{8.39}$$

$$M_1 : \eta = \gamma_1\xi_1 + \gamma_2\xi_2 + \gamma_3\xi_1^2 + \delta, \tag{8.40}$$

$$M_2 : \eta = \gamma_1\xi_1 + \gamma_2\xi_2 + \delta. \tag{8.41}$$

The corresponding linked models are:

$$M_{t01} : \eta = \gamma_1\xi_1 + \gamma_2\xi_2 + \gamma_3\xi_1^2 + (1-t)\{\gamma_4\xi_1\xi_2 + \gamma_5\xi_2^2\} + \delta,$$

$$M_{t02} : \eta = \gamma_1\xi_1 + \gamma_2\xi_2 + (1-t)\{\gamma_3\xi_1^2 + \gamma_4\xi_1\xi_2 + \gamma_5\xi_2^2\} + \delta,$$

$$M_{t12} : \eta = \gamma_1\xi_1 + \gamma_2\xi_2 + (1-t)\gamma_3\xi_1^2 + \delta.$$

Clearly, when $t = 0$, $M_{t01} = M_0$, $M_{t02} = M_0$ and $M_{t12} = M_1$; when $t = 1$, $M_{t01} = M_1$, $M_{t02} = M_2$ and $M_{t12} = M_2$. The log-likelihood functions corresponding to the linked models are:

$$\begin{aligned}\log p_{01}(\mathbf{Y}, \boldsymbol{\Omega}|\boldsymbol{\theta}, t) = -\frac{1}{2}\Bigg\{ & c^* + \sum_{i=1}^{n} \boldsymbol{\xi}_i^T \boldsymbol{\Phi}^{-1} \boldsymbol{\xi}_i \\ & + \sum_{i=1}^{n} (\mathbf{y}_i - \boldsymbol{\mu} - \boldsymbol{\Lambda}\boldsymbol{\omega}_i)^T \boldsymbol{\Psi}_\epsilon^{-1} (\mathbf{y}_i - \boldsymbol{\mu} - \boldsymbol{\Lambda}\boldsymbol{\omega}_i) \\ & + \sum_{i=1}^{n} [\eta_i - \gamma_1\xi_{i1} - \gamma_2\xi_{i2} - \gamma_3\xi_{i1}^2 - (1-t)(\gamma_4\xi_{i1}\xi_{i2} + \gamma_5\xi_{i2}^2)] \\ & \psi_\delta^{-1}[\eta_i - \gamma_1\xi_{i1} - \gamma_2\xi_{i2} - \gamma_3\xi_{i1}^2 - (1-t)(\gamma_4\xi_{i1}\xi_{i2} + \gamma_5\xi_{i2}^2)]\Bigg\}\end{aligned}$$

$$\begin{aligned}\log p_{02}(\mathbf{Y}, \boldsymbol{\Omega}|\boldsymbol{\theta}, t) = -\frac{1}{2}\Bigg\{ & c^* + \sum_{i=1}^{n} \boldsymbol{\xi}_i^T \boldsymbol{\Phi}^{-1} \boldsymbol{\xi}_i \\ & + \sum_{i=1}^{n} (\mathbf{y}_i - \boldsymbol{\mu} - \boldsymbol{\Lambda}\boldsymbol{\omega}_i)^T \boldsymbol{\Psi}_\epsilon^{-1} (\mathbf{y}_i - \boldsymbol{\mu} - \boldsymbol{\Lambda}\boldsymbol{\omega}_i) \\ & + \sum_{i=1}^{n} [\eta_i - \gamma_1\xi_{i1} - \gamma_2\xi_{i2} - (1-t)(\gamma_3\xi_{i1}^2 + \gamma_4\xi_{i1}\xi_{i2} + \gamma_5\xi_{i2}^2)]\end{aligned}$$

$$\times \psi_\delta^{-1}[\eta_i - \gamma_1\xi_{i1} - \gamma_2\xi_{i2} - (1-t)(\gamma_3\xi_{i1}^2 + \gamma_4\xi_{i1}\xi_{i2} + \gamma_5\xi_{i2}^2)]\Bigg\}$$

$$\log p_{12}(\mathbf{Y}, \mathbf{\Omega}|\boldsymbol{\theta}, t) = -\frac{1}{2}\Bigg\{c^* + \sum_{i=1}^n \boldsymbol{\xi}_i^T \mathbf{\Phi}^{-1}\boldsymbol{\xi}_i$$

$$+ \sum_{i=1}^n (\mathbf{y}_i - \boldsymbol{\mu} - \mathbf{\Lambda}\boldsymbol{\omega}_i)^T \mathbf{\Psi}_\epsilon^{-1}(\mathbf{y}_i - \boldsymbol{\mu} - \mathbf{\Lambda}\boldsymbol{\omega}_i)$$

$$+ \sum_{i=1}^n [\eta_i - \gamma_1\xi_{i1} - \gamma_2\xi_{i2} - (1-t)\gamma_3\xi_{i1}^2]\psi_\delta^{-1}$$

$$\times [\eta_i - \gamma_1\xi_{i1} - \gamma_2\xi_{i2} - (1-t)\gamma_3\xi_{i1}^2]\Bigg\}.$$

where c^* is a constant that is equal to $n[(p+q)\log(2\pi) + \log|\mathbf{\Psi}_\epsilon| + \log|\psi_\delta| + \log|\mathbf{\Phi}|]$. Derivatives of these functions with respect to t are equal to:

$$\frac{\mathrm{d}\log p_{01}(\mathbf{Y}, \mathbf{\Omega}|\boldsymbol{\theta}, t)}{\mathrm{d}t} = -\sum_{i=1}^n [\eta_i - \gamma_1\xi_{i1} - \gamma_2\xi_{i2} - \gamma_3\xi_{i1}^2 - (1-t)(\gamma_4\xi_{i1}\xi_{i2} + \gamma_5\xi_{i2}^2)]$$
$$\times \psi_\delta^{-1}(\gamma_4\xi_{i1}\xi_{i2} + \gamma_5\xi_{i2}^2),$$

$$\frac{\mathrm{d}\log p_{02}(\mathbf{Y}, \mathbf{\Omega}|\boldsymbol{\theta}, t)}{\mathrm{d}t} = -\sum_{i=1}^n [\eta_i - \gamma_1\xi_{i1} - \gamma_2\xi_{i2} - (1-t)(\gamma_3\xi_{i1}^2 + \gamma_4\xi_{i1}\xi_{i2} + \gamma_5\xi_{i2}^2)]$$
$$\times \psi_\delta^{-1}(\gamma_3\xi_{i1}^2 + \gamma_4\xi_{i1}\xi_{i2} + \gamma_5\xi_{i2}^2),$$

and

$$\frac{\mathrm{d}\log p_{12}(\mathbf{Y}, \mathbf{\Omega}|\boldsymbol{\theta}, t)}{\mathrm{d}t} = -\sum_{i=1}^n [\eta_i - \gamma_1\xi_{i1} - \gamma_2\xi_{i2} - (1-t)(\gamma_3\xi_{i1}^2)]\,\psi_\delta^{-1}\gamma_3\xi_{i1}^2.$$

These derivatives give $U(\mathbf{Y}, \mathbf{\Omega}, \boldsymbol{\theta}, t)$ for computing the logarithm Bayes factors, see Equations (8.35) and (8.36).

In the path sampling procedure, we take $S = 20$ grids in $[0, 1]$. Based on the previous analysis of this example, for each $t_{(s)}$, we take a burn-in phase of 2000 iterations in the MCMC algorithm, and 200 observations are collected after convergence for computing the logarithm Bayes factor. To give some idea about the sensitivity of the procedure with respect to prior inputs, three types of hyperparameters in the conjugate prior distributions are considered. The first type (Type I) is given by the hyperparameter values specified in Section 8.2.5, with some information provided by the auxiliary estimation. For Type II and Type III prior inputs, we take $\alpha_{0\epsilon k} = \alpha_{0\delta} = 10$, $\beta_{0\epsilon k} = \beta_{0\delta} = 10$, $\rho_0 = 8$, $\mathbf{H}_{0yk} = \mathbf{I}$

Table 8.13 Estimated log Bayes factors under different prior inputs.

	Prior inputs		
	I	II	III
$\log B_{01}$	0.30	0.37	0.43
$\log B_{02}$	2.15	1.93	1.84
$\log B_{12}$	1.85	1.55	1.40

and $\mathbf{H}_{0\omega k} = \mathbf{I}$, as before, but the prior inputs on $\mathbf{\Lambda}_{0yk}$, $\mathbf{\Lambda}_{0\omega k}$ and $\mathbf{R}_0$ in Type II and Type III are respectively equal to half and twice of the prior inputs given in Type I. The estimated logarithm Bayes factors are presented in Table 8.13. On the basis of the criterion for interpreting logarithm Bayes factors, both nonlinear models M_0 and M_1, are better than the linear model M_2; while the more complex model M_0 is barely better than M_1. As a result, either M_0 or M_1 is selected. The results also indicate that the estimated logarithm Bayes factors are not very sensitive to Type I, II and III prior inputs.

8.5.3 Model Comparison using DIC and WinBUGS

We use the artificial example discussed in Section 8.4.3 to illustrate model comparison of nonlinear SEMs using the DIC criterion that can be obtained from WinBUGS. The same random sample of observations simulated according to the settings described in Section 8.4.3 is considered. Let M_A be the true model that is defined by Equation (8.30) and (8.31), with the true values of Equation (8.32). Let M_B be a competing model, which is a linear SEM defined by the same measurement Equation (8.30) and the following structural equation:

$$\nu_i = \beta_1 x_i + \gamma_2 \xi_{i2} + \gamma_3 \xi_{i3}.$$

Using WinBUGS, we find that the DIC values corresponding to M_A and M_B are respectively equal to $DIC_A = 10\,848.9$, and $DIC_B = 11\,007.5$. As DIC_A is less than DIC_B, the true nonlinear SEM is selected. This model comparison result can be used to reject the null hypothesis $H_0 : \gamma_3 = \gamma_4 = \gamma_5 = 0$, where γ_3, γ_4 and γ_5 are the coefficients corresponding to the nonlinear terms of latent variables in Equation (8.31).

8.5.4 Remarks on Goodness-of-fit Assessment

For assessing the goodness-of-fit of a hypothesized linear SEM, we can compare it with the saturated model M_s which is defined by $\mathbf{y}_i = \boldsymbol{\mu} + \boldsymbol{\epsilon}_i$, where $\boldsymbol{\epsilon}_i$ is distributed as $N[\mathbf{0}, \boldsymbol{\Sigma}]$ with a generic covariance matrix $\boldsymbol{\Sigma}$ without any structure. However, for assessing NSEMs, the saturated model cannot be represented by M_s that is defined by the normal assumption. As a result, it is difficult to apply the classical likelihood ratio test. In the Bayesian approach, the goodness-of-fit of a posited NSEM can be assessed by the PP p-value (see Appendix 5.2 and Section 6.3.4). For continuous data, this statistic can be computed through Equation (A5.8) with the following χ^2 discrepancy variable

$$D(\mathbf{Y}^{\mathrm{rep}}|\boldsymbol{\theta}, \boldsymbol{\Omega}) = \sum_{i=1}^{n}(\mathbf{Y}_i^{\mathrm{rep}} - \boldsymbol{\mu} - \boldsymbol{\Lambda}\boldsymbol{\omega}_i)^T \boldsymbol{\Psi}_\epsilon^{-1}(\mathbf{Y}_i^{\mathrm{rep}} - \boldsymbol{\mu} - \boldsymbol{\Lambda}\boldsymbol{\omega}_i),$$

where $\mathbf{Y}_i^{\mathrm{rep}}$ is a replication of $\mathbf{Y}_i$. For mixed continuous and ordered categorical data, this statistic can be computed by the method as described in Section 6.3.4. Finally, the conclusion could be cross-validated with the residual plots.

REFERENCES

Bagozzi, R. P., Baumgartner, H. and Yi, Y. (1992) State versus action orientation and the theory of reasoned action: an application to coupon usage. *Journal of Consumer Research*, **18**, 505–517.

Bentler, P. M. (1983) Some contributions to efficient statistics for structural models: specification and estimation of moment structures. *Psychometrika*, **48**, 493–517.

Benter, P. M. (1992) *EQS: Structural Equation Program Manual.* Los Angeles: BMDP Statistical Software.

Bentler, P. M. and Dudgeon, P. (1996) Covariance structure analysis: statistical practice, theory and directions. *Annual Review of Psychology*, **47**, 541–570.

Besag, J., Green, P., Higdon, D. and Mengersen, K. (1995) Bayesian computation and stochastic systems. *Statistical Science*, **10**, 3–66.

Browne, M. W. (1984) Asymptotically distribution-free methods in the analysis of covariance structures. *British Journal of Mathematical and Statistical Psychology*, **37**, 62–83.

Busemeyer, J. R. and Jones, L. E. (1983) Analysis of multiplicative combination rules when the causal variables are measured with error. *Psychological Bulletin*, **93**, 549–562.

Etezadi-Amoli, J. and McDonald, R. P. (1983) A second generation nonlinear factor analysis. *Psychometrika*, **48**, 315–342.

Geman, S. and Geman, D. (1984) Stochastic relaxation, Gibbs distribution and the Bayesian restoration of images. *IEEE Transactions on Pattern Analysis and Machine Intelligence*, **6**, 721–741.

Gelman, A. (1996) Inference and monitoring convergence. In W. R. Gilks, S. Richardson and D. J. Spiegelhalter (eds), *Markov Chain Monte Carlo in Practice*, pp. 131–144. London: Chapman and Hall.

Gelman, A., Meng, X. L. and Stern, H. (1996) Posterior predictive assessment of model fitness via realized discrepancies. *Statistica Sinica*, **6**, 733–759.

Gelman, A., Roberts, G. O. and Gilks, W. R. (1995) Efficient Metropolis jumping rules. In J. M. Bernardo, J. O. Berger, A. P. Dawid and A. F. M. Smith (eds), *Bayesian Statistics 5*, pp. 599–607. Oxford: Oxford University Press.

Geyer, C. J. (1992) Practical Markov chain Monte Carlo. *Statistical Science*, 7, 473–511.

Gilks, W. R., Richardson, S. and Spiegelhalter, D. J. (1996) Introducing Markov chain Monte Carlo. In W. R. Gilks, S. Richardson and D. J. Spiegelhalter (eds), *Markov Chain Monte Carlo in Practice*, pp. 1–19. London: Chapman and Hall.

Hastings, W. K. (1970) Monte Carlo sampling methods using Markov chains and their application. *Biometrika*, **57**, 97–109.

Hu, L., Bentler, P. M. and Kano, Y. (1992) Can test statistics in covariance structure analysis be trusted *Psychological Bulletin*, **112**, 351–362.

Jaccard, J. and Wan, C. K. (1995) Measurement error in the analysis of interaction effects between continuous predictors using multiple regression: multiple indicator and structural equation approaches. *Psychological Bulletin*, **117**, 348–357.

Jonsson, F. Y. (1998) Modeling interaction and non-linear effects: a step by step LISREL example. In R. E. Schumacker and G. A. Marcoulides (eds), *Interaction and Nonlinear Effects in Structural Equation Models* pp. 17–42. Mahwah, NJ: Lawrence Erlbaum Associates, Publishers.

Jöreskog, K. G. and Sörbom, D. (1996) *LISREL 8: Structural Equation Modeling with the SIMPLIS Command Language*. Hove and London: Scientific Software International.

Jöreskog, K. G. and Yang, F. (1996) Nonlinear structural equation models: the Kenny–Judd model with interaction effects. In G. A. Marcoulides and R. E. Schumacker (eds), *Advanced Structural Equation Modeling Techniques*, pp. 57–88. Hillsdale, NJ: Lawrence Erlbaum Associates, Publishers.

Kenny, D. A. and Judd, C. M. (1984) Estimating the nonlinear and interactive effects of latent variables. *Psychological Bulletin*, **96**, 201–210.

Klein, A. and Moosbrugger, M. (2000) Maximum likelihood estimation of latent interaction effects with the LMS method. *Psychometrika*, **65**, 457–474.

Lee, S. Y. and Song, X. Y. (2003) Estimation and model comparison for a nonlinear latent variable model with fixed covariates. *Psychometrika*, **68**, 27–47.

Lee, S. Y. and Song, X. Y. (2004) Bayesian model comparison of nonlinear structural equation models with missing continuous and ordinal categorical data. *British Journal of Mathematical and Statistical Psychology*, **57**, 131–150.

Lee, S. Y. and Zhu, H. T. (2000) Statistical analysis of nonlinear structural equation models with continuous and polytomous data. *British Journal of Mathematical and Statistical Psychology*, **53**, 209–232.

Lee, S. Y. and Zhu, H. T. (2002) Maximum likelihood estimation of nonlinear structural equation models. sl Psychometrika, **67**, 189–210.

Lee, S. Y., Poon, W. Y. and Bentler, P. M. (1995) A two-stage estimation of structural equation models with continuous and polytomous variables. *British Journal of Mathematical and Statistical Psychology*, **48**, 339–358.

Lee, S. Y., Song, X. Y. and Poon, W. Y. (2004) Comparison of approaches in estimating interaction and quadratic effects of latent variables. *Multivariate Behavioral Research*, **39**, 37–67.

Lee, S. Y., Song, X. Y. and Tang, N. S. (2006) Bayesian methods for analyzing structural equation models with covariates, interactions and quadratic latent variables. *Structural Equation Modeling*, in press.

Li, F. *et al.* (1998) Approaches to testing interaction effects using structural equation modeling methodology. *Multivariate Behavioral Research*, **33**, 1–39.

Marsh, H. W., Wen, Z. and Hau, K. T. (2004) Structural equation models of latent interaction: evaluation of alternative estimation strategies and indicator construction. *Psychological Methods*. **9**, 275–300.

McDonald, R. P. (1962) A general approach to nonlinear factor analysis. *Psychometrika*, **27**, 123–157.

Metropolis, N. *et al.* (1953) Equations of state calculations by fast computing machine. *Journal of Chemical Physics*, **21**, 1087–1091.

Mooijaart, A. and Bentler, P. (1986) Random polynomial factor analysis. In Diday, E. *et al.* (eds), *Data Analysis and Informatics, IV*, pp. 241–250. Amsterdam: Elsevier Science Publishers B. V. (North-Holland).

Ping, R. A. (1996) Estimating latent variable interactions and quadratics: the state of this art. *Journal of Management*, **22**, 163–183.

Schermelleh-Engel, K., Klein, A. and Moosbrugger, H. (1998) Estimating nonlinear effects using a latent moderated structural equations approach. In R. E. Schumacker and G. A. Marcoulides (eds), *Interaction and Nonlinear Effects in Structural Equation Models*, pp. 203–238. Mahwah, NJ: Lawrence Erlbaum Associates, Publishers.

Schumacker, R. E. and Marcoulides, G. A. (1998) *Interaction and Nonlinear Effects in Structural Equation Models.* Mahwah, NJ: Lawrence Erlbaum Associates, Publishers.

Song, X. Y. and Lee, S. Y. (2006) Bayesian analysis of structural equation models with nonlinear covariates and latent variables. *Multivariate Behavioral Research*, **41**, 337–365.

Spiegelhalter, D. J., Thomas, A., Best, N. G. and Lunn, D. (2003) *WinBUGS User Manual. Version 1.4.* Cambridge, England: MRC Biostatistics Unit.

Wall, M. M. and Amemiya, Y. (2000) Estimation for polynomial structural equation models. *Journal of the American Statistical Association*, **95**, 929–940.

Wall, M. M. and Amemiya, Y. (2003) A method of moments technique for fitting interaction effects in structural equation models. *British Journal of Mathematical and Statistical Psychology*, **56**, 47–63.

World Values Study Group (1994) World Values Survey 1981–1984 and 1990–1993. ICPSR version. Ann Arbor, MI: Institute for Social Research (producer). Ann Arbor, MI: Inter-university Consortium for Political and Social Research (distributor).

Zhu, H. T. and Lee, S. Y. (1999) Statistical analysis of nonlinear factor analysis models. *British Journal of Mathematical and Statistical Psychology*, **52**, 225–242.

9

Two-level Nonlinear Structural Equation Models

9.1 INTRODUCTION

Statistical methods described in previous chapters for analyzing various SEMs assume that the available data are obtained from a random sample from a single population. However, in many applications, the data may exhibit two possible kinds of heterogeneity. The first kind is mixture data, which involve independent observations that come from one of the K populations with different distributions, and no information is available on which of the K populations an individual observation belongs to. Although K may be known or unknown, it is usually quite small. Mixture models are used to analyze this kind of heterogeneous data. Analysis of a mixture of SEMs will be postponed to the Chapter 11. The second kind of heterogeneous data are drawn from a number of different groups (clusters) with a known hierarchical structure. Examples may well be: the drawing of random samples of patients from within random samples of clinics or hospitals; individuals from within random samples of families; or students from within random samples of schools. In contrast to the mixture data, these hierarchically structured data usually involve a large number of G groups, and the group membership of each observation can be specified accurately. However, we allow that individuals within a group share certain common influential factors and hence lead to correlated observations. Hence, the assumption of independence among observed observations is violated when dealing with this kind of data. Clearly, ignoring the correlated structure of the

Structural Equation Modeling: A Bayesian Approach S-Y. Lee

data and analyzing them as observations from a single random sample give erroneous results. Moreover, it is also desirable to establish a meaningful model for the between-groups levels, and study the effects of the between-groups latent variables to the within-groups latent variables. Consequently, the need for developing two-level models that take into consideration the correlated structure of the data is well recognized in structural equation modeling. Statistical methods that are based on the ML or its related approaches have been developed, see McDonald and Goldstein (1989), Zhang and Lee (2001), Lee and Poon (1998), Lee and Tsang (1999), Rabe-Hesketh *et al.* (2004), Lee and Shi (2001) and Lee and Song (2005). Using a Bayesian approach, Song and Lee (2004) developed MCMC methods for analyzing two-level nonlinear models with continuous and ordered categorical data, and Lee and Tang (2006) consider two-level nonlinear SEMs with cross-level effects. For reasons stated in previous chapters, we will describe Bayesian methods for analyzing various two-level SEMs in this chapter.

To provide a comprehensive framework for analyzing two-level models, nonlinear structural equations are incorporated in the SEMs that are associated with within-groups and between-groups models. Moreover, the model can accommodate mixed type of continuous and ordered categorical data. In addition to Bayesian estimation, we will present a path sampling procedure to compute the Bayes factor for model comparison. The generality of the model is important for providing a comprehensive framework for model comparison of different kinds of SEMs. For example, we can compare a two-level nonlinear model with a two-level linear model. Again, the idea of data augmentation is utilized. Here, the observed data are augmented with various latent variables at both the levels, and the latent continuous random vectors that underly the ordered categorical variables. An algorithm that is based on the Gibbs sampler and the Metropolis–Hastings (MH) algorithm is described for estimation. Observations generated by this algorithm will be used in the path sampling procedure in computing Bayes factor. Although we concentrate on a two-level SEM, the methodology proposed can be extended to higher level SEMs. Finally, an application of WinBUGS to two-level nonlinear SEMs will be discussed.

9.2 A TWO-LEVEL NONLINEAR SEM WITH MIXED TYPE VARIABLES

Consider a collection of p-variate random vectors $\mathbf{u}_{gi}$, $i = 1, \ldots, N_g$, within groups $g = 1, \ldots, G$. The sample sizes N_g may differ from group to group so that the data set is unbalanced. At the first level we assume that, conditional on the group mean $\mathbf{v}_g$, random observations in each group satisfy the following measurement equation:

$$\mathbf{u}_{gi} = \mathbf{v}_g + \mathbf{v}_{gi} = \mathbf{v}_g + \boldsymbol{\Lambda}_{1g}\boldsymbol{\omega}_{1gi} + \boldsymbol{\epsilon}_{1gi}, \quad g = 1, \ldots, G, \; i = 1, \ldots, N_g, \tag{9.1}$$

where $\mathbf{\Lambda}_{1g}$ is a $p \times q_1$ matrix of factor loadings, $\boldsymbol{\omega}_{1gi}$ is a $q_1 \times 1$ random vector of latent factors, and $\boldsymbol{\epsilon}_{1gi}$ is a $p \times 1$ random vector of error measurements which is independent of $\boldsymbol{\omega}_{1gi}$ and is distributed as $N[\mathbf{0}, \mathbf{\Psi}_{1g}]$, where $\mathbf{\Psi}_{1g}$ is a diagonal matrix. Note that due to the existence of $\mathbf{v}_g$, $\mathbf{u}_{gi}$ and $\mathbf{u}_{gj}$ are not independent. Hence, in the two-level SEM, the usual assumption on the independence of the observations is violated. This induces some difficulty in the analysis. To account for the structure at the between-groups level, we assume that the group mean $\mathbf{v}_g$ satisfies the following factor analysis model:

$$\mathbf{v}_g = \boldsymbol{\mu} + \mathbf{\Lambda}_2\boldsymbol{\omega}_{2g} + \boldsymbol{\epsilon}_{2g}, \quad g = 1, \ldots, G, \tag{9.2}$$

where $\boldsymbol{\mu}$ is the vector of intercepts, $\mathbf{\Lambda}_2$ is a $p \times q_2$ matrix of factor loadings, $\boldsymbol{\omega}_{2g}$ is a $q_2 \times 1$ vector of latent variables and $\boldsymbol{\epsilon}_{2g}$ is a $p \times 1$ random vector of error measurements which is independent of $\boldsymbol{\omega}_{2g}$ and is distributed as $N[\mathbf{0}, \mathbf{\Psi}_2]$, where $\mathbf{\Psi}_2$ is a diagonal matrix. Moreover, the first- and second-level measurement errors are assumed to be independent. It follows from Equations (9.1) and (9.2) that

$$\mathbf{u}_{gi} = \boldsymbol{\mu} + \mathbf{\Lambda}_2\boldsymbol{\omega}_{2g} + \boldsymbol{\epsilon}_{2g} + \mathbf{\Lambda}_{1g}\boldsymbol{\omega}_{1gi} + \boldsymbol{\epsilon}_{1gi}. \tag{9.3}$$

For assessing the interrelationships among the latent variables, latent vectors $\boldsymbol{\omega}_{1gi}$ and $\boldsymbol{\omega}_{2g}$ are partitioned as $\boldsymbol{\omega}_{1gi} = (\boldsymbol{\eta}_{1gi}^T, \boldsymbol{\xi}_{1gi}^T)^T$ and $\boldsymbol{\omega}_{2g} = (\boldsymbol{\eta}_{2g}^T, \boldsymbol{\xi}_{2g}^T)^T$, respectively; where $\boldsymbol{\eta}_{1gi}(q_{11} \times 1)$, $\boldsymbol{\xi}_{1gi}(q_{12} \times 1)$, $\boldsymbol{\eta}_{2g}(q_{21} \times 1)$ and $\boldsymbol{\xi}_{2g}(q_{22} \times 1)$ are latent vectors, with $q_{j1} + q_{j2} = q_j$, for $j = 1, 2$. The distributions of $\boldsymbol{\xi}_{1gi}$ and $\boldsymbol{\xi}_{2g}$ are $N[\mathbf{0}, \mathbf{\Phi}_{1g}]$ and $N[\mathbf{0}, \mathbf{\Phi}_2]$, respectively. The following non-linear structural equations are incorporated in the between-groups and within-groups models of the proposed two-level model:

$$\boldsymbol{\eta}_{1gi} = \mathbf{\Pi}_{1g}\boldsymbol{\eta}_{1gi} + \mathbf{\Gamma}_{1g}\mathbf{H}_1(\boldsymbol{\xi}_{1gi}) + \boldsymbol{\delta}_{1gi}, \text{ and} \tag{9.4}$$

$$\boldsymbol{\eta}_{2g} = \mathbf{\Pi}_2\boldsymbol{\eta}_{2g} + \mathbf{\Gamma}_2\mathbf{H}_2(\boldsymbol{\xi}_{2g}) + \boldsymbol{\delta}_{2g}, \tag{9.5}$$

where $\mathbf{H}_1(\boldsymbol{\xi}_{1gi}) = (h_{11}(\boldsymbol{\xi}_{1gi}), \ldots, h_{1a}(\boldsymbol{\xi}_{1gi}))^T$ and $\mathbf{H}_2(\boldsymbol{\xi}_{2g}) = (h_{21}(\boldsymbol{\xi}_{2g}), \ldots, h_{2b}(\boldsymbol{\xi}_{2g}))^T$ are vector-valued functions with nonzero differentiable known functions h_{1k} and h_{2k}, and usually $a \geq q_{12}$ and $b \geq q_{22}$, $\mathbf{\Pi}_{1g}(q_{11} \times q_{11})$, $\mathbf{\Pi}_2(q_{21} \times q_{21})$, $\mathbf{\Gamma}_{1g}(q_{11} \times a)$ and $\mathbf{\Gamma}_2(q_{21} \times b)$ are unknown parameter matrices, $\boldsymbol{\delta}_{1gi}$ is a vector of error measurements which is distributed as $N[\mathbf{0}, \mathbf{\Psi}_{1g\delta}]$, $\boldsymbol{\delta}_{2g}$ is a vector of error measurements which is distributed as $N[\mathbf{0}, \mathbf{\Psi}_{2\delta}]$ and $\mathbf{\Psi}_{1g\delta}$ and $\mathbf{\Psi}_{2\delta}$ are diagonal matrices. Due to the nonlinearity induced by $\mathbf{H}_1$ and $\mathbf{H}_2$, the underlying distribution of $\mathbf{u}_{gi}$ is not normal. In the within-groups structural equation, we assume as usual that $\boldsymbol{\xi}_{1gi}$ and $\boldsymbol{\delta}_{1gi}$ are independent. Similarly, in the between-groups structural equation, we assume that $\boldsymbol{\xi}_{2g}$ and $\boldsymbol{\delta}_{2g}$ are independent. Moreover, we assume that the within-groups latent vectors $\boldsymbol{\eta}_{1gi}$ and $\boldsymbol{\xi}_{1gi}$ are independent of the between-groups latent vectors $\boldsymbol{\eta}_{2g}$ and $\boldsymbol{\xi}_{2g}$.

Hence, it follows from Equation (9.4) that $\boldsymbol{\eta}_{1gi}$ is independent of $\boldsymbol{\eta}_{2g}$ and $\boldsymbol{\xi}_{2g}$. That is, this two-level SEM does not accommodate the effects of the latent vectors in the between-groups level on the latent vectors in the within-groups level. However, in the within-groups model or in the between-groups model, nonlinear effects of the exogenous latent variables to the endogenous latent variables can be assessed through Equations (9.4) and (9.5), and the hierarchical structure of the data has been taken into account. As the functions h_{1k} in $\mathbf{H}_1(\boldsymbol{\xi}_{1gi})$ and h_{2k} in $\mathbf{H}_2(\boldsymbol{\xi}_{2g})$ are rather general, the common interaction and quadratic effects are their special cases. Practically, allowing nonlinear relationships such as interaction and quadratic terms among latent variables leads to models that more accurately represent reality. Furthermore, we assume that $\mathbf{I}_1 - \boldsymbol{\Pi}_{1g}$ and $\mathbf{I}_2 - \boldsymbol{\Pi}_2$ are nonsingular, and their determinants are respectively independent of the elements in $\boldsymbol{\Pi}_{1g}$ and $\boldsymbol{\Pi}_2$. The proposed two-level SEM is not identified without imposing the identification restrictions. The common method of fixing appropriate elements in $\boldsymbol{\Lambda}_{1g}, \boldsymbol{\Pi}_{1g}, \boldsymbol{\Gamma}_{1g}, \boldsymbol{\Lambda}_2, \boldsymbol{\Pi}_2$ and $\boldsymbol{\Gamma}_2$ at preassigned known values can be used to achieve an identified model. In a similar way to the analysis of single-level SEMs, there are no necessary and sufficient conditions to achieve identifiability, and the problem is approached on a problem-by-problem basis. In a similar way to the analysis of nonlinear SEMs, the choices of $\mathbf{H}_1(\boldsymbol{\xi}_{1gi})$ and $\mathbf{H}_2(\boldsymbol{\xi}_{2g})$ are not arbitrary (see discussion in Section 8.2).

To accommodate mixed ordered categorical and continuous variables, without loss of generality, we suppose that $\mathbf{u}_{gi} = (\mathbf{x}_{gi}^T, \mathbf{y}_{gi}^T)^T$, where $\mathbf{x}_{gi} = (x_{gi1}, \cdots, x_{gir})^T$ is an observable continuous random vector, and $\mathbf{y}_{gi} = (y_{gi1}, \cdots, y_{gis})^T$ an unobservable continuous random vector. In a similar way to the previous chapters, a threshold specification is used to model the observable ordered categorical vector $\mathbf{z} = (z_1, \cdots, z_s)^T$ with its underlying continuous vector $\mathbf{y} = (y_1, \cdots, y_s)^T$ as follows:

$$\mathbf{z} = \begin{bmatrix} z_1 \\ \vdots \\ z_s \end{bmatrix} \quad \text{if} \quad \begin{array}{c} \alpha_{1,z_1} < y_1 \leq \alpha_{1,z_1+1} \\ \vdots \\ \alpha_{s,z_s} < y_s \leq \alpha_{s,z_s+1}, \end{array} \tag{9.6}$$

where z_k is an integral value in $\{0, 1, \cdots, b_k\}$. In general, we let $\alpha_{k,0} = -\infty$, $\alpha_{b_k+1} = \infty$. For the kth variable, there are $b_k + 1$ categories which are defined by unknown thresholds α_{kj}. Dichotomous variables are treated as an ordered categorical variable with a single threshold that is fixed at zero. The link between a dichotomous variable with its underlying continous variable y is given by

$$d = 1 \text{ if } y > 0, \text{ and } d = 0 \text{ if } y \leq 0. \tag{9.7}$$

In general, the thresholds, mean and variance of an ordered categorical variable can be identified through the method given in Chapter 6; and the mean and variance of a dichotomous variable can be identified through the method given in Chapter 7.

Let $\boldsymbol{\theta}$ be the parameter vector that contains all the unknown structural parameters in $\boldsymbol{\Lambda}_{1g}, \boldsymbol{\Psi}_{1g}, \boldsymbol{\Pi}_{1g}, \boldsymbol{\Gamma}_{1g}, \boldsymbol{\Phi}_{1g}, \boldsymbol{\Psi}_{1g\delta}, \boldsymbol{\Lambda}_2, \boldsymbol{\Psi}_2, \boldsymbol{\Gamma}_2, \boldsymbol{\Phi}_2$ and $\boldsymbol{\Psi}_{2\delta}$, and $\boldsymbol{\alpha}$ be the parameter vector that contains all the unknown thresholds. The total number of unknown parameters in $\boldsymbol{\theta}$ and $\boldsymbol{\alpha}$ is usually large. In the following discussion, we assume that the two-level nonlinear model defined by $\boldsymbol{\theta}$ and $\boldsymbol{\alpha}$ is identified.

The above model subsumes a number of important models in the recent developments of SEMs. For instance, the models discussed in Shi and Lee (1998, 2000) are generalized in two aspects: one from linear models to nonlinear models and the other from single-level models to two-level models. The nonlinear SEMs proposed by Lee and Zhu (2000) are extended from single-level to two-level. Despite its generality, the proposed two-level SEM is defined by measurement and structural equations that describe the relationships among the observed and latent variables at both levels by conceptually simple regression models. Consider the following three major components of the proposed model and the data structure: (1) a two-level model for hierarchically structured data, (2) discrete natures of the data, and (3) a nonlinear structural equation in the within-group model. The first two components are important for achieving correct statistical results. The last component is essential for analyzing more complicated situations, as nonlinear terms of latent variables have been found to be useful in establishing a better model. The between-groups model is also defined with a nonlinear structural equation for generality. In practice, situations might not often arise that require all the aforementioned major components to fit the data. Still, the development is useful to provide a general framework for analyzing the large number of its submodels. This is particularly true from a model comparison perspective. For example, even if a linear model is better than a nonlinear model in fitting a data set, such a conclusion cannot be reached without the model comparison statistic under the more general nonlinear model framework. For practical situations where G is not large, it may not be worthwhile or practical to consider a complicated between-groups model. Moreover, most two-level SEMs in the literature assume that the within-groups parameters are invariant over groups.

9.3 BAYESIAN ESTIMATION

In this section, we consider the statistical analysis of the general two-level nonlinear SEM described in Section 9.2. We first note that due to the presence of $\mathbf{v}_g$ (see Equation (9.1)), $\mathbf{u}_{gi}$ and $\mathbf{u}_{gj}$ within the gth group are correlated. To cope with this dependence structure and to obtain independent observations in the traditional ML approach that focused on the observed data likelihood, one has to work with $\mathbf{u}_g^* = (\mathbf{u}_{g1}^{*T}, \ldots, \mathbf{u}_{gN_g}^{*T})^T$, where $\mathbf{u}_{gi}^* = (\mathbf{x}_{gi}^T, \mathbf{z}_{gi}^T)^T$ which is a mixed continuous and ordered categorical random observation. Because of the discrete nature of the ordered categorical data and the nonlinearities in the first- and second-level structural equations, $l_g(\mathbf{u}_g^*)$, the probability density function of the high dimensional random vector $\mathbf{u}_g^*$, involves high dimensional intractable integrals. The observed data likelihood is equal to $l_1(\mathbf{u}_1^*) \cdots l_G(\mathbf{u}_G^*)$. Maximizing this function through the classical numerical methods such as the Newton–Raphson algorithm is extremely complicated. Moreover, because $\mathbf{u}_1^*, \ldots, \mathbf{u}_G^*$ have different dimensions, they are not identically distributed. The ML theory that depends on the assumption of identically distributed observations may not be directly applicable to draw statistical conclusions. For example, the observed data log-likelihood should be used with great caution in testing the goodness-of-fit of the hypothesized model.

Motivated by its various advantages, we propose the Bayesian approach for analyzing the current two-level nonlinear SEM with mixed continuous, dichotomous and/or ordered categorical data. As we have discussed in previous chapters, recent MCMC methods in statistical computing for posterior simulation greatly enhance the applicability of the Bayesian inference. Our basic strategy is to augment the observed data with the latent data that come from the latent variables and/or latent measurements, then MCMC tools are applied to simulate observations in the posterior analysis.

9.3.1 Posterior Simulation and Bayesian Estimates

Let $\mathbf{X}_g = (\mathbf{x}_{g1}, \ldots, \mathbf{x}_{gN_g})$ and $\mathbf{X} = (\mathbf{X}_1, \ldots, \mathbf{X}_G)$ be the observed continuous data, and $\mathbf{Z}_g = (\mathbf{z}_{g1}, \ldots, \mathbf{z}_{gN_g})$ and $\mathbf{Z} = (\mathbf{Z}_1, \ldots, \mathbf{Z}_G)$ be the observed ordered categorical data. Let $\mathbf{Y}_g = (\mathbf{y}_{g1}, \ldots, \mathbf{y}_{gN_g})$ and $\mathbf{Y} = (\mathbf{Y}_1, \ldots, \mathbf{Y}_G)$ be the latent continuous measurements associated with $\mathbf{Z}_g$ and $\mathbf{Z}$, respectively. The observed data will be augmented with $\mathbf{Y}$ in the posterior analysis. Once $\mathbf{Y}$ is given, all the data are continuous and the problem will be easier to cope with. Let $\mathbf{V} = (\mathbf{v}_1, \ldots, \mathbf{v}_G)$ be the matrix of between-group latent variables. If $\mathbf{V}$ is observed, the model is reduced to the single-level multi-sample model. Moreover, let $\boldsymbol{\Omega}_{1g} = (\boldsymbol{\omega}_{1g1}, \ldots, \boldsymbol{\omega}_{1gN_g})$, $\boldsymbol{\Omega}_1 = (\boldsymbol{\Omega}_{11}, \ldots, \boldsymbol{\Omega}_{1G})$ and $\boldsymbol{\Omega}_2 = (\boldsymbol{\omega}_{21}, \ldots, \boldsymbol{\omega}_{2G})$ be the matrices of latent variables at the within-groups and between-groups levels. If these matrices are observed, the complicated nonlinear structural Equations (9.4) and (9.5) reduce to the regular simultaneous regression models. Difficulties due to the nonlinear relationships among the latent variables are greatly alleviated. Hence, problems associated with the complicated components of the model, such as the correlated structure of the observations induced by the two-level data, the discrete nature of the ordered categorical variables and the nonlinearity of the latent variables at both levels, can be handled by data augmentation. In the posterior analysis, the observed data $(\mathbf{X}, \mathbf{Z})$ will be augmented with $(\mathbf{Y}, \mathbf{V}, \boldsymbol{\Omega}_1, \boldsymbol{\Omega}_2)$, the hypothetically missing data matrices of latent measurements and variables. More specifically, we will consider the joint posterior distribution $[\boldsymbol{\theta}, \boldsymbol{\alpha}, \mathbf{Y}, \mathbf{V}, \boldsymbol{\Omega}_1, \boldsymbol{\Omega}_2|\mathbf{X}, \mathbf{Z}]$. The Gibbs sampler (Geman and Geman, 1984) will be used for generating a sequence of observations from this joint posterior distribution. Then the Bayesian solution is obtained by standard inferences on the basis of the generated sample of observations. In applying the Gibbs sampler, we iteratively sample from the following conditional distributions: $[\mathbf{V}|\boldsymbol{\theta}, \boldsymbol{\alpha}, \mathbf{Y}, \boldsymbol{\Omega}_1, \boldsymbol{\Omega}_2, \mathbf{X}, \mathbf{Z}]$, $[\boldsymbol{\Omega}_1|\boldsymbol{\theta}, \boldsymbol{\alpha}, \mathbf{Y}, \mathbf{V}, \boldsymbol{\Omega}_2, \mathbf{X}, \mathbf{Z}]$, $[\boldsymbol{\Omega}_2|\boldsymbol{\theta}, \boldsymbol{\alpha}, \mathbf{Y}, \mathbf{V}, \boldsymbol{\Omega}_1, \mathbf{X}, \mathbf{Z}]$, $[\boldsymbol{\alpha}, \mathbf{Y}|\boldsymbol{\theta}, \mathbf{V}, \boldsymbol{\Omega}_1, \boldsymbol{\Omega}_2, \mathbf{X}, \mathbf{Z}]$ and $[\boldsymbol{\theta}|\boldsymbol{\alpha}, \mathbf{Y}, \mathbf{V}, \boldsymbol{\Omega}_1, \boldsymbol{\Omega}_2, \mathbf{X}, \mathbf{Z}]$.

For the proposed two-level model, the conditional distribution $[\boldsymbol{\theta}|\boldsymbol{\alpha}, \mathbf{Y}, \mathbf{V}, \boldsymbol{\Omega}_1, \boldsymbol{\Omega}_2, \mathbf{X}, \mathbf{Z}]$ is further decomposed into components involving various structural parameters in the between-groups and within-groups models. These components are different under various special cases of the model. Some typical examples are:

(a) Models with different within-groups parameters across groups: in this case, the within-groups structural parameters $\boldsymbol{\theta}_{1g} = \{\boldsymbol{\Lambda}_{1g}, \boldsymbol{\Psi}_{1g}, \boldsymbol{\Pi}_{1g}, \boldsymbol{\Gamma}_{1g}, \boldsymbol{\Phi}_{1g}, \boldsymbol{\Psi}_{1g\delta}\}$ and threshold parameters $\boldsymbol{\alpha}_g$ associated with the gth group are different from those associated with the hth group, for $g \neq h$. Practically, G and N_g should not be too small for drawing valid statistical conclusions for the between-group model and the gth within-groups model.

(b) Models with some invariant within-groups parameters: in this case, parameters $\boldsymbol{\theta}_{1g}$ and/or $\boldsymbol{\alpha}_g$ associated with the gth group are equal to those associated with some other groups.

(c) Models with all invariant within-groups parameters: under this situation, $\boldsymbol{\theta}_{11} = \cdots = \boldsymbol{\theta}_{1G}$, and $\boldsymbol{\alpha}_1 = \cdots = \boldsymbol{\alpha}_G$.

Conditional distributions under various special cases are similar but different. Moreover, prior distributions of the parameters are also involved. On the basis of the reasoning given in previous chapters, conjugate type prior distributions are used. The non-informative distribution is used for the prior distribution of the thresholds. The conditional distributions of the components in $[\boldsymbol{\theta}|\boldsymbol{\alpha}, \mathbf{Y}, \mathbf{V}, \boldsymbol{\Omega}_1, \boldsymbol{\Omega}_2, \mathbf{X}, \mathbf{Z}]$ as well as other conditional distributions required by the Gibbs sampler are briefly discussed in Appendix 9.1. As we can see in this appendix, these conditional distributions are generalizations of those that are associated with a single-level model, and most of them are standard distributions such as normal, univariate truncated normal, gamma and inverted Wishart. Simulating observations from them requires little computing time. The MH algorithm will be used for simulating observations efficiently from the three complicated conditional distributions: $[\boldsymbol{\Omega}_1|\boldsymbol{\theta}, \boldsymbol{\alpha}, \mathbf{Y}, \mathbf{V}, \boldsymbol{\Omega}_2, \mathbf{X}, \mathbf{Z}]$, $[\boldsymbol{\Omega}_2|\boldsymbol{\theta}, \boldsymbol{\alpha}, \mathbf{Y}, \mathbf{V}, \boldsymbol{\Omega}_1, \mathbf{X}, \mathbf{Z}]$ and $[\boldsymbol{\alpha}, \mathbf{Y}|\boldsymbol{\theta}, \mathbf{V}, \boldsymbol{\Omega}_1, \boldsymbol{\Omega}_2, \mathbf{X}, \mathbf{Z}]$. Some technical details on the implementation of the MH algorithm are given in Appendix 9.2.

As in previous chapters, observations obtained from the posterior simulation can be used for various statistical inferences via some data analysis methods. Bayesian estimates of $\boldsymbol{\theta}$, $\boldsymbol{\alpha}$ and latent variables $\boldsymbol{\omega}_{2g}$ and $\boldsymbol{\omega}_{1gi}$ at both levels can be obtained easily via the corresponding sample means of the generated observations. Specifically, let $\{(\boldsymbol{\theta}^{(t)}, \boldsymbol{\alpha}^{(t)}, \boldsymbol{\Omega}_1^{(t)}, \boldsymbol{\Omega}_2^{(t)}); t = 1, \ldots, T^*\}$ be random observations generated from the joint posterior distribution $p(\boldsymbol{\theta}, \boldsymbol{\alpha}, \mathbf{Y}, \mathbf{V}, \boldsymbol{\Omega}_1, \boldsymbol{\Omega}_2|\mathbf{X}, \mathbf{Z})$, then joint Bayes estimates of $\boldsymbol{\theta}$, $\boldsymbol{\alpha}$, $\boldsymbol{\omega}_{2g}$ and $\boldsymbol{\omega}_{1gi}$ are obtained as follows:

$$\hat{\boldsymbol{\theta}} = \frac{1}{T^*}\sum_{t=1}^{T^*}\boldsymbol{\theta}^{(t)}, \ \hat{\boldsymbol{\alpha}} = \frac{1}{T^*}\sum_{t=1}^{T^*}\boldsymbol{\alpha}^{(t)}, \ \hat{\boldsymbol{\omega}}_{2g} = \frac{1}{T^*}\sum_{t=1}^{T^*}\boldsymbol{\omega}_{2g}^{(t)}, \ \hat{\boldsymbol{\omega}}_{1gi} = \frac{1}{T^*}\sum_{t=1}^{T^*}\boldsymbol{\omega}_{1gi}^{(t)} \quad (9.8)$$

where $\boldsymbol{\omega}_{2g}^{(t)}$ and $\boldsymbol{\omega}_{1gi}^{(t)}$ are from $\boldsymbol{\Omega}_2^{(t)}$ and $\boldsymbol{\Omega}_{1g}^{(t)}$, respectively. These joint Bayesian estimates tend to their corresponding posterior means in probability as T tends to infinity. Since we have a large sample of $\boldsymbol{\theta}$ from its posterior distribution, an estimate of $\text{Var}(\boldsymbol{\theta})$ can be obtained easily from the sample covariance matrix. Moreover, estimated residuals $\hat{\boldsymbol{\epsilon}}_{1gi}$, $\hat{\boldsymbol{\epsilon}}_{2g}$, $\hat{\boldsymbol{\delta}}_{1gi}$ and $\hat{\boldsymbol{\delta}}_{2g}$ can be obtained by means of the parameter estimates, the latent variable estimates, and the estimates of the unobserved continuous measurements that underly the dichotomous or ordered categorical data.

9.3.2 Simulation Studies

The objectives of the two simulation studies presented in this section are to reveal the performance of the Bayesian approach and its associated algorithm in recovering the true parameters, and roughly to examine the sensitivity of the parameter estimates to different prior inputs of the hyperparameters in the conjugate prior distributions. The first study is concerned with a model with invariant within-groups parameters, while the second study is focused on a different model with certain distinct parameters across some within-groups.

In the first simulation study, random vectors $\mathbf{u}_{gi}$ satisfied the two-level model defined by Equations (9.1) and (9.2) are considered with the following specifications: $\boldsymbol{\Lambda}_{1g} = \boldsymbol{\Lambda}_1$ and $\boldsymbol{\Psi}_{1g} = \boldsymbol{\Psi}_1$, where

$$\boldsymbol{\Lambda}_1^T = \begin{bmatrix} 1.0^* & 0.8 & 0.8 & 0^* & 0^* & 0^* & 0^* & 0^* & 0^* \\ 0^* & 0^* & 0^* & 1.0^* & 0.8 & 0.8 & 0^* & 0^* & 0^* \\ 0^* & 0^* & 0^* & 0^* & 0^* & 0^* & 1.0^* & 0.8 & 0.8 \end{bmatrix}$$

and $\boldsymbol{\Psi}_1 = 0.8\mathbf{I}_9$, $\boldsymbol{\mu} = (0, \ldots, 0)^T$; $\boldsymbol{\Lambda}_2$ has the same structure as $\boldsymbol{\Lambda}_1$ except the true values are all 0.6 instead of 0.8, and $\boldsymbol{\Psi}_2 = 0.4\mathbf{I}_9$. As usual, parameters with an asterisk are fixed at the preassigned values. On the basis of the structures in $\boldsymbol{\Lambda}_1$ and $\boldsymbol{\Lambda}_2$, there are three latent variables in the within-groups and between-groups models. These latent variables are denoted as $\{\eta_{1gi}, \xi_{1gi1}, \xi_{1gi2}\}$ and $\{\eta_{2g}, \xi_{2g1}, \xi_{2g2}\}$, respectively. The latent variables are respectively related by the following structural equations:

$$\eta_{1gi} = \gamma_{11}\xi_{1gi1} + \gamma_{12}\xi_{1gi2} + \gamma_{13}\xi_{1gi1}^2 + \gamma_{14}\xi_{1gi1}\xi_{1gi2} + \gamma_{15}\xi_{1gi2}^2 + \delta_{1gi}, \quad (9.9)$$

$$\eta_{2g} = \gamma_{21}\xi_{2g1} + \gamma_{22}\xi_{2g2} + \delta_{2g}, \quad (9.10)$$

and the following specifications: $\mathbf{\Gamma}_{1g} = \mathbf{\Gamma}_1 = (\gamma_{11}, \ldots, \gamma_{15}) = (0.5, 0.5, 0.8, 0.8, 0.8)$ and $\mathbf{\Gamma}_2 = (\gamma_{21}, \gamma_{22}) = (0.6, 0.6)$. True values of parameters in the covariance matrices $\mathbf{\Phi}_{1g} = \mathbf{\Phi}_1$, $\mathbf{\Psi}_{1g\delta} = \psi_{1\delta}$, $\mathbf{\Phi}_2$ and $\psi_{2\delta}$ are respectively given by $\phi_{1,11} = \phi_{1,22} = 1.0$, $\phi_{1,21} = 0.5$, $\psi_{1\delta} = 0.8$, $\phi_{2,11} = \phi_{2,22} = 1.0$, $\phi_{2,21} = 0.3$ and $\psi_{2\delta} = 0.7$. The between-groups structural equation is linear, and the within-groups structural equation is nonlinear with interaction and quadratic terms of latent variables. Random observations $\mathbf{u}_{gi}$ are simulated from $G = 180$ groups according to Equations (9.1) and (9.2) with the following unbalanced design: $N_g = 10$ for $g = 1, \ldots, 60$, $N_g = 15$ for $g = 61, \ldots, 120$, and $N_g = 20$ for $g = 121, \ldots, 180$. The last three continuous measurements are transformed to ordered categorical data according to the following true thresholds: $\boldsymbol{\alpha}_{g1} = \boldsymbol{\alpha}_{g2} = \boldsymbol{\alpha}_{g3} = \boldsymbol{\alpha}_1 = \boldsymbol{\alpha}_2 = \boldsymbol{\alpha}_3 = (-1.0^*, -0.6, 0.6, 1.0^*)$. In the analysis, ψ_{27}, ψ_{28} and ψ_{29} that correspond to the ordered categorical variables are fixed at 0.4. There are a total of 57 parameters in this model.

In the Bayesian estimation, prior distributions are selected from the conjugate families as given in Appendix 9.1. As the within-groups parameters are invariant over groups, their prior distributions are given by Equation (A9.8). To give a sensitivity analysis about the prior inputs of the hyperparameters in the prior distributions, we perturb the prior inputs as follows:

Type I: $\boldsymbol{\mu}_0 = 0.2\mathbf{J}_9$, where $\mathbf{J}_9$ is a 9×1 vector with elements 1.0, $\mathbf{\Lambda}_{01k}$, $\mathbf{\Lambda}^*_{01k}$, $\mathbf{\Lambda}_{02k}$ and $\mathbf{\Lambda}^*_{02k}$ are equal to half of the true population values, $\alpha_{01k} = \alpha_{02k} = \alpha_{01\delta k} = \alpha_{02\delta k} = 5$, $\beta_{01k} = \beta_{02k} = \beta_{01\delta k} = \beta_{02\delta k} = 4$, $\rho_{01} = \rho_{02} = 3$, $\mathbf{R}_{01}^{-1} = \mathbf{R}_{02}^{-1} = 5.0\mathbf{I}_2$, $\mathbf{H}_{01k}$, $\mathbf{H}^*_{01k}$, $\mathbf{H}_{02k}$, $\mathbf{H}^*_{02k}$ and $\mathbf{\Sigma}_0$ are equal to identity matrices of appropriate orders.

Type II: $\boldsymbol{\mu}_0 = 2.0\mathbf{J}_9$, $\mathbf{\Lambda}_{01k}$, $\mathbf{\Lambda}^*_{01k}$, $\mathbf{\Lambda}_{02k}$ and $\mathbf{\Lambda}^*_{02k}$ are equal to twice the true population values, α_{01k}, α_{02k}, $\alpha_{01\delta k}$, $\alpha_{02\delta k}$, β_{01k}, β_{02k}, $\beta_{01\delta k}$, $\beta_{02\delta k}$, ρ_{01} and ρ_{02} are equal to four times the values given in Type I, while the other hyperparameter values are the same as those given in Type I.

The Gibbs sampler coupled with the MH algorithm is used to produce Bayesian estimates under the two prior inputs in 100 replications. In the MH algorithm, we set $\sigma^2 = 0.68$ to give an approximate acceptance rate of 0.47. The convergence of the algorithm is monitored by the 'estimated potential scale reduction (EPSR)' values suggested by Gelman and Rubin (1992), and by plots of simulated sequences of the individual parameters. At convergence, the EPSR values should be less than 1.2 and the parallel sequences generated by different starting values should be mixed well together. We initially conduct a few test runs of the MCMC algorithm to decide the 'burn-in' phase and observe that it converged in less than 1000 iterations. Hence, we take a burn-in-phase of 1000 iterations and collected $T^* = 4000$ iterations after convergence to produce the

Bayesian estimates and the related statistics. The mean (MEAN) and the root mean squares (RMS) are computed on the basis of the results obtained from the 100 replications. Results are presented in Tables 9.1 and 9.2. These tables show close agreement between the Bayesian estimates under the different prior

Table 9.1 Bayesian estimates of the structural parameters (Str. Par.) and thresholds: first simulation study, Type I prior inputs.

Parameter	Within group		Parameter	Between group	
	MEAN	RMS		MEAN	RMS
Str. Par.			Str. Par.		
$\lambda_{1,21}=0.8$	0.800	0.007	$\lambda_{2,21}=0.6$	0.609	0.048
$\lambda_{1,31}=0.8$	0.800	0.007	$\lambda_{2,31}=0.6$	0.605	0.048
$\lambda_{1,52}=0.8$	0.800	0.024	$\lambda_{2,52}=0.6$	0.589	0.066
$\lambda_{1,62}=0.8$	0.801	0.024	$\lambda_{2,62}=0.6$	0.596	0.064
$\lambda_{1,83}=0.8$	0.796	0.038	$\lambda_{2,83}=0.6$	0.602	0.071
$\lambda_{1,93}=0.8$	0.800	0.035	$\lambda_{2,93}=0.6$	0.608	0.077
$\gamma_{11}=0.5$	0.487	0.064	$\gamma_{21}=0.6$	0.583	0.104
$\gamma_{12}=0.5$	0.486	0.058	$\gamma_{22}=0.6$	0.600	0.101
$\gamma_{13}=0.8$	0.810	0.057	$\psi_{21}=0.4$	0.416	0.080
$\gamma_{14}=0.8$	0.781	0.099	$\psi_{22}=0.4$	0.415	0.057
$\gamma_{15}=0.8$	0.797	0.089	$\psi_{23}=0.4$	0.414	0.048
$\psi_{11}=0.6$	0.603	0.028	$\psi_{24}=0.4$	0.396	0.065
$\psi_{12}=0.6$	0.601	0.025	$\psi_{25}=0.4$	0.398	0.050
$\psi_{13}=0.6$	0.606	0.023	$\psi_{26}=0.4$	0.415	0.051
$\psi_{14}=0.6$	0.602	0.024	$\psi_{\delta 2}=0.7$	0.706	0.125
$\psi_{15}=0.6$	0.603	0.021	$\phi_{2,11}=1.0$	1.067	0.155
$\psi_{16}=0.6$	0.607	0.022	$\phi_{2,12}=0.3$	0.322	0.109
$\psi_{17}=0.6$	0.612	0.047	$\phi_{2,22}=1.0$	1.094	0.215
$\psi_{18}=0.6$	0.606	0.034	$\mu_1=0.0$	−0.098	0.160
$\psi_{19}=0.6$	0.611	0.033	$\mu_2=0.0$	−0.064	0.104
$\psi_{\delta 1}=0.8$	0.791	0.032	$\mu_3=0.0$	−0.064	0.111
$\phi_{1,11}=1.0$	0.995	0.045	$\mu_4=0.0$	−0.067	0.115
$\phi_{1,12}=0.5$	0.501	0.037	$\mu_5=0.0$	−0.039	0.078
$\phi_{1,22}=1.0$	1.015	0.078	$\mu_6=0.0$	−0.038	0.084
			$\mu_7=0.0$	−0.053	0.107
			$\mu_8=0.0$	−0.041	0.079
			$\mu_9=0.0$	−0.037	0.077
Thresholds					
$\alpha_{72}=-0.6$	−0.597	0.023	$\alpha_{73}=0.6$	0.602	0.023
$\alpha_{82}=-0.6$	−0.600	0.020	$\alpha_{83}=0.6$	0.601	0.021
$\alpha_{92}=-0.6$	−0.600	0.021	$\alpha_{93}=0.6$	0.602	0.021

Table 9.2 Bayesian estimates of the structural parameters (Str. Par.) and thresholds: first simulation study, Type II prior inputs.

Parameter	Within group		Parameter	Between group	
	MEAN	RMS		MEAN	RMS
Str. Par.			Str. Par.		
$\lambda_{1,21} = 0.8$	0.801	0.007	$\lambda_{2,21} = 0.6$	0.611	0.048
$\lambda_{1,31} = 0.8$	0.800	0.007	$\lambda_{2,31} = 0.6$	0.607	0.046
$\lambda_{1,52} = 0.8$	0.804	0.025	$\lambda_{2,52} = 0.6$	0.612	0.068
$\lambda_{1,62} = 0.8$	0.804	0.024	$\lambda_{2,62} = 0.6$	0.621	0.066
$\lambda_{1,83} = 0.8$	0.810	0.039	$\lambda_{2,83} = 0.6$	0.628	0.083
$\lambda_{1,93} = 0.8$	0.814	0.038	$\lambda_{2,93} = 0.6$	0.637	0.090
$\gamma_{11} = 0.5$	0.506	0.065	$\gamma_{21} = 0.6$	0.596	0.106
$\gamma_{12} = 0.5$	0.517	0.061	$\gamma_{22} = 0.6$	0.633	0.113
$\gamma_{13} = 0.8$	0.793	0.064	$\psi_{21} = 0.4$	0.411	0.046
$\gamma_{14} = 0.8$	0.826	0.100	$\psi_{22} = 0.4$	0.413	0.044
$\gamma_{15} = 0.8$	0.812	0.084	$\psi_{23} = 0.4$	0.413	0.037
$\psi_{11} = 0.6$	0.608	0.028	$\psi_{24} = 0.4$	0.418	0.047
$\psi_{12} = 0.6$	0.603	0.024	$\psi_{25} = 0.4$	0.396	0.038
$\psi_{13} = 0.6$	0.609	0.023	$\psi_{26} = 0.4$	0.409	0.038
$\psi_{14} = 0.6$	0.607	0.025	$\psi_{\delta 2} = 0.7$	0.736	0.086
$\psi_{15} = 0.6$	0.604	0.020	$\phi_{2,11} = 1.0$	0.943	0.144
$\psi_{16} = 0.6$	0.610	0.023	$\phi_{2,12} = 0.3$	0.289	0.105
$\psi_{17} = 0.6$	0.620	0.044	$\phi_{2,22} = 1.0$	0.944	0.197
$\psi_{18} = 0.6$	0.612	0.033	$\mu_1 = 0.0$	0.099	0.163
$\psi_{19} = 0.6$	0.618	0.036	$\mu_2 = 0.0$	0.068	0.108
$\psi_{\delta 1} = 0.8$	0.799	0.061	$\mu_3 = 0.0$	0.066	0.117
$\phi_{1,11} = 1.0$	0.983	0.048	$\mu_4 = 0.0$	0.051	0.107
$\phi_{1,12} = 0.5$	0.491	0.038	$\mu_5 = 0.0$	0.037	0.078
$\phi_{1,22} = 1.0$	0.977	0.077	$\mu_6 = 0.0$	0.039	0.084
			$\mu_7 = 0.0$	0.065	0.115
			$\mu_8 = 0.0$	0.037	0.079
			$\mu_9 = 0.0$	0.042	0.081
Thresholds					
$\alpha_{72} = -0.6$	-0.596	0.023	$\alpha_{73} = 0.6$	0.602	0.022
$\alpha_{82} = -0.6$	-0.600	0.020	$\alpha_{83} = 0.6$	0.602	0.021
$\alpha_{92} = -0.6$	-0.600	0.022	$\alpha_{93} = 0.6$	0.603	0.021

inputs. Hence, under the given sample size, the proposed Bayesian estimation is not sensitive to these two different prior inputs. We also observe the close agreement between the true parameters and the means of Bayesian estimates, and rather small RMS.

In the second simulation, a model with noninvariant within-group parameters is considered. Here, $G = 230$. For $g = 1, \ldots, 150$, we took $N_g = 8$, and the following true parameter values:

$$\mathbf{\Lambda}_{1g}^T = \begin{bmatrix} 1.0^* & 1.2 & 0^* & 0^* & 0^* & 0^* \\ 0^* & 0^* & 1.0^* & 1.2 & 0^* & 0^* \\ 0^* & 0^* & 0^* & 0^* & 1.0^* & 1.2 \end{bmatrix}, \text{ and } \mathbf{\Psi}_{1g} = 0.8\mathbf{I}_6;$$

whilst for $g = 151, \ldots, 230$, we took $N_g = 10$,

$$\mathbf{\Lambda}_{1g}^T = \begin{bmatrix} 1.0^* & 0.6 & 0^* & 0^* & 0^* & 0^* \\ 0^* & 0^* & 1.0^* & 0.6 & 0^* & 0^* \\ 0^* & 0^* & 0^* & 0^* & 1.0^* & 0.6 \end{bmatrix}, \text{ and } \mathbf{\Psi}_{1g} = 0.36\mathbf{I}_6.$$

Here, there are three latent variables: η_{1g}, ξ_{1g1} and ξ_{1g2}. These latent variables are related by the following nonlinear structural equation

$$\eta_{1gi} = \gamma_{1g1}\xi_{1gi1} + \gamma_{1g2}\xi_{1gi2} + \gamma_{1g3}\xi_{1gi1}\xi_{1gi2} + \delta_{1gi}, \tag{9.11}$$

where for $g = 1, \ldots, 150$, the true values of the parameters are $\mathbf{\Gamma}_{1g} = (0.5, 0.5, 0.6)$, $\psi_{1\delta g} = 0.8$, $\phi_{1g,11} = \phi_{1g,22} = 1.0$ and $\phi_{1g,21} = 0.3$; while for $g = 151, \ldots, 230$, $\mathbf{\Gamma}_{1g} = (0.3, 0.3, 0.4)$, $\psi_{1\delta g} = 0.36$, $\mathbf{\Phi}_{1g}$ is equal to the matrix just given above. We consider a correlated factor analysis model with $\boldsymbol{\mu} = \mathbf{0}$ in the between-groups level. The structure of $\mathbf{\Lambda}_2$ is the same as the $\mathbf{\Lambda}_{1g}$ specified above except the true value of the free loading parameters is 0.8, the variances of the latent factors are all 0.7, the covariances among all pairs of factors are 0.15, and the unique variances are 0.3. The fourth and sixth continuous measurements of observations are transformed to ordered categorical through the following thresholds: $\boldsymbol{\alpha}_{g1} = \boldsymbol{\alpha}_1 = (-1.0^*, -0.5, 0.3, 1.0^*)$ and $\boldsymbol{\alpha}_{g2} = \boldsymbol{\alpha}_2 = (-1.0^*, -0.3, 0.5, 1.0^*)$. In the analysis, ψ_{24} and ψ_{26} are fixed at 0.3. There are a total of 55 unknown parameters.

Based on the results obtained from the first simulation about the sensitivity of prior inputs in Bayesian estimation, only one set of hyperparameter values was considered. These values are given by: $\boldsymbol{\mu}_0 = 0.0\mathbf{J}_6$, $\mathbf{\Lambda}_{01gk}$, $\mathbf{\Lambda}_{01gk}^*$, $\mathbf{\Lambda}_{02k}$ and $\mathbf{\Lambda}_{02k}^*$, are the true population values, $\mathbf{H}_{01gk}$, $\mathbf{H}_{01gk}^*$, $\mathbf{H}_{02k}$ and $\mathbf{H}_{02k}^*$ are identity matrices of appropriate orders, $\alpha_{01gk} = \alpha_{01g\delta k} = 6$, $\beta_{01gk} = \beta_{01g\delta k} = 4$, $\alpha_{02k} = \alpha_{02\delta k} = 10$, $\beta_{02k} = \beta_{02\delta k} = 3$, $\rho_{01g} = \rho_{02} = 6$, $\mathbf{R}_{01g}^{-1} = \mathbf{R}_{02}^{-1} = 5.0\mathbf{I}_2$. Again, 100 replications are completed via the MCMC algorithm. The σ^2 in the MH algorithm is taken to be 0.68, giving an approximate acceptance rate of 0.52. On the basis of the results obtained from some test runs, we took a burn-in-phase of 500 iterations and collected 3000 observations after convergence to produce the Bayesian solution. Results are presented in Table 9.3. Again, we observe close agreement between the true parameters and the means of Bayesian estimates, and small RMS.

Table 9.3 Bayesian estimates of the structural parameters (Str. Par.) and thresholds: second simulation study.

Within-group 1, $g = 1, \ldots, 150$			Within-group 2, $g = 151, \ldots, 230$		
Str. Par.	MEAN	RMS	Str. Par.	MEAN	RMS
$\lambda_{1g,21} = 1.2$	1.206	0.051	$\lambda_{1g,21} = 0.6$	0.596	0.044
$\lambda_{1g,42} = 1.2$	1.214	0.089	$\lambda_{1g,42} = 0.6$	0.628	0.051
$\lambda_{1g,63} = 1.2$	1.209	0.094	$\lambda_{1g,63} = 0.6$	0.626	0.050
$\gamma_{1g1} = 0.5$	0.500	0.053	$\gamma_{1g1} = 0.3$	0.305	0.038
$\gamma_{1g2} = 0.5$	0.496	0.058	$\gamma_{1g2} = 0.3$	0.306	0.048
$\gamma_{1g3} = 0.6$	0.608	0.056	$\gamma_{1g3} = 0.4$	0.420	0.056
$\psi_{1g1} = 0.8$	0.809	0.069	$\psi_{1g1} = 0.8$	0.809	0.040
$\psi_{1g2} = 0.8$	0.783	0.098	$\psi_{1g2} = 0.8$	0.783	0.024
$\psi_{1g3} = 0.8$	0.794	0.073	$\psi_{1g3} = 0.8$	0.794	0.070
$\psi_{1g4} = 0.8$	0.814	0.104	$\psi_{1g4} = 0.8$	0.814	0.035
$\psi_{1g5} = 0.8$	0.805	0.063	$\psi_{1g5} = 0.8$	0.805	0.068
$\psi_{1g6} = 0.8$	0.803	0.098	$\psi_{1g6} = 0.8$	0.803	0.031
$\psi_{\delta 1g} = 0.8$	0.777	0.086	$\psi_{\delta 1g} = 0.36$	0.366	0.042
$\phi_{1g,11} = 1.0$	1.004	0.087	$\phi_{1g,11} = 1.0$	0.956	0.094
$\phi_{1g,12} = 0.3$	0.305	0.046	$\phi_{1g,12} = 0.3$	0.291	0.047
$\phi_{1g,22} = 1.0$	0.998	0.089	$\phi_{1g,22} = 1.0$	0.965	0.091
Between-group					
$\mu_1 = 0.0$	−0.015	0.081	$\phi_{2,11} = 0.7$	0.690	0.116
$\mu_2 = 0.0$	−0.014	0.066	$\phi_{2,12} = .15$	0.137	0.070
$\mu_3 = 0.0$	−0.015	0.073	$\phi_{2,13} = .15$	0.142	0.074
$\mu_4 = 0.0$	−0.022	0.074	$\phi_{2,22} = 0.7$	0.717	0.108
$\mu_5 = 0.0$	−0.001	0.078	$\phi_{2,23} = .15$	0.140	0.064
$\mu_6 = 0.0$	0.000	0.068	$\phi_{2,33} = 0.7$	0.722	0.111
$\lambda_{2,21} = 0.8$	0.832	0.076	$\psi_{21} = 0.3$	0.318	0.045
$\lambda_{2,42} = 0.8$	0.812	0.092	$\psi_{22} = 0.3$	0.296	0.037
$\lambda_{2,63} = 0.8$	0.819	0.084	$\psi_{23} = 0.3$	0.313	0.050
			$\psi_{25} = 0.3$	0.309	0.050
Thresholds					
$\alpha_{42} = -0.5$	−0.498	0.025	$\alpha_{43} = 0.3$	0.299	0.028
$\alpha_{62} = -0.3$	−0.301	0.026	$\alpha_{63} = 0.5$	0.496	0.024

9.4 GOODNESS-OF-FIT AND MODEL COMPARISON

Assessing the goodness-of-fit of a hypothesized model is an important issue in SEM. In the Bayesian approach, the posterior predictive (PP) p-value (Gelman, Meng and Stern, 1996, see also Appendix 5.2) can be used as a goodness-of-fit assessment for a hypothesized two-level nonlinear SEM with the mixed type data. A brief description of the PP p-values in the context of the current model is given in Appendix 9.3. As the PP p-value is not suitable for comparing two competing models M_0 and M_1 (see Carlin and Louis, 1996), the following Bayes factor (see Kass and Raftery, 1995) is used for comparing M_0 and M_1:

$$B_{10} = \frac{p(\mathbf{X}, \mathbf{Z}|M_1)}{p(\mathbf{X}, \mathbf{Z}|M_0)}.$$

In general, it is well known that the computation of B_{10} is nontrivial. This is particularly true for the current nonlinear two-level model which includes a large number of parameters and latent measurements, and latent variables. Inspired by the good features of path sampling (Gelman and Meng, 1998), we will use it to compute B_{10}.

In the application of path sampling in computing B_{10}, we again use the data augmentation idea to augment $(\mathbf{X}, \mathbf{Z})$ with $(\mathbf{Y}, \mathbf{V}, \boldsymbol{\Omega}_1, \boldsymbol{\Omega}_2)$ in the analysis. Consider the following class of densities defined by a continuous parameter t in [0,1]:

$$p(\boldsymbol{\theta}, \boldsymbol{\alpha}, \mathbf{Y}, \mathbf{V}, \boldsymbol{\Omega}_1, \boldsymbol{\Omega}_2|\mathbf{X}, \mathbf{Z}, t) = \frac{p(\boldsymbol{\theta}, \boldsymbol{\alpha}, \mathbf{Y}, \mathbf{V}, \boldsymbol{\Omega}_1, \boldsymbol{\Omega}_2, \mathbf{X}, \mathbf{Z}|t)}{z(t)},$$

where $z(t) = p(\mathbf{X}, \mathbf{Z}|t)$. Let t in $[0, 1]$ be a parameter linking the competing models M_0 and M_1 such that for $a = 0, 1$, $z(a) = p(\mathbf{X}, \mathbf{Z}|t = a) = p(\mathbf{X}, \mathbf{Z}|M_a)$, then $B_{10} = z(1)/z(0)$. Taking logarithms and differentiating $z(t)$ with respect to t, it can be shown by reasoning similar to that in Section 5.4 that:

$$\log B_{10} = \frac{1}{2}\sum_{s=0}^{S}(t_{(s+1)} - t_{(s)})(\bar{U}_{(s+1)} + \bar{U}_{(s)}), \tag{9.12}$$

where $t_{(0)} = 0 < t_{(1)} < \cdots < t_{(S)} < t_{(S+1)} = 1$ are fixed grids in [0,1] and

$$\bar{U}_{(s)} = \frac{1}{J}\sum_{j=1}^{J} U(\boldsymbol{\theta}^{(j)}, \boldsymbol{\alpha}^{(j)}, \mathbf{Y}^{(j)}, \mathbf{V}^{(j)}, \boldsymbol{\Omega}_1^{(j)}, \boldsymbol{\Omega}_2^{(j)}, \mathbf{X}, \mathbf{Z}, t_{(s)}), \tag{9.13}$$

in which $\{(\boldsymbol{\theta}^{(j)}, \boldsymbol{\alpha}^{(j)}, \mathbf{Y}^{(j)}, \mathbf{V}^{(j)}, \boldsymbol{\Omega}_1^{(j)}, \boldsymbol{\Omega}_2^{(j)}) : \text{j=1}, \ldots, J\}$ is a sample of observations simulated from the joint posterior distribution $[\boldsymbol{\theta}, \boldsymbol{\alpha}, \mathbf{Y}, \mathbf{V}, \boldsymbol{\Omega}_1, \boldsymbol{\Omega}_2|\mathbf{X}, \mathbf{Z}, t_{(s)}]$, and

$$U(\boldsymbol{\theta}, \boldsymbol{\alpha}, \mathbf{Y}, \mathbf{V}, \boldsymbol{\Omega}_1, \boldsymbol{\Omega}_2, \mathbf{X}, \mathbf{Z}, t) = \mathrm{d}\log p(\mathbf{Y}, \mathbf{V}, \boldsymbol{\Omega}_1, \boldsymbol{\Omega}_2, \mathbf{X}, \mathbf{Z}|\boldsymbol{\theta}, \boldsymbol{\alpha}, t)/\mathrm{d}t, \tag{9.14}$$

where $p(\mathbf{Y}, \mathbf{V}, \boldsymbol{\Omega}_1, \boldsymbol{\Omega}_2, \mathbf{X}, \mathbf{Z}|\boldsymbol{\theta}, \boldsymbol{\alpha}, t)$ is the complete data likelihood. Note that this complete data likelihood is not complicated and obtaining the function U through differentiation is not difficult. Moreover, the program implemented in estimation can be used for simulating observations in Equation (9.13), hence there is little additional programming effort required. Usually, $S = 10$ grids is sufficient for providing a good approximation of the logarithm B_{10} for competing models which are not far apart. More grids are required for very different M_1 and M_0, and the issue should be approached on a problem-by-problem basis. In Equation (9.13), a value of $J = 2000$ is usually enough for most practical applications.

An important step in applying path sampling for computing logarithm B_{12} is to find a good path t in $[0, 1]$ to link the competing models M_1 and M_2. Because the two-level nonlinear SEM is rather complex, M_1 and M_2 can be quite different and finding a path to link them may require some insight. Two illustrative examples are discussed as follows.

Example1:

$$M_k: \quad \mathbf{u}_{gi} = \mathbf{v}_g + \boldsymbol{\Lambda}_1^k \boldsymbol{\omega}_{1gi} + \boldsymbol{\epsilon}_{1gi}, \quad k = 1, 2 \tag{9.15}$$

$$\mathbf{v}_g = \boldsymbol{\mu} + \boldsymbol{\Lambda}_2^k \boldsymbol{\omega}_{2g} + \boldsymbol{\epsilon}_{2g}, \tag{9.16}$$

$$\boldsymbol{\eta}_{1gi} = \boldsymbol{\Pi}_1^k \boldsymbol{\eta}_{1gi} + \boldsymbol{\Gamma}_1^k \mathbf{H}_1^k(\boldsymbol{\xi}_{1gi}) + \boldsymbol{\delta}_{1gi}, \tag{9.17}$$

$$\boldsymbol{\eta}_{2g} = \boldsymbol{\Pi}_2^k \boldsymbol{\eta}_{2g} + \boldsymbol{\Gamma}_2^k \mathbf{H}_2^k(\boldsymbol{\xi}_{2g}) + \boldsymbol{\delta}_{2g}, \tag{9.18}$$

where the superscript k in the parameters and functions is used for denoting the model M_k, $k = 1, 2$. M_1 and M_2 can be linked via the following M_{t12} with t in $[0, 1]$:

$$\begin{aligned}
M_{t12}: \quad \mathbf{u}_{gi} &= \mathbf{v}_g + \{t\boldsymbol{\Lambda}_1^1 + (1-t)\boldsymbol{\Lambda}_1^2\}\boldsymbol{\omega}_{1gi} + \boldsymbol{\epsilon}_{1gi}, \\
\mathbf{v}_g &= \boldsymbol{\mu} + \{t\boldsymbol{\Lambda}_2^1 + (1-t)\boldsymbol{\Lambda}_2^2\}\boldsymbol{\omega}_{2g} + \boldsymbol{\epsilon}_{2g}, \\
\boldsymbol{\eta}_{1gi} &= [t\boldsymbol{\Pi}_1^1 + (1-t)\boldsymbol{\Pi}_1^2]\boldsymbol{\eta}_{1gi} + [t\boldsymbol{\Gamma}_1^1\mathbf{H}_1^1(\boldsymbol{\xi}_{1gi}) + (1-t)\boldsymbol{\Gamma}_1^2\mathbf{H}_1^2(\boldsymbol{\xi}_{1gi})] + \boldsymbol{\delta}_{1gi}, \\
\boldsymbol{\eta}_{2g} &= [t\boldsymbol{\Pi}_2^1 + (1-t)\boldsymbol{\Pi}_2^2]\boldsymbol{\eta}_{2g} + [t\boldsymbol{\Gamma}_2^1\mathbf{H}_2^1(\boldsymbol{\xi}_{2g}) + (1-t)\boldsymbol{\Gamma}_2^2\mathbf{H}_2^2(\boldsymbol{\xi}_{2g})] + \boldsymbol{\delta}_{2g}.
\end{aligned}$$

When $t = 1$, M_{t12} reduces to M_1; when $t = 0$, M_{t12} reduces to M_2. From Equations (9.12) and (9.13), logarithm B_{12} can be computed. Hence, general models involving different matrix coefficients in the measurement equations, and different forms of nonlinear structural equations, can be easily compared.

Example 2: The competing models M_1 and M_2 have the following within-groups measurement and structural equations which are defined in a similar way to Equations (9.15) and (9.16):

$$\mathbf{u}_{gi} = \mathbf{v}_g + \mathbf{\Lambda}_1 \boldsymbol{\omega}_{1gi} + \boldsymbol{\epsilon}_{1gi}, \tag{9.19}$$

$$\boldsymbol{\eta}_{1gi} = \mathbf{\Pi}_1 \boldsymbol{\eta}_{1gi} + \mathbf{\Gamma}_1 \mathbf{H}_1(\boldsymbol{\xi}_{1gi}) + \boldsymbol{\delta}_{1gi}. \tag{9.20}$$

The difference between M_1 and M_2 is on the between-groups models. Let

$$M_1: \quad \mathbf{v}_g = \boldsymbol{\mu} + \mathbf{\Lambda}_2^1 \boldsymbol{\omega}_{2g} + \boldsymbol{\epsilon}_{2g},$$

where $\boldsymbol{\omega}_{2g}$ is distributed as $N[\mathbf{0}, \mathbf{\Phi}_2]$. Thus, the between-groups model in M_1 is a factor analysis model. In M_2, $\boldsymbol{\omega}_{2g} = (\boldsymbol{\eta}_{2g}^T, \boldsymbol{\xi}_{2g}^T)^T$, and the measurement and structural equations in the between-groups model are given as follows:

$$M_2: \quad \mathbf{v}_g = \boldsymbol{\mu} + \mathbf{\Lambda}_2^2 \boldsymbol{\omega}_{2g} + \boldsymbol{\epsilon}_{2g}, \tag{9.21}$$

$$\boldsymbol{\eta}_{2g} = \mathbf{\Pi}_2^2 \boldsymbol{\eta}_{2g} + \mathbf{\Gamma}_2^2 \mathbf{H}_2(\boldsymbol{\xi}_{2g}) + \boldsymbol{\delta}_{2g}. \tag{9.22}$$

The between-groups model in M_2 is an NSEM with a nonlinear structural equation. Note that M_1 and M_2 are non-nested. As there are two different models for $\boldsymbol{\omega}_{2g}$, it is rather difficult to directly link M_1 and M_2. This difficulty can be solved via an auxiliary model M_a which can be linked with both M_1 and M_2. We first compute $\log B_{1a}$ and $\log B_{2a}$, and then obtain $\log B_{12}$ via the following equation:

$$\log B_{12} = \log \frac{p(\mathbf{X}, \mathbf{Z}|M_1)/p(\mathbf{X}, \mathbf{Z}|M_a)}{p(\mathbf{X}, \mathbf{Z}|M_2)/p(\mathbf{X}, \mathbf{Z}|M_a)} = \log B_{1a} - \log B_{2a}. \tag{9.23}$$

For our current problem, one auxiliary model is M_a in which the measurement and structural equations of the within-groups model are given by Equations (9.19) and (9.20), while the between-groups model is defined by $\mathbf{v}_g = \boldsymbol{\mu} + \boldsymbol{\epsilon}_{2g}$. The link model M_{t1a} is defined by $M_{t1a}: \mathbf{u}_{gi} = \boldsymbol{\mu} + t\mathbf{\Lambda}_2^1 \boldsymbol{\omega}_{2g} + \boldsymbol{\epsilon}_{2g} + \mathbf{\Lambda}_1 \boldsymbol{\omega}_{1gi} + \boldsymbol{\epsilon}_{2gi}$, with within-groups structural equations given by Equation (9.20), where $\boldsymbol{\omega}_{2g}$ is distributed as $N[\mathbf{0}, \mathbf{\Phi}]$ and without a between-groups structural equation. Clearly, $t = 1$ and 0 corresponds to M_1 and M_a, respectively. Hence, $\log B_{1a}$ can be computed under this setting via the path sampling procedure. The link model M_{t2a} is defined by $M_{t2a}: \mathbf{u}_{gi} = \boldsymbol{\mu} + t\mathbf{\Lambda}_2^2 \boldsymbol{\omega}_{2g} + \boldsymbol{\epsilon}_{2g} + \mathbf{\Lambda}_1 \boldsymbol{\omega}_{1gi} + \boldsymbol{\epsilon}_{1gi}$, with the within-groups and between-groups structural equations respectively given by Equations (9.20) and (9.22). Clearly, $t = 1$ and 0 corresponds to M_2

and M_a. Hence, $\log B_{2a}$ can be obtained. Finally, $\log B_{12}$ can be obtained from $\log B_{1a}$ and $\log B_{2a}$ via Equation (9.23).

In general, just one auxiliary model may not be adequate to link two very different M_1 and M_2. However, based on the key idea of the above example, the difficulty can be solved by using more than one appropriate auxiliary model $M_a, M_b, \ldots$ in between M_1 and M_2. For example, suppose we use M_a and M_b to link M_1 and M_2, with M_a closer to M_1. Then

$$\frac{p(\mathbf{X},\mathbf{Z}|M_1)}{p(\mathbf{X},\mathbf{Z}|M_2)} = \frac{p(\mathbf{X},\mathbf{Z}|M_1)/p(\mathbf{X},\mathbf{Z}|M_a)}{p(\mathbf{X},\mathbf{Z}|M_2)/p(\mathbf{X},\mathbf{Z}|M_a)} \quad \text{and}$$
$$\frac{p(\mathbf{X},\mathbf{Z}|M_2)}{p(\mathbf{X},\mathbf{Z}|M_a)} = \frac{p(\mathbf{X},\mathbf{Z}|M_2)/p(\mathbf{X},\mathbf{Z}|M_b)}{p(\mathbf{X},\mathbf{Z}|M_a)/p(\mathbf{X},\mathbf{Z}|M_b)}; \tag{9.24}$$

hence, $\log B_{12} = \log B_{1a} + \log B_{ab} - \log B_{2b}$. Each logarithm Bayes factor can be computed via path sampling.

Similar to the goodness-of-fit assessment in the context of single-level nonlinear SEMs, it is rather difficult to find a saturated model for the two-level nonlinear SEMs. However, the goodness-of-fit of a proposed model can be assessed by means of the PP p-value, and the estimated residual plots.

9.5 AN APPLICATION: FILIPINA CSWs STUDY

As an illustration of the proposed methodology, we use a small portion of the data set in the study of Morisky *et al.* (1998) on the effects of establishment policies, knowledge and attitudes on condom use among Filipina commercial sex workers (CSWs). It has been argued that the nature of commercial sex work promotes the spread of AIDS and other sexually transmitted diseases; thus promotion of safer sexual practice among CSWs is important. The study of Morisky *et al.* (1998) concerned the development and preliminary findings from an AIDS preventative intervention for Filipina CSWs. The data set was collected from female CSWs in establishments (bars, night clubs, Karaoke TV and massage parlours) in cities of the Philippines. The whole questionnaire consisted of 134 items on areas of demographics knowledge, attitudes, beliefs, behaviors, self-efficacy for condom use and social desirability. Latent psychological determinants such as CSWs' risk behaviors, knowledge and attitudes associated with AIDS and condom use are important issues to be assessed. For instance, a basic concern is to explore whether linear relationships among these latent variables are sufficient, or is it better to incorporate nonlinear relationships in the model. The manifest variables that are used as indicators for latent quantities are measured in terms of ordered categorical and continuous scales. Moreover, as emphasized by Morisky *et al.* (1998), establishments' policies on their CSWs condom use practices exert a strong influence on CSWs. Hence, it is interesting to study the influence of the establishment by incorporating a between-group model for the data. As observations within each establishment are correlated, the usual assumption of independence in the standard single-level SEMs is violated. On the basis of the above considerations, it is desirable to develop a two-level nonlinear SEM in the context of mixed ordered categorical and continuous data.

Nine manifest variables, of which the seventh, eighth and ninth variables are continuous and the remaining are ordered categorical with a five-point scale, are selected. Questions corresponding to these variables are given in Appendix 9.4. For brevity, we delete those observations with missing entries in the analysis, and the remaining sample size is 755. There are 97 establishments. The numbers of individuals in establishments varied from 1 to 58 which gives an unbalanced data set. The sample means and standard deviations of the continuous variables are $\{2.442, 1.180, 0.465\}$ and $\{5.299, 2.208, 1.590\}$, respectively. The cell frequencies of the ordered categorical variables are ranged from 12 to 348. To unify scales of variables, the raw continuous data are standardized.

After some preliminary studies and based on the meanings of the questions corresponding to the manifest variables (see Appendix 9.4), in the measurement equations corresponding to the between-groups and within-groups models, we use the first three, the next three, and the last three manifest variables as indicators for latent factors that can be roughly interpreted as 'worry about AIDS', 'attitude to the risk of getting AIDS' and 'aggressiveness'. For the between-groups model, we propose a factor analysis model with the following specifications:

$$\mathbf{\Lambda}_2^T = \begin{bmatrix} 1.0^* & \lambda_{2,21} & \lambda_{2,31} & 0^* & 0^* & 0^* & 0^* & 0^* & 0^* \\ 0^* & 0^* & 0^* & 1.0^* & \lambda_{2,52} & \lambda_{2,62} & 0^* & 0^* & 0^* \\ 0^* & 0^* & 0^* & 0^* & 0^* & 0^* & 1.0^* & \lambda_{2,83} & \lambda_{2,93} \end{bmatrix},$$

$$\mathbf{\Phi}_2 = \begin{bmatrix} \phi_{2,11} & & \text{sym} \\ \phi_{2,21} & \phi_{2,22} & \\ \phi_{2,31} & \phi_{2,32} & \phi_{2,33} \end{bmatrix},$$

and $\mathbf{\Psi}_2 = \text{diag}(0.3, 0.3, 0.3, 0.3, 0.3, 0.3, \psi_{27}, \psi_{28}, \psi_{29})$, where the unique variances corresponding to the ordered categorical variables are fixed at 0.3. Although other structures for $\mathbf{\Lambda}_2$ can be considered, we choose this common form in confirmatory factor analysis (see, for example, Ansari, Jedidi and Dube, 2002; Lee and Zhu, 2000, among others) that gives nonoverlapping latent factors for clear interpretation. These latent factors are allowed to be correlated. For the within-groups model with the latent factors $\{\eta_{1gi}, \xi_{1gi1}, \xi_{1gi2}\}$, we considered invariant within-groups parameters such that $\mathbf{\Psi}_{1g} = \mathbf{\Psi}_1 = \text{diag}(\psi_{11}, \cdots, \psi_{19})$, and $\mathbf{\Lambda}_{1g} = \mathbf{\Lambda}_1$, where $\mathbf{\Lambda}_1$ has the same common structure as $\mathbf{\Lambda}_2$ with unknown loadings $\{\lambda_{1,21}, \lambda_{1,31}, \lambda_{1,52}, \lambda_{1,62}, \lambda_{1,83}, \lambda_{1,93}\}$. However, as the within-groups model is directly related to the CSWs, we wish to consider a more subtle model with a structural equation that accounts for relationships among the latent factors. To assess the interaction effect of the exogenous latent factors, the following structural equation for the latent variables is taken:

$$\eta_{1gi} = \gamma_{11}\xi_{1gi1} + \gamma_{12}\xi_{1gi2} + \gamma_{13}\xi_{1gi1}\xi_{1gi2} + \delta_{1gi}. \tag{9.25}$$

To identify the model with respect to ordered categorical variables via the common method (see, Lee, Poon and Bentler, 1995), α_{k1} and α_{k4}, $k = 1, \cdots, 6$ are fixed at $\alpha_{kj} = \Phi^{*^{-1}}(m_k)$, where Φ^* is the distribution function of $N[0, 1]$, and m_k is the observed cumulative marginal proportion of the categories with $z_{gk} < j$. There are a total of 48 parameters in this two-level nonlinear SEM.

In the Bayesian analysis, we need to specify hyperparameter values in the proper conjugate prior distributions of the unknown parameters. For situations where we have good prior information, for example from closely related data or the knowledge of experts, subjective hyperparameter values should be taken. Under the general situation without good prior information, alternative methods have to be used to fix the hyperparameters. Some Bayesian analyses of SEMs (see e.g. Ansari, Jedidi and Dube, 2002) used vague but proper priors with ad hoc hyperparameter values. Many kinds of data-dependent priors have appeared in Bayesian literature (see Raftery, 1996a,b; Richardson and Green, 1997; Pauler, Wakefield and Kass, 1999; Song and Lee, 2001; Zhu and Lee, 2001, among others). In this illustrative example, we use some data-dependent prior inputs, and ad hoc prior inputs that give rather vague but proper prior distributions. We emphasize that these prior inputs are used for the purpose of illustration only, we are not routinely recommending them for other substantive applications. The data-dependent prior inputs are obtained by conducting an auxiliary Bayesian estimation with proper vague conjugate prior distributions which gives estimates $\tilde{\mathbf{\Lambda}}_{01k}$, $\tilde{\mathbf{\Lambda}}^*_{01k}$, $\tilde{\mathbf{\Lambda}}_{02k}$ and $\tilde{\mathbf{\Lambda}}^*_{02k}$ for some hyperparameter values (according to the notation in Appendix 9.1). Then, results are obtained and compared on the basis of the following types of hyperparameter values:

(I): Hyperparameters $\mathbf{\Lambda}_{01k}$, $\mathbf{\Lambda}^*_{01k}$, $\mathbf{\Lambda}_{02k}$ and $\mathbf{\Lambda}^*_{02k}$ are equal to $\tilde{\mathbf{\Lambda}}_{01k}$, $\tilde{\mathbf{\Lambda}}^*_{01k}$, $\tilde{\mathbf{\Lambda}}_{02k}$ and $\tilde{\mathbf{\Lambda}}^*_{02k}$, respectively; $\mathbf{H}_{01k}$, $\mathbf{H}^*_{01k}$, $\mathbf{H}_{02k}$ and $\mathbf{H}^*_{02k}$ are equal to identity matrices of appropriate orders; $\alpha_{01k} = \alpha_{02k} = \alpha_{01\delta k} = \alpha_{02\delta k} = 10$, $\beta_{01k} = \beta_{02k} = \beta_{01\delta k} = \beta_{02\delta k} = 8$, $\rho_{01} = \rho_{02} = 6$, $\mathbf{R}_{01}^{-1} = 5.0\mathbf{I}_2$ and $\mathbf{R}_{02}^{-1} = 5.0\mathbf{I}_3$.

(II): Hyperparameter values in $\mathbf{\Lambda}_{01k}$, $\mathbf{\Lambda}^*_{01k}$, $\mathbf{\Lambda}_{02k}$ and $\mathbf{\Lambda}^*_{02k}$ are equal to the zeros, $\mathbf{H}_{01k}$, $\mathbf{H}^*_{01k}$, $\mathbf{H}_{02k}$ and $\mathbf{H}^*_{02k}$ are equal to 5.0 times the identity matrices of appropriate orders. Other hyperparameter values are equal to those given in (I). These prior inputs are not data-dependent.

Bayesian estimates are obtained by the proposed algorithm that involves the Gibbs sampler and the MH algorithm. The convergence of this algorithm is

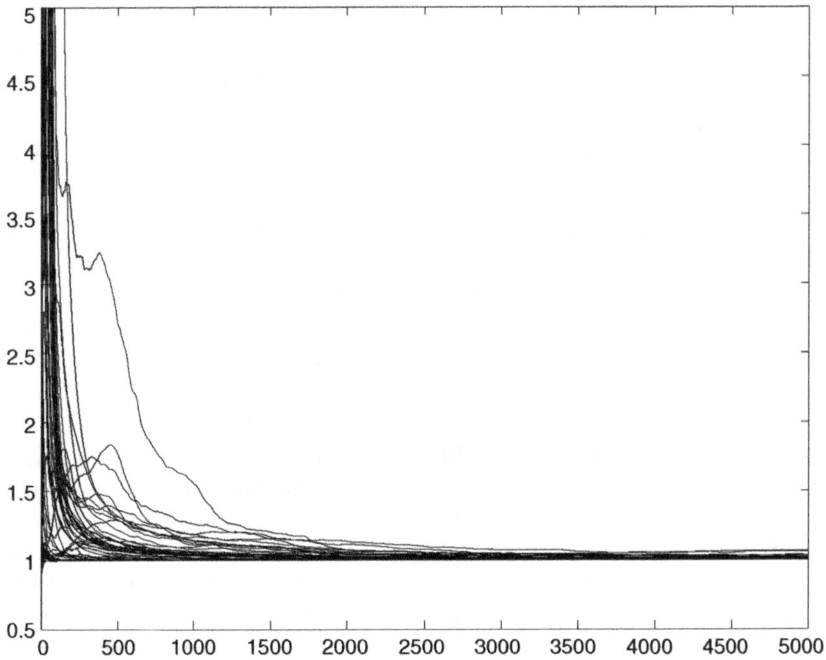

Figure 9.1 EPSR values against the number of iterations in the analysis of AIDS data. This figure is from Song and Lee (2004).

monitored by the (EPSR) values suggested by Gelman and Rubin (1992), and by plots of generated observations obtained with different starting values. To give some idea about the convergence of the MCMC method in analyzing this complex SEM, the EPSR values corresponding to the analysis with type (I) prior inputs are displayed in Figure 9.1. We observe that the algorithm converged in less than 2000 iterations. Hence, we take a burn-in-phase of 2000 iterations, and further collect 3000 observations to produce the Bayesian estimates and their standard error estimates. Results obtained under prior inputs (I) and (II) are reported in Tables 9.4 and 9.5. From these tables, we see that the estimates obtained under these different prior inputs are reasonably close. The PP p-values corresponding to these two sets of estimates are equal to 0.592 and 0.600, which indicate that the proposed model fits the sample data, and this statistic is quite robust to the selected prior inputs under the given sample size of 755.

In order to illustrate the proposed path sampling in computing the Bayes factor for model comparison, we compare this two-level nonlinear model with some non-nested models. Let M_1 be the two-level nonlinear SEM with the above specifications and the nonlinear structural Equation (9.25), and M_2 and

Table 9.4 Bayesian estimates of the structural parameters (Str. Par.) and thresholds under prior (I) for M_1: AIDS data.

	Within-group			Between-group	
	EST	SE		EST	SE
Str. Par.			Str. Par.		
$\lambda_{1,21}$	0.238	0.081	$\lambda_{2,21}$	1.248	0.218
$\lambda_{1,31}$	0.479	0.112	$\lambda_{2,31}$	0.839	0.189
$\lambda_{1,52}$	1.102	0.213	$\lambda_{2,52}$	0.205	0.218
$\lambda_{1,62}$	0.973	0.185	$\lambda_{2,62}$	0.434	0.221
$\lambda_{1,83}$	0.842	0.182	$\lambda_{2,83}$	0.159	0.209
$\lambda_{1,93}$	0.885	0.192	$\lambda_{2,93}$	0.094	0.164
γ_{11}	0.454	0.147	$\phi_{2,11}$	0.212	0.042
γ_{12}	−0.159	0.159	$\phi_{2,12}$	−0.032	0.032
γ_{13}	−0.227	0.382	$\phi_{2,13}$	0.008	0.037
$\phi_{1,11}$	0.216	0.035	$\phi_{2,22}$	0.236	0.054
$\phi_{1,12}$	−0.031	0.017	$\phi_{2,23}$	0.006	0.041
$\phi_{1,22}$	0.202	0.037	$\phi_{2,33}$	0.257	0.063
ψ_{11}	0.558	0.087	ψ_{27}	0.378	0.070
ψ_{12}	0.587	0.049	ψ_{28}	0.349	0.053
ψ_{13}	0.725	0.063	ψ_{29}	0.259	0.039
ψ_{14}	0.839	0.084			
ψ_{15}	0.691	0.085			
ψ_{16}	0.730	0.081			
ψ_{17}	0.723	0.056			
ψ_{18}	0.629	0.053			
$\psi_{\delta 1}$	0.460	0.080			
Thresholds					
α_{12}	−1.163	0.054	α_{13}	−0.751	0.045
α_{22}	−0.083	0.033	α_{23}	0.302	0.035
α_{32}	−0.985	0.045	α_{33}	−0.589	0.044
α_{42}	−0.406	0.035	α_{43}	0.241	0.029
α_{52}	−1.643	0.063	α_{53}	−0.734	0.027
α_{62}	−1.038	0.043	α_{63}	−0.118	0.025

This table and Tables 9.5–9.6 are taken from Song and Lee (2004).

M_3 be non-nested models with the same specifications except that the corresponding nonlinear structural equations are given by

$$M_2: \quad \eta_{1gi} = \gamma_{11}\xi_{1gi1} + \gamma_{12}\xi_{1gi2} + \gamma_{14}\xi_{1gi1}^2 + \delta_{1gi}, \tag{9.26}$$

$$M_3: \quad \eta_{1gi} = \gamma_{11}\xi_{1gi1} + \gamma_{12}\xi_{1gi2} + \gamma_{15}\xi_{1gi2}^2 + \delta_{1gi}. \tag{9.27}$$

Table 9.5 Bayesian estimates of the structural parameters (Str. Par.) and thresholds under prior (II) for M_1: AIDS data.

	Within-group			Between-group	
	EST	SE		EST	SE
Str. Par.			Str. Par.		
$\lambda_{1,21}$	0.239	0.080	$\lambda_{2,21}$	1.404	0.283
$\lambda_{1,31}$	0.495	0.119	$\lambda_{2,31}$	0.869	0.228
$\lambda_{1,52}$	1.210	0.284	$\lambda_{2,52}$	0.304	0.293
$\lambda_{1,62}$	1.083	0.250	$\lambda_{2,62}$	0.602	0.264
$\lambda_{1,83}$	0.988	0.215	$\lambda_{2,83}$	0.155	0.230
$\lambda_{1,93}$	0.918	0.197	$\lambda_{2,93}$	0.085	0.176
γ_{11}	0.474	0.157	$\phi_{2,11}$	0.196	0.043
γ_{12}	−0.232	0.165	$\phi_{2,12}$	−0.026	0.030
γ_{13}	−0.353	0.540	$\phi_{2,13}$	0.010	0.032
$\phi_{1,11}$	0.198	0.048	$\phi_{2,22}$	0.219	0.048
$\phi_{1,12}$	−0.022	0.015	$\phi_{2,23}$	0.008	0.037
$\phi_{1,22}$	0.181	0.035	$\phi_{2,33}$	0.251	0.060
ψ_{11}	0.562	0.100	ψ_{27}	0.376	0.068
ψ_{12}	0.587	0.049	ψ_{28}	0.350	0.055
ψ_{13}	0.715	0.066	ψ_{29}	0.256	0.039
ψ_{14}	0.849	0.091			
ψ_{15}	0.702	0.093			
ψ_{16}	0.697	0.079			
ψ_{17}	0.738	0.055			
ψ_{18}	0.601	0.056			
ψ_{19}	0.828	0.064			
$\psi_{\delta 1}$	0.460	0.077			
Thresholds					
α_{12}	−1.170	0.060	α_{13}	−0.755	0.053
α_{22}	−0.084	0.029	α_{23}	0.303	0.033
α_{32}	−0.986	0.043	α_{33}	−0.589	0.042
α_{42}	−0.406	0.035	α_{43}	0.242	0.029
α_{52}	−1.656	0.066	α_{53}	−0.738	0.028
α_{62}	−1.032	0.044	α_{63}	−0.119	0.026

To apply the path sampling in computing the Bayes factor for comparing M_1 and M_2, we link up M_1 and M_2 by M_t with the following structural equation:

$$\eta_{1gi} = \gamma_{11}\xi_{1gi1} + \gamma_{12}\xi_{1gi2} + t\gamma_{13}\xi_{1gi1}\xi_{1gi2} + (1-t)\gamma_{14}\xi_{1gi1}^2 + \delta_{1gi}. \quad (9.28)$$

Clearly, when t=1, $M_t = M_1$; when t=0, $M_t = M_2$. By differentiating the complete-data likelihood $p(\mathbf{Y}, \mathbf{V}, \boldsymbol{\Omega}_1, \boldsymbol{\Omega}_2, \mathbf{X}, \mathbf{Z}|\boldsymbol{\theta}, \boldsymbol{\alpha}, t)$ with respect to t, we obtain

$$U(\boldsymbol{\theta}, \boldsymbol{\alpha}, \mathbf{Y}, \mathbf{V}, \boldsymbol{\Omega}_1, \boldsymbol{\Omega}_2, \mathbf{X}, \mathbf{Z}, t) = \sum_{i=1}^{n} \left[\eta_{1gi} - \gamma_{11}\xi_{1gi1} - \gamma_{12}\xi_{1gi2} \right.$$
$$\left. - t\gamma_{13}\xi_{1gi1}\xi_{1gi2} - (1-t)\gamma_{14}\xi_{1gi1}^2\right]\psi_{1\delta}^{-1}\left[\gamma_{13}\xi_{1gi1}\xi_{1gi2} - \gamma_{14}\xi_{1gi1}^2\right]. \tag{9.29}$$

Consequently, $\log B_{12}$ can be computed by Equations (9.12) and (9.13) with a sample of observations simulated from the appropriate posterior distributions. The above procedure can be similarly used for computing $\log B_{13}$ and $\log B_{23}$.

In this example, we take 20 grids in [0,1] and $J = 1000$ in computing logarithm Bayes factors. The $\log B_{12}$, $\log B_{13}$ and $\log B_{23}$ under prior inputs (I, II) are respectively equal to (0.317, 0.018), (0.176, 0.131) and (−0.138, −0.203). Hence, the values of logarithm Bayes factors are reasonably close to the given different prior inputs. According to the criterion given in Kass and Raftery (1995) for comparing non-nested models, M_2 is slightly better than M_1 and M_3. To apply the procedure for comparing nested models, we further compare M_2 with a linear model M_0, and a more comprehensive model M_4. Competing models M_0 and M_4 have the same specifications as M_2, except the corresponding structural equations are given by:

$$M_0 : \eta_{1gi} = \gamma_{11}\xi_{1gi1} + \gamma_{12}\xi_{1gi2} + \delta_{1gi},$$
$$M_4 : \eta_{1gi} = \gamma_{11}\xi_{1gi1} + \gamma_{12}\xi_{1gi2} + \gamma_{13}\xi_{1gi1}\xi_{1gi2} + \gamma_{14}\xi_{1gi1}^2 + \gamma_{15}\xi_{1gi2}^2 + \delta_{1gi}.$$

Note that M_0 is nested in M_2, and M_2 is nested in M_4. Using the path sampling procedure, estimates of $\log B_{40}$ and $\log B_{42}$ under prior inputs (I, II) are found to be (1.181, 1.233) and (1.043, 1.071) respectively. Based on the criterion given in Kass and Raftery (1995), M_4 is slightly better than M_0 and M_2. The PP p-values corresponding to M_4 under prior inputs (I) and (II) are equal to 0.582 and 0.611, respectively. These two values are close and indicate the expected result that the selected model also fits the data. Bayesian estimates under M_4 and their standard error estimates are reported in Table 9.6. Results obtained under prior inputs (II) are similar. We also observe comparatively large variability corresponding to estimates of parameters in $\boldsymbol{\Lambda}_2$ in the between-groups model, and estimates of $\{\gamma_{13}, \gamma_{14}\}$ corresponding to the nonlinear terms of the latent variables. This phenomenon may be due to the small sample size at the between-groups level and the complicated nature of the parameters. Other straightforward interpretations are not discussed. Based on the proposed methodology, more complicated or other combinations of nonlinear terms can be similarly analyzed.

Table 9.6 Bayesian estimates of the structural parameters (Str. Par.) and thresholds under prior (I) for M_4: AIDS data.

	Within-group			Between-group	
	EST	SE		EST	SE
Str. Par.			Str. Par.		
$\lambda_{1,21}$	0.203	0.070	$\lambda_{2,21}$	1.261	0.233
$\lambda_{1,31}$	0.450	0.100	$\lambda_{2,31}$	0.842	0.193
$\lambda_{1,52}$	0.992	0.205	$\lambda_{2,52}$	0.189	0.227
$\lambda_{1,62}$	0.868	0.180	$\lambda_{2,62}$	0.461	0.209
$\lambda_{1,83}$	0.936	0.172	$\lambda_{2,83}$	0.157	0.230
$\lambda_{1,93}$	0.880	0.194	$\lambda_{2,93}$	0.074	0.167
γ_{11}	0.489	0.147	$\phi_{2,11}$	0.211	0.040
γ_{12}	−0.026	0.217	$\phi_{2,12}$	−0.029	0.033
γ_{13}	−0.212	0.265	$\phi_{2,13}$	0.010	0.035
γ_{14}	0.383	0.442	$\phi_{2,22}$	0.223	0.053
γ_{15}	−0.147	0.188	$\phi_{2,23}$	0.013	0.038
$\phi_{1,11}$	0.245	0.042	$\phi_{2,33}$	0.243	0.059
$\phi_{1,12}$	−0.029	0.020	ψ_{27}	0.377	0.068
$\phi_{1,22}$	0.186	0.031	ψ_{28}	0.351	0.055
ψ_{11}	0.546	0.093	ψ_{29}	0.258	0.040
ψ_{12}	0.591	0.047			
ψ_{13}	0.724	0.063			
ψ_{14}	0.826	0.081			
ψ_{15}	0.716	0.092			
ψ_{16}	0.731	0.077			
ψ_{17}	0.733	0.049			
ψ_{18}	0.610	0.048			
ψ_{19}	0.833	0.064			
$\psi_{\delta 1}$	0.478	0.072			
Thresholds					
α_{12}	−1.163	0.058	α_{13}	−0.753	0.048
α_{22}	−0.088	0.032	α_{23}	0.299	0.036
α_{32}	−0.980	0.046	α_{33}	−0.583	0.047
α_{42}	−0.407	0.034	α_{43}	0.244	0.028
α_{52}	−1.650	0.063	α_{53}	−0.734	0.027
α_{62}	−1.034	0.043	α_{63}	−0.118	0.026

To summarize, we have establish a two-level nonlinear SEM with three non-overlapping factors: 'worry about AIDS', 'attitude to the risk of getting AIDS' and 'aggressiveness' in the within-groups and between-groups covariance structures. The significance of the establishments' influence is reflected by relatively large estimates of some between-groups parameters.

9.6 TWO-LEVEL NONLINEAR SEMs WITH CROSS-LEVEL EFFECTS

The within-groups nonlinear structural equation of the within-groups model of the two-level nonlinear SEM (see Equation (9.4)) discussed in previous sections is capable of assessing the nonlinear effects of the within-groups latent variables in $\boldsymbol{\eta}_{1gi}$ and $\boldsymbol{\xi}_{1gi}$ to the within-groups endogenous latent variables in $\boldsymbol{\eta}_{1gi}$. It does not accommodate the cross-level effects of between-groups latent variables in $\boldsymbol{\eta}_{2g}$ to $\boldsymbol{\eta}_{1gi}$. However, in practice, cross-level effects may be important, see Rabe-Hesketh *et al.* (2004). For example, effects of the latent variables at the school level can have great influence on some of latent variables at the teacher level. Hence, it is desirable to develop two-level SEMs with cross-level effects.

The main purpose of this subsection is to introduce a two-level nonlinear SEM with cross-level effects. For brevity, we consider data that are continuous with the usual normality assumption and invariant parameter matrices over groups. Moreover, we combine the within-groups and between-groups error measurements at the measurement equation and focus on the structures of the latent variables.

9.6.1 The Model

We consider the following measurement equation to relate the observed variables with the latent variables at the within-groups and between-groups models (see Lee and Tang (2006)):

$$\mathbf{u}_{gi} = \boldsymbol{\mu} + \boldsymbol{\Lambda}_2\boldsymbol{\omega}_{2g} + \boldsymbol{\Lambda}_1\boldsymbol{\omega}_{1gi} + \boldsymbol{\epsilon}_{gi}, \quad g = 1, \ldots, G, \quad i = 1, \ldots, N_g, \quad (9.30)$$

where $\boldsymbol{\omega}_{2g}$ is a $(q_2 \times 1)$ random vector of latent factors with distribution $N[\mathbf{0}, \boldsymbol{\Phi}_2]$, $\boldsymbol{\epsilon}_{gi}$ is a $(p \times 1)$ random vector with distribution $N[\mathbf{0}, \boldsymbol{\Psi}_\epsilon]$, where $\boldsymbol{\Psi}_\epsilon$ is a diagonal matrix, and $\boldsymbol{\epsilon}_{gi}$ are independent of $\boldsymbol{\omega}_{2g}$ and $\boldsymbol{\omega}_{1gi}$. The definitions of the other quantities are the same as before. For brevity, we consider a factor analysis model at the between-groups level. For the within-groups model, we let $\boldsymbol{\omega}_{1gi} = (\boldsymbol{\eta}_{gi}^T, \boldsymbol{\xi}_{gi}^T)^T$ be a partition of $\boldsymbol{\omega}_{1gi}$. To simplify notation, we omit the subscript 1 in $\boldsymbol{\eta}_{gi}$ and $\boldsymbol{\xi}_{gi}$. We consider the following structural equation in the within-groups model:

$$\boldsymbol{\eta}_{gi} = \boldsymbol{\Gamma}\mathbf{H}(\boldsymbol{\xi}_{gi}, \boldsymbol{\omega}_{2g}) + \boldsymbol{\delta}_{gi}, \quad (9.31)$$

where $\boldsymbol{\eta}_{gi}(q_{11} \times 1)$ and $\boldsymbol{\xi}_{gi}(q_{12} \times 1)$ are latent subvectors of $\boldsymbol{\omega}_{1gi}$ and $\mathbf{H}(\boldsymbol{\xi}_{gi}, \boldsymbol{\omega}_{2g}) = [h_1(\boldsymbol{\xi}_{gi}, \boldsymbol{\omega}_{2g}), \ldots, h_m(\boldsymbol{\xi}_{gi}, \boldsymbol{\omega}_{2g})]^T$ is an $m \times 1$ nonzero vector-valued funtion with differential known functions $h_1, \ldots, h_m$, and $m \geq \max\{q_{12}, q_2\}$, $\boldsymbol{\Gamma}(q_{11} \times m)$ is the matrix of unknown coefficients, $\boldsymbol{\xi}_{gi}$ and $\boldsymbol{\delta}_{gi}$ are respectively distributed as $N[\mathbf{0}, \mathbf{\Phi}_1]$ and $N[\mathbf{0}, \mathbf{\Psi}_\delta]$, where $\mathbf{\Psi}_\delta$ is diagonal; and $\boldsymbol{\delta}_{gi}$ is independent of $\boldsymbol{\xi}_{gi}$ and $\boldsymbol{\omega}_{2g}$. The generality of the vector-valued function $\mathbf{H}(\boldsymbol{\xi}_{gi}, \boldsymbol{\omega}_{2g})$ accommodates nonlinear terms of the exogenous latent variables in $\boldsymbol{\omega}_{2g}$ and $\boldsymbol{\xi}_{gi}$ to predict the endogenous latent variables in $\boldsymbol{\eta}_{gi}$. A concrete example that is associated with $\boldsymbol{\eta}_{gi} = (\eta_{gi})$, $\boldsymbol{\xi}_{gi} = (\xi_{gi1}, \xi_{gi2})^T$ and $\boldsymbol{\omega}_{2g} = (\omega_{2g1}, \omega_{2g2}, \omega_{2g3})$ is given below:

$$\eta_{gi} = \gamma_1\xi_{gi1} + \gamma_2\xi_{gi2} + \gamma_3\xi_{gi1}\xi_{gi2} + \gamma_4\omega_{2g1} + \gamma_5\omega_{2g2} + \gamma_6\omega_{2g3} + \gamma_7\omega_{2g2}\omega_{2g3} + \delta_{gi}, \tag{9.32}$$

where $\mathbf{\Gamma} = (\gamma_1, \ldots, \gamma_7)$, $\mathbf{H}(\boldsymbol{\xi}_{gi}, \boldsymbol{\omega}_{2g}) = (\xi_{gi1}, \xi_{gi2}, \xi_{gi1}\xi_{gi2}, \omega_{2g1}, \omega_{2g2}, \omega_{2g3}, \omega_{2g2}\omega_{2g3})^T$. Note that linear and interaction terms of the exogenous latent variables at the within-groups and between-groups model are involved in Equation (9.32). If necessary, we can easily include other nonlinear terms. Let $\mathbf{\Lambda} = (\mathbf{\Lambda}_2, \mathbf{\Lambda}_1)$ and $\boldsymbol{\zeta}_{gi} = (\boldsymbol{\omega}_{2g}^T, \boldsymbol{\omega}_{1gi}^T)^T$, then Equation (9.30) can be rewritten as:

$$\mathbf{u}_{gi} = \boldsymbol{\mu} + \mathbf{\Lambda}\boldsymbol{\zeta}_{gi} + \boldsymbol{\epsilon}_{gi}. \tag{9.33}$$

It is assumed that for $g \neq h$, $\mathbf{u}_{gi}$ and $\mathbf{u}_{hj}$ are independent, for any i and j. However, due to the presence of $\boldsymbol{\omega}_{2g}$, the observed measurements $\mathbf{u}_{gi}$ and $\mathbf{u}_{gj}$ are correlated. Moreover, due to the presence of $\boldsymbol{\omega}_{2g}$ in the within-groups structural Equation (9.31), for $i \neq j$, $\boldsymbol{\eta}_{gi}$ and $\boldsymbol{\eta}_{gj}$ are dependent and hence $\boldsymbol{\omega}_{1gi}$ and $\boldsymbol{\omega}_{1gj}$ are dependent. Similarly, the within-groups latent vector $\boldsymbol{\omega}_{1gi}$ depends on the between-groups latent vector $\boldsymbol{\omega}_{2g}$. Hence, the usual assumption in the common two-level SEMs (Ansari and Jedidi, 2000; Lee and Shi, 2001; Song and Lee, 2004) about the independence of $\boldsymbol{\omega}_{2g}$ and $\boldsymbol{\omega}_{1gi}$ is violated. The covariances among the observed and latent variables become more complicated because of the various kinds of dependence, not only among the $\mathbf{u}_{gi}$ and $\mathbf{u}_{gj}$, but also among the $\boldsymbol{\omega}_{1gi}$, $\boldsymbol{\omega}_{1gj}$ and $\boldsymbol{\omega}_{2g}$. For instance, due to the complexity of $\mathbf{H}(\boldsymbol{\xi}_{gi}, \boldsymbol{\omega}_{2g})$ in Equation (9.31), the covariance matrix of $\boldsymbol{\omega}_{1gi}$ can be complicated; due to the correlated structure of $\boldsymbol{\omega}_{2g}$ and $\boldsymbol{\omega}_{1gi}$, their covariance is complicated; and the covariance matrix of $\mathbf{u}_{gi}$ can be very complicated. Moreover, as the covariance of $\mathbf{u}_{gi}$ and $\mathbf{u}_{gj}$ can be very complicated, the covariance matrix of $\mathbf{u}_g = (\mathbf{u}_{g1}^T, \ldots, \mathbf{u}_{gN_g}^T)^T$ can be very complicated. Therefore, the accommodation of the between-groups latent variables' effect to the within-groups endogenous latent variables $\boldsymbol{\eta}_{gi}$ further compounds the difficulty in analyzing the two-level nonlinear SEMs. As we can see, the difficulty can be solved by the technique of data augmentation. In the next section, we assume that the model defined by Equations (9.30) and (9.31) is identified.

9.6.2 Bayesian Analysis

Let $\mathbf{U} = (\mathbf{u}_1, \cdots, \mathbf{u}_G)$ be the overall observed data, $n = N_1 + \cdots + N_G$, and let $\mathbf{\Omega}_{1g}$, $\mathbf{\Omega}_1$ and $\mathbf{\Omega}_2$ be data matrices defined as before. Further, let $\boldsymbol{\theta}$ be the parameter vector that contains all unknown parameters in $\boldsymbol{\mu}, \boldsymbol{\Lambda}_1, \boldsymbol{\Lambda}_2, \boldsymbol{\Gamma}$, $\boldsymbol{\Phi}_2, \boldsymbol{\Phi}_1, \boldsymbol{\Psi}_\epsilon$ and $\boldsymbol{\Psi}_\delta$. Utilizing the key idea of data augmentation (Tanner and Wong, 1987), the joint posterior distribution of interest is $[\boldsymbol{\theta}, \boldsymbol{\Omega}_2, \boldsymbol{\Omega}_1|\mathbf{U}]$. The Bayesian estimates of the parameters and latent variables and the PP p-value can be obtained given a sufficiently large number of observations that are simulated from $[\boldsymbol{\theta}, \boldsymbol{\Omega}_2, \boldsymbol{\Omega}_1|\mathbf{U}]$. Moreover, the Bayes factor for model comparison can be similarly computed through Equations (9.12) and (9.13), with a slight modification on the derivative of the complete-data likelihood (see Equation (9.14)). Hence, the major task is to simulate observations from the joint posterior distribution by the Gibbs sampler coupled with the MH algorithm. The Gibbs sampler is implemented as follows. At the $(j+1)$th iteration with current value $(\boldsymbol{\theta}^{(j)}, \boldsymbol{\Omega}_2^{(j)}, \boldsymbol{\Omega}_1^{(j)})$, iteratively generate (a) $\boldsymbol{\Omega}_2^{(j+1)}$ from $[\boldsymbol{\Omega}_2|\boldsymbol{\theta}^{(j)}, \boldsymbol{\Omega}_1^{(j)}, \mathbf{U}]$, (b) $\boldsymbol{\Omega}_1^{(j+1)}$ from $[\boldsymbol{\Omega}_1|\boldsymbol{\theta}^{(j)}, \boldsymbol{\Omega}_2^{(j+1)}, \mathbf{U}]$, and (c) $\boldsymbol{\theta}^{(j+1)}$ from $p(\boldsymbol{\theta}|\boldsymbol{\Omega}_2^{(j+1)}, \boldsymbol{\Omega}_1^{(j+1)}, \mathbf{U})$. Brief derivations of the conditional distributions involved in the Gibbs sampler are presented in Appendix 9.5. The related MH algorithm is presented in Appendix 9.6. Convergence of the algorithm is monitored by the 'estimated potential scale reduction (EPSR)' values suggested by Gelman and Rubin (1992), or by the plots of parallel sequences of observations simulated via different starting values.

9.6.3 An Application

The Accelerated Schools for Quality Education (ASQE) Project is a huge project which was conducted for helping schools to achieve an internal cultural change in order to be self-reliant in attaining school-based goals in self-improvement. In this section, we focus on the particular issue about the causal relationships among the 'school values inventory', teachers 'job satisfaction', and their 'empowerment' in identifying and solving the schools' problems. Relationships among these latent variables at the school level and the teacher level are important in the cultivation of their own and their peers' skills in improving their teaching skills and practice. Based on the proposed two-level SEM that incorporates the effects of the between-groups (school level) latent variables to the within-groups (teacher level) latent variables, we can assess precise interrelationships among the latent variables in both levels.

To save space, we only present our results based on analyses of the data that were obtained in the year from September, 1998, to August, 1999. The data set is hierarchically structured with $n = 1555$ teachers nested in $G = 50$ schools. The data set is unbalanced with values of N_g ranged from 14 to 47. Three manifest variables (relating to questions: I proudly introduce my school as a worth-while working place to my friends; I find that my attitude of value is close to my school's attitude of value; and I can fully utilize my potential in my school work), u_{g1}, u_{g2} and u_{g3} that are related with respondents' job satisfaction are taken as indicators for the latent factor, 'job satisfaction'. These variables are measured via a seven-point scale. For brevity, they are treated as continuous. The manifest variables u_{g4}, u_{g5} and u_{g6} for the latent variable, 'school value inventory' are: (1) participation and collaboration, (2) collegiality, and (3) communication and consensus, which are respectively measured by the averages of seven, six and ten items in the questionnaire. The manifest variables u_{g7}, u_{g8} and u_{g9} for the latent factor, 'teachers empowerment' are: (1) decision making, (2) self efficacy, and (3) self autonomy, which are measured by the averages of four, four and five items in the questionnaire. The sample means and standard deviations of the manifest variables are {4.139, 4.553, 4.487, 2.406, 3.171, 3.468, 0.534, 0.381, 0.601} and {1.371, 1.187, 1.181, 0.848, 0.763, 0.728, 0.499, 0.486, 0.490}, respectively.

To establish an appropriate SEM for the school and teacher levels in the above data set, a two-level structural equation model with nine manifest variables and three latent variables is considered with the following specifications. We consider a factor analysis model for the between-groups model at the schools level. Although other structures for the factor loading matrix $\mathbf{\Lambda}_2$ can be used, based on the meaning of the questions that are associated with the manifest variables, we choose the following common structure for the factor loading matrix (see Jöreskog and Sörbom, 1996) that gives nonoverlapping latent factors with clear interpretation. These latent factors are allowed to be correlated. Specifically, we take

$$\mathbf{\Lambda}_2^T = \begin{bmatrix} 1.0 & \lambda_{2,21} & \lambda_{2,31} & 0.0 & 0.0 & 0.0 & 0.0 & 0.0 & 0.0 \\ 0.0 & 0.0 & 0.0 & 1.0 & \lambda_{2,52} & \lambda_{2,62} & 0.0 & 0.0 & 0.0 \\ 0.0 & 0.0 & 0.0 & 0.0 & 0.0 & 0.0 & 1.0 & \lambda_{2,83} & \lambda_{2,93} \end{bmatrix}, \quad \mathbf{\Phi}_2 = \begin{bmatrix} \phi_{2,11} & \phi_{2,12} & \phi_{2,13} \\ \phi_{2,12} & \phi_{2,22} & \phi_{2,23} \\ \phi_{2,13} & \phi_{2,23} & \phi_{2,33} \end{bmatrix},$$

where 1's and 0's in $\mathbf{\Lambda}_2$ are fixed parameters. Based on the meaning of the corresponding questions, the latent variables can be roughly interpreted as the influence of the schools on 'job satisfaction, ω_{2g1}', 'schools value inventory, ω_{2g2}', and 'teachers empowerment, ω_{2g3}' of the teachers. For the within-groups model at the teachers level, we also use the same factor loading structure of $\mathbf{\Lambda}_2$ for $\mathbf{\Lambda}_1$ (with unknown elements denoted by $\lambda_{1,ij}$) to relate the latent factors to the manifest variables. Again, there are three latent factors, η_{gi}, ξ_{gi1} and ξ_{gi2}, in the within-groups model. Similarly, based on the meaning of the corresponding questions, interpretations of η_{gi}, ξ_{gi1} and ξ_{gi2} are 'job satisfaction', 'schools value inventory', and 'teachers empowerment' that are directly related to the teachers. The variances and covariance of ξ_{gi1} and ξ_{gi2} are given by $\phi_{1,11}$, $\phi_{1,22}$ and $\phi_{1,12}$, respectively. As 'job satisfaction' of the teachers is an important factor in education, it is important to investigate its relationships with the other latent factors. Here, we study various within-groups structural equations with η_{gi} as the endogenous variable.

To address the questions on the importance of the exogenous latent variables, we compare competing models which have the same between-groups (school level) factor analysis model, and the same within-groups (teacher level) measurement equation, but with the following different within-groups structural equations:

$$\begin{aligned}
M_0 &: \eta_{gi} = \gamma_1\xi_{gi1} + \gamma_2\xi_{gi2} + \gamma_4\omega_{2g1} + \gamma_5\omega_{2g2} + \gamma_6\omega_{2g3} + \delta_{gi},\\
M_1 &: \eta_{gi} = \delta_{gi},\\
M_2 &: \eta_{gi} = \gamma_1\xi_{gi1} + \gamma_2\xi_{gi2} + \delta_{gi},\\
M_3 &: \eta_{gi} = \gamma_1\xi_{gi1} + \gamma_2\xi_{gi2} + \gamma_3\xi_{gi1}\xi_{gi2} + \gamma_4\omega_{2g1} + \gamma_5\omega_{2g2} + \gamma_6\omega_{2g3} + \delta_{gi},\\
M_4 &: \eta_{gi} = \gamma_1\xi_{gi1} + \gamma_2\xi_{gi2} + \gamma_3\xi_{gi1}\xi_{gi2} + \gamma_4\omega_{2g1} + \gamma_5\omega_{2g2} + \gamma_6\omega_{2g3}\\
&\qquad + \gamma_7\omega_{2g2}\omega_{2g3} + \delta_{gi}.
\end{aligned} \tag{9.34}$$

Model M_0 involves linear effects of the exogenous latent variables at both levels. Model M_1 corresponds to a simple model without any exogenous latent variables in the structural equation, and M_2 corresponds to a linear model without the effects of school level latent variables. Models M_3 and M_4 are nonlinear SEMs that involve both effects from the within-groups and between-groups exogenous latent variables to the endogenous latent variable in the structural equation. Model M_3 involves an interaction effect $\gamma_3\xi_{gi1}\xi_{gi2}$ of the teacher level latent variables, whereas M_4 further involves an additional interaction effect $\gamma_7\omega_{2g2}\omega_{2g3}$ of the school level latent variables. M_1 and M_2 are nested in M_0; while M_0 is nested in M_3 and M_4. Defining a path $t \in [0, 1]$ to link any two of the above models is straightforward. For example, M_0 and M_3 can be linked by:

$$M_{t03} : \eta_{gi} = \gamma_1\xi_{gi1} + \gamma_2\xi_{gi2} + t\gamma_3\xi_{gi1}\xi_{gi2} + \gamma_4\omega_{2g1} + \gamma_5\omega_{2g2} + \gamma_6\omega_{2g3} + \delta_{gi}.$$

Note that M_{t03} reduces to M_0 when $t = 0$, and reduces to M_3 when $t = 1$.

The logarithm Bayes factors for comparing the above models are computed via the path sampling procedure with the following Type I hyperparameters in the conjugate distributions: $\alpha_{0\epsilon k} = \alpha_{0\delta k} = 10$, $\beta_{0\epsilon k} = \beta_{0\delta k} = 4$, $\rho_{01} = \rho_{02} = 8$, $\mathbf{H}_{0\omega k} = \mathbf{H}_{0uk} = 0.25\mathbf{I}$, $\mathbf{R}_{01} = (\rho_{01} - q_{12} - 1)\tilde{\mathbf{\Phi}}_1$, $\mathbf{R}_{02} = (\rho_{02} - q_2 - 1)\tilde{\mathbf{\Phi}}_2$, $\mathbf{\Lambda}_{0k} = (\tilde{\mathbf{\Lambda}}_{2k}, \tilde{\mathbf{\Lambda}}_{1k})$ and $\mathbf{\Gamma}_{0k} = \tilde{\mathbf{\Gamma}}_k$, where $\tilde{\mathbf{\Lambda}}_{2k}$, $\tilde{\mathbf{\Lambda}}_{1k}$, $\tilde{\mathbf{\Gamma}}_k$, $\tilde{\mathbf{\Phi}}_1$ and $\tilde{\mathbf{\Phi}}_2$ are the Bayesian estimates obtained by an auxiliary estimation with noninformative prior distributions. In the MH algorithm, the variances in the proposal distributions are chosen such that the approximately average acceptance rates are 0.276 and 0.283, respectively. To monitor convergence, three parallel sequences of observations are generated for each parameter via different starting values, and the EPSR values of all parameters are computed. To give some idea about the convergent behavior of the MCMC methods in analyzing these kinds of complex

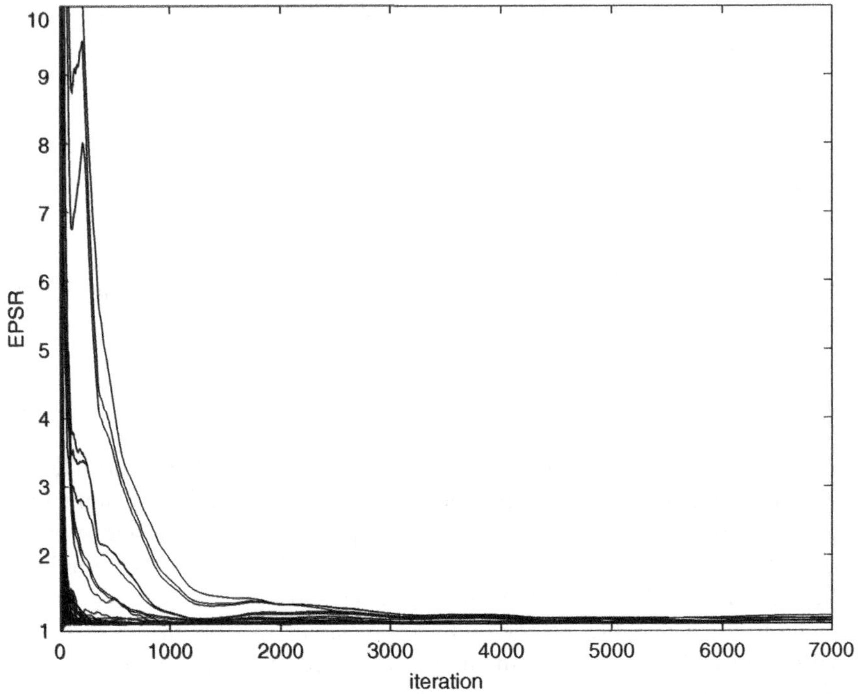

Figure 9.2 EPSR values of all parameters against iteration numbers in the ASQE example.

SEMs, plots of the EPSR values for all parameters against the iteration numbers in analyzing the selected model M_3 are presented in Figure 9.2. It can be seen that the EPSR values are less than 1.2 within 4000 iterations. To compute the Bayes factor, the number of grids is taken to be 10, and for each $t_{(s)}$ 6000 simulated observations are used to compute $\bar{U}_{(s)}$ after 4000 burn-in iterations. The estimated logarithm Bayes factors computed via the path sampling procedure are equal to $\widehat{\log B_{01}} = 882.31, \widehat{\log B_{02}} = 240.59, \widehat{\log B_{03}} = -4.018, \widehat{\log B_{04}} = -2.832$ and $\widehat{\log B_{43}} = -1.187$. Based on the criterion given in Table 5.1, M_0 is significantly better than M_1, and M_2, but not as good as M_4 and M_3. From $\widehat{\log B_{54}}$, M_3 is selected. The estimated PP p-value corresponding to M_3 is 0.512, which indicates that M_3 is a plausible model for fitting the data. As by-products, Bayesian estimates of the unknown parameters and their standard error estimates for the selected model M_3 are obtained from the observations simulated at $t = 1$ in the path sampling procedure. Results are reported in Table 9.7. Path diagrams of the structural equation and the measurement equation in the selected model M_3 are displayed in Figures 9.3 and 9.4, respectively.

Table 9.7 The Bayesian estimates and their standard errors under Type I prior inputs: ASQE example.

Para.	EST	SE	Para.	EST	SE	Para.	EST	SE
$\lambda_{2,21}$	0.518	0.029	$\lambda_{1,21}$	0.834	0.007	$\psi_{\epsilon 1}$	0.158	0.006
$\lambda_{2,31}$	0.760	0.024	$\lambda_{1,31}$	0.818	0.007	$\psi_{\epsilon 2}$	0.206	0.006
$\lambda_{2,52}$	0.828	0.094	$\lambda_{1,52}$	0.933	0.019	$\psi_{\epsilon 3}$	0.248	0.007
$\lambda_{2,62}$	2.904	0.128	$\lambda_{1,62}$	0.705	0.018	$\psi_{\epsilon 4}$	0.315	0.009
$\lambda_{2,83}$	0.839	0.051	$\lambda_{1,83}$	1.004	0.024	$\psi_{\epsilon 5}$	0.235	0.007
$\lambda_{2,93}$	0.640	0.054	$\lambda_{1,93}$	0.825	0.024	$\psi_{\epsilon 6}$	0.267	0.007
$\phi_{2,11}$	0.292	0.043	$\phi_{1,11}$	0.396	0.014	$\psi_{\epsilon 7}$	0.117	0.003
$\phi_{2,12}$	0.028	0.005	$\phi_{1,12}$	0.157	0.006	$\psi_{\epsilon 8}$	0.111	0.003
$\phi_{2,13}$	0.044	0.008	$\phi_{1,22}$	0.113	0.005	$\psi_{\epsilon 9}$	0.158	0.004
$\phi_{2,22}$	0.007	0.001	γ_1	0.533	0.049	ψ_δ	0.647	0.020
$\phi_{2,23}$	0.005	0.001	γ_2	1.856	0.101	μ_1	4.219	0.033
$\phi_{2,33}$	0.016	0.002	γ_3	−0.652	0.065	μ_2	4.644	0.022
			γ_4	−1.734	0.059	μ_3	4.557	0.026
			γ_5	1.561	0.230	μ_4	2.394	0.014
			γ_6	4.139	0.209	μ_5	3.161	0.012
						μ_6	3.437	0.020
						μ_7	0.520	0.012
						μ_8	0.369	0.011
						μ_9	0.592	0.010

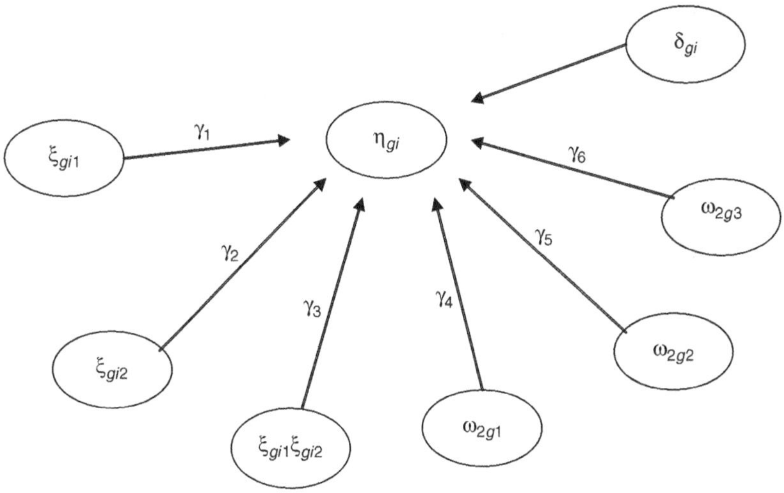

Figure 9.3 The path diagram corresponding to the within-groups structural equation: ASQE example.

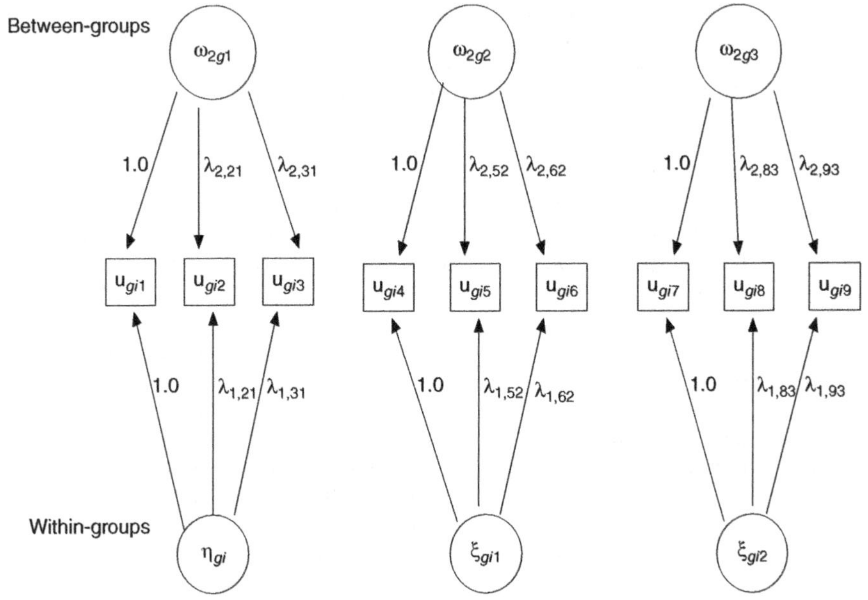

Figure 9.4 The path diagram corresponding to the measurement equations of the between-groups and within-groups models; for brevity, paths of error measurements are not displayed: ASQE example.

Comparing M_0 with M_1, and M_2, we conclude that a two-level model with a teacher level structural equation that includes effects of school level latent factors and the teacher level latent variables is significantly better than those that only involve the teacher level latent variables. This indicates that the cross-level effects are necessary for establishing an appropriate model. The model M_0 is further compared with more subtle models that contain nonlinear terms of latent variables from both levels in the within-groups structural equation. Finally, the two-level nonlinear SEM M_3 is selected. In M_3, a factor analysis model with three correlated latent factors is proposed for the between-groups model. An SEM is proposed for the within-groups model, which involves a structural equation which contains the linear effects of the school level latent factors ω_{2g1}, ω_{2g2} and ω_{2g3}, linear effects of the teacher level latent variables ξ_{gi1} and ξ_{gi2}, and an interaction effect of the teacher level latent variables, $\xi_{gi1}\xi_{gi2}$ to η_{gi}. We note from the SEs in Table 9.7 that the standard errors of some of the between-groups parameter estimates are slightly larger than those in the within-groups model. As G is smaller than $N_1 + \cdots + N_G$, this phenomenon is reasonable. From the Bayesian estimates of $\phi_{1,11}$, $\phi_{1,12}$ and $\phi_{1,22}$ given in Table 9.7, the estimated correlation between 'school value inventory' and 'teacher empowerment' at the teacher level is 0.742. As expected, these latent variables are highly correlated. From the Bayesian estimates of $\phi_{2,ij}$, we see that the corresponding correlation in the between-groups model is reduced to 0.473. This indicates a difference between these variables at the school and teacher levels. The estimated within-groups structural equation is:

$$\eta_{gi} = 0.533\xi_{gi1} + 1.856\xi_{gi2} - 0.652\xi_{gi1}\xi_{gi2} - 1.734\omega_{2g1} + 1.561\omega_{2g2} + 4.139\omega_{2g3} + \delta_{gi}.$$

We can see from the standard errors that the regression coefficients are significantly different from zero. Comparing the magnitudes of $\hat{\gamma}_1$, $\hat{\gamma}_2$ and $\hat{\gamma}_3$ with $\hat{\gamma}_4$, $\hat{\gamma}_5$ and $\hat{\gamma}_6$, we note that the impact of the effects from the latent variables in the school level is stronger than that from latent variables in the teacher level. Moreover, from $\hat{\gamma}_1(= 0.533)$ and $\hat{\gamma}_2(= 1.856)$, we see that the latent variables about 'school value inventory' and 'teacher empowerment' at the teacher level have positive effects on teachers' 'job satisfaction'. This is a logical finding. From $\hat{\gamma}_3(= -0.652)$, 'school value inventory' and 'teacher empowerment' at the teacher level have a negative interaction on teachers' 'job satisfaction'. The basic interpretation is that the additive effect of the linear random variables (ξ_{gi1}, ξ_{gi2}) in the structural equation are inadequate to account for their relationships with the latent variable 'job satisfaction, η_{gi}', and a negative interaction term of ξ_{gi1} and ξ_{gi2} has to be added. Depending on different situations, this negative interaction term has varying impact. For example, it indicates that for teachers with a positive feeling on both 'school value inventory' and 'teacher empowerment' (both ξ_{gi1} and ξ_{gi2} are positive), it would have a too strong additive positive effect on 'job satisfaction', and hence a negative adjustment by the interaction effect is necessary. From $\hat{\gamma}_5(= 1.561)$ and $\hat{\gamma}_6(= 4.139)$, the influences of the schools on teachers' 'empowerment' and 'school value inventory' also have a positive effect on the 'job satisfaction' of the teachers. Interestingly, we observe from $\hat{\gamma}_4(= -1.734)$ that the latent factor on influence of schools on teachers' 'job satisfaction' has rather strong negative effects on 'job satisfaction' of the teachers. It may indicate the presence of some conflicts between the school policy and the teachers' job satisfaction. This finding provides some insight to the administrators in deciding their school policy and their relationship with their teachers.

9.7 ANALYSIS OF TWO-LEVEL NONLINEAR SEMs USING WINBUGS

The software WinBUGS (Spiegelhalter, Thomas, Best and Lunn, 2003) can produce Bayesian estimates of the parameters in some two-level nonlinear SEMs. As we mentioned before, for models with ordered categorical data, it is hard to estimate the unknown thresholds and the unknown parameters in the structural model simultaneously by WinBUGS. Here, we fix the unknown thresholds

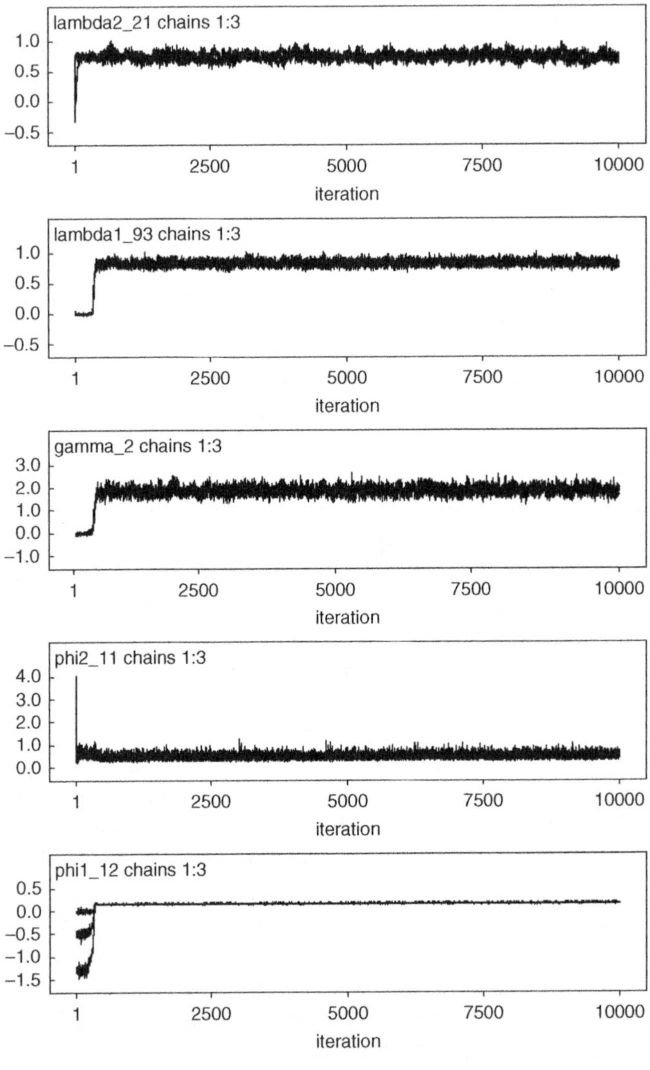

Figure 9.5 (a), (b), (c), (d) and (e) Plots of WinBUGS sequences corresponding to $\lambda_{2,21}$, $\lambda_{1,93}$, γ_2, $\phi_{2,11}$ and $\phi_{1,12}$.

at some preassigned values. Based on our experience in analyzing two-level SEMs with cross-level effects through WinBUGS, we find that the convergence of the MCMC chains corresponding to the parameters in $\mathbf{\Phi}_2$ is not good. This phenonmenon appears even in analyzing two-level linear SEMs. However, WinBUGS can be applied to analyze two-level nonlinear SEMs without cross-level effects. To demonstrate this, we apply WinBUGS to analyzing the ASQE data on the basis of the same two-level SEM and Type I prior inputs as in Section 9.6.3, except that the within-groups structural equation is now equal to

$$\eta_{gi} = \gamma_1 \xi_{gi1} + \gamma_2 \xi_{gi2} + \gamma_3 \xi_{gi1} \xi_{gi2} + \delta_{gi}. \qquad (9.35)$$

This structural equation involves an interaction term of the latent variables at the teacher level, but no cross-level effects. The DIC value corresponding to this model is 19 703.80. This value can be used for model comparison. Plots of three simulated sequences of observations obtained from WinBUGS are presented in Figure 9.5. Bayesian estimates are reported in Table 9.8. From estimates of the parameters and latent variables, Equations (9.30) and (9.31), we can obtain estimated residuals $\hat{\boldsymbol{\epsilon}}_{gi}$ and $\hat{\delta}_{gi}$, for $g = 1, \ldots, G, i = 1, \ldots, N_g$. These estimated residuals are useful for outlier analysis and for goodness-of-fit assessment. Some estimated residual plots, $\hat{\epsilon}_{gi4}$, $\hat{\epsilon}_{gi6}$ and $\hat{\delta}_{gi}$ versus the case numbers, are displayed in Figure 9.6. Plots of estimated residual $\hat{\delta}_{gi}$ versus $\hat{\xi}_{gi1}$ and $\hat{\xi}_{gi2}$ are presented in Figure 9.7, and plots of $\hat{\epsilon}_{gi4}$ versus $\hat{\xi}_{gi1}$, $\hat{\xi}_{gi2}$ and $\hat{\eta}_{gi}$ are presented in Figure 9.8. Other plots are similar. These plots indicate the corresponding measurement and structural equations of the proposed two-level model fit the data reasonably well. The WinBUGS codes and the data are given in the following website: http://www.wiley.com/go/lee_structural.

Table 9.8 The Bayesian estimates and their standard errors obtained through WinBUGS: ASQE example.

Para.	EST	SE	Para.	EST	SE	Para.	EST	SE
$\lambda_{2,21}$	0.738	0.060	$\lambda_{1,21}$	0.844	0.015	$\psi_{\epsilon 1}$	0.163	0.011
$\lambda_{2,31}$	0.706	0.041	$\lambda_{1,31}$	0.836	0.014	$\psi_{\epsilon 2}$	0.225	0.011
$\lambda_{2,52}$	0.688	0.162	$\lambda_{1,52}$	0.936	0.037	$\psi_{\epsilon 3}$	0.242	0.012
$\lambda_{2,62}$	2.040	0.247	$\lambda_{1,62}$	0.694	0.032	$\psi_{\epsilon 4}$	0.315	0.016
$\lambda_{2,83}$	0.754	0.083	$\lambda_{1,83}$	1.000	0.042	$\psi_{\epsilon 5}$	0.233	0.013
$\lambda_{2,93}$	0.545	0.095	$\lambda_{1,93}$	0.824	0.043	$\psi_{\epsilon 6}$	0.285	0.012
$\phi_{2,11}$	0.484	0.101	$\phi_{1,11}$	0.396	0.026	$\psi_{\epsilon 7}$	0.114	0.006
$\phi_{2,12}$	0.003	0.016	$\phi_{1,12}$	0.160	0.010	$\psi_{\epsilon 8}$	0.111	0.005
$\phi_{2,13}$	0.048	0.019	$\phi_{1,22}$	0.115	0.008	$\psi_{\epsilon 9}$	0.158	0.006
$\phi_{2,22}$	0.019	0.006	γ_1	0.508	0.089	ψ_{δ}	0.631	0.035
$\phi_{2,23}$	0.010	0.004	γ_2	1.858	0.176	μ_1	4.227	0.106
$\phi_{2,33}$	0.026	0.006	γ_3	−0.644	0.126	μ_2	4.629	0.081
						μ_3	4.562	0.077
						μ_4	2.395	0.030
						μ_5	3.164	0.024
						μ_6	3.445	0.044
						μ_7	0.523	0.026
						μ_8	0.373	0.021
						μ_9	0.595	0.018

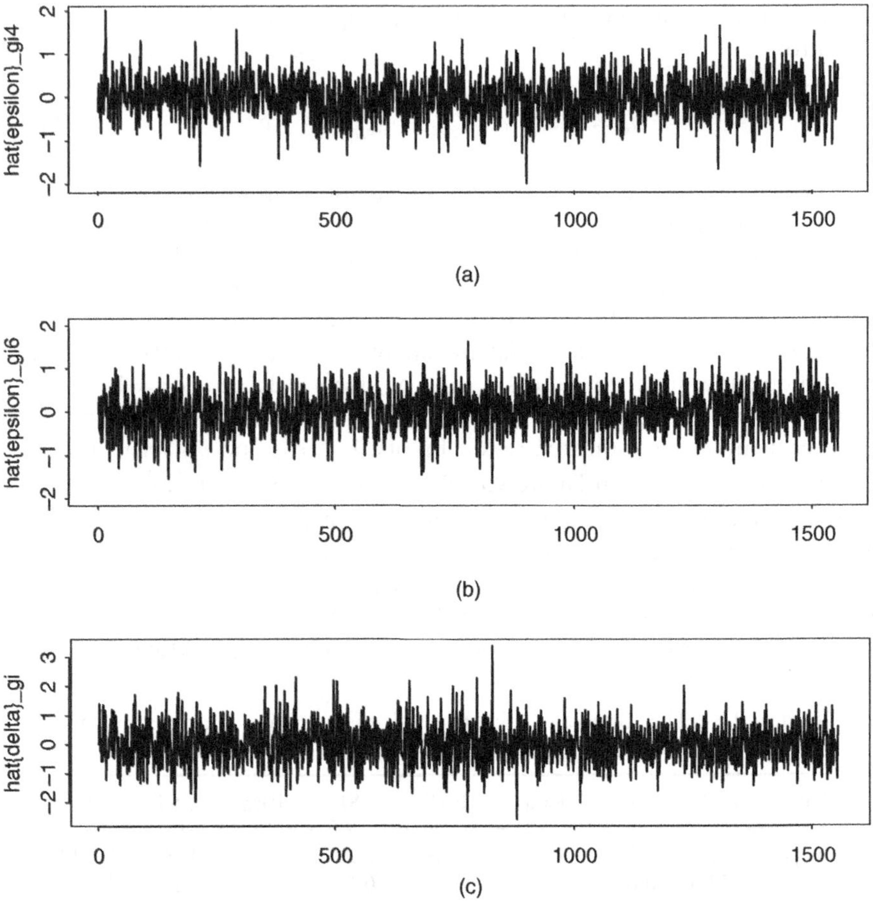

Figure 9.6 Estimated residual plots: (a) $\hat{\epsilon}_{gi4}$, (b) $\hat{\epsilon}_{gi6}$ and (c) $\hat{\delta}_{gi}$.

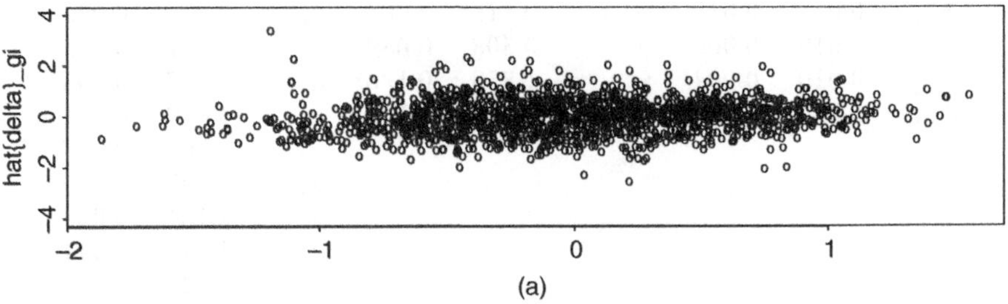

Figure 9.7 Plots of estimated residuals $\hat{\delta}_{gi}$ versus: (a) $\hat{\xi}_{gi1}$ and (b) $\hat{\xi}_{gi2}$.

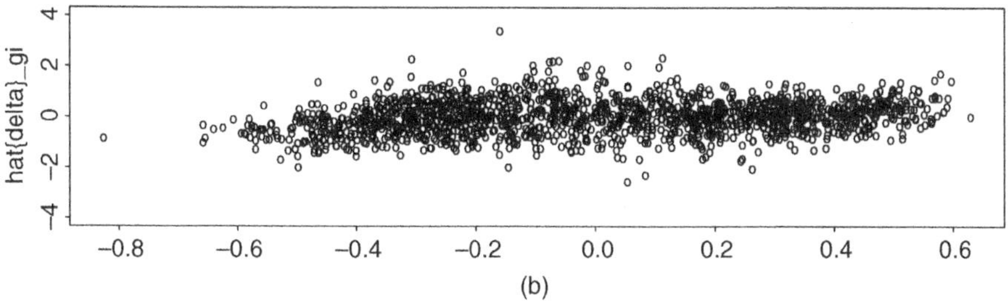

Figure 9.7 (Continued)

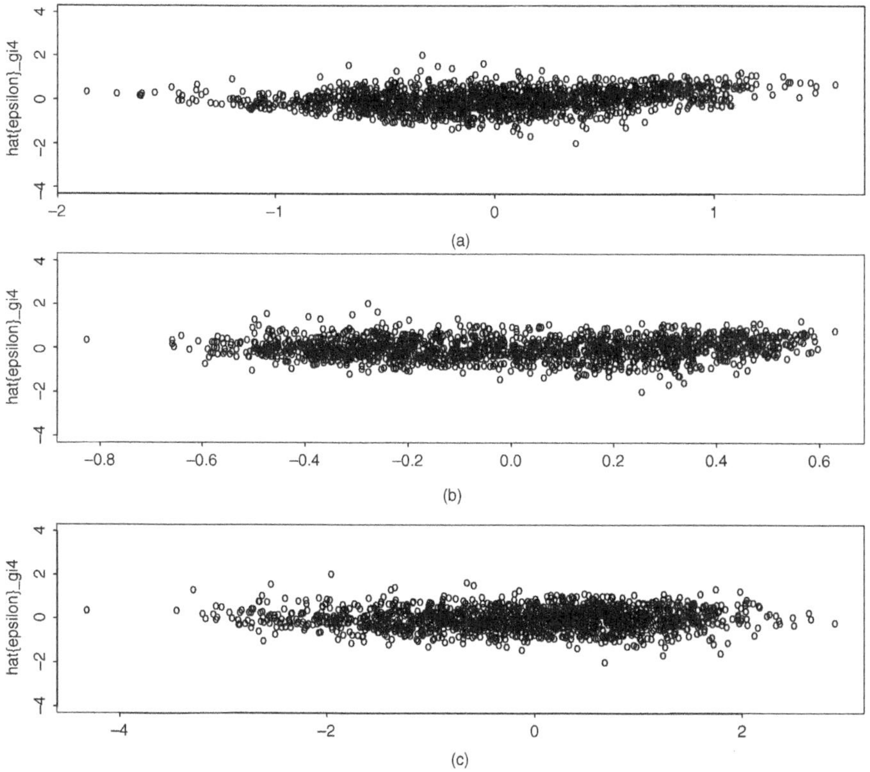

Figure 9.8 Plots of estimated residuals $\hat{\epsilon}_{gi4}$ versus: (a) $\hat{\xi}_{gi1}$, (b) $\hat{\xi}_{gi2}$, and (c) $\hat{\eta}_{gi}$.

APPENDIX 9.1: CONDITIONAL DISTRIBUTIONS: TWO-LEVEL NONLINEAR SEM

Owing to the complexity of the model, it is very tedious to derive all the conditional distributions required by the Gibbs sampler, hence only brief discussions are given. For brevity, we will use $p(\cdot|\cdot)$ to denote the conditional distribution if the context is clear. Moreover, we only consider the case that all parameters in $\mathbf{\Lambda}_{1g}, \mathbf{\Lambda}_2, \mathbf{\Pi}_{1g}, \mathbf{\Gamma}_{1g}, \mathbf{\Pi}_2$ and $\mathbf{\Gamma}_2$ are not fixed. Conditional distributions for the case with fixed parameters can be obtained by slight modifications as given in previous chapters.

$p(\mathbf{V}|\boldsymbol{\theta}, \boldsymbol{\alpha}, \mathbf{Y}, \mathbf{\Omega}_1, \mathbf{\Omega}_2, \mathbf{X}, \mathbf{Z})$: Since $\mathbf{v}_g$s are independent and not depending on $\boldsymbol{\alpha}$, this conditional distribution is equal to a product of $p(\mathbf{v}_g|\boldsymbol{\theta}, \mathbf{Y}_g, \mathbf{\Omega}_{1g}, \boldsymbol{\omega}_{2g}, \mathbf{X}_g, \mathbf{Z}_g)$ with $g = 1, \ldots, G$. For each gth term in this product,

$$p(\mathbf{v}_g|\boldsymbol{\theta}, \mathbf{Y}_g, \mathbf{\Omega}_{1g}, \boldsymbol{\omega}_{2g}, \mathbf{X}_g, \mathbf{Z}_g) \propto p(\mathbf{v}_g|\boldsymbol{\theta}, \boldsymbol{\omega}_{2g}) \prod_{i=1}^{N_g} p(\mathbf{u}_{gi}|\boldsymbol{\theta}, \mathbf{v}_g, \boldsymbol{\omega}_{1gi}) \tag{A9.1}$$

$$\propto \exp\left[-\frac{1}{2}\left\{\mathbf{v}_g^T(N_g\mathbf{\Psi}_{1g}^{-1} + \mathbf{\Psi}_2^{-1})\mathbf{v}_g - 2\mathbf{v}_g^T \right.\right.$$
$$\left.\left.\times\left[\mathbf{\Psi}_{1g}^{-1}\sum_{i=1}^{N_g}(\mathbf{u}_{gi} - \mathbf{\Lambda}_{1g}\boldsymbol{\omega}_{1gi}) + \mathbf{\Psi}_2^{-1}(\boldsymbol{\mu} + \mathbf{\Lambda}_2\boldsymbol{\omega}_{2g})\right]\right\}\right].$$

Hence, for each $\mathbf{v}_g$, its conditional distribution $p(\mathbf{v}_g|\cdot)$ is $N[\boldsymbol{\mu}_g^*, \mathbf{\Sigma}_g^*]$, where

$$\boldsymbol{\mu}_g^* = \mathbf{\Sigma}_g^*\left[\mathbf{\Psi}_{1g}^{-1}\sum_{i=1}^{N_g}(\mathbf{u}_{gi} - \mathbf{\Lambda}_{1g}\boldsymbol{\omega}_{1gi}) + \mathbf{\Psi}_2^{-1}(\boldsymbol{\mu} + \mathbf{\Lambda}_2\boldsymbol{\omega}_{2g})\right], \text{ and}$$
$$\mathbf{\Sigma}_g^* = (N_g\mathbf{\Psi}_{1g}^{-1} + \mathbf{\Psi}_2^{-1})^{-1}.$$

$p(\mathbf{\Omega}_1|\boldsymbol{\theta}, \boldsymbol{\alpha}, \mathbf{Y}, \mathbf{V}, \mathbf{\Omega}_2, \mathbf{X}, \mathbf{Z})$: Since $\boldsymbol{\omega}_{1gi}$ are mutually independent, $\mathbf{u}_{gi}$ is independent of $\mathbf{u}_{hj}$ for all $h \neq g$, and they are not depending on $\boldsymbol{\alpha}$ and $\mathbf{Y}$, we have

$$p(\mathbf{\Omega}_1|\cdot) = \prod_{g=1}^{G}\prod_{i=1}^{N_g} p(\boldsymbol{\omega}_{1gi}|\boldsymbol{\theta}, \mathbf{v}_g, \boldsymbol{\omega}_{2g}, \mathbf{u}_{gi})$$
$$\propto \prod_{g=1}^{G}\prod_{i=1}^{N_g} p(\mathbf{u}_{gi}|\boldsymbol{\theta}, \mathbf{v}_g, \boldsymbol{\omega}_{1gi})\, p(\boldsymbol{\eta}_{1gi}|\boldsymbol{\xi}_{1gi}, \boldsymbol{\theta})\, p(\boldsymbol{\xi}_{1gi}|\boldsymbol{\theta}).$$

It follows that $p(\boldsymbol{\omega}_{1gi}|\cdot)$ is proportional to

$$\exp\left[-\frac{1}{2}\{\boldsymbol{\xi}_{1gi}^T\mathbf{\Phi}_{1g}^{-1}\boldsymbol{\xi}_{1gi} + (\mathbf{u}_{gi} - \mathbf{v}_g - \mathbf{\Lambda}_{1g}\boldsymbol{\omega}_{1gi})^T\mathbf{\Psi}_{1g}^{-1}(\mathbf{u}_{gi} - \mathbf{v}_g - \mathbf{\Lambda}_{1g}\boldsymbol{\omega}_{1gi})\right.$$
$$\left. + [\boldsymbol{\eta}_{1gi} - \mathbf{\Pi}_{1g}\boldsymbol{\eta}_{1gi} - \mathbf{\Gamma}_{1g}\mathbf{H}_1(\boldsymbol{\xi}_{1gi})]^T\mathbf{\Psi}_{1g\delta}^{-1}[\boldsymbol{\eta}_{1gi} - \mathbf{\Pi}_{1g}\boldsymbol{\eta}_{1gi} - \mathbf{\Gamma}_{1g}\mathbf{H}_1(\boldsymbol{\xi}_{1gi})]\}\right]. \tag{A9.2}$$

$p(\boldsymbol{\Omega}_2|\boldsymbol{\theta}, \boldsymbol{\alpha}, \mathbf{Y}, \mathbf{V}, \boldsymbol{\Omega}_1, \mathbf{X}, \mathbf{Z})$: This distribution has a very similar form to $p(\boldsymbol{\Omega}_1|\cdot)$ and Equation (A9.2), and hence it is not presented.

$p(\boldsymbol{\alpha}, \mathbf{Y}|\boldsymbol{\theta}, \mathbf{V}, \boldsymbol{\Omega}_1, \boldsymbol{\Omega}_2, \mathbf{X}, \mathbf{Z})$: We only consider the case that all the thresholds corresponding to each within-groups are different. The other cases can be similarly derived. To deal with the situation with little or no information about these parameters, the following noninformative prior distribution is used:

$$p(\alpha_{gk}) = p(\alpha_{gk,2}, \ldots, \alpha_{gk,b_k-1}) \propto c, \quad g = 1, \ldots, G \; k = 1, \ldots, s.$$

where c is a constant. Now, since $(\boldsymbol{\alpha}_g, \mathbf{Y}_g)$ is independent of $(\boldsymbol{\alpha}_h, \mathbf{Y}_h)$ for $g \neq h$, and $\boldsymbol{\Psi}_{1g}$ is diagonal,

$$p(\boldsymbol{\alpha}, \mathbf{Y}|\cdot) = \prod_{g=1}^{G} p(\boldsymbol{\alpha}_g, \mathbf{Y}_g|\cdot) = \prod_{g=1}^{G} \prod_{k=1}^{s} p(\boldsymbol{\alpha}_{gk}, \mathbf{Y}_{gk}|\cdot), \tag{A9.3}$$

where $\mathbf{Y}_{gk} = (y_{g1k}, \ldots, y_{gN_gk})$. Let ψ_{1gk} be the kth diagonal element of $\boldsymbol{\Psi}_{1g}$, v_{gk} be the kth element of $\mathbf{v}_g$ and $\boldsymbol{\Lambda}_{1gk}^T$ be the kth row of $\boldsymbol{\Lambda}_{1g}$, and $I_A(y)$ be an indicator function with value 1 if y in A and zero otherwise, $p(\boldsymbol{\alpha}, \mathbf{Y}|\cdot)$ can be obtained from Equation (A9.3) and

$$p(\boldsymbol{\alpha}_{gk}, \mathbf{Y}_{gk}|\cdot) \propto \prod_{i=1}^{N_g} \phi[\psi_{1gk}^{-1/2}(y_{gik} - v_{gk} - \boldsymbol{\Lambda}_{1gk}^T \boldsymbol{\omega}_{1gi})] \mathrm{I}_{(\alpha_{gk,z_{gik}}, \alpha_{gk,z_{gik}+1}]}(y_{gik}), \tag{A9.4}$$

where ϕ is the probability density function of $N[0, 1]$.

$p(\boldsymbol{\theta}|\boldsymbol{\alpha}, \mathbf{Y}, \mathbf{V}, \boldsymbol{\Omega}_1, \boldsymbol{\Omega}_2, \mathbf{X}, \mathbf{Z})$: This conditional distribution is different under different special cases as discussed in Section 9.3.1. We first consider the situation with distinct within-groups parameters, that is $\boldsymbol{\theta}_{11} \neq \cdots \neq \boldsymbol{\theta}_{1G}$. Let $\boldsymbol{\theta}_2$ be the vector of unknown parameters in $\boldsymbol{\mu}$, $\boldsymbol{\Lambda}_2$ and $\boldsymbol{\Psi}_2$; and $\boldsymbol{\theta}_{2\omega}$ be the vector of unknown parameters in $\boldsymbol{\Pi}_2, \boldsymbol{\Gamma}_2, \boldsymbol{\Phi}_2$ and $\boldsymbol{\Psi}_{2\delta}$. These between-groups parameters are the same for each g. For the within-group parameters, let $\boldsymbol{\theta}_{1g}$ be the vector of unknown parameters in $\boldsymbol{\Lambda}_{1g}$ and $\boldsymbol{\Psi}_{1g}$ and $\boldsymbol{\theta}_{1g\omega}$ be the vector of unknown parameters in $\boldsymbol{\Pi}_{1g}, \boldsymbol{\Gamma}_{1g}, \boldsymbol{\Phi}_{1g}$ and $\boldsymbol{\Psi}_{1g\delta}$. It is natural to assume the prior distributions of these parameter vectors in different independent groups are independent of each other, and hence they can be treated separately.

For $\boldsymbol{\theta}_{1g}$, the following commonly used conjugate type prior distributions are used:

$$\psi_{1gk}^{-1} \stackrel{D}{=} Gamma[\alpha_{01gk}, \beta_{01gk}], \quad [\boldsymbol{\Lambda}_{1gk}|\psi_{1gk}] \stackrel{D}{=} N[\boldsymbol{\Lambda}_{01gk}, \psi_{1gk}\mathbf{H}_{01gk}], \quad k = 1, \ldots, p,$$

where $\boldsymbol{\Lambda}_{1gk}^T$ is the kth row of $\boldsymbol{\Lambda}_{1g}$ and $\alpha_{01gk}, \beta_{01gk}, \boldsymbol{\Lambda}_{01gk}$ and $\mathbf{H}_{01gk}$ are given hyperparameter values. For $k \neq h$, it is assumed that $(\psi_{1gk}, \boldsymbol{\Lambda}_{1gk})$ and $(\psi_{1gh}, \boldsymbol{\Lambda}_{1gh})$ are independent. Let $\mathbf{U}_g^* = \{\mathbf{u}_{gi} - \mathbf{v}_g;\; i = 1, \ldots, N_g\}$ and $\mathbf{U}_{gk}^{*T}$ be the kth row of $\mathbf{U}_g^*$, $\boldsymbol{\Sigma}_{1gk} = (\mathbf{H}_{01gk}^{-1} + \boldsymbol{\Omega}_{1g}\boldsymbol{\Omega}_{1g}^T)^{-1}$, $\mathbf{m}_{1gk} = \boldsymbol{\Sigma}_{1gk}(\mathbf{H}_{01gk}^{-1}\boldsymbol{\Lambda}_{01gk} + \boldsymbol{\Omega}_{1g}\mathbf{U}_{gk}^*)$, $\boldsymbol{\Omega}_{1g} = (\boldsymbol{\omega}_{1g1}, \ldots, \boldsymbol{\omega}_{1gN_g})$ and $\beta_{1gk} = \beta_{01gk} + \frac{1}{2}(\mathbf{U}_{gk}^{*T}\mathbf{U}_{gk}^* - \mathbf{m}_{1gk}^T\boldsymbol{\Sigma}_{1gk}^{-1}\mathbf{m}_{1gk} + \boldsymbol{\Lambda}_{01gk}^T\mathbf{H}_{01gk}^{-1}\boldsymbol{\Lambda}_{01gk})$, then it can be shown that

$$
\begin{aligned}
p(\psi_{1gk}^{-1}|\cdot) &\stackrel{D}{=} Gamma\ (2^{-1}N_g + \alpha_{01gk}, \beta_{1gk}) \text{ and} \\
p(\boldsymbol{\Lambda}_{1gk}|\psi_{1gk}^{-1}, \cdot) &\stackrel{D}{=} N\,[\mathbf{m}_{1gk}, \psi_{1gk}\boldsymbol{\Sigma}_{1gk}].
\end{aligned}
\tag{A9.5}
$$

For $\boldsymbol{\theta}_{1g\omega}$, it is assumed that $\boldsymbol{\Phi}_{1g}$ is independent of $(\boldsymbol{\Lambda}_{1g}^*, \boldsymbol{\Psi}_{1g\delta})$, where $\boldsymbol{\Lambda}_{1g}^* = (\boldsymbol{\Pi}_{1g}, \boldsymbol{\Gamma}_{1g})^T$. Also, $(\boldsymbol{\Lambda}_{1gk}^*, \psi_{1g\delta k})$ and $(\boldsymbol{\Lambda}_{1gh}^*, \psi_{1g\delta h})$ are independent, where $\boldsymbol{\Lambda}_{1gk}^{*T}$ and $\psi_{1g\delta k}$ are the kth row and diagonal element of $\boldsymbol{\Lambda}_{1g}^*$ and $\boldsymbol{\Psi}_{1g\delta}$ respectively. The associated prior distribution of $\boldsymbol{\Phi}_{1g}$ is: $p(\boldsymbol{\Phi}_{1g}^{-1}) \stackrel{D}{=} W_{q_{12}}[\mathbf{R}_{01g}, \rho_{01g}]$, where ρ_{01g} and the positive definite matrix $\mathbf{R}_{01g}$ are given hyperparameters. Moreover, the prior distribution of $\psi_{1g\delta k}$ and $\boldsymbol{\Lambda}_{1gk}^*$ are:

$$
\psi_{1g\delta k}^{-1} \stackrel{D}{=} Gamma\ [\alpha_{01g\delta k}, \beta_{01g\delta k}] \text{ and } [\boldsymbol{\Lambda}_{1gk}^*|\psi_{1g\delta k}] \stackrel{D}{=} N\,[\boldsymbol{\Lambda}_{01gk}^*, \psi_{1g\delta k}\mathbf{H}_{01gk}^*],
$$

where $\alpha_{01g\delta k}, \beta_{01g\delta k}, \boldsymbol{\Lambda}_{01gk}^*$ and $\mathbf{H}_{01gk}^*$ are given hyperparameters. Let $\boldsymbol{\Omega}_{1g}^* = \{\boldsymbol{\eta}_{1g1}, \ldots, \boldsymbol{\eta}_{1gN_g}\}$, $\boldsymbol{\Omega}_{1gk}^{*T}$ be the kth row of $\boldsymbol{\Omega}_{1g}^*$, $\boldsymbol{\Xi}_{1g} = \{\boldsymbol{\xi}_{1g1}, \ldots, \boldsymbol{\xi}_{1gN_g}\}$ and $\mathbf{H}_{1g}^* = \{\mathbf{H}_1^*(\boldsymbol{\omega}_{1g1}), \ldots, \mathbf{H}_1^*(\boldsymbol{\omega}_{1gN_g})\}$, in which $\mathbf{H}_1^*(\boldsymbol{\omega}_{1gi}) = (\boldsymbol{\eta}_{1gi}^T, \mathbf{H}_1(\boldsymbol{\xi}_{1gi})^T)^T$, $i = 1, \ldots, N_g$, it can be shown that for $k = 1, \ldots, q_{11}$,

$$
\begin{aligned}
p(\psi_{1g\delta k}|\cdot) &\stackrel{D}{=} Gamma\ [2^{-1}N_g + \alpha_{01g\delta k}, \beta_{1g\delta k}], \text{ and} \\
p(\boldsymbol{\Lambda}_{1gk}^*|\psi_{1g\delta k}^{-1}, \cdot) &\stackrel{D}{=} N\,[\mathbf{m}_{1gk}^*, \psi_{1g\delta k}\boldsymbol{\Sigma}_{1gk}^*],
\end{aligned}
\tag{A9.6}
$$

where $\boldsymbol{\Sigma}_{1gk}^* = (\mathbf{H}_{01gk}^{*-1} + \mathbf{H}_{1g}^*\mathbf{H}_{1g}^{*T})^{-1}$, $\mathbf{m}_{1gk}^* = \boldsymbol{\Sigma}_{1gk}^*(\mathbf{H}_{01gk}^{*-1}\boldsymbol{\Lambda}_{01gk}^* + \mathbf{H}_{1g}^*\boldsymbol{\Omega}_{1gk}^*)$ and $\beta_{1g\delta k} = \beta_{01g\delta k} + \frac{1}{2}(\boldsymbol{\Omega}_{1gk}^{*T}\boldsymbol{\Omega}_{1gk}^* - \mathbf{m}_{1gk}^{*T}\boldsymbol{\Sigma}_{1gk}^{*-1}\mathbf{m}_{1gk}^* + \boldsymbol{\Lambda}_{01gk}^{*T}\mathbf{H}_{01gk}^{*-1}\boldsymbol{\Lambda}_{01gk}^*)$. The conditional distribution relating to $\boldsymbol{\Phi}_{1g}$ is given by

$$
p(\boldsymbol{\Phi}_{1g}|\boldsymbol{\Xi}_{1g}) \stackrel{D}{=} IW_{q_{12}}[(\boldsymbol{\Xi}_{1g}\boldsymbol{\Xi}_{1g}^T + \mathbf{R}_{01g}^{-1}), N_g + \rho_{01g}].
\tag{A9.7}
$$

Conditional distributions involved in $\boldsymbol{\theta}_2$ are derived in the same way on the basis of the following independent conjugate type prior distributions: for $k = 1, \ldots, p$, and

$$
\begin{aligned}
&\boldsymbol{\mu} \stackrel{D}{=} N\,[\boldsymbol{\mu}_0, \boldsymbol{\Sigma}_0], \; \psi_{2k}^{-1} \stackrel{D}{=} Gamma\,[\alpha_{02k}, \beta_{02k}], \\
&[\boldsymbol{\Lambda}_{2k}|\psi_{2k}] \stackrel{D}{=} N\,[\boldsymbol{\Lambda}_{02k}, \psi_{2k}\mathbf{H}_{02k}].
\end{aligned}
$$

where $\mathbf{\Lambda}_{2k}^T$ is the kth row of $\mathbf{\Lambda}_2$; $\alpha_{02k}, \beta_{02k}, \boldsymbol{\mu}_0, \mathbf{\Sigma}_0, \mathbf{\Lambda}_{02k}$ and $\mathbf{H}_{02k}$ are given hyperparameters.

Similarly, conditional distributions involved in $\boldsymbol{\theta}_{2\omega}$ are derived on the basis of the following conjugate type distributions: for $k = 1, \ldots, q_{21}$,

$$\mathbf{\Phi}_2^{-1} \stackrel{D}{=} W_{q_{22}}[\mathbf{R}_{02}, \rho_{02}], \ \psi_{2\delta k}^{-1} \stackrel{D}{=} Gamma\,[\alpha_{02\delta k}, \beta_{02\delta k}], \text{ and}$$
$$[\mathbf{\Lambda}_{2k}^* | \psi_{2\delta k}] \stackrel{D}{=} N\,[\mathbf{\Lambda}_{02k}^*, \psi_{2\delta k}\mathbf{H}_{02k}^*],$$

where $\mathbf{\Lambda}_2^* = (\mathbf{\Pi}_2^T, \mathbf{\Gamma}_2^T)^T$ and $\mathbf{\Lambda}_{2k}^{*T}$ is the vector that contains the unknown parameters in the kth row of $\mathbf{\Lambda}_2^*$. As these conditional distributions are similar to those in Equations (A9.5)–(A9.7), they are not presented here.

In the situation where $\boldsymbol{\theta}_{11} = \cdots = \boldsymbol{\theta}_{1G}, (= \boldsymbol{\theta}_1)$, the prior distributions corresponding to components of $\boldsymbol{\theta}_1$ are not depending on g, and all the data in the within groups should be combined in deriving the conditional distributions for the estimation. Conditional distributions can be derived with the following conjugate type prior distributions: for $k = 1, \ldots, p$ and similar notations as above,

$$\psi_{1k}^{-1} \stackrel{D}{=} Gamma[\alpha_{01k}, \beta_{01k}], \ \mathbf{\Lambda}_{1k} | \psi_{1k}^{-1} \stackrel{D}{=} N\,[\mathbf{\Lambda}_{01k}, \psi_{1k}\mathbf{H}_{01k}],$$
$$[\mathbf{\Phi}_1^{-1}] \stackrel{D}{=} W_{q_{22}}[\mathbf{R}_{01}, \rho_{01}],$$
$$[\psi_{1\delta k}^{-1}] \stackrel{D}{=} Gamma[\alpha_{01\delta k}, \beta_{01\delta k}] \quad \text{and} \quad [\mathbf{\Lambda}_{1k}^* | \psi_{1\delta k}] \stackrel{D}{=} N\,[\mathbf{\Lambda}_{01k}^*, \psi_{1\delta k}\mathbf{H}_{01k}^*], \qquad \text{(A9.8)}$$

and the prior distributions and conditional distributions corresponding to structural parameters in the between-groups covariance matrix are the same as before.

APPENDIX 9.2: MH ALGORITHM: TWO-LEVEL NONLINEAR SEM

Simulating observations from the gamma, normal and inverted Wishart distributions is straightforward and fast. However, the conditional distributions $p(\mathbf{\Omega}_1|\cdot)$, $p(\mathbf{\Omega}_2|\cdot)$ and $p(\boldsymbol{\alpha}, \mathbf{Y}|\cdot)$ are complex and it is therefore necessary to implement the MH algorithm for efficient simulation of observations from these conditional distributions.

For the conditional distribution $p(\mathbf{\Omega}_1|\cdot)$, we need to simulate observations from the target density $p(\boldsymbol{\omega}_{1gi}|\cdot)$ as given in Equation (A9.2). Similar to the method of Zhu and Lee (1999) and Lee and Zhu (2000), we choose

$N[\cdot, \sigma_1^2 \mathbf{C}_{1g}]$ as the proposal distribution, where $\mathbf{C}_{1g}^{-1} = \mathbf{C}_{1g\omega}^{-1} + \mathbf{\Lambda}_{1g}^T \mathbf{\Psi}_{1g}^{-1} \mathbf{\Lambda}_{1g}$ and $\mathbf{C}_{1g\omega}^{-1}$ is given by

$$\mathbf{C}_{1g\omega}^{-1} = \begin{bmatrix} \mathbf{\Pi}_{1g0}^T \mathbf{\Psi}_{1g\delta}^{-1} \mathbf{\Pi}_{1g0} & -\mathbf{\Pi}_{1g0}^T \mathbf{\Psi}_{1g\delta}^{-1} \mathbf{\Gamma}_{1g} \mathbf{\Delta}_{1g} \\ -\mathbf{\Delta}_{1g}^T \mathbf{\Gamma}_{1g}^T \mathbf{\Psi}_{1g\delta}^{-1} \mathbf{\Pi}_{1g0} & \mathbf{\Phi}_{1g}^{-1} + \mathbf{\Delta}_{1g}^T \mathbf{\Gamma}_{1g}^T \mathbf{\Psi}_{1g\delta}^{-1} \mathbf{\Gamma}_{1g} \mathbf{\Delta}_{1g} \end{bmatrix},$$

where $\mathbf{\Pi}_{1g0} = \mathbf{I}_{q_{11}} - \mathbf{\Pi}_{1g}$ with an identity matrix $\mathbf{I}_{q_{11}}$ of order q_{11} and $\mathbf{\Delta}_{1g} = \partial \mathbf{H}_1(\boldsymbol{\xi}_{1gi})/\partial \boldsymbol{\xi}_{1gi}|_{\boldsymbol{\xi}_{1gi}=0}$. Let $p(\cdot|\boldsymbol{\xi}^*, \sigma_1^2 \mathbf{C}_{1g})$ be the density function corresponding to the proposal distribution $N[\boldsymbol{\xi}^*, \sigma_1^2 \mathbf{C}_{1g}]$, the MH algorithm is implemented as follows. At the mth iteration with a current value $\boldsymbol{\xi}_{1gi}^{(m)}$, a new candidate $\boldsymbol{\xi}_i^*$ is generated from $p(\cdot|\boldsymbol{\xi}_{1gi}^{(m)}, \sigma_1^2 \mathbf{C}_{1g})$ and accepting this new candidate with probability $\min[1, p(\boldsymbol{\xi}_i^*|\cdot)/p(\boldsymbol{\xi}_{1gi}^{(m)}|\cdot)]$. The variance σ_1^2 can be chosen such that the average acceptance rate is approximately 0.25 or more, see Gelman, Roberts, and Gilks (1995).

Observations from the conditional distribution $p(\mathbf{\Omega}_2|\cdot)$ with target density similar to Equation (A9.2) can be simulated via a similar MH algorithm as described above. To save space, details are not given.

An MH type algorithm that is similar to that described in Section 6.3.2 is necessary for simulating observations from the complex distribution $p(\boldsymbol{\alpha}, \mathbf{Y}|\cdot)$. Here, the target density is given in Equation (A9.4). According to the factorization recommended by Cowles (1996) (see also Lee and Zhu (2000)) the joint proposal density of $\boldsymbol{\alpha}_{gk}$ and $\mathbf{Y}_{gk}$ is constructed as

$$p(\boldsymbol{\alpha}_{gk}, \mathbf{Y}_{gk}|\cdot) = p(\boldsymbol{\alpha}_{gk}|\ \cdot) p(\mathbf{Y}_{gk}|\boldsymbol{\alpha}_{gk}\ \cdot). \tag{A9.9}$$

Then, the algorithm is implemented as follows. At the mth iteration with $(\boldsymbol{\alpha}_{gk}^{(m)}, \mathbf{Y}_{gk}^{(m)})$, the acceptance probability for a $(\boldsymbol{\alpha}_{gk}, \mathbf{Y}_{gk})$ as a new observation $(\boldsymbol{\alpha}_{gk}^{(m+1)}, \mathbf{Y}_{gk}^{(m+1)})$ is $\min\{1, R_{gk}\}$, where

$$R_{gk} = \frac{p(\boldsymbol{\alpha}_{gk}, \mathbf{Y}_{gk}|\boldsymbol{\theta}, \mathbf{Z}_{gk}, \mathbf{\Omega}_{1g}) p(\boldsymbol{\alpha}_{gk}^{(m)}, \mathbf{Y}_{gk}^{(m)}|\boldsymbol{\alpha}_{gk}, \mathbf{Y}_{gk}, \boldsymbol{\theta}, \mathbf{Z}_{gk}, \mathbf{\Omega}_{1g})}{p(\boldsymbol{\alpha}_{gk}^{(m)}, \mathbf{Y}_{gk}^{(m)}|\boldsymbol{\theta}, \mathbf{Z}_{gk}, \mathbf{\Omega}_{1g}) p(\boldsymbol{\alpha}_{gk}, \mathbf{Y}_{gk}|\boldsymbol{\alpha}_{gk}^{(m)}, \mathbf{Y}_{gk}^{(m)}, \boldsymbol{\theta}, \mathbf{Z}_{gk}, \mathbf{\Omega}_{1g})},$$

with $\mathbf{Z}_{gk} = (z_{gik}, \ldots, z_{gn_g k})$. To search for a new observation via the proposal density of Equation (A9.9), we first generate a vector of thresholds $(\alpha_{gk,2}, \ldots, \alpha_{gk,b_k-1})$ from the following truncated normal distribution

$$\alpha_{gk,z} \overset{D}{=} N[\alpha_{gk,z}^{(m)}, \sigma_{\alpha gk}^2] I_{(\alpha_{gk,z-1}, \alpha_{gk,z+1}^{(m)}]}(\alpha_{gk,z}), \quad z = 2, \ldots, b_k - 1,$$

where $\sigma^2_{\alpha gk}$ is a preassigned value to give an approximate acceptance rate 0.44 (see Cowles 1996). It follows from, Equation (A9.4) and the above result that

$$R_{gk} = \prod_{z=2}^{b_k-1} \frac{\Phi^*[(\alpha^{(m)}_{gk,z+1} - \alpha^{(m)}_{gk,z})/\sigma_{\alpha gk}] - \Phi^*[(\alpha_{gk,z-1} - \alpha^{(m)}_{gk,z})/\sigma_{\alpha gk}]}{\Phi^*[(\alpha_{gk,z+1} - \alpha_{gk,z})/\sigma_{\alpha gk}] - \Phi^*[(\alpha^{(m)}_{gk,z-1} - \alpha_{gk,z})/\sigma_{\alpha gk}]}$$

$$\prod_{i=1}^{N_g} \frac{\Phi^*[\psi_{1gk}^{-1/2}(\alpha_{gk,z_{gik}+1} - \nu_{gk} - \mathbf{\Lambda}^T_{1gk}\boldsymbol{\omega}_{1gi})] - \Phi^*[\psi_{1gk}^{-1/2}(\alpha_{gk,z_{gik}} - \nu_{gk} - \mathbf{\Lambda}^T_{1gk}\boldsymbol{\omega}_{1gi})]}{\Phi^*[\psi_{1gk}^{-1/2}(\alpha^{(m)}_{gk,z_{gik}+1} - \nu_{gk} - \mathbf{\Lambda}^T_{1gk}\boldsymbol{\omega}_{1gi})] - \Phi^*[\psi_{1gk}^{-1/2}(\alpha^{(m)}_{gk,z_{gik}} - \nu_{gk} - \mathbf{\Lambda}^T_{1gk}\boldsymbol{\omega}_{1gi})]},$$

where Φ^* is the distribution function of $N[0, 1]$. Since R_{gk} only depends on the old and new values of $\boldsymbol{\alpha}_{gk}$ but not on $\mathbf{Y}_{gk}$, it only needs to generate a new $\mathbf{Y}_{gk}$ for an accepted $\boldsymbol{\alpha}_{gk}$. This new $\mathbf{Y}_{gk}$ is simulated from the truncated normal distribution $p(\mathbf{Y}_{gk}|\boldsymbol{\alpha}_{gk}, \cdot)$ via the algorithm given in Roberts (1995).

APPENDIX 9.3: PP P-VALUE FOR TWO-LEVEL NSEM WITH MIXED CONTINUOUS AND ORDERED-CATEGORICAL VARIABLES

Suppose null hypothesis H_0 is that the proposed model defined in Equations (9.1) and (9.2) is plausible, the PP p-value is defined as

$$p_B(\mathbf{X}, \mathbf{Z}) = Pr[D(\mathbf{U}^{\text{rep}}|\boldsymbol{\theta}, \boldsymbol{\alpha}, \mathbf{Y}, \mathbf{V}, \mathbf{\Omega}_1, \mathbf{\Omega}_2) \geq D(\mathbf{U}|\boldsymbol{\theta}, \boldsymbol{\alpha}, \mathbf{Y}, \mathbf{V}, \mathbf{\Omega}_1, \mathbf{\Omega}_2)|\mathbf{X}, \mathbf{Z}, H_0],$$

where $\mathbf{U}^{\text{rep}}$ denotes a replication of $\mathbf{U} = \{\mathbf{u}_{gi}; i = 1, \ldots, N_g, g = 1, \ldots, G\}$, with $\mathbf{u}_{gi}$ satisfying the model defined by Equation (9.3) that involves structural parameters and latent variables satisfying equations (9.4) and (9.5) and $D(\cdot|\cdot)$ is a discrepancy variable. Here, the following χ^2 discrepancy variable is used:

$$D(\mathbf{U}^{\text{rep}}|\boldsymbol{\theta}, \boldsymbol{\alpha}, \mathbf{Y}, \mathbf{V}, \mathbf{\Omega}_1, \mathbf{\Omega}_2) = \sum_{g=1}^{G}\sum_{i=1}^{N_g}(\mathbf{u}^{\text{rep}}_{gi} - \mathbf{v}_g - \mathbf{\Lambda}_{1g}\mathbf{\Omega}_{1gi})^T \mathbf{\Psi}^{-1}_{1g}(\mathbf{u}^{\text{rep}}_{gi} - \mathbf{v}_g - \mathbf{\Lambda}_{1g}\mathbf{\Omega}_{1gi}),$$

which is distributed as $\chi^2(pn)$, where $n = N_1 + \cdots + N_G$, a χ^2 distribution with pn degrees of freedom. The PP p-value on the basis of this discrepancy variable is

$$P_B(\mathbf{X}, \mathbf{Z}) = \int Pr[\chi^2(pn) \geq D(\mathbf{U}|\boldsymbol{\theta}, \boldsymbol{\alpha}, \mathbf{Y}, \mathbf{V}, \mathbf{\Omega}_1, \mathbf{\Omega}_2)] \times$$
$$p(\boldsymbol{\theta}, \boldsymbol{\alpha}, \mathbf{Y}, \mathbf{V}, \mathbf{\Omega}_1, \mathbf{\Omega}_2|\mathbf{X}, \mathbf{Z}) d\boldsymbol{\theta} d\boldsymbol{\alpha} d\mathbf{Y} d\mathbf{V} d\mathbf{\Omega}_1 d\mathbf{\Omega}_2.$$

A Rao–Blackwellized type estimate of the PP p-value is equal to

$$\hat{P}_B(\mathbf{X}, \mathbf{Z})$$
$$= T^{-1} \sum_{t=1}^{T} Pr[\chi^2(pn) \geq D(\mathbf{U}|\boldsymbol{\theta}^{(t)}, \boldsymbol{\alpha}^{(t)}, \mathbf{Y}^{(t)}, \mathbf{V}^{(t)}, \boldsymbol{\Omega}_1^{(t)}, \boldsymbol{\Omega}_2^{(t)})]$$

where $D(\mathbf{U}|\boldsymbol{\theta}^{(t)}, \boldsymbol{\alpha}^{(t)}, \mathbf{Y}^{(t)}, \mathbf{V}^{(t)}, \boldsymbol{\Omega}_1^{(t)}, \boldsymbol{\Omega}_2^{(t)})$ is calculated at each iteration and the tail-area of a χ^2 distribution which can be obtained via standard statistical software. The hypothesized model is rejected if $\hat{P}_B(\mathbf{X}, \mathbf{Z})$ is not close to 0.5.

APPENDIX 9.4: QUESTIONS ASSOCIATED WITH THE MANIFEST VARIABLES

$u_g(1)$: How much of a threat do you think AIDS is to the health of people?
$u_g(2)$: What are the chances that you yourself might get AIDS?
$u_g(3)$: How worried are you about getting AIDS?
How great is the risk of getting AIDS or the AIDS virus from sexual intercourse with someone:
$u_g(4)$: who has the AIDS virus using a condom?
$u_g(5)$: whom you don't know very well without using a condom?
$u_g(6)$: who injects drugs?
$u_g(7)$: How often did you perform vaginal sex in the last 7 days?
$u_g(8)$: How often did you perform manual sex in the last 7 days?
$u_g(9)$: How often did you perform oral sex in the last 7 days?

APPENDIX 9.5: CONDITIONAL DISTRIBUTIONS: SEMs WITH CROSS-LEVEL EFFECTS

Let $\boldsymbol{\theta}_1$ be the structural parameter vector that contains all unknown parameters in $\boldsymbol{\Gamma}$ and $\boldsymbol{\Psi}_\delta$ that are involved in the structural Equation (9.31); let $\boldsymbol{\theta}_2$ be the structural parameter vector that contains all unknown parameters in $\boldsymbol{\mu}$, $\boldsymbol{\Lambda}_2$, $\boldsymbol{\Lambda}_1$ and $\boldsymbol{\Psi}_\epsilon$ that are involved in the measurement Equation (9.30); let $\boldsymbol{\theta}_3$ be the structural parameter vector that contains all unknown parameters in covariance matrices $\boldsymbol{\Phi}_2$ and $\boldsymbol{\Phi}_1$ of the latent vectors. Let $p(\boldsymbol{\theta})$ be some appropriate prior distribution of $\boldsymbol{\theta}$. Based on the definition of the model and the nature of the structural parameters, it is reasonable to consider prior distributions that satisfy $p(\boldsymbol{\theta}) = p(\boldsymbol{\theta}_1)p(\boldsymbol{\theta}_2)p(\boldsymbol{\theta}_3)$.

In deriving the conditional distribution $p(\mathbf{\Omega}_2|\mathbf{\Omega}_1, \boldsymbol{\theta}, \mathbf{U})$, it can be shown that

$$p(\mathbf{\Omega}_2|\mathbf{\Omega}_1, \boldsymbol{\theta}, \mathbf{U}) = \prod_{g=1}^{G} p(\boldsymbol{\omega}_{2g}|\mathbf{\Omega}_{1g}, \mathbf{u}_g, \boldsymbol{\theta}) \propto$$

$$\prod_{g=1}^{G}\left\{p(\boldsymbol{\omega}_{2g}|\boldsymbol{\theta}_3)\prod_{i=1}^{N_g}[p(\mathbf{u}_{gi}|\boldsymbol{\omega}_{2g}, \boldsymbol{\omega}_{1gi}, \boldsymbol{\theta}_2)p(\boldsymbol{\eta}_{gi}|\boldsymbol{\xi}_{gi}, \boldsymbol{\omega}_{2g}, \boldsymbol{\theta}_1)]\right\}.$$

Based on the definition of the model and assumptions, $p(\boldsymbol{\omega}_{2g}|\mathbf{\Omega}_{1g}, \mathbf{u}_g, \boldsymbol{\theta})$ is proportional to

$$\exp\left[-\frac{1}{2}\sum_{i=1}^{N_g}(\mathbf{u}_{gi} - \boldsymbol{\mu} - \mathbf{\Lambda}_2\boldsymbol{\omega}_{2g} - \mathbf{\Lambda}_1\boldsymbol{\omega}_{1gi})^T\mathbf{\Psi}_\epsilon^{-1}(\mathbf{u}_{gi} - \boldsymbol{\mu} - \mathbf{\Lambda}_2\boldsymbol{\omega}_{2g} - \mathbf{\Lambda}_1\boldsymbol{\omega}_{1gi})\right.$$
$$\left.-\frac{1}{2}\boldsymbol{\omega}_{2g}^T\mathbf{\Phi}_2^{-1}\boldsymbol{\omega}_{2g} - \frac{1}{2}\sum_{i=1}^{N_g}(\boldsymbol{\eta}_{gi} - \mathbf{\Gamma}\mathbf{H}(\boldsymbol{\xi}_{gi}, \boldsymbol{\omega}_{2g}))^T\mathbf{\Psi}_\delta^{-1}(\boldsymbol{\eta}_{gi} - \mathbf{\Gamma}\mathbf{H}(\boldsymbol{\xi}_{gi}, \boldsymbol{\omega}_{2g}))\right]. \quad \text{(A9.10)}$$

Consider the conditional distribution $p(\mathbf{\Omega}_1|\mathbf{\Omega}_2, \boldsymbol{\theta}, \mathbf{U})$. Note that given $\mathbf{U}$ and $\mathbf{\Omega}_2$, $\boldsymbol{\omega}_{1gi}$ are mutually independent for $g = 1, \ldots, G$ and $i = 1, \ldots, N_g$. Hence, it can be shown that

$$p(\mathbf{\Omega}_1|\mathbf{\Omega}_2, \boldsymbol{\theta}, \mathbf{U}) = \prod_{g=1}^{G}\prod_{i=1}^{N_g} p(\boldsymbol{\omega}_{1gi}|\mathbf{u}_{gi}, \boldsymbol{\theta}, \boldsymbol{\omega}_{2g}) \propto$$

$$\prod_{g=1}^{G}\prod_{i=1}^{N_g}[p(\mathbf{u}_{gi}|\boldsymbol{\omega}_{1gi}, \boldsymbol{\omega}_{2g}, \boldsymbol{\theta}_2)p(\boldsymbol{\eta}_{gi}|\boldsymbol{\xi}_{gi}, \boldsymbol{\omega}_{2g}, \boldsymbol{\theta}_1)p(\boldsymbol{\xi}_{gi}|\boldsymbol{\theta}_3)].$$

Note that given $\boldsymbol{\omega}_{1gi}$ and $\boldsymbol{\omega}_{2g}$, $\mathbf{u}_{gi}$ are mutually independent, and given $\boldsymbol{\xi}_{gi}$ and $\boldsymbol{\omega}_{2g}$, $\boldsymbol{\eta}_{gi}$ are mutually independent. Hence, it follows from the definition of the model and assumptions that $p(\boldsymbol{\omega}_{1gi}|\mathbf{u}_{gi}, \boldsymbol{\theta}, \boldsymbol{\omega}_{2g})$ is proportional to

$$\exp\left[-\frac{1}{2}(\mathbf{u}_{gi} - \boldsymbol{\mu} - \mathbf{\Lambda}_2\boldsymbol{\omega}_{2g} - \mathbf{\Lambda}_1\boldsymbol{\omega}_{1gi})^T\mathbf{\Psi}_\epsilon^{-1}(\mathbf{u}_{gi} - \boldsymbol{\mu} - \mathbf{\Lambda}_2\boldsymbol{\omega}_{2g} - \mathbf{\Lambda}_1\boldsymbol{\omega}_{1gi})\right.$$
$$\left.-\frac{1}{2}\boldsymbol{\xi}_{gi}^T\mathbf{\Phi}_1^{-1}\boldsymbol{\xi}_{gi} - \frac{1}{2}(\boldsymbol{\eta}_{gi} - \mathbf{\Gamma}\mathbf{H}(\boldsymbol{\xi}_{gi}, \boldsymbol{\omega}_{2g}))^T\mathbf{\Psi}_\delta^{-1}(\boldsymbol{\eta}_{gi} - \mathbf{\Gamma}\mathbf{H}(\boldsymbol{\xi}_{gi}, \boldsymbol{\omega}_{2g}))\right]. \quad \text{(A9.11)}$$

Hence, by utilizing the idea of data augmentation, and working with conditional distributions on the basis of a complete data set, the problems that are induced by the dependence of the observed and/or latent vectors can be solved.

Useful prior information of $\boldsymbol{\theta}$ can be incorporated in the conditional distribution of $p(\boldsymbol{\theta}|\boldsymbol{\Omega}_2, \boldsymbol{\Omega}_1, \mathbf{U})$ for achieving better results. Again, the commonly used conjugate prior distributions are used. Specifically, we use the following conjugate prior distribution for $\boldsymbol{\theta}_1$. For $k = 1, \ldots, r_1$,

$$\psi_{\delta k}^{-1} \stackrel{D}{=} Gamma[\alpha_{0\delta k}, \beta_{0\delta k}], \; [\boldsymbol{\Gamma}_k|\psi_{\delta k}] \stackrel{D}{=} N[\boldsymbol{\Gamma}_{0k}, \psi_{\delta k}\mathbf{H}_{0\omega k}], \tag{A9.12}$$

where $\psi_{\delta k}$ is the kth diagonal element of $\boldsymbol{\Psi}_\delta$, and $\boldsymbol{\Gamma}_k^T$ is the kth row vector of $\boldsymbol{\Gamma}$, $\alpha_{0\delta k}, \beta_{0\delta k}, \boldsymbol{\Gamma}_{0k}$ and the positive definite matrix $\mathbf{H}_{0\omega k}$ are hyperparameters whose values are assumed to be given from the prior information of previous studies or other sources. For $h \neq k$, it is assumed that $(\psi_{\delta k}, \boldsymbol{\Gamma}_k)$ and $(\psi_{\delta h}, \boldsymbol{\Gamma}_h)$ are independent. The following conjugate prior distributions for $\boldsymbol{\Psi}_\epsilon$ and $\boldsymbol{\Lambda} = (\boldsymbol{\Lambda}_2, \boldsymbol{\Lambda}_1)$ in $\boldsymbol{\theta}_2$ are used. For $k = 1, \ldots, p$

$$\boldsymbol{\mu} \stackrel{D}{=} N[\boldsymbol{\mu}_0, \boldsymbol{\Sigma}_0], \; \psi_{\epsilon k}^{-1} \stackrel{D}{=} Gamma[\alpha_{0\epsilon k}, \beta_{0\epsilon k}], \; [\boldsymbol{\Lambda}_k|\psi_{\epsilon k}] \stackrel{D}{=} N[\boldsymbol{\Lambda}_{0k}, \psi_{\epsilon k}\mathbf{H}_{0uk}], \tag{A9.13}$$

where $\psi_{\epsilon k}$ is the kth diagonal element of $\boldsymbol{\Psi}_\epsilon$, $\boldsymbol{\Lambda}_k^T$ is the kth row vector of $\boldsymbol{\Lambda}$, $\alpha_{0\epsilon k}, \beta_{0\epsilon k}, \boldsymbol{\Lambda}_{0k}$ and $\mathbf{H}_{0uk}$ are the given hyperparameters. For $h \neq k$, it is assumed that $(\psi_{\epsilon h}, \boldsymbol{\Lambda}_h)$ and $(\psi_{\epsilon k}, \boldsymbol{\Lambda}_k)$ are independent. The conjugate prior distributions for components in $\boldsymbol{\theta}_3$ are:

$$\boldsymbol{\Phi}_2 \stackrel{D}{=} IW_q[\mathbf{R}_{02}, \rho_{02}], \quad \boldsymbol{\Phi}_1 \stackrel{D}{=} IW_{r_2}[\mathbf{R}_{01}, \rho_{01}], \tag{A9.14}$$

where ρ_{02}, ρ_{01}, and the positive definite matrices $\mathbf{R}_{02}$ and $\mathbf{R}_{01}$ are the given hyperparameters.

Consider the components of the conditional distributions in relation to $p(\boldsymbol{\theta}_1|\boldsymbol{\Omega}_2, \boldsymbol{\Omega}_1, \mathbf{U})$. It follows from reasoning similar to that in Song and Lee (2004) that

$$\begin{aligned} &p(\psi_{\delta k}^{-1}|\boldsymbol{\Omega}_2, \boldsymbol{\Omega}_1, \mathbf{U}) \stackrel{D}{=} Gamma\,[\frac{n}{2} + \alpha_{0\delta k}, \beta_{\delta k}], \text{ and} \\ &p(\boldsymbol{\Gamma}_k|\boldsymbol{\Omega}_2, \boldsymbol{\Omega}_1, \mathbf{U}, \psi_{\delta k}^{-1}) \stackrel{D}{=} N[\boldsymbol{\nu}_{\delta k}, \psi_{\delta k}\boldsymbol{\Omega}_{\delta k}], \end{aligned} \tag{A9.15}$$

where $\boldsymbol{\Omega}_{\delta k} = [(\mathbf{H}_{0\omega k}^{-1} + \sum_{g=1}^{G}\sum_{i=1}^{N_g}\mathbf{H}(\boldsymbol{\xi}_{gi}, \boldsymbol{\omega}_{2g})\mathbf{H}(\boldsymbol{\xi}_{gi}, \boldsymbol{\omega}_{2g})^T)]^{-1}$, $\boldsymbol{\nu}_{\delta k} = \boldsymbol{\Omega}_{\delta k}[\mathbf{H}_{0\omega k}^{-1}\boldsymbol{\Gamma}_{0k} + \sum_{g=1}^{G}\sum_{i=1}^{N_g}\eta_{gik}\mathbf{H}(\boldsymbol{\xi}_{gi}, \boldsymbol{\omega}_{2g})]$, η_{gik} is the kth element of $\boldsymbol{\eta}_{gi}$ and $\beta_{\delta k} = \beta_{0\delta k} + (\sum_{g=1}^{G}\sum_{i=1}^{N_g}\eta_{gik}^2 - \boldsymbol{\nu}_{\delta k}^T\boldsymbol{\Omega}_{\delta k}^{-1}\boldsymbol{\nu}_{\delta k} + \boldsymbol{\Gamma}_{0k}^T\mathbf{H}_{0\omega k}^{-1}\boldsymbol{\Gamma}_{0k})/2$.

Consider the conditional distributions in relation to $p(\boldsymbol{\theta}_2|\boldsymbol{\Omega}_2, \boldsymbol{\Omega}_1, \mathbf{U})$. Let $\boldsymbol{\Omega}_{\epsilon k} = (\mathbf{H}_{0uk}^{-1} + \sum_{g=1}^{G}\sum_{i=1}^{N_g} \boldsymbol{\zeta}_{gi}\boldsymbol{\zeta}_{gi}^T)^{-1}$, $\boldsymbol{\nu}_{\epsilon k} = \boldsymbol{\Omega}_{\epsilon k}(\mathbf{H}_{0uk}^{-1}\boldsymbol{\Lambda}_{0k} + \sum_{g=1}^{G}\sum_{i=1}^{N_g} u_{gik}\boldsymbol{\zeta}_{gi})$ and $\beta_{\epsilon k} = \beta_{0\epsilon k} + (\sum_{g=1}^{G}\sum_{i=1}^{N_g} u_{gik}^2 - \boldsymbol{\nu}_{\epsilon k}^T\boldsymbol{\Omega}_{\epsilon k}^{-1}\boldsymbol{\nu}_{\epsilon k} + \boldsymbol{\Lambda}_{0k}^T\mathbf{H}_{0uk}^{-1}\boldsymbol{\Lambda}_{0k})/2$, where $\boldsymbol{\zeta}_{gi}$ is defined as in Equation (9.33). Similarly, it can be shown that

$$p(\psi_{\epsilon k}^{-1}|\boldsymbol{\Omega}_2, \boldsymbol{\Omega}_1, \mathbf{U}) \stackrel{D}{=} Gamma\,[\frac{n}{2} + \alpha_{0\epsilon k}, \beta_{\epsilon k}],$$

$$p(\boldsymbol{\Lambda}_k|\boldsymbol{\Omega}_2, \boldsymbol{\Omega}_1, \mathbf{U}, \psi_{\epsilon k}^{-1}) \stackrel{D}{=} N\,[\boldsymbol{\nu}_{\epsilon k}, \psi_{\epsilon k}\boldsymbol{\Omega}_{\epsilon k}],$$

$$p(\boldsymbol{\mu}|\mathbf{U}, \boldsymbol{\Psi}_\epsilon, \boldsymbol{\Omega}_1, \boldsymbol{\Lambda}) \stackrel{D}{=} N\,[(\boldsymbol{\Sigma}_0^{-1} + n\boldsymbol{\Psi}_\epsilon^{-1})^{-1}(n\boldsymbol{\Psi}_\epsilon^{-1}\bar{\mathbf{U}}^* + \boldsymbol{\Sigma}_0^{-1}\boldsymbol{\mu}_0), (\boldsymbol{\Sigma}_0^{-1} + n\boldsymbol{\Psi}_\epsilon^{-1})^{-1}], \tag{A9.16}$$

where $\bar{\mathbf{U}}^* = \sum_{g=1}^{G}\sum_{i=1}^{N_g}(\mathbf{u}_{gi} - \boldsymbol{\Lambda}\boldsymbol{\zeta}_{gi})/n$.

Finally, for $\boldsymbol{\theta}_3$, it can be shown that

$$p(\boldsymbol{\Phi}_2|\boldsymbol{\Omega}_2, \boldsymbol{\Omega}_1, \mathbf{U}) \stackrel{D}{=} IW_q\left(\sum_{g=1}^{G} \boldsymbol{\omega}_{2g}\boldsymbol{\omega}_{2g}^T + \mathbf{R}_{02}, G + \rho_{02}\right), \tag{A9.17}$$

$$p(\boldsymbol{\Phi}_1|\boldsymbol{\Omega}_2, \boldsymbol{\Omega}_1, \mathbf{U}) \stackrel{D}{=} IW_{r_2}\left(\sum_{g=1}^{G}\sum_{i=1}^{N_g} \boldsymbol{\xi}_{gi}\boldsymbol{\xi}_{gi}^T + \mathbf{R}_{01}, n + \rho_{01}\right). \tag{A9.18}$$

APPENDIX 9.6: THE MH ALGORITHM: SEMs WITH CROSS-LEVEL EFFECTS

This algorithm is implemented as follows. At the $(j+1)$th iteration with a current value $\boldsymbol{\omega}_{1gi}^{(j)}$, a new candidate $\boldsymbol{\omega}_{1gi}$ is generated from the proposal distribution $N\,[\boldsymbol{\omega}_{1gi}^{(j)}, \sigma^2\boldsymbol{\Sigma}_\omega]$, where $\boldsymbol{\Sigma}_\omega^{-1} = \boldsymbol{\Lambda}_1^T\boldsymbol{\Psi}_\epsilon^{-1}\boldsymbol{\Lambda}_1 + \boldsymbol{\Sigma}$ with

$$\boldsymbol{\Sigma} = \begin{pmatrix} \boldsymbol{\Psi}_\delta^{-1} & -\boldsymbol{\Psi}_\delta^{-1}\boldsymbol{\Gamma}\boldsymbol{\Delta} \\ -\boldsymbol{\Delta}^T\boldsymbol{\Gamma}^T\boldsymbol{\Psi}_\delta^{-1} & \boldsymbol{\Phi}_1^{-1} + \boldsymbol{\Delta}^T\boldsymbol{\Gamma}^T\boldsymbol{\Psi}_\delta^{-1}\boldsymbol{\Gamma}\boldsymbol{\Delta} \end{pmatrix},$$

where $\boldsymbol{\Delta} = \partial\mathbf{H}(\boldsymbol{\xi}_{gi}, \boldsymbol{\omega}_{2g})/\partial\boldsymbol{\xi}_{gi}^T|_{\boldsymbol{\xi}_{gi}=\mathbf{0}}$ and σ^2 is chosen such that the average acceptance rate is approximately 0.25 or more (see Gelman, Roberts and Gilks, 1995). The acceptance probability is

$$\min\left[1, \frac{p(\boldsymbol{\omega}_{1gi}|\boldsymbol{u}_{gi}, \boldsymbol{\theta}, \boldsymbol{\omega}_{2g})}{p(\boldsymbol{\omega}_{1gi}^{(j)}|\boldsymbol{u}_{gi}, \boldsymbol{\theta}, \boldsymbol{\omega}_{2g})}\right].$$

Similarly, the MH algorithm for sampling $\boldsymbol{\omega}_{2g}$ from $p(\boldsymbol{\omega}_{2g}|\boldsymbol{\Omega}_{1g}, \boldsymbol{u}_g, \boldsymbol{\theta})$ is as follows. At the $(j+1)$th iteration with a current value $\boldsymbol{\omega}_{2g}^{(j)}$, a new candidate $\boldsymbol{\omega}_{2g}$ is generated from the proposal distribution $N[\boldsymbol{\omega}_{2g}^{(j)}, \sigma^2\boldsymbol{\Sigma}]$, where $\boldsymbol{\Sigma} = N_g\boldsymbol{\Lambda}_2^T\boldsymbol{\Psi}_\epsilon^{-1}\boldsymbol{\Lambda}_2 + \boldsymbol{\Phi}_2^{-1} + \sum_{i=1}^{N_g}\boldsymbol{\Delta}_{2i}^T\boldsymbol{\Gamma}^T\boldsymbol{\Psi}_\delta^{-1}\boldsymbol{\Gamma}\boldsymbol{\Delta}_{2i}$ with $\boldsymbol{\Delta}_{2i} = \partial\mathbf{H}(\boldsymbol{\xi}_{gi}, \boldsymbol{\omega}_{2g})/\partial\boldsymbol{\omega}_{2g}^T|_{\boldsymbol{\omega}_{2g}=\mathbf{0}}$ and the choice of σ^2 is similar. The acceptance probability is

$$\min\left[1, \frac{p(\boldsymbol{\omega}_{2g}|\boldsymbol{\Omega}_{1g}, \boldsymbol{u}_g, \boldsymbol{\theta})}{p(\boldsymbol{\omega}_{2g}^{(j)}|\boldsymbol{\Omega}_{1g}, \boldsymbol{u}_g, \boldsymbol{\theta})}\right].$$

REFERENCES

Ansari, A. and Jedidi, K. (2000) Bayesian factor analysis for multilevel binary observations. *Psychometrika*, **65**, 475–498.

Ansari, A., Jedidi, K. and Dube, L. (2002) Heterogeneous factor analysis models: a Bayesian approach. *Psychometrika*, **67**, 49–78.

Carlin, B. P. and Louis, T. A. (1996) *Bayes and Empirical Bayes Methods for Data Analysis*. London: Chapman and Hall.

Cowles, M. K. (1996) Accelerating Monte Carlo Markov chain convergence for cumulative-link generalized linear models. *Statistics and Computing*, **6**, 101–111.

Gelman, A. and Meng, X. L. (1998) Simulating normalizing constants: from importance sampling to bridge sampling to path sampling. *Statistical Science*, **13**, 163–185.

Gelman, A. and Rubin, D. B. (1992) Inference from iterative simulation using multiple sequences. *Statistical Science*, **7**, 457–472.

Gelman, A., Meng, X. L. and Stern, H. (1996) Posterior predictive assessment of model fitness via realized discrepancies. *Statistica Sinica*, **6**, 733–759.

Gelman, A., Roberts, G. O. and Gilks, W. R. (1995) Efficient Metropolis jumping rules. In J. M. Bernardo, J. O. Berger, A. P. Dawid and A. F. M. Smith (eds), *Bayesian Statistics 5*, pp. 599–607. Oxford: Oxford University Press.

Geman, S. and Geman, D. (1984) Stochastic relaxation, Gibbs distribution and the Bayesian restoration of images. *IEEE Transactions on Pattern Analysis and Machine Intelligence*, **6**, 721–741.

Jöreskog, K. G. and Sörborn, D. (1996) *LISREL 8: Structural Equation Modeling with the SIMPLIS Command Language*. Hove and London: Scientific Software International.

Kass, R. E. and Raftery, A. E. (1995) Bayes factors. *Journal of the American Statistical Association*, **90**, 773–795.

Lee, S. Y. and Poon, W. Y. (1998) Analysis of two-level structural equation models via EM type algorithms. *Statistica Sinica*, **8**, 749–766.

Lee, S. Y. and Shi, J. Q. (2001) Maximum likelihood estimation of two-level latent variable models with mixed continuous and polytomous data. *Biometrics*, **57**, 787–794.

Lee, S. Y. and Song, X. Y. (2005) Maximum likelihood analysis of a two-level nonlinear structural equation model with fixed covariates. *Journal of Educational and Behavioral Statistics*, **30**, 1–26.

Lee, S. Y. and Tang, N. S. (2006) Bayesian analysis of two-level nonlinear structural equation models with cross-level effects. (Unpublished manuscript).

Lee, S. Y. and Tsang, S. Y. (1999) Constrained maximum likelihood estimation of two-level covariance structure model via EM type algorithms. *Psychometrika*, **64**, 435–450.

Lee, S. Y. and Zhu, H. T. (2000) Statistical analysis of nonlinear structural equation models with continuous and polytomous data. *British Journal of Mathematical and Statistical Psychology*, **53**, 209–232.

Lee, S. Y., Poon, W. Y. and Bentler, P. M. (1995) A two-stage estimation of structural equation models with continuous and polytomous variables. *British Journal of Mathematical and Statistical Psychology*, **48**, 339–358.

McDonald, R. P. and Goldstein, H. (1989) Balanced versus unbalanced designs for linear structural relations in two-level data. *The British Journal of Mathematical and Statistical Psychology*, **42**, 215–232.

Morisky, D. E. *et al.* (1998) The effects of establishment practices, knowledge and attitudes on condom use among Filipina sex workers. *AIDS Care*, **10**, 213–220.

Pauler, D. A., Wakefield, J. C. and Kass, R. E. (1999) Bayes factor and approximations for variance component models. *Journal of the American Statistical Association*, **94**, 1242–1253.

Rabe-Hesketh, S., Skrondal, A. and Pickles, A. (2004) Generalized multilevel structural equation modeling. *Psychometrika*, **69**, 167–190.

Raftery, A. E. (1996a) Approximate Bayes factors and accounting for model uncertainty in generalised linear models. *Biometrika*, **83**, 251–266.

Raftery, A. E. (1996b) Hypothesis testing and model selection. In W. R. Gilks, S. Richardson and D.J. Spieglhalter (eds), *Practical Markov Chain Monte Carlo*, pp. 163–188. London: Chapman and Hall.

Richardson, S. and Green, P. J. (1997) On Bayesian analysis of mixtures with an unknown number of components (with discussion). *Journal of the Royal Statistical Society, Series B*, **59**, 731–792.

Roberts, C. P. (1995) Simulation of truncated normal variables. *Statistics and Computing*, **5**, 121–125.

Shi, J. Q. and Lee, S. Y. (1998) Bayesian sampling-based approach for factor analysis model with continuous and polytomous data. *British Journal of Mathematical and Statistical Psychology*, **51**, 233–252.

Shi, J. Q. and Lee, S. Y. (2000) Latent variable models with mixed continuous and polytomous data. *Journal of the Royal Statistical Society, Series B*, **62**, 77–87.

Song, X. Y. and Lee, S. Y. (2001) Bayesian estimation and test for factor analysis model with continuous and polytomous data in several populations. *British Journal of Mathematical and Statistical Psychology*, **54**, 237–263.

Song, X. Y. and Lee, S. Y. (2004) Bayesian analysis of two-level nonlinear structural equation models with continuous and polytomous data. *British Journal of Mathematical and Statistical Psychology*, **57**, 29–52.

Spiegalhalter, D. J., Thomas, A., Best, N. G. and Lunn, D. (2003) *WinBUGS User Manual, Version 1.4*, Cambridge, England: MRC Biostatistics Unit.

Tanner, M. A. and Wong, W. H. (1987) The calculation of posterior distributions by data augmentation(with discussion). *Journal of the American Statistical Association*, **82**, 528–550.

Zhang, W. and Lee, S. Y. (2001) Asymptotic theory of two-level structural equation models with constraints. *Statistica Sinica*, **11**, 135–145.

Zhu, H. T. and Lee, S. Y. (1999) Statistical analysis of nonlinear factor analysis models. *British Journal of Mathematical and Statistical Psychology*, **52**, 225–242.

Zhu, H. T. and Lee, S. Y. (2001) A Bayesian analysis of finite mixtures in the LISREL model. *Psychometrika*, **66**, 133–152.

10 Multisample Analysis of Structural Equation Models

10.1 INTRODUCTION

The two-level SEMs are used to analyze data with hierarchical structures which usually involve a large number of groups, and observations within each group are correlated. The main purposes of two-level SEMs are to take into account the correlated structure of the data, and to establish a between-groups model. Multisample data are coming from a comparatively smaller number of groups (populations). The number of observations within each group is usually large, and it is assumed that observations within each group are independent. As the within-group observations are independent, the multisample data do not have a hierarchical structure. In multisample analysis of SEMs, each group is associated with a hypothesized model of interest. One main objective is to investigate the similarities or differences among the models in the different groups. As a result, the statistical inferences emphasized in analyzing multisample SEMs are different from those in analyzing two-level SEMs.

Analysis of multiple samples is a major topic in structural equation modeling. This kind of analysis is useful for investigating the behaviors of different groups of employees, different cultures, different treatment groups, etc.. The main interest is the testing of hypotheses about the different kinds of invariances among the models in different groups. The traditional approach for testing

Structural Equation Modeling: A Bayesian Approach S-Y. Lee

invariance applied the likelihood ratio test on the basis of the asymptotic chi-square distribution (see Bollen (1989)). For more complex SEMs, it is rather difficult to evaluate the observed-data likelihood at the ML estimates to obtain the test statistics; moreover, the asymptotic distribution of the likelihood ratio test statistic is unknown and hard to derive. As we will see later in this chapter, this important issue can be formulated as a model comparison problem, and can be effectively addressed by the Bayes factor or DIC in a Bayesian approach. Another advantage of the Bayesian model comparison through Bayes factor is that non-nested models (hypotheses) can be compared, hence it is not necessary to follow a hierarchy of hypotheses to assess the invariance for the SEMs in different groups.

Bayesian methods for analyzing mutlisample SEMs are presented in this chapter. We will emphasize nonlinear SEMs with ordered categorical (or dichotomous) variables, although the general ideas can be applied to SEMs with other settings. The rest of the chapter is organized as follows: Section 10.2 presents the model; the Bayesian approach for estimation and model comparison (hypothesis testing) are discussed in Section 10.3; Section 10.4 presents the results that are obtained from the analyses of two applications. One is about job attitude, emotion and benefit attitude, while the other is about quality of life, and analyzed via WinBUGS. Technical details are given in the appendix.

10.2 THE MULTISAMPLE NONLINEAR STRUCTURAL EQUATION MODEL

We first extend the standard SEMs to a multisample nonlinear SEM. Consider G independent groups of individuals that may represent different populations of employees, different cultures, etc.. For $g = 1, \ldots, G$, and $i = 1, \ldots, N_g$, let $\mathbf{v}_i^{(g)}$ be the $p \times 1$ random vector of manifest variables that correspond to the ith observation (subject) in the gth group. In contrast to two-level SEMs, for $i = 1, \ldots, N_g$ in the gth group, $\mathbf{v}_i^{(g)}$ are assumed to be independent. For each $g = 1, \ldots, G$, $\mathbf{v}_i^{(g)}$ is related to latent variables in a $q \times 1$ random vector $\boldsymbol{\omega}_i^{(g)}$ by the following measurement equation:

$$\mathbf{v}_i^{(g)} = \boldsymbol{\mu}^{(g)} + \boldsymbol{\Lambda}^{(g)}\boldsymbol{\omega}_i^{(g)} + \boldsymbol{\epsilon}_i^{(g)}, \tag{10.1}$$

where $\boldsymbol{\mu}^{(g)}$ is the vector of the intercepts, $\boldsymbol{\Lambda}^{(g)}$ is the parameter matrix of regression coefficients that reflect the relation of manifest variables in $\mathbf{v}_i^{(g)}$ with the latent variables in $\boldsymbol{\omega}_i^{(g)}$, and $\boldsymbol{\epsilon}_i^{(g)}$ is a random vector of the measurement errors. It is assumed that $\boldsymbol{\omega}_i^{(g)}$ and $\boldsymbol{\epsilon}_i^{(g)}$ are independent, and the distribution of $\boldsymbol{\epsilon}_i^{(g)}$ is $N[\mathbf{0}, \boldsymbol{\Psi}_\epsilon^{(g)}]$, where $\boldsymbol{\Psi}_\epsilon^{(g)}$ is a diagonal covariance matrix. Let $\boldsymbol{\omega}_i^{(g)} = (\boldsymbol{\eta}_i^{(g)T}, \boldsymbol{\xi}_i^{(g)T})^T$, where $\boldsymbol{\eta}_i^{(g)}$ represents the $(q_1 \times 1)$ vector of endogenous latent variables, and $\boldsymbol{\xi}_i^{(g)}$ represents the $(q_2 \times 1)$ vector of the exogenous latent variables. Note that it is naturally assumed that q_1 and q_2 are independent of g, that is they are the same for each group. To assess the effects of the nonlinear terms of latent variables in $\boldsymbol{\xi}_i^{(g)}$ to $\boldsymbol{\eta}_i^{(g)}$, we consider a nonlinear SEM with the following nonlinear structural equation:

$$\boldsymbol{\eta}_i^{(g)} = \boldsymbol{\Pi}^{(g)}\boldsymbol{\eta}_i^{(g)} + \boldsymbol{\Gamma}^{(g)}\mathbf{H}(\boldsymbol{\xi}_i^{(g)}) + \boldsymbol{\delta}_i^{(g)} = \boldsymbol{\Lambda}_\omega^{(g)}\mathbf{H}^*(\boldsymbol{\omega}_i^{(g)}) + \boldsymbol{\delta}_i^{(g)}, \tag{10.2}$$

where $\Pi^{(g)}$ and $\boldsymbol{\Gamma}^{(g)}$are unknown parameter matrices, and $\boldsymbol{\delta}_i^{(g)}$ is the random vector of error measurements that is independent of $\boldsymbol{\xi}_i^{(g)}$; $\boldsymbol{\Lambda}_\omega^{(g)} = (\boldsymbol{\Pi}^{(g)}, \boldsymbol{\Gamma}^{(g)})$, $\mathbf{H}^*(\boldsymbol{\omega}_i^{(g)}) = (\boldsymbol{\eta}_i^{(g)T}, \mathbf{H}(\boldsymbol{\xi}_i^{(g)})^T)^T$ and $\mathbf{H}(\boldsymbol{\xi}_i^{(g)}) = (h_1(\boldsymbol{\xi}_i^{(g)}), \ldots, h_a(\boldsymbol{\xi}_i^{(g)}))^T$, in which the hs are nonzero and known differentiable functions that include polynomials. It is assumed that the vector-valued functions $\mathbf{H}$ and $\mathbf{H}^*$ do not depend on g. However, different groups can have different linear or nonlinear terms of $\boldsymbol{\xi}_i^{(g)}$ by defining appropriate $\mathbf{H}$ (or $\mathbf{H}^*$) and assigning zero values to appropriate elements in $\boldsymbol{\Gamma}^{(g)}$. For example, let $\boldsymbol{\omega}_i^{(g)} = (\eta_i^{(g)}, \xi_{i1}^{(g)}, \xi_{i2}^{(g)})^T$, for $g = 1, 2$. Suppose that the structural equations in the first and second groups are given by

$$\begin{aligned}
\eta_i^{(1)} &= \gamma_1^{(1)}\xi_{i1}^{(1)} + \gamma_2^{(1)}\xi_{i2}^{(1)} + \gamma_3^{(1)}\xi_{i1}^{(1)}\xi_{i2}^{(1)} + \delta_i^{(1)}, \\
\eta_i^{(2)} &= \gamma_1^{(2)}\xi_{i1}^{(2)} + \gamma_2^{(2)}\xi_{i2}^{(2)} + \gamma_4^{(2)}\xi_{i1}^{(2)}\xi_{i1}^{(2)} + \delta_i^{(2)}.
\end{aligned} \tag{10.3}$$

In defining these structural equations, we consider $\mathbf{H}(\xi_i^{(g)}) = (\xi_{i1}^{(g)}, \xi_{i2}^{(g)}, \xi_{i1}^{(g)}\xi_{i2}^{(g)}, \xi_{i1}^{(g)}\xi_{i1}^{(g)})^T$, $\boldsymbol{\Gamma}^{(1)} = (\gamma_1^{(1)}, \gamma_2^{(1)}, \gamma_3^{(1)}, 0)$ and $\boldsymbol{\Gamma}^{(2)} = (\gamma_1^{(2)}, \gamma_2^{(2)}, 0, \gamma_4^{(2)})$. In practice, models in different groups usually have the same linear and nonlinear terms of $\boldsymbol{\xi}_i^{(g)}$. For handling very complicated different nonlinear functions in the models, this assumption can be relaxed with minor modifications. In the same way as for models in previous chapters, it is assumed that $\mathbf{I} - \boldsymbol{\Pi}^{(g)}$ is a nonsingular matrix, the determinant of which is not dependent on elements in $\boldsymbol{\Pi}^{(g)}$. It is further assumed that $\boldsymbol{\xi}_i^{(g)}$ is distributed as $N[\mathbf{0}, \boldsymbol{\Phi}^{(g)}]$, and $\boldsymbol{\delta}_i^{(g)}$ is distributed with $N[\mathbf{0}, \boldsymbol{\Psi}_\delta^{(g)}]$, where $\boldsymbol{\Psi}_\delta^{(g)}$ is a diagonal covariance matrix.

To handle the ordered categorical outcomes, suppose that $\mathbf{v}_i^{(g)} = (\mathbf{x}_i^{(g)T}, (y_i^{(g)T})^T$, where $\mathbf{x}_i^{(g)}$ is an observable $(r \times 1)$ subvector of continuous responses, while $\mathbf{y}_i^{(g)}$ is an $(s \times 1)$ subvector of unobservable continuous responses, the information of which is reflected by an observable ordered categorical vector $\mathbf{z}_i^{(g)}$. In a generic sense, an ordered categorical variable $z_m^{(g)}$ is defined with its underlying latent continuous random variable $y_m^{(g)}$ by:

$$z_m^{(g)} = a \text{ if } \alpha_{m,z_m}^{(g)} \le y_m^{(g)} < \alpha_{m,z_m+1}^{(g)}, \ a = 1, \ldots, b_m, \ m = 1, \ldots, s, \tag{10.4}$$

where $\{-\infty = \alpha_{m,1}^{(g)} < \alpha_{m,2}^{(g)} < \cdots < \alpha_{m,b_m}^{(g)} < \alpha_{m,b_m+1}^{(g)} = \infty\}$ is the set of threshold parameters that define the categories, and b_m is the number of categories for the ordered categorical variable $z_m^{(g)}$. Note that for each ordered categorical variable, the number of thresholds is the same for each group. However, the categories can be unequally spaced under this formulation.

To tackle the identification problem, we have to pay attention to the following issues. There are two kinds of indeterminacies in the multisample SEMs with ordered categorical variables. First, the SEM that is defined by Equations (10.1) and (10.2) is not identified. This indeterminacy can be solved by the common method of fixing appropriate elements in $\mathbf{\Lambda}^{(g)}$, $\mathbf{\Pi}^{(g)}$ and/or $\mathbf{\Gamma}^{(g)}$ at preassigned values. The other indeterminacy is induced by the ordered categorical variables. Consider an ordered categorical variable $z_m^{(g)}$ that is defined by a set of thresholds $\alpha_{m,k}^{(g)}$ and an underlying latent continuous variable $y_m^{(g)}$ with mean and variance $\mu_m^{(g)}$, and $\sigma_m^{2(g)}$. The indeterminacy is caused by the fact that $\alpha_{m,k}^{(g)}$, $\mu_m^{(g)}$ and $\sigma_m^{2(g)}$ are not simultaneously estimable. For a given group g, a common method to solve this identification problem with respect to each mth ordered categorical variable that corresponds to the gth group is to fix $\alpha_{m,2}^{(g)}$ and $\alpha_{m,b_m}^{(g)}$ at preassigned values (Lee, Poon and Bentler, 1995; Shi and Lee, 2000; Lee, Song, Skevington and Hao, 2005). For example, we may fix $\alpha_{m,2}^{(g)} = \Phi^{*-1}(f_{m,2}^{(g)})$ and $\alpha_{m,b_m}^{(g)} = \Phi^{*-1}(f_{m,b_m}^{(g)})$, where Φ^* is the cumulative function of $N[0, 1]$, $f_{m,2}^{(g)}$ and $f_{m,b_m}^{(g)}$ are the frequencies of the first category and the cumulative frequencies of categories with $z_m^{(g)} < b_m$. For analyzing multisample models with interest in group comparisons, it is important to impose conditions for identifying the ordered categorical variables such that the underlying latent continuous variables have the same scale among the groups. To achieve this, we can choose the first group as the reference group and identify its ordered categorical variables by fixing both end thresholds as above. Then, for any m, and $g \neq 1$, we impose the following restrictions (see Lee, Poon and Bentler,1989),

$$\alpha_{m,k}^{(g)} = \alpha_{m,k}^{(1)}, \qquad k = 1, \ldots, b_m, \tag{10.5}$$

on the thresholds for every ordered categorical variable $z_m^{(g)}$. Under these identification conditions, the unknown parameters in the groups are interpreted in a relative sense, compared over groups. Note that when a different reference group is used, relations over groups are unchanged. Hence, the statistical inferences are unaffected by the choice of the reference group. Clearly, the compatibility of the groups is reflected by the differences of the parameter estimates. Note that for dichotomous variables the threshold in each group is fixed at zero, and $z_m^{(g)}$ is either equal to '0' or '1'. In this case, it is not necessary to set any constraints on the thresholds across groups.

10.3 BAYESIAN ANALYSIS OF MULTISAMPLE NONLINEAR SEMs

In this section, we will describe Bayesian estimation and model comparison in the context of multisample nonlinear SEMs with ordered categorical variables. Again, the general strategy that utilizes the idea of data augmentation and MCMC tools is used. Theoretically, as a multisample SEM is a special case of the two-level SEM, some conditional distributions required in the Gibbs sampler can be obtained from the results in Chapter 9. However, as specific constraints among the parameters in different groups are imposed, it is necessary to pay more attention in specifying the corresponding prior distributions. In the same way as the model comparison in two-level SEMs, it requires some insight in applying the path sampling procedure.

10.3.1 Bayesian Estimation

Let $\boldsymbol{\theta}^{(g)}$ be the unknown parameter vector in the identified model and let $\boldsymbol{\alpha}^{(g)}$ be the vector of unknown thresholds that correspond to the gth group. In multisample analysis, a certain type of parameter in $\boldsymbol{\theta}^{(g)}$ is often hypothesized to be invariant over the group models. For example, restrictions on the thresholds, and the following constraints $\boldsymbol{\Lambda}^{(1)} = \cdots = \boldsymbol{\Lambda}^{(G)}$, $\boldsymbol{\Phi}^{(1)} = \cdots = \boldsymbol{\Phi}^{(G)}$ and/or $\boldsymbol{\Gamma}^{(1)} = \cdots = \boldsymbol{\Gamma}^{(G)}$, are often imposed. Hence, in the analysis we allow common parameters in $\boldsymbol{\theta}^{(1)}, \ldots, \boldsymbol{\theta}^{(G)}$. Let $\boldsymbol{\theta}$ be the vector that contains all unknown distinct parameters in $\boldsymbol{\theta}^{(1)}, \ldots, \boldsymbol{\theta}^{(G)}$, and let $\boldsymbol{\alpha}$ be the vector that contains all the unknown thresholds, the Bayesian estimates of $\boldsymbol{\theta}$ and $\boldsymbol{\alpha}$ are obtained by the Gibbs sampler.

Let $\mathbf{X}^{(g)} = (\mathbf{x}_1^{(g)}, \ldots, \mathbf{x}_{N_g}^{(g)})$ and $\mathbf{X} = (\mathbf{X}^{(1)}, \ldots, \mathbf{X}^{(G)})$ be the observed continuous data, and $\mathbf{Z}^{(g)} = (\mathbf{z}_1^{(g)}, \ldots, \mathbf{z}_{N_g}^{(g)})$ and $\mathbf{Z} = (\mathbf{Z}^{(1)}, \ldots, \mathbf{Z}^{(G)})$ be the observed ordered categorical data. Let $\mathbf{Y}^{(g)} = (\mathbf{y}_1^{(g)}, \ldots, \mathbf{y}_{N_g}^{(g)})$ and $\mathbf{Y} = (\mathbf{Y}^{(1)}, \ldots, \mathbf{Y}^{(G)})$ be the latent continuous measurement associated with $\mathbf{Z}^{(g)}$ and $\mathbf{Z}$, respectively. The observed data will be augmented with $\mathbf{Y}$ in the posterior analysis. Once $\mathbf{Y}$ is given, all the data are continuous and the problem will be easier to cope with. Moreover, let $\boldsymbol{\Omega}^{(g)} = (\boldsymbol{\omega}_1^{(g)}, \ldots, \boldsymbol{\omega}_{N_g}^{(g)})$ and $\boldsymbol{\Omega} = (\boldsymbol{\Omega}^{(1)}, \ldots, \boldsymbol{\Omega}^{(G)})$ be the matrices of latent variables. Note that when $\boldsymbol{\Omega}$ is observed, the measurement equation and the nonlinear structural equation reduce to the regular simultaneous regression model. Difficulties due to the nonlinear relationships among the latent variables are greatly alleviated. Hence, problems associated with the complicated components of the model can be handled by data augmentation. In the posterior analysis, the observed data $(\mathbf{X}, \mathbf{Z})$ will be augmented with $(\mathbf{Y}, \boldsymbol{\Omega})$. More specifically, we will consider the joint posterior distribution $[\boldsymbol{\theta}, \boldsymbol{\alpha}, \mathbf{Y}, \boldsymbol{\Omega}|\mathbf{X}, \mathbf{Z}]$. The Gibbs sampler (Geman and Geman, 1984) will be used in generating a sequence of observations from this joint posterior distribution. Then the Bayesian solution is obtained by standard inferences on the basis of the generated sample of observations. In applying the Gibbs sampler, we iteratively sample observations from the following conditional distributions: $[\boldsymbol{\Omega}|\boldsymbol{\theta}, \boldsymbol{\alpha}, \mathbf{Y}, \mathbf{X}, \mathbf{Z}]$, $[\boldsymbol{\alpha}, \mathbf{Y}|\boldsymbol{\theta}, \boldsymbol{\Omega}, \mathbf{X}, \mathbf{Z}]$ and $[\boldsymbol{\theta}|\boldsymbol{\alpha}, \mathbf{Y}, \boldsymbol{\Omega}, \mathbf{X}, \mathbf{Z}]$. As in our previous treatments of the thresholds, we consider non-informative prior to $\boldsymbol{\alpha}$, so that the corresponding prior distribution is proportional to a constant. The first two conditional distributions, which can be derived by similar reasoning in Chapter 9, Appendix 9.1, are presented in Appendix 10.1. The conditional distribution $[\boldsymbol{\theta}|\boldsymbol{\alpha}, \mathbf{Y}, \boldsymbol{\Omega}, \mathbf{X}, \mathbf{Z}]$ is further decomposed into components involving various structural parameters in the different group models. These components are different under various hypotheses of interest or various competing models. Some examples of non-nested competing models (or hypotheses) are:

$$
\begin{aligned}
M_A &: \text{No constraints.} \quad M_1 : \boldsymbol{\mu}^{(1)} = \cdots = \boldsymbol{\mu}^{(G)}. \quad M_2 : \boldsymbol{\Lambda}^{(1)} = \cdots = \boldsymbol{\Lambda}^{(G)} \\
M_3 &: \boldsymbol{\Lambda}_\omega^{(1)} = \cdots = \boldsymbol{\Lambda}_\omega^{(G)}. \quad M_4 : \boldsymbol{\Phi}^{(1)} = \cdots = \boldsymbol{\Phi}^{(G)} \qquad (10.6) \\
M_5 &: \boldsymbol{\Psi}_\epsilon^{(1)} = \cdots = \boldsymbol{\Psi}_\epsilon^{(G)}. \quad M_6 : \boldsymbol{\Psi}_\delta^{(1)} = \cdots = \boldsymbol{\Psi}_\delta^{(G)}.
\end{aligned}
$$

The components in the conditional distribution $[\boldsymbol{\theta}|\boldsymbol{\alpha}, \mathbf{Y}, \boldsymbol{\Omega}, \mathbf{X}, \mathbf{Z}]$ and the specification of prior distributions are slightly different under different M_k defined above. First, prior distributions for nonconstrained parameters in different groups are naturally assumed to be independent. In estimating the unconstrained parameters, we need to specify its own prior distribution, and the data in the corresponding group are used. For constrained parameters across groups, only one prior distribution for these constrained parameters is necessary, and all the data in the groups should be combined in the estimation (see Song and Lee (2001)). In contrast to the Bayesian analysis of other single group SEMs, we may not wish to take a joint prior distribution for the factor loading matrix and the unique variance of the error measurement. In the gth group model, let $\psi_{\epsilon k}^{(g)}$ and $\boldsymbol{\Lambda}_k^{(g)T}$ be the kth diagonal element of $\boldsymbol{\Psi}_\epsilon^{(g)}$ and the kth row of $\boldsymbol{\Lambda}^{(g)}$, respectively. In multisample analysis, if those parameters are not invariant over groups, their joint prior distribution could be taken as: $\psi_{\epsilon k}^{(g)-1} \stackrel{D}{=} Gamma[\alpha_{0\epsilon k}^{(g)}, \beta_{0\epsilon k}^{(g)}]$, $[\boldsymbol{\Lambda}_k^{(g)}|\psi_{\epsilon k}^{(g)}] \stackrel{D}{=} N[\boldsymbol{\Lambda}_{0k}^{(g)}, \psi_{\epsilon k}^{(g)}\mathbf{H}_{0yk}^{(g)}]$, where $\alpha_{0\epsilon k}^{(g)}, \beta_{0\epsilon k}^{(g)}, \boldsymbol{\Lambda}_{0k}^{(g)}$ and $\mathbf{H}_{0yk}^{(g)}$ are hyperparameters. This kind of joint prior distribution may cause problems under the constrained situation where $\boldsymbol{\Lambda}^{(1)} = \cdots = \boldsymbol{\Lambda}^{(G)} = \boldsymbol{\Lambda}$ and $\boldsymbol{\Psi}_\epsilon^{(1)} \neq \cdots \neq \boldsymbol{\Psi}_\epsilon^{(G)}$, because it is difficult to select $p(\boldsymbol{\Lambda}_k|\cdot)$ based on a set of different $\psi_{\epsilon k}^{(g)}$. Hence, for convenience, and following the suggestion of Song and Lee (2001), we select independent prior distributions for $\boldsymbol{\Lambda}^{(g)}$ and $\boldsymbol{\Psi}_\epsilon^{(g)}$, such that $p(\boldsymbol{\Lambda}^{(g)}, \boldsymbol{\Psi}_\epsilon^{(g)}) = p(\boldsymbol{\Lambda}^{(g)})p(\boldsymbol{\Psi}_\epsilon^{(g)})$, for $g = 1, 2, \ldots, G$. Under this kind of prior distribution, the prior distribution of $\boldsymbol{\Lambda}$ under the constraint $\boldsymbol{\Lambda}^{(1)} = \cdots = \boldsymbol{\Lambda}^{(G)} = \boldsymbol{\Lambda}$ is given by: $\boldsymbol{\Lambda}_k \stackrel{D}{=} N[\boldsymbol{\Lambda}_{0k}, \mathbf{H}_{0yk}]$, which is independent of $\psi_{\epsilon k}^{(g)}$. Under the situation $\boldsymbol{\Lambda}^{(1)} \neq \cdots \neq \boldsymbol{\Lambda}^{(G)}$, the prior distribution of each $\boldsymbol{\Lambda}_k^{(g)}$ is given by $N[\boldsymbol{\Lambda}_{0k}^{(g)}, \mathbf{H}_{0yk}^{(g)}]$. The prior distribution of $\psi_{\epsilon k}^{(g)-1}$

is again $Gamma[\alpha_{0\epsilon k}^{(g)}, \beta_{0\epsilon k}^{(g)}]$. Similarly, we select the prior distributions for $\mathbf{\Lambda}_{\omega}^{(g)}$ and $\mathbf{\Psi}_{\delta}^{(g)}$ such that $p(\mathbf{\Lambda}_{\omega}^{(g)}, \mathbf{\Psi}_{\delta}^{(g)}) = p(\mathbf{\Lambda}_{\omega}^{(g)})p(\mathbf{\Psi}_{\delta}^{(g)})$, for $g = 1, \ldots, G$. Let $\mathbf{\Lambda}_{\omega k}^{T}$ and $\mathbf{\Lambda}_{\omega k}^{(g)^T}$ be the kth rows of $\mathbf{\Lambda}_{\omega}$ and $\mathbf{\Lambda}_{\omega}^{(g)}$, respectively. The prior distributions of $\mathbf{\Lambda}_{\omega}$, and $\mathbf{\Lambda}^{(g)}$ are given by:

(i) if $\mathbf{\Lambda}_{\omega}^{(1)} = \cdots = \mathbf{\Lambda}_{\omega}^{(G)} = \mathbf{\Lambda}_{\omega}, \ \mathbf{\Lambda}_{\omega k} \stackrel{D}{=} N[\mathbf{\Lambda}_{0\omega k}, \mathbf{H}_{0\omega k}]$,

(ii) if $\mathbf{\Lambda}_{\omega}^{(1)} \neq \cdots \neq \mathbf{\Lambda}_{\omega}^{(G)}, \quad \mathbf{\Lambda}_{\omega k}^{(g)} \stackrel{D}{=} N[\mathbf{\Lambda}_{0\omega k}^{(g)}, \mathbf{H}_{0\omega k}^{(g)}]$,

with the hyperparameters $\mathbf{\Lambda}_{0\omega k}$, $\mathbf{\Lambda}_{0\omega k}^{(g)}$, $\mathbf{H}_{0\omega k}$ and $\mathbf{H}_{0\omega k}^{(g)}$. The prior distribution of $\psi_{\delta k}^{(g)-1}$ is $Gamma[\alpha_{0\delta k}^{(g)}, \beta_{0\delta k}^{(g)}]$, with hyperparamters $\alpha_{0\delta k}^{(g)}$, and $\beta_{0\delta k}^{(g)}$. The prior distributions of $\boldsymbol{\mu}$ and $\mathbf{\Phi}$ are given by:

$$\boldsymbol{\mu}^{(g)} \stackrel{D}{=} N[\boldsymbol{\mu}_0^{(g)}, \mathbf{\Sigma}_0^{(g)}]; \ \mathbf{\Phi}^{(g)-1} \stackrel{D}{=} W_{q_2}[\mathbf{R}_0^{(g)}, \rho_0^{(g)}], \quad g = 1, \ldots, G;$$

where $\boldsymbol{\mu}_0^{(g)}$, $\mathbf{\Sigma}_0^{(g)}$, $\mathbf{R}_0^{(g)}$ and $\rho_0^{(g)}$ are hyperparameters. Similar adjustments are taken under various combinations of constraints. Based on the above understanding, and the reasoning given in Chapter 9, the conditional distribution $[\boldsymbol{\theta}|\boldsymbol{\alpha}, \mathbf{Y}, \mathbf{\Omega}, \mathbf{X}, \mathbf{Z}]$ under various competing models can be obtained. Some results are given in Appendix 10.1.

10.3.2 Bayesian Model Comparison

In analyzing multisample SEMs, one important statistical inference beyond estimation is on testing whether some type of parameters are invariant over the groups. In Bayesian analysis, each hypothesis of interest is associated with a model, and the problem is approached through model comparison. For instance, the competing models of interest could be those given in (10.6), or any combinations of those models, which specify certain kinds of constraints on some parameters among groups. In contrast to the traditional approach using the likelihood ratio test, it is not necessary to proceed Bayesian model comparison with a sequence of hierarchical models. For example, depending on the interest of a substantive problem, we can compare any two nonnested models M_k and M_h, as given in (10.6); or compare any M_k with any combination of the models given in (10.6).

Let M_h and M_k be any two competing multisample SEMs. The following Bayes factor is used to compare M_h and M_k:

$$B_{kh} = \frac{p(\mathbf{X}, \mathbf{Z}|M_k)}{p(\mathbf{X}, \mathbf{Z}|M_h)}, \tag{10.7}$$

where $(\mathbf{X}, \mathbf{Z})$ is the observed data set. Again, this Bayes factor is computed by a path sampling (Gelman and Meng, 1998) procedure. Let t be a path in $[0, 1]$ to link M_h and M_k, and $t_{(0)} = 0 < t_{(1)} < \cdots < t_{(S)} < t_{(S+1)} = 1$ be fixed grids in $[0, 1]$. Let $p(\mathbf{Y}, \boldsymbol{\Omega}, \mathbf{X}, \mathbf{Z}|\boldsymbol{\theta}, \boldsymbol{\alpha}, t)$ be the complete-data likelihood, $\boldsymbol{\theta}$ is the vector of unknown parameters in the linked model and

$$U(\boldsymbol{\theta}, \boldsymbol{\alpha}, \mathbf{Y}, \boldsymbol{\Omega}, \mathbf{X}, \mathbf{Z}, t) = \mathrm{d}\log p(\mathbf{Y}, \boldsymbol{\Omega}, \mathbf{X}, \mathbf{Z}|\boldsymbol{\theta}, \boldsymbol{\alpha}, t)/\mathrm{d}t.$$

It can be shown by reasoning similar to that used in previous chapters that

$$\widehat{\log B_{kh}} = \frac{1}{2}\sum_{s=0}^{S}(t_{(s+1)} - t_{(s)})(\bar{U}_{(s+1)} + \bar{U}_{(s)}), \tag{10.8}$$

where

$$\bar{U}_{(s)} = \frac{1}{J}\sum_{j=1}^{J} U(\boldsymbol{\theta}^{(j)}, \boldsymbol{\alpha}^{(j)}, \mathbf{Y}^{(j)}, \boldsymbol{\Omega}^{(j)}, \mathbf{X}, \mathbf{Z}, t_{(s)}), \tag{10.9}$$

in which $\{\boldsymbol{\theta}^{(j)}, \boldsymbol{\alpha}^{(j)}, \mathbf{Y}^{(j)}, \boldsymbol{\Omega}^{(j)} : \text{j=1,2,...,J}\}$ is a sample of observations simulated from the joint posterior distribution $[\boldsymbol{\theta}, \boldsymbol{\alpha}, \mathbf{Y}, \boldsymbol{\Omega}|\mathbf{X}, \mathbf{Z}, t_{(s)}]$.

To find a good path to link the competing models M_h and M_k is a crucial step in the path sampling procedure for computing $\log B_{kh}$. In multisample analysis, it may not be easy to find a path that directly links M_h and M_k. Under this situation, the use of an auxiliary model M_a which can be linked with both M_h and M_k is helpful. We have used this technique in analyzing two-level SEMs (see Example 2 in Chapter 9, Section 9.4). An illustrative example for multisample analysis is given below.

In this illustrative example, $G = 2$. For each g, the model is defined by Equations (10.1) and (10.2) with $\boldsymbol{\Pi}^{(g)} = \mathbf{0}$. For $g = 1, 2$, $i = 1, \ldots, N_g$, suppose that the competing models M_1 and M_2 are defined as:

$$\begin{aligned}
M_1: \quad & \mathbf{v}_i^{(g)} = \boldsymbol{\mu}^{(g)} + \boldsymbol{\Lambda}\boldsymbol{\omega}_i^{(g)} + \boldsymbol{\epsilon}_i^{(g)} \\
& \boldsymbol{\eta}_i^{(g)} = \boldsymbol{\Gamma}^{(g)}\mathbf{H}(\boldsymbol{\xi}_i^{(g)}) + \boldsymbol{\delta}_i^{(g)}; \\
M_2: \quad & \mathbf{v}_i^{(g)} = \boldsymbol{\mu}^{(g)} + \boldsymbol{\Lambda}^{(g)}\boldsymbol{\omega}_i^{(g)} + \boldsymbol{\epsilon}_i^{(g)} \\
& \boldsymbol{\eta}_i^{(g)} = \boldsymbol{\Gamma}\mathbf{H}(\boldsymbol{\xi}_i^{(g)}) + \boldsymbol{\delta}_i^{(g)}.
\end{aligned}$$

In M_1, $\boldsymbol{\Lambda}$ is invariant over the two groups; whilst in M_2, $\boldsymbol{\Gamma}$ is invariant over the groups. Note that in both models, the form of the nonlinear terms is the same.

Due to the constraints imposed on the parameters, it is quite difficult to find a path t in $[0, 1]$ that directly links M_1 and M_2. This difficulty can be solved through the use of the following auxiliary model M_a:

$$M_a: \quad \mathbf{v}_i^{(g)} = \boldsymbol{\mu}^{(g)} + \boldsymbol{\epsilon}_i^{(g)}, g = 1, 2; i = 1, \ldots, N_g.$$

The linked model M_{ta} for linking M_1 and M_a is defined as: for $g = 1, 2; i = 1, \ldots, N_g$

$$\begin{aligned} M_{ta1}: \quad & \mathbf{v}_i^{(g)} = \boldsymbol{\mu}^{(g)} + t\boldsymbol{\Lambda}\boldsymbol{\omega}_i^{(g)} + \boldsymbol{\epsilon}_i^{(g)}, \\ & \boldsymbol{\eta}_i^{(g)} = \boldsymbol{\Gamma}^{(g)}\mathbf{H}(\boldsymbol{\xi}_i^{(g)}) + \boldsymbol{\delta}_i^{(g)}. \end{aligned}$$

When $t = 1$, M_{ta1} reduces to M_1; and when $t = 0$, M_{ta1} reduces to M_a. The parameter vector $\boldsymbol{\theta}$ in M_{ta1} contains $\boldsymbol{\mu}^{(1)}, \boldsymbol{\mu}^{(2)}, \boldsymbol{\Lambda}, \boldsymbol{\Psi}_\epsilon^{(1)}, \boldsymbol{\Psi}_\epsilon^{(2)}, \boldsymbol{\Gamma}^{(1)}, \boldsymbol{\Gamma}^{(2)}, \boldsymbol{\Phi}^{(1)}, \boldsymbol{\Phi}^{(2)}, \boldsymbol{\Psi}_\delta^{(1)}$ and $\boldsymbol{\Psi}_\delta^{(2)}$. The linked model M_{ta2} for linking M_2 and M_a is defined as:

$$\begin{aligned} M_{ta2}: \quad & \mathbf{v}_i^{(g)} = \boldsymbol{\mu}^{(g)} + t\boldsymbol{\Lambda}^{(g)}\boldsymbol{\omega}_i^{(g)} + \boldsymbol{\epsilon}_i^{(g)}, \\ & \boldsymbol{\eta}_i^{(g)} = \boldsymbol{\Gamma}\mathbf{H}(\boldsymbol{\xi}_i^{(g)}) + \boldsymbol{\delta}_i^{(g)}. \end{aligned}$$

Clearly, when $t = 1$ and 0, M_{ta2} reduces to M_2 and M_a, respectively. The parameter vector in M_{ta2} contains $\boldsymbol{\mu}^{(1)}, \boldsymbol{\mu}^{(2)}, \boldsymbol{\Lambda}^{(1)}, \boldsymbol{\Lambda}^{(2)}, \boldsymbol{\Psi}_\epsilon^{(1)}, \boldsymbol{\Psi}_\epsilon^{(2)}, \boldsymbol{\Gamma}, \boldsymbol{\Phi}^{(1)}, \boldsymbol{\Phi}^{(2)}, \boldsymbol{\Psi}_\delta^{(1)}$ and $\boldsymbol{\Psi}_\delta^{(2)}$. We first compute $\log B_{1a}$ and $\log B_{2a}$, and then obtain $\log B_{12}$ via the following equation:

$$\log B_{12} = \log \frac{p(\mathbf{X}, \mathbf{Z}|M_1)/p(\mathbf{X}, \mathbf{Z}|M_a)}{p(\mathbf{X}, \mathbf{Z}|M_2)/p(\mathbf{X}, \mathbf{Z}|M_a)} = \log B_{1a} - \log B_{2a}. \tag{10.10}$$

The above method can be applied to link other competing models. Moreover, the approach presented in Chapter 9, Section 9.4, can be applied to situations where just one auxiliary model is not adequate to link the two very different competing models M_1 and M_2. Similarly, we use more than one auxiliary model $M_a, M_b, \ldots$ to link M_1 and M_2. For example, suppose we use M_a and M_b to link M_1 and M_2, with M_a closer to M_1, and M_b closer to M_2. Then

$$\frac{p(\mathbf{X}, \mathbf{Z}|M_1)}{p(\mathbf{X}, \mathbf{Z}|M_2)} = \frac{p(\mathbf{X}, \mathbf{Z}|M_1)/p(\mathbf{X}, \mathbf{Z}|M_a)}{p(\mathbf{X}, \mathbf{Z}|M_2)/p(\mathbf{X}, \mathbf{Z}|M_a)}, \quad \text{and}$$

$$\frac{p(\mathbf{X}, \mathbf{Z}|M_2)}{p(\mathbf{X}, \mathbf{Z}|M_a)} = \frac{p(\mathbf{X}, \mathbf{Z}|M_2)/p(\mathbf{X}, \mathbf{Z}|M_b)}{p(\mathbf{X}, \mathbf{Z}|M_a)/p(\mathbf{X}, \mathbf{Z}|M_b)};$$

hence, $\log B_{12} = \log B_{1a} + \log B_{ab} - \log B_{2b}$. Each logarithm Bayes factor is computed via path sampling.

Model comparison can also be conducted with DIC that is available via WinBUGS, see Section 10.4.2. In a similar way to the goodness-of-fit assessment in the context of single group nonlinear SEMs, it is quite difficult to find a saturated model for the multisample nonlinear SEMs. However, the goodness-of-fit of a proposed multisample model can be assessed by the estimated residual plots, or the PP p-value that can be obtained through straightforward modification of the derivation presented in Chapter 9, Appendix 9.3.

10.4 NUMERICAL ILLUSTRATIONS

10.4.1 Analysis of Multisample Management Data

Assessing employee job attitudes, benefits and emotion is an interesting issue that has received much attention in organizational and management research. In this section, we use the proposed Bayesian approach to analyze a multisample nonlinear SEM with these latent constructs. The aims are to study the nonlinear effects of the exogenous latent variables of 'benefit attitude' and 'emotion' to the endogenous latent variable of 'job attitude', and assess various hypotheses for comparing the nonlinear SEMs between two populations.

A portion of the Inter-University Consortium for Political and Social Research (ICPSR) data set collected in the project World Value Survey 1981–1984 and 1990–1993 (World Value Study Group, 1994) was analyzed. In this illustration, the data obtained from the UK ($g = 1$) and Canada ($g = 2$) with nine manifest variables (variables 252, 253, 254, 89, 91, 93, 99, 102 and 103, see Appendix 1.1) were used. After deleting observations with missing entries, the sample sizes are $N_1 = 1412$ and $N_2 = 1686$. It is clear from Appendix 1.1 that the first three manifest variables, which were measured via a 10-point Likert scale and hence treated as continuous for convenience, are indicators of the latent variable 'job attitude'; the next three dichotomous manifest variables are indicators of the latent variable 'emotion', and that the last three dichotomous manifest variables are indicators of the latent variable 'benefit attitude'. The dichotomous variables are coded with 1 for 'yes, or mentioned', and 0 for 'no, or not mentioned'. On the basis of the clear meaning associated with the indicators of the latent variables, the factor loading matrices are taken to be of the following nonoverlapping structure for better interpretation:

$$\Lambda^{(g)T} = \begin{bmatrix} 1.0^* & \lambda_{21}^{(g)} & \lambda_{31}^{(g)} & 0^* & 0^* & 0^* & 0^* & 0^* & 0^* \\ 0^* & 0^* & 0^* & 1.0^* & \lambda_{52}^{(g)} & \lambda_{62}^{(g)} & 0^* & 0^* & 0^* \\ 0^* & 0^* & 0^* & 0^* & 0^* & 0^* & 1.0^* & \lambda_{83}^{(g)} & \lambda_{93}^{(g)} \end{bmatrix}, \quad g = 1, 2. \tag{10.11}$$

The following nonlinear structural equation was considered for each group to assess the interaction of 'emotion, $\xi_1^{(g)}$', and 'benefit attitude, $\xi_2^{(g)}$', with 'job attitude, $\eta^{(g)}$':

$$\eta_i^{(g)} = \gamma_1^{(g)}\xi_{i1}^{(g)} + \gamma_2^{(g)}\xi_{i2}^{(g)} + \gamma_3^{(g)}\xi_{i1}^{(g)}\xi_{i2}^{(g)} + \delta_i^{(g)}. \tag{10.12}$$

The covariance matrix of the error measurements in the measurement equation is set equal to $\mathbf{\Psi}_\epsilon^{(g)}$, with diagonal elements $(\psi_{\epsilon 1}^{(g)}, \psi_{\epsilon 2}^{(g)}, \psi_{\epsilon 3}^{(g)}, 1.0^*, 1.0^*, 1.0^*, 1.0^*, 1.0^*, 1.0^*)$. Here, the variances of the error measurements that correspond to the dichotomous variables are fixed at 1.0 for achieving an identified model. Other unknown parameters in this model are $\phi_{11}^{(g)}, \phi_{22}^{(g)}, \phi_{12}^{(g)}$, the variances and covariances of the latent variables $(\xi_1^{(g)}, \xi_2^{(g)})$ in $\mathbf{\Phi}^{(g)}$, and $\psi_\delta^{(1)}$ and $\psi_\delta^{(2)}$, the variances of $\delta_i^{(g)}$ in the structural Equation (10.12). In the following Bayesian analysis, some prior inputs of the conjugate prior distributions, such as $\boldsymbol{\mu}_0^{(g)}, \mathbf{\Lambda}_{0k}^{(g)}, \mathbf{R}_0^{(g)}$ and $\mathbf{\Lambda}_{0\omega k}^{(g)}$, are taken from the Bayesian estimates of the parameters obtained by an initial estimation, based on the noninformative prior distributions with the following ad hoc prior inputs: $\mathbf{\Sigma}_0^{(g)} = \mathbf{I}, \mathbf{H}_{0yk}^{(g)} = \mathbf{I}, \mathbf{H}_{0\omega k}^{(g)} = \mathbf{I}, \alpha_{0\epsilon k}^{(1)} = \alpha_{0\delta 1}^{(1)} = 10, \beta_{0\epsilon k}^{(1)} = \beta_{0\delta 1}^{(1)} = 8, \alpha_{0\epsilon k}^{(2)} = \alpha_{0\delta 1}^{(2)} = 20, \beta_{0\epsilon k}^{(2)} = \beta_{0\delta 1}^{(2)} = 16, \rho_0^{(1)} = 20$ and $\rho_0^{(2)} = 4$.

In this multisample analysis, we employed the Bayes factor to study the relations of the parameter matrices in the nonlinear SEMs of Canada and the UK, and for finding a better model. We consider the following non-nested models M_1, M_2, M_3, M_4, M_5 and M_6, which are associated with the following hypotheses: H_1: no constraints; $H_2 : \mathbf{\Lambda}^{(1)} = \mathbf{\Lambda}^{(2)}$; $H_3 : \mathbf{\Gamma}^{(1)} = \mathbf{\Gamma}^{(2)}$; $H_4 : \mathbf{\Phi}^{(1)} = \mathbf{\Phi}^{(2)}$; $H_5 : \mathbf{\Psi}_\epsilon^{(1)} = \mathbf{\Psi}_\epsilon^{(2)}$; and $H_6 : \psi_\delta^{(1)} = \psi_\delta^{(2)}$. The logarithm Bayes factors for comparing these non-nested models are computed by the path sampling procedure as described in Section 10.3.2. In the computation, we selected $S = 15$, took 2000 burn-in iterations and further collected $J = 2000$ observations in computing $\overline{U}_{(s)}$ in Equation (10.9) . We obtained $\widehat{\log B_{21}} = -1600$. Note that as the sample sizes N_1 and N_2 are large, the magnitude of the estimated logarithm Bayes factor is large. This reveals the fact that large sample sizes provide clear evidence of selecting the better model. Based on the interpretation of $\widehat{\log B_{21}}$, M_1 with no constraints is better than M_2 with the constraint $\mathbf{\Lambda}^{(1)} = \mathbf{\Lambda}^{(2)}$. What is the interpretation of this result? From the structure of $\mathbf{\Lambda}^{(g)}$ (see Equation (10.11)) and the corresponding indicators (see Appendix 1.1), we have a rather clear interpretation of the latent variables $\eta^{(g)}, \xi_1^{(g)}$ and $\xi_2^{(g)}$ as 'job attitude', 'emotion' and 'benefit attitude'. Based on Equation (10.11), we see that the associations between the latent variables and their respective indicators are clearly indicated by the corresponding elements in $\mathbf{\Lambda}^{(g)}$. The rejection of M_2 reveals that the associations in group 1 (UK) are different from those in group 2 (Canada). The common multisample analysis of structural equation models via the likelihood ratio test based on p-value is to stop if the hypothesis $\mathbf{\Lambda}^{(1)} = \mathbf{\Lambda}^{(2)}$ is rejected. The reason for ending further analysis may be due to the belief that rejection of $\mathbf{\Lambda}^{(1)} = \mathbf{\Lambda}^{(2)}$ implies that the scale of measurements of the latent variables in group 1 are different from the scale of measurements of the latent variables in group 2. However, based on the likelihood ratio test with p-value, the conclusion that H_2 is not rejected does not imply that H_2 is true, and it does not imply that the scales of these two groups are the same. Hence, the justification that is just based on the result of hypothesis testing for continuing the analysis is not very strong. Moreover, for dichotomous variables, as the underlying continuous responses are not observed, it is difficult to draw a conclusion on the scales of the latent variables on the basis of the hypothesis testing result. From the non-overlapping structure of $\mathbf{\Lambda}^{(g)}$, these latent variables can be clearly interpreted as the latent constructs in relation to job attitude, emotion and benefit attitude, although the associations of these latent constructs and their indicators are not the same in groups 1 and 2. Hence, we think that it is desirable to conduct a further analysis to obtain a deeper understanding of these latent constructs. This additional analysis involves comparisons of M_1 and M_2 with models M_3, M_4, M_5 and M_6 that are associated with H_3, H_4, H_5 and H_6. The logarithm Bayes factors computed by the path sampling procedure for comparing these models are reported in Table 10.1. We again found that the magnitudes of these logarithm Bayes factors are large. It can be observed from Table 10.1 that M_5 associated with the constraint $\mathbf{\Psi}_\epsilon^{(1)} = \mathbf{\Psi}_\epsilon^{(2)}$, and M_6 associated with the constraint $\psi_\delta^{(1)} = \psi_\delta^{(2)}$, are better than M_1 with no constraints, whilst M_2, M_3 and M_4 with the other constraints are worse than M_1. As $\widehat{\log B_{65}} = 910$, M_6 is better than M_5.

The interpretations of the above model comparison results in the context of the two SEMs for Canada and UK are: the factor loading matrices are different, regression coefficients in the structural equation of the exogenous latent variables to the endogenous variables are different, the covariance matrices of the exogenous latent variables are different and the covariance matrices of the error measurements in the measurement equation are different; however, the variances of the residuals in the structural equations are equal. Basically, we can conclude that the measurement and structural equations of these two groups are different.

Table 10.1 Estimated log Bayes factors: $\log B_{jk}$.

	Model k					
Model j	M_1	M_2	M_3	M_4	M_5	M_6
M_1	–					
M_2	−1.60	–				
M_3	−5.16	−3.56	–			
M_4	−1.56	0.04	3.60	–		
M_5	0.11	1.71	5.27	1.67	–	
M_6	1.01	2.61	6.17	2.57	0.91	–

Note: $\widehat{\log B_{jk}} = (j, k)$ entry $\times 10^3$; hence $\widehat{\log B_{21}} = -1600$ and $\widehat{\log B_{65}} = 910$.

Table 10.2 Bayesian estimates and standard error estimates under M_6.

	UK		CANADA			UK		CANADA	
PAR	EST	SE	EST	SE	PAR	EST	SE	EST	SE
μ_1	5.313	0.068	3.936	0.067	γ_1	0.404	0.104	0.347	0.116
μ_2	5.493	0.074	4.831	0.069	γ_2	0.193	0.114	0.293	0.092
μ_3	3.949	0.063	3.109	0.056	γ_3	−0.277	0.204	0.035	0.126
μ_4	−0.566	0.051	−0.761	0.060	ϕ_{11}	0.827	0.133	0.747	0.209
μ_5	−1.225	0.096	−1.444	0.130	ϕ_{12}	0.052	0.042	0.077	0.042
μ_6	−1.132	0.048	−1.054	0.050	ϕ_{22}	0.605	0.102	0.885	0.141
μ_7	0.627	0.047	0.944	0.055	$\psi_{\epsilon1}$	3.814	0.275	4.268	0.313
μ_8	0.259	0.042	0.601	0.050	$\psi_{\epsilon2}$	5.753	0.294	7.136	0.279
μ_9	−0.269	0.027	−0.021	0.022	$\psi_{\epsilon3}$	4.269	0.217	3.886	0.190
λ_{21}	0.879	0.068	0.633	0.065	ψ_δ	3.077	0.290	3.077	0.290
λ_{31}	0.747	0.059	0.648	0.061					
λ_{52}	1.260	0.237	1.374	0.288					
λ_{62}	0.420	0.082	0.681	0.129					
λ_{83}	0.945	0.139	0.968	0.136					
λ_{93}	1.197	0.219	0.972	0.161					

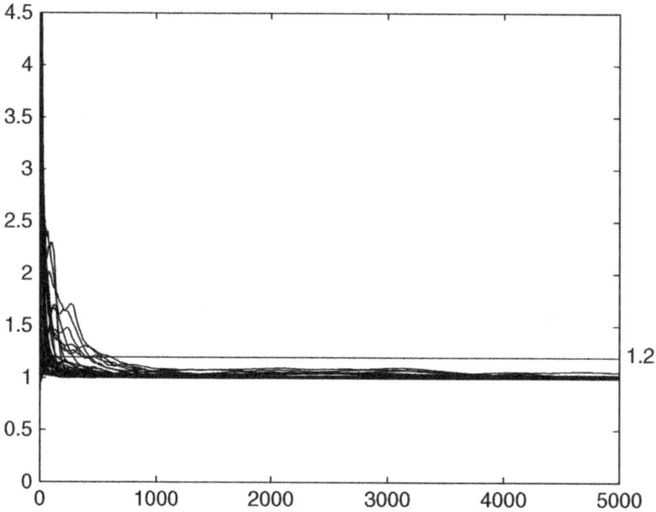

Figure 10.1 EPSR values against the number of iterations in the ICPSR data under M_6.

The Bayesian estimates of the unknown parameters in M_6 are presented in Table 10.2. To reveal the convergence of the MCMC algorithm in analyzing multisample SEMs, the EPSR values, and sequences of generated observations corresponding to some parameters in the two groups, are presented in Figures 10.1, 10.2 and 10.3, respectively. We first interpret the Bayesian

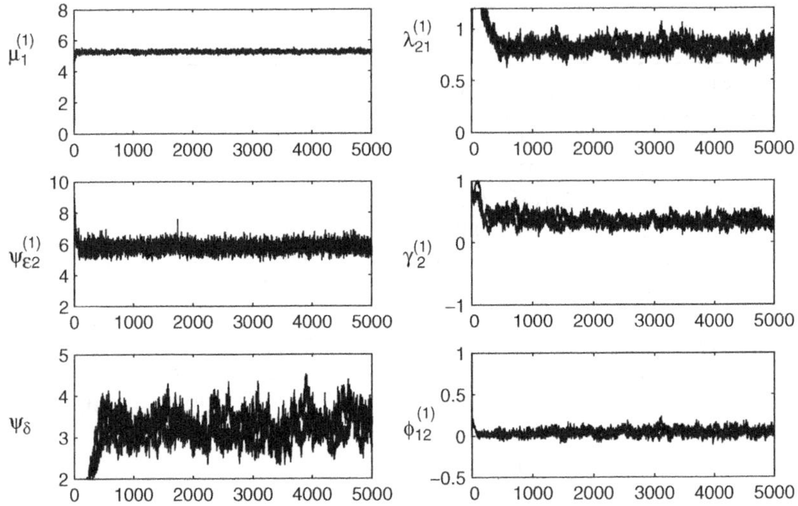

Figure 10.2 Plots of parallel sequences of $\mu_1^{(1)}, \lambda_{21}^{(1)}, \psi_{\epsilon 2}^{(1)}, \gamma_2^{(1)}, \psi_\delta^{(1)}$ and $\phi_{12}^{(1)}$ in the ICPSR data under M_6.

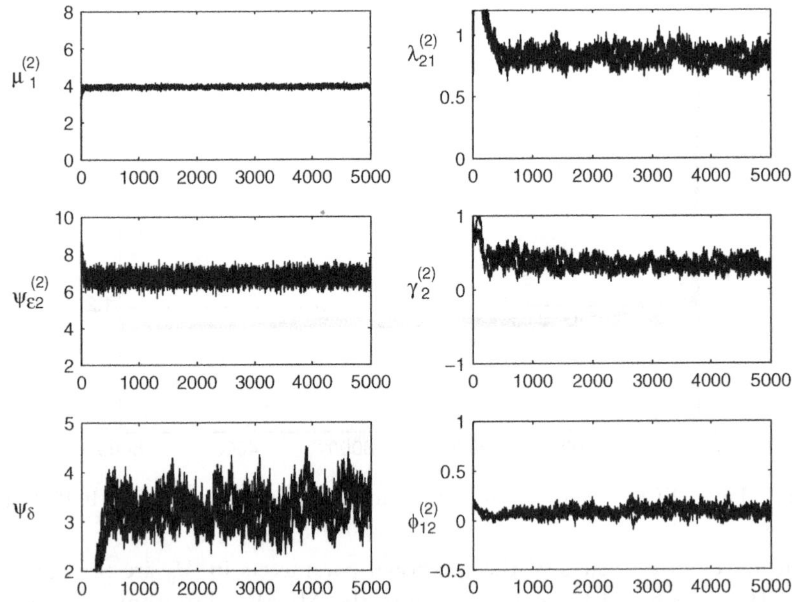

Figure 10.3 Plots of parallel sequences of $\mu_1^{(2)}, \lambda_{21}^{(2)}, \psi_{\epsilon 2}^{(2)}, \gamma_2^{(2)}, \psi_\delta^{(2)}$ and $\phi_{12}^{(2)}$ in the ICPSR data under M_6.

estimates in relation to UK. Note that for nonlinear SEMs, the intercept $\boldsymbol{\mu}^{(g)}$ is not equal to the mean of $\mathbf{v}^{(g)}$. Equation (8.4) in Chapter 8 can be used in the same way here to evaluate the mean of $\mathbf{v}^{(g)}$. From the specifications (see Equation (10.11)) of the nonlinear SEM in analyzing this example, $E(\xi_1^{(1)}) = E(\xi_2^{(1)}) = 0$ and $\hat{E}(\xi_1^{(1)}\xi_2^{(1)}) = \hat{\phi}_{12}^{(1)} = 0.053$, so we can compute the estimate of $E(v_k)$ for every $k = 1, \ldots, 9$. For $k = 1, 2$ and 3 we have $\hat{E}(\mathrm{v}_1^{(1)}) = \hat{\mu}_1^{(1)} + \hat{\gamma}_3^{(1)}\hat{E}(\xi_{i1}^{(1)}\xi_{i2}^{(1)}) = 5.314 + 1.0 \times (-0.277) \times 0.053 = 5.299$, and similarly $\hat{E}(\mathrm{v}_2^{(1)}) = 5.481$ and $\hat{E}(\mathrm{v}_3^{(1)}) = 3.938$. For $k = 4, \ldots, 9$, $\hat{E}(\mathrm{v}_k^{(1)}) = \hat{\mu}_k$. Hence, $\hat{E}(\mathrm{v}_k^{(1)})$ and $\hat{\mu}_k^{(1)}$ are close or equal to each other. The interpretation of the estimated means $\hat{E}(\mathrm{v}_1^{(1)})$, $\hat{E}(\mathrm{v}_2^{(1)})$ and $\hat{E}(\mathrm{v}_3^{(1)})$ that correspond to the indicators of 'job attitude' is standard, because it is based on a 10-point scale that is treated as continuous. From $\hat{E}(\mathrm{v}_k^{(1)})$, $k = 4, 5, 6$ that correspond to the dichotomous indicators, we know that the mean scores of the items that served as indicators for 'emotion' shift to the left. Hence, it seems that on the average, the British do have good emotions. From $\hat{E}(\mathrm{v}_k^{(1)})$, $k = 7, 8, 9$, it seems that they are more concerned about 'good pay' and 'good job security', but less concerned about 'good chances for promotion'. All of the factor loading estimates are quite large. This indicates a strong association between each of the latent variables and their respective indicators. The estimated nonlinear structural equation is given by:

$$\eta_i^{(1)} = 0.404\ \xi_{i1}^{(1)} + 0.193\ \xi_{i2}^{(1)} - 0.277\ \xi_{i1}^{(1)}\xi_{i2}^{(1)}. \tag{10.13}$$

Before giving the interpretations of this nonlinear structural equation, we note again from the scales of the dichotomous indicators in relation to 'emotion, ξ_1' and 'benefit attitude, ξ_2' that a comparatively larger (positive) value of ξ_1 implies that an individual has worse emotions, and a comparatively larger (positive) value of ξ_2 implies that an individual is more concerned about benefits. Moreover, a comparatively large (positive) value of η implies a comparatively bad job attitude. With the above understanding of the latent variables, it follows from $\hat{\gamma}_1^{(1)} = 0.404$ and $\hat{\gamma}_2^{(1)} = 0.193$ that worse emotions, and more concern about benefits, imply a worse job attitude. From $\hat{\gamma}_3^{(1)} = -0.277$, 'emotion, $\xi_1^{(1)}$' and 'benefit attitude, $\xi_2^{(1)}$' have an interaction effect on 'job attitude'. The basic interpretation is that the 'additive' effect of the linear latent variables of 'emotion, $\xi_1^{(1)}$', and 'benefit attitude, $\xi_2^{(1)}$' in the structural equation is inadequate to account for their relationships with the latent variable 'job attitude, $\eta^{(1)}$' and an interaction term of $\xi_1^{(1)}$ and $\xi_2^{(1)}$ has to be added. Depending on various situations, this interaction term (with a negative sign) has different effects. For example, it can indicate that: (i) for employees with good emotions (negative $\xi_{i1}^{(1)}$), less concern about benefits (negative $\xi_{i2}^{(1)}$) would have a positive effect on job attitude ($-0.277\ \xi_{i1}^{(1)}\xi_{i2}^{(1)}$ would decrease the value of $\eta_i^{(1)}$), whereas more concern about benefits (positive $\xi_{i2}^{(1)}$) would have a negative effect on job attitude ($-0.277\ \xi_{i1}^{(1)}\xi_{i2}^{(1)}$ would increase the value of $\eta_i^{(1)}$). These interpretations may provide useful insights for administrators to improve the management of their organizations. From $\hat{\phi}_{11}^{(1)}$, $\hat{\phi}_{12}^{(1)}$ and $\hat{\phi}_{22}^{(1)}$, the estimate of the correlation of 'emotion' and 'benefit attitude' is equal to 0.075. As expected, the correlation of these two latent variables is small. The interpretations of the Bayesian estimates of unknown parameters in the nonlinear SEM that corresponds to Canada are similar.

The following similarities and differences are observed in comparing the Bayesian estimates of these two countries. (i) Although the magnitudes are different, the mean behavior of manifest variables v_1 to v_9 (corresponding to questions 1–9) is similar in the sense that $\mu_k^{(1)}$ and $\mu_k^{(2)}$ have the same sign. (ii) The factor loading estimates in the Canada model are also quite large, thus indicating a strong association between the indicators (manifest variables) and their latent variables. (iii) The interaction effect of 'emotion' and 'benefit attitude' in the Canada model is very minor. (iv) As shown by the model comparison result, the variances of the residual in the structural equations of the models in the two countries are equal. (v) In general, the standard error estimates of the Bayesian estimates are quite small. This indicates a nice feature of the Bayesian application to this real example.

10.4.2 Analysis of Multisample Quality of Life Data via WinBUGS

Analysis of single group quality of life (QOL) data has been considered in Chapter 6, Section 6.6. Here, we describe the Bayesian methods in analyzing multisample QOL data. As the latent constructs of QOL can be naturally regarded as latent variables that are reflected by the related items (observed variables) in the questionnaire, multisample factor analysis and multisample structural equation models have been used in analyzing QOL data. For instance, Power *et al.* (1999) applied a multisample model to investigate whether the WHOQOL instrument is structurally comparable in different cultures, and Meuleners, Lee, Binns and Lower (2003) applied a LISREL model to analyze the QOL for adolescents. However, the above cited work, as well as most applications of the factor analysis model to QOL, are based on the assumption that the data are continuous and normally distributed.

The WHOQOL-BREF (Power *et al.*, 1999) instrument was taken from the WHOQOL-100 instrument by selecting 26 ordered categorical items out of 100 original items. The observations were taken from 15 international field centers, one of which is China, and the rest are Western countries, such as the UK, Italy and Germany. The first two items are the overall QOL and general health, the next seven items address physical health, the next six items address psychological health, the three items that follow are for social relationships and the last eight items address the environment. All of the items are measured with a five-point scale (1 = 'not at all/very dissatisfied'; 2 = 'a little/dissatisfied'; 3 = 'moderate/neither'; 4 = 'very much/satisfied'; and 5 = 'extremely/very satisfied').

To illustrate the Bayesian methodology, we use synthetic two-sample data that mimic the QOL study with the same items as mentioned above for each sample, see also Chapter 6, Section 6.6. Hence, we consider a two-sample SEM on the basis of a simulated data set of randomly drawn observations from two populations. The sample sizes are $N_1 = 338$ and $N_2 = 247$. We note that several items are seriously skew to the right. Treating these discrete data as coming from a normal distribution may lead to a misleading conclusion.

We apply the multisample SEM as defined in Equations (10.1) and (10.2) with $G = 2$ to analyze the data. In the Bayesian analysis, we identify the ordered categorical variables by the method described in Section 10.2, using the first group ($g = 1$) as the reference group. Based on the meaning of the questions, we use the following non-overlapping $\mathbf{\Lambda}^{(g)}$ for clear interpretation of latent variables: For $g = 1, 2$

$$\mathbf{\Lambda}^{(g)T} = \begin{bmatrix} 1 & \lambda_{2,1}^{(g)} & 0 & 0 & \cdots & 0 & 0 & 0 & \cdots & 0 & 0 & 0 & 0 & 0 & 0 & \cdots & 0 \\ 0 & 0 & 1 & \lambda_{4,2}^{(g)} & \cdots & \lambda_{9,2}^{(g)} & 0 & 0 & \cdots & 0 & 0 & 0 & 0 & 0 & 0 & \cdots & 0 \\ 0 & 0 & 0 & 0 & \cdots & 0 & 1 & \lambda_{11,3}^{(g)} & \cdots & \lambda_{15,3}^{(g)} & 0 & 0 & 0 & 0 & 0 & \cdots & 0 \\ 0 & 0 & 0 & 0 & \cdots & 0 & 0 & 0 & \cdots & 0 & 1 & \lambda_{17,4}^{(g)} & \lambda_{18,4}^{(g)} & 0 & 0 & \cdots & 0 \\ 0 & 0 & 0 & 0 & \cdots & 0 & 0 & 0 & \cdots & 0 & 0 & 0 & 0 & 1 & \lambda_{20,5}^{(g)} & \cdots & \lambda_{26,5}^{(g)} \end{bmatrix}, \tag{10.14}$$

where 1s and 0s are fixed parameters. In a similar way to the analysis presented in Section 6.6, the latent variables in $\boldsymbol{\omega}_i^{(g)T} = (\eta_i^{(g)}, \xi_{i1}^{(g)}, \xi_{i2}^{(g)}, \xi_{i3}^{(g)}, \xi_{i4}^{(g)})$ are interpreted as 'health related QOL, η', 'physical health, ξ_1', 'psychological health, ξ_2', 'social relationship, ξ_3' and 'environment, ξ_4'. The measurement equation in the model is given by

$$\mathbf{v}_i^{(g)} = \boldsymbol{\mu}^{(g)} + \mathbf{\Lambda}^{(g)} \boldsymbol{\omega}_i^{(g)} + \boldsymbol{\epsilon}_i^{(g)}, \quad g = 1, 2, \tag{10.15}$$

with $\mathbf{\Lambda}^{(g)}$ defined as above. The following structural equation is used to assess the effects of the latent constructs in $\boldsymbol{\xi}_i^{(g)}$ to the health related QOL, $\eta_i^{(g)}$:

$$\eta_i^{(g)} = \gamma_1^{(g)} \xi_{i1}^{(g)} + \gamma_2^{(g)} \xi_{i2}^{(g)} + \gamma_3^{(g)} \xi_{i3}^{(g)} + \gamma_4^{(g)} \xi_{i4}^{(g)} + \delta_i^{(g)}. \tag{10.16}$$

In the Bayesian analysis, the prior inputs of the hyperparameters in the conjugate prior distributions given at the end of Section 10.3.1 are taken as: $\alpha_{0\epsilon k}^{(g)} = \alpha_{0\delta k}^{(g)} = 10$, $\beta_{0\epsilon k}^{(g)} = \beta_{0\delta k}^{(g)} = 8$, elements in $\mathbf{\Lambda}_{0k}^{(g)}$ are taken as 0.8, elements in $\mathbf{\Lambda}_{0\omega k}^{(g)}$ are taken as 0.6, $\mathbf{H}_{0yk}^{(g)}$ and $\mathbf{H}_{0\omega k}^{(g)}$ are diagonal matrices with diagonal elements 0.25, $\mathbf{R}_0^{(g)} = 8\mathbf{I}_4$ and $\rho_0^{(g)} = 30$. Note that the prior distributions and the prior inputs are appropriately modified for handling situations with constraints.

Three multisample models M_1, M_2 and M_3 that are respectively associated with following hypotheses are considered: H_1 : no constraints $H_2 : \mathbf{\Lambda}^{(1)} = \mathbf{\Lambda}^{(2)}$ and $H_3 :\ \mathbf{\Lambda}^{(1)} = \mathbf{\Lambda}^{(2)}, \mathbf{\Phi}^{(1)} = \mathbf{\Phi}^{(2)}$. The software WinBUGS was applied to obtain the Bayesian results. In the analysis, the number of burn-in iterations was 10 000 and 10 000 observations were collected after convergence to produce the results. The DIC values corresponding to M_1, M_2 and M_3 are equal to 32 302.6, 32 321.7 and 32 341.9, respectively. Hence, model M_1 with the smallest DIC value is selected. Note that one may consider other competing models if desirable. The Bayesian estimates and standard error estimates produced by WinBUGS under M_1 are presented in Table 10.3. The

Table 10.3 Bayesian estimates of unknown parameters in the two-group SEM with no constraints

	Group			Group			Group	
PAR	$g=1$	$g=2$	PAR	$g=1$	$g=2$	PAR	$g=1$	$g=2$
μ_1	0.021	−0.519	$\lambda_{2,1}$	0.859	0.804	γ_1	0.847	0.539
μ_2	0.001	0.059	$\lambda_{4,2}$	0.952	0.754	γ_2	0.334	0.139
μ_3	0.002	−0.240	$\lambda_{5,2}$	1.112	1.016	γ_3	0.167	0.026
μ_4	0.009	−0.300	$\lambda_{6,2}$	1.212	0.976	γ_4	−0.068	0.241
μ_5	−0.004	−0.188	$\lambda_{7,2}$	0.820	0.805	ψ_1	0.400	0.248
μ_6	0.008	−0.382	$\lambda_{8,2}$	1.333	1.123	ψ_2	0.422	0.268
μ_7	−0.002	−0.030	$\lambda_{9,2}$	1.203	0.961	ψ_3	0.616	0.584
μ_8	0.008	−0.070	$\lambda_{11,3}$	0.799	0.827	ψ_4	0.628	0.445
μ_9	0.003	0.108	$\lambda_{12,3}$	0.726	0.987	ψ_5	0.462	0.214
μ_{10}	0.004	−0.358	$\lambda_{13,3}$	0.755	0.669	ψ_6	0.401	0.184
μ_{11}	0.003	−0.286	$\lambda_{14,3}$	1.011	0.762	ψ_7	0.709	0.253
μ_{12}	0.001	−0.087	$\lambda_{15,3}$	0.874	0.719	ψ_8	0.271	0.202
μ_{13}	0.004	−0.373	$\lambda_{17,4}$	0.273	0.627	ψ_9	0.393	0.191
μ_{14}	0.003	0.031	$\lambda_{18,4}$	0.954	0.961	ψ_{10}	0.471	0.288
μ_{15}	0.002	0.079	$\lambda_{20,5}$	0.804	1.108	ψ_{11}	0.654	0.262
μ_{16}	0.012	−0.404	$\lambda_{21,5}$	0.772	0.853	ψ_{12}	0.707	0.428
μ_{17}	0.000	0.037	$\lambda_{22,5}$	0.755	0.815	ψ_{13}	0.698	0.348
μ_{18}	0.010	−0.596	$\lambda_{23,5}$	0.723	0.672	ψ_{14}	0.453	0.269
μ_{19}	0.005	−0.183	$\lambda_{24,5}$	0.984	0.647	ψ_{15}	0.575	1.137
μ_{20}	0.004	−0.543	$\lambda_{25,5}$	0.770	0.714	ψ_{16}	0.462	0.267
μ_{21}	0.003	−0.571	$\lambda_{26,5}$	0.842	0.761	ψ_{17}	0.962	0.297
μ_{22}	0.002	−0.966	ϕ_{11}	0.450	0.301	ψ_{18}	0.522	0.301
μ_{23}	0.001	−0.220	ϕ_{12}	0.337	0.279	ψ_{19}	0.530	0.559
μ_{24}	0.017	−1.151	ϕ_{13}	0.211	0.162	ψ_{20}	0.679	0.565
μ_{25}	−0.001	−0.837	ϕ_{14}	0.299	0.207	ψ_{21}	0.708	0.392
μ_{26}	0.007	−0.982	ϕ_{22}	0.579	0.537	ψ_{22}	0.714	0.386
			ϕ_{23}	0.390	0.251	ψ_{23}	0.736	0.493
			ϕ_{24}	0.393	0.290	ψ_{24}	0.577	0.451
			ϕ_{33}	0.599	0.301	ψ_{25}	0.719	0.482
			ϕ_{34}	0.386	0.210	ψ_{26}	0.670	0.408
			ϕ_{44}	0.535	0.386			
			ψ_δ	0.246	0.234			

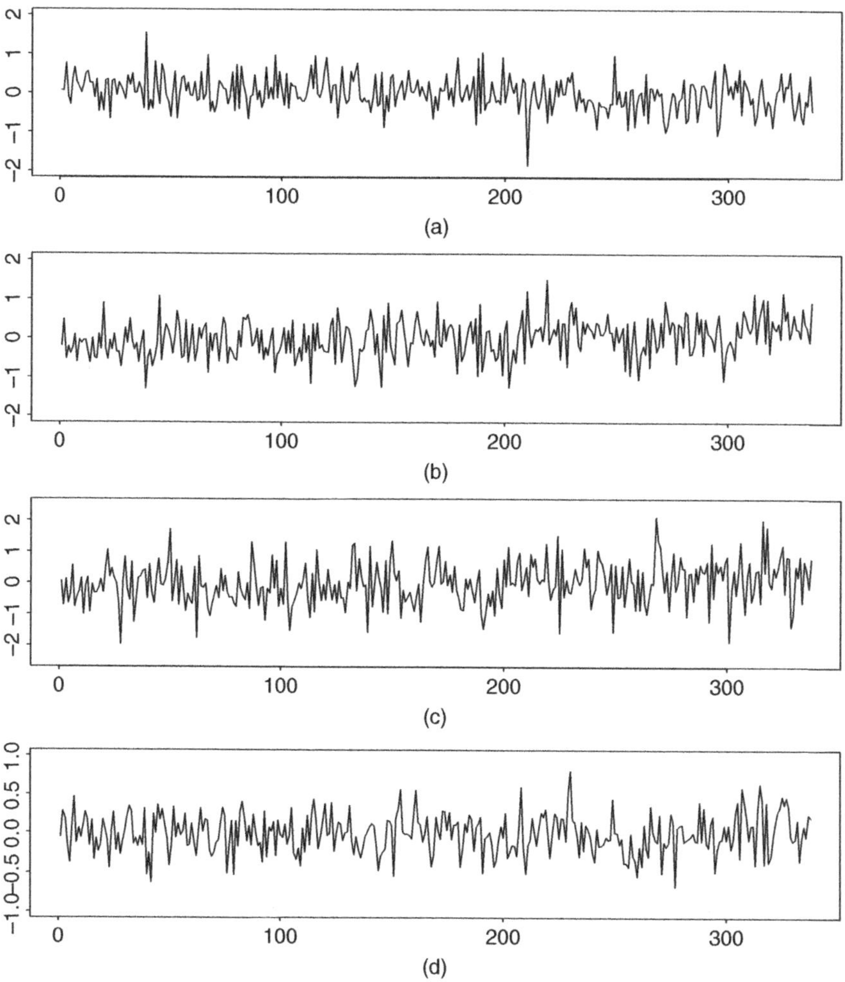

Figure 10.4 Estimated residual plots versus case numbers (a) $\hat{\epsilon}_{i1}^{(1)}$, (b) $\hat{\epsilon}_{i2}^{(1)}$, (c) $\hat{\epsilon}_{i3}^{(1)}$ and (d) $\hat{\delta}_{i}^{(1)}$.

estimated residuals $\hat{\boldsymbol{\epsilon}}_i^{(g)}$ and $\hat{\boldsymbol{\delta}}_i^{(g)}$ under M_1 can be obtained in the same way as in previous chapters. For group one, some estimated residual plots, $\hat{\epsilon}_{i1}^{(1)}$, $\hat{\epsilon}_{i2}^{(1)}$, $\hat{\epsilon}_{i3}^{(1)}$ and/or $\hat{\delta}_i^{(1)}$ versus case numbers, are displayed in Figure 10.4. Some estimated residual plots of $\hat{\epsilon}_{i1}^{(1)}$ and $\hat{\delta}_i^{(1)}$ versus $\hat{\xi}_{i1}^{(1)}$, $\hat{\xi}_{i2}^{(1)}$, $\hat{\xi}_{i3}^{(1)}$, $\hat{\xi}_{i4}^{(1)}$ and/or $\hat{\eta}_i^{(1)}$ are presented in Figures 10.5 and 10.6, respectively. Other plots are similar. Interpretations of these plots are the same as before. The WinBUGS codes in analyzing models M_1, M_2 and M_3 are given in the website: http://www.wiley.com/go/lee_structural.

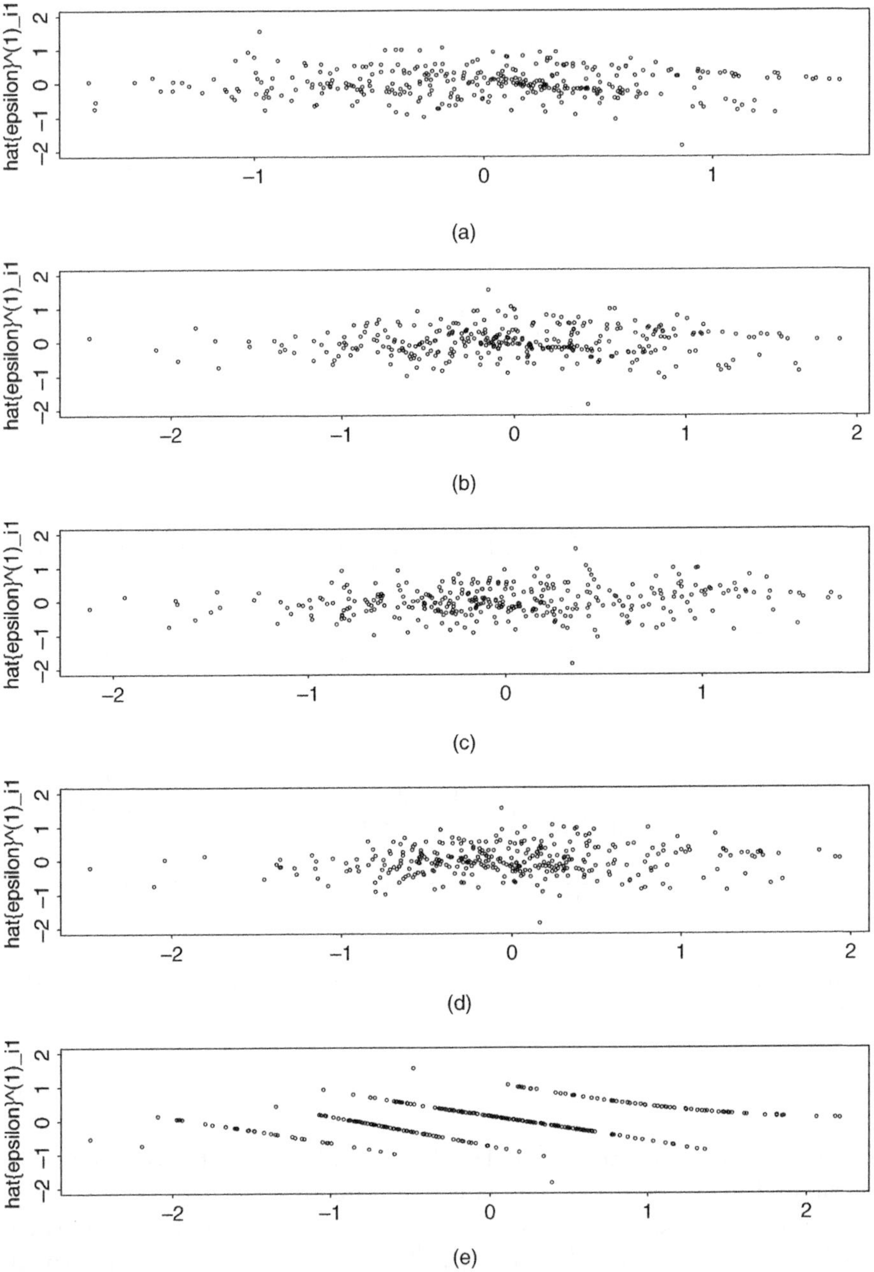

Figure 10.5 Plots of estimated residuals $\hat{\epsilon}_{i1}^{(1)}$ versus (a) $\hat{\xi}_{i1}^{(1)}$, (b) $\hat{\xi}_{i2}^{(1)}$, (c) $\hat{\xi}_{i3}^{(1)}$, (d) $\hat{\xi}_{i4}^{(1)}$ and (e) $\hat{\eta}_i^{(1)}$.

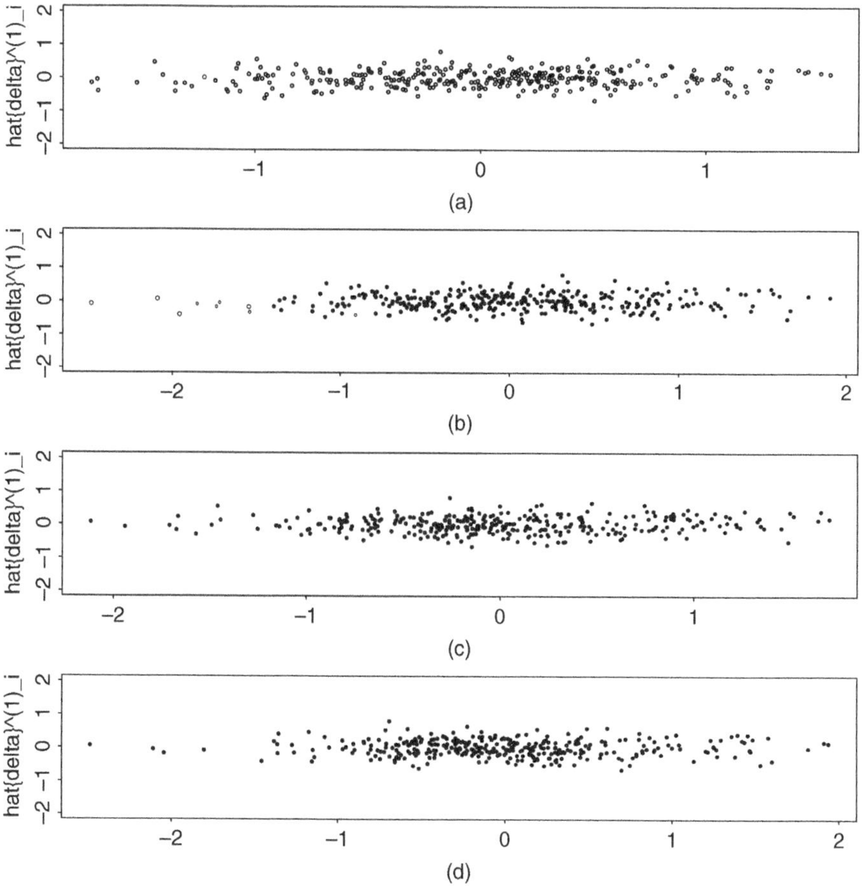

Figure 10.6 Plots of estimated residuals $\hat{\delta}_i^{(1)}$ versus (a) $\hat{\xi}_{i1}^{(1)}$, (b) $\hat{\xi}_{i2}^{(1)}$, (c) $\hat{\xi}_{i3}^{(1)}$ and (d) $\hat{\xi}_{i4}^{(1)}$.

APPENDIX 10.1: CONDITIONAL DISTRIBUTIONS: MULTISAMPLE SEMs

The conditional distributions $[\boldsymbol{\Omega}|\boldsymbol{\theta}, \boldsymbol{\alpha}, \mathbf{Y}, \mathbf{X}, \mathbf{Z}]$, $[\mathbf{Y}, \boldsymbol{\alpha}|\boldsymbol{\theta}, \mathbf{X}, \mathbf{Z}, \boldsymbol{\Omega}]$ and $[\boldsymbol{\theta}|\boldsymbol{\alpha}, \mathbf{Y}, \boldsymbol{\Omega}, \mathbf{X}, \mathbf{Z}]$ that are required in the implementation of the Gibbs sampler are presented in this appendix. Note that the results on the first two conditional distributions are natural extensions of those given in Chapter 6, but they can be regarded as the special cases of those given in Chapter 9. Also note that we allow common parameters in $\boldsymbol{\theta}$ according to the constraints under the competing models.

Conditional distribution of $[\boldsymbol{\Omega}|\boldsymbol{\theta}, \boldsymbol{\alpha}, \mathbf{Y}, \mathbf{X}, \mathbf{Z}]$ can be obtained as below:

$$p[\boldsymbol{\Omega}|\boldsymbol{\theta}, \boldsymbol{\alpha}, \mathbf{Y}, \mathbf{X}, \mathbf{Z}] = \prod_{g=1}^{G} \prod_{i=1}^{N_g} p(\boldsymbol{\omega}_i^{(g)}|\mathbf{v}_i^{(g)}, \boldsymbol{\theta}^{(g)}),$$

where

$$p(\boldsymbol{\omega}_i^{(g)}|\mathbf{v}_i^{(g)}, \boldsymbol{\theta}^{(g)}) \propto \exp\left\{-\frac{1}{2}\left[(\mathbf{v}_i^{(g)} - \boldsymbol{\mu}^{(g)} - \boldsymbol{\Lambda}^{(g)}\boldsymbol{\omega}_i^{(g)})^T \boldsymbol{\Psi}_\epsilon^{(g)-1}(\mathbf{v}_i^{(g)} - \boldsymbol{\mu}^{(g)} - \boldsymbol{\Lambda}^{(g)}\boldsymbol{\omega}_i^{(g)}) + (\boldsymbol{\eta}_i^{(g)} - \boldsymbol{\Lambda}_\omega^{(g)}\mathbf{H}^*(\boldsymbol{\omega}_i^{(g)}))^T \boldsymbol{\Psi}_\delta^{(g)^{-1}}(\boldsymbol{\eta}_i^{(g)} - \boldsymbol{\Lambda}_\omega^{(g)}\mathbf{H}^*(\boldsymbol{\omega}_i^{(g)})) + \boldsymbol{\xi}_i^{(g)T}\boldsymbol{\Phi}^{(g)^{-1}}\boldsymbol{\xi}_i^{(g)}\right]\right\}. \tag{A10.1}$$

Since the conditional distribution of (A10.1) is not standard, the Metropolis–Hasting (MH) algorithm can be used to draw random observations from this distribution.

Under the multisample situation, the notation in the conditional distribution $[\boldsymbol{\alpha}, \mathbf{Y}|\boldsymbol{\theta}, \boldsymbol{\Omega}, \mathbf{X}, \mathbf{Z}]$ is very tedious. The derivation is similar to the two-level case as given in Chapter 9, Appendix 9.1, Equations (A9.3) and (A9.4). As $(\boldsymbol{\alpha}^{(g)}, \mathbf{Y}^{(g)})$ is independent with $(\boldsymbol{\alpha}^{(h)}, \mathbf{Y}^{(h)})$ for $g \neq h$, and $\boldsymbol{\Psi}_\delta^{(g)}$ is diagonal,

$$p(\boldsymbol{\alpha}, \mathbf{Y}|\cdot) = \prod_{g=1}^{G} p(\boldsymbol{\alpha}^{(g)}, \mathbf{Y}^{(g)}|\cdot) = \prod_{g=1}^{G}\prod_{k=1}^{s} p(\boldsymbol{\alpha}_k^{(g)}, \mathbf{Y}_k^{(g)}|\cdot), \tag{A10.2}$$

where $\mathbf{Y}_k^{(g)} = [y_{1k}^{(g)}, \ldots, y_{N_g k}^{(g)}]$. Let $\psi_{\epsilon k}^{(g)}$ be the kth diagonal element of $\boldsymbol{\Psi}_\epsilon^{(g)}$, $\mu_k^{(g)}$ be the kth element of $\boldsymbol{\mu}^{(g)}$, $\boldsymbol{\Lambda}_k^{(g)T}$ be the kth row of $\boldsymbol{\Lambda}^{(g)}$ and $I_A(y)$ be an indicator function with value 1 if y in A and zero otherwise; $p(\boldsymbol{\alpha}, \mathbf{Y}|\cdot)$ can be obtained from Equation (A10.2) and

$$p(\boldsymbol{\alpha}_k^{(g)}, \mathbf{Y}_k^{(g)}|\cdot) \propto \prod_{i=1}^{N_g} \phi[\psi_{\epsilon k}^{(g)^{-1/2}}(y_{ik}^{(g)} - \mu_k^{(g)} - \boldsymbol{\Lambda}_k^{(g)T}\boldsymbol{\omega}_i^{(g)})] I_{[\alpha_{k,z_{ik}}, \alpha_{k,z_{ik}+1})}(y_{ik}^{(g)}), \tag{A10.3}$$

where ϕ is the probability density function of $N[0, 1]$. Note that in Equation (A10.3) the superscript '(g)' in the threshold is suppressed to simplify notation.

For dichotomous data, no thresholds are involved in the model. The conditional distribution of interest is $p[\mathbf{Y}|\boldsymbol{\theta}, \boldsymbol{\Omega}, \mathbf{X}, \mathbf{Z}]$. Based on the definition and the properties of the data, let $y_{ik}^{(g)}$ be the kth element of $\mathbf{y}_i^{(g)}$ and $z_{ik}^{(g)}$ be its corresponding dichotomous observations, $k = 1, \ldots, s$, so it can be shown that

$$p(\mathbf{Y}|\boldsymbol{\theta}, \boldsymbol{\Omega}, \mathbf{X}, \mathbf{Z}) = \prod_{g=1}^{G} \prod_{i=1}^{N_g} \prod_{k=1}^{s} p(y_{ik}^{(g)}|\cdot),$$

$$p(y_{ik}^{(g)}|\cdot) \stackrel{D}{=} \begin{cases} N(\mu_k^{(g)} + \boldsymbol{\Lambda}_k^{(g)^T} \boldsymbol{\omega}_i^{(g)}, \psi_{\epsilon k}^{(g)}) I_{(-\infty,0)}(y_{ik}^{(g)}), & z_{ik}^{(g)} = 0; \\ N(\mu_k^{(g)} + \boldsymbol{\Lambda}_k^{(g)^T} \boldsymbol{\omega}_i^{(g)}, \psi_{\epsilon k}^{(g)}) I_{[0,\infty)}(y_{ik}^{(g)}), & z_{ik}^{(g)} = 1. \end{cases} \quad \text{(A10.4)}$$

Under the prior distributions of components in $\boldsymbol{\theta}$ as given in Section 10.3.1, the conditional distribution $[\boldsymbol{\theta}|\boldsymbol{\alpha}, \mathbf{Y}, \boldsymbol{\Omega}, \mathbf{X}, \mathbf{Z}]$ is presented. Note that as $\mathbf{Y}$ is given, the model is defined with continuous data; hence, the conditional distribution is independent of $\boldsymbol{\alpha}$ and $\mathbf{Z}$.

The conditional distribution of some components in $\boldsymbol{\theta}^{(g)}$, $g = 1, \ldots, G$, under the situation without any parameter constraints are given as follows. Let $\boldsymbol{\Lambda}_k^{(g)^T}$ be the kth row of $\boldsymbol{\Lambda}^{(g)}$, $\psi_{\epsilon k}^{(g)}$ be the kth diagonal element of $\boldsymbol{\Psi}_\epsilon^{(g)}$, $\mathbf{V}_k^{*(g)T}$ be the kth row of $\mathbf{V}^{*(g)} = (\mathbf{v}_1^{(g)} - \boldsymbol{\mu}^{(g)}, \ldots, \mathbf{v}_{N_g}^{(g)} - \boldsymbol{\mu}^{(g)})$ and $\boldsymbol{\Omega}_2^{(g)} = (\boldsymbol{\xi}_1^{(g)}, \ldots, \boldsymbol{\xi}_{N_g}^{(g)})$. It can be shown that:

$$\begin{aligned} &[\boldsymbol{\mu}^{(g)}|\boldsymbol{\Lambda}^{(g)}, \boldsymbol{\Psi}_\epsilon^{(g)}, \mathbf{Y}, \boldsymbol{\Omega}, \mathbf{X}] \stackrel{D}{=} N(\mathbf{a}_\mu^{(g)}, \mathbf{A}_\mu^{(g)}); \\ &[\boldsymbol{\Lambda}_k^{(g)}|\boldsymbol{\Psi}_\epsilon^{(g)}, \boldsymbol{\mu}^{(g)}, \mathbf{Y}, \boldsymbol{\Omega}, \mathbf{X}] \stackrel{D}{=} N(\mathbf{a}_k^{(g)}, \mathbf{A}_k^{(g)}); \\ &[\psi_{\epsilon k}^{(g)-1}|\boldsymbol{\Lambda}_k^{(g)}, \boldsymbol{\mu}^{(g)}, \mathbf{Y}, \boldsymbol{\Omega}, \mathbf{X}] \stackrel{D}{=} Gamma(N_g/2 + \alpha_{0\epsilon k}^{(g)}, \beta_{\epsilon k}^{(g)}); \\ &[\boldsymbol{\Phi}^{(g)}|\boldsymbol{\Omega}_2^{(g)}] \stackrel{D}{=} IW_{q_2}[(\mathbf{R}_0^{(g)-1} + \boldsymbol{\Omega}_2^{(g)} \boldsymbol{\Omega}_2^{(g)^T}), N_g + \rho_0^{(g)}]; \end{aligned} \quad \text{(A10.5)}$$

in which

$$\begin{aligned} \mathbf{a}_\mu^{(g)} &= \mathbf{A}_\mu^{(g)} \left[\boldsymbol{\Sigma}_0^{(g)-1} \boldsymbol{\mu}_0^{(g)} + N_g \boldsymbol{\Psi}_\epsilon^{(g)-1} (\bar{\mathbf{v}}^{(g)} - \boldsymbol{\Lambda}^{(g)} \bar{\boldsymbol{\omega}}^{(g)})\right], \mathbf{A}_\mu^{(g)} = \left(\boldsymbol{\Sigma}_0^{(g)-1} + N_g \boldsymbol{\Psi}_\epsilon^{(g)-1}\right)^{-1}, \\ \mathbf{a}_k^{(g)} &= \mathbf{A}_k^{(g)} \left[\mathbf{H}_{0yk}^{(g)-1} \boldsymbol{\Lambda}_{0k}^{(g)} + \psi_{\epsilon k}^{(g)-1} \boldsymbol{\Omega}^{(g)} \mathbf{V}_k^{*(g)}\right], \mathbf{A}_k^{(g)} = \left[\psi_{\epsilon k}^{(g)-1} \boldsymbol{\Omega}^{(g)} \boldsymbol{\Omega}^{(g)T} + \mathbf{H}_{0yk}^{(g)-1}\right]^{-1} \\ \beta_{\epsilon k}^{(g)} &= \beta_{0\epsilon k}^{(g)} + \frac{1}{2}\left[\boldsymbol{\Lambda}_k^{(g)T} (\boldsymbol{\Omega}_k^{(g)} \boldsymbol{\Omega}_k^{(g)T}) \boldsymbol{\Lambda}_k^{(g)} - 2\boldsymbol{\Lambda}_k^{(g)T} \boldsymbol{\Omega}^{(g)} \mathbf{V}_k^{*(g)} + \mathbf{V}_k^{*(g)T} \mathbf{V}_k^{*(g)}\right], \end{aligned} \quad \text{(A10.6)}$$

with $\bar{\mathbf{v}}^{(g)} = \sum_{i=1}^{N_g} \mathbf{v}_i^{(g)}/N_g$ and $\bar{\boldsymbol{\omega}}^{(g)} = \sum_{i=1}^{N_g} \boldsymbol{\omega}_i^{(g)}/N_g$ being the means of $\mathbf{v}_i^{(g)}$ and $\boldsymbol{\omega}_i^{(g)}$ within the gth group.

As we mentioned, slight modifications are required to handle models with parameter constraints, see Section 10.3.1. Under the constraints $\mathbf{\Lambda}_k^{(1)} = \cdots = \mathbf{\Lambda}_k^{(G)} = \mathbf{\Lambda}_k$, the conjugate prior distribution of $\mathbf{\Lambda}_k$ is $N[\mathbf{\Lambda}_{0k}, \mathbf{H}_{0yk}]$ and the conditional distribution is

$$[\mathbf{\Lambda}_k \,|\psi_{\epsilon k}^{(1)}, \ldots, \psi_{\epsilon k}^{(G)}, \boldsymbol{\mu}^{(1)}, \ldots, \boldsymbol{\mu}^{(G)}, \mathbf{Y}, \mathbf{\Omega}, \mathbf{X}] \overset{D}{=} N[\mathbf{a}_k, \mathbf{A}_k], \qquad \text{(A10.7)}$$

where $\mathbf{a}_k = \mathbf{A}_k[\mathbf{H}_{0yk}^{-1}\mathbf{\Lambda}_{0k} + \sum_{g=1}^{G} \psi_{\epsilon k}^{(g)-1}\,\mathbf{\Omega}^{(g)}\mathbf{V}_k^{*(g)}]$, and $\mathbf{A}_k = [\sum_{g=1}^{G} \psi_{\epsilon k}^{(g)-1}\mathbf{\Omega}^{(g)}\mathbf{\Omega}^{(g)T} + \mathbf{H}_{0yk}^{-1}]^{-1}$. Under the constraints $\psi_{\epsilon k}^{(1)} = \cdots = \psi_{\epsilon k}^{(G)} = \psi_{\epsilon k}$, the conjugate prior distribution of $\psi_{\epsilon k}$ is $Gamma(\alpha_{0\epsilon k}, \beta_{0\epsilon k})$, and the conditional distribution is

$$[\psi_{\epsilon k}^{-1} \mid \mathbf{\Lambda}_k^{(1)}, \ldots, \mathbf{\Lambda}_k^{(G)}, \boldsymbol{\mu}^{(1)}, \ldots, \boldsymbol{\mu}^{(G)}, \mathbf{Y}, \mathbf{\Omega}, \mathbf{X}] \overset{D}{=} Gamma(N^*/2 + \alpha_{0\epsilon k}, \beta_{\epsilon k}), \qquad \text{(A10.8)}$$

where $N^* = N_1 + \cdots + N_G$,

$$\beta_{\epsilon k} = \beta_{0\epsilon k} + \tfrac{1}{2}\sum_{g=1}^{G}\left[\mathbf{\Lambda}_k^{(g)T}(\mathbf{\Omega}^{(g)}\mathbf{\Omega}^{(g)T})\mathbf{\Lambda}_k^{(g)} - 2\mathbf{\Lambda}_k^{(g)T}\mathbf{\Omega}^{(g)}\,\mathbf{V}_k^{*(g)} + \mathbf{V}_k^{*(g)T}\mathbf{V}_k^{*(g)}\right].$$

Under the constraints $\mathbf{\Phi}^{(1)} = \cdots = \mathbf{\Phi}^{(G)} = \mathbf{\Phi}$, the conjugate prior distribution of $\mathbf{\Phi}^{-1}$ is $W_{q_2}[\mathbf{R}_0, \rho_0]$ and the conditional distribution is

$$[\mathbf{\Phi} \mid \mathbf{\Omega}_2^{(1)}, \cdots, \mathbf{\Omega}_2^{(G)}] \overset{D}{=} IW_{q_2}\left[\left(\mathbf{R}_0^{-1} + \sum_{g=1}^{G}\mathbf{\Omega}_2^{(g)}\mathbf{\Omega}_2^{(g)T}\right),\ N^* + \rho_0\right]. \qquad \text{(A10.9)}$$

The conditional distributions of $\mathbf{\Lambda}_k^{(g)}$ and $\psi_{\delta k}^{(g)}$ are similar, and hence not presented.

As the conditional distributions involved in Equation (A10.5) or (A10.7)–(A10.9) are standard distributions, drawing observations from them is straightforward. Simulating observations from the conditional distributions that are given in Equation (A10.4) or (A10.3) involves the univariate truncated normal distribution, and this is done by the inverse distribution method proposed by Devroye (1985). A Metropolis–Hasting (MH) algorithm is used to simulate observations from the complex conditional distributions of Equations (A10.1) and (10.3).

REFERENCES

Bollen, K. A. (1989) *Structural Equations with Latent Variables.* New York: John Wiley & Sons, Inc..

Devroye, L. (1985) *Non-Uniform Random Variate Generation.* New York: Springer Verlag.

Gelman, A. and Meng, X. L. (1998) Simulating normalizing constant: from importance sampling to bridge sampling to path sampling. *Statistical Science*, **13**, 163–185.

Geman, S. and Geman, D. (1984) Stochastic relaxation, Gibbs distributions and the Bayesian restoration of images. *IEEE Transactions on Pattern Analysis and Machine Intelligence*, **6**, 721–741.

Lee, S. Y., Poon, W. Y. and Bentler, P. M. (1989) Simultaneous analysis of multivariate polychoric correlation model in several groups. *Psychometrika*, **54**, 63–73.

Lee, S. Y., Poon, W. Y. and Bentler, P. M. (1995) A two-stage estimation of structural equation models with continuous and polytomous variables. *British Journal of Mathematical and Statistical Psychology*, **48**, 339–358.

Lee, S. Y., Song, X. Y., Skevington, S. and Hao, Y. T. (2005) Application of structural equation models to quality of life. *Structural Equation Modeling: A Multidisciplinary Journal*, **12**, 435–453.

Meuleners, L. B., Lee, A. H., Binns, C. W. and Lower, A. (2003) Quality of life for adolescents: assessing measurement properties using structural equation modeling. *Quality of Life Research*, **12**, 283–290.

Power, M., Bullingen, M., Hazper, A. and WHOQOL Group (1999) The World Health Organization WHOQOL-100: tests of the universality of quality of life in 15 different cultural groups worldwide. *Health Psychology*, **18**(5), 495–505.

Shi, J. Q. and Lee, S. Y. (2000) Latent variable models with mixed continuous and polytomous data. *Journal of the Royal Statistical Society, Series B*, **62**, 77–87.

Song, X. Y. and Lee, S. Y. (2001) Bayesian estimation and test for factor analysis model with continuous and polytomous data in several populations. *British Journal of Mathematical and Statistical Psychology*, **54** 237–263.

World Values Study Group (1994) World Values Survey, 1981–1984 and 1990–1993. ICPSR version. Ann Arbor, MI: Institute for Social Research (producer). Ann Arbo, MI: Inter-university Consortium for Political and Social Research (distribution).

11

Finite Mixtures in Structural Equation Models

11.1 INTRODUCTION

In behavioral science such as psychology, sociology, education, etc., heterogeneity of population is inevitable and is an important concern. The result would be seriously distorted if analyzing a heterogeneous population as a homogenous population. Ordinary multiple group methods should not be used, unless the membership of each independent observation can be specified accurately. Recently, finite mixtures have been applied to structural equation models in order to deal with heterogeneous populations.

In general, a finite mixture model (see Redner and Walker, 1984; Titterington, Smith and Markov, 1985) arises with a population which is a mixture of K components with associative probability densities $\{f_k, k = 1, \ldots, K\}$ and mixing proportions $\{\pi_k, k = 1, \ldots, K\}$. Mixture models arise in many fields, including behavioral, medical, economics and environmental sciences. They have been used in modeling heterogeneity, handling outliers (Pettit and Smith, 1985) and density estimation (Roeder and Wasserman, 1997). It is well recognized that statistical analysis of mixture models is not straightforward. For estimation with a fixed number of components K, numerous methods have been proposed. Examples are the method of moments (Lindsay and Basak, 1993), Bayesian methods with MCMC techniques (Diebolt and Robert, 1994; Robert, 1996), and the ML method (Hathaway, 1985). For mixture models with K treated as random, Richardson and Green (1997) developed a full Bayesian analysis on the basis of the reversible jump MCMC

Structural Equation Modeling: A Bayesian Approach S-Y. Lee

method introduced by Green (1995). For the challenging problem of testing the number of components, the classical likelihood-based inference encountered serious difficulties. Due to some non-regular problems, some standard asymptotic properties associated with the ML estimation and the likelihood ratio test are not valid. To deal with problems involved in the likelihood ratio test, some bootstrap methods have been proposed (see Feng and McCulloch, 1996; McLachlan, 1987). However, as pointed out by Richardson and Green (1997), the Bayes paradigm is particularly suitable for analyzing mixture models with an unknown K.

In the field of SEM, Arminger and Stein (1997) used a two-stage method to analyze finite mixtures of conditional distributions with covariance structures. Jedidi, Jagpal and DeSarbo (1997a) analyzed the finite mixtures of multivariate regression and simultaneous equation models, while Jedidi, Jagpal and DeSarbo (1997b) considered the estimation of a general finite mixtures of structural equation models and gave a brief discussion on the problem of model selection via the Bayesian information criterion (BIC) with a fixed number of components. Yung (1997) investigated finite mixtures of confirmatory factor analysis models. He proposed an approximated scoring algorithm and an EM algorithm to solve the likelihood equation. Dolan and van der Maas (1998) applied a quasi-Newton algorithm to finite mixtures and inferred the estimation by changing the degree of separation and the sample size. Arminger, Stein and Wittenberg (1999) discussed ML analysis for mixtures of conditional mean- and covariance-structure models, and three estimation strategies on the basis of the EM algorithm were established. They also pointed out the difficulty of the usual likelihood ratio test for testing the number of unknown components, briefly discussed an ad hoc test and proposed a procedure that uses the parametric bootstrap to construct an estimate of the distribution of the likelihood ratio test under the null hypothesis. The test is based on the unrestricted parameter estimates of the variances/covariances in the covariance matrices. Detailed statistical properties and justifications associated with these tests have not been established.

Zhu and Lee (2001) proposed a Bayesian analysis to finite mixtures in the LISREL model, using the idea of augmenting the observed data with latent variables and allocation variables. Joint Bayesian estimates of the mixing proportions, structural parameters and latent variables were obtained via some MCMC methods. Lee and Song (2003b) developed a path sampling procedure to compute the observed data log-likelihood, for evaluating the BIC in selecting the appropriate number of components for a mixture SEM with missing data. A Bayesian approach to analyze mixtures of SEMs with an unknown number of components has been developed by Lee and Song (2002). They formulated the problem as a model selection problem for selecting one of the two mixture SEMs with different numbers of components. Their approach is based on the Bayes factor (Berger, 1985; Kass and Raftery, 1995), which is computed via a path sampling (Gelman and Meng, 1998) procedure. Spiegelhalter, Thomas,

Best and Lunn (2003) pointed out that DIC may not be appropriate for model comparison in the context of mixture models. Hence, for mixture models, WinBUGS does not give DIC results.

For a mixture SEM with K components, it is well-known that the model is invariant with respect to permutation of the labels $k = 1, \ldots, K$. Thus, the model is not identified, and adoption of a unique labelling for identifiability is important. In the literature, a common method is to use some constraints on the components of the mean vector to force a unique labelling. In a Bayesian approach, arbitrary constraints may not be able to solve the problem. In this chapter, we will apply the permutation sampler (Frühwirth-Schnatter, 2001) to find the appropriate identifiability constraints. We start with the description of finite mixtures in structural equation models, then move to the Bayesian estimation of the model and the Bayesian model comparison. The methodologies will be illustrated with examples. The application of WinBUGS in obtaining some Bayesian results is presented in Section 11.4.1.

11.2 FINITE MIXTURES IN SEMs

Let $\mathbf{y}_i$ be a $p \times 1$ random vector corresponding to the ith observation in a random sample of size n, and suppose that its distribution is given by the following probability density function:

$$f(\mathbf{y}_i|\boldsymbol{\theta}) = \sum_{k=1}^{K} \pi_k f_k(\mathbf{y}_i|\boldsymbol{\mu}_k, \boldsymbol{\theta}_k), \quad i = 1, \ldots, n, \tag{11.1}$$

where K is a given integer, π_k is the unknown mixing proportion such that $\pi_k > 0$ and $\pi_1 + \cdots + \pi_K = 1.0$, $f_k(\mathbf{y}|\boldsymbol{\mu}_k, \boldsymbol{\theta}_k)$ is the multivariate normal density function with an unknown mean vector $\boldsymbol{\mu}_k$ and a general covariance structure $\boldsymbol{\Sigma}_k = \boldsymbol{\Sigma}_k(\boldsymbol{\theta}_k)$ that depends on an unknown parameter vector $\boldsymbol{\theta}_k$, and $\boldsymbol{\theta}$ is the parameter vector that contains all unknown parameters in π_k, $\boldsymbol{\mu}_k$ and $\boldsymbol{\theta}_k$, $k = 1, \ldots, K$. As the LISREL type model (Jöreskog and Sörbom, 1996) is a very popular model, it will be used here as a representative model for the random vector $\mathbf{y}_i$ conditional on the kth component. For the kth component, the measurement equation of the model is given by:

$$\mathbf{y}_i = \boldsymbol{\mu}_k + \boldsymbol{\Lambda}_k \boldsymbol{\omega}_{ki} + \boldsymbol{\epsilon}_{ki}, \tag{11.2}$$

where $\boldsymbol{\mu}_k$ is the mean vector, $\boldsymbol{\Lambda}_k$ is the $p \times q$ factor loading matrix, $\boldsymbol{\omega}_{ki}$ is a random vector of latent variables and $\boldsymbol{\epsilon}_{ki}$ is a random vector of residuals which is distributed according to $N[\mathbf{0}, \boldsymbol{\Psi}_k]$, where $\boldsymbol{\Psi}_k$ is a diagonal matrix. It is assumed that $\boldsymbol{\omega}_{ki}$ and $\boldsymbol{\epsilon}_{ki}$ are independent. Let $\boldsymbol{\omega}_{ki} = (\boldsymbol{\eta}_{ki}^T, \boldsymbol{\xi}_{ki}^T)^T$ be a partition of $\boldsymbol{\omega}_{ki}$

into an endogenous latent vector $\boldsymbol{\eta}_{ki}$ and an exogenous latent vector $\boldsymbol{\xi}_{ki}$. The structural equation of the model, which describes the relationships among the latent variables, is defined as

$$\boldsymbol{\eta}_{ki} = \boldsymbol{\Pi}_k \boldsymbol{\eta}_{ki} + \boldsymbol{\Gamma}_k \boldsymbol{\xi}_{ki} + \boldsymbol{\delta}_{ki}, \tag{11.3}$$

where $\boldsymbol{\eta}_{ki}$ and $\boldsymbol{\xi}_{ki}$ are $q_1 \times 1$ and $q_2 \times 1$ subvectors of $\boldsymbol{\omega}_{ki}$ respectively, $\boldsymbol{\delta}_{ki}$ is a random vector of residuals that is independent of $\boldsymbol{\xi}_{ki}$, $\boldsymbol{\Pi}_k$ and $\boldsymbol{\Gamma}_k$ are unknown parameter matrices such that $\boldsymbol{\Pi}_{0k}^{-1} = (\mathbf{I} - \boldsymbol{\Pi}_k)^{-1}$ exists and $|\boldsymbol{\Pi}_{0k}|$ is independent of elements in $\boldsymbol{\Pi}_k$. Let the distributions of $\boldsymbol{\xi}_{ki}$ and $\boldsymbol{\delta}_k$ be $N[\mathbf{0}, \boldsymbol{\Phi}_k]$ and $N[\mathbf{0}, \boldsymbol{\Psi}_{\delta k}]$, respectively, where $\boldsymbol{\Psi}_{\delta k}$ is a diagonal matrix. The parameter vector $\boldsymbol{\theta}_k$ in the kth component contains the free unknown parameters in $\boldsymbol{\Lambda}_k$, Π_k, $\boldsymbol{\Gamma}_k$, $\boldsymbol{\Pi}_k$, $\boldsymbol{\Psi}_{\delta k}$ and $\boldsymbol{\Psi}_k$. The covariance structure of $\boldsymbol{\omega}_{ki}$ is given by

$$\boldsymbol{\Sigma}_{\omega k} = \begin{bmatrix} \boldsymbol{\Pi}_{0k}^{-1}(\boldsymbol{\Gamma}_k \boldsymbol{\Phi}_k \boldsymbol{\Gamma}_k^T + \boldsymbol{\Psi}_{\delta k})(\boldsymbol{\Pi}_{0k}^{-1})^T & \boldsymbol{\Pi}_{0k}^{-1}\boldsymbol{\Gamma}_k \boldsymbol{\Phi}_k \\ \boldsymbol{\Phi}_k \boldsymbol{\Gamma}_k^T (\boldsymbol{\Pi}_{0k}^{-1})^T & \boldsymbol{\Phi}_k \end{bmatrix}, \tag{11.4}$$

and $\boldsymbol{\Sigma}_k(\boldsymbol{\theta}_k) = \boldsymbol{\Lambda}_k \boldsymbol{\Sigma}_{\omega k} \boldsymbol{\Lambda}_k^T + \boldsymbol{\Psi}_k$. Any of these unknown parameter matrices can be invariant across components. However, it is important to assign a different $\boldsymbol{\mu}_k$ in the measurement Equation (11.2) of each component, in order to analyze effectively data from the heterogenous populations that are different by their mean vectors. We do not formulate mixtures of SEMs with different mean vectors in the latent vector $\boldsymbol{\omega}_k$ (or $\boldsymbol{\xi}_k$) and allow the same $\boldsymbol{\mu}_k = \boldsymbol{\mu}$ for the manifest vector $\mathbf{y}$. The main reason is that for the kth component under such formulation, $E(\mathbf{y}) = \boldsymbol{\mu} + \boldsymbol{\Lambda}_k E(\boldsymbol{\omega}_k)$; hence, $E(\mathbf{y})$ is not directly related to $E(\boldsymbol{\omega}_k)$, but affected by $\boldsymbol{\Lambda}_k$ which may be greatly influenced by the variation of population through $\boldsymbol{\Sigma}_k$.

As the mixture model defined in Equation (11.1) is invariant with respect to the permutation of labels $k = 1, \ldots, K$, adoption of a unique labelling for identifiability is important. Roeder and Wasserman (1997) and Zhu and Lee (2001) proposed imposing the ordering $\mu_{1,1} < \cdots < \mu_{K,1}$ for solving the label switching problem (jumping between the various labelling subspace), where $\mu_{k,1}$ is the first element of the mean vector $\boldsymbol{\mu}_k$. This method works fine if $\mu_{1,1} < \cdots < \mu_{K,1}$ are well separated. However, if $\mu_{1,1} < \cdots < \mu_{K,1}$ are close to each other, it may not be able to eliminate the label switching and may give incorrect results. Hence, it is important to find an appropriate identifiability constraint. Here, the random permutation sampler that is developed by Frühwirth-Schnatter (2001) will be applied for finding the suitable identifiability constraints.

Moreover, for each $k = 1, \ldots K$, structural parameters in the covariance matrix $\boldsymbol{\Sigma}_k$ corresponding to the model defined by Equations (11.2) and (11.3) are not identified. This problem is solved by the common method in structural equation modeling by fixing appropriate elements in $\boldsymbol{\Lambda}_k$, Π_k and/or $\boldsymbol{\Gamma}_k$ at preassigned values that are chosen on a problem-by-problem basis. For clear presentation of the Bayesian method, we assume that all the unknown parameters in the model are identified.

The next section discusses Bayesian estimation with the permutation sampler, and Bayesian classification. A simulation study and some illustrative examples, including one that is analyzed through WinBUGS, are presented in Section 11.4. Bayesian model comparison is addressed in Section 11.5. Technical details are given in Appendices 11.1 and 11.2.

11.3 BAYESIAN ESTIMATION AND CLASSIFICATION

Let $\boldsymbol{\theta}_{yk}$ be the vector of unknown parameters in $\boldsymbol{\Lambda}_k$ and $\boldsymbol{\Psi}_k$, and let $\boldsymbol{\theta}_{\omega k}$ be the vector of unknown parameters in $\boldsymbol{\Pi}_k$, $\boldsymbol{\Gamma}_k$, $\boldsymbol{\Phi}_k$ and $\boldsymbol{\Psi}_{\delta k}$. Let $\boldsymbol{\mu}$, $\boldsymbol{\pi}$, $\boldsymbol{\theta}_y$ and $\boldsymbol{\theta}_\omega$ be the vectors that contain the unknown parameters in $\{\boldsymbol{\mu}_1, \ldots, \boldsymbol{\mu}_K\}$, $\{\pi_1, \ldots, \pi_K\}$, $\{\boldsymbol{\theta}_{y1}, \ldots, \boldsymbol{\theta}_{yK}\}$ and $\{\boldsymbol{\theta}_{\omega 1}, \ldots, \boldsymbol{\theta}_{\omega K}\}$, respectively; and let $\boldsymbol{\theta} = (\boldsymbol{\mu}, \boldsymbol{\pi}, \boldsymbol{\theta}_y, \boldsymbol{\theta}_\omega)$ be the overall parameter vector. Inspired by Zhu and Lee (2001) and other works in finite mixture models, we introduce a group label w_i for the ith observation $\mathbf{y}_i$ as a latent allocation variable, and assume that it is independently drawn from the following distribution:

$$p(w_i = k) = \pi_k, \quad \text{for} \quad k = 1, \ldots, K. \tag{11.5}$$

Moreover, let $\mathbf{Y} = (\mathbf{y}_1, \ldots, \mathbf{y}_n)$ be the observed data matrix, $\boldsymbol{\Omega} = (\boldsymbol{\omega}_1, \ldots, \boldsymbol{\omega}_n)$ with $\boldsymbol{\omega}_i = \boldsymbol{\omega}_{ki}$ if it is in the kth component, and $\mathbf{W} = (w_1, \ldots, w_n)$ be the matrix of allocation variables.

In a standard Bayesian analysis, we need to evaluate the posterior distribution of the unknown parameters, $p(\boldsymbol{\theta}|\mathbf{Y})$. Due to the nature of the mixture model, this posterior distribution is complicated. However, if $\mathbf{W}$ is observed, the component of every $\mathbf{y}_i$ can be identified and the mixture model becomes the more familiar multiple group model. In addition, if $\boldsymbol{\Omega}$ is observed, the underlying SEM will become the linear simultaneous equation model which is comparatively easy to handle. Hence, the observed data $\mathbf{Y}$ are augmented with the latent quantities $\boldsymbol{\Omega}$ and $\mathbf{W}$ in the posterior analysis. In the following sections, we will concentrate on analyzing $p(\boldsymbol{\theta}, \boldsymbol{\Omega}, \mathbf{W}|\mathbf{Y})$, the posterior distribution of $(\boldsymbol{\theta}, \boldsymbol{\Omega}, \mathbf{W})$ given $\mathbf{Y}$.

The label switching problem has to be solved in the posterior analysis. For general mixture models with K components, the unconstrained parameter space contains $K!$ subspaces, each one corresponding to a different way to label the states. In the current mixture of SEM, the likelihood is invariant to relabelling the states. If the prior distributions of $\boldsymbol{\pi}$ and other parameters in $\boldsymbol{\theta}$ are also invariant, the unconstrained posterior is invariant to relabelling the states and identical on all labelling subspaces. This induces a multimodal posterior, and a serious problem in Bayesian estimation.

We will use the MCMC approach proposed by Frühwirth-Schnatter (2001) to deal with the above label switching problem. In this approach, an unidentified model is first estimated by sampling from the unconstrained posterior using the random permutation sampler, where each sweep is concluded by a random permutation of the current labelling of the components. The random permutation sampler delivers a sample that explores the whole unconstrained parameter space and jumps between the various labelling subspace in a balanced fashion. As pointed out by Frühwirth-Schnatter (2001), although the model is unidentified, the output of the random permutation sampler can be used to estimate unknown parameters that are invariant to relabelling the states, and can be explored to find suitable identifiability constraints. Then, the model is reestimated by sampling from the posterior distribution under the imposed identifiability constraints, again using the permutation sampler. The implementation of the permutation sampler in relation to the mixtures of SEMs, and the method of selecting the identifiability constraint are briefly described in Appendices 11.1 and 11.2, respectively.

The Bayesian estimates of $\boldsymbol{\theta}$ and $\boldsymbol{\Omega}$ will be obtained by computing the posterior means of $\boldsymbol{\theta}$ and $\boldsymbol{\Omega}$ in the posterior distribution of $[\boldsymbol{\theta}, \boldsymbol{\Omega}, \mathbf{W}|\mathbf{Y}]$. The main task is to simulate a sufficient large sample of observations from this posterior distribution, then the Bayesian estimates can be approximated by the sample means. Similar to many Bayesian analyses of SEMs, the Gibbs sampler (Geman and Geman, 1984) is used to generate a sequence of observations from $p(\boldsymbol{\theta}, \boldsymbol{\Omega}, \mathbf{W}|\mathbf{Y})$. The basic algorithm of this sampler is briefly given as below. At the rth iteration with current values $\boldsymbol{\theta}^{(r)}$, $\boldsymbol{\Omega}^{(r)}$ and $\mathbf{W}^{(r)}$:

Step (a): Generate $(\mathbf{W}^{(r+1)}, \boldsymbol{\Omega}^{(r+1)})$ from $p(\boldsymbol{\Omega}, \mathbf{W}|\mathbf{Y}, \boldsymbol{\theta}^{(r)})$;
Step (b): Generate $\boldsymbol{\theta}^{(r+1)}$ from $p(\boldsymbol{\theta}|\mathbf{Y}, \boldsymbol{\Omega}^{(r+1)}, \mathbf{W}^{(r+1)})$.
Step (c): Reorder the label through the permutation sampler to make the identifiability fulfill.

As $p(\boldsymbol{\Omega}, \mathbf{W}|\mathbf{Y}, \boldsymbol{\theta}) = p(\mathbf{W}|\mathbf{Y}, \boldsymbol{\theta})p(\boldsymbol{\Omega}|\mathbf{Y}, \mathbf{W}, \boldsymbol{\theta})$, Step (a) can be further decomposed into the following two steps:

Step (a1): Generate $\mathbf{W}^{(r+1)}$ from $p(\mathbf{W}|\mathbf{Y}, \boldsymbol{\theta}^{(r)})$;
Step (a2): Generate $\boldsymbol{\Omega}^{(r+1)}$ from $p(\boldsymbol{\Omega}|\mathbf{Y}, \boldsymbol{\theta}^{(r)}, \mathbf{W}^{(r+1)})$.

Simulating observations $(\mathbf{W}, \boldsymbol{\Omega})$ through Steps (a1) and (a2) is more efficient than using Step (a).

Conditional distributions required for implementing the Gibbs sampler are discussed below. As we have mentioned before, the finite mixtures model becomes a multisample model with given $\mathbf{W}$. Thus the conditional distributions associated with the Gibbs sampler can be derived without much difficulty.

We first consider the conditional distribution associated with Step (a1). As w_i are independent,

$$p(\mathbf{W}|\mathbf{Y}, \boldsymbol{\theta}) = \prod_{i=1}^{n} p(w_i|\mathbf{y}_i, \boldsymbol{\theta}). \tag{11.6}$$

Moreover,

$$p(w_i = k|\mathbf{y}_i, \boldsymbol{\theta}) = \frac{p(w_i = k, \mathbf{y}_i|\boldsymbol{\theta})}{p(\mathbf{y}_i|\boldsymbol{\theta})} = \frac{p(w_i = k|\boldsymbol{\pi})p(\mathbf{y}_i|w_i = k, \boldsymbol{\mu}_k, \boldsymbol{\theta}_k)}{p(\mathbf{y}_i|\boldsymbol{\theta})}$$
$$= \frac{\pi_k f_k(\mathbf{y}_i|\boldsymbol{\mu}_k, \boldsymbol{\theta}_k)}{p(\mathbf{y}_i|\boldsymbol{\theta})}, \qquad (11.7)$$

where $f_k(\mathbf{y}_i|\boldsymbol{\mu}_k, \boldsymbol{\theta}_k)$ is the probability density function of $N[\boldsymbol{\mu}_k, \boldsymbol{\Sigma}_k(\boldsymbol{\theta}_k)]$. Hence, the conditional distribution of $\mathbf{W}$ given $\mathbf{Y}$ and $\boldsymbol{\theta}$ can be derived from Equations (11.6) and (11.7). It can be seen from Equation (11.7) that drawing observations from $p(\mathbf{W}|\mathbf{Y}, \boldsymbol{\theta}^{(r)})$ is not difficult.

Consider the conditional distribution involved in Step (a2). Because $\boldsymbol{\omega}_i$ are mutually independent with given w_i, we have

$$\prod_{i=1}^{n} p(\boldsymbol{\omega}_i|\mathbf{y}_i, w_i, \boldsymbol{\theta}) = p(\boldsymbol{\Omega}|\mathbf{Y}, \boldsymbol{\theta}, \mathbf{W}) \propto p(\mathbf{Y}|\boldsymbol{\Omega}, \mathbf{W}, \boldsymbol{\theta}_y)p(\boldsymbol{\Omega}|\mathbf{W}, \boldsymbol{\theta}_\omega)$$
$$= \prod_{i=1}^{n} p(\mathbf{y}_i|\boldsymbol{\omega}_i, w_i, \boldsymbol{\theta}_y)p(\boldsymbol{\omega}_i|w_i, \boldsymbol{\theta}_\omega). \qquad (11.8)$$

Let $\mathbf{C}_k = \boldsymbol{\Sigma}_{\omega k}^{-1} + \boldsymbol{\Lambda}_k^T\boldsymbol{\Psi}_k^{-1}\boldsymbol{\Lambda}_k$; where $\boldsymbol{\Sigma}_{\omega k}$ is the covariance matrix of $\boldsymbol{\omega}_{ki}$ in the kth component (see Equation (11.4)). Moreover, as the conditional distribution of $\boldsymbol{\omega}_i$ given $\boldsymbol{\theta}$ and $w_i = k$ is $N[\mathbf{0}, \boldsymbol{\Sigma}_{\omega k}]$, and the conditional distribution of $\mathbf{y}_i$ given $\boldsymbol{\omega}_i, \boldsymbol{\mu}, \boldsymbol{\theta}_y$, and $w_i = k$ is $N[\boldsymbol{\mu}_k + \boldsymbol{\Lambda}_k\boldsymbol{\omega}_{ki}, \boldsymbol{\Psi}_k]$, it can be shown (see Lindley and Smith (1972), pp. 4 and 5) that

$$p(\boldsymbol{\omega}_i|\mathbf{y}_i, w_i = k, \boldsymbol{\theta}) \stackrel{D}{=} N[\mathbf{C}_k^{-1}\boldsymbol{\Lambda}_k^T\boldsymbol{\Psi}_k^{-1}(\mathbf{y}_i - \boldsymbol{\mu}_k), \mathbf{C}_k^{-1}]. \qquad (11.9)$$

The conditional distribution of $p(\boldsymbol{\Omega}|\mathbf{Y}, \boldsymbol{\theta}, \mathbf{W})$ can be obtained from Equations (11.8) and (11.9). Drawing observations from this familiar normal distribution is fast. Hence Step (a2) of the Gibbs sampler can be completed.

We now consider the conditional distribution $p(\boldsymbol{\theta}|\mathbf{Y}, \boldsymbol{\Omega}, \mathbf{W})$ that is required in Step (b) of the Gibbs sampler. This conditional distribution is quite complicated. However, the difficulty can be reduced by assuming the following mild conditions on the prior distribution of $\boldsymbol{\theta}$. We assume that the prior distribution of the mixing proportion $\boldsymbol{\pi}$ is independent of the prior distributions of $\boldsymbol{\mu}, \boldsymbol{\theta}_y$ and $\boldsymbol{\theta}_\omega$. Like many Bayesian analyses in SEMs, the prior distribution of the mean vector $\boldsymbol{\mu}$ can be taken to be independent of the prior distributions of the parameters $\boldsymbol{\theta}_y$ and $\boldsymbol{\theta}_\omega$ in the covariance structures. Moreover, when $\boldsymbol{\Omega}$ is given, the parameters in $\boldsymbol{\theta}_{yk} = \{\boldsymbol{\Lambda}_k, \boldsymbol{\Psi}_k\}$ are the parameters involved in the linear regression model with the manifest variables in $\mathbf{y}$ (see Equation (11.2)), and the parameters in $\boldsymbol{\theta}_{\omega k} = \{\boldsymbol{\Pi}_k, \boldsymbol{\Gamma}_k, \boldsymbol{\Phi}_k, \boldsymbol{\Psi}_k\}$ are the parameters involved in the other simultaneous equation model with the latent variables (see Equation (11.3)). Hence, we assume that the prior distributions of $\boldsymbol{\theta}_y$ and $\boldsymbol{\theta}_\omega$ are independent. As a result, $p(\boldsymbol{\theta}) = p(\boldsymbol{\pi}, \boldsymbol{\mu}, \boldsymbol{\theta}_y, \boldsymbol{\theta}_\omega) = p(\boldsymbol{\pi})p(\boldsymbol{\mu})p(\boldsymbol{\theta}_y)p(\boldsymbol{\theta}_\omega)$. Moreover, from the definition of the model and the properties of $\mathbf{W}, \boldsymbol{\Omega}$ and $\boldsymbol{\theta}$, we have $p(\mathbf{W}|\boldsymbol{\theta}) = p(\mathbf{W}|\boldsymbol{\pi})$, $p(\boldsymbol{\Omega}, \mathbf{Y}|\mathbf{W}, \boldsymbol{\theta}) = p(\mathbf{Y}|\boldsymbol{\Omega}, \mathbf{W}, \boldsymbol{\mu}, \boldsymbol{\theta}_y)p(\boldsymbol{\Omega}|\mathbf{W}, \boldsymbol{\theta}_\omega)$ and $p(\boldsymbol{\Omega}|\mathbf{W}, \boldsymbol{\theta}_\omega) = p(\boldsymbol{\Omega}|\boldsymbol{\theta}_\omega)$. Consequently, the joint conditional distribution of $\boldsymbol{\theta} = (\boldsymbol{\pi}, \boldsymbol{\mu}, \boldsymbol{\theta}_y, \boldsymbol{\theta}_\omega)$ can be expressed as

$$
\begin{aligned}
p(\boldsymbol{\theta}|\mathbf{W}, \boldsymbol{\Omega}, \mathbf{Y}) &= p(\boldsymbol{\pi}, \boldsymbol{\mu}, \boldsymbol{\theta}_y, \boldsymbol{\theta}_\omega|\mathbf{W}, \boldsymbol{\Omega}, \mathbf{Y}) \\
&\propto p(\boldsymbol{\pi})p(\boldsymbol{\mu})p(\boldsymbol{\theta}_y)p(\boldsymbol{\theta}_\omega)p(\mathbf{W}, \boldsymbol{\Omega}, \mathbf{Y}|\boldsymbol{\theta}) \\
&\propto p(\boldsymbol{\pi})p(\boldsymbol{\mu})p(\boldsymbol{\theta}_y)p(\boldsymbol{\theta}_\omega)p(\mathbf{W}|\boldsymbol{\theta})p(\boldsymbol{\Omega}, \mathbf{Y}|\boldsymbol{\theta}, \mathbf{W}) \\
&\propto p(\boldsymbol{\pi})p(\boldsymbol{\mu})p(\boldsymbol{\theta}_y)p(\boldsymbol{\theta}_\omega)p(\mathbf{W}|\boldsymbol{\pi})p(\boldsymbol{\Omega}|\boldsymbol{\theta}_\omega)p(\mathbf{Y}|\mathbf{W}, \boldsymbol{\Omega}, \boldsymbol{\mu}, \boldsymbol{\theta}_y) \\
&= [p(\boldsymbol{\pi})p(\mathbf{W}|\boldsymbol{\pi})][p(\boldsymbol{\mu})p(\boldsymbol{\theta}_y)p(\mathbf{Y}|\mathbf{W}, \boldsymbol{\Omega}, \boldsymbol{\mu}, \boldsymbol{\theta}_y)][p(\boldsymbol{\theta}_\omega)p(\boldsymbol{\Omega}|\boldsymbol{\theta}_\omega)]. \qquad (11.10)
\end{aligned}
$$

Using this result, the marginal densities $p(\boldsymbol{\pi}|\cdot)$, $p(\boldsymbol{\mu}, \boldsymbol{\theta}_y|\cdot)$ and $p(\boldsymbol{\theta}_\omega|\cdot)$ can be treated separately.

The prior distribution of $\boldsymbol{\pi}$ is taken to be the symmetric Dirichlet distribution, that is, $\boldsymbol{\pi} \sim D(\alpha, \ldots, \alpha)$ with the probability density function given by

$$
p(\boldsymbol{\pi}) = \frac{\Gamma(K\alpha)}{\Gamma(\alpha)^K} \pi_1^\alpha \cdots \pi_K^\alpha,
$$

where $\Gamma(\cdot)$ is the Gamma function. Since $p(\mathbf{W}|\boldsymbol{\pi}) \propto \prod_{k=1}^K \pi_k^{n_k}$, it follows from Equation (11.10) that the full conditional distribution for $\boldsymbol{\pi}$ remains Dirichlet in the following form:

$$
p(\boldsymbol{\pi}|\cdot) \propto p(\boldsymbol{\pi})p(\mathbf{W}|\boldsymbol{\pi}) \propto \prod_{k=1}^K \pi_k^{n_k+\alpha}, \qquad (11.11)
$$

where n_k is the total number of i such that $w_i = k$. Thus $p(\boldsymbol{\pi}|\cdot)$ is distributed as $D(\alpha + n_1, \ldots, \alpha + n_K)$.

Let $\mathbf{Y}_k$ and $\boldsymbol{\Omega}_k$ be the respective submatrices of $\mathbf{Y}$ and $\boldsymbol{\Omega}$, such that all the ith columns with $w_i \neq k$ are deleted. It is natural to assume that for $k \neq h$, $(\boldsymbol{\mu}_k, \boldsymbol{\theta}_{yk}, \boldsymbol{\theta}_{\omega k})$ and $(\boldsymbol{\mu}_h, \boldsymbol{\theta}_{yh}, \boldsymbol{\theta}_{\omega h})$ are independent. Hence, given $\mathbf{W}$, we have

$$
p(\boldsymbol{\mu}, \boldsymbol{\theta}_y, \boldsymbol{\theta}_\omega|\mathbf{Y}, \boldsymbol{\Omega}, \mathbf{W}) \propto \prod_{k=1}^K p(\boldsymbol{\mu}_k)p(\boldsymbol{\theta}_{yk})p(\boldsymbol{\theta}_{\omega k})p(\mathbf{Y}_k|\boldsymbol{\Omega}_k, \boldsymbol{\mu}_k, \boldsymbol{\theta}_{yk})p(\boldsymbol{\Omega}_k|\boldsymbol{\theta}_{\omega k}), \qquad (11.12)
$$

and we can treat the product in Equation (11.12) separately for each k. With $\mathbf{W}$ given, the original complicated problem of finite mixtures reduces to a much simpler multisample problem. Here, for brevity, we assume that there are no cross-group constraints, and the analysis can be carried out separately with each individual sample. Situations with cross-group constraints can be handled in a similar way to multiple groups analysis.

For mixture models, Diebolt and Robert (1994), and Roeder and Wasserman (1997) pointed out that using fully noninformative prior distributions may lead to improper posterior distributions. Thus most existing Bayesian analyses on mixtures of the normal distribution used the conjugate type prior distributions (see Roeder and Wasserman, 1997). This type of prior distribution for various components of $\boldsymbol{\theta}$ is adopted here. Let $\boldsymbol{\Lambda}_{\omega k} = (\mathbf{\Pi}_k, \mathbf{\Gamma}_k)$, for $m = 1, \ldots, p$ and $l = 1, \ldots, q_1$, for convenience, we take

$$
\begin{aligned}
&[\boldsymbol{\Lambda}_{km}|\psi_{km}] \stackrel{D}{=} N[\boldsymbol{\Lambda}_{0km}, \psi_{km}\mathbf{H}_{0ykm}], \ \psi_{km}^{-1} \stackrel{D}{=} Gamma[\alpha_{0\epsilon k}, \beta_{0\epsilon k}],\\
&[\boldsymbol{\Lambda}_{\omega kl}|\psi_{\delta kl}] \stackrel{D}{=} N[\boldsymbol{\Lambda}_{0\omega kl}, \psi_{\delta kl}\mathbf{H}_{0\omega kl}], \ \psi_{\delta kl}^{-1} \stackrel{D}{=} Gamma[\alpha_{0\delta k}, \beta_{0\delta k}], \qquad (11.13)\\
&\boldsymbol{\mu}_k \stackrel{D}{=} N\,\boldsymbol{\mu}_0, \boldsymbol{\Sigma}_0, \ \boldsymbol{\Phi}_k^{-1} \stackrel{D}{=} W_{q_2}[\mathbf{R}_0, \rho_0],
\end{aligned}
$$

where ψ_{km} and $\psi_{\delta kl}$ are the mth diagonal element of $\boldsymbol{\Psi}_k$ and the lth diagonal element of $\boldsymbol{\Psi}_{\delta k}$, respectively; $\boldsymbol{\Lambda}_{km}^T$ and $\boldsymbol{\Lambda}_{\omega kl}^T$ are vectors that contain unknown parameters in the mth row of $\boldsymbol{\Lambda}_k$ and the lth row of $\boldsymbol{\Lambda}_{\omega k}$, respectively; $\boldsymbol{\mu}_0, \alpha_{0\epsilon k}, \beta_{0\epsilon k}, \boldsymbol{\Lambda}_{0km}, \alpha_{0\delta k}, \beta_{0\delta k}, \boldsymbol{\Lambda}_{0\omega kl}$ and ρ_0, and positive definite matrices $\boldsymbol{\Sigma}_0, \mathbf{H}_{0ykm}, \mathbf{H}_{0\omega kl}$ and $\mathbf{R}_0$ are hyperparameters whose values are assumed to be given, and $W_{q_2}[\cdot,\cdot]$ denotes the Wishart distribution of dimension q_2. Moreover, we also assume that $(\psi_{km}, \boldsymbol{\Lambda}_{km})$ is independent of $(\psi_{kh}, \boldsymbol{\Lambda}_{kh})$ for $m \neq h$, and $(\psi_{\delta kl}, \boldsymbol{\Lambda}_{\omega kl})$ is independent of $(\psi_{\delta kh}, \boldsymbol{\Lambda}_{\omega kh})$ for $l \neq h$.

Let $\boldsymbol{\Omega}_{1k}$ and $\boldsymbol{\Omega}_{2k}$ be submatrices of $\boldsymbol{\Omega}_k$ that respectively contain the q_1 rows of η_k and the remaining q_2 rows of $\boldsymbol{\xi}_k$. It can be shown by similar derivations as in previous chapters that the conditional distributions of components of $\boldsymbol{\theta}_k$ are the following familiar normal, Gamma and inverted Wishart distributions:

$$
\begin{aligned}
[\boldsymbol{\mu}_k|\mathbf{Y}_k, \boldsymbol{\Omega}_k, \boldsymbol{\Lambda}_k, \boldsymbol{\Psi}_k) &\stackrel{D}{=} N[(\boldsymbol{\Sigma}_0^{-1} + n_k\boldsymbol{\Psi}_k^{-1})^{-1}(n_k\boldsymbol{\Psi}_k^{-1}\bar{\mathbf{Y}}_k + \boldsymbol{\Sigma}_0^{-1}\boldsymbol{\mu}_0),\\
&\qquad (\boldsymbol{\Sigma}_0^{-1} + n_k\boldsymbol{\Psi}_k^{-1})^{-1}],\\
[\boldsymbol{\Lambda}_{km}|\mathbf{Y}_k, \boldsymbol{\Omega}_k, \mu_{km}, \psi_{km}^{-1}] &\stackrel{D}{=} N[\mathbf{a}_{ykm}, \psi_{km}\mathbf{A}_{ykm}],\\
[\psi_{km}^{-1}|\mathbf{Y}_k, \boldsymbol{\Omega}_k, \mu_{km}] &\stackrel{D}{=} Gamma[n_k/2 + \alpha_{0\epsilon k}, \beta_{\epsilon km}],\\
[\boldsymbol{\Lambda}_{\omega kl}|\mathbf{Y}_k, \boldsymbol{\Omega}_k, \psi_{\delta kl}^{-1}] &\stackrel{D}{=} N[\mathbf{a}_{\delta kl}, \psi_{\delta kl}\mathbf{A}_{\omega kl}],\\
[\psi_{\delta kl}^{-1}|\mathbf{Y}_k, \boldsymbol{\Omega}_k] &\stackrel{D}{=} Gamma[n_k/2 + \alpha_{0\delta k}, \beta_{\delta kl}],\\
[\boldsymbol{\Phi}_k|\boldsymbol{\Omega}_{2k}] &\stackrel{D}{=} IW_{q_2}[\boldsymbol{\Omega}_{2k}\boldsymbol{\Omega}_{2k}^T + \mathbf{R}_0^{-1}, n_k + \rho_0],
\end{aligned}
$$

where $\bar{\mathbf{Y}}_k = \sum_{i:w_i=k}(\mathbf{y}_i - \mathbf{\Lambda}_k\boldsymbol{\omega}_{ki})/n_k$, and $\sum_{i:w_i=k}$ denotes the summation with respect to those i such that $w_i = k$, and

$$\mathbf{a}_{ykm} = \mathbf{A}_{ykm}(\mathbf{H}_{0ykm}^{-1}\mathbf{\Lambda}_{0km} + \mathbf{\Omega}_k\tilde{\mathbf{Y}}_{km}^T),\ \mathbf{A}_{ykm} = (\mathbf{H}_{0ykm}^{-1} + \mathbf{\Omega}_k\mathbf{\Omega}_k^T)^{-1},$$

$$\beta_{\epsilon km} = \beta_{0\epsilon k} + 2^{-1}[\tilde{\mathbf{Y}}_{km}\tilde{\mathbf{Y}}_{km}^T - \mathbf{a}_{ykm}^T\mathbf{A}_{ykm}^{-1}\mathbf{a}_{ykm} + \mathbf{\Lambda}_{0km}^T\mathbf{H}_{0ykm}^{-1}\mathbf{\Lambda}_{0km}],$$

$$\mathbf{a}_{\delta kl} = \mathbf{A}_{\omega kl}(\mathbf{H}_{0\omega kl}^{-1}\mathbf{\Lambda}_{0\omega kl} + \mathbf{\Omega}_k\mathbf{\Omega}_{1kl}^T), \mathbf{A}_{\omega kl} = (\mathbf{H}_{0\omega kl}^{-1} + \mathbf{\Omega}_k\mathbf{\Omega}_k^T)^{-1},$$

$$\beta_{\delta kl} = \beta_{0\delta k} + 2^{-1}[\mathbf{\Omega}_{1kl}\mathbf{\Omega}_{1kl}^T - \mathbf{a}_{\delta kl}^T\mathbf{A}_{\omega kl}^{-1}\mathbf{a}_{\delta kl} + \mathbf{\Lambda}_{0\omega kl}^T\mathbf{H}_{0\omega kl}^{-1}\mathbf{\Lambda}_{0\omega kl}],$$

in which $\tilde{\mathbf{Y}}_{km}$ is the mth row of $\tilde{\mathbf{Y}}_k$ which is a matrix whose columns are equal to the columns of $\mathbf{Y}_k$ minus $\boldsymbol{\mu}_k$, and $\mathbf{\Omega}_{1kl}$ is the lth row of $\mathbf{\Omega}_{1k}$. The computational burden in simulating observations from these conditional distributions is light. As the situation with fixed parameters can be handled in the same way as in Zhu and Lee (2001), details are not presented here.

The conditional distributions given above are familiar and simple distributions. The computational burden required in simulating observations from these distributions is not heavy. For brevity, we do not introduce any model to the mean vector. Extension of the method discussed here to models with mean structures is straightforward.

Let $\{(\boldsymbol{\theta}^{(t)}, \mathbf{\Omega}^{(t)}, \mathbf{W}^{(t)}), t = 1, \ldots, T^*\}$ be the observations of $(\boldsymbol{\theta}, \mathbf{\Omega}, \mathbf{W})$ generated by the Gibbs sampler from the posterior distribution $[\boldsymbol{\theta}, \mathbf{\Omega}, \mathbf{W}|\mathbf{Y}]$. The Bayesian estimates of $\boldsymbol{\theta}$ and $\mathbf{\Omega}$ can easily be obtained via the corresponding sample means of the generated observations. For example,

$$\hat{\boldsymbol{\theta}} = T^{*-1}\sum_{t=1}^{T^*}\boldsymbol{\theta}^{(t)} \quad \text{and } \hat{\mathbf{\Omega}} = T^{*-1}\sum_{t=1}^{T^*}\mathbf{\Omega}^{(t)}. \tag{11.14}$$

These Bayesian estimates are consistent estimates of the corresponding posterior means. It is rather difficult to derive analytic forms for the covariance matrices $\text{Var}(\boldsymbol{\theta}|\mathbf{Y})$ and $\text{Var}(\boldsymbol{\omega}_i|\mathbf{Y})$. However, their consistent estimates can be obtained as follows:

$$\widehat{\text{Var}(\boldsymbol{\theta}|\mathbf{Y})} = (T^* - 1)^{-1}\sum_{t=1}^{T^*}(\boldsymbol{\theta}^{(t)} - \hat{\boldsymbol{\theta}})(\boldsymbol{\theta}^{(t)} - \hat{\boldsymbol{\theta}})^T, \tag{11.15}$$

$$\widehat{\text{Var}(\boldsymbol{\omega}_{ki}|\mathbf{Y})} = (T^* - 1)^{-1}\sum_{t=1}^{T^*}(\boldsymbol{\omega}_{ki}^{(t)} - \hat{\boldsymbol{\omega}}_{ki})(\boldsymbol{\omega}_{ki}^{(t)} - \hat{\boldsymbol{\omega}}_{ki})^T,\ i = 1, \ldots, n.$$

Standard error estimates of $\boldsymbol{\theta}$ can be obtained through the square roots of $\widehat{\text{Var}(\boldsymbol{\theta}|\mathbf{Y})}$. Other statistical inferences on $\boldsymbol{\theta}$ or $\boldsymbol{\omega}_{ki}$ can be achieved based on the simulated observations as well (see, for example, Besag, Green, Higdon and

Mengersen, 1995; Gilks, Richardson and Spiegelhalter, 1996). For $k = 1, 2, \ldots, K$ estimated residuals can be obtained from $\hat{\boldsymbol{\omega}}_{ki}$ and $\hat{\boldsymbol{\theta}}_k$ through Equations (11.2) and (11.3) as follows:

$$\hat{\boldsymbol{\epsilon}}_{ki} = \mathbf{y}_i - \hat{\boldsymbol{\mu}}_k - \hat{\boldsymbol{\Lambda}}_k \hat{\boldsymbol{\omega}}_{ki}, \text{ and}$$
$$\hat{\boldsymbol{\delta}}_{ki} = \hat{\boldsymbol{\eta}}_{ki} - \hat{\boldsymbol{\Pi}}_k \hat{\boldsymbol{\eta}}_{ki} - \hat{\boldsymbol{\Gamma}}_k \hat{\boldsymbol{\xi}}_{ki},$$

where the index i belongs to the set $\{i : \hat{w}_i = k\}$.

In addition to their role in facilitating estimation, the allocation variables in $\mathbf{W}$ also form a coherent basis for Bayesian classification of the observations. Classification can either be addressed on a within-sample basis or a predictive basis. Using the 'percentage correctly classified' loss function (see Richardson and Green, 1997; Zhu and Lee, 2001), Bayesian classification of an existing observation $\mathbf{y}_i$ and a new observation $\mathbf{y}^*$ are respectively given by

$$\hat{w}_i = \text{argmax}_k\{\Pr(w_i = k|\mathbf{Y})\} \quad \text{and} \quad \hat{w}^* = \text{argmax}_k\{\Pr(w^* = k|\mathbf{Y}, \mathbf{y}^*)\}.$$

The posterior probabilities $\{\Pr(w_i = k|\mathbf{y}_i); k = 1, \ldots, K\}$ can be approximated via the sample mean of the observations generated by the Gibbs sampler:

$$\Pr(w_i = k|\mathbf{Y}) \approx T^{*-1} \sum_{t=1}^{T^*} I(w_i^{(t)} = k),$$

where I is an indicator function. Consider the predictive classification in classifying a new observation $\mathbf{y}^*$. Let the corresponding allocation variable be w^*, the Bayes classification needs to evaluate $\Pr(w^* = k|\mathbf{Y}, \mathbf{y}^*)$. Inclusion of the additional $\mathbf{y}^*$ theoretically changes the posterior distributions, and the simulation process should be rerun for each new $\mathbf{y}^*$. This is obviously not practical. Hence, we use the following approximation

$$\Pr(w^* = k|\mathbf{Y}, \mathbf{y}^*) = \int p(w^* = k|\boldsymbol{\theta}, \mathbf{y}^*) p(\boldsymbol{\theta}|\mathbf{Y}, \mathbf{y}^*) d\boldsymbol{\theta} \approx \int p(w^* = k|\boldsymbol{\theta}, \mathbf{y}^*) p(\boldsymbol{\theta}|\mathbf{Y}) d\boldsymbol{\theta},$$

and estimate the last integral by the following sample average of the generated observations from the Gibbs sampler procedure:

$$\Pr(w^* = k|\mathbf{Y}, \mathbf{y}^*) \approx T^{*-1} \sum_{t=1}^{T^*} \left[\pi_k^{(t)} f_k(\mathbf{y}^*|\boldsymbol{\mu}_k^{(t)}, \boldsymbol{\theta}_k^{(t)}) / \sum_{h=1}^{K} \pi_h^{(t)} f_h(\mathbf{y}^*|\boldsymbol{\mu}_h^{(t)}, \boldsymbol{\theta}_h^{(t)}) \right].$$

The associated computing burden is light.

11.4 EXAMPLES AND SIMULATION STUDY

11.4.1 Analysis of an Artificial Example

An important issue in analyzing mixture SEMs is the separation of the components. Yung (1997) and Dolan and van der Maas (1998) pointed out that some statistical results they achieved cannot be trusted when the separation is poor. Yung (1997) considered the following measure of separation: $d_{kh} = \max_{l\in\{k,h\}}\{(\boldsymbol{\mu}_k - \boldsymbol{\mu}_h)^T\boldsymbol{\Sigma}_l^{-1}(\boldsymbol{\mu}_k - \boldsymbol{\mu}_h)\}^{1/2}$ and recommended that d_{kh} should be about 3.8 or over. In view of this, an objective of this artificial example is to investigate the performance of the proposed Bayesian approach in analyzing a mixture of SEMs with two components which are not well separated. Another objective is to demonstrate the random permutation sampler for finding suitable identifiability constraints. Random observations are simulated from a mixture SEM with two components defined by Equations (11.1), (11.2) and (11.3). The model for each $k = 1, 2$ involves nine manifest variables which are indicators for three latent variables η_k, ξ_{k1} and ξ_{k2}. The structure of the factor loading matrix in each component is

$$\boldsymbol{\Lambda}_k^T = \begin{bmatrix} 1.0 & \lambda_{k,21} & \lambda_{k,31} & 0 & 0 & 0 & 0 & 0 & 0 \\ 0 & 0 & 0 & 1.0 & \lambda_{k,52} & \lambda_{k,62} & 0 & 0 & 0 \\ 0 & 0 & 0 & 0 & 0 & 0 & 1.0 & \lambda_{k,83} & \lambda_{k,93} \end{bmatrix},$$

where the 1s and 0s are fixed parameters for achieving an identified covariance structure, whilst the others are distinct unknown parameters. In the kth component, the structural equation is given by: $\eta_{ki} = \gamma_{k,1}\xi_{ki1} + \gamma_{k,2}\xi_{ki2} + \delta_{ki}$, where $\gamma_{k,1}$ and $\gamma_{k,2}$ are unknown parameters. The true population values are given by: $\pi_1 = \pi_2 = 0.5$, $\boldsymbol{\mu}_1 = (0.0, 0.0, 0.0, 0.0, 0.0, 1.0, 1.0, 1.0, 1.0)^T$, $\boldsymbol{\mu}_2 = (0.0, 0.0, 0.0, 0.5, 1.5, 0.0, 1.0, 1.0, 1.0)^T$; $\lambda_{1,21} = \lambda_{1,31} = \lambda_{1,83} = \lambda_{1,93} = 0.4$, $\lambda_{1,52} = \lambda_{1,62} = 0.8$, $\lambda_{2,21} = \lambda_{2,31} = \lambda_{2,83} = \lambda_{2,93} = 0.8$, $\lambda_{2,52} = \lambda_{2,62} = 0.4$, $\gamma_{1,1} = 0.2$, $\gamma_{1,2} = 0.7$, $\gamma_{2,1} = 0.7$, $\gamma_{2,2} = 0.2$, $\phi_{1,11} = \phi_{1,22} = \phi_{2,11} = \phi_{2,22} = 1.0$, $\phi_{1,12} = \phi_{2,12} = 0.3$, $\psi_{11} = \cdots = \psi_{19} = \psi_{21} = \cdots = \psi_{29} = \psi_{\delta 11} = \psi_{\delta 21} = 0.5$. In this two-component mixture SEM, the total number of unknown parameters is 62. The separation d_{12} is equal to 2.56, which is less than the suggested value in Yung (1997).

Based on the above settings, we simulate 400 observations from each component and the total sample size is 800. We focus on $\boldsymbol{\mu}_1$ (or $\boldsymbol{\mu}_2$) in finding a suitable identifiability constraint. The first step is to apply the random permutation sampler to produce an MCMC sample from the unconstrained posterior with size 5000 after a burn-in phase of 500 simulations. This random permutation sampler delivers a sample that explores a whole unconstrained parameter space and jumps between the various labelling subspaces in a balanced fashion. For a mixture of SEMs with two components, we just have 2! labelling subspaces. In the random permutation sampler, after each sweep the 1s and 2s are permuted

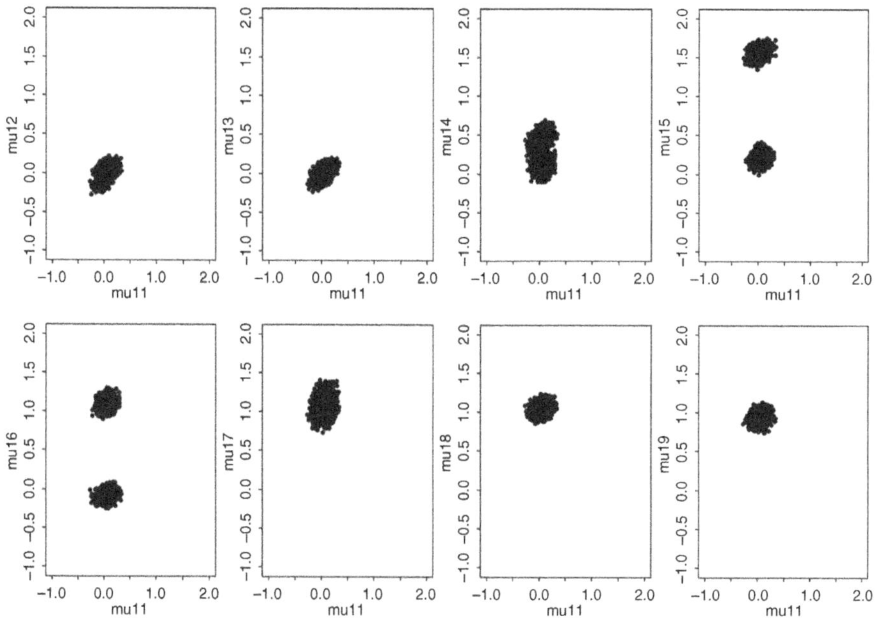

Figure 11.1 Scatterplots of MCMC output for components of $\boldsymbol{\mu}_1$.

randomly; that is, with probability 0.5, the 1s stay as 1s, and with probability 0.5 they become 2s. The output can be explored to find a suitable identifiability constraint. Based on the reasoning given in Appendix 11.2, it suffices to consider the parameters in $\boldsymbol{\mu}_1$. To search for an appropriate identifiability constraint, we look at scatterplots of $\mu_{1,1}$ versus $\mu_{1,l}$, $l = 2, \ldots, 9$, for obtaining information on aspects of the states that are most different. These scatterplots are presented in Figure 11.1. These plots clearly indicate that the two most significant differences between the two components are sampled values corresponding to $\mu_{1,5}$ and $\mu_{1,6}$. If permutation sampling is based on the constraint $\mu_{1,5} < \mu_{2,5}$ or $\mu_{1,6} > \mu_{2,6}$, the label switching will not appear.

Bayesian estimates are obtained by using the permutation sampler with the identifiability constraint $\mu_{1,5} < \mu_{2,5}$. Values of hyperparameters in the conjugate prior distributions (see Equation 11.13) are taken as: for $m = 1, \ldots, 9$, $\mu_{0,m}$ is equal to the sample mean $\bar{y}_m$, $\boldsymbol{\Sigma}_0 = 10^2\mathbf{I}$, elements in $\boldsymbol{\Lambda}_{0km}$ and $\boldsymbol{\Lambda}_{0\omega kl}$ (which only involves the γs) are taken to be true parameter values, $\mathbf{H}_{0ykm}$ and $\mathbf{H}_{0\omega kl}$ are the identity matrices, $\alpha_{0\epsilon k} = \alpha_{0\delta k} = 10$, $\beta_{0\epsilon k} = \beta_{0\delta k} = 8$, $\rho_0 = 6$ and $\mathbf{R}_0^{-1} = 5\mathbf{I}$. The α in the Dirichlet distribution of $\boldsymbol{\pi}$ is taken as 1. We conduct a few test runs and find that the algorithm converged in less than 500 iterations. Based on this finding, Bayesian estimates are obtained using a burn-in phase of 500 iterations and a total of 2000 observations are collected after the burn-in phase. Results are reported in Table 11.1. We observe that the Bayesian estimates are pretty close to their true parameter values.

Table 11.1 Bayesian estimates in the artificial example.

Component 1			Component 2		
Par	EST	SE	Par	EST	SE
$\pi_1 = 0.5$	0.504	0.021	$\pi_2 = 0.5$	0.496	0.028
$\mu_{1,1} = 0.0$	0.070	0.082	$\mu_{2,1} = 0.0$	0.030	0.084
$\mu_{1,2} = 0.0$	0.056	0.046	$\mu_{2,2} = 0.0$	−0.056	0.061
$\mu_{1,3} = 0.0$	0.036	0.045	$\mu_{2,3} = 0.0$	−0.014	0.061
$\mu_{1,4} = 0.0$	0.108	0.075	$\mu_{2,4} = 0.5$	0.430	0.071
$\mu_{1,5} = 0.0$	0.209	0.061	$\mu_{2,5} = 1.5$	1.576	0.057
$\mu_{1,6} = 1.0$	1.101	0.062	$\mu_{2,7} = 0.0$	−0.084	0.052
$\mu_{1,7} = 1.0$	1.147	0.112	$\mu_{2,7} = 1.0$	0.953	0.110
$\mu_{1,8} = 1.0$	1.033	0.046	$\mu_{2,8} = 1.0$	1.041	0.056
$\mu_{1,9} = 1.0$	0.974	0.046	$\mu_{2,9} = 1.0$	0.941	0.057
$\lambda_{1,21} = 0.4$	0.322	0.052	$\lambda_{2,21} = 0.8$	0.851	0.060
$\lambda_{1,31} = 0.4$	0.411	0.052	$\lambda_{2,31} = 0.8$	0.810	0.060
$\lambda_{1,52} = 0.8$	0.712	0.060	$\lambda_{2,52} = 0.4$	0.498	0.067
$\lambda_{1,62} = 0.8$	0.695	0.066	$\lambda_{2,62} = 0.4$	0.480	0.064
$\lambda_{1,83} = 0.4$	0.386	0.075	$\lambda_{2,83} = 0.8$	0.738	0.077
$\lambda_{1,93} = 0.4$	0.428	0.079	$\lambda_{2,93} = 0.8$	0.826	0.083
$\gamma_{1,1} = 0.2$	0.236	0.104	$\gamma_{2,1} = 0.7$	0.817	0.104
$\gamma_{1,2} = 0.7$	0.740	0.074	$\gamma_{2,2} = 0.2$	0.210	0.074
$\phi_{1,11} = 1.0$	1.017	0.121	$\phi_{2,11} = 1.0$	0.820	0.118
$\phi_{1,12} = 0.3$	0.249	0.090	$\phi_{2,12} = 0.3$	0.283	0.074
$\phi_{1,22} = 1.0$	0.900	0.185	$\phi_{2,22} = 1.0$	0.982	0.163
$\psi_{11} = 0.5$	0.535	0.092	$\psi_{21} = 0.5$	0.588	0.080
$\psi_{12} = 0.5$	0.558	0.046	$\psi_{22} = 0.5$	0.489	0.053
$\psi_{13} = 0.5$	0.510	0.045	$\psi_{23} = 0.5$	0.565	0.057
$\psi_{14} = 0.5$	0.483	0.067	$\psi_{24} = 0.5$	0.620	0.085
$\psi_{15} = 0.5$	0.492	0.056	$\psi_{25} = 0.5$	0.556	0.056
$\psi_{16} = 0.5$	0.554	0.063	$\psi_{26} = 0.5$	0.507	0.050
$\psi_{17} = 0.5$	0.696	0.126	$\psi_{27} = 0.5$	0.569	0.108
$\psi_{18} = 0.5$	0.563	0.052	$\psi_{28} = 0.5$	0.578	0.062
$\psi_{19} = 0.5$	0.566	0.053	$\psi_{29} = 0.5$	0.508	0.065
$\psi_{\delta 11} = 0.5$	0.549	0.094	$\psi_{\delta 21} = 0.5$	0.549	0.082

This artificial data set has also been analyzed by using WinBUGS (Spiegelhalter, Thomas, Best and Lunn, 2003). We first conduct an initial Bayesian estimation without the identifiability constraints to identify the appropriate identifiability constraint $\mu_{1,5} < \mu_{2,5}$ from the output as before. The model is then reestimated with this identifiability constraint, and three starting values of the parameters that are obtained from the sample mean, the fifth, and 95th percentile of the corresponding simulated samples. Bayesian estimates are obtained by using the permutation samples with the identifiability constraint, and the hyperparameter values given above. Results are presented in Table 11.2.

Table 11.2 Bayesian estimates in the artificial example via WinBUGS.

Component 1			Component 2		
Par	EST	SE	Par	EST	SE
$\pi_1 = 0.5$	0.503	0.027	$\pi_2 = 0.5$	0.498	0.027
$\mu_{1,1} = 0.0$	0.035	0.068	$\mu_{2,1} = 0.0$	0.028	0.070
$\mu_{1,2} = 0.0$	0.050	0.046	$\mu_{2,2} = 0.0$	−0.043	0.062
$\mu_{1,3} = 0.0$	0.034	0.046	$\mu_{2,3} = 0.0$	−0.008	0.062
$\mu_{1,4} = 0.0$	0.124	0.067	$\mu_{2,4} = 0.5$	0.435	0.066
$\mu_{1,5} = 0.0$	0.192	0.060	$\mu_{2,5} = 1.5$	1.590	0.056
$\mu_{1,6} = 1.0$	1.089	0.061	$\mu_{2,7} = 0.0$	−0.065	0.053
$\mu_{1,7} = 1.0$	1.068	0.067	$\mu_{2,7} = 1.0$	0.961	0.066
$\mu_{1,8} = 1.0$	1.024	0.046	$\mu_{2,8} = 1.0$	1.053	0.056
$\mu_{1,9} = 1.0$	0.974	0.047	$\mu_{2,9} = 1.0$	0.916	0.057
$\lambda_{1,21} = 0.4$	0.355	0.047	$\lambda_{2,21} = 0.8$	0.850	0.053
$\lambda_{1,31} = 0.4$	0.428	0.047	$\lambda_{2,31} = 0.8$	0.817	0.053
$\lambda_{1,52} = 0.8$	0.723	0.062	$\lambda_{2,52} = 0.4$	0.507	0.065
$\lambda_{1,62} = 0.8$	0.741	0.068	$\lambda_{2,62} = 0.4$	0.495	0.064
$\lambda_{1,83} = 0.4$	0.403	0.058	$\lambda_{2,83} = 0.8$	0.741	0.061
$\lambda_{1,93} = 0.4$	0.385	0.061	$\lambda_{2,93} = 0.8$	0.837	0.064
$\gamma_{1,1} = 0.2$	0.208	0.071	$\gamma_{2,1} = 0.7$	0.873	0.103
$\gamma_{1,2} = 0.7$	0.740	0.094	$\gamma_{2,2} = 0.2$	0.152	0.068
$\phi_{1,11} = 1.0$	0.953	0.115	$\phi_{2,11} = 1.0$	0.785	0.110
$\phi_{1,12} = 0.3$	0.278	0.071	$\phi_{2,12} = 0.3$	0.305	0.067
$\phi_{1,22} = 1.0$	0.932	0.133	$\phi_{2,22} = 1.0$	0.988	0.117
$\psi_{11} = 0.5$	0.496	0.072	$\psi_{21} = 0.5$	0.519	0.059
$\psi_{12} = 0.5$	0.536	0.043	$\psi_{22} = 0.5$	0.512	0.052
$\psi_{13} = 0.5$	0.504	0.042	$\psi_{23} = 0.5$	0.566	0.054
$\psi_{14} = 0.5$	0.517	0.067	$\psi_{24} = 0.5$	0.620	0.076
$\psi_{15} = 0.5$	0.492	0.056	$\psi_{25} = 0.5$	0.536	0.052
$\psi_{16} = 0.5$	0.542	0.063	$\psi_{26} = 0.5$	0.518	0.051
$\psi_{17} = 0.5$	0.636	0.093	$\psi_{27} = 0.5$	0.533	0.066
$\psi_{18} = 0.5$	0.536	0.045	$\psi_{28} = 0.5$	0.574	0.054
$\psi_{19} = 0.5$	0.586	0.048	$\psi_{29} = 0.5$	0.487	0.054
$\psi_{\delta 11} = 0.5$	0.479	0.077	$\psi_{\delta 21} = 0.5$	0.510	0.072

Some estimated residual plots for the first component, $\hat{\epsilon}_{1i1}$, $\hat{\epsilon}_{1i4}$, $\hat{\epsilon}_{1i7}$ and $\hat{\delta}_{1i}$ versus case numbers, are displayed in Figure 11.2. Estimated residual plots of $\hat{\delta}_{1i}$ versus $\hat{\xi}_{1i1}$ and $\hat{\xi}_{1i2}$; and plots of $\hat{\epsilon}_{1i1}$ versus $\hat{\xi}_{1i1}$, $\hat{\xi}_{1i2}$ and $\hat{\eta}_{1i}$ are presented in Figures 11.3 and 11.4, respectively. The estimated residual plots associated with the second component are similar. These plots lie within two parallel horizontal lines that are centered at zero and no linear or quadratic trends are detected. This roughly indicates that the proposed measurement and structural equations are adequate. As WinBUGS does not give the DIC value, these estimated residual plots provide useful information for goodness-of-fit assessment. The

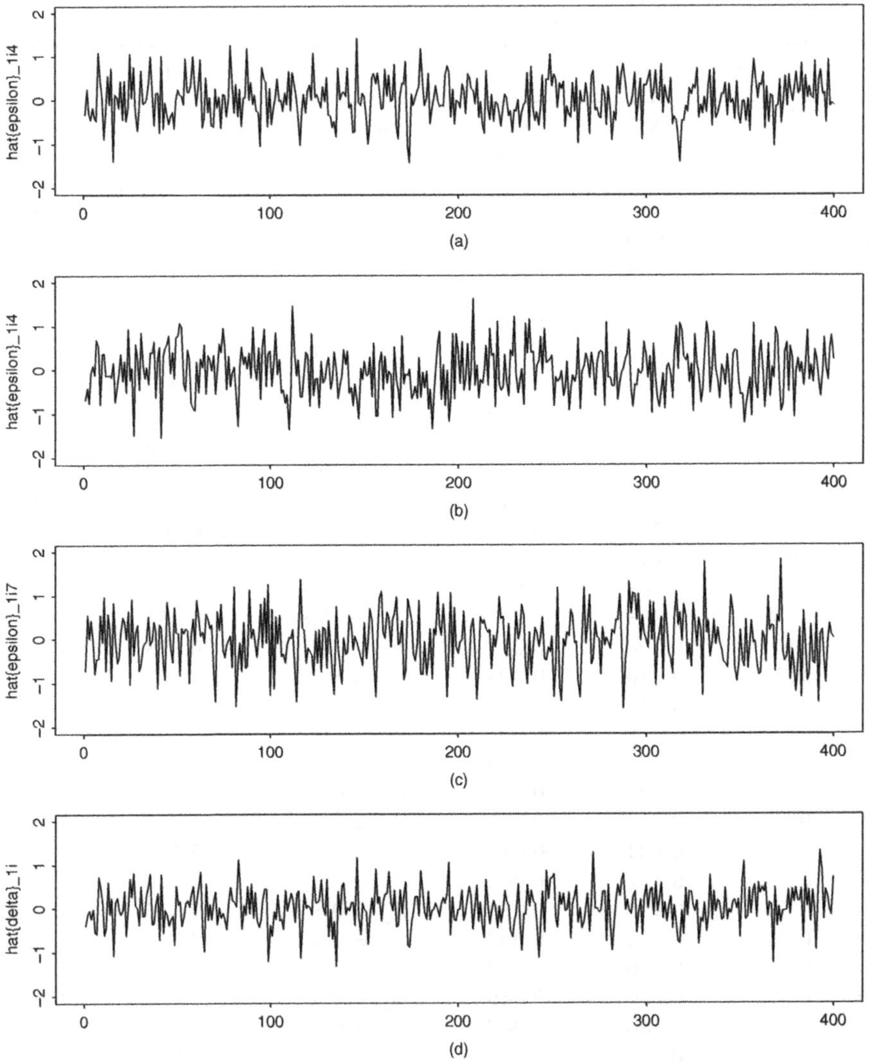

Figure 11.2 Estimated residual plots (a) $\hat{\epsilon}_{1i1}$, (b) $\hat{\epsilon}_{1i4}$, (c) $\hat{\epsilon}_{1i7}$ and (d) $\hat{\delta}_{1i}$.

WinBUGS codes under the identifiability constraint $\mu_{1,5} < \mu_{2,5}$, and the data are given in the following website: http://www.wiley.com/go/lee_structural.

11.4.2 A Simulation Study

Results of a simulation study will be presented to give some ideas on the performance of the proposed Bayesian approach for analyzing finite mixtures

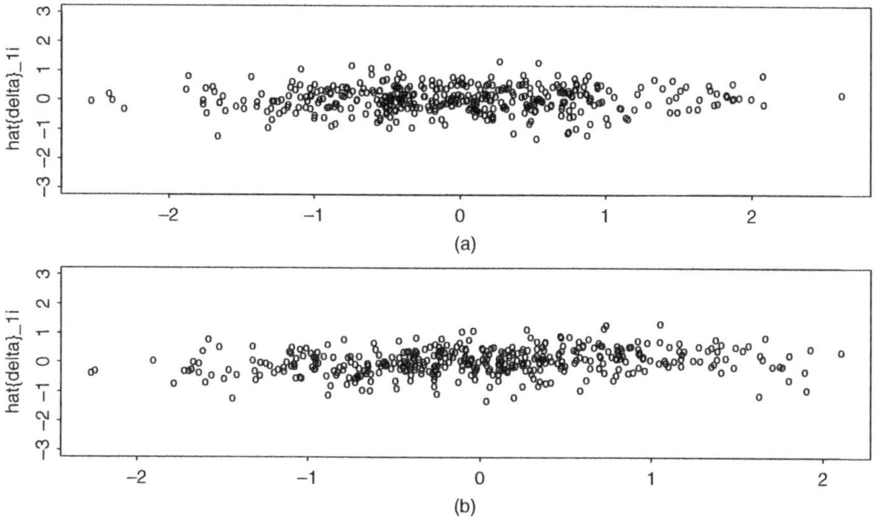

Figure 11.3 Plots of estimated residuals $\hat{\delta}_{1i}$ versus (a) $\hat{\xi}_{1i1}$ and (b) $\hat{\xi}_{1i2}$.

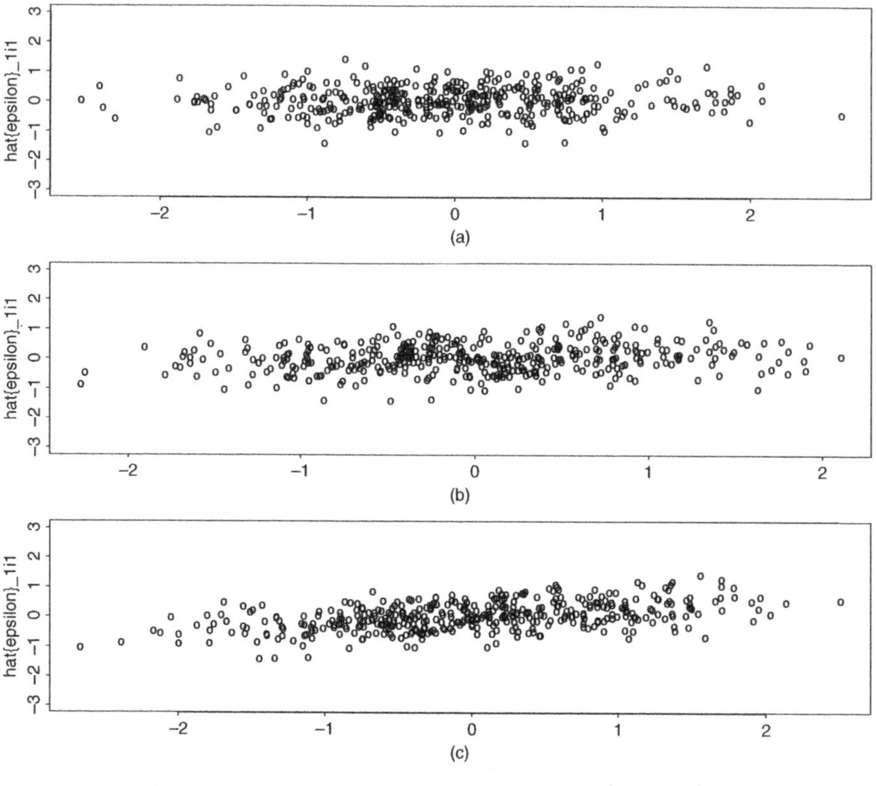

Figure 11.4 Plots of estimated residuals $\hat{\epsilon}_{1i1}$ versus (a) $\hat{\xi}_{1i1}$, (b) $\hat{\xi}_{1i2}$ and (c) $\hat{\eta}_{1i}$.

of SEMs. The data set is generated from a mixture of two SEMs defined in Equations (11.1), (11.2) and (11.3). Each model involves six manifest variables which are related with three latent factors $\boldsymbol{\eta}_k = (\eta_{k1}, \eta_{k2})$, and ξ_k for $k = 1, 2$. In these two components, the population values of the elements in $\boldsymbol{\Lambda}_1$, $\boldsymbol{\Lambda}_2$, $\boldsymbol{\Pi}_1$ and $\boldsymbol{\Pi}_2$ are taken as:

$$\boldsymbol{\Lambda}_1^T = \boldsymbol{\Lambda}_2^T = \begin{bmatrix} 1.0^* & 0.8 & 0.0^* & 0.0^* & 0.0^* & 0.0^* \\ 0.0^* & 0.0^* & 1.0^* & 0.8 & 0.0^* & 0.0^* \\ 0.0^* & 0.0^* & 0.0^* & 0.0^* & 1.0^* & 0.8 \end{bmatrix},$$

$$\boldsymbol{\Pi}_1 = \begin{bmatrix} 0^* & 0^* \\ 0.5 & 0^* \end{bmatrix}, \boldsymbol{\Pi}_2 = \begin{bmatrix} 0^* & 0^* \\ -0.5 & 0^* \end{bmatrix}.$$

In the analysis, parameters with asterisks are treated as fixed known parameters. The true population values of the other unknown parameters are given by: $\Phi_1 = \Phi_2 = (1.0)$, $\boldsymbol{\mu}_1 = 0.0 \times \mathbf{J}_6$, $\boldsymbol{\mu}_2 = 2.0 \times \mathbf{J}_6$, $\psi_{1m} = \psi_{2m} = 0.8$ for all $m = 1, \ldots, 6$, $\psi_{\delta 1l} = \psi_{\delta 2l} = 1.0$ for all $l = 1, 2$, $\boldsymbol{\Gamma}_1 = [0.6, 0.6]^T, \boldsymbol{\Gamma}_2 = [0.6, -0.6]^T$, where $\mathbf{J}_6$ is a 6×1 vector with all elements equal to 1. The following two designs with different mixing proportions are considered: $\{\pi_1 = 0.5, \pi_2 = 0.5\}$ and $\{\pi_1 = 0.3, \pi_2 = 0.7\}$. For each design, we have a mixture of SEMs with two components and 40 unknown parameters. Sample sizes $n = 400$ and 800 were selected and 100 replications were completed for each combination.

Two sets of Bayesian estimates are obtained using the conjugate prior distributions with different prior inputs: (I) Estimates based on conjugate priors with hyperparameters $\{\boldsymbol{\Lambda}_{0km}, \boldsymbol{\Lambda}_{0\omega kl}\}$ fixed at the true values; $\alpha = 1$, $\boldsymbol{\mu}_0 = \bar{\mathbf{y}}$, $\boldsymbol{\Sigma}_0 = \mathbf{S}_y/2$, $\rho_0 = 5$, $\mathbf{R}_0^{-1} = (5.0)$, $\alpha_{0\epsilon k} = \alpha_{0\delta k} = 10$, $\beta_{0\epsilon k} = \beta_{0\delta k} = 8$; $\mathbf{H}_{0ykm} = \mathbf{I}$ and $\mathbf{H}_{0\omega kl} = \mathbf{I}$ for all k, m and l, where $\bar{\mathbf{y}}$ and $\mathbf{S}_y$ are the sample mean and the sample covariance matrix of the simulated data. This can be regarded as a situation with accurate prior information. (II) Estimates based on the following values of hyperparameters: $\{\boldsymbol{\Lambda}_{0km}, \boldsymbol{\Lambda}_{0\omega kl}\}$ equal to 2.0 times the true values; other hyperparameter values are fixed at the same values as in (I). Starting values of the unknown parameters are given by: $\pi_1 = \pi_2 = 0.5$, $\phi_1 = \phi_2 = (2.0)$, $\boldsymbol{\mu}_1 = 0.0 \times \mathbf{J}_6$, $\boldsymbol{\mu}_2 = 3.0 \times \mathbf{J}_6$, $\psi_{1m} = \psi_{2m} = 1.2$ for $m = 1, \ldots, 6$, $\psi_{\delta 1l} = \psi_{\delta 2l} = 1.2$ for $l = 1, 2$, $\boldsymbol{\Gamma}_1 = (1.0, 1.0)^T, \boldsymbol{\Gamma}_2 = (1.0, -1.0)^T$ and 0.0 for all the unknown parameters in $\boldsymbol{\Lambda}_1$ and $\boldsymbol{\Lambda}_2$.

Based on the exploration of the MCMC outputs of the random permutation sampler with an unconstrained model, we observe that any element in $\boldsymbol{\mu}_k$ can be used to define the identifiability constraint. In this simulation study we use $\mu_{1,1} < \mu_{2,1}$. A few test runs are conducted initially as a pilot study to obtain some idea about the number of the Gibbs sampler iterations required in achieving convergence. In all these runs, the Gibbs sampler converges in about 500 to 1000 iterations, where the ESPR values (Gelman and Rubin, 1992) are less than 1.2. Hence, for the 100 replications in the simulation, random observations are collected after 1000 iterations. Then a total of an additional 3000 observations are collected to produce the Bayesian estimates and their standard error estimates via Equations (11.14) and (11.15). Based on the 100 replications, the mean and the standard deviations (SD) of the estimates, as well as the mean of the standard error estimate (SE) are computed. Moreover, the bias which is the difference between the true parameter and the mean of the corresponding estimates, and the root mean squares (RMS) between the estimates and the true values based on the 100 replications, are computed. The results are reported in Tables 11.3 to 11.6. We have the following findings from these tables. (i) Bayesian estimates are reasonably accurate. (ii) As expected, increasing the sample size improves the accuracy of the estimates. (iii) Bayesian estimates with more accurate priors are better, but the differences are not significant. Hence, it seems that the requirement of accurate hyperparameter values is not crucial in the Bayesian estimation under sample sizes $n = 400$ and 800. (iv) In most cases, the SE values are slightly smaller than the SD values. As pointed out by Dolan and van der Maas (1998), this minor difference may be due to the difference in the model that featured in the simulation study. However, the SE and SD values are quite close to each other, this indicates that the standard error estimates produced by the proposed procedure are close to the standard deviation of the estimates.

All computations are performed using a Sun Enterprise 4000 server. For $n = 400$, the average computing time over the four designs with different π and prior inputs is about 292 minutes for 100 replications, while for $n = 800$, the corresponding average computer time is about 640 minutes.

11.4.3 An Example on 'Job' and 'Homelife'

A small portion of the ICPSR data set collected in the project World Values Survey 1981–1984 and 1990–1993 (World Value Study Group, 1994) is analyzed in this example. The whole data set was collected in 45 societies around the world on broad topics such as work, the meaning and purpose of life, family life and contemporary social issues. As an illustration of our proposed method, only the data obtained from the UK with a sample size of 1484 are used. Eight variables in the original data set (variables 116, 117, 180, 132, 96, 255, 254 and 252) that related with respondents' job and homelife are taken as manifest variables in $\mathbf{y}$. A description of these variables in the questionnaire is given in Appendix 1.1. After deleting the cases with missing data the sample size is 824. The data set is analyzed with a mixture SEM with two components. For each component, there are three latent variables which can be roughly interpreted as 'job satisfaction, η', 'homelife, ξ_1' and 'job attitude, ξ_2'. For $k = 1, 2$, the

Table 11.3 Summary statistics for Bayesian estimates (I) for model with $\pi_1 = 0.5$.

	$n = 400$				$n = 800$			
Par	Bias	SE	SD	RMS	Bias	SE	SD	RMS
$\pi_1 = 0.5$	0.01	0.05	0.04	0.05	0.01	0.03	0.03	0.03
$\pi_2 = 0.5$	0.01	0.04	0.04	0.05	0.01	0.03	0.03	0.03
$\Phi_1 = 1.0$	0.09	0.24	0.25	0.25	−0.07	0.17	0.18	0.18
$\Phi_2 = 1.0$	0.08	0.21	0.21	0.22	−0.06	0.15	0.16	0.16
$\mu_{1,1} = 0.0$	0.03	0.16	0.15	0.16	0.01	0.11	0.11	0.11
$\mu_{1,2} = 0.0$	0.05	0.14	0.13	0.15	0.01	0.10	0.10	0.10
$\mu_{1,3} = 0.0$	0.05	0.19	0.18	0.20	0.01	0.14	0.13	0.14
$\mu_{1,4} = 0.0$	0.06	0.17	0.16	0.18	0.01	0.13	0.12	0.13
$\mu_{1,5} = 0.0$	0.02	0.14	0.13	0.14	−0.00	0.11	0.10	0.11
$\mu_{1,6} = 0.0$	0.04	0.14	0.13	0.14	0.00	0.09	0.09	0.09
$\mu_{2,1} = 2.0$	0.02	0.13	0.12	0.13	0.00	0.08	0.09	0.08
$\mu_{2,2} = 2.0$	0.01	0.10	0.11	0.10	0.00	0.09	0.08	0.08
$\mu_{2,3} = 2.0$	−0.02	0.14	0.14	0.14	0.00	0.09	0.09	0.09
$\mu_{2,4} = 2.0$	−0.01	0.12	0.11	0.12	0.00	0.08	0.08	0.08
$\mu_{2,5} = 2.0$	0.03	0.11	0.12	0.12	0.00	0.08	0.08	0.08
$\mu_{2,6} = 2.0$	0.02	0.10	0.11	0.10	0.00	0.08	0.08	0.08
$\psi_{\delta 11} = 1.0$	−0.09	0.13	0.18	0.16	−0.09	0.12	0.14	0.16
$\psi_{\delta 12} = 1.0$	−0.09	0.13	0.18	0.16	−0.08	0.15	0.15	0.16
$\psi_{\delta 21} = 1.0$	−0.11	0.14	0.18	0.18	−0.08	0.12	0.14	0.15
$\psi_{\delta 22} = 1.0$	−0.12	0.15	0.18	0.19	−0.07	0.14	0.15	0.15
$\psi_{11} = 0.8$	0.06	0.12	0.15	0.14	0.06	0.10	0.12	0.12
$\psi_{12} = 0.8$	−0.03	0.09	0.12	0.10	−0.02	0.08	0.10	0.09
$\psi_{13} = 0.8$	0.06	0.13	0.16	0.15	0.03	0.09	0.13	0.09
$\psi_{14} = 0.8$	−0.01	0.10	0.13	0.10	−0.02	0.08	0.10	0.08
$\psi_{15} = 0.8$	0.09	0.11	0.14	0.14	0.06	0.11	0.12	0.13
$\psi_{16} = 0.8$	−0.03	0.09	0.11	0.09	−0.01	0.09	0.09	0.08
$\psi_{21} = 0.8$	0.05	0.12	0.15	0.13	0.05	0.11	0.12	0.13
$\psi_{22} = 0.8$	0.00	0.10	0.12	0.10	0.00	0.09	0.10	0.08
$\psi_{23} = 0.8$	0.03	0.12	0.16	0.13	0.04	0.12	0.13	0.11
$\psi_{24} = 0.8$	0.01	0.09	0.13	0.09	0.00	0.08	0.11	0.08
$\psi_{25} = 0.8$	0.06	0.12	0.15	0.14	0.04	0.10	0.11	0.10
$\psi_{26} = 0.8$	0.00	0.10	0.12	0.10	−0.02	0.08	0.09	0.08
$\lambda_{1,21} = 0.8$	0.07	0.09	0.12	0.11	0.06	0.08	0.09	0.10
$\lambda_{1,42} = 0.8$	0.04	0.08	0.10	0.08	0.02	0.06	0.07	0.07
$\lambda_{1,63} = 0.8$	0.11	0.15	0.16	0.18	0.08	0.13	0.12	0.15
$\lambda_{2,21} = 0.8$	0.03	0.09	0.11	0.09	0.04	0.08	0.08	0.09
$\lambda_{2,42} = 0.8$	0.01	0.07	0.08	0.07	0.02	0.05	0.06	0.06
$\lambda_{2,63} = 0.8$	0.08	0.12	0.14	0.14	0.07	0.09	0.10	0.12
$\Pi_{1,21} = 0.5$	0.01	0.16	0.16	0.16	0.04	0.12	0.11	0.12
$\gamma_{1,11} = 0.6$	0.06	0.19	0.18	0.20	0.03	0.12	0.12	0.12
$\gamma_{1,21} = 0.6$	0.06	0.24	0.23	0.24	0.04	0.15	0.15	0.16
$\Pi_{2,21} = -0.5$	−0.02	0.17	0.18	0.17	−0.02	0.12	0.11	0.12
$\gamma_{2,11} = 0.6$	0.10	0.15	0.15	0.17	0.03	0.11	0.10	0.11
$\gamma_{2,21} = -0.6$	−0.11	0.21	0.23	0.23	−0.04	0.14	0.14	0.15

This table and Table 11.4 are taken from Zhu and Lee (2001).

Table 11.4 Summary statistics for Bayesian estimates (II) with $\pi_1 = 0.5$.

	$n = 400$				$n = 800$			
Par	Bias	SE	SD	RMS	Bias	SE	SD	RMS
$\pi_1 = 0.5$	0.01	0.05	0.04	0.05	0.01	0.04	0.03	0.04
$\pi_2 = 0.5$	−0.01	0.05	0.04	0.05	−0.01	0.04	0.03	0.04
$\Phi_1 = 1.0$	−0.14	0.23	0.25	0.27	−0.08	0.20	0.19	0.21
$\Phi_2 = 1.0$	−0.18	0.18	0.20	0.26	−0.12	0.16	0.15	0.20
$\mu_{1,1} = 0.0$	0.04	0.13	0.15	0.14	0.01	0.13	0.11	0.13
$\mu_{1,2} = 0.0$	0.05	0.14	0.14	0.15	0.02	0.12	0.10	0.12
$\mu_{1,3} = 0.0$	0.06	0.20	0.19	0.21	0.03	0.15	0.13	0.15
$\mu_{1,4} = 0.0$	0.06	0.18	0.17	0.19	0.04	0.13	0.12	0.13
$\mu_{1,5} = 0.0$	−0.01	0.14	0.14	0.14	0.01	0.10	0.10	0.10
$\mu_{1,6} = 0.0$	0.03	0.14	0.13	0.14	0.02	0.10	0.09	0.10
$\mu_{2,1} = 2.0$	0.00	0.13	0.13	0.13	0.02	0.09	0.09	0.09
$\mu_{2,2} = 2.0$	−0.00	0.11	0.12	0.11	0.01	0.08	0.08	0.08
$\mu_{2,3} = 2.0$	−0.03	0.12	0.14	0.13	−0.02	0.10	0.09	0.10
$\mu_{2,4} = 2.0$	−0.03	0.11	0.12	0.12	−0.02	0.07	0.08	0.07
$\mu_{2,5} = 2.0$	0.05	0.13	0.12	0.14	0.02	0.08	0.08	0.08
$\mu_{2,6} = 2.0$	0.01	0.10	0.11	0.10	0.03	0.08	0.08	0.09
$\psi_{\delta 11} = 1.0$	−0.11	0.16	0.18	0.20	−0.07	0.12	0.15	0.14
$\psi_{\delta 12} = 1.0$	−0.12	0.15	0.18	0.19	−0.06	0.13	0.15	0.14
$\psi_{\delta 21} = 1.0$	−0.11	0.13	0.18	0.17	−0.09	0.11	0.14	0.14
$\psi_{\delta 22} = 1.0$	−0.09	0.14	0.20	0.17	−0.09	0.14	0.16	0.17
$\psi_{11} = 0.8$	0.08	0.12	0.15	0.15	0.06	0.10	0.13	0.12
$\psi_{12} = 0.8$	−0.02	0.10	0.13	0.10	0.00	0.08	0.10	0.08
$\psi_{13} = 0.8$	0.08	0.13	0.16	0.15	0.04	0.10	0.13	0.11
$\psi_{14} = 0.8$	−0.03	0.10	0.13	0.10	−0.01	0.08	0.10	0.08
$\psi_{15} = 0.8$	0.11	0.12	0.15	0.17	0.08	0.12	0.12	0.14
$\psi_{16} = 0.8$	−0.04	0.10	0.12	0.11	−0.02	0.08	0.09	0.08
$\psi_{21} = 0.8$	0.06	0.11	0.15	0.12	0.04	0.11	0.12	0.12
$\psi_{22} = 0.8$	−0.03	0.09	0.13	0.10	−0.01	0.08	0.10	0.08
$\psi_{23} = 0.8$	0.02	0.11	0.16	0.12	0.03	0.11	0.13	0.11
$\psi_{24} = 0.8$	−0.00	0.09	0.13	0.09	−0.01	0.09	0.10	0.09
$\psi_{25} = 0.8$	0.09	0.11	0.15	0.14	0.08	0.11	0.11	0.13
$\psi_{26} = 0.8$	−0.02	0.11	0.12	0.12	−0.01	0.09	0.09	0.09
$\lambda_{1,21} = 0.8$	0.09	0.11	0.13	0.14	0.04	0.09	0.10	0.10
$\lambda_{1,42} = 0.8$	0.03	0.08	0.10	0.09	0.03	0.06	0.07	0.07
$\lambda_{1,63} = 0.8$	0.16	0.19	0.18	0.24	0.09	0.11	0.13	0.14
$\lambda_{2,21} = 0.8$	0.07	0.10	0.11	0.13	0.04	0.07	0.08	0.08
$\lambda_{2,42} = 0.8$	0.03	0.08	0.09	0.08	0.02	0.05	0.06	0.05
$\lambda_{2,63} = 0.8$	0.13	0.14	0.16	0.20	0.09	0.10	0.11	0.14
$\Pi_{1,21} = 0.5$	0.02	0.15	0.17	0.15	0.02	0.11	0.12	0.11
$\gamma_{1,11} = 0.6$	0.10	0.17	0.19	0.20	0.06	0.12	0.13	0.13
$\gamma_{1,21} = 0.6$	0.15	0.23	0.24	0.27	0.06	0.15	0.16	0.16
$\Pi_{2,21} = -0.5$	0.02	0.17	0.18	0.17	0.01	0.12	0.12	0.12
$\gamma_{2,11} = 0.6$	0.12	0.17	0.18	0.21	0.06	0.13	0.12	0.14
$\gamma_{2,21} = -0.6$	−0.14	0.20	0.26	0.25	−0.12	0.16	0.17	0.20

Table 11.5 Summary statistics for Bayesian estimates (I) for model with $\pi_1 = 0.3$.

	$n = 400$				$n = 800$			
Par	Bias	SE	SD	RMS	Bias	SE	SD	RMS
$\pi_1 = 0.3$	0.00	0.04	0.04	0.04	0.01	0.03	0.03	0.03
$\pi_2 = 0.7$	0.00	0.04	0.04	0.04	0.01	0.03	0.03	0.03
$\Phi_1 = 1.0$	−0.19	0.27	0.28	0.32	0.08	0.21	0.23	0.23
$\Phi_2 = 1.0$	−0.08	0.17	0.18	0.19	0.03	0.14	0.13	0.14
$\mu_{1,1} = 0.0$	−0.01	0.19	0.21	0.19	0.04	0.16	0.15	0.17
$\mu_{1,2} = 0.0$	0.02	0.18	0.19	0.18	0.06	0.14	0.14	0.15
$\mu_{1,3} = 0.0$	0.02	0.27	0.26	0.27	0.06	0.19	0.19	0.20
$\mu_{1,4} = 0.0$	0.02	0.22	0.22	0.21	0.06	0.18	0.17	0.19
$\mu_{1,5} = 0.0$	−0.05	0.17	0.19	0.18	0.03	0.14	0.14	0.14
$\mu_{1,6} = 0.0$	−0.01	0.17	0.18	0.17	0.05	0.14	0.13	0.15
$\mu_{2,1} = 2.0$	0.00	0.09	0.10	0.09	0.01	0.06	0.07	0.06
$\mu_{2,2} = 2.0$	0.00	0.08	0.09	0.08	0.01	0.05	0.06	0.05
$\mu_{2,3} = 2.0$	−0.01	0.10	0.11	0.10	−0.01	0.07	0.08	0.07
$\mu_{2,4} = 2.0$	−0.01	0.11	0.09	0.11	−0.01	0.07	0.07	0.07
$\mu_{2,5} = 2.0$	0.01	0.08	0.09	0.09	0.01	0.06	0.07	0.06
$\mu_{2,6} = 2.0$	0.01	0.09	0.08	0.09	0.01	0.06	0.06	0.06
$\psi_{\delta 11} = 1.0$	−0.12	0.14	0.21	0.18	−0.08	0.16	0.18	0.18
$\psi_{\delta 12} = 1.0$	−0.13	0.15	0.21	0.20	−0.10	0.14	0.18	0.17
$\psi_{\delta 21} = 1.0$	−0.08	0.15	0.16	0.17	−0.04	0.11	0.13	0.12
$\psi_{\delta 22} = 1.0$	−0.05	0.16	0.18	0.17	−0.06	0.11	0.14	0.12
$\psi_{11} = 0.8$	0.07	0.12	0.18	0.14	0.05	0.10	0.14	0.11
$\psi_{12} = 0.8$	0.02	0.11	0.15	0.11	−0.03	0.11	0.12	0.11
$\psi_{13} = 0.8$	0.05	0.11	0.18	0.12	0.06	0.12	0.15	0.13
$\psi_{14} = 0.8$	0.00	0.11	0.15	0.11	−0.02	0.09	0.12	0.10
$\psi_{15} = 0.8$	0.09	0.13	0.16	0.15	0.08	0.13	0.14	0.15
$\psi_{16} = 0.8$	0.00	0.11	0.15	0.11	−0.01	0.09	0.11	0.09
$\psi_{21} = 0.8$	0.05	0.11	0.14	0.12	0.03	0.09	0.11	0.10
$\psi_{22} = 0.8$	0.01	0.09	0.11	0.09	0.00	0.08	0.08	0.08
$\psi_{23} = 0.8$	0.03	0.10	0.14	0.11	0.02	0.10	0.11	0.10
$\psi_{24} = 0.8$	0.01	0.10	0.12	0.10	−0.01	0.08	0.10	0.08
$\psi_{25} = 0.8$	0.05	0.11	0.13	0.12	0.02	0.08	0.10	0.08
$\psi_{26} = 0.8$	0.00	0.09	0.10	0.09	0.00	0.07	0.08	0.07
$\lambda_{1,21} = 0.8$	0.06	0.13	0.16	0.14	0.08	0.10	0.12	0.12
$\lambda_{1,42} = 0.8$	0.02	0.11	0.13	0.11	0.04	0.07	0.09	0.08
$\lambda_{1,63} = 0.8$	0.16	0.17	0.23	0.24	0.12	0.11	0.16	0.17
$\lambda_{2,21} = 0.8$	0.04	0.09	0.09	0.10	0.02	0.07	0.07	0.07
$\lambda_{2,42} = 0.8$	0.00	0.06	0.07	0.06	0.00	0.05	0.05	0.05
$\lambda_{2,63} = 0.8$	0.06	0.11	0.12	0.13	0.02	0.08	0.08	0.08
$\Pi_{1,21} = 0.5$	0.08	0.21	0.22	0.22	0.01	0.15	0.15	0.15
$\gamma_{1,11} = 0.6$	0.10	0.23	0.26	0.25	0.18	0.16	0.17	0.18
$\gamma_{1,21} = 0.6$	0.03	0.27	0.31	0.26	0.17	0.20	0.21	0.21
$\Pi_{2,21} = -0.5$	−0.01	0.14	0.14	0.14	0.01	0.09	0.09	0.09
$\gamma_{2,11} = 0.6$	0.05	0.14	0.13	0.15	0.03	0.09	0.09	0.09
$\gamma_{2,21} = -0.6$	−0.07	0.16	0.18	0.16	−0.04	0.12	0.12	0.13

Table 11.6 Summary statistics for Bayesian estimates (II) with $\pi_1 = 0.3$.

	$n = 400$				$n = 800$			
Par	Bias	SE	SD	RMS	Bias	SE	SD	RMS
$\pi_1 = 0.3$	0.02	0.04	0.04	0.04	0.01	0.03	0.03	0.03
$\pi_2 = 0.7$	−0.02	0.04	0.04	0.04	−0.01	0.03	0.03	0.03
$\Phi_1 = 1.0$	−0.17	0.26	0.31	0.30	−0.17	0.23	0.23	0.29
$\Phi_2 = 1.0$	−0.12	0.16	0.18	0.20	−0.05	0.13	0.13	0.14
$\mu_{1,1} = 0.0$	0.08	0.17	0.21	0.19	0.03	0.15	0.15	0.15
$\mu_{1,2} = 0.0$	0.09	0.18	0.20	0.20	0.05	0.15	0.14	0.16
$\mu_{1,3} = 0.0$	0.13	0.25	0.27	0.28	0.05	0.21	0.19	0.22
$\mu_{1,4} = 0.0$	0.12	0.20	0.24	0.23	0.06	0.18	0.17	0.19
$\mu_{1,5} = 0.0$	0.02	0.18	0.19	0.18	0.01	0.14	0.14	0.14
$\mu_{1,6} = 0.0$	0.09	0.16	0.19	0.19	0.04	0.13	0.13	0.14
$\mu_{2,1} = 2.0$	0.01	0.10	0.10	0.10	0.01	0.08	0.07	0.08
$\mu_{2,2} = 2.0$	0.00	0.08	0.09	0.08	−0.01	0.06	0.06	0.07
$\mu_{2,3} = 2.0$	−0.01	0.09	0.11	0.10	0.01	0.08	0.08	0.08
$\mu_{2,4} = 2.0$	−0.01	0.09	0.10	0.09	0.00	0.07	0.07	0.07
$\mu_{2,5} = 2.0$	0.02	0.10	0.09	0.10	0.02	0.07	0.07	0.07
$\mu_{2,6} = 2.0$	0.01	0.08	0.09	0.08	0.00	0.06	0.06	0.06
$\psi_{\delta 11} = 1.0$	−0.13	0.14	0.21	0.19	−0.10	0.12	0.18	0.16
$\psi_{\delta 12} = 1.0$	−0.10	0.14	0.22	0.18	−0.09	0.16	0.18	0.18
$\psi_{\delta 21} = 1.0$	−0.09	0.12	0.16	0.15	−0.04	0.11	0.13	0.12
$\psi_{\delta 22} = 1.0$	−0.09	0.15	0.17	0.18	−0.04	0.12	0.14	0.13
$\psi_{11} = 0.8$	0.08	0.12	0.18	0.15	0.08	0.10	0.15	0.13
$\psi_{12} = 0.8$	−0.00	0.10	0.15	0.10	−0.01	0.11	0.12	0.11
$\psi_{13} = 0.8$	0.05	0.11	0.18	0.12	0.05	0.13	0.15	0.14
$\psi_{14} = 0.8$	−0.04	0.09	0.14	0.10	−0.01	0.10	0.12	0.10
$\psi_{15} = 0.8$	0.13	0.13	0.17	0.18	0.10	0.11	0.14	0.15
$\psi_{16} = 0.8$	−0.02	0.10	0.15	0.10	−0.02	0.10	0.12	0.10
$\psi_{21} = 0.8$	0.05	0.11	0.14	0.12	0.02	0.10	0.11	0.10
$\psi_{22} = 0.8$	−0.01	0.10	0.11	0.10	0.01	0.07	0.08	0.07
$\psi_{23} = 0.8$	−0.00	0.09	0.14	0.09	0.02	0.09	0.12	0.09
$\psi_{24} = 0.8$	0.01	0.10	0.11	0.10	−0.01	0.07	0.09	0.07
$\psi_{25} = 0.8$	0.08	0.11	0.13	0.13	0.04	0.09	0.10	0.10
$\psi_{26} = 0.8$	−0.02	0.08	0.11	0.09	0.01	0.08	0.08	0.08
$\lambda_{1,21} = 0.8$	0.09	0.11	0.16	0.14	0.07	0.10	0.13	0.12
$\lambda_{1,42} = 0.8$	0.04	0.10	0.11	0.11	0.03	0.08	0.09	0.08
$\lambda_{1,63} = 0.8$	0.23	0.19	0.25	0.30	0.20	0.15	0.19	0.25
$\lambda_{2,21} = 0.8$	0.05	0.09	0.10	0.11	0.02	0.06	0.07	0.07
$\lambda_{2,42} = 0.8$	0.01	0.06	0.07	0.06	0.01	0.04	0.05	0.04
$\lambda_{2,63} = 0.8$	0.10	0.11	0.13	0.15	0.04	0.08	0.09	0.09
$\Pi_{1,21} = 0.5$	0.21	0.22	0.27	0.30	0.09	0.18	0.18	0.20
$\gamma_{1,11} = 0.6$	0.03	0.21	0.23	0.21	0.02	0.14	0.16	0.14
$\gamma_{1,21} = 0.6$	0.22	0.37	0.35	0.43	0.12	0.23	0.23	0.26
$\Pi_{2,21} = -0.5$	0.06	0.13	0.14	0.15	0.04	0.10	0.09	0.11
$\gamma_{2,11} = 0.6$	0.01	0.13	0.14	0.13	−0.01	0.09	0.10	0.09
$\gamma_{2,21} = -0.6$	−0.11	0.17	0.19	0.21	−0.05	0.13	0.13	0.14

specification of the parameter matrices in the model formulation are given by: $\mathbf{\Pi}_k = \mathbf{0}$, $\mathbf{\Psi}_{\delta k} = \psi_{\delta k}$, $\mathbf{\Gamma}_k = (\gamma_{k,1}, \gamma_{k,2})$,

$$\mathbf{\Lambda}_k^T = \begin{bmatrix} 1.0^* & \lambda_{k,21} & 0.0^* & 0.0^* & 0.0^* & 0.0^* & 0.0^* & 0.0^* \\ 0.0^* & 0.0^* & 1.0^* & \lambda_{k,42} & \lambda_{k,52} & 0.0^* & 0.0^* & 0.0^* \\ 0.0^* & 0.0^* & 0.0^* & 0.0^* & 0.0^* & 1.0^* & \lambda_{k,73} & \lambda_{k,83} \end{bmatrix}, \quad \mathbf{\Phi}_k = \begin{bmatrix} \phi_{k,11} & \phi_{k,12} \\ \phi_{k,21} & \phi_{k,22} \end{bmatrix} \tag{11.16}$$

and $\mathbf{\Psi}_k = \text{diag}(\psi_{k1}, \cdots, \psi_{k8})$. To identify the model, elements in $\mathbf{\Lambda}_k$ with an asterisk are fixed. The total number of unknown parameters is 62.

Bayesian estimates of the structural parameters and estimates of the factor scores are obtained via the Gibbs sampler. The following hyperparameters are selected: $\alpha = 1$, $\boldsymbol{\mu}_0 = \bar{\mathbf{y}}$, $\mathbf{\Sigma}_0 = \mathbf{S}_y/2.0$, $\rho_0 = 5$ and $\mathbf{R}_0^{-1} = 5\mathbf{I}_2$; $\alpha_{0\epsilon k} = \alpha_{0\delta k} = \beta_{0\epsilon k} = \beta_{0\delta k} = 6$ for all k; $\mathbf{H}_{0ykm} = \mathbf{I}$ and $\mathbf{H}_{0\omega kl} = \mathbf{I}$, $\mathbf{\Lambda}_{0km} = \tilde{\mathbf{\Lambda}}_{0km}$ and $\mathbf{\Lambda}_{0\omega kl} = \tilde{\mathbf{\Lambda}}_{0\omega kl}$ for all k, m and l, where $\tilde{\mathbf{\Lambda}}_{0km}$ and $\tilde{\mathbf{\Lambda}}_{0\omega kl}$ are obtained by some initial estimation with noninformative prior distributions. We first use MCMC samples simulated by the random permutation sampler to find an identifiability constraint. We find that $\mu_{1,1} < \mu_{2,1}$ is a suitable one. Under this constraint, Bayesian estimates are obtained by the procedure described in Section 11.3. Based on different starting values of the parameters, three parallel sequences of observations are generated and the EPSR values are calculated. Figure 11.5 presents the plots of the EPSR values against the iteration numbers. We observe that the EPSR values of the

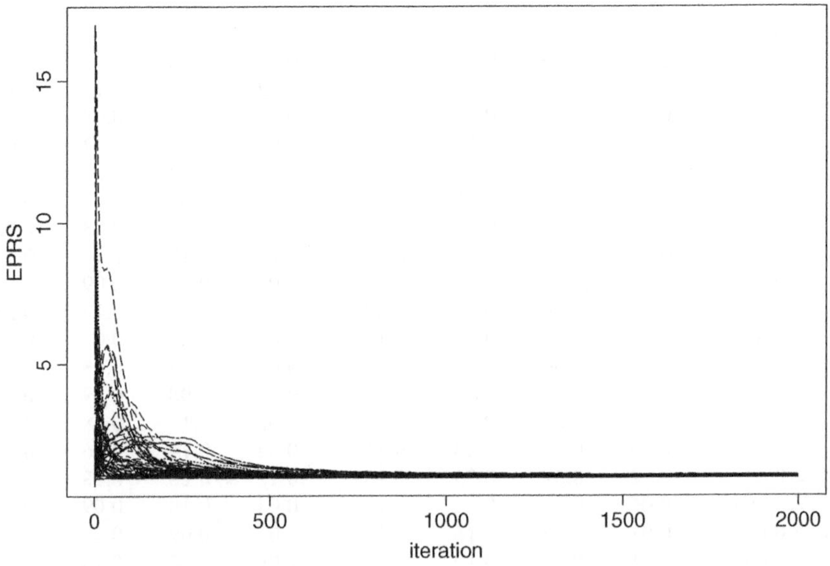

Figure 11.5 EPSR values of all parameters from three parallel runs in the ICPSR example. This table is taken from Zhu and Lee (2001).

parameters at the starting points were quite large, this indicates that the starting values are far away from the solution. The Gibbs sampler algorithm converged within 1000 iterations. After the convergence of the Gibbs sampler, a total of 1000 observations with a spacing of 10 are collected for analysis. The Bayesian estimates of the structural parameters and their standard error estimates are reported in Table 11.7. From this table, it can be seen that there are at least two components which have different sets of Bayesian parameter estimates.

Table 11.7 Bayesian estimates and standard error estimates of the ICPSR example.

Parameter	Component 1		Component 2	
	EST	SE	EST	SE
π_k	0.56	0.03	0.44	0.03
$\mu_{k,1}$	6.91	0.11	8.09	0.09
$\mu_{k,2}$	6.30	0.14	7.90	0.14
$\mu_{k,3}$	5.87	0.14	7.83	0.11
$\mu_{k,4}$	7.83	0.10	8.70	0.07
$\mu_{k,5}$	7.10	0.11	8.07	0.09
$\mu_{k,6}$	5.41	0.14	4.01	0.15
$\mu_{k,7}$	4.06	0.13	3.61	0.14
$\mu_{k,8}$	5.59	0.14	4.61	0.14
$\lambda_{k,11}$	1*	–	1*	–
$\lambda_{k,21}$	0.49	0.11	0.86	0.13
$\lambda_{k,32}$	1*	–	1*	–
$\lambda_{k,42}$	1.30	0.17	0.94	0.10
$\lambda_{k,52}$	1.58	0.20	1.02	0.11
$\lambda_{k,63}$	1*	–	1*	–
$\lambda_{k,73}$	2.05	0.44	0.98	0.07
$\lambda_{k,83}$	1.08	0.27	0.74	0.08
$\gamma_{k,1}$	0.68	0.14	0.77	0.11
$\gamma_{k,2}$	−0.02	0.15	−0.09	0.04
$\phi_{k,11}$	1.18	0.26	0.90	0.18
$\phi_{k,21}$	−0.12	0.09	−0.28	0.15
$\phi_{k,22}$	0.92	0.30	4.30	0.52
ψ_{k1}	1.56	0.65	0.56	0.11
ψ_{k2}	6.92	0.50	2.80	0.34
ψ_{k3}	4.86	0.37	1.35	0.18
ψ_{k4}	2.51	0.27	0.45	0.07
ψ_{k5}	1.29	0.27	0.55	0.08
ψ_{k6}	6.31	0.50	1.25	0.35
ψ_{k7}	2.43	0.76	1.07	0.23
ψ_{k8}	6.39	0.57	3.15	0.40
$\psi_{\delta k}$	3.38	0.72	0.70	0.12

This table is taken from Zhu and Lee (2001).

11.5 BAYESIAN MODEL COMPARISON OF MIXTURE SEMs

In previous sections, we discussed the Bayesian estimation for a finite mixture of SEMs with a known number of components. In the analysis, K is given and each π_k is greater than zero. The objective of this section is to consider the Bayesian model selection problem for selecting one of the two mixtures of SEMs with a different number of components. An approach based on the Bayes factor (Berger, 1985) will be developed. For mixtures of SEMs that involve a large number of unknown parameters and latent variables, an efficient procedure for computing the Bayes factor is important. Here, an algorithm based on path sampling (Gelman and Meng, 1998; Lee and Song, 2003a) is described. In the implementation, random observations are needed to sample from some appropriate conditional distributions. The Gibbs sampler (Geman and Geman, 1984) developed in previous sections in the estimation can be used for this purpose.

The underlying finite mixtures of SEMs is defined again by Equations (11.1), (11.2) and (11.3), except that K is not fixed and π_k are non-negative component probabilities that sum to 1.0. Note that some π_k may be equal to zero. When using the likelihood ratio test, some unknown parameters may be on the boundary of the parameter space, and this causes serious difficulty in developing the test statistics for testing the hypothesis about the number of components. On the contrary, the Bayesian approach for model comparison with the Bayes factor does not have this problem.

11.5.1 A Model Selection Procedure through Path Sampling

Let M_1 be a mixture SEM with K components, and M_0 be a mixture SEM with c components, where $c < K$. The Bayes factor for selection between M_0 and M_1 is defined by

$$B_{10} = \frac{P(\mathbf{Y}|M_1)}{P(\mathbf{Y}|M_0)}, \tag{11.17}$$

based on the given data set $\mathbf{Y}$. In computing the Bayes factor through path sampling, the observed data set $\mathbf{Y}$ is augmented with the matrix of latent variables $\boldsymbol{\Omega}$ and $\mathbf{W}$. Based on similar reasoning and derivation to that given in previous chapters, $\log B_{10}$ can be estimated as follows. Let

$$U(\mathbf{Y}, \boldsymbol{\Omega}, \boldsymbol{\theta}, t) = \frac{\mathrm{d}}{\mathrm{d}t}\log\{p(\mathbf{Y}, \boldsymbol{\Omega}|\boldsymbol{\theta}, t)\},$$

where $p(\mathbf{Y}, \boldsymbol{\Omega}|\boldsymbol{\theta}, \mathbf{t})$ is the complete-data log-likelihood function. Then,

$$\widehat{\log B_{10}} = \frac{1}{2}\sum_{s=0}^{S}(t_{(s+1)} - t_{(s)})(\overline{U}_{(s+1)} + \overline{U}_{(s)}), \tag{11.18}$$

where $t_{(s)}$ are fixed grids $\{t_{(s)}\}_{s=0}^{S}$ in $[0, 1]$ such that $t_{(0)} = 0 < t_{(1)} < t_{(2)} < \cdots < t_{(S)} < t_{(S+1)} = 1$, and $\overline{U}_{(s)}$ is the average of the $U(\mathbf{Y}, \boldsymbol{\Omega}, \boldsymbol{\theta}, t)$ on the basis of all simulation draws for which $t = t_{(s)}$. That is,

$$\overline{U}_{(s)} = J^{-1}\sum_{j=1}^{J} U(\mathbf{Y}, \boldsymbol{\Omega}^{(j)}, \boldsymbol{\theta}^{(j)}, t_{(s)}), \tag{11.19}$$

in which $\{(\boldsymbol{\Omega}^{(j)}, \boldsymbol{\theta}^{(j)}), j = 1, \ldots, J\}$ are simulated observations drawn from $p(\boldsymbol{\Omega}, \mathbf{W}, \boldsymbol{\theta}|\mathbf{Y}, t_{(s)})$.

The implementation of path sampling is straightforward. Drawing observations $\{(\boldsymbol{\Omega}^{(j)}, \boldsymbol{\theta}^{(j)}) : j = 1, \ldots, J\}$ from the posterior distribution $p(\boldsymbol{\Omega}, \mathbf{W}, \boldsymbol{\theta}|\mathbf{Y}, t_{(s)})$ is the major task in the proposed procedure. This posterior distribution is complicated and difficult to cope with. Similar to estimation, we further utilize the idea of data augmentation to augment the observed data $\mathbf{Y}$ with the latent matrix $\mathbf{W}$ of the allocation variables. As a consequence, the Gibbs sampler described in Section 11.3 can be applied to simulate observations from $p(\boldsymbol{\Omega}, \mathbf{W}, \boldsymbol{\theta}|\mathbf{Y}, t_{(s)})$. However, note that in computing Bayes factors, the inclusion of an identifiability constraint in simulating $\{(\boldsymbol{\Omega}^{(j)}, \boldsymbol{\theta}^{(j)}), j = 1, \ldots, J\}$ is not necessary. As the likelihood is invariant to relabelling the states, the inclusion of such a constraint will not change the values of $U(\mathbf{Y}, \boldsymbol{\Omega}, \boldsymbol{\theta}, t)$. As a result, the logarithm Bayes factors estimated by Equations (11.18) and (11.19) will not be changed.

Finding a good path t in [0,1] to link competing models M_1 and M_0 is an important step in applying path sampling. An illustrative example is given as follows. Consider the following competing models:

$$M_1: \quad [\mathbf{y}|\boldsymbol{\theta}, \boldsymbol{\pi}] \stackrel{D}{=} \sum_{k=1}^{K} \pi_k f_k(\mathbf{y}|\boldsymbol{\mu}_k, \boldsymbol{\Sigma}_k), \tag{11.20}$$

corresponding to a model of K components with positive component probabilities π_k, and

$$M_0: \quad [\mathbf{y}|\boldsymbol{\theta}, \boldsymbol{\pi}^*] \stackrel{D}{=} \sum_{k=1}^{c} \pi_k^* f_k(\mathbf{y}|\boldsymbol{\mu}_k, \boldsymbol{\Sigma}_k), \tag{11.21}$$

corresponding to a model of c components with positive component probabilities π_k^*, where $1 \leq c < K$. To apply path sampling for computing log B_{10}, these competing models are linked up by a path $t \in [0, 1]$ as follows:

$$\begin{aligned} M_t: \quad [\mathbf{y}|\boldsymbol{\theta}, \boldsymbol{\pi}, t] \stackrel{D}{=} &[\pi_1 + (1-t)a_1(\pi_{c+1} + \cdots + \pi_K)]f_1(\mathbf{y}|\boldsymbol{\mu}_1, \boldsymbol{\Sigma}_1) + \cdots \\ &+ [\pi_c + (1-t)a_c(\pi_{c+1} + \cdots + \pi_K)]f_c(\mathbf{y}|\boldsymbol{\mu}_c, \boldsymbol{\Sigma}_c) \\ &+ t\pi_{c+1}f_{c+1}(\mathbf{y}|\boldsymbol{\mu}_{c+1}, \boldsymbol{\Sigma}_{c+1}) + \cdots + t\pi_K f_K(\mathbf{y}|\boldsymbol{\mu}_K, \boldsymbol{\Sigma}_K), \end{aligned} \tag{11.22}$$

where $a_1, \cdots, a_c$ are given positive weights such that $a_1 + \cdots + a_c = 1$. Clearly, when $t = 1$, M_t reduces to M_1; when $t = 0$, M_t reduces to M_0 with $\pi_k^* = \pi_k + a_k(\pi_{c+1} + \cdots + \pi_K)$, $k = 1, \ldots, c$. The weights $a_1, \ldots, a_c$ represent the increases of the corresponding component probabilities from a K component SEM to a c component SEM. A natural and simple suggestion for practical applications is to take $a_k = c^{-1}$.

The complete-data log-likelihood function can be written as

$$\begin{aligned} \log p(\mathbf{Y}, \boldsymbol{\Omega}|\boldsymbol{\theta}, t) = \sum_{i=1}^n \log \Bigg\{ &\sum_{k=1}^c [\pi_k + (1-t)a_k \sum_{h=c+1}^K \pi_h] \\ &\times f_k(\mathbf{y}_i, \boldsymbol{\omega}_{ki}|\boldsymbol{\mu}_k, \boldsymbol{\Sigma}_k) + \sum_{k=c+1}^K t\pi_k f_k(\mathbf{y}_i, \boldsymbol{\omega}_{ki}|\boldsymbol{\mu}_k, \boldsymbol{\Sigma}_k) \Bigg\}. \end{aligned} \tag{11.23}$$

By differentiation with respect to t, we have

$$U(\mathbf{Y}, \boldsymbol{\Omega}, \boldsymbol{\theta}, t) = \tag{11.24}$$

$$\sum_{i=1}^n \frac{-\sum_{h=c+1}^K \pi_h \sum_{k=1}^c a_k f_k(\mathbf{y}_i, \boldsymbol{\omega}_{ki}|\boldsymbol{\mu}_k, \boldsymbol{\Sigma}_k) + \sum_{k=c+1}^K \pi_k f_k(\mathbf{y}_i, \boldsymbol{\omega}_{ki}|\boldsymbol{\mu}_k, \boldsymbol{\Sigma}_k)}{\sum_{k=1}^c \left[\pi_k + (1-t)a_k \sum_{h=c+1}^K \pi_h\right] f_k(\mathbf{y}_i, \boldsymbol{\omega}_{ki}|\boldsymbol{\mu}_k, \boldsymbol{\Sigma}_k) + \sum_{k=c+1}^K t\pi_k f_k(\mathbf{y}_i, \boldsymbol{\omega}_{ki}|\boldsymbol{\mu}_k, \boldsymbol{\Sigma}_k)},$$

where

$$\begin{aligned} f_k(\mathbf{y}_i, \boldsymbol{\omega}_{ki}|\boldsymbol{\mu}_k, \boldsymbol{\Sigma}_k) = &(2\pi)^{-p/2}|\boldsymbol{\Psi}_k|^{-1/2} \\ &\times \exp\left[-\frac{1}{2}(\mathbf{y}_i - \boldsymbol{\mu}_k - \boldsymbol{\Lambda}_k\boldsymbol{\omega}_{ki})^T \boldsymbol{\Psi}_k^{-1}(\mathbf{y}_i - \boldsymbol{\mu}_k - \boldsymbol{\Lambda}_k\boldsymbol{\omega}_{ki})\right] \\ &\times (2\pi)^{-q_1/2}|\mathbf{I}_{q_1} - \boldsymbol{\Pi}_k||\boldsymbol{\Psi}_{\delta k}|^{-1/2} \end{aligned}$$

$$\times \exp\left[-\frac{1}{2}(\boldsymbol{\eta}_{ki} - \boldsymbol{\Lambda}_{\omega k}\boldsymbol{\omega}_{ki})^T \boldsymbol{\Psi}_{\delta k}^{-1}(\boldsymbol{\eta}_{ki} - \boldsymbol{\Lambda}_{\omega k}\boldsymbol{\omega}_{ki})\right]$$
$$\times (2\pi)^{-q_2/2}|\boldsymbol{\Phi}_k|^{-1/2}\exp\left(-\frac{1}{2}\boldsymbol{\xi}_{ki}^T\boldsymbol{\Phi}_k^{-1}\boldsymbol{\xi}_{ki}\right),$$

and $\boldsymbol{\Lambda}_{\omega k} = (\boldsymbol{\Pi}_k, \boldsymbol{\Gamma}_k)$. Thus, the Bayes factor can be estimated via Equation (11.18) with

$$\bar{U}_{(s)} = J^{-1}\sum_{j=1}^{J} U(\mathbf{Y}, \boldsymbol{\Omega}^{(j)}, \boldsymbol{\theta}^{(j)}, t_{(s)}), \quad (11.25)$$

where $\{(\boldsymbol{\Omega}^{(j)}, \boldsymbol{\theta}^{(j)}) : j = 1, \ldots, J\}$ are observations drawn from $p(\boldsymbol{\Omega}, \boldsymbol{\theta}|\mathbf{Y}, t_{(s)})$.

11.5.2 A Simulation Study

In this section, results obtained from a simulation study will be presented to illustrate the accuracy of the results obtained from the path sampling procedure, and the sensitivity to prior inputs, for model selection in finite mixtures of SEMs. The true model is a mixture SEM with two components defined in Equations (11.2) and (11.3). In each component, there are six manifest variables, which are related to three latent factors in $\boldsymbol{\eta}_k = (\eta_{k1}, \eta_{k2})$, and $\boldsymbol{\xi}_k = \xi_k$. The specifications and population values of the elements in $\boldsymbol{\Lambda}_1$, $\boldsymbol{\Lambda}_2$, $\boldsymbol{\Pi}_1$ and $\boldsymbol{\Pi}_2$ are taken as:

$$\boldsymbol{\Lambda}_1^T = \boldsymbol{\Lambda}_2^T = \begin{pmatrix} 1^* & 0.8 & 0^* & 0^* & 0^* & 0^* \\ 0^* & 0^* & 1^* & 0.8 & 0^* & 0^* \\ 0^* & 0^* & 0^* & 0^* & 1^* & 0.8 \end{pmatrix}, \boldsymbol{\Pi}_1 = \begin{pmatrix} 0^* & 0^* \\ 0.5 & 0^* \end{pmatrix}, \boldsymbol{\Pi}_2 = \begin{pmatrix} 0^* & 0^* \\ -0.5 & 0^* \end{pmatrix},$$

where parameters with an asterisk are fixed at the preassigned values. The true population values of the other unknown parameters are given by: $\boldsymbol{\Phi}_1 = 1.0$, $\boldsymbol{\Phi}_2 = 1.0$, $\mu_{1,m} = 0.0$, $\mu_{2,m} = 2.0$, $\psi_{1m} = \psi_{2m} = 0.64$, for $m = 1, \ldots, 6$, $\psi_{\delta 1l} = \psi_{\delta 2l} = 0.8$ for $l = 1, 2$, $\boldsymbol{\Gamma}_1 = [0.5, 0.5]^T$, $\boldsymbol{\Gamma}_2 = [0.5, -0.5]^T$ and $\pi_1 = \pi_2 = 0.5$. Note that the mean vectors, $\boldsymbol{\Pi}$ and $\boldsymbol{\Gamma}$, of these two components are different. The separation $d_{12} = \max_{k\in\{1,2\}}[(\boldsymbol{\mu}_1 - \boldsymbol{\mu}_2)^T\boldsymbol{\Sigma}_k^{-1}(\boldsymbol{\mu}_1 - \boldsymbol{\mu}_2)]$ of these two components is equal to 4.128.

Using $n = 400$ and the above specifications, a random sample of observations is generated to illustrate the sensitivity of the proposed procedure for computing the logarithm Bayes factor with respect to prior distributions. Since there are many parameters involved in the prior distributions, it would be very tedious to provide a detailed analysis on the impact of each type of them. Hence, we concentrate on the important hyperparameters and fix the less important ones at the following given values: $\mathbf{H}_{0ykm} = \mathbf{I}$, $\mathbf{H}_{0\omega kl} = \mathbf{I}$, for all k, m and l, $\rho_0 = 8$ and $\mathbf{R}_0 = 5.0$, where $\mathbf{I}$ denotes an identity matrix of appropriate order. Based on the nature of the parameters, we divide the remaining more important hyperparameters into three groups: $\{\alpha\}$, $\{\boldsymbol{\mu}_0, \boldsymbol{\Sigma}_0, \boldsymbol{\Lambda}_{0km}, \boldsymbol{\Lambda}_{0\omega kl}\}$ and $\{(\alpha_{0\epsilon k}, \beta_{0\epsilon k}), (\alpha_{0\delta k}, \beta_{0\delta k})\}$. In the sensitivity analysis, we follow the suggestion of Kass and Raftery (1995) to perturb alternatively the hyperparameters in these groups. We take $\alpha = 1$ and $\alpha = 2$ in the symmetric Dirichlet distribution, the prior distribution of $\boldsymbol{\pi}$. For each value of α, the following different types of hyperparameter values in the other groups are considered.

Type I: Fix $\alpha_{0\epsilon k} = \alpha_{0\delta k} = 6$ and $\beta_{0\epsilon k} = \beta_{0\delta k} = 4$ for each component. Hyperparameters in the remaining groups are given by:

(a) For each $k = 1, 2$, $m = 1, \ldots, 6$, $l = 1, 2$, $\{\boldsymbol{\Lambda}_{0km}, \boldsymbol{\Lambda}_{0\omega kl}\}$ are equal to the true population values for all k, m and l; $\boldsymbol{\mu}_0 = \bar{\mathbf{y}}$, $\boldsymbol{\Sigma}_0 = \mathbf{S}_y/2$, where $\bar{\mathbf{y}}$ and $\mathbf{S}_y$ are the sample mean and the sample covariance matrix of the simulated data.

(b) $\{\boldsymbol{\Lambda}_{0km}, \boldsymbol{\Lambda}_{0\omega kl}\}$, $\boldsymbol{\mu}_0$ and $\boldsymbol{\Sigma}_0$ are set equal to half the values given in (a).

(c) $\{\boldsymbol{\Lambda}_{0km}, \boldsymbol{\Lambda}_{0\omega kl}\}$, $\boldsymbol{\mu}_0$ and $\boldsymbol{\Sigma}_0$ are set equal to twice the values given in (a).

Type II: For each $k = 1, 2$, $m = 1, \ldots, 6$, $l = 1, 2$, fix $\{\boldsymbol{\Lambda}_{0km}, \boldsymbol{\Lambda}_{0\omega kl}\}$ at the true values for all k, m and l; $\boldsymbol{\mu}_0 = \bar{\mathbf{y}}$, and $\boldsymbol{\Sigma}_0 = \mathbf{S}_y/2$. Hyperparameters in the remaining group are taken as:

(a) $\alpha_{0\epsilon k} = \alpha_{0\delta k} = 6$, $\beta_{0\epsilon k} = \beta_{0\delta k} = 4$ for each component.

(b) $\alpha_{0\epsilon k} = \alpha_{0\delta k} = 10$, $\beta_{0\epsilon k} = \beta_{0\delta k} = 8$ for each component.

The path sampling procedure with 20 grids in [0,1] is applied to estimate the logarithm Bayes factors for comparing M_1, M_2 and M_3, where M_k is a mixture of SEM with k components. In simulating the required observations, we first conduct a few test runs in a pilot study to obtain some idea about the convergence of the Gibbs sampler. We find that in all cases, the Gibbs sampler converged quickly within 100 iterations. For each $t_{(s)}$ in every replication, $J = 1000$ observations are collected after discarding 100 burn-in iterations. The logarithm Bayes factors for different types of hyperparameters are estimated via Equations (11.18) and (11.19). Results are presented in Table 11.8. All the estimated $\log B_{21}$ are significantly larger than 3.0 and the estimated $\log B_{32}$ are less than zero. These results lead to the conclusion that M_2 is much better than M_1 and it is also better than M_3. On the basis of the criterion for interpreting the logarithm Bayes factor, it is concluded that a two-component SEM should be selected. This conclusion is consistent with the true situation.

To obtain more understanding about the sensitivity of the Bayes factor to prior inputs, we further conduct a study on the basis of some changes on $\alpha_{0\epsilon k}$, $\alpha_{0\delta k}$, $\beta_{0\epsilon k}$ and $\beta_{0\delta k}$ under the Type I hyperparameter values. Here, we take

Table 11.8 The estimated logarithms of Bayes factors.

Type		I			II	
		(a)	(b)	(c)	(a)	(b)
$\alpha = 1$	$\log B_{21}$	12.21	9.39	11.05	12.21	10.79
	$\log B_{32}$	−0.42	−0.37	−0.46	−0.42	−0.51
$\alpha = 2$	$\log B_{21}$	13.19	11.80	11.41	13.20	12.21
	$\log B_{32}$	−0.73	−0.66	−0.89	−0.73	−1.00

Table 11.9 The estimated logarithms of Bayes factors under different prior inputs and S.

		$(S = 20)$			$(S = 10)$		
		(a)	(b)	(c)	(a)	(b)	(c)
$\alpha = 1$	$\log B_{21}$	12.43	10.94	11.24	11.33	9.34	9.87
	$\log B_{32}$	−0.32	−0.32	−0.40	−0.33	−0.35	−0.43
$\alpha = 2$	$\log B_{21}$	13.35	14.19	12.60	11.86	10.21	12.30
	$\log B_{32}$	−0.60	−0.60	−0.71	−0.63	−0.64	−0.18

$\alpha_{0\epsilon 1} = \alpha_{0\delta 1} = 7$, $\alpha_{0\epsilon 2} = \alpha_{0\delta 2} = 5$, $\beta_{0\epsilon 1} = \beta_{0\delta 1} = 4$ and $\beta_{0\epsilon 2} = \beta_{0\delta 2} = 3$, while the other hyperparameter values under (a) and (b) are the same as before. Moreover, to provide some idea about the Bayes factor estimates under different choices on the number of grids in $[0, 1]$, we consider $S = 20$ and $S = 10$. The estimated logarithm of Bayes factors are reported in Table 11.9. We observe that the estimates given in Table 11.9 are close to those given in columns under Type I of Table 11.8. From the $\widehat{\log B_{21}}$ and $\widehat{\log B_{32}}$ values under various choices of hyperparameters, it is observed that under the given sample size, the Bayes factor for analyzing mixtures of SEMs is not very sensitive to these prior inputs.

11.5.3 An Illustrative Example

The same portion of the ICPSR data set as described in Section 11.4.3 is reanalyzed to illustrate the path sampling procedure. We wish to find out whether there are some mixture models that are better than the mixture SEM with two components proposed in Section 11.4.3. For each component, the specifications of the model are the same as before, see descriptions around Equation (11.16). However, the number of components, K, is not fixed.

The hyperparameter values are selected as follows. First $\alpha = 1$, $\boldsymbol{\mu}_0 = \bar{\mathbf{y}}$, $\boldsymbol{\Sigma}_0 = \mathbf{S}_y/2$, $\rho_0 = 6$ and $\mathbf{R}_0^{-1} = 5\mathbf{I}$, $\mathbf{H}_{0ykm} = \mathbf{I}$ and $\mathbf{H}_{0\omega kl} = \mathbf{I}$ are selected for each k; $m = 1, \ldots, p$, and $l = 1, \ldots, q_1$. Moreover, $\{\alpha_{0\epsilon k}, \beta_{0\epsilon k}\}$ and $\{\alpha_{0\delta k}, \beta_{0\delta k}\}$ are selected such that the means and standard deviations of the prior distributions associated with ψ_{km} and $\psi_{\delta kl}$ are equal to 5.0. Finally, we take $\boldsymbol{\Lambda}_{0km} = \tilde{\boldsymbol{\Lambda}}_{0km}$ and $\boldsymbol{\Lambda}_{0\omega kl} = \tilde{\boldsymbol{\Lambda}}_{0\omega kl}$ for all k, m and l, where $\tilde{\boldsymbol{\Lambda}}_{0km}$ and $\tilde{\boldsymbol{\Lambda}}_{0\omega kl}$ are the corresponding Bayesian estimates obtained through a single component model with noninformative prior distributions. Again, for each $t_{(s)}$, the convergence of the Gibbs sampler algorithm is monitored by parallel sequences of the generated observations from very different starting values. We observe that the algorithm converged quickly within 200 iterations. Since the convergence behaviors are similar to those in the simulation study, they are not presented. A total of $J = 1000$ additional observations are collected after a burn-in phase of 200 iterations for computing $\bar{U}_{(s)}$ in Equation (11.19), and then the logarithms of Bayes factors are estimated via Equation (11.18), using 20 fixed grids in [0,1]. Let M_k denote the mixture model with k components, and the estimated logarithms of Bayes factors are equal to: $\widehat{\log B_{21}} = 75.055$, $\widehat{\log B_{32}} = 4.381$, $\widehat{\log B_{43}} = -0.824$ and $\widehat{\log B_{53}} = -1.395$. According to the criterion of the logarithm Bayes factor, the one-component model is significantly worse than the two-component model which is significantly worse than the three-component model; while the three-component model is better than the four-component and five-component models. Hence, it can be concluded that a mixture model with three components should be chosen. Although the two-component model suggested in Section 11.5.2 is a plausible model, it does not give as strong a support of evidence as the three-component model.

In estimation, based on the MCMC samples simulated by the random permutation sampler, we find $\mu_{1,1} < \mu_{2,1}$ is a suitable identifiability constraint. Bayesian estimates of the selected three-component mixture model obtained under the selected constraint are presented in Table 11.10, together with the corresponding standard error estimates. For parameters directly associated with manifest variables y_1 to y_5, we from this table that their Bayesian estimates under component two are close to those under component three, but these estimates are quite different from those under component one. In contrast, for parameters directly associated with manifest variables y_6 to y_8, estimates under component one are close to those under component three, but these estimates are quite different from those under component two. Hence, it is reasonable to select a three-component model for this data set. For completeness, estimate the separations of these components and find that they are equal to $d_{12} = 2.257$, $d_{13} = 2.590$ and $d_{23} = 2.473$. These results indicate that the proposed procedure is able to select the appropriate three-component model whose components are not well separated.

Table 11.10 Bayesian estimates of parameters and standard errors for the selected model with three components in analyzing the ICPSR data set.

	Component 1		Component 2		Component 3	
PAR	EST	SE	EST	SE	EST	SE
π_k	0.51	0.03	0.23	0.03	0.26	0.03
$\mu_{k,1}$	6.75	0.13	8.05	0.15	8.25	0.15
$\mu_{k,2}$	5.95	0.12	7.53	0.21	8.63	0.17
$\mu_{k,3}$	5.76	0.18	7.67	0.18	7.87	0.14
$\mu_{k,4}$	7.74	0.13	8.65	0.12	8.77	0.12
$\mu_{k,5}$	7.05	0.12	8.06	0.12	8.04	0.11
$\mu_{k,6}$	5.50	0.25	2.70	0.18	5.20	0.15
$\mu_{k,7}$	4.12	0.23	2.60	0.16	4.35	0.14
$\mu_{k,8}$	5.66	0.24	3.08	0.23	5.94	0.15
$\lambda_{k,21}$	0.31	0.12	1.10	0.21	0.66	0.08
$\lambda_{k,42}$	1.38	0.18	0.84	0.13	0.87	0.16
$\lambda_{k,52}$	1.67	0.19	0.92	0.15	1.10	0.18
$\lambda_{k,73}$	2.15	0.31	0.98	0.09	1.94	0.22
$\lambda_{k,83}$	0.88	0.22	0.97	0.11	0.64	0.20
$\gamma_{k,1}$	0.62	0.16	0.52	0.15	0.69	0.14
$\gamma_{k,2}$	0.01	0.11	-0.37	0.12	-0.12	0.14
$\phi_{k,11}$	1.07	0.21	1.30	0.33	0.81	0.20
$\phi_{k,21}$	-0.13	0.13	-0.59	0.20	0.07	0.07
$\phi_{k,22}$	1.10	0.49	1.45	0.38	1.57	0.21
ψ_{k1}	1.08	0.12	0.87	0.19	0.56	0.35
ψ_{k2}	6.79	0.17	2.02	0.45	0.83	0.46
ψ_{k3}	4.75	0.38	1.71	0.40	1.17	0.37
ψ_{k4}	2.54	0.13	0.45	0.08	0.69	0.27
ψ_{k5}	1.11	0.16	0.55	0.09	0.71	0.23
ψ_{k6}	5.75	0.55	0.55	0.13	4.71	0.46
ψ_{k7}	1.36	0.35	0.60	0.11	1.13	0.47
ψ_{k8}	6.10	0.52	1.05	0.50	4.52	0.47
$\psi_{\delta k}$	3.80	0.13	0.63	0.15	0.68	0.51

APPENDIX 11.1: THE PERMUTATION SAMPLER

Let $\boldsymbol{\psi} = (\boldsymbol{\Omega}, \mathbf{W}, \boldsymbol{\theta})$, the permutation sampler for generating $\boldsymbol{\psi}$ from the posterior $p(\boldsymbol{\psi}|\mathbf{Y})$ is implemented as follows.

(1) First generate $\tilde{\boldsymbol{\psi}}$ from the unconstrained posterior $p(\boldsymbol{\psi}|\mathbf{Y})$ using standard Gibbs sampling steps.

(2) Select some permutation $\rho(1), \ldots, \rho(K)$ of the current labelling of the states and define $\boldsymbol{\psi} = \rho(\tilde{\boldsymbol{\psi}})$ from $\tilde{\boldsymbol{\psi}}$ by reordering the labelling through this permutation, $(\boldsymbol{\theta}_1, \ldots, \boldsymbol{\theta}_K) := (\boldsymbol{\theta}_{\rho(1)}, \ldots, \boldsymbol{\theta}_{\rho(K)})$ and $\mathbf{W} = (w_1, \ldots, w_n) := (\rho(w_1), \ldots, \rho(w_n))$.

One application of permutation sampling is the random permutation sampler, where each sweep of the MCMC chain is concluded by relabelling the states through a random permutation of $\{1, \cdots, K\}$. This method delivers a sample that explores the whole unconstrained parameter space and jumps between the various labelling subspaces in a balanced fashion. Another application of the permutation sampler is the permutation sampling under identifiability constraints. A common way to include an identifiability constraint is to use a permutation sampler, where the permutation is selected in such a way that the identifiability constraint is fulfilled.

APPENDIX 11.2: SEARCHING FOR IDENTIFIABILITY CONSTRAINTS

For $k = 1, \ldots, K$, let $\boldsymbol{\theta}_k$ denote the parameter vector corresponding to the kth component. According to the suggestion by Frühwirth-Schnatter (2001), the MCMC output of the random permutation sampler can be explored to find a suitable identifiability constraint. It is sufficient to consider only the parameters in $\boldsymbol{\theta}_1$, because a balanced sample from the unconstrained posterior will contain the same information for all parameters in $\boldsymbol{\theta}_k$ with $k > 1$. As the random permutation sampler jumps between the various labelling subspaces, part of the values sampled for $\boldsymbol{\theta}_1$ will belong to the first state, part will belong to the second state, and so on. To differ for various states, it is most useful to consider bivariate scatterplots of $\boldsymbol{\theta}_{1,i}$ versus $\boldsymbol{\theta}_{1,l}$ for possible combinations of i and l. Jumping between the labelling subspaces produces groups in these scatterplots that correspond to different states. By describing the difference between the various groups geometrically, identification of a unique labelling subspace through conditions on the state-specific parameters is attempted. If the values sampled for a certain component of $\boldsymbol{\theta}$ differ markedly between the groups when jumping between the labelling subspaces, then an order condition on this component could be used to separate the labelling subspaces, while if the values sampled for a certain component of $\boldsymbol{\theta}$ hardly differ between the states when jumping between the labelling subspaces, then this component will be a poor candidate for separating the labelling subspaces.

REFERENCES

Arminger, G. and Stein, P. (1997) Finite mixtures of covariance structure models with regressors. *Sociological Methods and Research*, **26**, 148–182.

Arminger, G., Stein, P. and Wittenberg (1999) Mixtures of conditional mean- and covariance-structure models. *Psychometrika*, **64**, 475–494.

Berger, J. O. (1985) *Statistical Decision Theory and Bayesian Analysis*. New York: Springer-Verlag.

Besag, J., Green, P., Higdon, D. and Mengersen, K. (1995) Bayesian computation and stochastic systems. *Statistical Science*, **10**, 3–66.

Diebolt, J. and Robert, C. P. (1994) Estimation of finite mixture distributions through Bayesian sampling. *Journal of the Royal Statistical Society, Series B*, **56**, 363–375.

Dolan, C. V. and van der Maas, J. J. L. (1998) Fitting multivariate normal mixtures subject to structural equation modeling. *Psychometrika*, **63**, 227–253.

Feng, Z. D. and McCulloch, C. E. (1996) Using bootstrap likelihood ratios in finite mixture models. *Journal of the Royal Statistical Society, Series B*, **58**, 609–617.

Frühwirth-Schnatter, S. (2001) Markov chain Monte Carlo estimation of classical and dynamic switching and mixture models. *Journal of the American Statistical Association*, **96**, 194–208.

Gelman, A. and Meng, X. L. (1998) Simulating normalizing constants: from importance sampling to bridge sampling to path sampling. *Statistical Science*, **13**, 163–185.

Gelman, A. and Rubin, D. B. (1992) Inference from iterative simulation using multiple sequences. *Statistical Science*, **7**, 457–472.

Geman, S. and Geman, D. (1984) Stochastic relaxation, Gibbs distribution and the Bayesian restoration of images. *IEEE Transactions on Pattern Analysis and Machine Intelligence*, **6**, 721–741.

Gilks, W. R., Richardson, S. and Spiegelhalter, D. J. (1996) Introducing Markov chain Monte Carlo. In W.R. Gilks, S. Richardson and D. J. Spiegelhalter (eds), *Markov Chain Monte Carlo in Practice*, pp. 1–19. London: Chapman and Hall.

Green, P. J. (1995) Reversible jump Markov chain Monte Carlo computation and Bayesian model determination. *Biometrika*, **82**, 711–732.

Hathaway, R. J. (1985) A constrained formulation of maximum-likelihood estimation for normal mixture distributions. *The Annals of Statistics*, **13**, 795–800.

Jedidi, K., Jagpal, H. S. and DeSarbo, W. S. (1997a) Finite-mixture structural equation models for response-based segmentation and unobserved heterogeneity. *Marketing Science*, **16**, 39–59.

Jedidi, K., Jagpal, H. S. and DeSarbo, W. S. (1997b) STEMM: a general finite mixture structural equation model. *Journal of Classification*, **14**, 23–50.

Jöreskog, K. G. and Sörbom, D. (1996) *LISREL 8: Structural Equation Modeling with the SIMPLIS Command Language*. Hove and London: Scientific Software International.

Kass, R. E. and Raftery, A. E. (1995) Bayes factors. *Journal of the American Statistical Association*, **90**, 773–795.

Lee, S. Y. and Song, X. Y. (2002) Bayesian selection on the number of factors in a factor analysis model. *Behaviormetrika*, **29**, 23–39.

Lee, S. Y. and Song, X. Y. (2003a) Bayesian model selection for mixtures of structural equation models with an unknown number of components. *British Journal of Mathematical and Statistical Psychology*, **56**, 145–65.

Lee, S. Y. and Song, X. Y. (2003b) Maximum likelihood estimation and model comparison for mixtures of structural equation models with ignorable missing data. *Journal of Classification*, **20**, 221–255.

Lindley, D. V. and Smith, A. F. M. (1972) Bayes estimates for the linear model (with discussion). *Journal of the Royal Statistical Society, Series B*, **34**, 1–42.

Lindsay, B. G. and Basak, P. (1993) Multivariate normal mixtures: a fast consistent method of moments. *Journal of the American Statistical Association*, **88**, 468–476.

McLachlan, G. J. (1987) On bootstrapping the likelihood ratio test statistics for the number of components in a normal mixture. *Applied Statistics*, **36**, 318–324.

Pettit, L. I. and Smith, A. F. M. (1985) Outliers and influential observations in linear models. In J. M. Bernardo *et al.* (eds), *Bayesian Statistics*, **2** pp. 473–494. Amsterdam: North-Holland.

Redner, R. A. and Walker, H. F. (1984) Mixture densities, maximum likelihood and the EM algorithm. *SIAM Review*, **26**, 195–239.

Richardson, S. and Green, P. J. (1997) On Bayesian analysis of mixtures with an unknown number of components (with discussion). *Journal of the Royal Statistical Society, Series B*, **59**, 731–792.

Robert, C. (1996) Mixtures of distributions: Inference and estimation. In W. R. Gilks, S. Richardson and D. J. Spiegelhalter (eds), *Practical Markov Chain Monte Carlo*, pp. 163–188. London: Chapman and Hall.

Roeder, K. and Wasserman, L. (1997) Practical Bayesian density estimation using mixtures of normals. *Journal of the American Statistical Association*, **92**, 894–902.

Spiegelhalter, D. J., Thomas, A., Best, N. G. and Lunn, D. (2003) *WinBUGS User Manual. Version 1.4.* Cambridge, England: MRC Biostatistics Unit.

Titterington, D. M., Smith, A. F. M. and Markov, U. E. (1985) *Statistical Analysis of Finite Mixture Distributions.* Chichester: John Wiley & Sons, Ltd.

World Values Study Group (1994) World Values Survey 1981–1984 and 1990–1993. ICPSR version. Ann Arbor, MI: Institute for Social Research (producer). Ann Arbor, MI: Inter-university Consortium for Political and Social Research (distributor).

Yung, Y. F. (1997) Finite mixtures in confirmatory factor-analysis models. *Psychometrika*, **62**, 297–330.

Zhu, H. T. and Lee, S. Y. (2001) A Bayesian analysis of finite mixtures in the LISREL model. *Psychometrika*, **66**, 133–152.

12

Structural Equation Models with Missing Data

12.1 INTRODUCTION

Missing data are very common in substantive research. For example, respondents in a household survey may refuse to report income, individuals in an opinion survey may refuse to express their attitudes toward some sensitive or embarrassing questions. Moreover, it is common to have missing individuals at one or more time points in longitudinal studies, or some results are missing because of mechanical breakdowns in a psychological experiment. Clearly, the effect of missing data needs to be taken into account for better statistical inferences. In statistics, analysis of missing data has a long history, it continuously receives a lot of attention and is still an active area of research (see, for example, Afifi and Elashoff (1969), Little and Rubin (1987), and Ibrahim, Chen and Lipsitz (2001), among many others).

In structural equation modeling, much attention has been given to the analysis of models in the presence of missing data. Some historical approaches to missing data problems – such as the listwise deletion, and filling in the missing values by mean estimates or least square estimates – although not without value, tend to have an ad hoc character and will encounter serious difficulty in achieving theoretical properties. More statistically rigorous methods have been proposed. The earlier contributions were mainly focused on standard SEMs and were developed through the multigroup analysis of a covariance structure analysis (CSA) approach. The basic technique treated observations that belong to the same missing pattern as an independent group, obtained the sample

Structural Equation Modeling: A Bayesian Approach S-Y. Lee

covariance matrix of each independent group, and then analyzed the sample covariance matrices through the multigroup methods in the CSA approach. This approach would encounter theoretical difficulty if some missing patterns just have a small number of observations so that the corresponding sample covariance matrices may be singular. It is computationally tedious if the number of missing patterns is large. Moreover, it is difficult to extend this approach to handle more complicated SEMs or missing data that are missing with a nonignorable missing mechanism. Motivated by the deficiency of the multi-group CSA approach, methods that are focused on the raw observations have been developed. For example, in analyzing standard SEMs, Arbuckle (1996) proposed the full information ML method which maximizes the case wise likelihood of the observed continuous data. Recently, through the utilization of the data augmentation idea and MCMC methods in statistical computing, Bayesian methods for analyzing missing data in the context of more complex SEMs have been developed. For instance, Song and Lee (2002) and Lee and Song (2004) developed Bayesian methods for analyzing linear and nonlinear SEMs with mixed continuous and ordered categorical variables. While the above mentioned contributions in SEM are established on the basis of ignorable missing data that are missing at random (MAR), Lee and Tang (2006) developed Bayesian methods for analyzing nonlinear SEMs with nonignorable missing data.

The main objective of this chapter is to introduce the Bayesian approach for analyzing SEMs with ignorable missing data that are missing at random (MAR), and nonignorable missing data that are missing according to a nonignorable missing mechanism. As described in Little and Rubin (1987), if the probability of response depends on the fully observed data but not on the missing data, the corresponding missing data can be regarded as MAR. For nonignorable missing data, the probability of response depends not only on the observed data but also on the missing data, according to a nonignorable missing model. In the development of the Bayesian approach, we again emphasize the useful strategy that combines the idea of data augmentation and application of MCMC methods. We will show that Bayesian methods for analyzing complex SEMs with fully observed data can be extended to handle missing data with a large number of missing patterns without much theoretical and practical difficulty. In this chapter, we regard observations with missing entries as partially observed data.

We will present a general Bayesian framework for analyzing general SEMs with missing data that are MAR, which includes Bayesian estimation, and model comparison via the Bayes factor. This general framework will be applied to nonlinear SEMs with missing continuous and ordered categorical data, and to mixtures of SEMs. We assess the effect of missing data on estimation and show that the proposed Bayesian approach that uses both the fully and partially observed data produces accurate estimates. Moreover, we investigate the impact of ignoring the partially observed data on model selection. In the analysis of mixtures of SEMs, we give an example to illustrate that ignoring the partially observed data will give different estimates and model comparison results. Then, we present Bayesian methods to analyze nonlinear SEMs with missing data that are missing with a nonignorable mechanism. Finally, we demonstrate the use of the software WinBUGS to obtain the Bayesian solutions.

12.2 A GENERAL FRAMEWORK FOR SEMs WITH MISSING DATA THAT ARE MAR

Consider a general SEM of interest, which could be the standard model, a nonlinear model, a multilevel model or a finite mixture of SEMs. Let $\mathbf{V} = (\mathbf{v}_1, \ldots, \mathbf{v}_n)$ be a matrix of random vectors with $\mathbf{v}_i = (\mathbf{x}_i^T, \mathbf{y}_i^T)^T$, in which $\mathbf{x}_i$ and $\mathbf{y}_i$ are vectors of manifest continuous variables whose exact measurements are observable and unobservable, respectively. Let $\mathbf{Z} = (\mathbf{z}_1, \ldots, \mathbf{z}_n)$ be observable ordered categorical data (or dichotomous data) that correspond to $\mathbf{Y} = (\mathbf{y}_1, \ldots, \mathbf{y}_n)$. Suppose $\mathbf{v}_i$ follows a SEM with real and/or augmented latent variables $\boldsymbol{\Omega} = (\boldsymbol{\Omega}_1, \boldsymbol{\Omega}_2)$, where $\boldsymbol{\Omega}_1 = (\boldsymbol{\omega}_{11}, \ldots, \boldsymbol{\omega}_{1n})$ contains latent variables in the structural equation model such as the factor scores, and $\boldsymbol{\Omega}_2 = (\boldsymbol{\omega}_{21}, \ldots, \boldsymbol{\omega}_{2n})$ contains external latent quantities such as the allocation variables in a finite mixture of SEMs.

To deal with the missing data that are MAR, let $\mathbf{X}_{\text{obs}}$ and $\mathbf{X}_{\text{mis}}$ be the observed and missing data sets corresponding to the continuous data $\mathbf{X} = (\mathbf{x}_1, \ldots, \mathbf{x}_n)$; $\mathbf{Z}_{\text{obs}}$ and $\mathbf{Z}_{\text{mis}}$ be the observed and missing data sets corresponding to the ordered categorical data $\mathbf{Z}$; $\mathbf{Y}_{\text{obs}}$ and $\mathbf{Y}_{\text{mis}}$ be the hypothetical observed and missing data sets of $\mathbf{Y}$ corresponding $\mathbf{Z}_{\text{obs}}$ and $\mathbf{Z}_{\text{mis}}$, respectively. Moreover, we let $\mathbf{V}_{\text{obs}} = \{\mathbf{X}_{\text{obs}}, \mathbf{Y}_{\text{obs}}\}$ and $\mathbf{V}_{\text{mis}} = \{\mathbf{X}_{\text{mis}}, \mathbf{Y}_{\text{mis}}\}$. The main goal is to develop Bayesian methods for estimating the unknown parameter vector $\boldsymbol{\theta}^*$ of the model, and comparing competitive models on the basis of the observed data $\{\mathbf{X}_{\text{obs}}, \mathbf{Z}_{\text{obs}}\}$

We first consider the Bayesian estimation by investigating the following posterior distribution of $\boldsymbol{\theta}^*$ with given $\mathbf{X}_{\text{obs}}$ and $\mathbf{Z}_{\text{obs}}$,

$$p(\boldsymbol{\theta}^* | \mathbf{X}_{\text{obs}}, \mathbf{Z}_{\text{obs}}) \propto p(\mathbf{X}_{\text{obs}}, \mathbf{Z}_{\text{obs}} | \boldsymbol{\theta}^*) p(\boldsymbol{\theta}^*),$$

where $p(\mathbf{X}_{\text{obs}}, \mathbf{Z}_{\text{obs}} | \boldsymbol{\theta}^*)$ is the observed data likelihood and $p(\boldsymbol{\theta}^*)$ is the prior density of $\boldsymbol{\theta}^*$. Owing to the complexities of the model and the data, $p(\boldsymbol{\theta}^* | \mathbf{X}_{\text{obs}}, \mathbf{Z}_{\text{obs}})$ is usually very complicated. In a similar way to the Bayesian analyses discussed in previous chapters in solving this problem, we utilize the idea of data augmentation (Tanner and Wong, 1987), and then perform the posterior simulation with the MCMC methods. For the current situation, it is natural to augment the observed data $\{\mathbf{X}_{\text{obs}}, \mathbf{Z}_{\text{obs}}\}$ with the latent and missing quantities $\{\boldsymbol{\Omega}, \mathbf{X}_{\text{mis}}, \mathbf{Y}_{\text{mis}}, \mathbf{Y}_{\text{obs}}\} = \{\boldsymbol{\Omega}, \mathbf{V}_{\text{mis}}, \mathbf{Y}_{\text{obs}}\}$. As $\mathbf{Y}_{\text{mis}}$ is included, it is not necessary to augment with the corresponding $\mathbf{Z}_{\text{mis}}$. A sufficiently large number of random observations will be simulated from the joint posterior distribution $[\boldsymbol{\theta}^*, \boldsymbol{\Omega}, \mathbf{V}_{\text{mis}}, \mathbf{Y}_{\text{obs}} | \mathbf{X}_{\text{obs}}, \mathbf{Z}_{\text{obs}}]$, so that its distribution can be approximated adequately from the empirical distribution of the generated observations. This task can be similarly completed by a hybrid algorithm that combines the Gibbs sampler (Geman and Geman, 1984), and the MH algorithm (Metropolis *et al.*, 1953; Hastings, 1970) as before. Bayesian estimates of parameters in $\boldsymbol{\theta}^*$ and the standard error estimates can be obtained through the sample of simulated observations, $\{(\boldsymbol{\theta}^{*(t)}, \boldsymbol{\Omega}^{(t)}, \mathbf{V}_{\text{mis}}^{(t)}, \mathbf{Y}_{\text{obs}}^{(t)}), t = 1, \ldots, T^*\}$, that are drawn from the joint posterior distribution.

The following conditional distributions: $p(\boldsymbol{\theta}^*|\boldsymbol{\Omega}, \mathbf{V}_{\text{mis}}, \mathbf{Y}_{\text{obs}}, \mathbf{X}_{\text{obs}}, \mathbf{Z}_{\text{obs}})$, $p(\boldsymbol{\Omega}|\boldsymbol{\theta}^*, \mathbf{V}_{\text{mis}}, \mathbf{Y}_{\text{obs}}, \mathbf{X}_{\text{obs}}, \mathbf{Z}_{\text{obs}})$, $p(\mathbf{V}_{\text{mis}}|\boldsymbol{\theta}^*, \boldsymbol{\Omega}, \mathbf{Y}_{\text{obs}}, \mathbf{X}_{\text{obs}}, \mathbf{Z}_{\text{obs}})$ and $p(\mathbf{Y}_{\text{obs}}|\boldsymbol{\theta}^*, \boldsymbol{\Omega}, \mathbf{V}_{\text{mis}}, \mathbf{X}_{\text{obs}}, \mathbf{Z}_{\text{obs}})$, are required in the implementation of the Gibbs sampler. With given $\mathbf{Y} = (\mathbf{Y}_{\text{mis}}, \mathbf{Y}_{\text{obs}})$ and $\mathbf{V} = (\mathbf{V}_{\text{mis}}, \mathbf{V}_{\text{obs}})$, the conditional distributions corresponding to $\boldsymbol{\theta}^*$ and $\boldsymbol{\Omega}$ can be derived in exactly the same way as in the situation with fully observed data. Similarly, with $\mathbf{V}_{\text{mis}}$, $\mathbf{X}_{\text{obs}}$ and $\mathbf{Z}_{\text{obs}}$ given, the conditional distribution corresponding to $\mathbf{Y}_{\text{obs}}$ can be derived as in previous chapters. We only need to pay more attention to the conditional distribution corresponding to $\mathbf{V}_{\text{mis}}$. Under the usual assumption that $\mathbf{v}_1, \ldots, \mathbf{v}_n$ are mutually independent, it follows that:

$$p(\mathbf{V}_{\text{mis}}|\boldsymbol{\theta}^*, \boldsymbol{\Omega}, \mathbf{Y}_{\text{obs}}, \mathbf{X}_{\text{obs}}, \mathbf{Z}_{\text{obs}}) = \prod_{i=1}^{n} p(\mathbf{v}_{\text{mis}i}|\boldsymbol{\theta}^*, \boldsymbol{\Omega}, \mathbf{y}_{\text{obs}i}, \mathbf{x}_{\text{obs}i}, \mathbf{z}_{\text{obs}i}) \tag{12.1}$$

where $\mathbf{v}_{\text{mis}i} = (\mathbf{x}_{\text{mis}i}, \mathbf{y}_{\text{mis}i})$ is the ith data point in the random sample of size n. The individual $\mathbf{v}_{\text{mis}i}$ can be separately simulated from the conditional distribution in Equation (12.1), hence the simulation is simple and is not affected by the missing patterns. Moreover, it will be seen in the subsections of this chapter that even for complex SEMs, the conditional distribution $p(\mathbf{v}_{\text{mis}i}|\boldsymbol{\theta}^*, \boldsymbol{\Omega}, \mathbf{y}_{\text{obs}i}, \mathbf{x}_{\text{obs}i}, \mathbf{z}_{\text{obs}i})$ is usually simple. Consequently, the computational burden for simulating $\mathbf{V}_{\text{mis}}$ from its conditional distribution is light. Note that the statistical properties of the Bayesian estimates obtained do not depend on the sample sizes within the missing patterns.

To address the model comparison problem, we let M_0 and M_1 be two competing models and consider the computation of the following Bayes factor

$$B_{10} = \frac{p(\mathbf{X}_{\text{obs}}, \mathbf{Z}_{\text{obs}}|M_1)}{p(\mathbf{X}_{\text{obs}}, \mathbf{Z}_{\text{obs}}|M_0)}.$$

The $\log B_{10}$ can be similarly computed by path sampling (Gelman and Meng, 1998). Consider the following class of densities defined by a continuous parameter t in $[0, 1]$:

$$p(\boldsymbol{\theta}^*, \boldsymbol{\Omega}, \mathbf{V}_{\text{mis}}, \mathbf{Y}_{\text{obs}}|\mathbf{X}_{\text{obs}}, \mathbf{Z}_{\text{obs}}, t) = p(\boldsymbol{\theta}^*, \boldsymbol{\Omega}, \mathbf{V}_{\text{mis}}, \mathbf{Y}_{\text{obs}}, \mathbf{X}_{\text{obs}}, \mathbf{Z}_{\text{obs}}|t)/z(t).$$

where $z(t) = p(\mathbf{X}_{\text{obs}}, \mathbf{Z}_{\text{obs}}|t)$. Let t be a parameter linking the competing models M_0 and M_1 such that $z(1) = p(\mathbf{X}_{\text{obs}}, \mathbf{Z}_{\text{obs}}|t = 1) = p(\mathbf{X}_{\text{obs}}, \mathbf{Z}_{\text{obs}}|M_1)$ and $z(0) = p(\mathbf{X}_{\text{obs}}, \mathbf{Z}_{\text{obs}}|t = 0) = p(\mathbf{X}_{\text{obs}}, \mathbf{Z}_{\text{obs}}|M_0)$, then $B_{10} = z(1)/z(0)$. Based on the reasoning given in Chapter 5, it can be shown in the same way that the logarithm of Bayes factor can be estimated as follows:

$$\widehat{\log B_{10}} = \frac{1}{2}\sum_{s=0}^{S}(t_{(s+1)} - t_{(s)})(\overline{U}_{(s+1)} + \overline{U}_{(s)}), \tag{12.2}$$

where $t_{(0)} = 0 < t_{(0)}, \ldots, t_{(S)} < t_{(S+1)} = 1$, which are fixed grids in $[0, 1]$, and

$$\overline{U}_{(s)} = J^{-1}\sum_{j=1}^{J} U(\boldsymbol{\theta}^{*(j)}, \boldsymbol{\Omega}^{(j)}, \mathbf{V}_{\text{mis}}^{(j)}, \mathbf{Y}_{\text{obs}}^{(j)}, \mathbf{X}_{\text{obs}}, \mathbf{Z}_{\text{obs}}, t_{(s)}), \tag{12.3}$$

in which $\{(\boldsymbol{\theta}^{*(j)}, \boldsymbol{\Omega}^{(j)}, \mathbf{V}_{\text{mis}}^{(j)}, \mathbf{Y}_{\text{obs}}^{(j)}), j = 1, \ldots, J\}$ is a sample of observations simulated from $p(\boldsymbol{\theta}^*, \boldsymbol{\Omega}, \mathbf{V}_{\text{mis}}, \mathbf{Y}_{\text{obs}}|\mathbf{X}_{\text{obs}}, \mathbf{Z}_{\text{obs}}, t_{(s)})$ and

$$U(\boldsymbol{\theta}^*, \boldsymbol{\Omega}, \mathbf{V}_{\text{mis}}, \mathbf{Y}_{\text{obs}}, \mathbf{X}_{\text{obs}}, \mathbf{Z}_{\text{obs}}, t) = \frac{\mathrm{d}}{\mathrm{d}t}\log p(\boldsymbol{\Omega}, \mathbf{V}_{\text{mis}}, \mathbf{Y}_{\text{obs}}, \mathbf{X}_{\text{obs}}, \mathbf{Z}_{\text{obs}}|\boldsymbol{\theta}^*, t),$$

where $p(\boldsymbol{\Omega}, \mathbf{V}_{\text{mis}}, \mathbf{Y}_{\text{obs}}, \mathbf{X}_{\text{obs}}, \mathbf{Z}_{\text{obs}}|\boldsymbol{\theta}^*, t)$ is the complete data likelihood. As we have a program to simulate $(\boldsymbol{\theta}^*, \boldsymbol{\Omega}, \mathbf{V}_{\text{mis}}, \mathbf{Y}_{\text{obs}})$ in the Bayesian estimation, the implementation of the path sampling procedure for computing the logarithm Bayes factor is straightforward. Basically, by focusing on the raw observations, and conducting the Bayesian approach with data augmentation and MCMC methods, the methodologies in analyzing SEMs with fully observed data can be generalized to handle missing data that are MAR without too many new derivations nor too much additional programming work. It just needs to deal with one additional simple component in the Gibbs sampler (see Equation (12.1), and concrete examples in the following sections).

12.3 NONLINEAR SEM WITH MISSING CONTINUOUS AND ORDERED CATEGORICAL DATA

We first consider the analysis of a nonlinear SEM with missing continuous and ordered categorical data. As this model is rather general, the results presented here can be applied to many special cases, such as a linear SEM with missing continuous and ordered categorical data, or a nonlinear SEM with missing continuous data. This nonlinear SEM is the same as that defined in Section 8.3; however, for completeness it is briefly described here.

Consider $p \times 1$ random vectors $\mathbf{v}_1, \ldots, \mathbf{v}_n$, which are identically and independently distributed, and satisfy the following measurement equation:

$$\mathbf{v}_i = \boldsymbol{\mu} + \boldsymbol{\Lambda}\boldsymbol{\omega}_i + \boldsymbol{\epsilon}_i, \quad i = 1, \ldots, n, \tag{12.4}$$

where $\boldsymbol{\mu}$ is a vector of intercepts, $\boldsymbol{\Lambda}$ is an unknown parameter matrix, $\boldsymbol{\omega}_i$ is a $q \times 1$ vector of latent variables, $\boldsymbol{\epsilon}_i$ is a vector of error measurements with distribution $N(\mathbf{0}, \boldsymbol{\Psi}_\epsilon)$, $\boldsymbol{\Psi}_\epsilon = \text{diag}(\psi_{\epsilon 1}, \ldots, \psi_{\epsilon p})$ and $\boldsymbol{\omega}_i$ and $\boldsymbol{\epsilon}_i$ are independent. The latent vector $\boldsymbol{\omega}_i$ is partitioned into subvectors $(\boldsymbol{\eta}_i^T, \boldsymbol{\xi}_i^T)^T$ which satisfy the following nonlinear structural equation:

$$\boldsymbol{\eta}_i = \boldsymbol{\Pi}\boldsymbol{\eta}_i + \boldsymbol{\Gamma}\mathbf{H}(\boldsymbol{\xi}_i) + \boldsymbol{\delta}_i, \tag{12.5}$$

where $\boldsymbol{\eta}_i$ and $\boldsymbol{\xi}_i$ are $q_1 \times 1$ and $q_2 \times 1$ latent random vectors, respectively, $\mathbf{H}(\boldsymbol{\xi}) = (h_1(\boldsymbol{\xi}), \ldots, h_m(\boldsymbol{\xi}))^T$ is a vector-valued function with differentiable functions $h_1, \ldots, h_m$ and $m \geq q_2$, and $\boldsymbol{\Pi}$ and $\boldsymbol{\Gamma}$ are unknown parameter matrices. Moreover, it is assumed that $|\mathbf{I}_{q_1} - \boldsymbol{\Pi}|$ is independent of elements in $\boldsymbol{\Pi}$, $\boldsymbol{\xi}_i$ and $\boldsymbol{\delta}_i$ are independently distributed as $N[\mathbf{0}, \boldsymbol{\Phi}]$ and $N[\mathbf{0}, \boldsymbol{\Psi}_\delta]$, respectively; and $\boldsymbol{\Psi}_\delta = \text{diag}(\psi_{\delta 1}, \ldots, \psi_{\delta q_1})$. Let $\boldsymbol{\Lambda}_\omega = (\boldsymbol{\Pi}, \boldsymbol{\Gamma})$ and $\mathbf{G}(\boldsymbol{\omega}_i) = (\boldsymbol{\eta}_i^T, \mathbf{H}(\boldsymbol{\xi}_i)^T)^T$, then the above structural equation can be written as:

$$\boldsymbol{\eta}_i = \boldsymbol{\Lambda}_\omega \mathbf{G}(\boldsymbol{\omega}_i) + \boldsymbol{\delta}_i. \tag{12.6}$$

Without loss of generality, suppose $\mathbf{v}_i = (\mathbf{x}_i^T, \mathbf{y}_i^T)^T$, where $\mathbf{x}_i$ is an $r \times 1$ vector of observed measurements and $\mathbf{y}_i$ is an $s \times 1$ vector of unobserved measurements. The information of $\mathbf{y} = (y_1, \ldots, y_s)^T$ is given by an observable ordinal categorical vector $\mathbf{z}$ such that

$$\mathbf{z} = \begin{pmatrix} z_1 \\ \vdots \\ z_s \end{pmatrix} \quad \text{if} \quad \begin{matrix} \alpha_{1,z_1} < y_1 \leq \alpha_{1,z_1+1} \\ \vdots \\ \alpha_{s,z_s} < y_s \leq \alpha_{s,z_s+1} \end{matrix}, \tag{12.7}$$

where z_k is an integral value in the set $\{0, 1, \ldots, b_k\}$ for $k = 1, \ldots, s$, $\alpha_{k,0} = -\infty$ and $\alpha_{k,b_k+1} = \infty$. Hence, for the kth variable, there are $b_k + 1$ categories which are defined by unknown threshold parameters $\alpha_{k,j}$. Here, we assume the model is identified after imposing appropriate conditions as describe in Chapter 6.

To deal with the missing data problem, let $\mathbf{x}_i = \{\mathbf{x}_{\text{obs}i}, \mathbf{x}_{\text{mis}i}\}$ and $\mathbf{z}_i = \{\mathbf{z}_{\text{obs}i}, \mathbf{z}_{\text{mis}i}\}$ where $\mathbf{x}_{\text{obs}i}$ and $\mathbf{z}_{\text{obs}i}$ represent the observed data, while $\mathbf{x}_{\text{mis}i}$ and $\mathbf{z}_{\text{mis}i}$ represent the missing data. For a fully observed $\{\mathbf{x}_i, \mathbf{z}_i\}$ data point, $\mathbf{x}_{\text{mis}i}$ and $\mathbf{z}_{\text{mis}i}$ are empty. Let $\mathbf{y}_i = \{\mathbf{y}_{\text{obs}i}, \mathbf{y}_{\text{mis}i}\}$ represent the latent continuous measurements, where $\mathbf{y}_{\text{obs}i}$ and $\mathbf{y}_{\text{mis}i}$ are respectively corresponding to $\mathbf{z}_{\text{obs}i}$ and $\mathbf{z}_{\text{mis}i}$.

Let $\mathbf{v}_{\text{obs}i} = \{\mathbf{x}_{\text{obs}i}, \mathbf{y}_{\text{obs}i}\}$ and $\mathbf{v}_{\text{mis}i} = \{\mathbf{x}_{\text{mis}i}, \mathbf{y}_{\text{mis}i}\}$, thus $\mathbf{v}_i = \{\mathbf{v}_{\text{obs}i}, \mathbf{v}_{\text{mis}i}\}$. If there are no missing data in the whole data set, then it is not necessary to define $\mathbf{x}_{\text{mis}i}, \mathbf{z}_{\text{mis}i}, \mathbf{y}_{\text{mis}i}$ and $\mathbf{v}_{\text{mis}i}$, and the observed data set is $\{(\mathbf{x}_i, \mathbf{z}_i), i = 1, \ldots, n\}$ as in Section 8.3. Here, we require the additional notation due to the presence of missing data. In this situation, we keep in mind that each data point is composed of its observed part and missing part, and the Bayesian methods are developed on the basis of the observed data set $\{(\mathbf{x}_{\text{obs}i}, \mathbf{z}_{\text{obs}i}); i = 1, \ldots, n\}$.

We first consider Bayesian estimation of the unknown parameters in $\boldsymbol{\theta}^* = (\boldsymbol{\theta}, \boldsymbol{\alpha})$, where $\boldsymbol{\theta}$ contains all the structural parameters of the model, whilst $\boldsymbol{\alpha}$ contains all the thresholds parameters. Let $p(\boldsymbol{\theta}, \boldsymbol{\alpha})$ be the prior density of $\boldsymbol{\theta}$ and $\boldsymbol{\alpha}$, $\mathbf{X}_{\text{obs}} = \{\mathbf{x}_{\text{obs}i}; i = 1, \ldots, n\}$ and $\mathbf{Z}_{\text{obs}} = \{\mathbf{z}_{\text{obs}i}; i = 1, \ldots, n\}$ be observed continuous and ordered categorical data, respectively. The joint posterior density of $\boldsymbol{\alpha}$ and $\boldsymbol{\theta}$ given $\mathbf{X}_{\text{obs}}$ and $\mathbf{Z}_{\text{obs}}$ is $p(\boldsymbol{\theta}, \boldsymbol{\alpha}|\mathbf{X}_{\text{obs}}, \mathbf{Z}_{\text{obs}})$. We again apply the idea of data augmentation to the posterior analysis. As before, we have to augment the observed data with the matrix of latent variables $\boldsymbol{\Omega} = \{\boldsymbol{\omega}_i, i = 1, \ldots, n\}$, and unobservable continuous measurements that underline the observed and missing ordered categorical data. Due to the presence of missing entries in the observable continuous measurements we additionally need to augment with these missing entries in the posterior analysis. More specifically, we let $\mathbf{X}_{\text{mis}} = \{\mathbf{x}_{\text{mis}i}; i = 1, \ldots, n\}$, $\mathbf{Y}_{\text{obs}} = \{\mathbf{y}_{\text{obs}i}; i = 1, \ldots, n\}$ and $\mathbf{Y}_{\text{mis}} = \{\mathbf{y}_{\text{mis}i}; i = 1, \ldots, n\}$ be the latent data. Further, we let $\mathbf{V}_{\text{obs}} = \{\mathbf{v}_{\text{obs}i}; i = 1, \ldots, n\}$, $\mathbf{V}_{\text{mis}} = \{\mathbf{v}_{\text{mis}i}; i = 1, \ldots, n\}$, $\mathbf{X} = \{\mathbf{X}_{\text{obs}}, \mathbf{X}_{\text{mis}}\}$, $\mathbf{Y} = \{\mathbf{Y}_{\text{obs}}, \mathbf{Y}_{\text{mis}}\}$ and $\mathbf{V} = \{\mathbf{V}_{\text{obs}}, \mathbf{V}_{\text{mis}}\}$. The observed data $\{\mathbf{X}_{\text{obs}}, \mathbf{Z}_{\text{obs}}\}$ are augmented with the missing quantities $\{\boldsymbol{\Omega}, \mathbf{X}_{\text{mis}}, \mathbf{Y}_{\text{mis}}, \mathbf{Y}_{\text{obs}}\} = \{\boldsymbol{\Omega}, \mathbf{V}_{\text{mis}}, \mathbf{Y}_{\text{obs}}\}$ in the posterior analysis. A sufficiently large number of random observations will be simulated from the joint posterior distribution $[\boldsymbol{\theta}, \boldsymbol{\Omega}, \mathbf{V}_{\text{mis}}, \boldsymbol{\alpha}, \mathbf{Y}_{\text{obs}}|\mathbf{X}_{\text{obs}}, \mathbf{Z}_{\text{obs}}]$ so that its distribution can be approximated adequately by the empirical distribution of the generated observations. The following Gibbs sampler (Geman and Geman, 1984) will be used. At the jth iteration with current values $\boldsymbol{\theta}^{(j)}$, $\boldsymbol{\Omega}^{(j)}$, $\mathbf{V}_{\text{mis}}^{(j)}$, $\boldsymbol{\alpha}^{(j)}$ and $\mathbf{Y}_{\text{obs}}^{(j)}$, we iteratively generate:

$$
\begin{aligned}
\boldsymbol{\theta}^{(j+1)} &\text{ from } p(\boldsymbol{\theta}|\mathbf{X}_{\text{obs}}, \mathbf{Z}_{\text{obs}}, \boldsymbol{\Omega}^{(j)}, \mathbf{V}_{\text{mis}}^{(j)}, \boldsymbol{\alpha}^{(j)}, \mathbf{Y}_{\text{obs}}^{(j)}),\\
\boldsymbol{\Omega}^{(j+1)} &\text{ from } p(\boldsymbol{\Omega}|\mathbf{X}_{\text{obs}}, \mathbf{Z}_{\text{obs}}, \boldsymbol{\theta}^{(j+1)}, \mathbf{V}_{\text{mis}}^{(j)}, \boldsymbol{\alpha}^{(j)}, \mathbf{Y}_{\text{obs}}^{(j)}),\\
\mathbf{V}_{\text{mis}}^{(j+1)} &\text{ from } p(\mathbf{V}_{\text{mis}}|\mathbf{X}_{\text{obs}}, \mathbf{Z}_{\text{obs}}, \boldsymbol{\theta}^{(j+1)}, \boldsymbol{\Omega}^{(j+1)}, \boldsymbol{\alpha}^{(j)}, \mathbf{Y}_{\text{obs}}^{(j)}),\\
(\boldsymbol{\alpha}^{(j+1)}, \mathbf{Y}_{\text{obs}}^{(j+1)}) &\text{ from } p(\boldsymbol{\alpha}, \mathbf{Y}_{\text{obs}}|\mathbf{X}_{\text{obs}}, \mathbf{Z}_{\text{obs}}, \boldsymbol{\theta}^{(j+1)}, \boldsymbol{\Omega}^{(j+1)}, \mathbf{V}_{\text{mis}}^{(j+1)}).
\end{aligned}
\tag{12.8}
$$

Recall that once $\mathbf{y}_{\text{mis}i}$ is given, it is not necessary to simulate $\mathbf{z}_{\text{mis}i}$. Thus, the missing data set $\{\mathbf{z}_{\text{mis}i}; i = 1, \ldots, n\}$ is not involved. As before, convergence of the Gibbs sampler is monitored by the 'estimated potential scale reduction (EPSR)' values corresponding to the parameters, and/or by inspecting several parallel sequences of observations generated from different starting values.

Conditional distribution corresponding to $\boldsymbol{\theta}$, $\boldsymbol{\Omega}$ and $(\boldsymbol{\alpha}, \mathbf{Y}_{\text{obs}})$ with given $\mathbf{V}_{\text{mis}}$ can be derived in exactly the same way with fully observed data $\mathbf{V} = (\mathbf{X}, \mathbf{Y})$ as described in Chapter 8. More specifically, based on the same conjugate prior distributions of $\boldsymbol{\theta}$ as given in Equations (8.5) and (8.9), the conditional distributions corresponding to components in $\boldsymbol{\theta}$ are given by Equations (8.6), (8.7), (8.8) and (8.10). The conditional distribution of $\boldsymbol{\Omega}$ is given by Equation (8.11). Compared with the Gibbs sampler (see Equation (8.21)) for analyzing the model without missing data, the only additional task to implement the Gibbs sampler of Equation (12.8) just involves the conditional distribution corresponding to $\mathbf{V}_{\text{mis}}$.

For $i = 1, \ldots, n$, since $\mathbf{v}_i$ are mutually independent, $\mathbf{v}_{\text{mis}i}$ are also mutually independent. Since $\boldsymbol{\Psi}_\epsilon$ is diagonal, $\mathbf{v}_{\text{mis}i}$ is independent with $\mathbf{v}_{\text{obs}i} = (\mathbf{x}_{\text{obs}i}, \mathbf{y}_{\text{obs}i})$. Let p_i be the dimension of $\mathbf{v}_{\text{mis}i}$, it follows from Equation (12.4) that

$$p(\mathbf{V}_{\text{mis}i}|\mathbf{X}_{\text{obs}}, \mathbf{Z}_{\text{obs}}, \boldsymbol{\Omega}, \boldsymbol{\theta}, \boldsymbol{\alpha}, \mathbf{Y}_{\text{obs}}) = \prod_{i=1}^{n} p(\mathbf{v}_{\text{mis}i}|\boldsymbol{\theta}, \boldsymbol{\omega}_i), \text{ and}$$

$$[\mathbf{v}_{\text{mis}i}|\boldsymbol{\theta}, \boldsymbol{\omega}_i] \stackrel{D}{=} N[\boldsymbol{\mu}_{\text{mis}i} + \boldsymbol{\Lambda}_{\text{mis}i}\boldsymbol{\omega}_i, \boldsymbol{\Psi}_{\epsilon\text{mis}i}], \tag{12.9}$$

where $\boldsymbol{\mu}_{\text{mis}i}$ is a $p_i \times 1$ subvector of $\boldsymbol{\mu}$ with elements corresponding to observed components deleted, $\boldsymbol{\Lambda}_{\text{mis}i}$ is a $p_i \times q$ submatrix of $\boldsymbol{\Lambda}$ with rows corresponding to observed components deleted and $\boldsymbol{\Psi}_{\epsilon\text{mis}i}$ is a $p_i \times p_i$ submatrix of $\boldsymbol{\Psi}_\epsilon$ with the appropriate rows and columns deleted. Note that even though the form of $\mathbf{V}_{\text{mis}}$ is complicated with many distinct missing patterns, its conditional distribution only involves a product of very simple normal distributions. The computational burden for simulating $\mathbf{V}_{\text{mis}}$ is light.

Simulating observations from $p(\boldsymbol{\theta}|\boldsymbol{\Omega}, \mathbf{V}_{\text{mis}}, \boldsymbol{\alpha}, \mathbf{Y}_{\text{obs}}, \mathbf{X}_{\text{obs}}, \mathbf{Z}_{\text{obs}})$ is straightforward, because they are the familiar normal, Gamma and inverted Wishart distributions. In a similar way to the analysis of the model with fully observed data, conditional densities involved in $p(\boldsymbol{\Omega}|\boldsymbol{\theta}, \mathbf{V}_{\text{mis}}, \boldsymbol{\alpha}, \mathbf{Y}_{\text{obs}}, \mathbf{X}_{\text{obs}}, \mathbf{Z}_{\text{obs}})$ and $p(\boldsymbol{\alpha}, \mathbf{Y}_{\text{obs}}|\boldsymbol{\theta}, \boldsymbol{\Omega}, \mathbf{Z}_{\text{obs}})$ are nonstandard and complex. The well-known Metropolis–Hastings (MH) algorithm (Metropolis *et al.*, 1953; Hastings, 1970) is again used to simulate observations from these conditional distributions. As the implementation of the MH algorithm is very similar to that described in Sections 8.2.3 and 6.3.2, the discussion is not repeated here. Bayesian estimates and their standard error estimates are similarly obtained via the sample mean and the sample covariance matrix as before. Moreover, estimated residuals $\hat{\boldsymbol{\epsilon}}_i$ and $\hat{\boldsymbol{\delta}}_i$ can be obtained in a similar way to previous chapters.

We now consider the issue on model comparison of two competing nonlinear SEMs M_0 and M_1 on the basis of the observed data $(\mathbf{X}_{\text{obs}}, \mathbf{Z}_{\text{obs}})$. Again, the Bayes factor B_{10} is used, and a procedure for computing this statistic is developed via path sampling (Gelman and Meng, 1998). Based on the general framework given in Section 12.2, the logarithm Bayes factor is estimated by Equation (12.2) with

$$\bar{U}_{(s)} = J^{-1} \sum_{j=1}^{J} U(\boldsymbol{\theta}^{(j)}, \boldsymbol{\Omega}^{(j)}, \mathbf{V}_{\text{mis}}^{(j)}, \boldsymbol{\alpha}^{(j)}, \mathbf{Y}_{\text{obs}}^{(j)}, \mathbf{X}_{\text{obs}}, \mathbf{Z}_{\text{obs}}, t_{(s)}), \tag{12.10}$$

in which $\{(\boldsymbol{\theta}^{(j)}, \boldsymbol{\Omega}^{(j)}, \mathbf{V}_{\text{mis}}^{(j)}, \boldsymbol{\alpha}^{(j)}, \mathbf{Y}_{\text{obs}}^{(j)}), j = 1, \ldots, J\}$ is a sample of observations simulated from $p(\boldsymbol{\theta}, \boldsymbol{\Omega}, \mathbf{V}_{\text{mis}}, \boldsymbol{\alpha}, \mathbf{Y}_{\text{obs}} | \mathbf{X}_{\text{obs}}, \mathbf{Z}_{\text{obs}}, t_{(s)})$, and

$$U(\boldsymbol{\theta}, \boldsymbol{\Omega}, \mathbf{V}_{\text{mis}}, \boldsymbol{\alpha}, \mathbf{Y}_{\text{obs}}, \mathbf{X}_{\text{obs}}, \mathbf{Z}_{\text{obs}}, t) = \frac{\text{d}}{\text{d}t} \log p(\boldsymbol{\Omega}, \mathbf{V}_{\text{mis}}, \mathbf{Y}_{\text{obs}}, \mathbf{X}_{\text{obs}}, \mathbf{Z}_{\text{obs}} | \boldsymbol{\theta}, \boldsymbol{\alpha}, t). \tag{12.11}$$

Defining a linked model with a good choice of continuous path to link the competing models M_0 and M_1 is crucial in the path sampling procedure. This is done on a problem-by-problem basis. As an illustration, we consider the following nonlinear SEMs:

$$\begin{aligned} M_0 : \quad & \mathbf{v}_i = \boldsymbol{\mu}_0 + \boldsymbol{\Lambda}_0 \boldsymbol{\omega}_i + \boldsymbol{\epsilon}_i, \;\; \boldsymbol{\eta}_i = \boldsymbol{\Lambda}_{\omega_0} \mathbf{G}_0(\boldsymbol{\omega}_i) + \boldsymbol{\delta}_i, \;\; i = 1, \ldots, n; \\ M_1 : \quad & \mathbf{v}_i = \boldsymbol{\mu}_1 + \boldsymbol{\Lambda}_1 \boldsymbol{\omega}_i + \boldsymbol{\epsilon}_i, \;\; \boldsymbol{\eta}_i = \boldsymbol{\Lambda}_{\omega_1} \mathbf{G}_1(\boldsymbol{\omega}_i) + \boldsymbol{\delta}_i, \;\; i = 1, \ldots, n, \end{aligned} \tag{12.12}$$

where $\{\boldsymbol{\mu}_0, \boldsymbol{\Lambda}_0, \boldsymbol{\Lambda}_{\omega_0}, \mathbf{G}_0(\boldsymbol{\omega})\}$ and $\{\boldsymbol{\mu}_1, \boldsymbol{\Lambda}_1, \boldsymbol{\Lambda}_{\omega_1}, \mathbf{G}_1(\boldsymbol{\omega})\}$ are two sets of parameters and functions of $\boldsymbol{\omega}$ that associate with M_0 and M_1 respectively. In general, some components in one set may be equal to or different from the corresponding components in the other set. These models can be linked up by t in [0,1] as below:

$$\begin{aligned} M_t : \quad & \mathbf{v}_i = (1-t)[\boldsymbol{\mu}_0 + \boldsymbol{\Lambda}_0 \boldsymbol{\omega}_i] + t[\boldsymbol{\mu}_1 + \boldsymbol{\Lambda}_1 \boldsymbol{\omega}_i] + \boldsymbol{\epsilon}_i \\ & \boldsymbol{\eta}_i = (1-t)\boldsymbol{\Lambda}_{\omega_0} \mathbf{G}_0(\boldsymbol{\omega}_i) + t \boldsymbol{\Lambda}_{\omega_1} \mathbf{G}_1(\boldsymbol{\omega}_i) + \boldsymbol{\delta}_i, \;\; i = 1, \ldots, n. \end{aligned}$$

Clearly, when $t = 0$, $M_t = M_0$ and when $t = 1$, $M_t = M_1$. It follows from the definition of the model that $\log p(\boldsymbol{\Omega}, \mathbf{V}_{\text{mis}}, \mathbf{Y}_{\text{obs}}, \mathbf{X}_{\text{obs}}, \mathbf{Z}_{\text{obs}} | \boldsymbol{\theta}, \boldsymbol{\alpha}, t)$ is equal to

$$\sum_{i=1}^{n} \left\{ C - \frac{1}{2} [\boldsymbol{\epsilon}_i(t)^T \boldsymbol{\Psi}_{\epsilon}^{-1} \boldsymbol{\epsilon}_i(t) + \boldsymbol{\delta}_i(t)^T \boldsymbol{\Psi}_{\delta}^{-1} \boldsymbol{\delta}_i(t)] \right\}, \tag{12.13}$$

where C is a constant independent of t, $\boldsymbol{\epsilon}_i(t) = \mathbf{v}_i - (1-t)(\boldsymbol{\mu}_0 + \boldsymbol{\Lambda}_0 \boldsymbol{\omega}_i) - t(\boldsymbol{\mu}_1 + \boldsymbol{\Lambda}_1 \boldsymbol{\omega}_i)$ and $\boldsymbol{\delta}_i(t) = \boldsymbol{\eta}_i - (1-t)\boldsymbol{\Lambda}_{\omega_0} \mathbf{G}_0(\boldsymbol{\omega}_i) - t \boldsymbol{\Lambda}_{\omega_1} \mathbf{G}_1(\boldsymbol{\omega}_i)$. In $p(\boldsymbol{\Omega}, \mathbf{V}_{\text{mis}}, \mathbf{Y}_{\text{obs}}, \mathbf{X}_{\text{obs}}, \mathbf{Z}_{\text{obs}} | \boldsymbol{\theta}, \boldsymbol{\alpha}, t)$, $\boldsymbol{\theta}$ is the parameter vector in the linked model.

It contains all the common and distinct parameters in $\boldsymbol{\mu}_0$, $\boldsymbol{\mu}_1$, $\boldsymbol{\Lambda}_0$, $\boldsymbol{\Lambda}_1$, $\boldsymbol{\Lambda}_{\omega_0}$ and $\boldsymbol{\Lambda}_{\omega_1}$, $\boldsymbol{\Phi}$, $\boldsymbol{\Psi}_\epsilon$ and $\boldsymbol{\Psi}_\delta$. Differentiating the above log-likelihood function with respect to t, we have

$$U(\boldsymbol{\theta}, \boldsymbol{\Omega}, \mathbf{V}_{\text{mis}}, \boldsymbol{\alpha}, \mathbf{Y}_{\text{obs}}, \mathbf{X}_{\text{obs}}, \mathbf{Z}_{\text{obs}}, t) =$$
$$\boldsymbol{\epsilon}_i(t)^T \boldsymbol{\Psi}_\epsilon^{-1}[\boldsymbol{\mu}_1 + \boldsymbol{\Lambda}_1\boldsymbol{\omega}_i - \boldsymbol{\mu}_0 - \boldsymbol{\Lambda}_0\boldsymbol{\omega}_i] + \boldsymbol{\delta}_i(t)^T \boldsymbol{\Psi}_\delta^{-1}[\boldsymbol{\Lambda}_{\omega_1}\mathbf{G}_1(\boldsymbol{\omega}_i) - \boldsymbol{\Lambda}_{\omega_0}\mathbf{G}_0(\boldsymbol{\omega}_i)]. \quad (12.14)$$

The Bayes factor is computed via Equations (12.10) and (12.11) with a sample $\{(\boldsymbol{\theta}^{(j)}, \boldsymbol{\Omega}^{(j)}, \mathbf{V}_{\text{mis}}^{(j)}, \boldsymbol{\alpha}^{(j)}, \mathbf{Y}_{\text{obs}}^{(j)}), j = 1, \ldots, J\}$ simulated by the hybrid algorithm. Note that in the Bayesian approach, the presence of missing data does not induce much difficulty in computing the logarithm Bayes factor.

12.3.1 A Simulation Study

The main purpose is to illustrate the empirical performance of the proposed Bayesian approach and to study the impact of missing data in estimation and model comparison. Random observations are generated from an NSEM defined by Equations (12.4) and (12.5) with six manifest variables that are related with latent variables η, ξ_1 and ξ_2. The specifications in Equation (12.4) are

$$\boldsymbol{\mu} = (0, \ldots, 0)^T, \quad \text{and} \quad \boldsymbol{\Lambda}^T = \begin{bmatrix} 1 & \lambda_{21} & 0 & 0 & 0 & 0 \\ 0 & 0 & 1 & \lambda_{42} & 0 & 0 \\ 0 & 0 & 0 & 0 & 1 & \lambda_{63} \end{bmatrix}, \quad (12.15)$$

where 1s and 0s in $\boldsymbol{\Lambda}$ are fixed, λ_{21}, λ_{42} and λ_{63} are unknown parameters with true values 0.8, 0.7 and 0.8, respectively. The nonlinear structural equation is given by

$$M_1 : \eta = \gamma_1\xi_1 + \gamma_2\xi_2 + \gamma_3\xi_1^2 + \delta,$$

with true values of γ_1, γ_2 and γ_3 equal to 0.6, 0.6 and 0.3, respectively. True variances of ϵ_{ik} and δ are all equal to 0.5, $\phi_{11} = \phi_{22} = 1.0$ and $\phi_{12} = 0.2$. Continuous measurements corresponding to the last two variables are transformed to ordered categorical data via true thresholds $\boldsymbol{\alpha}_1 = \boldsymbol{\alpha}_2 = (-1.0, -0.6, 0.6, 1.0)$, where -1.0 and 1.0 are fixed.

In each replication, a complete data set with 500 random observations is generated. Then MAR missing data are created as follows. (1) 100 fully observed data points are randomly selected, and the sample mean of the first four variables, $\bar{x}_1$, $\bar{x}_2$, $\bar{x}_3$ and $\bar{x}_4$ are computed. (2) For each element, x_1, x_2, x_3 and x_4, and in each and every of the remaining 400 observations, we generate randomly an observation u from the uniform distribution on [0, 1] to decide to whether the element is missing or not. More specifically, we randomly generate four independent observations u_1, u_2, u_3 and u_4 from the uniform distribution on [0, 1], then x_1 is deleted if $x_1+\bar{x}_1+\bar{x}_2 > u_1 - 1.0$, otherwise x_2 is deleted if $x_2+\bar{x}_1+\bar{x}_2 > u_2 - 1.0$; also x_3 is deleted if $x_3+\bar{x}_3 > u_3$, otherwise x_4 is deleted if $x_4+\bar{x}_4 > u_4$. In the created missing data sets, all entries relating to y_1 and y_2 are retained and about two-thirds of the observations contain one or more missing entries in $x_1, \ldots, x_4$. The number of fully observed data points in each replication can be different. For example, we see from Table 12.1 that the number of fully observed data in replication 1 is 179, and that number in replication 10 is 150.

We will compare the above NSEM (M_1) to a linear SEM which is defined with the same measurement equation as specified, and the following linear structural equation:

$$M_0 : \eta = \gamma_1\xi_1 + \gamma_2\xi_2 + \delta.$$

To give some rough idea about the sensitivity of the results to prior inputs, three types of hyperparameter values in the conjugate prior distributions of $\boldsymbol{\theta}$ (see Equations (8.5) and (8.9)) are considered. For $k = 1, \ldots, 6$, we fixed $\mathbf{H}_{0yk} = \mathbf{I}$ and $\mathbf{H}_{0\omega k} = \mathbf{I}$ for all types of prior inputs. Then in prior inputs Type I, we take $\rho_0 = 10$, $\mathbf{R}_0 = 4\mathbf{I}$, $\alpha_{0\epsilon k} = \alpha_{0\delta k} = 8$, $\beta_{0\epsilon k} = \beta_{0\delta k} = 10$ for all k; and other values in $\boldsymbol{\Lambda}_{0k}$ and $\boldsymbol{\Lambda}_{0\omega k}$ equal to the true parameter values. Hyperparameter values in prior inputs Type II and Type III are respectively equal to half and twice those given in Type I. On the basis of the fully observed data, and the

Table 12.1 Estimated $\log B_{10}$ in the 10 replications. 'FOD' stands for fully observed data, and 'FBA' stands for the proposed Bayesian approach that uses all fully observed and incomplete data, and 'LDA' stands for the listwise approach.

Rep.#	# of FOD	FBA II	FBA I	FBA III	LDA II	LDA I	LDA III
1	179	8.201	6.805	5.809	0.283	0.279	0.349
2	156	5.777	5.234	4.439	0.205	0.251	0.173
3	172	7.427	6.350	4.739	0.849	0.998	1.129
4	144	3.005	2.868	2.163	0.190	0.172	0.188
5	165	5.280	4.694	3.660	0.351	0.292	0.312
6	161	5.035	4.121	3.378	0.093	0.129	0.246
7	175	3.880	3.543	2.816	0.300	0.316	0.252
8	193	6.730	5.530	4.611	0.692	0.733	0.862
9	184	8.550	7.241	6.303	0.484	0.450	0.383
10	150	8.280	7.365	5.773	0.325	0.287	0.231

Note: Tables 12.1–12.4 are taken from Lee and Song (2004).

fully and partially observed data with missing entries, estimates of the logarithm Bayes factors for comparing M_1 and M_0 are obtained via the path sampling procedure with $S = 20$ and $J = 1000$. Results obtained from 10 replications are reported in Table 12.1. Based on the criterion for interpreting the Bayes factor, in most cases, results produced by the Bayesian approach that uses all the fully and partially observed data (FBA) clearly suggest the correct model is M_1; but results obtained from the listwise deletion approach (LDA) that only uses fully observed data cannot provide a definite suggestion. It seems that the conclusions are not changed with the above different prior inputs.

Results of simulation studies to investigate the accuracy of Bayesian estimates for mixed continuous and order categorical data have been reported in Song and Lee (2002), and Lee and Song (2004) in the context of linear and nonlinear SEMs, respectively. The conclusions are that Bayesian estimates are accurate, and are significantly better than the listwise deletion approach under some missing patterns and sample sizes.

12.3.2 An Illustrative Example

To illustrate the methodology further, a portion of the data set obtained from a study (Morisky *et al.*, 1998) of the effects of establishment policies, knowledge and attitudes on condom use among Filipino commercial sex workers (CSWs) is analyzed. As the nature of commercial sex work promotes the spread of AIDS and other sexually transmitted diseases, promotion of safer sexual practice among CSWs is an important issue. The primary concerns are on the developments and findings from an AIDS preventative intervention for Filipino CSWs. The data set was collected from female CSWs in the cities of the Philippines. The entire questionnaire consisted of 134 items, covering the areas of demographics knowledge, attitudes, beliefs, behaviors, self-efficacy for condom use and social desirability. In our illustrative example, only six manifest variables ($v_1, \ldots, v_6$) are selected. Variables v_1 and v_2 are related to the 'worry about getting AIDS', v_3 and v_4 are related to the 'number of times of vaginal sex in the last seven days' and the 'average weekly money (in pesos) earned as an entertainer', while v_5 and v_6 are about the 'attitudes of getting AIDS from sexual intercourse using a condom'. Variables v_3 and v_4 are continuous, while the others are ordered categorical measured with a five-point scale. Since the respondents and nonrespondents with the same values of recorded variables do not differ systematically on the values of variables missing for the nonrespondents, we assume that the missing values are missing at random and the missing mechanism is ignorable. After deleting obvious outliers, the data set contains 1080 observations, only 754 of them are fully observed. The missing patterns are presented in Table 12.2. Note that some missing patterns just have a very small number of observations. These missing patterns cannot be analyzed by the multisample method with the covariance structure analysis approach. To unify scales of the continuous variables, the corresponding

Table 12.2 Missing patterns and their sample sizes: AIDS data set, '×' and 'o' indicate missing and observed entries, respectively.

Pattern	Sample size	Manifest variables 1	2	3	4	5	6	Pattern	Sample size	Manifest variables 1	2	3	4	5	6
1	784	o	o	o	o	o	o	11	7	×	o	o	o	×	o
2	100	×	o	o	o	o	o	12	7	×	o	o	o	o	×
3	57	o	×	o	o	o	o	13	9	o	×	o	o	×	o
4	6	o	o	×	o	o	o	14	3	o	×	o	o	o	×
5	4	o	o	o	×	o	o	15	1	o	×	o	×	o	o
6	25	o	o	o	o	×	o	16	1	o	×	×	o	o	o
7	26	×	o	o	o	o	×	17	4	o	×	o	o	×	×
8	17	×	×	o	o	o	o	18	2	×	o	o	o	×	×
9	23	o	o	o	o	×	×	19	1	o	o	×	o	×	×
10	2	×	o	×	o	o	o	20	1	×	×	o	o	×	×

raw continuous data are standardized. The sample means and standard deviations of the continuous variables are $\{1.58, 1203.74\}$ and $\{1.84, 1096.32\}$, respectively. The cell frequencies of each individual ordinal categorical variable range from 21 to 709 (see Morisky *et al.* (1998) for other descriptive statistics).

To identify parameters associated with the ordered categorical variables α_{11}, α_{14}, α_{21}, α_{24}, α_{31}, α_{34}, α_{41} and α_{44} are fixed at -0.478, 1.034, -1.420, 0.525, -0.868, 0.559, -2.130 and -0.547, respectively. These fixed values are selected via $\alpha_{kh} = \Phi^{*^{-1}}(f_k)$, where Φ^* is the distribution function of $N[0, 1]$, and f_k are observed cumulative marginal proportions of the categories with $z_k < h$. Based on the meanings of the questions corresponding to the selected manifest variables, the data set is analyzed by a model with three latent variables η, ξ_1 and ξ_2 and the measurement equation as specified in Equations (12.4) and (12.15).

Competing models associated with the same measurement equation but the following different structural equations are considered for illustration purposes:

$$
\begin{aligned}
M_1 &: \ \eta = \gamma_1\xi_1 + \gamma_2\xi_2 + \delta, \\
M_2 &: \ \eta = \gamma_1\xi_1 + \gamma_2\xi_2 + \gamma_3\xi_1^2 + \delta, \\
M_3 &: \ \eta = \gamma_1\xi_1 + \gamma_2\xi_2 + \gamma_4\xi_1\xi_2 + \delta, \ \text{and} \\
M_4 &: \ \eta = \gamma_1\xi_1 + \gamma_2\xi_2 + \gamma_5\xi_2^2 + \delta.
\end{aligned}
$$

Note that M_1 is nested in M_2, M_3 and M_4, whilst M_2, M_3 and M_4 are nonnested. Estimated logarithm Bayes factors are obtained by the proposed path sampling procedure with $S = 20$ and $J = 1000$. Assuming that we have no prior information from other sources, we conduct an initial Bayesian estimation based on M_1 with noninformative priors in order to obtain prior inputs of some hyperparameters. Again, three types of prior inputs are considered. Prior inputs in Type I are the same as those given by Type I in the simulation study except that the true values are replaced by the Bayesian estimates obtained from the initial estimation, and the prior distribution of γ_3, γ_4 and γ_5 is a normal distribution with mean zero and a large variance. Prior inputs in Type II and Type III are obtained the same way as in the simulation study.

We are interested in comparing the linear model M_1 with the nonlinear models. It is easy to construct a path to link the competing models. For example, the linked model M_t for M_1 and M_2 is $M_t : \eta = \gamma_1\xi_1 + \gamma_2\xi_2 + t\gamma_3\xi_1^2 + \delta$. Hence, when $t = 0$, $M_t = M_0$ and when $t = 1$, $M_t = M_1$. Results obtained on the basis of the FBA and LDA approaches are presented in the top part of Table 12.3. The estimated log Bayes factors are not very sensitive to Type I, Type II and Type III prior inputs. In comparing M_2 and M_1, $\widehat{\log B_{21}}$ obtained from the FBA approach clearly recommends the nonlinear model M_2, while $\widehat{\log B_{21}}$ obtained from the LDA approach provides no suggestion. From $\widehat{\log B_{31}}$ and $\widehat{\log B_{41}}$, the other nonlinear models are not significantly better than M_1. Hence, M_2 is the best model from $M_1, \ldots, M_4$.

To compare M_2 with more complex models, we consider the following models with more complicated structural equations:

$$M_5 : \eta = \gamma_1\xi_1 + \gamma_2\xi_2 + \gamma_3\xi_1^2 + \gamma_4\xi_1\xi_2 + \delta,$$
$$M_6 : \eta = \gamma_1\xi_1 + \gamma_2\xi_2 + \gamma_3\xi_1^2 + \gamma_5\xi_2^2 + \delta.$$

Here, M_2 is nested in M_5 and M_6. The estimated logarithm Bayes factors are presented in the bottom part of Table 12.3. We see that the more complex models are not significantly better than M_2; hence the simpler model M_2 is selected. We compute and find that the PP p-value (Gelman, Meng and Stern,

Table 12.3 Estimated log Bayes factors under different prior inputs.

	LDA			FBA		
Priors	I	II	III	I	II	III
$\widehat{\log B_{21}}$	0.551	0.864	0.418	2.303	2.325	2.097
$\widehat{\log B_{31}}$	0.296	0.296	0.186	0.340	0.411	0.354
$\widehat{\log B_{41}}$	0.811	0.871	0.846	0.780	0.766	0.714
$\widehat{\log B_{52}}$	0.419	0.496	0.336	0.406	0.466	0.395
$\widehat{\log B_{62}}$	0.364	0.344	0.307	0.489	0.442	0.414

1996) corresponding to M_2 is 0.572. This indicates that M_2 fits the data well. In almost all cases, the EPSR values in monitoring the convergence of the hybrid algorithm for drawing observations from the posterior distribution are less than 1.2 after about 500 iterations. For completeness, estimates of unknown parameters in M_2 obtained by the FBA approach on the basis of different prior inputs are reported in Table 12.4.

Based on the results obtained, an NSEM has been chosen. Its specification about Λ in the measurement equation suggests that there are three non-overlapping latent factors η, ξ_1 and ξ_2, which can be roughly interpreted as 'worry about AIDS', 'aggressiveness' of CSWs and 'attitude to the risk of getting AIDS'. Based on the Type I prior inputs, these latent factors are related by the following nonlinear structural equation: $\eta = 0.545\xi_1 - 0.033\xi_2 - 0.226\xi_1^2 + \delta$, estimated with Type I prior inputs. Thus, 'aggressiveness' of the CSWs has both linear and quadratic effects on 'worry about AIDS'. Plotting the quadratic curve of η against ξ_1, we find that the maximum of η is roughly at $\xi_1 = 1.2$, and η decreases as ξ_1 moves away from both directions at 1.2. This indicates that the 'more aggressive' CSWs are not afraid of getting AIDS; on the other hand, the 'less aggressive' CSWs are not worried about getting AIDS. From the model comparison results, the model with the quadratic term of 'attitude to the risk of getting AIDS' or the corresponding interaction term with 'aggressiveness' is not as good. Thus, these nonlinear relationships are not important, and it is not necessary to consider the more complicated model that involves both the interaction and quadratic terms.

Table 12.4 Bayesian estimates of parameters in M_2, under different prior inputs.

	Bayesian estimate				Bayesian estimate		
Par	I	II	III	Par	I	II	III
λ_{21}	0.228	0.226	0.226	$\psi_{\epsilon 1}$	0.593	0.510	0.660
λ_{42}	0.353	0.318	0.402	$\psi_{\epsilon 2}$	0.972	0.973	0.975
λ_{63}	0.358	0.276	0.405	$\psi_{\epsilon 3}$	0.519	0.433	0.586
				$\psi_{\epsilon 4}$	0.943	0.943	0.940
γ_1	0.544	0.505	0.538	$\psi_{\epsilon 5}$	0.616	0.545	0.716
γ_2	−0.033	−0.051	−0.076	$\psi_{\epsilon 6}$	1.056	1.086	1.048
γ_3	−0.226	−0.153	−0.223				
				α_{12}	−0.030	−0.031	−0.015
ϕ_{11}	0.508	0.592	0.442	α_{13}	0.340	0.345	0.359
ϕ_{12}	−0.029	−0.028	−0.460	α_{22}	−0.961	−0.964	−0.966
ϕ_{22}	0.394	0.479	0.305	α_{23}	−0.620	−0.623	−0.627
				α_{32}	−0.394	−0.399	−0.399
ψ_δ	0.663	0.668	0.659	α_{33}	0.257	0.255	0.256
				α_{42}	−1.604	−1.613	−1.604
				α_{43}	−0.734	−0.736	−0.738

12.4 MIXTURE OF SEMs WITH MISSING DATA

In this section, we consider the mixture of SEMs with missing data and an unknown number of components. As one does not know the component memberships of the observations, it is not possible to handle missing data in mixtures of SEMs by replacement with the mean estimate or the least square estimate. We will present a Bayesian approach to solve the problem, and give an example to reveal the impact of ignoring the incomplete data on model comparison. The mixture model has been discussed in Chapter 11, Section 11.2, hence, it is just briefly described here for completeness.

A K component mixture SEM for a $p \times 1$ random vector $\mathbf{y}_i$ is defined as follows:

$$f(\mathbf{y}_i|\boldsymbol{\theta}) = \sum_{k=1}^{K} \pi_k f_k(\mathbf{y}_i|\boldsymbol{\mu}_k, \boldsymbol{\theta}_k), \quad i = 1, \ldots, n. \tag{12.16}$$

For the kth component,

$$\mathbf{y}_i = \boldsymbol{\mu}_k + \boldsymbol{\Lambda}_k \boldsymbol{\omega}_{ki} + \boldsymbol{\epsilon}_{ki}, \tag{12.17}$$

$$\boldsymbol{\eta}_{ki} = \boldsymbol{\Pi}_k \boldsymbol{\eta}_{ki} + \boldsymbol{\Gamma}_k \boldsymbol{\xi}_{ki} + \boldsymbol{\delta}_{ki}, \tag{12.18}$$

where $\boldsymbol{\epsilon}_k \stackrel{D}{=} N[\mathbf{0}, \boldsymbol{\Psi}_k]$, $\boldsymbol{\xi}_k$ and $\boldsymbol{\delta}_k$ are independently distributed as $N(\mathbf{0}, \boldsymbol{\Phi}_k)$ and $N(\mathbf{0}, \boldsymbol{\Psi}_{\delta k})$, respectively. The definitions of the random vectors and parameter matrices, as well as the assumptions of this mixture SEM are exactly the same as those given in Equations (11.1), (11.2) and (11.3). Recall that as the mixture model is invariant with respect to permutation of labels $k = 1, \ldots, K$, adoption of a unique labelling for identifiability is important. The MCMC approach proposed by Frühwirth-Schnatter (2001) as described in Chapter 11 will also be used here to deal with the label switching problem. Moreover, for each $k = 1, \ldots, K$, the covariance matrix Σ_k is identified by fixing appropriate elements in $\boldsymbol{\Lambda}_k$, $\boldsymbol{\Pi}_k$ and/or $\boldsymbol{\Gamma}_k$ at preassigned values that are chosen on a problem-by-problem basis. Let $\boldsymbol{\pi} = (\pi_1, \ldots, \pi_K)$, and $\boldsymbol{\theta}^*$ be the parameter vector that contains all unknown parameters in $\boldsymbol{\mu}_k$ and $\boldsymbol{\theta}_k$, $k = 1, \ldots, K$. In the terminology of Chapter 11, $\boldsymbol{\theta} = (\boldsymbol{\theta}^*, \boldsymbol{\pi})$.

To deal with the missing data problem, let $\mathbf{y}_i = \{\mathbf{y}_{\text{obs}i}, \mathbf{y}_{\text{mis}i}\}$, where $\mathbf{y}_{\text{obs}i}$ represents the observed elements of $\mathbf{y}_i$, whilst $\mathbf{y}_{\text{mis}i}$ represents the missing entries. Bayesian analysis of the current mixture of SEMs will be studied on the basis of the observed data set $\mathbf{Y}_{\text{obs}} = \{\mathbf{y}_{\text{obs}i}; i = 1, \ldots, n\}$. Let $\mathbf{Y}_{\text{mis}} = \{\mathbf{y}_{\text{mis}i}; i = 1, \ldots, n\}$ be the collection of missing data, $\mathbf{Y} = (\mathbf{Y}_{\text{obs}}, \mathbf{Y}_{\text{mis}})$, $\boldsymbol{\Omega}_1 = \{\boldsymbol{\omega}_1, \ldots, \boldsymbol{\omega}_n\}$ be the matrix of latent variables and $\mathbf{W} = (w_1, \ldots, w_n)$ be the matrix of allocation variables (see Equation (11.5)). In the posterior analysis, the observed data $\mathbf{Y}_{\text{obs}}$ is augmented with $\mathbf{Y}_{\text{mis}}$, $\boldsymbol{\Omega}_1$ and $\mathbf{W}$. The Gibbs sampler (Geman and Geman, 1984) is applied to simulate observations from the joint posterior distribution

$p(\mathbf{Y}_{\text{mis}}, \boldsymbol{\Omega}_1, \mathbf{W}, \boldsymbol{\pi}, \boldsymbol{\theta}^*|\mathbf{Y}_{\text{obs}})$ as follows. At the $(j+1)$th iteration with a current $\boldsymbol{\theta}^{*(j)}$, $\boldsymbol{\pi}^{(j)}$, $\mathbf{Y}_{\text{mis}i}^{(j)}$, $\boldsymbol{\Omega}_1^{(j)}$ and $\mathbf{W}^{(j)}$,

$$\begin{aligned}
&\text{Step (a) : generate } (\mathbf{Y}_{\text{mis}}^{(j+1)}, \boldsymbol{\Omega}_1^{(j+1)}, \mathbf{W}^{(j+1)}) \text{ from } p(\mathbf{Y}_{\text{mis}}, \boldsymbol{\Omega}_1, \mathbf{W}|\mathbf{Y}_{\text{obs}}, \boldsymbol{\theta}^{*(j)}, \boldsymbol{\pi}^{(j)});\\
&\text{Step (b) : generate } (\boldsymbol{\theta}^{*(j+1)}, \boldsymbol{\pi}^{(j+1)}) \text{ from } p(\boldsymbol{\theta}^*, \boldsymbol{\pi}|\mathbf{Y}_{\text{obs}}, \mathbf{Y}_{\text{mis}}^{(j+1)}, \boldsymbol{\Omega}_1^{(j+1)}, \mathbf{W}^{(j+1)}).
\end{aligned} \tag{12.19}$$

Step (a) can be further broken down into the following three steps:

$$\begin{aligned}
&\text{Step (a1)}: \quad \text{generate } \mathbf{W}^{(j+1)} \text{ from } p(\mathbf{W}|\boldsymbol{\theta}^{*(j)}, \boldsymbol{\pi}^{(j)}, \mathbf{Y}_{\text{mis}}^{(j)}, \mathbf{Y}_{\text{obs}});\\
&\text{Step (a2)}: \quad \text{generate } \boldsymbol{\Omega}_1^{(j+1)} \text{ from } p(\boldsymbol{\Omega}_1|\boldsymbol{\theta}^{*(j)}, \boldsymbol{\pi}^{(j)}, \mathbf{Y}_{\text{mis}}^{(j)}, \mathbf{W}^{(j+1)}, \mathbf{Y}_{\text{obs}});\\
&\text{Step (a3)}: \quad \text{generate } \mathbf{Y}_{\text{mis}}^{(j+1)} \text{ from } p(\mathbf{Y}_{\text{mis}}|\boldsymbol{\theta}^{*(j)}, \boldsymbol{\pi}^{(j)}, \boldsymbol{\Omega}_1^{(j+1)}, \mathbf{W}^{(j+1)}, \mathbf{Y}_{\text{obs}}).
\end{aligned}$$

Note that these steps are very similar to those given in Section 11.3, under the situation without missing data. The prior distributions of the parameters in $\boldsymbol{\theta}^*$ and $\boldsymbol{\pi}$ are taken from the conjugate prior distributions and the Dirichlet distribution as given in Section 11.3. As $\mathbf{Y} = (\mathbf{Y}_{\text{mis}}, \mathbf{Y}_{\text{obs}})$ is given, the conditional distributions corresponding to $\boldsymbol{\theta}^*$ and $\boldsymbol{\pi}$ in Step (b), and those corresponding to $\mathbf{W}$ and $\boldsymbol{\Omega}_1$, in Steps (a1) and (a2) can be obtained from the expressions given as before.

The remaining conditional distribution $p(\mathbf{Y}_{\text{mis}}|\boldsymbol{\theta}^*, \boldsymbol{\pi}, \boldsymbol{\Omega}_1, \mathbf{W}, \mathbf{Y}_{\text{obs}})$ in Step (a3) is derived below. For $i = 1, \ldots, n$, as $\mathbf{y}_i$ are independent given $\boldsymbol{\Omega}_1$, $\mathbf{y}_{\text{mis}i}$ are also conditionally independent. As $\boldsymbol{\Psi}_\epsilon$ is diagonal; and given w_i, $\boldsymbol{\omega}_i$, and $\boldsymbol{\theta}^*$; $\mathbf{y}_{\text{mis}i}$ is independent of $\mathbf{y}_{\text{obs}i}$. Moreover, with given $\mathbf{W}_i$, we know the component membership of $\mathbf{y}_{\text{mis}i}$; hence, $\boldsymbol{\pi}$ is irrelevant. Let p_i be the dimension of $\mathbf{y}_{\text{mis}i}$, it follows that

$$p(\mathbf{Y}_{\text{mis}}|\boldsymbol{\theta}^*, \boldsymbol{\pi}, \mathbf{W}, \mathbf{Y}_{\text{obs}}, \boldsymbol{\Omega}_1) = \prod_{i=1}^{n} p(\mathbf{y}_{\text{mis}i}|\boldsymbol{\theta}^*, \boldsymbol{\omega}_i, w_i), \quad \text{and}$$

$$[\mathbf{y}_{\text{mis}i}|\boldsymbol{\theta}^*, \boldsymbol{\omega}_i, w_i = k] \stackrel{D}{=} N[\boldsymbol{\mu}_{\text{mis}i,k} + \boldsymbol{\Lambda}_{\text{mis}i,k}\boldsymbol{\omega}_{ki}, \boldsymbol{\Psi}_{\epsilon\text{mis}i,k}],$$

where $\boldsymbol{\mu}_{\text{mis}i,k}$ is a $p_i \times 1$ subvector of $\boldsymbol{\mu}_k$ with elements corresponding to observed components deleted, $\boldsymbol{\Lambda}_{\text{mis}i,k}$ is the corresponding $p_i \times q$ submatrix of $\boldsymbol{\Lambda}_k$, and $\boldsymbol{\Psi}_{\epsilon\text{mis}i,k}$ is the corresponding $p_i \times p_i$ submatrix of $\boldsymbol{\Psi}_{\epsilon k}$, with the appropriate rows and/or columns deleted. Thus, even though the form of $\mathbf{Y}_{\text{mis}}$ is complicated with many distinct missing patterns, its conditional distribution only involves a product of very simple normal distributions. The computational burden for simulating $\mathbf{Y}_{\text{mis}}$ is light.

Let M_0 and M_1 be two mixtures of SEMs with different members of components. The choice between M_0 and M_1 is based on the Bayes factor that is defined by

$$B_{10} = \frac{p(\mathbf{Y}_{\text{obs}}|M_1)}{p(\mathbf{Y}_{\text{obs}}|M_0)}. \tag{12.20}$$

Path sampling can be used again to compute $\log B_{10}$, by utilizing the simulated observations $\{\mathbf{Y}_{\text{mis}}^{(j)}, \boldsymbol{\Omega}_1^{(j)}, \boldsymbol{\theta}^{*(j)}, \boldsymbol{\pi}^{(j)}\}$ from the joint posterior distribution, see Equation (11.18). To explicitly demonstrate how to apply the above procedure for Bayesian model selection in the context of a mixture of SEMs with missing data, let

$$M_1: \quad [\mathbf{y}_{\text{obs}i}|\boldsymbol{\theta}^*, \boldsymbol{\pi}] \sim \sum_{k=1}^{K} \pi_k f_k(\mathbf{y}_{\text{obs}i}|\boldsymbol{\mu}_k, \boldsymbol{\Sigma}_k), \tag{12.21}$$

corresponding to a model of K components, and

$$M_0: \quad [\mathbf{y}_{\text{obs}i}|\boldsymbol{\theta}^*, \boldsymbol{\pi}^*] \sim \sum_{k=1}^{c} \pi_k^* f_k(\mathbf{y}_{\text{obs}i}|\boldsymbol{\mu}_k, \boldsymbol{\Sigma}_k), \tag{12.22}$$

corresponding to a model of c components with component probabilities π_k^*, where $1 \leq c < K$. Note here the density in each component is based on the observed data $\mathbf{y}_{\text{obs}i}$. To apply the proposed method for computing $\log B_{10}$, these competing models are linked up by a path with $t \in [0, 1]$:

$$\begin{aligned} M_t: \quad &[\mathbf{y}_{\text{obs}i}|\boldsymbol{\theta}^*, \boldsymbol{\pi}, t] \sim [\pi_1 + (1-t)a_1(\pi_{c+1} + \cdots + \pi_K)]f_1(\mathbf{y}_{\text{obs}i}|\boldsymbol{\mu}_1, \boldsymbol{\Sigma}_1) + \cdots \\ &+ [\pi_c + (1-t)a_c(\pi_{c+1} + \cdots + \pi_K)]f_c(\mathbf{y}_{\text{obs}i}|\boldsymbol{\mu}_c, \boldsymbol{\Sigma}_c) + t\pi_{c+1}f_{c+1}(\mathbf{y}_{\text{obs}i}|\boldsymbol{\mu}_{c+1}, \boldsymbol{\Sigma}_{c+1}) \\ &+ \cdots + t\pi_K f_K(\mathbf{y}_{\text{obs}i}|\boldsymbol{\mu}_K, \boldsymbol{\Sigma}_K), \end{aligned} \tag{12.23}$$

where $a_1, \ldots, a_c$ are given non-negative weights such that $a_1 + \cdots + a_c = 1$. Clearly, when $t = 1$, M_t reduces to M_1; when $t = 0$, M_t reduces to M_0 with $\pi_k^* = \pi_k + a_k(\pi_{c+1} + \cdots + \pi_K)$, $k = 1, \ldots, c$. The complete data log-likelihood function $\log p(\mathbf{Y}_{\text{obs}}, \mathbf{Y}_{\text{mis}}, \boldsymbol{\Omega}_1|\boldsymbol{\theta}^*, \boldsymbol{\pi}, t)$ is given by Equation (11.23), and its derivative is given in Equation (11.24).

12.4.1 An Illustrative Example

To illustrate the path sampling procedure in model comparison, a real example on the basis of a small portion of ICPSR data set collected in the project World

Values Survey 1981–1984 and 1990–1993 (World Values Study Group, 1994) is analyzed. Based on the data obtained from the UK, six variables (variables 180, 96, 62, 176, 116 and 117, see Appendix 1.1) that are related to respondents' job and homelife are taken as manifest variables in $\mathbf{y} = (y_1, \ldots, y_6)^T$. These variables were measured via a 10-point scale; for convenience, they are treated as continuous in this example. There are 1483 random observations, many of them are with missing entries and some of the sample sizes within the missing patterns are very small. The missing patterns are presented in Table 12.5. We observe that there are only 196 fully observed data. From the questions associated with the manifest variables, it is natural to consider a measurement model Equation (12.17) with three latent variables: η, ξ_1 and ξ_2, such that the first two manifest variables are indicators for η, the third and fourth manifest variables are indicators for ξ_1, and the remaining manifest variables are indicators for ξ_2. More specifically, the following specifications on the parameter matrices of component are used: $\mathbf{\Pi} = \mathbf{0}$, $\mathbf{\Gamma} = (\gamma_1, \gamma_2)$, $\mathbf{\Psi}_\delta = \psi_\delta$,

$$\mathbf{\Lambda}^T = \begin{pmatrix} 1.0^* & \lambda_{21} & 0^* & 0^* & 0^* & 0^* \\ 0^* & 0^* & 1.0^* & \lambda_{42} & 0^* & 0^* \\ 0^* & 0^* & 0^* & 0^* & 1.0^* & \lambda_{63} \end{pmatrix}, \qquad \mathbf{\Phi} = \begin{pmatrix} \phi_{11} & \phi_{12} \\ \phi_{21} & \phi_{22} \end{pmatrix}, \tag{12.24}$$

and $\mathbf{\Psi} = \text{diag}(\psi_1, \ldots, \psi_6)$. The latent variables can be roughly interpreted as 'job satisfaction, η', 'homelife, ξ_1', and 'job attitude, ξ_2'.

This data set is analyzed on the basis of a mixture of SEMs with an unknown number of components. The formulation of the model in every component is taken to be the same as in Equation (12.24). The path sampling procedure is used to select a model with the most appropriate number of components. In

Table 12.5 Missing patterns and their sample sizes: ICPSR data set, '×' and 'o' indicate missing and observed entries, respectively.

		Manifest variables								Manifest variables					
Pattern	Sample size	1	2	3	4	5	6	Pattern	Sample size	1	2	3	4	5	6
1	196	o	o	o	o	o	o	9	3	o	×	×	o	×	×
2	630	o	o	×	o	o	o	10	2	×	o	×	o	o	o
3	515	o	o	×	o	×	×	11	2	×	o	×	o	×	×
4	106	o	o	o	o	×	×	12	2	o	o	o	o	×	×
5	10	o	o	×	×	o	o	13	1	o	o	o	o	o	×
6	7	o	o	×	×	×	×	14	1	o	×	o	o	×	×
7	4	o	×	×	o	o	o	15	1	×	o	o	o	×	×
8	3	o	o	×	o	o	×								

the implementation of the Gibbs sampler, the following hyperparameter values in the conjugate prior distributions are used. For $m = 1, \ldots, p$; $l = 1, \ldots, q$,

$$\begin{aligned}
&\boldsymbol{\mu}_0 = \bar{\mathbf{y}}, \text{ where } \bar{\mathbf{y}} \text{ is obtained on the basis of the 196 fully observed data,}\\
&\boldsymbol{\Sigma}_0 = 9^2\mathbf{I};\\
&\boldsymbol{\Lambda}_{0km} = \mathbf{0}, \boldsymbol{\Gamma}_{0kl} = \mathbf{0}, \mathbf{H}_{0ykm} = \mathbf{I}, \mathbf{H}_{0\omega kl} = \mathbf{I};\\
&(\alpha_{0\epsilon k}, \beta_{0\epsilon k}) = (\alpha_{0\delta k}, \beta_{0\delta k}) = (2, 30);\ \text{ and } \rho_0 = 20, \mathbf{R}_0^{-1} = 5\mathbf{I}.
\end{aligned} \tag{12.25}$$

For the prior distribution of $\boldsymbol{\pi}$, the value of the hyperparameter α in the symmetric Dirichlet distribution is taken to be 1. Again, for each $t_{(s)}$, the convergence of the Gibbs sampler algorithm is monitored by ESPR values associated with different starting values. We observe that the algorithm converged quickly within 500 iterations. The logarithm Bayes factors were estimated by using 20 fixed grids in [0,1], and a total of $J = 1000$ additional observations were simulated by the Gibbs sampler after a burn-in phase of 500 iterations. Let M_k denote the mixture model with k components, the logarithm Bayes factor estimates obtained via FBA are equal to $\widehat{\log B_{21}} = 61.42$ and $\widehat{\log B_{32}} = -0.96$, whilst the estimates obtained via LDA are equal to $\widehat{\log B_{21}} = -0.77$ and $\widehat{\log B_{32}} = -0.46$, respectively. According to the criterion for interpreting the log Bayes factor, a two component model is selected by FBA, whilst a one component model is selected by LDA. Clearly, the conclusions are different.

To cross-validate the selection results obtained via the logarithm Bayes factors with respect to prior inputs, the following hyperparameter values in the conjugate distributions are considered:

Type A: Same prior inputs as given in Equation (12.25), except $(\alpha_{0\epsilon k}, \beta_{0\epsilon k}) = (\alpha_{0\delta k}, \beta_{0\delta k})$ are perturbed as below. I: (2,30), II: (2,20) and III: (2,40).

Type B: Same prior inputs as given in Equation (12.25), except $\boldsymbol{\Sigma}_0$ is perturbed as below. I: $\boldsymbol{\Sigma}_0 = 9^2\mathbf{I}$, II: $\boldsymbol{\Sigma}_0 = 3^2\mathbf{I}$ and III: $\boldsymbol{\Sigma}_0 = (9/2)^2\mathbf{I}$.

Type C: Same prior inputs as given in Equation (12.25), except ρ_0 is perturbed as below. I: $\rho_0 = 20$, II: $\rho_0 = 10$ and III: $\rho_0 = 30$.

The estimated logarithm Bayes factors based on these prior inputs are reported in Table 12.6. We observe that under all different prior inputs, the $\log B_{21}$ estimates obtained from FBA are very large. These results clearly give the same conclusion on selecting a two component mixture SEM. From the $\log B_{32}$ estimates, it is obvious that a two component model is preferable to a three component model. Finally, the $\log B_{21}$ estimates obtained by LDA suggest a misleading conclusion to select a one component model. In Bayesian estimation, we find from the

Table 12.6 The estimated logarithms of Bayes factors under different prior inputs.

	Type	A			B			C		
		I	II	III	I	II	III	I	II	III
FBA	$\log B_{21}$	61.42	45.93	73.31	61.42	64.86	65.03	61.42	59.54	61.11
	$\log B_{32}$	−0.96	−0.97	−0.94	−0.96	−0.93	−0.90	−0.96	−0.90	−0.94
LDA	$\log B_{21}$	−0.77	−0.94	0.28	−0.77	−0.08	0.24	−0.774	−0.72	−0.84

Table 12.7 Bayesian estimates of parameters in the two component model in analyzing the ICPSR data set.

PAR	Estimate in Component 1	Estimate in Component 2	PAR	Estimate in Component 1	Estimate in Component 2
π	0.317	0.683	ϕ_{11}	0.984	1.259
μ_1	6.879	8.843	ϕ_{12}	−0.124	0.185
μ_2	5.975	8.177	ϕ_{22}	0.603	0.816
μ_3	1.913	2.520	$\psi_{\epsilon 1}$	2.812	0.620
μ_4	5.157	5.448	$\psi_{\epsilon 2}$	3.937	0.965
μ_5	5.770	8.151	$\psi_{\epsilon 3}$	2.423	1.352
μ_6	5.146	7.735	$\psi_{\epsilon 4}$	4.820	2.535
λ_{21}	0.535	0.996	$\psi_{\epsilon 5}$	5.929	1.342
λ_{42}	2.349	2.375	$\psi_{\epsilon 6}$	6.305	3.185
λ_{63}	2.308	1.082	ψ_{δ}	2.603	0.764
γ_1	0.268	0.182			
γ_2	−0.794	0.366			

MCMC samples simulated by the random permutation sampler that $\mu_{1,1} < \mu_{2,1}$ is a suitable identifiability constraint. By using the permutation sampler with this identifiability constraint, Bayesian estimates of the selected mixture model with two components are obtained. Results are presented in Table 12.7. We observe that parameters of μ_1, μ_2, μ_3, μ_5, μ_6, λ_{21}, λ_{63}, γ_2, ϕ_{12}, ψ_1, ψ_2, ψ_3, ψ_4, ψ_5, ψ_6 and ψ_δ under component 1 and component 2 are quite different. This indicates that a two component model is more reasonable.

12.5 NONLINEAR SEMs WITH NONIGNORABLE MISSING DATA

In addition to the ignorable missing data that are MAR, many missing data in behavioral, medical, social and psychological research are nonignorable in the sense that the reason for missing depends on the observed data and the missing data themselves. For example, the side effects of the treatment may make the patients worse and thereby affect patients' participation. Nonignorable missing data are more difficult to handle than the ignorable missing data, and have received considerable attention in statistics (see Diggle and Kenward (1994) and Ibrahim, Chen and Lipsitz (2001) among others). In the field of SEM, limited work has been done on analyzing nonignorable missing data. In this section, we present a Bayesian approach (see Lee and Tang, 2006) for analyzing a nonlinear SEM with nonignorable missing data. For brevity, we focus on continuous data. However, the Baysian development can be extended to other SEMs or to dichotomous and/or ordered categorical data, based on the key ideas presented in this section and previous chapters. Again, the idea of data augmentation and the MCMC tools will be used in the Bayesian analysis. Bayesian estimates of the unknown parameters, and the Bayes factor will be computed. Although Ibrahim, Chen and Lipsitz (2001) pointed out that the parametric form of the assumed missing mechanism itself is not 'testable' from the data, the Bayes factor provides a useful statistic for comparing different missing data models. Moreover, in the context of a given nonignorable missing data model, the Bayes factor can be used to select a better NSEM for fitting the data.

12.5.1 The Model and the Nonignorable Missing Mechanism

For each $p \times 1$ random vector $\mathbf{v}_i = (\mathrm{v}_{i1}, \ldots, \mathrm{v}_{ip})^T$ in the data set $\mathbf{V} = (\mathbf{v}_1, \ldots, \mathbf{v}_n)$, we define a missing indicator $\boldsymbol{r}_i = (r_{i1}, \ldots, r_{ip})^T$ such that $r_{ij} = 1$ if v_{ij} is missing, and $r_{ij} = 0$ if v_{ij} is observed. Let $\boldsymbol{r} = (\boldsymbol{r}_1, \ldots, \boldsymbol{r}_n)$; and let $\mathbf{V}_{\text{mis}}$ and $\mathbf{V}_{\text{obs}}$ be the missing and observed data, respectively. If the distribution of $\boldsymbol{r}$ is independent of $\mathbf{V}_{\text{mis}}$, the missing mechanism is defined to be MAR; otherwise the missing mechanism is nonignorable (see Little and Rubin, 1987). For a nonignorable missing mechanism, it is necessary to investigate the conditional distribution of $\boldsymbol{r}$ given $\mathbf{V}$, $[\boldsymbol{r}|\mathbf{V}, \boldsymbol{\varphi}]$, where $\boldsymbol{\varphi}$ is a parameter vector. If this distribution does not contain unknown parameters in $\boldsymbol{\varphi}$, the missing mechanism is called nonignorable and known, otherwise it is called nonignorable and unknown. An example of a nonignorable and known mechanism is censored data with a known censoring point. For analyzing missing data with a nonignorable and unknown mechanism, the basic issues are to specify a reasonable model for $\boldsymbol{r}$ given $\mathbf{V}$, and then develop statistical methods for analyzing this model together with the model in relation to $\mathbf{V}$.

Let $\mathbf{v}_1, \ldots, \mathbf{v}_n$ be independent random observations, and let $\boldsymbol{\omega}_1, \ldots, \boldsymbol{\omega}_n$ be the corresponding latent vectors in the hypothesized nonlinear SEM as defined by Equations (12.4), (12.5) and (12.6). In this section, we consider the situation where the manifest vector $\mathbf{v}_i$ is incompletely observed with an nonignorable mechanism. Let $\mathbf{v}_i = (\mathbf{v}_{\text{obs}i}^T, \mathbf{v}_{\text{mis}i}^T)^T$, where $\mathbf{v}_{\text{obs}i}$ is a $p_{1i} \times 1$ vector of observed manifest variables, $\mathbf{v}_{\text{mis}i}$ is a $p_{2i} \times 1$ vector of missing components, and $p_{1i} + p_{2i} = p$. Here, we assume an arbitrary pattern of missing data in $\mathbf{v}_i$, and thus $\mathbf{v}_i = (\mathbf{v}_{\text{obs}i}^T, \mathbf{v}_{\text{mis}i}^T)^T$ may represent some permutation of the indices of the original $\mathbf{v}_i$. Let $[\boldsymbol{r}_i|\mathbf{v}_i, \boldsymbol{\omega}_i, \boldsymbol{\varphi}]$ be the conditional distribution of $\boldsymbol{r}_i$ given $\mathbf{v}_i$

and $\boldsymbol{\omega}_i$ with a probability density function $p(\boldsymbol{r}_i|\mathbf{v}_i, \boldsymbol{\omega}_i, \boldsymbol{\varphi})$ that is related to the nonignorable missing mechanism. Let $\boldsymbol{\theta}$ be the structural parameter vector that contains all unknown distinct parameters in $\boldsymbol{\mu}$, $\boldsymbol{\Lambda}$, $\boldsymbol{\Lambda}_\omega$, $\boldsymbol{\Psi}_\epsilon$, $\boldsymbol{\Psi}_\delta$ and $\boldsymbol{\Phi}$. Our main interest is in the posterior inferences about $\boldsymbol{\theta}$ and $\boldsymbol{\varphi}$ based on the missing data indicator $\boldsymbol{r}$ and the observed data $\mathbf{V}_{\text{obs}} = \{\mathbf{v}_{\text{obs}1}, \ldots, \mathbf{v}_{\text{obs}n}\}$. According to the definition of the model, the joint posterior density of $\boldsymbol{\theta}$ and $\boldsymbol{\varphi}$ based on $\mathbf{V}_{\text{obs}}$ and $\boldsymbol{r}$ is given by:

$$p(\boldsymbol{\theta}, \boldsymbol{\varphi}|\mathbf{V}_{\text{obs}}, \boldsymbol{r}) \propto \left\{\prod_{i=1}^{n} \int_{\boldsymbol{\omega}_i, \mathbf{v}_{\text{mis}i}} p(\mathbf{v}_i|\boldsymbol{\omega}_i, \boldsymbol{\theta}) p(\boldsymbol{r}_i|\mathbf{v}_i, \boldsymbol{\omega}_i, \boldsymbol{\varphi}) p(\boldsymbol{\omega}_i|\boldsymbol{\theta}) \mathrm{d}\boldsymbol{\omega}_i \mathrm{d}\mathbf{v}_{\text{mis}i}\right\} \times p(\boldsymbol{\theta}, \boldsymbol{\varphi}), \tag{12.26}$$

where $p(\boldsymbol{\theta}, \boldsymbol{\varphi})$ denotes the joint prior distribution of $\boldsymbol{\theta}$ and $\boldsymbol{\varphi}$. In general, the integral in Equation (12.26) does not have a closed form and its dimension is equal to the sum of the dimensions of $\boldsymbol{\omega}_i$ and $\mathbf{v}_{\text{mis}i}$.

We now consider the selection of a model for the nonignorable missing mechanism. Theoretically, any general model can be taken. However, one must be careful not to use too complicated or too large a model, since it can easily become unidentifiable. Moreover, too complex a model will also induce difficulty in deriving the corresponding conditional distributions of the missing responses given the observed data, and inefficient sampling from those conditional distributions. Since the observations are independent,

$$p(\boldsymbol{r}|\mathbf{V}, \boldsymbol{\Omega}, \boldsymbol{\varphi}) = \prod_{i=1}^{n} p(\boldsymbol{r}_i|\mathbf{v}_i, \boldsymbol{\omega}_i, \boldsymbol{\varphi}),$$

where $\boldsymbol{\Omega} = (\boldsymbol{\omega}_1, \ldots, \boldsymbol{\omega}_n)$. As the covariance matrix of the error measurement, $\boldsymbol{\epsilon}_i$, is diagonal, it follows that when $\boldsymbol{\omega}_i$ is given, the components of $\mathbf{v}_i$ are independent. Hence, for $j \neq \ell$, it is reasonable to assume that the conditional distributions of $\boldsymbol{r}_{ij}$ and $\boldsymbol{r}_{il}$ given $\boldsymbol{\omega}_i$ are independent. Under this assumption, we propose the following binomial model for the nonignorable missing mechanism (see Ibrahim, Chen and Lipsitz, 2001; Lee and Tang, 2006):

$$p(\boldsymbol{r}|\mathbf{V}, \boldsymbol{\Omega}, \boldsymbol{\varphi}) = \prod_{i=1}^{n}\prod_{j=1}^{p} [\text{pr}(r_{ij} = 1|\mathbf{v}_i, \boldsymbol{\omega}_i, \boldsymbol{\varphi})]^{r_{ij}} [1 - \text{pr}(r_{ij} = 1|\mathbf{v}_i, \boldsymbol{\omega}_i, \boldsymbol{\varphi})]^{1-r_{ij}}. \tag{12.27}$$

Ibrahim, Chen and Lipsitz (2001) pointed out that since r_{ij} is binary, one can use a sequence of logistic regressions for modeling $\text{pr}(r_{ij} = 1|\mathbf{v}_i, \boldsymbol{\omega}_i, \boldsymbol{\varphi})$ in Equation (12.27). They also pointed out that this model has the potential for reducing the number of parameters in the missing data mechanism, yields correlation structures between the r_{ij}s, allows more flexibility in specifying the missing data model, and facilitates efficient sampling from the conditional distribution of the missing response given the observed data. Following the suggestion given in Ibrahim, Chen and Lipsitz (2001), the following logistic regression model is used:

$$
\begin{aligned}
m(\mathbf{v}_i, \boldsymbol{\omega}_i, \boldsymbol{\varphi}) &= \text{logit}[\text{pr}(r_{ij} = 1|\mathbf{v}_i, \boldsymbol{\omega}_i, \boldsymbol{\varphi})] \\
&= \varphi_0 + \varphi_1 \mathrm{v}_{i1} + \cdots + \varphi_p \mathrm{v}_{ip} + \varphi_{p+1}\omega_{i1} + \cdots + \varphi_{p+q}\omega_{iq} = \boldsymbol{\varphi}^T \boldsymbol{d}_i,
\end{aligned} \tag{12.28}
$$

where $\boldsymbol{d}_i = (1, \mathrm{v}_{i1}, \ldots, \mathrm{v}_{ip}, \omega_{i1}, \ldots, \omega_{iq})^T$ and $\boldsymbol{\varphi} = (\varphi_0, \varphi_1, \ldots, \varphi_{p+q})^T$. For the normal random effects model: $\mathbf{v}_i = \boldsymbol{X}_i\boldsymbol{\beta} + \boldsymbol{Z}_i\boldsymbol{b}_i + \boldsymbol{\epsilon}_i$, where $\boldsymbol{X}_i$ and $\boldsymbol{Z}_i$ are covariates, $\boldsymbol{\beta}$ is an unknown parameter vector and $\boldsymbol{b}_i$ is the vector of latent random effects with distribution $N[\mathbf{0}, \boldsymbol{\Phi}]$; Ibrahim, Chen and Lipsitz (2001) suggested the use of the logistic regression that depends on $\mathbf{v}_i$ but not $\boldsymbol{b}_i$. They pointed out this is a reasonable assumption in practice, since models for $[\boldsymbol{r}_i|\mathbf{v}_i, \boldsymbol{\varphi}]$ can be chosen to resemble and closely approximate a model for $[\boldsymbol{r}_i|\mathbf{v}_i, \boldsymbol{\varphi}, \boldsymbol{b}_i]$ that might include $\boldsymbol{b}_i$. Considering the similarity between the measurement equation of our NSEM and the above normal random effect model, and based on the above arguments of Ibrahim, Chen and Lipsitz (2001), it is desirable to adopt the following special case of Equation (12.28) which does not depend on $\boldsymbol{\omega}_i$:

$$
m(\mathbf{v}_i, \boldsymbol{\omega}_i, \boldsymbol{\varphi}) = \text{logit}[\text{pr}(\boldsymbol{r}_{ij} = 1|\mathbf{v}_i, \boldsymbol{\omega}_i, \boldsymbol{\varphi})] = \varphi_0 + \varphi_1 \mathrm{v}_{i1} + \cdots + \varphi_p \mathrm{v}_{ip}. \tag{12.29}
$$

Moreover, it follows from the measurement equation and the basic concepts of latent variables and their indicators in SEMs that the characteristics of $\boldsymbol{\omega}_i$ are revealed by the manifest variables in $\mathbf{v}_i$. Hence, in practice, even if we think the nonignorable missing data depend on $\boldsymbol{\omega}_i$, the missing mechanism can still be adequately accounted for by the model given in Equation (12.29) that only depends on $\mathbf{v}_i$. However, for generality, the Bayesian approach will be developed on the basis of the more general model of Equation (12.28). As Equation (12.29) is equivalent to specifying $\varphi_{p+1} = \cdots = \varphi_{p+q} = 0$ in Equation (12.28), the modifications of the general Bayesian development in addressing this special case are straightforward. We do not recommend the routine use of Equations (12.28) or (12.29) for modeling the nonignorable missing mechanism in every practical application. Other missing mechanism models may be preferable for situations where one is certain about the specific form for the missing mechanism. However, the proposed model specified in Equations (12.28) or (12.29) is a reasonable one, and is useful for sensitivity analysis of the estimates with respect to missing data with different missing mechanisms.

12.5.2 Bayesian Analysis of the Model

In this and the following sections, let $\boldsymbol{\Lambda}_k^T$ and $\boldsymbol{\Lambda}_{\omega k}^T$ be the kth row vectors of $\boldsymbol{\Lambda}$ and $\boldsymbol{\Lambda}_\omega$, respectively; $\psi_{\epsilon k}$ and $\psi_{\delta k}$ be the kth diagonal elements of $\boldsymbol{\Psi}_\epsilon$ and $\boldsymbol{\Psi}_\delta$, respectively. Let $\mathbf{V}_{\text{mis}} = \{\mathbf{v}_{\text{mis}}1, \ldots, \mathbf{v}_{\text{mis}}n\}$ be the set of missing values associated with the manifest variables. Again, we use the useful strategy that combines the idea of data augmentation and MCMC methods in the Bayesian analysis. The observed data $\mathbf{V}_{\text{obs}}$ and the missing data indicator $\boldsymbol{r}$ are augmented with the missing quantities $\{\mathbf{V}_{\text{mis}}, \boldsymbol{\Omega}\}$ in the posterior analysis. The Gibbs sampler (Geman and Geman, 1984) is used to generate a sequence of random observations from the joint posterior distribution $[\boldsymbol{\Omega}, \mathbf{V}_{\text{mis}}, \boldsymbol{\theta}, \boldsymbol{\varphi}|\mathbf{V}_{\text{obs}}, \boldsymbol{r}]$, and then the Bayesian estimates are obtained from the observations in this generated sequence. In this algorithm, observations $\{\boldsymbol{\Omega}, \mathbf{V}_{\text{mis}}, \boldsymbol{\theta}, \boldsymbol{\varphi}\}$ are sampled iteratively from the following conditional distributions: $p(\boldsymbol{\Omega}|\mathbf{V}_{\text{obs}}, \mathbf{V}_{\text{mis}}, \boldsymbol{\theta}, \boldsymbol{\varphi}, \boldsymbol{r}) = p(\boldsymbol{\Omega}|\mathbf{V}, \boldsymbol{\theta}, \boldsymbol{\varphi}, \boldsymbol{r})$, $p(\mathbf{V}_{\text{mis}}|\mathbf{V}_{\text{obs}}, \boldsymbol{\Omega}, \boldsymbol{\theta}, \boldsymbol{\varphi}, \boldsymbol{r})$, $p(\boldsymbol{\varphi}|\mathbf{V}_{\text{obs}}, \mathbf{V}_{\text{mis}}, \boldsymbol{\Omega}, \boldsymbol{\theta}, \boldsymbol{r}) = p(\boldsymbol{\varphi}|\mathbf{V}, \boldsymbol{\Omega}, \boldsymbol{\theta}, \boldsymbol{r})$, and $p(\boldsymbol{\theta}|\mathbf{V}_{\text{obs}}, \mathbf{V}_{\text{mis}}, \boldsymbol{\Omega}, \boldsymbol{\varphi}, \boldsymbol{r}) = p(\boldsymbol{\theta}|\mathbf{V}, \boldsymbol{\Omega})$. Note that because of the data augmentation, the last conditional distribution does not depend on $\boldsymbol{r}$, and can be obtained from the expressions in Chapter 8.

Consider the conditional distribution $p(\boldsymbol{\Omega}|\mathbf{V}, \boldsymbol{\theta}, \boldsymbol{\varphi}, \boldsymbol{r})$. Note that when $\mathbf{V}_{\text{mis}}$ is given, the underlying model with missing data reduces to an NSEM with fully observed data. Thus, it follows from a derivation similar to that in Chapter 8 that

$$
\begin{aligned}
p(\boldsymbol{\Omega}|\mathbf{V}, \boldsymbol{\theta}, \boldsymbol{\varphi}, \boldsymbol{r}) &= \prod_{i=1}^{n} p(\boldsymbol{\omega}_i|\mathbf{v}_i, \boldsymbol{\theta}, \boldsymbol{\varphi}, \boldsymbol{r}_i) \\
&\propto \prod_{i=1}^{n} p(\mathbf{v}_i|\boldsymbol{\omega}_i, \boldsymbol{\theta}) p(\boldsymbol{\eta}_i|\boldsymbol{\xi}_i, \boldsymbol{\theta}) p(\boldsymbol{\xi}_i|\boldsymbol{\theta}) p(\boldsymbol{r}_i|\mathbf{v}_i, \boldsymbol{\omega}_i, \boldsymbol{\varphi}),
\end{aligned}
$$

where $p(\boldsymbol{\omega}_i|\mathbf{v}_i, \boldsymbol{\theta}, \boldsymbol{\varphi}, \boldsymbol{r}_i)$ is proportional to

$$
\begin{aligned}
\exp\Bigg\{ &-\frac{1}{2}\boldsymbol{\xi}_i^T \boldsymbol{\Phi}^{-1}\boldsymbol{\xi}_i - \frac{1}{2}(\mathbf{v}_i - \boldsymbol{\mu} - \boldsymbol{\Lambda}\boldsymbol{\omega}_i)^T \boldsymbol{\Psi}_\epsilon^{-1}(\mathbf{v}_i - \boldsymbol{\mu} - \boldsymbol{\Lambda}\boldsymbol{\omega}_i) - \frac{1}{2}[\boldsymbol{\eta}_i - \boldsymbol{\Lambda}_\omega \boldsymbol{G}(\boldsymbol{\omega}_i)]^T \\
&\times \boldsymbol{\Psi}_\delta^{-1}[\boldsymbol{\eta}_i - \boldsymbol{\Lambda}_\omega \boldsymbol{G}(\boldsymbol{\omega}_i)] + \left(\sum_{j=1}^{p} r_{ij}\right)\boldsymbol{\varphi}^T \boldsymbol{d}_i - p\log[1 + \exp(\boldsymbol{\varphi}^T \boldsymbol{d}_i)]\Bigg\}. \qquad (12.30)
\end{aligned}
$$

To derive $p(\mathbf{V}_{\text{mis}}|\mathbf{V}_{\text{obs}}, \boldsymbol{\Omega}, \boldsymbol{\theta}, \boldsymbol{\varphi}, \boldsymbol{r})$, we note that $\boldsymbol{r}_i$ is independent of $\boldsymbol{\theta}$. Moreover, as $\boldsymbol{\Psi}_\epsilon$ is diagonal, $\mathbf{v}_{\text{mis}i}$ is independent of $\mathbf{v}_{\text{obs}i}$. It can be shown that:

$$
\begin{aligned}
p(\mathbf{V}_{\text{mis}}|\mathbf{V}_{\text{obs}}, \boldsymbol{\Omega}, \boldsymbol{\theta}, \boldsymbol{\varphi}, \boldsymbol{r}) &= \prod_{i=1}^{n} p(\mathbf{v}_{\text{mis}i}|\mathbf{v}_{\text{obs}i}, \boldsymbol{\omega}_i, \boldsymbol{\theta}, \boldsymbol{\varphi}, \boldsymbol{r}_i) \\
&\propto \prod_{i=1}^{n} p(\mathbf{v}_{\text{mis}i}|\boldsymbol{\omega}_i, \boldsymbol{\theta}) p(\boldsymbol{r}_i|\mathbf{v}_i, \boldsymbol{\omega}_i, \boldsymbol{\varphi}),
\end{aligned}
$$

and $p(\mathbf{v}_{\text{mis}i}|\mathbf{v}_{\text{obs}i}, \boldsymbol{\omega}_i, \boldsymbol{\theta}, \boldsymbol{\varphi}, \boldsymbol{r}_i)$ is proportional to

$$\frac{\exp\left[-\frac{1}{2}(\mathbf{v}_{\text{mis}i}-\boldsymbol{\mu}_{\text{mis}i}-\boldsymbol{\Lambda}_{\text{mis}i}\boldsymbol{\omega}_i)^T\boldsymbol{\Psi}_{\epsilon\text{mis}i}^{-1}(\mathbf{v}_{\text{mis}i}-\boldsymbol{\mu}_{\text{mis}i}-\boldsymbol{\Lambda}_{\text{mis}i}\boldsymbol{\omega}_i)+\left(\sum_{j=1}^{p} r_{ij}\right)\boldsymbol{\varphi}^T\boldsymbol{d}_i\right]}{[1+\exp(\boldsymbol{\varphi}^T\boldsymbol{d}_i)]^p}, \tag{12.31}$$

where $\boldsymbol{\mu}_{\text{mis}i}$ is a $p_{2i}\times 1$ subvector of $\boldsymbol{\mu}$ with its elements corresponding to missing components of $\mathbf{v}_i$, $\boldsymbol{\Lambda}_{\text{mis}i}$ is a $p_{2i}\times q$ submatrix of $\boldsymbol{\Lambda}$ with its rows corresponding to missing mponents of $\mathbf{v}_i$, and $\boldsymbol{\Psi}_{\epsilon\text{mis}i}$ i is a $p_{2i}\times p_{2i}$ submatrix of $\boldsymbol{\Psi}_\epsilon$ with the rows and columns corresponding to missing components of $\mathbf{v}_i$.

Finally, we consider the conditional distribution of $\boldsymbol{\varphi}$ given $\mathbf{V}, \boldsymbol{\Omega}, \boldsymbol{\theta}$ and $\boldsymbol{r}$. Let $p(\boldsymbol{\varphi})$ be the prior density of $\boldsymbol{\varphi}$, such that $\varphi \overset{D}{=} N(\boldsymbol{\varphi}_0, \mathbf{H}_{0\varphi})$, where $\boldsymbol{\varphi}_0$ and $\mathbf{H}_{0\varphi}$ are the hyperparameters whose values are assumed to be given by the prior information. Since the distribution of $\boldsymbol{r}$ only involves $\mathbf{V}$, $\boldsymbol{\Omega}$ and $\boldsymbol{\varphi}$, and it is assumed that the prior distribution of $\boldsymbol{\varphi}$ is independent of the prior distribution of $\boldsymbol{\theta}$, we have

$$p(\boldsymbol{\varphi}|\mathbf{V}, \boldsymbol{\Omega}, \boldsymbol{\theta}, \boldsymbol{r}) \propto p(\boldsymbol{r}|\mathbf{V}, \boldsymbol{\Omega}, \boldsymbol{\varphi})p(\boldsymbol{\varphi}).$$

Thus, it follows from Equations (12.27) and (12.28) that $p(\boldsymbol{\varphi}|\mathbf{V}, \boldsymbol{\Omega}, \boldsymbol{\theta}, \boldsymbol{r})$ is proportional to

$$\frac{\exp\left[\sum_{i=1}^{n}\left(\sum_{j=1}^{p} r_{ij}\right)\boldsymbol{\varphi}^T\boldsymbol{d}_i-\frac{1}{2}(\boldsymbol{\varphi}-\boldsymbol{\varphi}_0)^T\mathbf{H}_{0\varphi}^{-1}(\boldsymbol{\varphi}-\boldsymbol{\varphi}_0)\right]}{\prod_{i=1}^{n}[1+\exp(\boldsymbol{\varphi}^T\boldsymbol{d}_i)]^p}. \tag{12.32}$$

This completes the derivation of the full conditional distributions that are required in the implementation of the Gibbs sampler. The conditional distributions $p(\boldsymbol{\omega}_i|\mathbf{v}_i, \boldsymbol{\theta}, \boldsymbol{\varphi}, \boldsymbol{r}_i)$, $p(\mathbf{v}_{\text{mis}i}|\mathbf{v}_{\text{obs}i}, \boldsymbol{\omega}_i, \boldsymbol{\theta}, \boldsymbol{\varphi}, \boldsymbol{r}_i)$ and $p(\boldsymbol{\varphi}|\mathbf{V}, \boldsymbol{\Omega}, \boldsymbol{\theta}, \boldsymbol{r})$ are nonstandard. Some details of the MH algorithm for simulating observations form these conditional distributions are presented in Appendix 12.1

We again propose the well-known Bayes factor for model comparisons. Let M_0 and M_1 be two competing models, the Bayes factor (see Kass and Raftery, 1995) is defined as

$$B_{10}=\frac{p(\mathbf{V}_{\text{obs}}, \boldsymbol{r}|M_1)}{p(\mathbf{V}_{\text{obs}}, \boldsymbol{r}|M_0)}, \tag{12.33}$$

where

$$p(\mathbf{V}_{\text{obs}}, \boldsymbol{r}|M_k)=\int p(\mathbf{V}_{\text{obs}}, \boldsymbol{r}|\boldsymbol{\theta}_k, \boldsymbol{\varphi}_k)p(\boldsymbol{\theta}_k, \boldsymbol{\varphi}_k)\mathrm{d}\boldsymbol{\theta}_k\mathrm{d}\boldsymbol{\varphi}_k, \quad k=1,0 \tag{12.34}$$

is the marginal density of M_k with parameter vectors $\boldsymbol{\theta}_k$ and $\boldsymbol{\varphi}_k$, and $p(\boldsymbol{\theta}_k, \boldsymbol{\varphi}_k)$ is the prior density of $\boldsymbol{\theta}_k$ and $\boldsymbol{\varphi}_k$. As $p(\mathbf{V}_{\text{obs}}, \boldsymbol{r}|M_k)$ is difficult to evaluate, computation of the Bayes factor is a very challenging problem (see DiCiccio, Kass, Raftery and Wasserman, 1997). In a similar way to the treatment of other models, the following path sampling (Gelman and Meng, 1998) procedure is used to evaluate the logarithm Bayes factor.

We consider the following class of densities defined with a continuous parameter $t \in [0, 1]$:

$$p(\boldsymbol{\theta}, \boldsymbol{\varphi}, \boldsymbol{\Omega}, \mathbf{V}_{\text{mis}}|\mathbf{V}_{\text{obs}}, \boldsymbol{r}, t) = p(\boldsymbol{\theta}, \boldsymbol{\varphi}, \boldsymbol{\Omega}, \mathbf{V}_{\text{mis}}, \mathbf{V}_{\text{obs}}, \boldsymbol{r}|t)/z(t),$$

where

$$z(t) = p(\mathbf{V}_{\text{obs}}, \boldsymbol{r}|t) = \int p(\boldsymbol{\theta}, \boldsymbol{\varphi}, \boldsymbol{\Omega}, \mathbf{V}_{\text{mis}}, \mathbf{V}_{\text{obs}}, \boldsymbol{r}|t) d\boldsymbol{\theta} d\boldsymbol{\varphi} d\boldsymbol{\Omega} d\mathbf{V}_{\text{mis}},$$

and t is in [0,1] for linking the competing models M_0 and M_1 such that for $k = 0, 1$, $z(k) = p(\mathbf{V}_{\text{obs}}, \boldsymbol{r}|t = k) = p(\mathbf{V}_{\text{obs}}, \boldsymbol{r}|M_k)$. Let $U(\boldsymbol{\theta}, \boldsymbol{\varphi}, \boldsymbol{\Omega}, \mathbf{V}_{\text{mis}}, \mathbf{V}_{\text{obs}}, \boldsymbol{r}, t) = d \log p(\boldsymbol{\Omega}, \mathbf{V}_{\text{mis}}, \mathbf{V}_{\text{obs}}, \boldsymbol{r}|\boldsymbol{\theta}, \boldsymbol{\varphi}, t)/dt$, where $p(\boldsymbol{\Omega}, \mathbf{V}_{\text{mis}}, \mathbf{V}_{\text{obs}}, \boldsymbol{r}|\boldsymbol{\theta}, \boldsymbol{\varphi}, t)$ is the complete data likelihood function at t. On the basis of a natural assumption that the prior of $(\boldsymbol{\theta}, \boldsymbol{\varphi})$ is independent of t, it can be shown by similar reasoning to that in the previous chapters that

$$\log B_{10} = \log \frac{z(1)}{z(0)} = \int_0^1 E_{\boldsymbol{\theta}, \boldsymbol{\varphi}, \boldsymbol{\Omega}, \mathbf{V}_{\text{mis}}}[U(\boldsymbol{\theta}, \boldsymbol{\varphi}, \boldsymbol{\Omega}, \mathbf{V}_{\text{mis}}, \mathbf{V}_{\text{obs}}, \boldsymbol{r}, t)] dt,$$

where $E_{\boldsymbol{\theta}, \boldsymbol{\varphi}, \boldsymbol{\Omega}, \mathbf{V}_{\text{mis}}}$ is the expectation with respect to the distribution $p(\boldsymbol{\theta}, \boldsymbol{\varphi}, \boldsymbol{\Omega}, \mathbf{V}_{\text{mis}}|\mathbf{V}_{\text{obs}}, \boldsymbol{r}, t)$. Let $t_{(0)} = 0 < t_{(1)} < t_{(2)} < \cdots < t_{(S)} < t_{(S+1)} = 1$ be fixed and ordered grids; then, $\log B_{10}$ is estimated by

$$\widehat{\log B_{10}} = \frac{1}{2} \sum_{s=0}^{S} (t_{(s+1)} - t_{(s)})(\bar{U}_{(s+1)} + \bar{U}_s),$$

where

$$\bar{U}_{(s)} = J^{-1} \sum_{j=1}^{J} U(\boldsymbol{\theta}^{(j)}, \boldsymbol{\varphi}^{(j)}, \boldsymbol{\Omega}^{(j)}, \mathbf{V}_{\text{mis}}^{(j)}, \mathbf{V}_{\text{obs}}, \boldsymbol{r}, t_{(s)}),$$

and $\{(\boldsymbol{\theta}^{(j)}, \boldsymbol{\varphi}^{(j)}, \boldsymbol{\Omega}^{(j)}, \mathbf{V}_{\text{mis}}^{(j)}), j = 1, \ldots, J\}$ are observations that are simulated from $p(\boldsymbol{\theta}, \boldsymbol{\varphi}, \boldsymbol{\Omega}, \mathbf{V}_{\text{mis}}|\mathbf{V}_{\text{obs}}, \boldsymbol{r}, t_{(s)})$.

12.5.3 An Illustrative Real Example

To give an illustration of the proposed methodology, a small portion of the ICPSR data set collected by the World Values Survey 1981–1984 and 1990–1993 (World Values Study Group, 1994) is analyzed in this example (see also Lee and Tang (2006)). Here, eight variables of the original data set (variables 116, 117, 252, 253, 254, 296, 298 and 314, see Appendix 1.1) are taken as manifest variables in $\mathbf{v} = (v_1, \ldots, v_8)^T$. These variables are measured on a 10-point scale and for convenience they are treated as continuous. We choose the data corresponding to females in Russia, who either answered question 116 or 117, or both. Under this choice, most of the data were obtained from working females. There are 712 random observations in the data set in which there are only 451 (63.34%) fully observed cases. The missing data are rather complicated, with 69 different missing patterns. Considering that the questions are either related to personal attitudes or related to personal morality, the corresponding missing data are better treated as nonignorable. To roughly unify the scales, the raw data are standardized using the sample mean and standard deviation obtained from the fully observed data.

Based on the meanings of the questions corresponding to the manifest variables, we propose an NSEM with the following specifications. For the measurement equation, we consider $\boldsymbol{\mu} = (\mu_1, \ldots, \mu_8)^T$, and

$$\boldsymbol{\Lambda}^T = \begin{bmatrix} 1.0^* & \lambda_{21} & 0.0^* & 0.0^* & 0.0^* & 0.0^* & 0.0^* & 0.0^* \\ 0.0^* & 0.0^* & 1.0^* & \lambda_{42} & \lambda_{52} & 0.0^* & 0.0^* & 0.0^* \\ 0.0^* & 0.0^* & 0.0^* & 0.0^* & 0.0^* & 1.0^* & \lambda_{73} & \lambda_{83} \end{bmatrix},$$

which corresponds to latent variables η, ξ_1 and ξ_2. As before, we choose the structure that gives nonoverlapping latent factors on the basis of the meaning of the questions, and for clear interpretation. The latent variable η can be roughly interpreted as 'job satisfaction', and the latent variables ξ_1 and ξ_2 can be roughly interpreted as 'job attitude' and 'morality (in relation to money)', respectively. We first consider the following model M_1, which involves an encompassing structural equation with all second-order terms of ξ_{i1} and ξ_{i2}:

$$M_1 : \eta_i = \gamma_1\xi_{i1} + \gamma_2\xi_{i2} + \gamma_3\xi_{i1}\xi_{i2} + \gamma_4\xi_{i1}^2 + \gamma_5\xi_{i2}^2 + \delta_i.$$

Although the Bayes factor can be applied to compare many other missing data models, the following three models are considered for assessing the missing data in this illustrative example:

$$M_a : \quad \text{logit}[\text{pr}(r_{ij} = 1|\mathbf{v}_i, \boldsymbol{\omega}_i, \boldsymbol{\varphi})] = \varphi_0 + \varphi_1 v_{i1} + \cdots + \varphi_8 v_{i8},$$

$$M_b : \quad \text{logit}[\text{pr}(r_{ij} = 1|\mathbf{v}_i, \boldsymbol{\omega}_i, \boldsymbol{\varphi})] = \varphi_0 + \varphi_1\eta_i + \varphi_2\xi_{i1} + \varphi_3\xi_{i2},$$

$$M_M : \quad \text{MAR}.$$

Note that M_a involves all the manifest variables, while M_b involves all the latent variables.

The logarithm Bayes factors for comparing the above models M_a, M_b and M_M under the NSEM M_1 are computed via the path sampling procedure. The prior inputs in the conjugate prior distributions are selected as before via an auxilary estimation. The number of grids in the path sampling procedure for computing all the logarithm Bayes factors is taken to be 10, and for each $t_{(s)}$ 5000 simulated observations collected after 5000 burn-in iterations are used to compute $\bar{U}_{(s)}$. To give some idea about the convergent behaviors in simulating the observations, plots of EPSR values (Gelman, 1996) of all parameters against iterations in analyzing the encompassing model M_1 with M_a are displayed in Figure 12.1. In addition, plots of three parallel sequences generated from different starting values of some parameters (others are similar) are presented in Figure 12.2. The estimated log Bayes factors are equal to $\widehat{\log B}^{1}_{ab} = 47.34$ and $\widehat{\log B}^{1}_{aM} = 43.85$, where the superscript of $\widehat{\log B}$ indicates the NSEM M_1, and subscripts of $\widehat{\log B}$ indicate the competing models for the missing mechanism. Clearly, from the estimated logarithm Bayes factors, the data give strong evidence to support the missing data model M_a, which is in the form of the proposed model Equation (12.29), for modeling the nonignorable missing data.

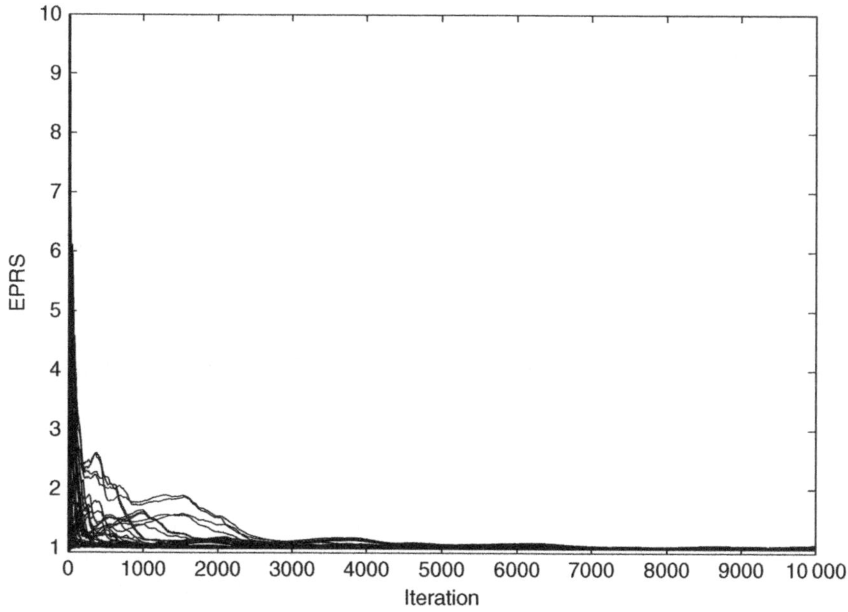

Figure 12.1 EPSR values of all parameters against iteration numbers in the real example with encompassing model M_1 and missing data model M_a. This figure is taken from Lee and Tang (2006).

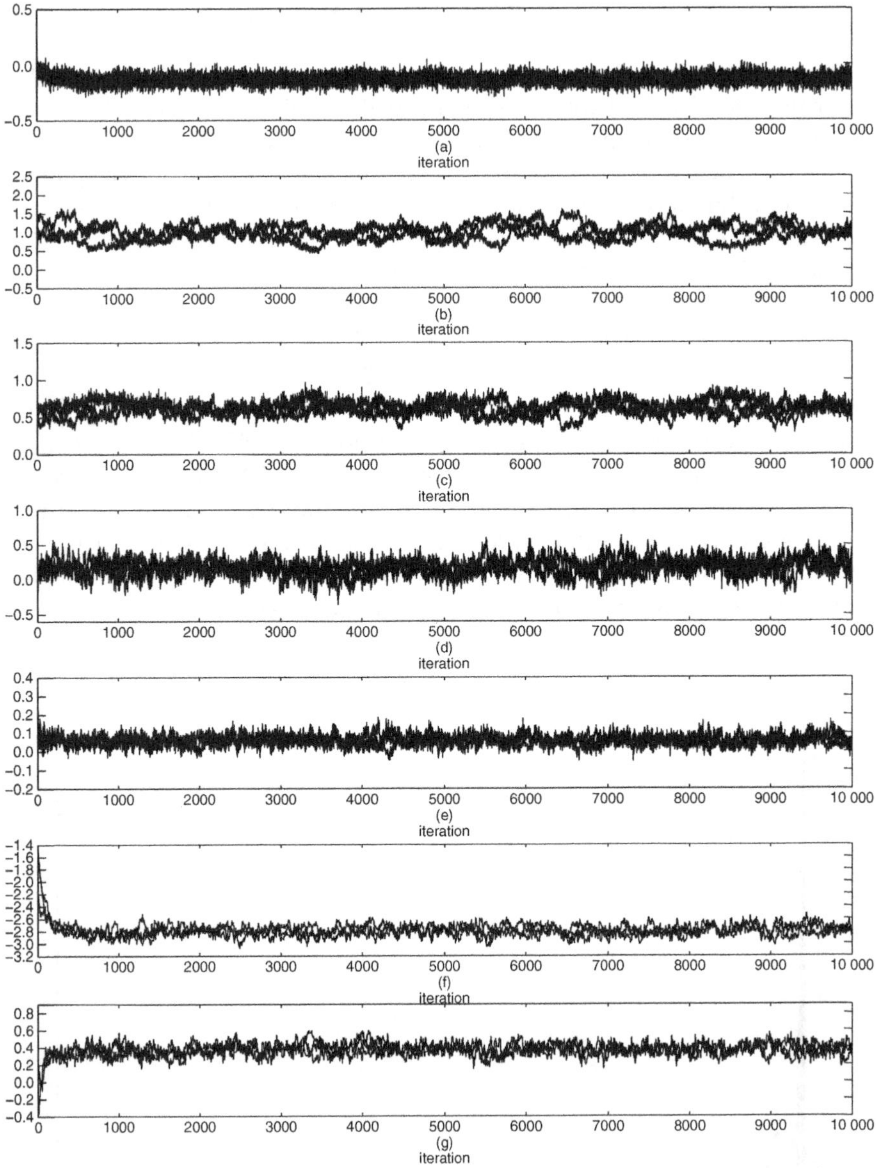

Figure 12.2 (a), (b), (c), (d), (e), (f) and (g) are plots of three parallel sequences corresponding to different starting values of μ_5, λ_{21}, $\psi_{\epsilon 2}$, γ_2, ϕ_{12}, φ_0 and φ_3 against iteration numbers in the real example with the encompassing model M_1, and missing data model M_a. This figure is taken from Lee and Tang (2006).

In addition to the encompassing NSEM M_1, we also consider the following NSEMs:

$$
\begin{aligned}
M_2: &\quad \eta_i = \gamma_1 \xi_{i1} + \gamma_2 \xi_{i2} + \gamma_3 \xi_{i1}^2 + \delta_i, \\
M_3: &\quad \eta_i = \gamma_1 \xi_{i1} + \gamma_2 \xi_{i2} + \gamma_3 \xi_{i1} \xi_{i2} + \delta_i, \\
M_4: &\quad \eta_i = \gamma_1 \xi_{i1} + \gamma_2 \xi_{i2} + \gamma_3 \xi_{i2}^2 + \delta_i.
\end{aligned}
$$

Under each of the above models, we compare the missing mechanism models M_a, M_b and M_M. The estimated logarithm Bayes factors are reported in Table 12.8, for example, $\widehat{\log B}_{ab}^2 = 48.56$ and $\widehat{\log B}_{aM}^3 = 44.04$. It is clear from the results in this table that for every M_2, M_3 and M_4, the data strongly support M_a. For each case, Bayesian estimates obtained under M_a are different from those obtained under MAR. To save space, these estimates are not reported. Based on the above comparison results, the model M_a is used for comparing NSEMs M_1, M_2, M_3 and M_4. In this model comparison, the estimated log Bayes factors are equal to $\widehat{\log B}_{12} = -1.29$, $\widehat{\log B}_{32} = -2.59$ and $\widehat{\log B}_{42} = -1.51$. These results give evidence that the data support the NSEM M_2.

The Bayesian estimates and their standard error estimates of the unknown parameters in the selected model M_2 are presented in the left columns of Table 12.9. We observe that the estimates of the coefficients $\varphi_0, \varphi_2, \varphi_3, \varphi_4, \varphi_5, \varphi_6$ and φ_8 are significantly different from zero. This result indicates that the nonignorable missing data model for accounting the nature of the missing data is necessary. The factor loading estimates indicate significant associations between the latent variables and their indicators. From $\hat{\phi}_{11}, \hat{\phi}_{12}$ and $\hat{\phi}_{22}$, the estimate of the correlation between ξ_1 and ξ_2 is 0.163. This estimate indicates that 'job attitude, ξ_1', and 'morality, ξ_2', is weakly correlated. The estimated nonlinear structural equation is equal to

$$\eta_i = -0.103\xi_{i1} + 0.072\xi_{i2} + 0.306\xi_{i1}^2.$$

Table 12.8 The estimated log Bayes factors: $\log B_{ab}^r$ and $\log B_{aM}^r$, $r = 2, 3, 4$.

SEM	$\log B_{ab}^r$	$\log B_{aM}^r$
M_2	48.56	46.66
M_3	50.40	44.04
M_4	50.38	44.69

Note: Tables 12.8 and 12.9 are taken from Lee and Tang (2006).

Table 12.9 Bayesian estimates and their standard error estimates of M_2 with M_a.

PAR	Our method EST	Our method SE	WinBUGS EST	WinBUGS SE	PAR	Our method EST	Our method SE	WinBUGS EST	WinBUGS SE
φ_0	−2.791	0.043	−2.794	0.076	μ_1	−0.135	0.038	−0.139	0.065
φ_1	0.038	0.033	0.040	0.059	μ_2	−0.136	0.032	−0.129	0.058
φ_2	−0.280	0.037	−0.280	0.068	μ_3	0.018	0.023	0.015	0.039
φ_3	0.370	0.036	0.365	0.073	μ_4	0.004	0.023	0.005	0.041
φ_4	−0.265	0.041	−0.262	0.083	μ_5	−0.129	0.026	−0.139	0.045
φ_5	−0.455	0.070	−0.502	0.126	μ_6	−0.046	0.023	−0.040	0.041
φ_6	−0.405	0.073	−0.341	0.154	μ_7	0.053	0.026	0.045	0.046
φ_7	0.059	0.056	0.013	0.134	μ_8	0.144	0.026	0.141	0.045
φ_8	0.332	0.038	0.323	0.061	λ_{21}	0.917	0.129	0.830	0.168
					λ_{42}	0.307	0.060	0.317	0.123
					λ_{52}	0.328	0.068	0.320	0.119
					λ_{73}	1.244	0.122	0.955	0.203
					λ_{83}	0.455	0.071	0.388	0.114
					$\psi_{\epsilon 1}$	0.544	0.067	0.508	0.096
					$\psi_{\epsilon 2}$	0.637	0.060	0.673	0.080
					$\psi_{\epsilon 3}$	0.493	0.058	0.492	0.111
					$\psi_{\epsilon 4}$	0.935	0.033	0.932	0.059
					$\psi_{\epsilon 5}$	0.907	0.039	0.922	0.068
					$\psi_{\epsilon 6}$	0.640	0.042	0.548	0.095
					$\psi_{\epsilon 7}$	0.612	0.051	0.714	0.086
					$\psi_{\epsilon 8}$	1.065	0.040	1.065	0.069
					γ_1	−0.103	0.047	−0.103	0.085
					γ_2	0.072	0.052	0.044	0.081
					γ_3	0.306	0.083	0.317	0.139
					ϕ_{11}	0.459	0.057	0.459	0.113
					ϕ_{12}	0.062	0.016	0.071	0.033
					ϕ_{22}	0.316	0.041	0.405	0.096
					ψ_δ	0.413	0.056	0.463	0.105

The interpretation of this equation in relation to the effect of the exogenous latent variables ξ_{i1} and ξ_{i2} to the endogenous latent variable η_i is similar to the interpretation of other nonlinear structural equations.

12.6 ANALYSIS OF SEMs WITH MISSING DATA VIA WINBUGS

The software WinBUGS (Spiegelhalter, Thomas, Best and Lunn, 2003) can be used to produce Bayesian solutions, such as Bayesian estimates, standard error estimates and estimated residuals, for SEMs with missing data that are ignorably missing with MAR or missing with a nonignorable missing mechanism.

As nonignorable missing data subsume missing data that are MAR, we focus on a discussion about nonignorable missing data. In the context of a specific SEM, in addition to the WinBUGS codes that are required to specify that SEM, we require codes that relate to the nonignorable missing data, which involve the definition of the missing model, the prior distribution in relation to the parameters in the missing model, etc..

The real example given in Section 12.5.3 has been reanalyzed by using WinBUGS with the same settings – for example, the same model structure and same prior inputs. Results obtained on the basis of M_2 with M_a are reported in the right-hand columns of Table 12.9. The DIC value corresponding to this model is 16961.2. We observe that the estimates obtained from WinBUGS are close to our estimates that are presented in the left-hand columns of Table 12.9. However, the numerical standard error (SE) estimates produced by this general software are larger than those that are produced by our tailor-made program for the specific SEM. Estimated residual plots, $\hat{\epsilon}_{i1}$, $\hat{\epsilon}_{i4}$ and $\hat{\delta}_i$ versus case numbers, are displayed in Figure 12.3. Some estimated residual plots of $\hat{\delta}_i$ versus $\hat{\xi}_{i1}$ and $\hat{\xi}_{i2}$, and plots of $\hat{\epsilon}_{i4}$ versus $\hat{\xi}_{i1}$, $\hat{\xi}_{i2}$ and $\hat{\eta}_i$, are presented in Figures 12.4 and 12.5, respectively. These plots roughly indicate

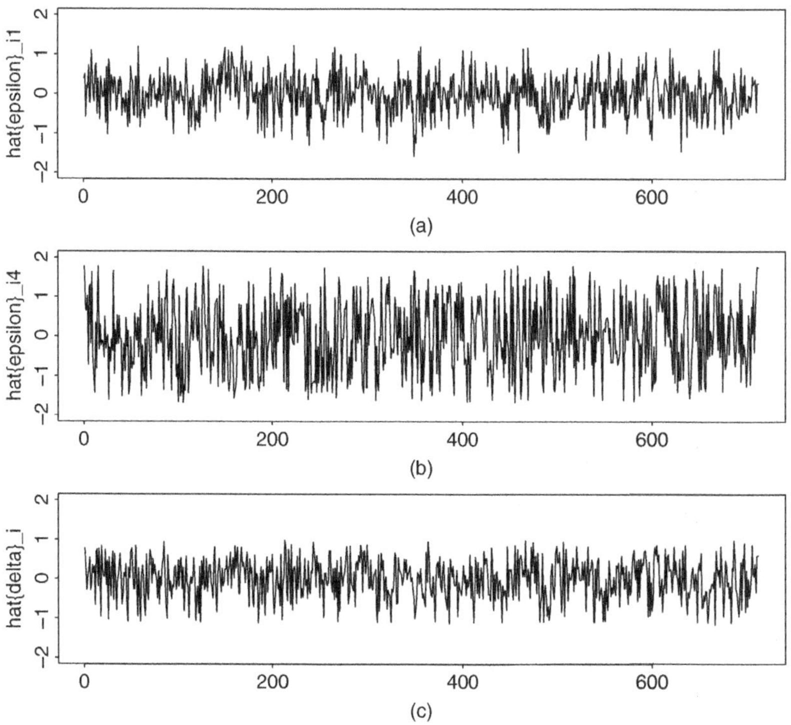

Figure 12.3 Estimated residual plots (a) $\hat{\epsilon}_{i1}$, (b) $\hat{\epsilon}_{i4}$ and (c) $\hat{\delta}_i$.

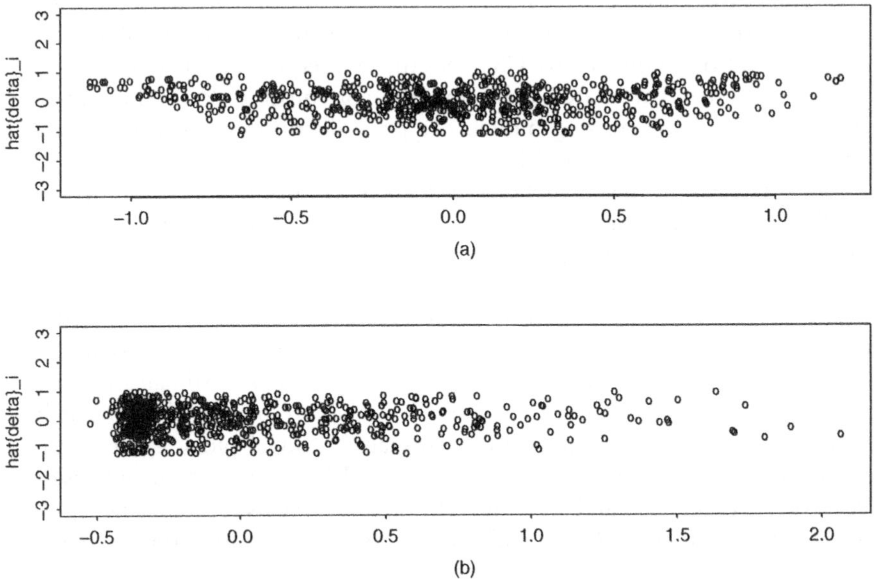

Figure 12.4 Plots of estimated residuals $\hat{\delta}_i$ versus (a) $\hat{\xi}_{i1}$ and (b) $\hat{\xi}_{i2}$.

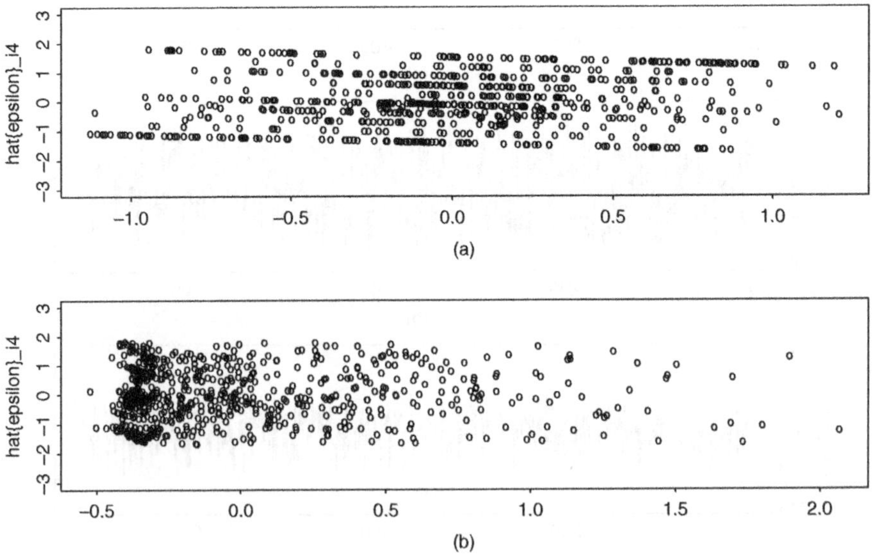

Figure 12.5 Plots of estimated residuals $\hat{\epsilon}_{i4}$ versus (a) $\hat{\xi}_{i1}$, (b) $\hat{\xi}_{i2}$ and (c) $\hat{\eta}_i$.

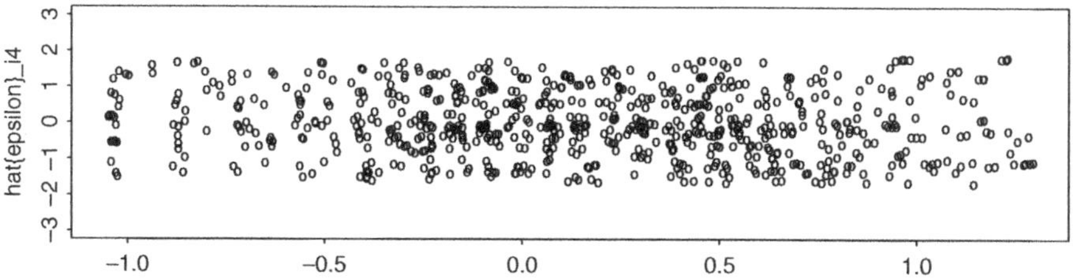

(c)

Figure 12.5 (Continued)

that the measurement equation and the structural equation are adequate. From Figure 12.5(b) we observe that the empirical distribution of $\hat{\xi}_{i2}$ is skewed to the left. For this complicated nonlinear SEM with nonignorable missing data, estimates of latent variables should be used with great caution. The WinBUGS codes and data are given at the following website: http://www.wiley.com/go/lee_structural.

APPENDIX 12.1: IMPLEMENTATION OF THE MH ALGORITHM

Simulating observations from $p(\boldsymbol{\omega}_i|\mathbf{v}_i, \boldsymbol{r}_i, \boldsymbol{\theta}, \boldsymbol{\varphi})$ is based on the reasoning in Roberts (1996). Specially, we choose $N(\mathbf{0}, \sigma_\omega^2 \boldsymbol{\Omega}_\omega)$ as the proposal distribution, where $\boldsymbol{\Omega}_\omega^{-1} = \boldsymbol{\Sigma}_\omega + \boldsymbol{\Lambda}^T \boldsymbol{\Psi}_\epsilon^{-1} \boldsymbol{\Lambda}$,

$$\boldsymbol{\Sigma}_\omega = \begin{pmatrix} \boldsymbol{\Pi}_0^T \boldsymbol{\Psi}_\delta^{-1} \boldsymbol{\Pi}_0 & -\boldsymbol{\Pi}_0^T \boldsymbol{\Psi}_\delta^{-1} \boldsymbol{\Gamma}\boldsymbol{\Delta} \\ -\boldsymbol{\Delta}^T \boldsymbol{\Gamma}^T \boldsymbol{\Psi}_\delta^{-1} \boldsymbol{\Pi}_0 & \boldsymbol{\Phi}^{-1} + \boldsymbol{\Delta}^T \boldsymbol{\Gamma}^T \boldsymbol{\Psi}_\delta^{-1} \boldsymbol{\Gamma}\boldsymbol{\Delta} \end{pmatrix} + \frac{p \exp(\varphi_0 + \boldsymbol{\varphi}_I^T \mathbf{v}_i)}{(1 + \exp(\varphi_0 + \boldsymbol{\varphi}_I^T \mathbf{v}_i))^2} \boldsymbol{\varphi}_{II} \boldsymbol{\varphi}_{II}^T,$$

with $\boldsymbol{\Pi}_0 = \mathbf{I} - \boldsymbol{\Pi}$, $\boldsymbol{\Delta} = \partial \boldsymbol{H}(\boldsymbol{\xi}_i)/\partial \boldsymbol{\xi}_i^T |_{\boldsymbol{\xi}_i = \mathbf{0}}$, $\boldsymbol{\varphi}_I = (\varphi_1, \ldots, \varphi_p)^T$, and $\boldsymbol{\varphi}_{II} = (\varphi_{p+1}, \ldots, \varphi_{p+q})^T$. The MH algorithm is implemented as follows. At the $(j+1)$th iteration with a current value $\boldsymbol{\omega}_i^{(j)}$, a new candidate $\boldsymbol{\omega}_i$ is generated from $N(\boldsymbol{\omega}_i^{(j)}, \sigma_\omega^2 \boldsymbol{\Omega}_\omega)$ and accepted with probability:

$$\min\left[1, \frac{p(\boldsymbol{\omega}_i|\mathbf{v}_i, \boldsymbol{r}_i, \boldsymbol{\theta}, \boldsymbol{\varphi})}{p(\boldsymbol{\omega}_i^{(j)}|\mathbf{v}_i, \boldsymbol{r}_i, \boldsymbol{\theta}, \boldsymbol{\varphi})}\right].$$

The variance σ_ω^2 can be chosen such that the average acceptance rate is approximately 0.25 or more (see Gelman, Roberts and Gilks, 1995).

To sample $\mathbf{v}_{mi}$ and $\boldsymbol{\varphi}$ from $p(\mathbf{v}_{mi}|\mathbf{v}_{oi}, \boldsymbol{\omega}_i, \boldsymbol{\theta}, \boldsymbol{\varphi}, \boldsymbol{r}_i)$ and $p(\boldsymbol{\varphi}|\mathbf{V}, \boldsymbol{\Omega}, \boldsymbol{r}, \boldsymbol{\theta})$, we respectively choose $N(\mathbf{0}, \sigma_{\mathrm{v}}^2\boldsymbol{\Omega}_{\mathrm{v}mi})$ and $N(\mathbf{0}, \sigma_{\varphi}^2\boldsymbol{\Omega}_{\varphi})$ as their corresponding proposal distributions, where

$$\boldsymbol{\Omega}_{\mathrm{v}mi}^{-1} = \boldsymbol{\Psi}_{\epsilon mi}^{-1} + \frac{p\exp(\varphi_0 + \sum_{l\in\bar{D}} \varphi_l \mathrm{v}_{il} + \boldsymbol{\varphi}_{II}^T\boldsymbol{\omega}_i)}{\left(1+\exp\left[\varphi_0 + \sum_{l\in\bar{D}} \varphi_l \mathrm{v}_{il} + \boldsymbol{\varphi}_{II}^T\boldsymbol{\omega}_i)\right]\right)^2}\boldsymbol{\varphi}_{\mathrm{m}}\boldsymbol{\varphi}_{\mathrm{m}}^T,$$

$$\boldsymbol{\Omega}_{\varphi}^{-1} = \frac{p}{4}\sum_{i=1}^{n} \boldsymbol{d}_i\boldsymbol{d}_i^T + \mathbf{H}_{0\varphi}^{-1},$$

in which $\boldsymbol{\varphi}_m$ is a vector that contains the elements of $\boldsymbol{\varphi}$ corresponding to $\mathbf{v}_{mi}$, $\boldsymbol{\Psi}_{\epsilon mi}$ is a submatrix of $\boldsymbol{\Psi}_{\epsilon}$ corresponding to $\mathbf{v}_{mi}$, $\bar{D}$ is the set of indexes corresponding to $\mathbf{v}_{oi}$ and σ_{v}^2 and σ_{φ}^2 are chosen as before.

REFERENCES

Afifi, A. A. and Elashoff, R. M. (1969) Multivariate two sample tests with dichotomous and continuous variables. I: The location model. *The Annals of Mathematical Statistics*, **40**, 290–298.

Arbuckle, J. L. (1996) Full information estimation in the presence of incomplete data. In G. A. Marcoulides and R. E. Schumacker (eds), *Advanced Structural Equation Modeling*. Mahwah, NJ: Lawrence Erlbaum Publishers.

DiCiccio, T. J., Kass, R. E., Raftery, A. and Wasserman, L. (1997) Computing Bayes factors by combining simulation and asymptotic approximations. *Journal of the American Statistical Association*, **92**, 903–915.

Diggle, P. and Kenward, M. G. (1994) Informative drop-out in longitudinal data analysis (with discussion). *Applied Statistics*, **43**, 49–93.

Frühwirth-Schnatter, S. (2001) Markov chain Monte Carlo estimation of classical and dynamic switching and mixture models. *Journal of the American Statistical Association*, **96**, 194–208.

Gelman, A. (1996). Inference and monitoring convergence. In W.R. Gilks, S. Richardson and D.J. Spiegelhalter (eds), *Markov Chain Monte Carlo in Practice*, pp. 131–144. London: Chapman and Hall.

Gelman, A. and Meng, X. L. (1998) Simulating normalizing constants: from importance sampling to bridge sampling to path sampling. *Statistical Science*, **13**, 163–185.

Gelman, A., Meng, X. L. and Stern, H. (1996) Posterior predictive assessment of model fitness via realized discrepancies. *Statistica Sinica*, **6**, 733–759.

Gelman, A., Roberts, G. and Gilks, W. (1995) Efficient Metropolis jumping rules . In J.M. Bernardo, J. O. Berger, A. P. David and A. F. M. Smith (eds), *Bayesian Statistics* (5th edn). Oxford: Oxford University Press.

Geman, S. and Geman, D. (1984) Stochastic relaxation, Gibbs distribution and the Bayesian restoration of images. *IEEE Transactions on Pattern Analysis and Machine Intelligence*, **6**, 721–741.

Hastings, W. K. (1970) Monte Carlo sampling methods using Markov chains and their application. *Biometrika*, **57**, 97–109.

Ibrahim, J. G., Chen, M. H. and Lipsitz, S. R. (2001) Missing responses in generalised linear mixed models when the missing data mechanism is nonignorable. *Biometrika*, **88**, 551–564.

Kass, R. E. and Raftery, A. E. (1995) Bayes factors. *Journal of the American Statistical Association*, **90**, 773–795.

Lee, S. Y. and Song, X. Y. (2004) Bayesian model comparison of nonlinear latent variable models with missing continuous and ordinal categorical data. *British Journal of Mathematical and Statistical Psychology*, **57**, 131–150.

Lee, S. Y. and Tang, N. S. (2006) Bayesian analysis of nonlinear structural equation models with nonignorable missing data. *Psychometrika*, in press.

Little, R. J. A. and Rubin, D. B. (1987) *Statistical Analysis with Missing Data*. New York: John Wiley & Sons, Inc.

Metropolis, N. *et al.* (1953) Equations of state calculations by fast computing machine. *Journal of Chemical Physics*, **21**, 1087–1091.

Morisky, D. E. *et al.* (1998) The effects of establishment practices, knowledge and attitudes on condom use among Filipina sex workers. *AIDS Care*, **10**, 213–220.

Roberts, G. O. (1996) Markov chain concepts related to sampling algorithms. In W. R. Gilks, S. Richardson and D. J. Spiegelhalter (eds), *Markov Chain Monte Carlo in Practice*, pp. 45–58. London: Chapman and Hall.

Song, X. Y. and Lee, S. Y. (2002) Analysis of structural equation model with ignorable missing continuous and polytomous data. *Psychometrika*, **67**, 261–288.

Spiegelhalter, D. J., Thomas, A., Best, N. G. and Lunn, D. (2003) *WinBUGS User Manual, Version 1.4.* Cambridge, England: MRC Biostatistics Unit.

Tanner, M. A. and Wong, W. H. (1987) The calculation of posterier distribution by data augmentation (with discussion). *Journal of the American Statistical Association*, **86**, 79–86.

World Values Study Group (1994) World Values Survey, 1981–1984 and 1990–1993. ICPSR version. Ann Arbor, MI: Institute for Social Research (producer). Ann Arbor, MI: Inter-university Consortium for Political and Social Research (distributor).

13
Structural Equation Models with Exponential Family of Distributions

13.1 INTRODUCTION

The standard and the complex SEMs discussed in previous chapters are developed under the crucial assumption that the conditional distribution of the manifest variables, given the latent variables, is normal. For example, consider the nonlinear SEM defined by Equations (8.1) and (8.2) in Chapter 8. Although we do not assume that $\mathbf{y}$ is normal, we do assume that $\mathbf{y}|\boldsymbol{\omega}$ is normal, or that $\boldsymbol{\epsilon}$ is normal. In this chapter, we generalize the distribution of $\mathbf{y}|\boldsymbol{\omega}$ from normal to the exponential family of distributions (EFDs). This family is very general, it includes discrete distributions such as binomial and Poisson, and continuous distributions such as normal, exponential and gamma, as special cases. Some common distributions in the univariate exponential family are presented in Table 13.1, see also Wedel and Kamakura (2001). The generalization from the normal distribution to EFDs is not only for theoretical interest, but also has significant practical value, particularly in analyzing more general discrete data, which include but are not limited to unordered binary data.

The problem of treating non-normal continuous data has received a considerable amount of attention in the field of SEM. Based on the traditional covariance structure analysis approach, general robust methods against the normal assumption have been developed with the sample covariance matrix (see Browne (1987), Shapiro and Browne (1987), Kano, Berkane and Bentler (1993) and Yuan and Bentler (1997), among others). In this chapter, we present a Bayesian approach that is developed on the basis of raw observations under a slightly

Table 13.1 Some common distributions in the univariate exponential family.

Distribution	Density function	Range
Discrete		
Binomial, $B(K, Pr)$	$\binom{K}{y} Pr^y(1-Pr)^{K-y}$	$[0, K]$
Poisson, $P(\mu)$	$e^{-\mu}\mu^y/y!$	$(0, \infty)$
Continuous		
Normal, $N(\mu, \sigma^2)$	$(2\pi\sigma^2)^{-\frac{1}{2}}\exp[-(y-\mu)^2/2\sigma^2]$	$(-\infty, \infty)$
Gamma, $G(\alpha, \beta)$	$(1/y\Gamma(\beta))(y\beta/\alpha)^\beta \exp(-\beta y/\alpha)$	$(0, \infty)$

more restrictive assumption that the non-normal data are coming from the EFDs. As the methodologies are tailor-made for this family of distributions, they are expected to be more effective for data that satisfy this distributional assumption. Moveover, the Bayesian approach can be applied to more subtle SEMs and complex data structures.

The exponential family of distributions has been extensively used in many areas of statistics, particularly in relation to latent variable models, such as the generalized linear models (McCullagh and Nelder, 1989) and generalized linear mixed models (Booth and Hobert, 1999). In contrast to SEMs, the main objective of these latent variable models is to assess the effects of covariates on the manifest variables, and their latent variables are usually used to model the random effects. They have been extensively applied to longitudinal studies where nonignorable missing data are frequently encountered.

Motivated by the above consideration, in this chapter we will concentrate on a nonlinear SEM that can accommodate covariates, variables from the exponential family of distributions, ordered categorical variables and missing data with a nonignorable mechanism. We will show that the strategy that combines the data augmentation and MCMC methods is again useful in developing the Bayesian methodologies. The model will be described in Section 13.2, Bayesian methods are presented in Section 13.3, and results of a simulation study are reported in Section 13.4. Section 13.5 demonstrates these methods with a real example. The application of WinBUGS will be illustrated with an artificial example in Section 13.6 and Section 13.7 concludes with a discussion.

13.2 THE SEM FRAMEWORK WITH EXPONENTIAL FAMILY OF DISTRIBUTIONS

13.2.1 The Model

The aim of this section is to describe a nonlinear SEM with fixed covariates on the basis of the exponential family of distributions (EFDs) (see Song and

Lee (2006)). For $i = 1, \ldots, n$, let $\mathbf{y}_i = (y_{i1}, \ldots, y_{ip})^T$ be a vector of manifest variables measured on each of the n independently distributed individuals. For brevity, we assume that the dimension of $\mathbf{y}_i$ is the same for every i, however, this assumption can be relaxed without much difficulty. We wish to identify the relationship between the manifest variables in $\mathbf{y}_i$ and the related latent vector $\boldsymbol{\omega}_i = (\omega_{i1}, \ldots, \omega_{iq})^T$ with fixed covariates. For $k = 1, \ldots, p$, we assume that the conditional distribution of y_{ik} given $\boldsymbol{\omega}_i$ is independent and comes from the following exponential family with a canonical parameter ϑ_{ik} (Sammel, Ryan and Legler, 1997):

$$p_k(y_{ik}|\boldsymbol{\omega}_i) = \exp\{[y_{ik}\vartheta_{ik} - b(\vartheta_{ik})]/\psi_{\epsilon k} + c_k(y_{ik}, \psi_{\epsilon k})\}, \tag{13.1}$$
$$E(y_{ik}|\boldsymbol{\omega}_i) = \dot{b}(\vartheta_{ik}), \quad \text{and} \quad Var(y_{ik}|\boldsymbol{\omega}_i) = \psi_{\epsilon k}\ddot{b}(\vartheta_{ik}),$$

where $b(\cdot)$ and $c_k(\cdot)$ are specific differentiable functions with the dots denoting the derivatives, and $\vartheta_{ik} = g_k(\mu_{ik})$ with a link function g_k. Let $\boldsymbol{\vartheta}_i = (\vartheta_{i1}, \ldots, \vartheta_{ip})^T$, $\mathbf{x}_{ik}(m_k \times 1)$ be vectors of fixed covariates, and

$$\mathbf{X}_i = \begin{bmatrix} \mathbf{x}_{i1}^T & \mathbf{0} & \cdots & \mathbf{0} \\ \mathbf{0} & \mathbf{x}_{i2}^T & \cdots & \mathbf{0} \\ \vdots & \vdots & \vdots & \vdots \\ \mathbf{0} & \mathbf{0} & \mathbf{0} & \mathbf{x}_{ip}^T \end{bmatrix}$$

be a $p \times m$ matrix, where $m = m_1 + \cdots + m_p$. We propose the model $\boldsymbol{\vartheta}_i = \mathbf{X}_i\mathbf{A} + \boldsymbol{\Lambda}\boldsymbol{\omega}_i$ for assessing the relation of $\boldsymbol{\vartheta}_i$ with $\mathbf{X}_i$ and $\boldsymbol{\omega}_i$, where $\mathbf{A} = (\mathbf{a}_1^T, \ldots, \mathbf{a}_p^T)^T$ is an $m \times 1$ vector with $\mathbf{a}_k$ being an $m_k \times 1$ vector of unknown parameters, and $\boldsymbol{\Lambda} = (\boldsymbol{\Lambda}_1, \ldots, \boldsymbol{\Lambda}_p)^T$ is a matrix of unknown parameters. Clearly, for $k = 1, \ldots, p$,

$$\vartheta_{ik} = \mathbf{x}_{ik}^T\mathbf{a}_k + \boldsymbol{\Lambda}_k^T\boldsymbol{\omega}_i. \tag{13.2}$$

Note that Equation (13.2) can accommodate an intercept μ_k by taking a component of $\mathbf{x}_{ik}$ as 1 and defining the corresponding component of $\mathbf{a}_k$ as μ_k. In situations where $\boldsymbol{\omega}_i$ is distributed according to $N[\mathbf{0}, \boldsymbol{\Phi}^*]$, in which $\boldsymbol{\Phi}^*$ is an unknown covariance matrix, the model defined by Equation (13.2) with EFDs subsumes some useful latent variable models in biostatistics and psychometrics. For instance, the model can be regarded as a factor analysis model with fixed covariates, and thus is a generalization of the normal factor analysis model with covariates (Sammel and Ryan, 1996) to a factor analysis model with covariate based on variables from EFDs. Moreover, it extends the factor analysis model with EFDs (Wedel and Kamakura, 2001) to a factor analysis model with EFDs and covariates.

Equation (13.2) can be viewed as a 'measurement' model. Its main purpose is to identify the latent variables via the corresponding manifest variables in $\mathbf{y}$, with the help of the fixed covariates in $\mathbf{X}_i$. Another basic goal of the SEM is to assess how the latent variables affect each other, by incorporating a structural model. Let $\boldsymbol{\omega}_i = (\boldsymbol{\eta}_i^T, \boldsymbol{\xi}_i^T)^T$, where $\boldsymbol{\eta}_i$ $(q_1 \times 1)$ is the endogenous latent vector and $\boldsymbol{\xi}_i$ $(q_2 \times 1)$ the exogenous latent vector. Inspired by the strong demand for investigating the effects of fixed covariates (Sammel and Ryan, 1996; Lee and Song, 2004) and the nonlinear effects of the latent variables in $\boldsymbol{\xi}_i$ to $\boldsymbol{\eta}_i$ (Schumacker and Marcoulides, 1998), we propose the following nonlinear structural model with fixed covariates:

$$\boldsymbol{\eta}_i = \mathbf{B}\mathbf{c}_i + \boldsymbol{\Pi}\boldsymbol{\eta}_i + \boldsymbol{\Gamma}\mathbf{H}(\boldsymbol{\xi}_i) + \boldsymbol{\delta}_i, \tag{13.3}$$

where $\mathbf{c}_i$ is a vector of fixed covariates, $\mathbf{H}(\boldsymbol{\xi}_i) = (h_1(\boldsymbol{\xi}_i), \cdots, h_l(\boldsymbol{\xi}_i))^T$, $h_j(\boldsymbol{\xi}_i)$ is a nonzero differentiable function of $\boldsymbol{\xi}_i$, $\boldsymbol{\delta}_i$ is an error term and $\mathbf{B}, \boldsymbol{\Pi}$, and $\boldsymbol{\Gamma}$ are unknown matrices. Fixed covariates in $\mathbf{c}_i$ may or may not be equal to those in $\mathbf{x}_{ik}$. The distributions of $\boldsymbol{\xi}_i$ and $\boldsymbol{\delta}_i$ are $N[\mathbf{0}, \boldsymbol{\Phi}]$ and $N[\mathbf{0}, \boldsymbol{\Psi}_\delta]$, respectively, and $\boldsymbol{\xi}_i$ and $\boldsymbol{\delta}_i$ are uncorrelated. For computing efficiency and stability, the covariance matrix $\boldsymbol{\Psi}_\delta$ is assumed to be diagonal. We assume that, similar to many other nonlinear SEMs in previous chapters, $\mathbf{I} - \boldsymbol{\Pi}$ is nonsingular and its determinant is independent of elements in $\boldsymbol{\Pi}$. Let $\boldsymbol{\Lambda}_\omega = (\mathbf{B}, \boldsymbol{\Pi}, \boldsymbol{\Gamma})$ and $\mathbf{G}(\boldsymbol{\omega}_i) = (\mathbf{c}_i^T, \boldsymbol{\eta}_i^T, \mathbf{H}(\boldsymbol{\xi}_i)^T)^T$, then Equation (13.3) can be rewritten as $\boldsymbol{\eta}_i = \boldsymbol{\Lambda}_\omega \mathbf{G}(\boldsymbol{\omega}_i) + \boldsymbol{\delta}_i$.

In behavioral, medical and psychological research, a lot of data are ordered categorical. To accommodate these kind of data, we allow any component y of $\mathbf{y}$ to be unobservable, and its information is given by an observable ordered categorical variable z as follows: $z = k$ if $\alpha_k < y \le \alpha_{k+1}$, for $k = 0, \cdots, b-1$, where $\{-\infty = \alpha_0 < \alpha_1 < \cdots < \alpha_{b-1} < \alpha_b = \infty\}$ is the set of thresholds that defines the categories. As the distribution of $\mathbf{y}$ is defined with the EFDs, the ordered categorical variables are incorporated into the exponential family of distributions. Note that the underlying distribution of the ordered categorical variable is from EFDs rather than the more restrictive normal distribution as in the previous chapters. This model can be identified through the methods given in the previous chapters. For instance, for each ordered categorical variable which is not identifiable, we fix α_1 and α_{b-1} at preassigned values; and the SEM can be identified by the common practice of restricting the appropriate elements in $\boldsymbol{\Lambda}$ and $\boldsymbol{\Lambda}_\omega$ to fixed known values. Dichotomous variables can be analyzed as a special case of ordered categorical variables with some modifications.

13.2.2 The Nonignorable Missing Mechanism

Missing data are very common in substantive research. For generality, we consider missing data with a nonignorable missing mechanism. Nonignorable missing data are more difficult to handle than ignorable missing data. As in Chapter 12, nonignorable missing data are analyzed with a missing mechanism using logistic regression.

To accommodate missing data, we define for each $\mathbf{y}_i$ a missing indicator vector $\mathbf{r}_i$, such that for $k = 1, \ldots, p$, $r_{ik} = 1$ if y_{ik} is missing and $r_{ik} = 0$ if y_{ik} is observed. Let $\mathbf{y}_i = (\mathbf{y}_{\text{obs}i}^T, \mathbf{y}_{\text{mis}i}^T)^T$, where $\mathbf{y}_{\text{obs}i}$ is a $p_{i1} \times 1$ vector of observed manifest variables and $\mathbf{y}_{\text{mis}i}$ is a $p_{i2} \times 1$ vector of the missing components in $\mathbf{y}_i : p_{i1} + p_{i2} = p$. We assume that there is an arbitrary pattern ofs missing data in $\mathbf{y}_i$, and thus that $(\mathbf{y}_{\text{obs}i}^T, \mathbf{y}_{\text{mis}i}^T)^T$ may represent the original $\mathbf{y}_i$ with a permutation of the indices. To account for the nonignorable missing mechanism, $\mathbf{r}_i$ is treated as random. In specifying a model in relation to the conditional distribution $[\mathbf{r}_i|\mathbf{y}_i, \boldsymbol{\omega}_i, \boldsymbol{\varphi}]$ with a parameter vector $\boldsymbol{\varphi}$, it is not worthwhile using too complicated or large a model, because it can easily become unidentifiable. Moreover, a model that is too complex will also make it difficult to derive the corresponding conditional distribution of the missing responses given the observed data, and will induce inefficient sampling of the conditional distributions in the MCMC algorithm.

Let $\mathbf{Y} = (\mathbf{y}_1, \ldots, \mathbf{y}_n)$, $\boldsymbol{\Omega} = (\boldsymbol{\omega}_1, \ldots, \boldsymbol{\omega}_n)$ and $\mathbf{r} = (\mathbf{r}_1, \ldots, \mathbf{r}_n)$. As the observations are conditionally independent, $p(\mathbf{r}|\mathbf{Y}, \boldsymbol{\Omega}, \boldsymbol{\varphi}) = p(\mathbf{r}_1|\mathbf{y}_1, \boldsymbol{\omega}_1, \boldsymbol{\varphi}) \times \cdots \times p(\mathbf{r}_n|\mathbf{y}_n, \boldsymbol{\omega}_n, \boldsymbol{\varphi})$. Based on the suggestion of Ibrahim, Chen and Lipsitz (2001), one possible model for the missing data mechanism is the following binomial model, which assumes independence between r_{ik}:

$$p(\mathbf{r}_i|\mathbf{y}_i, \boldsymbol{\omega}_i, \boldsymbol{\varphi}) = \prod_{k=1}^{p} [\text{pr}(r_{ik} = 1|\mathbf{y}_i, \boldsymbol{\omega}_i, \boldsymbol{\varphi})]^{r_{ik}} [1 - \text{pr}(r_{ik} = 1|\mathbf{y}_i, \boldsymbol{\omega}_i, \boldsymbol{\varphi})]^{1-r_{ik}}. \tag{13.4}$$

The independent assumption can be relaxed by the more general multinomial distribution that specifies the joint distribution of $\mathbf{r}_i$ (see Ibrahim, Chen and Lipsitz, 2001). In this chapter, the $\text{pr}(r_{ik} = 1|\mathbf{y}_i, \boldsymbol{\omega}_i, \boldsymbol{\varphi})$ in Equation (13.4) is modeled by the following logistic regression:

$$\begin{aligned} \text{logit}[\text{pr}(r_{ik} = 1|\mathbf{y}_i, \boldsymbol{\omega}_i, \boldsymbol{\varphi})] &= \varphi_0 + \varphi_1 y_{i1} + \cdots + \varphi_p y_{ip} + \varphi_{p+1}\omega_{i1} \\ &\quad + \cdots + \varphi_{p+q}\omega_{iq} = \boldsymbol{\varphi}^T \mathbf{d}_i, \end{aligned} \tag{13.5}$$

where $\boldsymbol{\varphi} = (\varphi_0, \cdots, \varphi_{p+q})^T$ and $\mathbf{d}_i = (1, y_{i1}, \ldots, y_{ip}, \omega_{i1}, \ldots, \omega_{iq})^T$. As any parameters in the sets $\{\varphi_1, \ldots, \varphi_p\}$ or $\{\varphi_{p+1}, \ldots, \varphi_{p+q}\}$ can be fixed to zero, the nonignorable missing mechanism defined in Equation (13.5) is rather flexible. It can handle the special cases in which $\mathbf{r}_i$ just depends on a subset of entries in $\mathbf{y}_i$, or a subset of entries in $\boldsymbol{\omega}_i$, or both. However, we do not recommend using Equation (13.5) routinely for modeling the nonignorable mechanism in every practical application. Other mechanisms may be preferable for situations where one is certain about its specific form. Finally, Ibrahim, Chen and Lipsitz (2001) mentioned that the parametric form of the missing mechanism itself is not 'testable' from the data.

13.3 A BAYESIAN APPROACH

Again, the Bayesian methods are developed by the useful strategy that combines data augmentation and some MCMC methods. In this section, we present the full conditional distributions in the implementation of the Gibbs sampler for simulating observations of the parameters and the latent variables from their joint posterior distribution. These generated observations are used to obtain the Bayesian estimates and their standard error estimates, and to compute the Bayes factor for model comparison. Under the special case that the missing data are MAR, we do not need to specify the missing mechanism model, and the related components of $\mathbf{r}_i$ and $\boldsymbol{\varphi}$ are not involved.

13.3.1 Prior Distributions

Proper conjugate prior distributions are taken for various matrices of the unknown regression coefficients, as well as the variance and covariance matrices of the latent variables and the error term. More specifically, let $\boldsymbol{\Psi}_\epsilon$ be the diagonal covariance matrix of the error measurements that correspond to the ordered categorical variables; for $k = 1, \ldots, p$, or $k = 1, \ldots, q_1$

$$\begin{aligned} &\mathbf{a}_k \stackrel{D}{=} N[\mathbf{a}_{0k}, \mathbf{H}_{0k}],\ \psi_{\epsilon k}^{-1} \stackrel{D}{=} Gamma[\alpha_{0\epsilon k}, \beta_{0\epsilon k}],\ [\boldsymbol{\Lambda}_k|\psi_{\epsilon k}] \stackrel{D}{=} N[\boldsymbol{\Lambda}_{0k}, \psi_{\epsilon k}\mathbf{H}_{0yk}], \\ &\boldsymbol{\Phi}^{-1} \stackrel{D}{=} W_{q_2}[\mathbf{R}_0, \rho_0],\ \psi_{\delta k}^{-1} \stackrel{D}{=} Gamma[\alpha_{0\delta k}, \beta_{0\delta k}],\ [\boldsymbol{\Lambda}_{\omega k}|\psi_{\delta k}] \stackrel{D}{=} N[\boldsymbol{\Lambda}_{0\omega k}, \psi_{\delta k}\mathbf{H}_{0\omega k}], \\ &\boldsymbol{\varphi} \stackrel{D}{=} N[\boldsymbol{\varphi}_0, \mathbf{H}_{0\varphi}], \end{aligned} \tag{13.6}$$

where $\psi_{\epsilon k}$ and $\psi_{\delta k}$ are the kth diagonal elements of $\boldsymbol{\Psi}_\epsilon$ and $\boldsymbol{\Psi}_\delta$, respectively; $\boldsymbol{\Lambda}_k$ and $\boldsymbol{\Lambda}_{\omega k}$ are the kth rows of $\boldsymbol{\Lambda}$ and $\boldsymbol{\Lambda}_\omega$, respectively; $\mathbf{a}_{0k}$, $\alpha_{0\epsilon k}$, $\beta_{0\epsilon k}$, $\boldsymbol{\Lambda}_{0k}$, $\boldsymbol{\varphi}_0$, $\alpha_{0\delta k}$, $\beta_{0\delta k}$, $\boldsymbol{\Lambda}_{0\omega k}$, ρ_0, and positive definite matrices $\mathbf{H}_{0k}, \mathbf{H}_{0\varphi}, \mathbf{H}_{0yk}, \mathbf{H}_{0\omega k}$ and $\mathbf{R}_0$ are hyperparameters whose values are assumed to be given by the prior information, and $W[\cdot, \cdot]$ denotes Wishart distribution. For $k \neq l$, it is assumed that prior distributions of $(\psi_{\epsilon k}, \boldsymbol{\Lambda}_k)$ and $(\psi_{\epsilon l}, \boldsymbol{\Lambda}_l)$, $(\psi_{\delta k}, \boldsymbol{\Lambda}_{\omega k})$ and $(\psi_{\delta l}, \boldsymbol{\Lambda}_{\omega l})$, as well as $\mathbf{a}_k$ and $\mathbf{a}_l$, are independent.

13.3.2 Full Conditional Distributions

We first consider the situation in which components of $\mathbf{y}_{\text{obs}i}$ in $\mathbf{y}_i$ are neither dichotomous nor ordered categorical, but can be directly observed.

Let $\mathbf{Y} = \{\mathbf{Y}_{\text{obs}}, \mathbf{Y}_{\text{mis}}\}$, where $\mathbf{Y}_{\text{obs}} = \{y_{\text{obs}i}, i = 1, \ldots, n\}$ is the observed data set and $\mathbf{Y}_{\text{mis}} = \{y_{\text{mis}i}, i = 1, \ldots, n\}$ is the missing data set. Based on the idea of data augmentation, we focus on the joint posterior distribution $[\boldsymbol{\Omega}, \mathbf{Y}_{\text{mis}}, \boldsymbol{\theta}, \boldsymbol{\varphi} | \mathbf{Y}_{\text{obs}}, r]$, where $\boldsymbol{\theta}$ is the parameter vector that contains all the unknown parameters in the model. The Gibbs sampler (Geman and Geman, 1984) is used for simulating observations of the posterior distribution $[\boldsymbol{\Omega}, \mathbf{Y}_{\text{mis}}, \boldsymbol{\theta}, \boldsymbol{\varphi} | \mathbf{Y}_{\text{obs}}, r]$. The required full conditional distributions are given as follows.

It can be shown by the definition of the proposed SEM that the full conditional distribution of $\boldsymbol{\Omega}$ is given by

$$\begin{aligned} p(\boldsymbol{\Omega}|\mathbf{Y}, \mathbf{r}, \boldsymbol{\theta}, \boldsymbol{\varphi}) &= \prod_{i=1}^{n} p(\boldsymbol{\omega}_i | \mathbf{y}_i, \mathbf{r}_i, \boldsymbol{\theta}, \boldsymbol{\varphi}) \\ &\propto \prod_{i=1}^{n} p(\mathbf{y}_i|\boldsymbol{\omega}_i, \boldsymbol{\theta}) p(\boldsymbol{\eta}_i|\boldsymbol{\xi}_i, \boldsymbol{\theta})\, p(\boldsymbol{\xi}_i|\boldsymbol{\theta}) p(\mathbf{r}_i|\mathbf{y}_i, \boldsymbol{\omega}_i, \boldsymbol{\varphi}), \end{aligned}$$

where $p(\boldsymbol{\omega}_i|\mathbf{y}_i, \mathbf{r}_i, \boldsymbol{\theta}, \boldsymbol{\varphi})$ is proportional to

$$\begin{aligned} \exp \Bigg\{ & \sum_{k=1}^{p} [y_{ik}\vartheta_{ik} - b(\vartheta_{ik})]/\psi_{\epsilon k} + \left(\sum_{k=1}^{p} r_{ik} \right) \boldsymbol{\varphi}^T \mathbf{d}_i - p \log[1 + \exp(\boldsymbol{\varphi}^T \mathbf{d}_i)] \\ & - \frac{1}{2} [\boldsymbol{\xi}_i^T \boldsymbol{\Phi}^{-1} \boldsymbol{\xi}_i \\ & + (\boldsymbol{\eta}_i - \mathbf{B}\mathbf{c}_i - \boldsymbol{\Pi}\boldsymbol{\eta}_i - \boldsymbol{\Gamma}\mathbf{H}(\boldsymbol{\xi}_i))^T \ \boldsymbol{\Psi}_\delta^{-1} (\boldsymbol{\eta}_i - \mathbf{B}\mathbf{c}_i - \boldsymbol{\Pi}\boldsymbol{\eta}_i - \boldsymbol{\Gamma}\mathbf{H}(\boldsymbol{\xi}_i))] \Bigg\}. \end{aligned} \quad (13.7)$$

For every i, once $\boldsymbol{\omega}_i$ is given, $\mathbf{y}_{\text{mis}i}$ is independent of $\mathbf{y}_{\text{obs}i}$. It follows from the definition of the model and Equation (13.5) that the full conditional distribution of $\mathbf{Y}_m$ is given by

$$p(\mathbf{Y}_{\text{mis}}|\boldsymbol{\Omega}, \mathbf{r}, \boldsymbol{\theta}, \boldsymbol{\varphi}, \mathbf{Y}_{\text{obs}}) = \prod_{i=1}^{n} p(\mathbf{y}_{\text{mis}i}|\boldsymbol{\omega}_i, \mathbf{r}_i, \boldsymbol{\theta}, \boldsymbol{\varphi}, \mathbf{y}_{\text{obs}i}),$$

where

$$\begin{aligned} p(\mathbf{y}_{\text{mis}i}|\boldsymbol{\omega}_i, \mathbf{r}_i, \boldsymbol{\theta}, \boldsymbol{\varphi}, \mathbf{y}_{\text{obs}i}) \propto \exp \Bigg\{ & \sum_{\{k: r_{ik}=1\}} \{[y_{ik}\vartheta_{ik} - b(\vartheta_{ik})]/\psi_{\epsilon k} + c_k(y_{ik}, \psi_{\epsilon k})\} \\ & + \left(\sum_{k=1}^{p} r_{ik} \right) \boldsymbol{\varphi}^T \mathbf{d}_i - p \log[1 + \exp(\boldsymbol{\varphi}^T \mathbf{d}_i)] \Bigg\}. \end{aligned} \quad (13.8)$$

Under the conjugate prior distributions given in Equation (13.6), it can be shown that the full conditional distributions of the components of $\boldsymbol{\theta}$ are given by

$$p(\mathbf{a}_k|\mathbf{Y}, \boldsymbol{\Omega}, \boldsymbol{\Lambda}_k, \psi_{\epsilon k}^{-1}) \propto \exp\left[\sum_{i=1}^{n} \frac{y_{ik}\vartheta_{ik} - b(\vartheta_{ik})}{\psi_{\epsilon k}} - \frac{1}{2}(\mathbf{a}_k - \mathbf{a}_{0k})^T \mathbf{H}_{0k}^{-1}(\mathbf{a}_k - \mathbf{a}_{0k})\right], \tag{13.9}$$

$$p(\psi_{\epsilon k}|\mathbf{Y}, \boldsymbol{\Omega}, \mathbf{a}_k, \boldsymbol{\Lambda}_k) \propto \psi_{\epsilon k}^{-(\frac{n}{2}+\alpha_{0\epsilon k}-1)} \exp\left[\sum_{i=1}^{n}\left[\frac{y_{ik}\vartheta_{ik} - b(\vartheta_{ik})}{\psi_{\epsilon k}} + c_k(y_{ik}, \psi_{\epsilon k})\right] - \frac{\beta_{0\epsilon k}}{\psi_{\epsilon k}}\right], \tag{13.10}$$

$$p(\boldsymbol{\Lambda}_k|\mathbf{Y}, \boldsymbol{\Omega}, \mathbf{a}_k, \psi_{\epsilon k}^{-1}) \propto \exp\left[\sum_{i=1}^{n} \frac{y_{ik}\vartheta_{ik} - b(\vartheta_{ik})}{\psi_{\epsilon k}} - \frac{1}{2}\psi_{\epsilon k}^{-1}(\boldsymbol{\Lambda}_k - \boldsymbol{\Lambda}_{0k})^T \mathbf{H}_{0yk}^{-1}(\boldsymbol{\Lambda}_k - \boldsymbol{\Lambda}_{0k})\right], \tag{13.11}$$

$$p(\psi_{\delta k}|\boldsymbol{\Omega}, \boldsymbol{\Lambda}_{\omega k}) \stackrel{D}{=} Gamma[n/2 + \alpha_{0\delta k}, \beta_{\delta k}], \tag{13.12}$$

$$p(\boldsymbol{\Lambda}_{\omega k}|\boldsymbol{\Omega}, \psi_{\delta k}^{-1}) \stackrel{D}{=} N[\mathbf{a}_{\omega k}, \psi_{\delta k}\mathbf{A}_{\omega k}], \tag{13.13}$$

$$p(\boldsymbol{\Phi}|\boldsymbol{\Omega}) \stackrel{D}{=} IW_{q_2}[\boldsymbol{\Omega}_2\boldsymbol{\Omega}_2^T + \mathbf{R}_0^{-1}, n + \rho_0], \tag{13.14}$$

where $\mathbf{A}_{\omega k} = (\mathbf{H}_{0\omega k}^{-1} + \mathbf{G}\mathbf{G}^T)^{-1}$, $\mathbf{a}_{\omega k} = \mathbf{A}_{\omega k}(\mathbf{H}_{0\omega k}^{-1}\boldsymbol{\Lambda}_{0\omega k} + \mathbf{G}\boldsymbol{\Omega}_{1k})$ and $\beta_{\delta k} = \beta_{0\delta k} + (\boldsymbol{\Omega}_{1k}^T\boldsymbol{\Omega}_{1k} - \mathbf{a}_{\omega k}^T\mathbf{A}_{\omega k}^{-1}\mathbf{a}_{\omega k} + \boldsymbol{\Lambda}_{0\omega k}^T\mathbf{H}_{0\omega k}^{-1}\boldsymbol{\Lambda}_{0\omega k})/2$, in which $\mathbf{G} = (\mathbf{G}(\boldsymbol{\omega}_1), \ldots, \mathbf{G}(\boldsymbol{\omega}_n))$, $\boldsymbol{\Omega}_1 = (\boldsymbol{\eta}_1, \ldots, \boldsymbol{\eta}_n)$, $\boldsymbol{\Omega}_2 = (\boldsymbol{\xi}_1, \ldots, \boldsymbol{\xi}_n)$ and $\boldsymbol{\Omega}_{1k}^T$ is the kth row of $\boldsymbol{\Omega}_1$.

It can be derived from Equation (13.5) that the full conditional distribution of $\boldsymbol{\varphi}$ is equal to

$$p(\boldsymbol{\varphi}|\mathbf{Y}, \boldsymbol{\Omega}, \mathbf{r}, \boldsymbol{\theta}) \propto p(\mathbf{r}|\mathbf{Y}, \boldsymbol{\Omega}, \boldsymbol{\varphi})p(\boldsymbol{\varphi}) \propto p(\boldsymbol{\varphi}) \prod_{i=1}^{n}\prod_{k=1}^{p} p(r_{ik}|y_{ik}, \boldsymbol{\omega}_i, \boldsymbol{\varphi})$$

$$\propto \exp\left\{\sum_{i=1}^{n}(\sum_{k=1}^{p} r_{ik})\boldsymbol{\varphi}^T\mathbf{d}_i - p\sum_{i=1}^{n}\log[1 + \exp(\boldsymbol{\varphi}^T\mathbf{d}_i)] - \frac{1}{2}(\boldsymbol{\varphi} - \boldsymbol{\varphi}_0)^T\mathbf{H}_{0\varphi}^{-1}(\boldsymbol{\varphi} - \boldsymbol{\varphi}_0)\right\}. \tag{13.15}$$

To handle the ordered categorical data, let $\mathbf{Y}_k$ be the kth row of $\mathbf{Y}$ that is not directly observable. Let $\mathbf{z}_k$ be the corresponding ordered categorical vector that includes n_k observable components after discarding the missing entries, and let $\boldsymbol{\alpha}_k = (\alpha_{k,1}, \ldots, \alpha_{k,b_k-1})$. It is natural to assume that the prior distribution of $\boldsymbol{\alpha}_k$

is independent of the prior distribution of $\boldsymbol{\theta}$. To deal with a general situation in which there is little or no information about the thresholds, the following noninformative prior distribution is used: $p(\boldsymbol{\alpha}_k) = p(\alpha_{k,1}, \ldots, \alpha_{k,b_k-1}) \propto c$ for $\alpha_{k,1} < \cdots < \alpha_{k,b_k-1}$, where c is a constant. Moreover, it is assumed that $\boldsymbol{\alpha}_k$ and $\boldsymbol{\alpha}_l$ are independent for $k \neq l$. It can be shown by a derivation similar to that in Chapter 6, Section 6.3.1 that

$$p(\boldsymbol{\alpha}_k, \mathbf{y}_k | \mathbf{z}_k, \boldsymbol{\Omega}, \boldsymbol{\theta}) = p(\boldsymbol{\alpha}_k | \mathbf{z}_k, \boldsymbol{\Omega}, \boldsymbol{\theta}) p(\mathbf{y}_k | \boldsymbol{\alpha}_k, \mathbf{z}_k, \boldsymbol{\Omega}, \boldsymbol{\theta})$$

$$\propto \prod_{i=1}^{n_k} \exp\{[y_{ik}\vartheta_{ik} - b(\vartheta_{ik})]/\psi_{\epsilon k} + c_k(y_{ik}, \psi_{\epsilon k})\} \, I_{(\alpha_{k,z_{ik}}, \alpha_{k,z_{ik}+1}]}(y_{ik}), \tag{13.16}$$

where $I_A(y)$ is an index function which takes 1 if $y \in A$, and 0 otherwise. The treatment of dichotomous variables is similar – see Chapter 7. Note that for the missing entries, the underlying continuous values can be simulated by Equation (13.8).

The Gibbs sampler algorithm proceeds by sampling $\boldsymbol{\omega}_i$, $\mathbf{y}_{\text{mis}i}$, $(\boldsymbol{\alpha}_k, \mathbf{y}_k)$, $\boldsymbol{\theta}$ and $\boldsymbol{\varphi}$ from conditional distributions of Equations (13.7) to (13.16), respectively. The simulation of observations from the standard distributions involved in Equations (13.12), (13.13) and (13.14) is straightforward. However, various forms of the Metropolis–Hastings (MH) algorithm (Metropolis *et al.*, 1953; Hastings, 1970) are required to simulate observations from the remaining complex conditional distributions. Due to the complexity of the proposed model and data structures, the implementations are not straightforward. Some details are given in Appendix 13.1.

13.3.3 Model Comparison

In this section, we use $\mathbf{D}_{\text{obs}}$ to denote the observed data, which include the directly observable data and the ordered categorical data, and use $\mathbf{D}_{\text{mis}}$ to denote unobservable data, which include missing data, latent variables and unobserved data that underlie the ordered categorical data. Moreover, let $\boldsymbol{\theta}_* = (\boldsymbol{\theta}, \boldsymbol{\alpha}, \boldsymbol{\varphi})$ be the overall unknown parameter vector, where $\boldsymbol{\alpha}$ is a vector that includes all unknown thresholds. Suppose that $\mathbf{D}_{\text{obs}}$ has arisen under one of the two competing models M_0 and M_1. For $l = 0, 1$, let $p(\mathbf{D}_{\text{obs}}|M_l)$ be the probability density of $\mathbf{D}_{\text{obs}}$ under M_l. Recall that the Bayes factor is defined by $B_{10} = p(\mathbf{D}_{\text{obs}}|M_1)/p(\mathbf{D}_{\text{obs}}|M_0)$. In a similar way to the previous chapters, a path sampling procedure is presented to compute the Bayes factor for model comparison.

Utilizing the idea of data augmentation, $\mathbf{D}_{\text{obs}}$ is augmented with $\mathbf{D}_{\text{mis}}$ in the computation. Let t be a continuous parameter in [0, 1] to link the competing models M_0 and M_1, let $p(\mathbf{D}_{\text{mis}}, \mathbf{D}_{\text{obs}}, \mathbf{r}|\boldsymbol{\theta}_*, t)$ be the complete data likelihood and $U(\boldsymbol{\theta}_*, \mathbf{D}_{\text{mis}}, \mathbf{D}_{\text{obs}}, \mathbf{r}, t) = \mathrm{d}\log p(\mathbf{D}_{\text{mis}}, \mathbf{D}_{\text{obs}}, \mathbf{r}|\boldsymbol{\theta}_*, t)/\mathrm{d}t$. It can be shown by similar reasoning to that in the previous chapters that

$$\log B_{10} = \int_0^1 E_{\boldsymbol{\theta}_*, \mathbf{D}_{\text{mis}}}[U(\boldsymbol{\theta}_*, \mathbf{D}_{\text{mis}}, \mathbf{D}_{\text{obs}}, \mathbf{r}, t)]\, \mathrm{d}t,$$

where $E_{\boldsymbol{\theta}_*, \mathbf{D}_{\text{mis}}}$ is the expectation with respect to the distribution $p(\boldsymbol{\theta}_*, \mathbf{D}_{\text{mis}}|\mathbf{D}_{\text{obs}}, \mathbf{r}, t)$. This integral can be numerically evaluated with the trapezoidal rule. Specifically, let S different fixed grids $\{t_{(s)}\}_{s=0}^{S}$ be ordered as $t_{(0)} = 0 < t_{(1)} < t_{(2)} < \cdots < t_{(S)} < t_{(S+1)} = 1$. Then, $\log B_{10}$ can be computed as

$$\widehat{\log B_{10}} = \frac{1}{2}\sum_{s=0}^{S}(t_{(s+1)} - t_{(s)})(\bar{U}_{(s+1)} + \bar{U}_{(s)}), \tag{13.17}$$

where $\bar{U}_{(s)}$ is the average of the $U(\boldsymbol{\theta}_*, \mathbf{D}_{\text{mis}}, \mathbf{D}_{\text{obs}}, \mathbf{r}, t)$ on the basis of all simulation draws for which $t = t_{(s)}$, that is, $\bar{U}_{(s)} = J^{-1}\sum_{j=1}^{J} U(\boldsymbol{\theta}_*^{(j)}, \mathbf{D}_{\text{mis}}^{(j)}, \mathbf{D}_{\text{obs}}, \mathbf{r}, t_{(s)})$, in which $\{(\boldsymbol{\theta}_*^{(j)}, \mathbf{D}_{\text{mis}}^{(j)}); j = 1, \ldots, J\}$ are observations simulated from the joint conditional distribution $p(\boldsymbol{\theta}_*, \mathbf{D}_{\text{mis}}|\ \mathbf{D}_{\text{obs}}, \mathbf{r}, t_{(s)})$.

13.4 A SIMULATION STUDY

A Bayesian approach for analyzing SEMs with dichotomous variables was presented in Chapter 7. A dichotomous variable is an ordered categorical variable that is defined by two categories with a threshold of zero. It is usually coded with '0' and '1', and the probability of observing '0' and '1' is decided by an underlying normal distribution with a fixed threshold. Another kind of discrete variable which also has two categories and is usually coded with '0' and '1' is the binary variable. Although they have same coding, binary variables are different from dichotomous variables. Binary variables are unordered, they do not associate with thresholds and the probabilities of observing '0' and '1' are decided by a binomial distribution rather than the normal distribution. Hence, the methods for analyzing dichotomous variables should not be directly applied to analyze binary variables. Given a variable with binary codings '0' and '1' it is important to decide whether it is an ordered dichotomous variable or an unordered binary variable. Clearly, binary variables are also common in behavioral, medical and social research.

Results obtained from a simulation study are presented here to illustrate the empirical performance of the Bayesian approach in analyzing binary and ordered categorical data. For brevity, we only consider ignorable missing data that are MAR. A data set $\mathbf{Y} = \{\mathbf{y}_i, i = 1, \ldots, n\}$ was generated with $\mathbf{y}_i = (y_{i1}, \ldots, y_{i9})^T$. For $k = 1, 2, 3$, the distribution of y_{ik} is binomial, $B(1, p_{ik})$, that is $y_{ik} \propto$

$\exp\left[y_{ik}\vartheta_{ik} - \log(1+e^{\vartheta_{ik}})\right]$ with $b(\vartheta_{ik}) = \log(1+e^{\vartheta_{ik}})$ and $\vartheta_{ik} = \log\left[p_{ik}/(1-p_{ik})\right]$. Note that for each of these binomial variables, $\psi_{\epsilon k} = 1.0$ is treated as a fixed parameter. For $k = 4, 5, 6$, the observations y_{i4}, y_{i5} and y_{i6} are simulated from a multivariate normal distribution. For $k = 7, 8, 9$ the underlying latent continuous measurements corresponding to y_{i7}, y_{i8} and y_{i9} are first simulated from a multivariate normal distribution; then they are transformed to ordered categorical observations z_{i7}, z_{i8} and z_{i9} via the same thresholds $(-1.0^*, -0.5, 0.5, 1.0^*)$, where -1.0^* and 1.0^* with an asterisk are treated as fixed for identification. The 'measurement equation' (see Equation (13.2)): $\vartheta_{ik} = \mu_k + \mathbf{x}_{ik}^T\mathbf{e}_k + \mathbf{\Lambda}_k^T\boldsymbol{\omega}_i$, is defined by an intercept μ_k, a (3×1) vector of fixed covariates $\mathbf{x}_{ik}$ that is simulated from $N[\mathbf{0}, \mathbf{I}]$, and a (3×1) vector of latent variables $\boldsymbol{\omega}_i = (\eta_i, \xi_{i1}, \xi_{i2})^T$. The specifications of the corresponding matrices of the unknown coefficients are given by: $\boldsymbol{\mu} = (0.0, 0.0, 0.0, 0.0, 0.0, 0.0, 0.0, 0.0, 0.0)^T$,

$$\mathbf{A}^T = \begin{bmatrix} 1.0 & 1.0 & 1.0 & 1.0 & 1.0 & 1.0 & 1.0 & 1.0 & 1.0 \\ 0.0^* & 0.0^* & 0.0^* & 0.0^* & 0.0^* & 0.0^* & 1.0 & 1.0 & 1.0 \\ 1.0 & 1.0 & 1.0 & 0.0^* & 0.0^* & 0.0^* & 0.0^* & 0.0^* & 0.0^* \end{bmatrix}, \quad (13.18)$$

$$\mathbf{\Lambda}^T = \begin{bmatrix} 1.0^* & 0.8 & 0.8 & 0.0^* & 0.0^* & 0.0^* & 0.0^* & 0.0^* & 0.0^* \\ 0.0^* & 0.0^* & 0.0^* & 1.0^* & 0.8 & 0.8 & 0.0^* & 0.0^* & 0.0^* \\ 0.0^* & 0.0^* & 0.0^* & 0.0^* & 0.0^* & 0.0^* & 1.0^* & 0.8 & 0.8 \end{bmatrix}, \quad (13.19)$$

where the values given in $\boldsymbol{\mu}$, $\mathbf{A}$ and $\mathbf{\Lambda}$ are the true population values and as usual the entries with an asterisk denote parameters that are fixed at that values. The true population values for the diagonal elements of $\mathbf{\Psi}_\epsilon$ are $(1.0^*, 1.0^*, 1.0^*, 1.0, 1.0, 1.0, 1.0, 1.0, 1.0)$. The structural equation (see Equation (13.3)) is defined by

$$\eta_i = b_1 c_{i1} + b_2 c_{i2} + \gamma_1 \xi_{i1} + \gamma_2 \xi_{i2} + \gamma_3 \xi_{i1}\xi_{i2} + \delta_i, \quad (13.20)$$

where c_{i1} and c_{i2} are covariates that are simulated from $N[0, 1]$, the true values of $b_1, b_2, \gamma_1, \gamma_2$ and γ_3 are 1.0, 1.0, 0.6, 0.6 and -0.4, respectively. The true values of ϕ_{11}, ϕ_{12} and ϕ_{22} are 1.0, 0.3 and 1.0, respectively; while the true value of ψ_δ is 0.36. There are a total of 51 unknown parameters in this model.

A random sample of size $n = 500$ was simulated; 50 observations out of the 500 observations contain missing entries that are missing at random. Simulation results were obtained on the basis of 100 replications. The prior inputs of the hyperparameters in the conjugate prior distributions are taken as: $\alpha_{0\epsilon k} = 9, \beta_{0\epsilon k} = 8, \alpha_{0\delta 1} = 9, \beta_{0\delta 1} = 3, \rho_0 = 6, \mathbf{R}_0 = 6\mathbf{I}, \mathbf{H}_{0k}$ and $\mathbf{H}_{0yk}$ are identity matrices, $\mathbf{H}_{0\omega 1} = 1.0$ and elements in $\mathbf{\Lambda}_{0k}$ and $\mathbf{\Lambda}_{0\omega 1}$ are taken to be the true values. These prior inputs represent a situation with good prior information. Based on some preliminary analyses for checking convergence, we took 3000 burn-in iterations and collected 3000 observations to obtain the Bayesian

Table 13.2 Performance of the Bayesian estimates with $n = 500$ in the simulation study.

Par	BIAS	SD	RMS	Par	BIAS	SD	RMS	Par	BIAS	SD	RMS
μ_1	0.025	0.143	0.133	a_{11}	0.004	0.150	0.142	λ_{21}	0.034	0.133	0.127
μ_2	0.016	0.135	0.128	a_{13}	0.042	0.147	0.155	λ_{31}	0.047	0.136	0.142
μ_3	0.023	0.134	0.150	a_{21}	0.035	0.150	0.137	λ_{52}	0.011	0.087	0.094
μ_4	0.009	0.065	0.071	a_{23}	0.019	0.147	0.128	λ_{62}	0.004	0.086	0.087
μ_5	0.001	0.059	0.065	a_{31}	0.045	0.150	0.196	λ_{83}	0.037	0.124	0.123
μ_6	0.002	0.059	0.059	a_{33}	0.034	0.148	0.154	λ_{93}	0.015	0.123	0.109
μ_7	0.002	0.078	0.086	a_{41}	0.001	0.065	0.063	b_1	0.011	0.118	0.119
μ_8	0.003	0.072	0.072	a_{51}	0.001	0.059	0.059	b_2	0.006	0.118	0.140
μ_9	0.013	0.071	0.075	a_{61}	0.016	0.059	0.062	γ_1	0.009	0.142	0.137
$\psi_{\epsilon4}$	0.015	0.112	0.128	a_{71}	0.020	0.089	0.090	γ_2	0.031	0.148	0.172
$\psi_{\epsilon5}$	0.002	0.091	0.097	a_{72}	0.018	0.089	0.085	γ_3	0.036	0.147	0.167
$\psi_{\epsilon6}$	0.009	0.090	0.086	a_{81}	0.001	0.083	0.081	α_{72}	0.004	0.063	0.060
$\psi_{\epsilon7}$	0.020	0.165	0.144	a_{82}	0.009	0.083	0.086	α_{73}	0.002	0.063	0.069
$\psi_{\epsilon8}$	0.008	0.140	0.120	a_{91}	0.003	0.084	0.078	α_{82}	0.007	0.061	0.059
$\psi_{\epsilon9}$	0.016	0.141	0.124	a_{92}	0.020	0.084	0.089	α_{83}	0.004	0.061	0.067
ψ_δ	0.014	0.107	0.060	ϕ_{11}	0.019	0.142	0.143	α_{92}	0.003	0.061	0.049
				ϕ_{12}	0.003	0.080	0.083	α_{93}	0.003	0.061	0.070
				ϕ_{22}	0.023	0.192	0.195				

solution. The absolute bias (BIAS) of the mean of Bayesian estimates and the true values, the standard deviation (SD) of the Bayesian estimates and the root mean squares (RMS) of the Bayesian estimates are reported in Table 13.2. It can be seen that under the given sample size and prior inputs, the empirical performance of the Bayesian approach is acceptable. For instance, all the 'BIAS' values are less than 0.05, and most of the 'SD' and 'RMS' values are less than 0.15. Note that the Bayesian estimates corresponding to $\mu_1, \mu_2, \mu_3, \lambda_{21}$ and λ_{31}, and some elements in $\mathbf{E}$ that are associated with the binary manifest variables are not as good as others. Finally, we expect slightly worse performances under situations with less accurate prior inputs.

13.5 A REAL EXAMPLE: A COMPLIANCE STUDY OF PATIENTS

The Bayesian methodology is used to assess the effects of some exogenous latent variables on patient nonadherence to medication in a study which was conducted by the Department of Medicine and Therapeutics, the Department of Community and Family Medicine, and the Department of Pharmacy at the Chinese University of Hong Kong (Chan, 2001). See Chapter 7, Section 7.2.1 for a background of this study, and Song and Lee (2006) for a similar analysis. A total of 849 ethnic Chinese patients diagnosed with hypertension were randomly selected from hospitals and clinics to serve as subjects for the study. In this example, we are interested in nine manifest variables, $(y_1, \ldots, y_9)$. A translation of the corresponding items in the questionnaires from Chinese to English is presented in Appendix 13.2. The first three binary variables are indicators of the latent variable of 'patient nonadherence, η', the next three binary variables are indicators of the latent variable of 'knowledge of medication, ξ_1', and the last three ordered categorical variables with a five-point scale are indicators of the latent variable of '(dis)satisfaction with the physician, ξ_2'. As an illustration, three continuous fixed covariates, 'age of the patient', 'years of having hypertension', 'years of using antihypertensive drugs', and a binary fixed covariate of whether the patient 'can read the instructions on the labels', are incorporated in the first six manifest variables in the measurement model Equation (13.2). In the presence of some missing data, the frequency of 'yes' answers corresponding to the binary variables ranged from 69 to 651, and the cell frequency of each ordered categorical variable ranged from 15 to 508. In the analysis of this data set, for each $k = 1, \ldots, 6$, the distribution of y_{ik} is naturally taken to be a binomial distribution $B(1, p_{ik})$, such that

$$y_{ik} \propto \exp\left[y_{ik}\vartheta_{ik} - \log(1 + e^{\vartheta_{ik}})\right], \quad \text{and} \tag{13.21}$$

$$\vartheta_{ik} = \log \frac{p_{ik}}{1 - p_{ik}} = \mathbf{x}_{ik}^T \mathbf{a}_k + \boldsymbol{\Lambda}_k^T \boldsymbol{\omega}_i, \tag{13.22}$$

where $\boldsymbol{\omega}_i = (\eta_i, \xi_{i1}, \xi_{i2})^T$ and Λ_k^T is the kth row of the following $\boldsymbol{\Lambda}$:

$$\boldsymbol{\Lambda}^T = \begin{bmatrix} 1 & \lambda_{21} & \lambda_{31} & 0 & 0 & 0 & 0 & 0 & 0 \\ 0 & 0 & 0 & 1 & \lambda_{52} & \lambda_{62} & 0 & 0 & 0 \\ 0 & 0 & 0 & 0 & 0 & 0 & 1 & \lambda_{83} & \lambda_{93} \end{bmatrix},$$

where $\lambda_{ij}s$ are the unknown factor loading parameters, and 1s and 0s are fixed in the estimation to achieve an identified model. For $k = 1, \ldots, 6$, $\psi_{\epsilon k} = 1.0$, which are treated as fixed parameters. For $k = 7, 8, 9$, the elements in $\mathbf{a}_k$ are fixed at 0, and the unobservable variable y_{ik} that corresponds to the observable ordered categorical variable is assumed to have a normal distribution $N(\boldsymbol{\Lambda}_k^T \boldsymbol{\omega}_i, \psi_{\epsilon k})$ in the EFDs. To identify parameters that are associated with ordered categorical variables, the thresholds α_{71}, α_{74}, α_{81}, α_{84}, α_{91} and α_{94} are fixed via the same method presented in previous chapters in identifying ordered categorical variables. The endogenous latent variable of patient nonadherence (η) is regressed on the exogenous latent variables of knowledge of medication (ξ_1) and the (dis)satisfaction with the physician (ξ_2), and two binary fixed covariates that are related to whether the patient is taking other long-term medication (c_1) and the existence of side effects (c_2). The following nonlinear structural model with the fixed covariates is considered:

$$\eta = b_1 c_1 + b_2 c_2 + \gamma_1 \xi_1 + \gamma_2 \xi_2 + \gamma_3 \xi_1 \xi_2 + \delta. \tag{13.23}$$

In this formulation of Equation (13.3), $\mathbf{B} = (b_1, b_2)$, $\mathbf{\Pi} = \mathbf{0}$ and $\mathbf{\Gamma} = (\gamma_1, \gamma_2, \gamma_3)$. Other unknown parameters are the variances and covariance of ξ_1 and ξ_2, (ϕ_{11}, ϕ_{22} and ϕ_{12}), the variances of ordered categorical variables $\psi_{\epsilon k}$ for $k = 7, 8, 9$, and the variance of δ, ψ_δ. In this example, we use the logistic regression for the nonignorable missing data mechanism logit $[\mathrm{pr}(r_{ik} = 1|\mathbf{y}_i, \boldsymbol{\omega}_i, \boldsymbol{\varphi})] = \varphi_0 + \varphi_1 y_{i1} + \cdots + \varphi_9 y_{i9}$. There are a total of 58 unknown parameters.

To specify the conjugate prior distributions, we use the following data-dependent prior inputs. We first conducted an auxiliary Bayesian estimation with a noninformative prior to obtain the prior inputs of the hyperparameter values for $\mathbf{a}_{0k}$, $\mathbf{\Lambda}_{0k}$ and $\mathbf{\Lambda}_{0\omega k}$ in the conjugate prior distributions in the actual estimation. Consequently, the values of the hyperparameters are given by $\alpha_{0\epsilon k} = 10$, $\beta_{0\epsilon k} = 5$, $\alpha_{0\delta k} = 10$, $\beta_{0\delta k} = 5$, $\rho_0 = 8$ and $\mathbf{H}_{0yk}$, $\mathbf{H}_{0\omega k}$, $\mathbf{H}_{0k}$ and $\mathbf{H}_{0\varphi}$ are assigned to be $0.25\mathbf{I}$, where $\mathbf{I}$ is an identity matrix with appropriate dimensions, $\mathbf{a}_{0k} = \tilde{\mathbf{a}}_k$, $\mathbf{\Lambda}_{0k} = \tilde{\mathbf{\Lambda}}_k$, $\mathbf{\Lambda}_{0\omega k} = \tilde{\mathbf{\Lambda}}_{\omega k}$ and $\mathbf{R}_0 = 2\mathbf{I}$, where $\tilde{\mathbf{a}}_k$, $\tilde{\mathbf{\Lambda}}_k$ and $\tilde{\mathbf{\Lambda}}_{\omega k}$ are estimates obtained from the auxiliary estimation. The Gibbs sampler algorithm with the MH algorithm is used to generate samples from the joint posterior distribution of the parameters and the latent quantities. Three chains with different initial values are run. To give some information about the convergence of the algorithm, plots of these chains for some randomly chosen parameter values against the iteration numbers are displayed in Figure 13.1. Moreover, the 'estimated potential scale reduction (EPSR)' values (Gelman and Rubin, 1992) of the parameters on the basis of the different initial values are displayed in Figure 13.2. We observe that the algorithm converged after about 5000 iterations. The Bayesian inferences are based on 10 000 observations that are collected after convergence of the Gibbs sampler. The Bayesian estimates and standard error estimates are obtained from the sample mean and the sample covariance matrix of the simulated observations.

We first conduct a model comparison of the encompassing model (M_0) as defined above with the competing models M_1 and M_2, which are defined by the same measurement model and the following structural models: $M_1 : \eta = b_1 c_1 + b_2 c_2 + \gamma_1 \xi_1 + \gamma_2 \xi_2 + \delta$ and $M_2 : \eta = \gamma_1 \xi_1 + \gamma_2 \xi_2 + \gamma_3 \xi_1 \xi_2 + \delta$. The interaction effect of the exogenous latent variables is not included in M_1, while the fixed covariates (c_1 and c_2) are not included in M_2. To assess the impact of the fixed covariates in the measurement equation, we consider an additional competing model M_3, which is defined by the structural model given in Equation (13.23) and the measurement model that without any fixed covariate, $\vartheta_{ik} = \mathbf{\Lambda}_k^T \boldsymbol{\omega}_i$. Models M_1, M_2 and M_3 are nested in M_0. It is easy to construct a path to link the competing models with the encompassing model. For example, the linked model M_t for M_0 and M_1 is defined with Equations (13.21) and (13.22), and the structural model $\eta = b_1 c_1 + b_2 c_2 + \gamma_1 \xi_1 + \gamma_2 \xi_2 + (1 - t)\gamma_3 \xi_1 \xi_2 + \delta$. When $t = 0$, M_t reduces to M_0, and when $t = 1$, M_t reduces to M_1. In the path sampling procedure, we take $S = 20$ and 1000 observations collected after convergence of the Gibbs sampler for each $t_{(s)}$. The computed logarithm Bayes factors, $\log B_{01}$, $\log B_{02}$ and $\log B_{03}$,

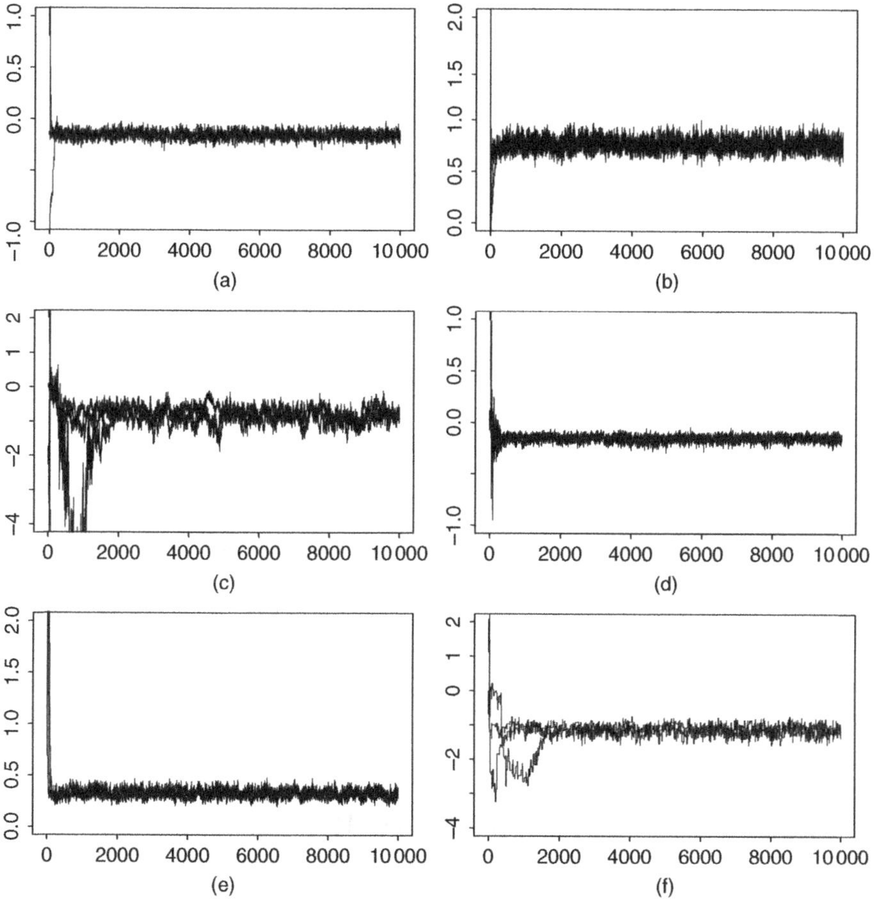

Figure 13.1 (a), (b), (c), (d), (e) and (f) are plots of three parallel sequences corresponding to different starting values of α_{72}, λ_{93}, γ_1, ϕ_{12}, $\psi_{\epsilon 7}$ and a_{41} against iterations in the analysis of the real data.

are equal to 1.23, 18.52 and 211.65, respectively. According to the criterion that is based on two times the logarithm Bayes factor (Kass and Raftery, 1995), the models without fixed covariates perform significantly worse. Moreover, M_0 is better than M_1, which does not involve the interaction effect. The PP p-value (see Gelman, Stern and Meng, 1996) for the goodness-of-fit testing of M_0 is 0.42, which indicates that M_0 fits the data well. To reveal the fit of the structural model Equation (13.23), the estimated residuals $\hat{\delta}_i = \hat{\eta}_i - \hat{b}_1 c_{i1} - \hat{b}_2 c_{i2} - \hat{\gamma}_1 \hat{\xi}_{i1} - \hat{\gamma}_2 \hat{\xi}_{i2} - \hat{\gamma}_3 \hat{\xi}_{i1} \hat{\xi}_{i2}$ were computed. Plots of $\hat{\delta}_i$ versus the case numbers $\hat{\xi}_{i1}$ and $\hat{\xi}_{i2}$ are displayed in Figure 13.3. These plots lie within two parallel horizontal lines that are centered at zero, and no linear or quadratic

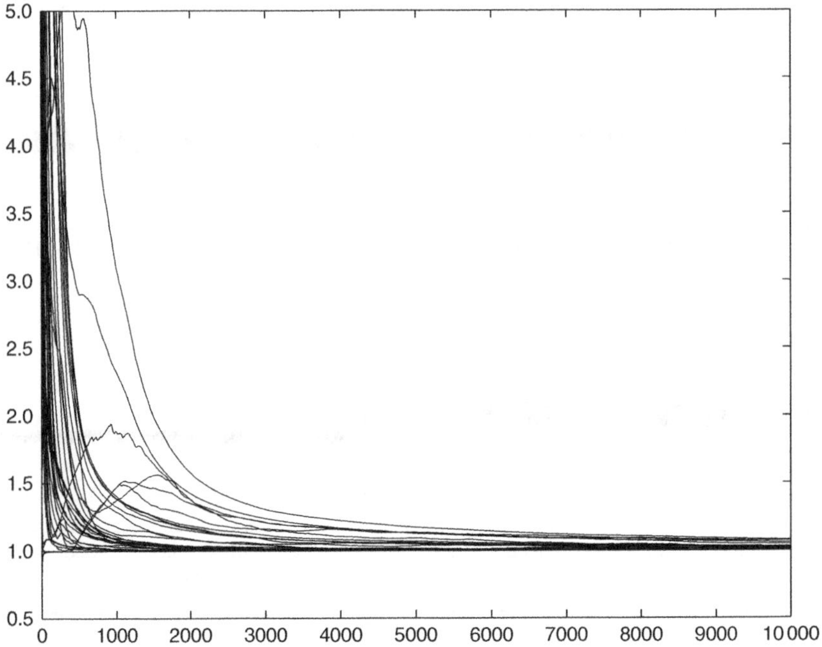

Figure 13.2 EPSR values against the number of iterations in the analysis of the real data.

trends are observed. This roughly reveals that the proposed structural model is adequate.

The Bayesian estimates of the parameters in M_0 are presented in Table 13.3, together with their standard error estimates. It can be seen from the estimates of φ_k that except for φ_1 and φ_2, all other φs are significantly different from zero. This confirms the demand for a nonignorable mechanism for the missing data. As expected, some elements in $\hat{\mathbf{a}}_k$, $\hat{b}_1$ and $\hat{b}_2$ are significantly different from zero, indicating the significant effects of the fixed covariates. Except for $\hat{\lambda}_{31}$, the loading estimates in $\mathbf{\Lambda}$ are high, indicating a strong association between the latent variables and their respective items. It can be calculated from $\hat{\phi}_{11}$, $\hat{\phi}_{12}$ and $\hat{\phi}_{22}$ that the estimated correlation between ξ_1 and ξ_2 is -0.337. This indicates the expected phenomenon that better 'knowledge of medication' and '(dis)satisfaction with the physician' are negatively correlated. The estimated structural model is $\eta = -0.89c_1 + 0.34c_2 - 0.79\xi_1 + 0.31\xi_2 + 0.68\xi_1\xi_2$. We have the following interpretations. (i) From the estimates of the coefficients of the covariates, it seems that 'taking other long-term medication' has a negative effect on 'nonadherence', whereas 'existence of side effects' has a positive effect. (ii) Better 'knowledge of medication' has a negative effect on 'nonadherence'. (iii) '(Dis)satisfaction with the physician' has a positive effect on 'nonadherence'. The reasonable findings in (i), (ii) and (iii) provide concrete suggestions for improving the medication scheme; for example, in allocating resources to give

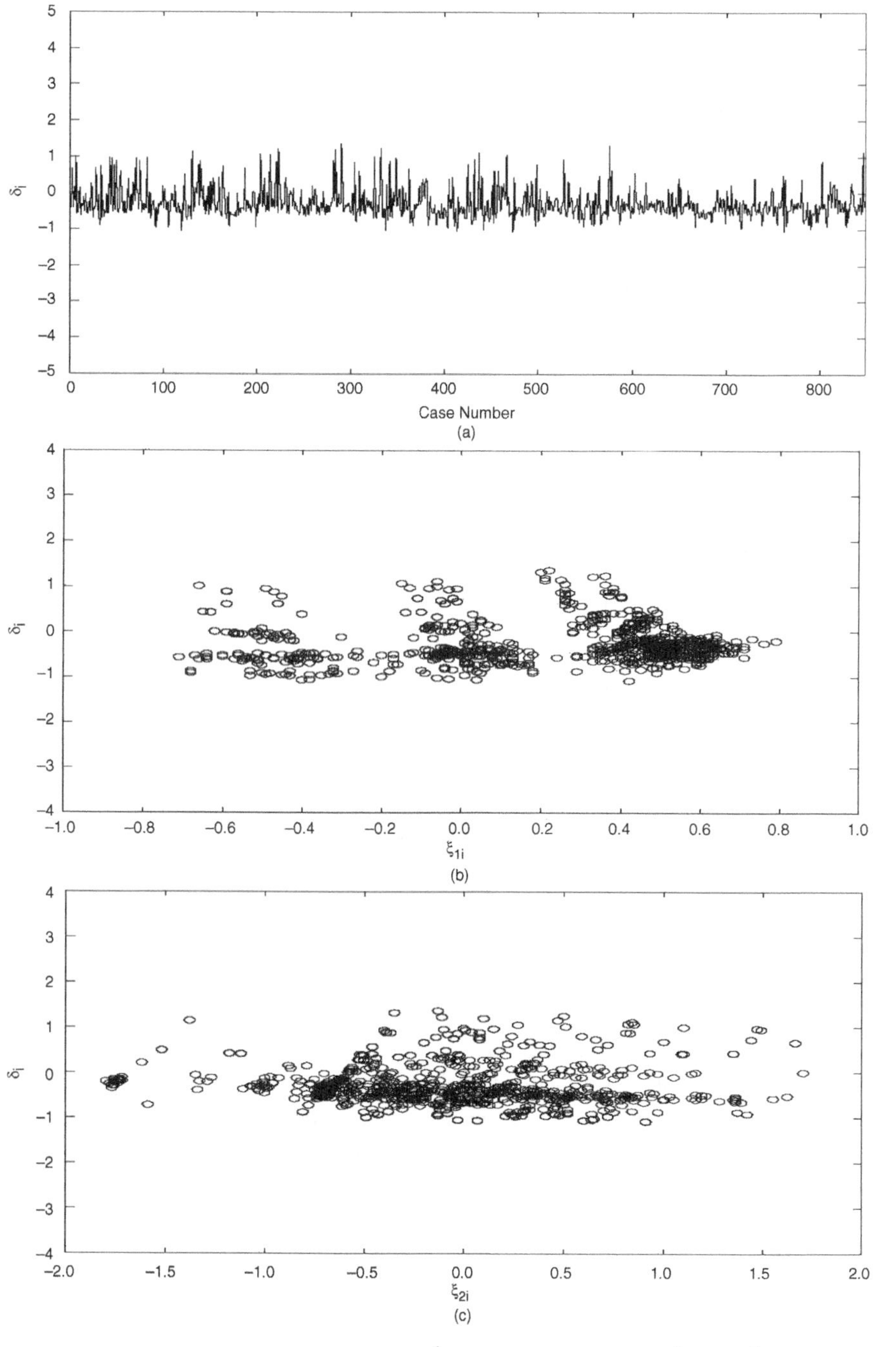

Figure 13.3 Plots of residuals $\hat{\delta}_i$ versus case numbers, $\hat{\xi}_{i1}$ and $\hat{\xi}_{i2}$.

Table 13.3 Bayesian estimates and standard error estimates of the real example.

Par	EST	SE	Par	EST	SE	Par	EST	SE
a_{11}	0.07	0.100	λ_{21}	2.60	0.373	α_{72}	−0.15	0.039
a_{12}	−0.31	0.171	λ_{31}	0.17	0.088	α_{73}	0.46	0.039
a_{13}	−0.30	0.166	λ_{52}	4.02	0.393	α_{82}	−0.19	0.039
a_{14}	−0.13	0.128	λ_{62}	3.83	0.389	α_{83}	0.27	0.037
a_{15}	−0.44	0.186	λ_{83}	0.92	0.062	α_{92}	0.56	0.045
a_{16}	−0.26	0.260	λ_{93}	0.77	0.059	α_{93}	1.14	0.045
a_{21}	−0.07	0.286	$\psi_{\epsilon 7}$	0.32	0.039	φ_0	−4.78	0.263
a_{22}	0.04	0.319	$\psi_{\epsilon 8}$	0.43	0.041	φ_1	−0.08	0.292
a_{23}	−0.12	0.075	$\psi_{\epsilon 9}$	0.60	0.047	φ_2	0.11	0.301
a_{24}	−0.13	0.136	b_1	−0.89	0.134	φ_3	−1.23	0.290
a_{25}	0.22	0.130	b_2	0.34	0.143	φ_4	−0.94	0.275
a_{26}	0.55	0.100	γ_1	−0.79	0.209	φ_5	−1.11	0.261
a_{31}	−0.14	0.078	γ_2	0.31	0.110	φ_6	−0.64	0.259
a_{32}	−0.50	0.143	γ_3	0.68	0.266	φ_7	0.47	0.251
a_{33}	0.18	0.143	ϕ_{11}	0.29	0.057	φ_8	−0.46	0.259
a_{34}	0.27	0.088	ϕ_{12}	−0.14	0.028	φ_9	0.40	0.231
a_{35}	0.17	0.139	ϕ_{22}	0.69	0.068			
a_{36}	0.08	0.210	ψ_δ	0.83	0.222			
a_{41}	−1.15	0.220						
a_{42}	−3.17	0.222						
a_{43}	−0.05	0.124						
a_{44}	−0.31	0.221						
a_{45}	2.54	0.206						
a_{46}	1.95	0.184						

better education and consultation to patients about the disease, and to maintain a good 'physician–patients' relationship. (iv) The interaction term of 'knowledge of medication' and '(dis)satisfaction with the physician' has a significant effect on 'nonadherence'. Depending on the situation, this interaction has various impacts. For example, it indicates that less knowledge of medication (negative ξ_1) and a bad relationship with the doctor (positive ξ_2) would have a very strong additive effect on nonadherence, which would need to be reduced by the interaction effect. Interpretations of the less important parameters in $\boldsymbol{\alpha}$, ψ_δ, $\psi_{\epsilon k}$ and $\boldsymbol{\varphi}$ are not discussed.

The robustness of the above results to the choice of prior inputs has been evaluated by repeating the analysis using reasonable alternative prior inputs. Moreover, the logarithm Bayes factors have been recomputed using different values of S and J. The same model comparison conclusion, and similar Bayesian estimates and standard error estimates, are obtained.

13.6 BAYESIAN ANALYSIS OF AN ARTIFICIAL EXAMPLE USING WINBUGS

To illustrate the application of WinBUGS (Spiegelhalter, Thomas, Best and Lunn, 2003) in analyzing the nonlinear SEMs with fixed covariates in the context of unordered binary data, the following artificial example is used. A data set $\mathbf{Y} = \{\mathbf{y}_i, i = 1, \ldots, n\}$ was generated with $\mathbf{y}_i = (y_{i1}, \ldots, y_{i9})^T$. For $k = 1, \ldots, 9$, the distribution of y_{ik} is binomial, $B(5, p_{ik})$, that is $y_{ik} \propto \exp\left[y_{ik}\vartheta_{ik} - 5\log(1+e^{\vartheta_{ik}})\right]$, $b(\vartheta_{ik}) = \log(1+e^{\vartheta_{ik}})$ and $\vartheta_{ik} = \log\left[p_{ik}/(1-p_{ik})\right] = \mu_k + \boldsymbol{\Lambda}_k^T \boldsymbol{\xi}_i$. Note that for these binomial variables, $\psi_{\epsilon k} = 1.0$ are treated as fixed parameters. The nonlinear structural equation is given by $\eta_i = bc_i + \gamma_1\xi_{i1} + \gamma_2\xi_{i2} + \gamma_3\xi_{i1}\xi_{i2} + \delta_i$. The fixed covariates c_i are generated from $B(1, 0.5)$. The specifications of the matrices $\boldsymbol{\Lambda}$ and $\boldsymbol{\Phi}$ are:

$$\boldsymbol{\Lambda}^T = \begin{bmatrix} 1.0^* & \lambda_{21} & \lambda_{31} & 0.0^* & 0.0^* & 0.0^* & 0.0^* & 0.0^* & 0.0^* \\ 0.0^* & 0.0^* & 0.0^* & 1.0^* & \lambda_{52} & \lambda_{62} & 0.0^* & 0.0^* & 0.0^* \\ 0.0^* & 0.0^* & 0.0^* & 0.0^* & 0.0^* & 0.0^* & 1.0^* & \lambda_{83} & \lambda_{93} \end{bmatrix} \text{ and } \boldsymbol{\Phi} = \begin{bmatrix} \phi_{11} & \phi_{12} \\ \phi_{12} & \phi_{22} \end{bmatrix},$$

where the 1s and 0s are treated as known parameters, whilst λ_{21}, λ_{31}, λ_{52}, λ_{62}, λ_{83}, λ_{93}, ϕ_{11}, ϕ_{12} and ϕ_{22} are unknown parameters. The true population values of the unknown parameters are: $\boldsymbol{\mu} = (\mu_1, \cdots, \mu_9)^T = (0.8, \cdots, 0.8)^T$, $\lambda_{21} = \lambda_{31} = \lambda_{83} = \lambda_{93} = 0.6$, $\lambda_{52} = \lambda_{62} = 0.7$, $\boldsymbol{\Gamma} = (\gamma_1, \gamma_2, \gamma_3) = (0.5, 0.5, 0.5)$, $b = 0.6$, $\psi_\delta = 0.7$, $(\phi_{11}, \phi_{12}, \phi_{22}) = (1.0, 0.5, 1.0)$.

Two independent data sets with $n = 500$ random observations are generated based on the above settings. The first data set contains fully observed data without any missing data and the second data set contains some nonignorable missing data. We use the following logistic regression to generate the nonignorable missing data in the second data set:

$$\text{logit}[\text{pr}(r_{ik} = 1|\mathbf{y}_i, \boldsymbol{\omega}_i, \boldsymbol{\varphi})] = \varphi_0 + \varphi_1 y_{i1} + \cdots + \varphi_9 y_{i9}, \tag{13.24}$$

where the true values of parameter vector $\boldsymbol{\varphi}$ are given by $\varphi_0 = -4.0$, $\varphi_1 = \cdots = \varphi_6 = 0.5$, $\varphi_7 = -1.5$ and $\varphi_8 = \varphi_9 = -1.0$. In this created missing data set the number of full observations is 397. That is, about 20.6% of the total number of observations have missing entries.

We use WinBUGS to analyze these two data sets. Hyperparameters of the conjugate prior distributions are taken as: $\alpha_{0\delta 1} = 10$, $\beta_{0\delta 1} = 8$, $\rho_0 = 8$, $\mathbf{H}_{0yk} = 0.25\mathbf{I}$, $\mathbf{H}_{0\omega 1} = 0.25$, $\mathbf{H}_{0k} = 0.25\mathbf{I}$, $\mathbf{H}_{0\varphi} = 0.25\mathbf{I}$, $\mathbf{R}_0 = 5\boldsymbol{\Phi}_0$, where $\boldsymbol{\Phi}_0$ is the true value of $\boldsymbol{\Phi}$; $\boldsymbol{\mu}_0$, b_0, $\boldsymbol{\Lambda}_{0k}$, $\boldsymbol{\Lambda}_{0\omega 1}$ and $\boldsymbol{\varphi}_0$ are all equal to the true values of the corresponding parameters. In the analysis of the second data set, the missing model specified in Equation (13.24) is used to cope with the nonignorable missing data. Plots of sequences of observations that correspond to some parameters for the first and second data sets are displayed in Figures 13.4 and 13.5, respectively.

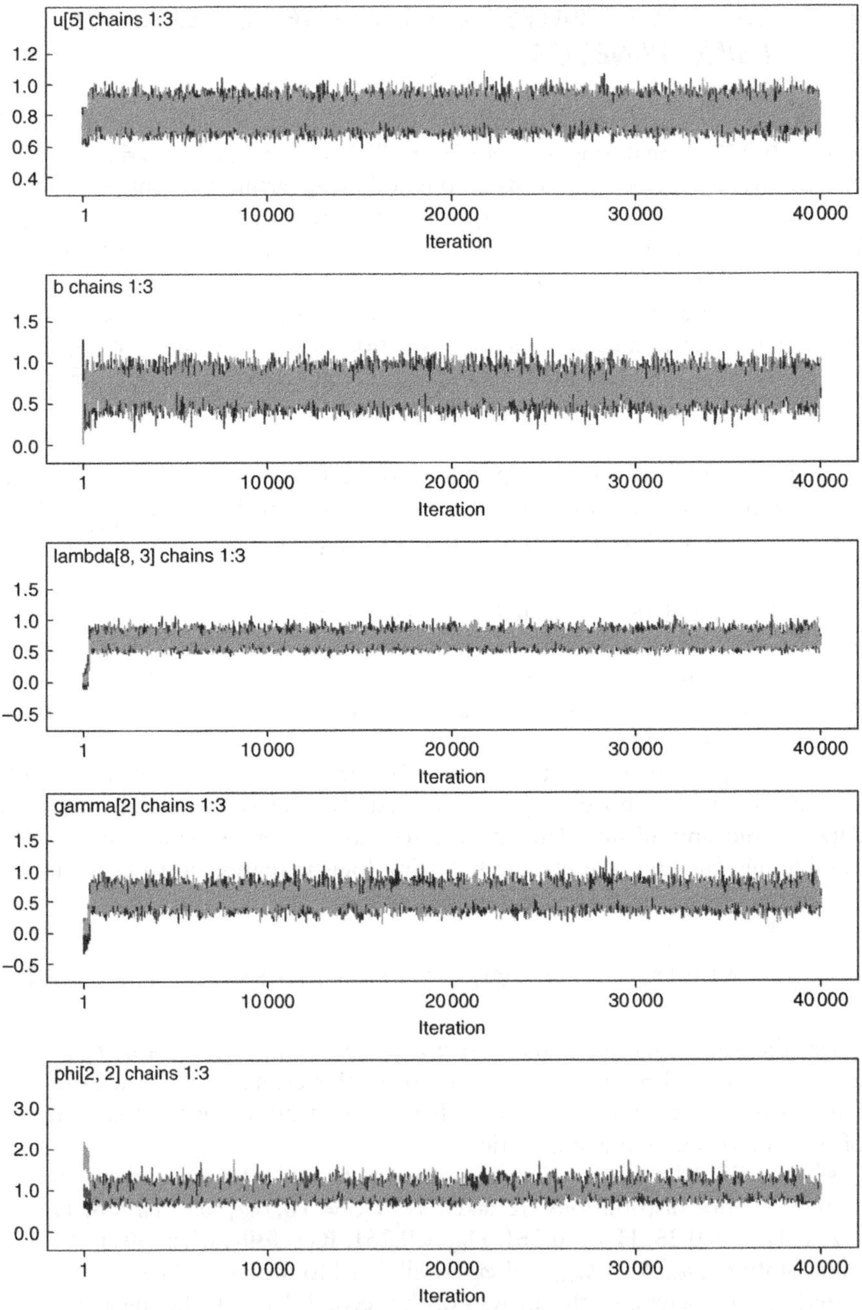

Figure 13.4 Three chains of observation corresponding to μ_5, b, λ_{83}, γ_2 and ϕ_{22} generated by different initial values in the analysis of the first data set.

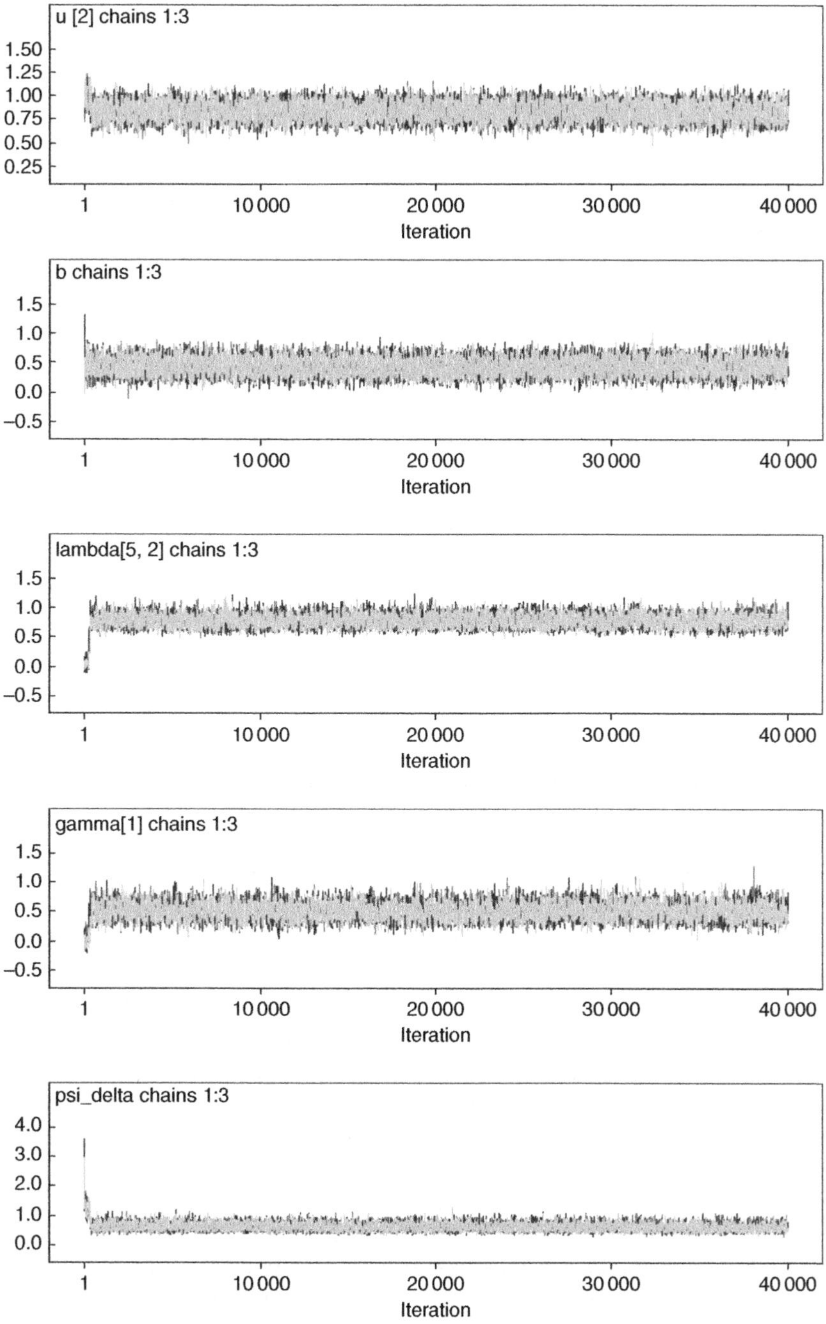

Figure 13.5 Three chains of observation corresponding to μ_2, b, λ_{52}, γ_1 and ψ_δ generated by different initial values in the analysis of the second data set.

Table 13.4 Bayesian estimates and standard error estimates in the artificial example.

Without missing data				With nonignorable missing data							
Par	True value	EST	SE	Par	True value	EST	SE	Par	True value	EST	SE
μ_1	0.80	0.79	0.10	μ_1	0.80	0.82	0.11	φ_0	−4.00	−3.96	0.40
μ_2	0.80	0.76	0.07	μ_2	0.80	0.83	0.07	φ_1	0.50	0.40	0.12
μ_3	0.80	0.77	0.07	μ_3	0.80	0.78	0.07	φ_2	0.50	0.53	0.13
μ_4	0.80	0.73	0.06	μ_4	0.80	0.82	0.07	φ_3	0.50	0.47	0.12
μ_5	0.80	0.81	0.06	μ_5	0.80	0.87	0.06	φ_4	0.50	0.41	0.11
μ_6	0.80	0.82	0.06	μ_6	0.80	0.86	0.06	φ_5	0.50	0.42	0.13
μ_7	0.80	0.85	0.07	μ_7	0.80	0.97	0.07	φ_6	0.50	0.47	0.13
μ_8	0.80	0.89	0.05	μ_8	0.80	0.85	0.05	φ_7	−1.50	−1.21	0.12
μ_9	0.80	0.85	0.05	μ_9	0.80	0.93	0.05	φ_8	−1.00	−1.07	0.12
b	0.60	0.69	0.12	b	0.60	0.56	0.13	φ_9	−1.00	−0.82	0.11
λ_{21}	0.60	0.59	0.06	λ_{21}	0.60	0.52	0.06				
λ_{31}	0.60	0.60	0.06	λ_{31}	0.60	0.57	0.06				
λ_{52}	0.70	0.82	0.09	λ_{52}	0.70	0.76	0.08				
λ_{62}	0.70	0.85	0.10	λ_{62}	0.70	0.77	0.08				
λ_{83}	0.60	0.67	0.08	λ_{83}	0.60	0.64	0.08				
λ_{93}	0.60	0.59	0.07	λ_{93}	0.60	0.51	0.08				
γ_1	0.50	0.68	0.13	γ_1	0.50	0.67	0.12				
γ_2	0.50	0.58	0.12	γ_2	0.50	0.51	0.14				
γ_3	0.50	0.55	0.13	γ_3	0.50	0.65	0.13				
ϕ_{11}	1.00	0.73	0.11	ϕ_{11}	1.00	1.08	0.15				
ϕ_{12}	0.50	0.43	0.07	ϕ_{12}	0.50	0.59	0.09				
ϕ_{22}	1.00	0.94	0.15	ϕ_{22}	1.00	0.91	0.15				
ψ_δ	0.70	0.75	0.12	ψ_δ	0.70	0.80	0.14				

These plots indicate that the algorithm converged in less than 5000 iterations. To be conservative, we discard the first 20 000 burn-in iterations, and collect 20 000 further observations for obtaining the Bayesian results. The Bayesian estimates and their standard error estimates are presented in Table 13.4. It seems that the Bayesian estimates are acceptable, and the standard error estimates are reasonable. For the first data set, the DIC value corresponding to the model is 12 604.5. Plots of estimated residuals $\hat{\delta}_i$ versus case numbers $\hat{\xi}_{i1}$ and $\hat{\xi}_{i2}$ in the analyses of this data set are presented in Figure 13.6. For the second set with nonignorable missing data, WinBUGS does not give the DIC value. To assess the goodness-of-fit roughly, we examine the estimated residual plots of $\hat{\delta}_i$ versus the case numbers $\hat{\xi}_{i1}$ and $\hat{\xi}_{i2}$. These plots are presented in Figure 13.7. From these estimated residual plots, we see that the measurement and structural equations fit the data reasonably well. The WinBUGS codes and data are given in the following website: http://www.wiley.com/go/lee_structural.

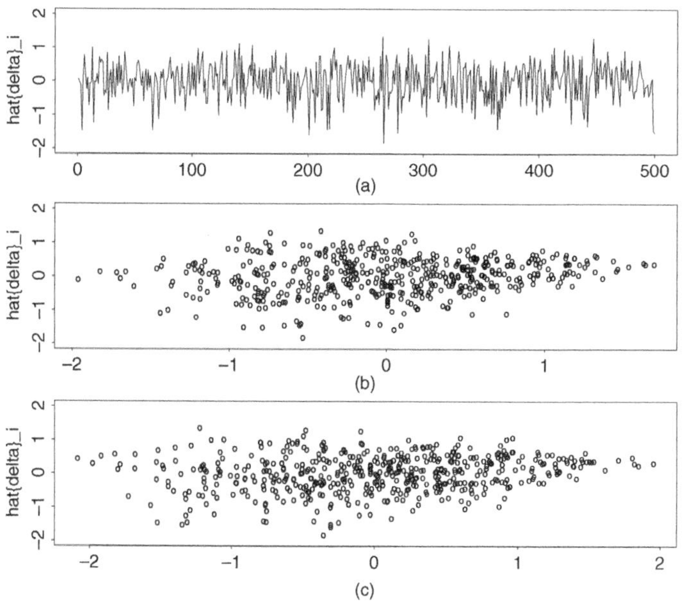

Figure 13.6 Plots of estimated residuals $\hat{\delta}_i$ versus (a) case numbers, (b) $\hat{\xi}_{i1}$ and (c) $\hat{\xi}_{i2}$ in the analysis of the first data set.

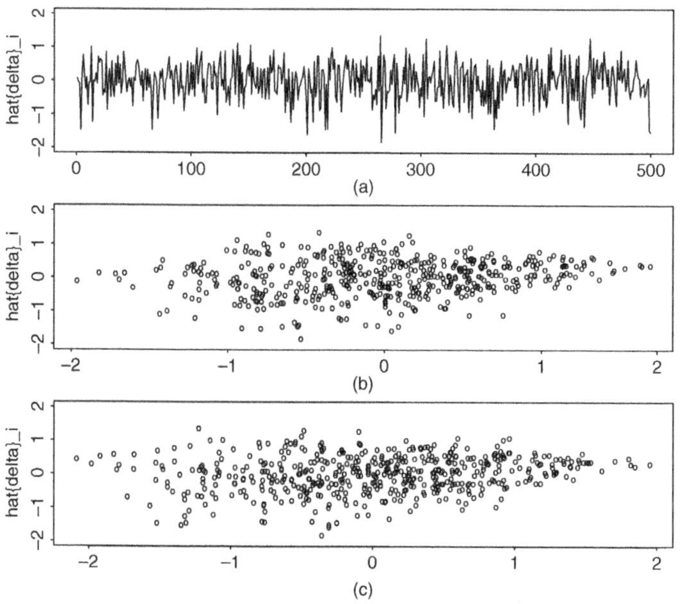

Figure 13.7 Plots of estimated residuals $\hat{\delta}_i$ versus (a) case numbers, (b) $\hat{\xi}_{i1}$ and (c) $\hat{\xi}_{i2}$ in the analysis of the second data set.

13.7 DISCUSSION

Based on the Bayesian framework presented in this chapter, general mixed non-normal and ordered categorical data that are common in practical applications can be analyzed through a nonlinear SEM with fixed covariates and EFDs. After the accommodation of missing data with a nonignorable missing mechanism, a comprehensive model is provided for analyzing complex continuous and discrete data in substantive research. We discuss the application of the Gibbs sampler and the MH algorithm for obtaining the Bayesian estimates and their standard error estimates. For model comparison, we present a path sampling procedure for computing the Bayes factor, which can be used to compare different SEMs or different models for the nonignorable missing mechanism. The freely available software WinBUGS is able to produce the Bayesian solution for most special cases of the framework. However, for some complex situations, for example the model considered in Section 13.6 with nonignorable missing data, WinBUGS does not provide the DIC value. However, this software is able to produce estimated residual plots which can be used to assess the goodness-of-fit roughly.

As the binomial distribution belonged to EFDs, unordered binary data can be analyzed through the methodology presented here. We mentioned in Chapter 7 that it is incorrect to treat unordered binary variables as ordered binary variables. Here, we point out that it is also incorrect to treat ordered binary variables as unordered binary variables. The methodologies presented in Chapter 7 should be used to analyze ordered binary data.

The proposed model framework and the key formulae in the Bayesian methods have a high potential for applications to the following useful models. Consider the latent variable model developed in Sammel, Ryan and Legler (1997) that concentrated on the mean structure of the manifest variables. An equivalent representation is defined as: for $i = 1, \ldots, n$, $\mathbf{y}_i$ and $\boldsymbol{\omega}_i$ satisfy the expressions given in Equation (13.1), the vector of canonical parameters $\boldsymbol{\vartheta}_i$ is related to $\boldsymbol{\omega}_i$ and $\mathbf{X}_i$ by the model defined by Equation (13.2), and $\boldsymbol{\omega}_i$ is related to fixed covariates $\mathbf{c}_i$ by the linear model $\boldsymbol{\omega}_i = \mathbf{B}\mathbf{c}_i + \boldsymbol{\delta}_i$. Comparing this linear model with Equation (13.3), we see that it is a special case of our nonlinear structural equation (with $\boldsymbol{\omega}_i = \boldsymbol{\eta}_i$, $\boldsymbol{\Pi} = \mathbf{0}$ and $\boldsymbol{\Gamma} = \mathbf{0}$), and relationships among the latent variables in $\boldsymbol{\omega}_i$ are not assessed. The Bayesian framework in this chapter provides methods to analyze these relationships. The binomial probit model is defined by $\mathbf{y}_i = \mathbf{X}_i\mathbf{A} + \boldsymbol{\epsilon}_i$, where $\mathbf{y}_i$ is a vector of binomial variables, $\mathbf{X}_i$ are covariates and $\boldsymbol{\epsilon}_i$ is a vector of error measurement with distribution $N[\mathbf{0}, \boldsymbol{\Psi}_\epsilon]$. This model, which is a special case of the model presented in this chapter, has wide applications in medicine, for example to gene-variable selection (Lee *et al.*, 2003; Zhou, Wang and Dougherty, 2003). Moreover, the multivariate probit model (Bock and Gibbons, 1996) as discussed in Chapter 7 is also a special case of the presented model. Finally, to accommodate a $p_i \times q$ matrix of fixed covariates, $\mathbf{C}_i^* = (\mathbf{c}_{i1}^{*T}, \ldots, \mathbf{c}_{ip_i}^{*T})^T$, which links to $\boldsymbol{\omega}_i$, we can allow the dimension of $\mathbf{y}_i$ to be p_i, which varies with i, and modify the 'measurement' model of Equation (13.2) to

$$\vartheta_{ik} = \mathbf{x}_{ik}^T \mathbf{e}_k + \mathbf{c}_{ik}^* \boldsymbol{\omega}_i, \quad k = 1, \ldots, p_i. \tag{13.25}$$

The model defined by Equations (13.1) and (13.25) is a generalized linear mixed model (GLMM). Hence, the nonlinear SEM that is defined by Equations (13.1), (13.25) and (13.3) generalizes the GLMM to a model with a more general covariance structure. Bayesian methods for analyzing this model can be established by essentially the same procedure as that proposed in Section 13.3. For example, the full conditional distributions can be obtained from those presented in Section 13.3.1, with $\boldsymbol{\Lambda}_k^T$ be replaced by $\mathbf{c}_{ik}^*$.

The Bayesian framework with EFDs presented in this chapter is a generalization of two types of method. First, it generalizes SEMs, which can be regarded as well-known methods in social–psychological research, to model with variables in EFDs and subtle mean structure with covariates. Secondly, it generalizes some widely used latent variable models in biostatistics, such as the multivariate probit model, and possibly GLMMs, to models with subtle covariance structures.

APPENDIX 13.1: IMPLEMENTATION OF THE MH ALGORITHMS

In simulating observations from $p(\boldsymbol{\omega}_i|\mathbf{y}_i, \mathbf{r}_i, \boldsymbol{\theta}, \boldsymbol{\varphi})$, we choose $N[\cdot, \sigma_\omega^2 \boldsymbol{\Omega}_\omega]$ as the proposal distribution, where $\boldsymbol{\Omega}_\omega^{-1} = \boldsymbol{\Sigma}_\omega + \boldsymbol{\Sigma}_\varphi + \boldsymbol{\Lambda}^T \boldsymbol{\Psi}_\omega \boldsymbol{\Lambda}$, in which

$$\boldsymbol{\Sigma}_\omega = \begin{bmatrix} \boldsymbol{\Pi}_0^T \boldsymbol{\Psi}_\delta^{-1} \boldsymbol{\Pi}_0 & -\boldsymbol{\Pi}_0^T \boldsymbol{\Psi}_\delta^{-1} \boldsymbol{\Gamma} \boldsymbol{\Delta} \\ -\boldsymbol{\Delta}^T \boldsymbol{\Gamma}^T \boldsymbol{\Psi}_\delta^{-1} \boldsymbol{\Pi}_0 & \boldsymbol{\Phi}^{-1} + \boldsymbol{\Delta}^T \boldsymbol{\Gamma}^T \boldsymbol{\Psi}_\delta^{-1} \boldsymbol{\Gamma} \boldsymbol{\Delta} \end{bmatrix},$$

$$\boldsymbol{\Sigma}_\varphi = \frac{p \exp(\varphi_0 + \boldsymbol{\varphi}_y^T \mathbf{y}_i)}{[1 + \exp(\varphi_0 + \boldsymbol{\varphi}_y^T \mathbf{y}_i)]^2} \boldsymbol{\varphi}_\omega \boldsymbol{\varphi}_\omega^T,$$

with $\boldsymbol{\Pi}_0 = \mathbf{I} - \boldsymbol{\Pi}$, $\boldsymbol{\Delta} = \partial \mathbf{H}(\boldsymbol{\xi}_i)/\partial \boldsymbol{\xi}_i^T|_{\boldsymbol{\xi}_i=0}$, $\boldsymbol{\varphi}_y = (\varphi_1, \ldots, \varphi_p)^T$, $\boldsymbol{\varphi}_\omega = (\varphi_{p+1}, \ldots, \varphi_{p+q})^T$ and $\boldsymbol{\Psi}_\omega = \text{diag}(\ddot{b}(\vartheta_{i1})/\psi_{\epsilon 1}, \ldots, \ddot{b}(\vartheta_{ip})/\psi_{\epsilon p})|_{\boldsymbol{\omega}_i=\mathbf{0}}$.

In simulating observations from the conditional distributions $p(\mathbf{y}_{\text{mis}i}|\boldsymbol{\omega}_i, \mathbf{r}_i, \boldsymbol{\theta}, \mathbf{y}_{\text{obs}i})$, $p(\mathbf{a}_k|\mathbf{Y}, \boldsymbol{\Omega}, \boldsymbol{\Lambda}_k, \psi_{\epsilon k})$, $p(\psi_{\epsilon k}|\mathbf{Y}, \boldsymbol{\Omega}, \mathbf{a}_k, \boldsymbol{\Lambda}_k)$, $p(\boldsymbol{\Lambda}_k|\mathbf{Y}, \boldsymbol{\Omega}, \mathbf{a}_k, \psi_{\epsilon k})$ and $p(\boldsymbol{\varphi}|\mathbf{Y}, \boldsymbol{\Omega}, \mathbf{r}, \boldsymbol{\theta})$, the proposal distributions are $N[\cdot, \sigma_y^2 \Omega_{ymi}]$, $N[\cdot, \sigma_a^2 \boldsymbol{\Omega}_{ak}]$, $N[\cdot, \sigma_\psi^2 \boldsymbol{\Omega}_{\psi k}]$, $N[\cdot, \sigma_\lambda^2 \boldsymbol{\Omega}_{\lambda k}]$ and $N[\cdot, \sigma_\varphi^2 \boldsymbol{\Omega}_\varphi]$, respectively, where

$$\boldsymbol{\Omega}_{ymi}^{-1} = \sum_{\{k: r_{ik}=1\}} \left. \frac{\partial^2 c_k(y_{ik}, \psi_{\epsilon k})}{\partial \mathbf{y}_{\text{mis}i} \partial \mathbf{y}_{\text{mis}i}} \right|_{y_{ik}=0} + \frac{p \exp(\varphi_0 + \sum_{\ell \in \bar{D}} \varphi_\ell y_{i\ell} + \boldsymbol{\varphi}_\omega^T \boldsymbol{\omega}_i)}{(1 + \exp(\varphi_0 + \sum_{\ell \in \bar{D}} \varphi_\ell y_{i\ell} + \boldsymbol{\varphi}_\omega^T \boldsymbol{\omega}_i))^2} \boldsymbol{\varphi}_m \boldsymbol{\varphi}_m^T,$$

where $\boldsymbol{\varphi}_m$ is a vector that contains the elements of $\boldsymbol{\varphi}_y$ corresponding to $\mathbf{y}_{\text{mis}i}$, $\bar{D}$ is the set of indexes corresponding to $\mathbf{y}_{0i}$.

$$\boldsymbol{\Omega}_{ak}^{-1} = \sum_{i=1}^{n} \ddot{b}(\vartheta_{ik})\mathbf{x}_{ik}\mathbf{x}_{ik}^T/\psi_{\epsilon k}\Big|_{\mathbf{a}_k=\mathbf{0}} + \mathbf{H}_{0k}^{-1},$$

$$\Omega_{\psi k}^{-1} = 1 - n/2 - \alpha_{0\epsilon k} - 2\sum_{i=1}^{n}[y_{ik}\vartheta_{ik} - b(\vartheta_{ik})] - \ddot{c}_k(y_{ik}, \psi_{\epsilon k})\Big|_{\psi_{\epsilon k}=1} + 2\beta_{0\epsilon k},$$

$$\boldsymbol{\Omega}_{\lambda k}^{-1} = \sum_{i=1}^{n} \ddot{b}(\vartheta_{ik})\boldsymbol{\omega}_i\boldsymbol{\omega}_i^T\Big|_{\boldsymbol{\Lambda}_k=\mathbf{0}} + \psi_{\epsilon k}^{-1}\mathbf{H}_{0yk}^{-1} \quad \text{and} \quad \boldsymbol{\Omega}_{\varphi}^{-1} = \frac{p}{4}\sum_{i=1}^{n}\mathbf{d}_i\mathbf{d}_i^T + \mathbf{H}_{0\varphi}^{-1}.$$

To improve efficiency, we respectively use $N[\boldsymbol{\mu}_{ymi}, \boldsymbol{\Omega}_{ymi}]$, $N[\boldsymbol{\mu}_{ak}, \boldsymbol{\Omega}_{ak}]$, $N[\mu_{\psi k}, \Omega_{\psi k}]$, $N[\boldsymbol{\mu}_{\lambda k}, \boldsymbol{\Omega}_{\lambda k}]$ and $N[\boldsymbol{\mu}_{\varphi}, \boldsymbol{\Omega}_{\varphi}]$ as initial proposal distributions in the first few iterations, where

$$\begin{aligned}\boldsymbol{\mu}_{ymi} &= \sum_{\{k: r_{ik}=1\}} \frac{\partial\{y_{ik}\vartheta_{ik}/\psi_{\epsilon k} + c_k(y_{ik}, \psi_{\epsilon k})\}}{\partial \mathbf{y}_{\text{mis}i}}\Bigg|_{y_{ik}=0} \\ &\quad + \left[\sum_{k=1}^{p} r_{ik} - \frac{p\exp(\varphi_0 + \sum_{\ell\in\bar{D}}\varphi_\ell y_{i\ell} + \boldsymbol{\varphi}_\omega^T\boldsymbol{\omega}_i)}{1 + \exp(\varphi_0 + \sum_{\ell\in\bar{D}}\varphi_\ell y_{i\ell} + \boldsymbol{\varphi}_\omega^T\boldsymbol{\omega}_i)}\right]\boldsymbol{\varphi}_m\end{aligned}$$

$$\boldsymbol{\mu}_{ak} = \sum_{i=1}^{n}\left[y_{ik} - \dot{b}(\vartheta_{ik})|_{\mathbf{a}_k=\mathbf{0}}\right]\frac{\mathbf{x}_{ik}}{\psi_{\epsilon k}} + \mathbf{H}_{0k}^{-1}\mathbf{a}_{0k},$$

$$\mu_{\psi k} = 1 - n/2 - \alpha_{0\epsilon k} - \sum_{i=1}^{n}[y_{ik}\vartheta_{ik} - b(\vartheta_{ik})] + \dot{c}_k(y_{ik}, \psi_{\epsilon k})\Big|_{\psi_{\epsilon k}=1} + \beta_{0\epsilon k},$$

$$\boldsymbol{\mu}_{\lambda k} = \sum_{i=1}^{n}\left[y_{\psi_{ik}} - \dot{b}(\vartheta_{ik})|_{\boldsymbol{\Lambda}_k=\mathbf{0}}\right]\frac{\boldsymbol{\omega}_i}{\psi_{\epsilon k}} + \mathbf{H}_{0yk}^{-1}\boldsymbol{\Lambda}_{0k} \quad \text{and}$$

$$\boldsymbol{\mu}_{\varphi} = \sum_{i=1}^{n}\left[\sum_{k=1}^{p} r_{ik} - \frac{p}{2}\right]\mathbf{d}_i + \mathbf{H}_{0\varphi}^{-1}\boldsymbol{\varphi}_0.$$

A multivariate version of the MH algorithm is used to simulate observations from $p(\boldsymbol{\alpha}_k, \mathbf{y}_k^* | \mathbf{z}_k, \boldsymbol{\Omega}, \boldsymbol{\theta})$ (see Equation (13.16)). Following Cowles (1996), for the joint proposal distribution of $\boldsymbol{\alpha}_k$ and $\mathbf{y}_k^*$ given $\mathbf{z}_k$, $\boldsymbol{\Omega}$ and $\boldsymbol{\theta}$ can be constructed according to the factorization $p(\boldsymbol{\alpha}_k, \mathbf{y}_k^*|\mathbf{z}_k, \boldsymbol{\Omega}, \boldsymbol{\theta}) = p(\boldsymbol{\alpha}_k|\mathbf{z}_k, \boldsymbol{\Omega}, \boldsymbol{\theta})p(\mathbf{y}_k^*|\boldsymbol{\alpha}_k, \mathbf{z}_k, \boldsymbol{\Omega}, \boldsymbol{\theta})$. At the jth iteration, we generate a candidate vector of thresholds $(\alpha_{k,1}, \ldots, \alpha_{k,b_k-1})$ from the following univariate truncated normal distribution

$$\alpha_{k,m} \sim N[\alpha_{k,m}^{(j)}, \sigma_{\alpha_k}^2]I_{(\alpha_{k,m-1}, \alpha_{k,m+1}^{(j)}]}(\alpha_{k,m}), \quad \text{for} \quad m = 1, \ldots, b_k - 1,$$

where $\alpha_{k,m}^{(j)}$ is the current value of $\alpha_{k,m}$ and $\sigma_{\alpha_k}^2$ is chosen to obtain an average acceptance rate of approximately 0.25 or greater. The acceptance probability for a candidate vector $(\boldsymbol{\alpha}_k, \mathbf{y}_k^*)$ as a new observation $(\boldsymbol{\alpha}_k^{(j+1)}, \mathbf{y}_k^{*(j+1)})$ is $\min\{1, R_k\}$, where

$$R_k = \frac{p(\boldsymbol{\alpha}_k, \mathbf{y}_k^* | \mathbf{z}_k, \boldsymbol{\Omega}, \boldsymbol{\theta}) p(\boldsymbol{\alpha}_k^{(j)}, \mathbf{y}_k^{*(j)} | \boldsymbol{\alpha}_k, \mathbf{y}_k^*, \mathbf{z}_k, \boldsymbol{\Omega}, \boldsymbol{\theta})}{p(\boldsymbol{\alpha}_k^{(j)}, \mathbf{y}_k^{*(j)} | \mathbf{z}_k, \boldsymbol{\Omega}, \boldsymbol{\theta}) p(\boldsymbol{\alpha}_k, \mathbf{y}_k^* | \boldsymbol{\alpha}_k^{(j)}, \mathbf{y}_k^{*(j)}, \mathbf{z}_k, \boldsymbol{\Omega}, \boldsymbol{\theta})}.$$

For an accepted $\boldsymbol{\alpha}_k$, a new $\mathbf{y}_k^*$ is simulated from the following univariate truncated distribution:

$$p(y_{ik} | \boldsymbol{\alpha}_k, z_{ik}, \boldsymbol{\omega}_i, \boldsymbol{\theta}) \stackrel{D}{=} \exp\{[y_{ik}\vartheta_{ik} - b(\vartheta_{ik})]/\psi_{\epsilon k} + c_k(y_{ik}, \psi_{\epsilon k})\} I_{(\alpha_{k,z_{ik}}, \alpha_{k,z_{ik}+1}]}(y_{ik}),$$

where y_{ik} and z_{ik} are the ith components of $\mathbf{y}_k^*$ and $\mathbf{z}_k$, respectively, and $I_A(y)$ is an indicator function which takes 1 if y is in A and zero otherwise.

APPENDIX 13.2

y_1: Did you have unused drugs remaining? (No, '0'/Yes, '1')
y_2: Did you stop, decrease or increase the dosage? (No, '0'/Yes, '1')
y_3: Did you forget to take drugs? (No, '0'/Yes, '1')
y_4: Do you know that you have hypertension? (No, '0'/Yes, '1')
y_5: Do you know the reason for taking drugs? (No, '0'/Yes, '1')
y_6: Do you know the reason for taking drugs for a long time? (No, '0'/Yes, '1')
y_7: Your physician consults your opinion in deciding the medication method. (Strongly agree, '1'/agree, '2'/no opinion, '3'/disagree, '4'/strongly disagree, '5')
y_8: You are free to discuss your important opinion with your physician. (Strongly agree, '1'/agree, '2'/no opinion, '3'/disagree, '4'/strongly disagree, '5')
y_9: Your physician listens carefully to you. (Strongly agree, '1'/agree, '2'/no opinion, '3'/disagree, '4'/strongly disagree, '5')

REFERENCES

Bock, R. D. and Gibbons, R. D. (1996) High dimensional multivariate probit analysis. *Biometrics*, **52**, 1183–193.

Booth, J. G. and Hobert, J. P. (1999) Maximum generalized linear mixed model likelihoods with an automated Monte Carlo EM algorithm. *Journal of the Royal Statistical Society, Series B*, **6**, 265–285.

Browne, M. W. (1987) Robustness of statistical inference in factor analysis and related models. *Biometrika*, **74**, 375–384.

Chan, G. M. C. (2001) *The Effects of Treatment Compliance on Clinical Outcomes in Patients with Chronic Diseases.* Ph.D. Thesis, Department of Medicine and Therapeutics, the Chinese University of Hong Kong.

Cowles, M. K. (1996) Accelerating Markov chain Monte Carlo convergence for cumulative-link generalized linear models. *Statistics and Computing*, **6**, 101–111.

Gelman, A. and Rubin, D. B. (1992) Inference from iterative simulation using multiple sequences. *Statistical Science*, 7, 457–472.

Gelman, A., Meng, X. L. and Stern, H. (1996) Posterior predictive assessment of model fitness via realized discrepancies. *Statistica Sinica*, **6**, 733–807.

Geman, S. and Geman, D. (1984) Stochastic relaxation, Gibbs distributions and the Bayesian restoration of images. *IEEE Transactions on Pattern Analysis and Machine Intelligence*, **6**, 721–741.

Hastings, W. K. (1970) Monte Carlo sampling methods using Markov chains and their application. *Biometrika*, **57**, 97–100.

Ibrahim, J. G., Chen, M. H. and Lipsitz, S. R. (2001) Missing responses in generalized linear mixed models when the missing data mechanism is nonignorable. *Biometrics*, **88**, 551–564.

Kano, Y., Berkane, M. and Bentler, P. M. (1993) Statistical inference based on pseudo-maximum likelihood estimators in elliptical populations. *Journal of the American Statistical Association*, **88**, 135–143.

Kass, R. E. and Raftery, A. E. (1995) Bayes factors. *Journal of the American Statistical Association*, **90**, 773–795.

Lee, S. Y. and Song, X. Y. (2004) Maximum likelihood analysis of a general latent variable model with hierarchically mixed data. *Biometrics*, **60**, 624–636.

Lee, K. E. *et al.* (2003) Gene selection: a Bayesian variable selection approach. *Bioinformatics*, **19**, 90–97.

McCullagh, P. and Nelder, J. A. (1989) *Generalized Linear Models* (2nd edn.). London: Chapman and Hall.

Metropolis, N. *et al.* (1953) Equations of state calculations by fast computing machine. *Journal of Chemical Physics*, **21**, 1087–1091.

Sammel, M. D. and Ryan, L. M. (1996) Latent variables with fixed effects. *Biometrics*, **52**, 220–43.

Sammel, M. D., Ryan, L. M. and Legler, J. M. (1997) Latent variable models for mixed discrete and continuous outcomes. *Journal of the Royal Statistical Society, Series B*, **59**, 667–678.

Schumacker, R. E. and Marcoulides, G. A. (1998) *Interaction and Nonlinear Effects in Structural Equation Models*. Mahwah, N. J: Lawrence Erlbaum Associates.

Shapiro, A. and Browne, M. W. (1987) Analysis of covariance structures under elliptical distributions. *Journal of American Statistical Association*, **82**, 1092–1097.

Song, X. Y. and Lee, S. Y. (2006) Bayesian analysis of latent variable models with nonignorable missing outcomes from exponential family. *Statistics in Medicine*, in press.

Spiegelhalter, D. J., Thomas, A., Best, N. G. and Lunn, D. (2003). *WinBUGS User Manual, Version 1.4*. Cambridge, England: MRC Biostatistics Unit.

Wedel, M. and Kamakura, W. (2001) Factor analysis with (mixed) observed and latent variables in the exponential family. *Psychometrika*, **55**, 515–530.

Yuan, K. H. and Bentler, P. M. (1997) Mean and covariance structures analysis: theoretical and practical improvements. *Journal of the American Statistical Association*, **92**, 767–774.

Zhou, X. B., Wang, X. D. and Dougherty, E. R. (2003) Missing-value estimation using linear and non-linear regression with Bayesian gene selection. *Bioinformatics*, **19**, 2302–2307.

14
Conclusion

In this book, we have introduced the Bayesian approach coupled with the data augmentation algorithm and MCMC methods to analyze SEMs. We have discussed the basic factor analysis model, the LISREL type models, as well as more complex models such as models with ordered categorical variables, and/or dichotomous variables, nonlinear models, multilevel models, multi-sample models, mixture models, models with missing data, models with variables from the exponential family of distributions, and some of their special cases and combinations. We have concentrated on obtaining the Bayesian estimates of the unknown parameters and their standard error estimates, and computing the Bayes factor for model comparison. Moreover, we have shown that estimates of latent variables and their standard error estimates can be produced as by-products, and we have also demonstrated the use of the latent variable estimates to conduct residual analysis through the estimated residual plots. The Bayesian methodologies in analyzing the above complex SEMs have been demonstrated by real examples from behavioral, educational, medical, social and psychological research. Almost all of these examples cannot be effectively handled by the existing commercial software in SEM.

As we have mentioned and shown throughout this book, the Bayesian approach has the following advantages. (i) It can directly incorporate genuine prior knowledge in the analysis for obtaining better results. (ii) Except possibly for dichotomous or binary variables, the associated sampling-based Bayesian methods give reliable estimates and standard error estimates even with small sample sizes. (iii) Compared with the classical methods, it gives better estimates of the latent variables. (iv) Samples that are simulated from the posterior distributions of parameters and latent variables are not only useful in Bayesian estimation and model comparison, but also in residual and outlier analyses. Moreover, statistical analyses such as construction of confidence intervals based

Structural Equation Modeling: A Bayesian Approach S-Y. Lee

on these samples do not rely on asymptotic theory. (v) The Bayes factor or DIC that is closely related to a Bayesian approach gives a more flexible statistic for model comparison/selection than the classical likelihood ratio test in the ML approach (see Kass and Raftery, 1995).

So far, even in the context of standard SEMs with normal data, latent variable estimates have not been used much in practice. One of the reasons may be due to the problems of the Bartlett's method or the regression type method (Lawley and Maxwell, 1971). For example, the estimates produced by the above methods depend on the parameter estimates without taking into account the sampling errors, and the sampling distribution of the estimates is complicated. As the latent variable estimates produced by the Bayesian approach do not have these problems, we hope to see more applications of them to real research in the future. Due to the nature of the data, latent variable estimates obtained from dichotomous data or binary data may be inaccurate. These estimates should be treated with great caution.

The Bayesian methodologies for the aforementioned complex SEMs are developed through a commonly used strategy in the literature of statistics. That is, we first utilize the idea of data augmentation (Tanner and Wong, 1987) to augment the observed data with the latent quantities which cause the difficulties in the analysis. The latent quantities may be the latent variables in the model, the unobservable continuous measurements that underly the dichotomous or the ordered categorical variables, the allocation variables in the mixture models, and/or missing data. It has been shown that after data augmentation, the posterior analysis based on the complete data set can be handled. Usually, a sufficiently large number of observations are simulated from the joint posterior distribution of the parameters and the augmented latent quantities to obtain the Bayesian solutions. This task is done by applying tools in statistical computing, namely the Gibbs sampler and the MH algorithm in estimation, and path sampling in computing the Bayes factor for model comparison. These tools are well-known in statistics, and conceptually simple, although the derivations of the full conditional distributions and the implementations of the algorithms in the context of various complex SEMs are not straightforward. Recently, we have demonstrated that this useful strategy is also effective in developing Bayesian robust methods (Lee and Xia, 2006b). Based on our experience with this strategy, we strongly recommend it to readers for handling complicated problems.

Dichotomous, ordered categorical and/or missing data are often encountered in behavioral, educational, medical and social sciences; it is also well-recognized that two-level models should be used to cope with the hierarchical data, mixture models should be used to handle heterogeneity, and nonlinear models are necessary to analyze the important interaction or quadratic effects of the exogenous latent variables to the endogenous latent variables. In fact, the developments of the rigorous methods through the Bayesian approach are motivated by the need of these techniques for conducting correct statistical inferences under the real complex situations. Unfortunately, applications of these new techniques to substantive research are rather limited. In our opinion the main reason for this phenonemon is that the common software in SEMs cannot provide satisfactory solutions for the complex models or data structures, and applied researchers encounter difficulty in implementing the technically involved computer programs. Thanks to the development of WinBUGS (Speighalter Thomas, Best and Lunn, 2003), this difficulty could be overcome. As we have shown, most of the complex SEMs discussed in this book can be analyzed through WinBUGS. Recently, Sturtz, Ligges and Gelman (2005) developed an R2WinBUGS package, which provides convenient functions to call WinBUGS from *R* (*R* Development Core Team, 2004) for further analyses. The Bayes factor and the PP p-values could be obtained by saving the output in WinBUGS and reading it into *R*. Given the advantages of the Bayesian approach, and the feasibility of WinBUGS in analyzing complicated models and data, we strongly recommend this general software to applied researchers. We think that it is worthwhile to give time and effort to know more about this useful software, despite its slower convergence than our tailor-made programs.

The developments of Bayesian statistical methods as well as Bayesian computing tools has been very rapid in recent years. In the process of writing this book, we realized that some of our content could be improved by these recent developments. With regard to the Bayesian methods, the partial posterior predictive (PPP) p-value (Bayarri and Berger, 2000) is an improvement over the PP p-value (Gelman, Meng and Stern, 1996) for overcoming the problem of 'double use' of the data. In the context of SEMs with nonignorable missing data, Lee and Tang (2006) proposed an algorithm for computing the PPP p-value, and showed that the additional computation is not difficult. Moreover, it may be desirable to present the highest posterior density (HPD) interval (see Chen, Shao and Ibrahim, 2000) of a parameter to reveal the variability of its estimate, as was done in Lee and Song (2003) and Song and Lee (2004). Computationally, the efficiency of the proposed data augmentation scheme and the MCMC methods may be improved by the more recent developments in statistical computing, such as Meng and van Dyk (1999), van Dyk and Meng (2001) and Gelman (2004). We may incorporate these and other possible improvements in a revised edition.

There are a few common assumptions in formulating the SEMs in this book. First, we usually assume that $\boldsymbol{\Psi}_\epsilon$ is diagonal. This assumption can be partially relaxed by extending $\boldsymbol{\Psi}_\epsilon$ to be a diagonal block matrix of the form

$$\boldsymbol{\Psi}_\epsilon = \begin{bmatrix} \boldsymbol{\Psi}_{\epsilon 1} & & \mathbf{0} \\ & \ddots & \\ \mathbf{0} & & \boldsymbol{\Psi}_{\epsilon K} \end{bmatrix},$$

where $\boldsymbol{\Psi}_{\epsilon k}$ is an unknown positive definite matrix. Based on the conjugate prior distribution of $\boldsymbol{\Psi}_{\epsilon k}$, which is a similar Wishart distribution, the corresponding conditional distribution $p(\boldsymbol{\Psi}_{\epsilon k}|\cdot)$ required by the Gibbs sampler can be derived. For dichotomous or ordered categorical data, the conditional distribution of latent continuous variables would be related to a multivariate truncated normal distribution rather than a univariate one, and simulating observations is more complicated. The idea given in Chapter 7, Section 7.3, can be used but details have to be worked out. The extension of $\boldsymbol{\Psi}_\epsilon$ to an arbitrary patterned matrix is nontrivial mainly because it may not be possible to use the Wishart distribution as its prior distribution, and the full conditional distribution of the whole parameter matrix Λ is more complicated.

Secondly, we usually assume that the matrix of coefficients $\boldsymbol{\Pi}$ in the structural equation satisfies the condition that $|\mathbf{I}-\boldsymbol{\Pi}|$ is independent of the elements in $\boldsymbol{\Pi}$. Let the conjugate prior distribution be $N[\boldsymbol{\Pi}_o, \mathbf{H}_{o\pi}]$, it follows from Lee and Zhu (2000) that the required conditional distribution of $\boldsymbol{\Pi}$ in the Gibbs sampler is proportional to

$$\exp\left[n\log|\mathbf{I}-\boldsymbol{\Pi}| - \frac{1}{2}(\boldsymbol{\Pi}-\boldsymbol{\Pi}_o)^T\mathbf{H}_{o\pi}^{-1}(\boldsymbol{\Pi}-\boldsymbol{\Pi}_o) \right.$$
$$\left. -\frac{1}{2}\sum_{i=1}^{n}(\boldsymbol{\eta}_i-\boldsymbol{\Pi}\boldsymbol{\eta}_i-\boldsymbol{\Gamma}\boldsymbol{\xi}_i)^T\boldsymbol{\Psi}_\delta^{-1}(\boldsymbol{\eta}_i-\boldsymbol{\Pi}\boldsymbol{\eta}_i-\boldsymbol{\Gamma}\boldsymbol{\xi}_i)\right].$$

If $|\mathbf{I}-\boldsymbol{\Pi}|$ is independent of elements in $\boldsymbol{\Pi}$, this conditional distribution is essentially a normal distribution. If $|\mathbf{I}-\boldsymbol{\Pi}|$ depends on $\boldsymbol{\Pi}$, this conditional distribution is not normal. However, an MH type of algorithm can be applied to simulate the required observations.

Thirdly, we assume the distributions of $\boldsymbol{\xi}_i$, $\boldsymbol{\epsilon}_i$ and $\boldsymbol{\delta}_i$ are multivariate normal distributions; although for certain SEMs, we allow that the distribution of the manifest variables is non-normal. An interesting topic for further research is about the robustness of the Bayesian methods against this assumption. For dichotomous and ordered categorical data with a threshold specification, we expect the problem due to violation of this assumption may not be serious. However, as latent variables play the most important role in both the measurement equation and the structural equation, it is desirable to develop semiparametric Bayesian methods to relax the normality assumption of $\boldsymbol{\xi}_i$.

Inspired by the pioneer work of Cook (1977, 1986) on case-deletion and local influence measures, as well as their wide applications to various statistical models, it is interesting to identify the influential observations in relation to the Bayesian methods. In view of this, a further research topic is to develop Bayesian case-deletion measures and/or Bayesian local influence measures. Moreover, it is also desirable to investigate the impact of outliers on the Bayesian methods. Recently, in the context of an ML approach and a nonlinear SEM with missing data, Lee and Xia (2006a) proposed a robust method to handle outliers, by incorporating stochastic weights in the covariance matrices of $\boldsymbol{\xi}_i$, $\boldsymbol{\epsilon}_i$ and $\boldsymbol{\delta}_i$. The effect of the stochastic weights is to reduce the influence of the outliers. Theoretically, the idea of stochastic weights can be applied to the Bayesian approach. However, many details have to be worked out.

Unordered categorical variables with a multinomial distribution are involved in much important substantive research. For instance, unordered categorical observations are associated with genotype variables in genetic studies, and the selection model in economics. Development of SEMs with this kind of discrete data has great practical value. Recently, analysis of longitudinal data has received a great deal of attention. With some appropriate specifications, the factor analysis model has been applied to analyze latent curve models (see, for example, Blozis (2004)). An interesting research topic would be to develop more subtle models and Bayesian methods for analyzing multivariate longitudinal data. More generally, it is useful to establish dynamic SEMs for analyzing time-series data, or functional data (see Ramsay and Silverman, 1997).

REFERENCES

Bayarri M. J. and Berger J. O. (2000) P values for composite null models. *Journal of American Statistical Association*, **95**, 1127–1142.

Blozis, S. A. (2004) Structured latent curve models for the study of change in multivariate repeated measures. *Psychologial Methods*, **9**, 334–353.

Chen, M. H., Shao, Q. M. and Ibrahim, J. G. (2000) *Monte Carlo Methods in Bayesian Computation*. New York: Springer-Verlag.

Cook, R. D. (1977) Detection of influential observations in linear regression. *Technometrics*, **19**, 15–18.

Cook, R. D. (1986) Assessment of local influence (with discussion). *Journal of the Royal Statistical Society, Series B*, **48**, 133–169.

Gelman, A. (2004) Parameterization and Bayesian modeling. *Journal of American Statistical Association*, **99**, 537–545.

Gelman, A., Meng, X. L. and Stern, H. (1996) Posterior predictive assessment of model fitness via realized discrepancies. *Statistica Sinica*, **6**, 733–807.

Kass, R. E. and Raftery, A. E. (1995) Bayes factors. *Journal of the American Statistical Association*, **90**, 773–795.

Lawley, D. N. and Maxwell, A. E. (1971) *Factor Analysis as a Statistical Method* (2nd edn). New York: American Elsevier.

Lee, S. Y. and Song, X. Y. (2003) Model comparison of nonlinear structural equation models with fixed covariates. *Psychometrika*, **68**, 27–47.

Lee, S. Y. and Tang, N. S. (2006) Bayesian analysis of nonlinear structural equation models with nonignorable missing data. *Psychometrika*, in press.

Lee, S. Y. and Xia, Y. M. (2006a) Maximum likelihood methods in treating outliers and symmetrically heavy-tailed distributions for nonlinear structural equation models with missing data. *Psychometrika*, in press.

Lee, S. Y. and Xia, Y. M. (2006b) A Bayesian robust approach for latent variable models with ignorable missing data. Submitted manuscript.

Lee, S. Y. and Zhu, H. T. (2000) Statistical analysis of nonlinear structural equation models with continuous and polyfomous data. *British Journal of Mathematical and Statistical Psychology*, **53**, 209–232.

Meng, X. L. and van Dyk, D. A. (1999) Seeking efficient data augmentation scheme via conditional and marginal augmentation. *Biometrika*, **86**, 301–320.

R Development Core Team (2004) *R: A Language and Environment for Statistical Computing*. Vienna, Austria: R Foundation for Statistical Computing.

Ramsay, J. O. and Silverman, B. W. (1997) *Functional Data Analysis*. New York: Springer-Verlag.

Song, X. Y. and Lee, S. Y. (2004) Bayesian analysis of two-level nonlinear structural equation models with continuous and polytomous data. *British Journal of Mathematical and Statistical Psychology*, **57**, 29–52.

Spiegelhalter, D. J., Thomas, A., Best, N. G. and Lunn, D. (2003) *WinBugs User Manual. Version 1.4*. Cambridge, England: MRC Biostatistics Unit.

Sturtz, S., Ligges, U. and Gelman, A. (2005) R2WinBUGS: a package for running WinBUGS from *R*. *Journal of Statistical Software*, **12**, 1–17.

Tanner, M. A. and Wong, W. H. (1987) The calculation of posterior distributions by data augmentation(with discussion). *Journal of the American Statistical Association*, **82**, 528–550.

Van Dyk, D. A. and Meng, X. L. (2001) The art of data augmentation (with discussion). *Journal of Computational and Graphical Statistics*, **10**, 1–111.

Index

Structural Equation Modeling: A Bayesian Approach S-Y. Lee

WILEY SERIES IN PROBABILITY AND STATISTICS

ESTABLISHED BY WALTER A. SHEWHART AND SAMUEL S. WILKS

Editors
David J. Balding, Peter Bloomfield, Noel A. C. Cressie, Nicholas I. Fisher, Iain M. Johnstone, J. B. Kadane, Geert Molenberghs, Louise M. Ryan, David W. Scott, Adrian F. M. Smith, Sanford Weisberg
Editors Emeriti
Vic Barnett, J. Stuart Hunter, David G. Kendall

The ***Wiley Series in Probability and Statistics*** is well established and authoritative. It covers many topics of current research interest in both pure and applied statistics and probability theory. Written by leading statisticians and institutions, the titles span both state-of-the-art developments in the field and classical methods.

Reflecting the wide range of current research in statistics, the series encompasses applied, methodological and theoretical statistics, ranging from applications and new techniques made possible by advances in computerized practice to rigorous treatment of theoretical approaches.

This series provides essential and invaluable reading for all statisticians, whether in academia, industry, government, or research.

ABRAHAM and LEDOLTER Statistical Methods for Forecasting
AGRESTI Analysis of Ordinal Categorical Data
AGRESTI An Introduction to Categorical Data Analysis
AGRESTI Categorical Data Analysis, Second Edition
ALTMAN, GILL, and McDONALD Numerical Issues in Statistical Computing for the Social Scientist
AMARATUNGA and CABRERA Exploration and Analysis of DNA Microarray and Protein Array Data
ANDĚL Mathematics of Chance
ANDERSON An Introduction to Multivariate Statistical Analysis, Third Edition
ANDERSON The Statistical Analysis of Time Series
ANDERSON, AUQUIER, HAUCK, OAKES, VANDAELE, and WEISBERG Statistical Methods for Comparative Studies
ANDERSON and LOYNES The Teaching of Practical Statistics
ARMITAGE and DAVID (editors) Advances in Biometry
ARNOLD, BALAKRISHNAN, and NAGARAJA Records
ARTHANARI and DODGE Mathematical Programming in Statistics
BAILEY The Elements of Stochastic Processes with Applications to the Natural Sciences
BALAKRISHNAN and KOUTRAS Runs and Scans with Applications
BALAKRISHNAN and NG Precedence-Type Tests and Applications
BARNETT Comparative Statistical Inference, Third Edition
BARNETT Environmental Statistics: Methods & Applications
BARNETT and LEWIS Outliers in Statistical Data, Third Edition
BARTOSZYNSKI and NIEWIADOMSKA-BUGAJ Probability and Statistical Inference
BASILEVSKY Statistical Factor Analysis and Related Methods: Theory and Applications
BASU and RIGDON Statistical Methods for the Reliability of Repairable Systems
BATES and WATTS Nonlinear Regression Analysis and Its Applications
BECHHOFER, SANTNER, and GOLDSMAN Design and Analysis of Experiments for Statistical Selection, Screening, and Multiple Comparisons

* Now available in a lower priced paperback edition in the Wiley Classics Library.

BELSLEY Conditioning Diagnostics: Collinearity and Weak Data in Regression
BELSLEY, KUH, and WELSCH Regression Diagnostics: Identifying Influential Data and Sources of Collinearity
BENDAT and PIERSOL Random Data: Analysis and Measurement Procedures, Third Edition
BERNARDO and SMITH Bayesian Theory
BERRY, CHALONER, and GEWEKE Bayesian Analysis in Statistics and Econometrics: Essays in Honor of Arnold Zellner
BHAT and MILLER Elements of Applied Stochastic Processes, Third Edition
BHATTACHARYA and JOHNSON Statistical Concepts and Methods
BHATTACHARYA and WAYMIRE Stochastic Processes with Applications
BIEMER, GROVES, LYBERG, MATHIOWETZ, and SUDMAN Measurement Errors in Surveys
BILLINGSLEY Convergence of Probability Measures, Second Edition
BILLINGSLEY Probability and Measure, Third Edition
BIRKES and DODGE Alternative Methods of Regression
BLISCHKE and MURTHY (editors) Case Studies in Reliability and Maintenance
BLISCHKE and MURTHY Reliability: Modeling, Prediction, and Optimization
BLOOMFIELD Fourier Analysis of Time Series: An Introduction, Second Edition
BOLLEN Structural Equations with Latent Variables
BOLLEN and CURRAN Latent Curve Models: A Structural Equation Perspective
BOROVKOV Ergodicity and Stability of Stochastic Processes
BOULEAU Numerical Methods for Stochastic Processes
BOX Bayesian Inference in Statistical Analysis
BOX R. A. Fisher, the Life of a Scientist
BOX and DRAPER Empirical Model-Building and Response Surfaces
BOX and DRAPER Evolutionary Operation: A Statistical Method for Process Improvement
BOX, HUNTER, and HUNTER Statistics for Experimenters: An Introduction to Design, Data Analysis, and Model Building
BOX, HUNTER, and HUNTER Statistics for Experimenters: Design, Innovation and Discovery, Second Edition
BOX and LUCE~NO Statistical Control by Monitoring and Feedback Adjustment
BRANDIMARTE Numerical Methods in Finance: A MATLAB-Based Introduction
BROWN and HOLLANDER Statistics: A Biomedical Introduction
BRUNNER, DOMHOF, and LANGER Nonparametric Analysis of Longitudinal Data in Factorial Experiments
BUCKLEW Large Deviation Techniques in Decision, Simulation, and Estimation
CAIROLI and DALANG Sequential Stochastic Optimization
CASTILLO, HADI, BALAKRISHNAN and SARABIA Extreme Value and Related Models with Applications in Engineering and Science
CHAN Time Series: Applications to Finance
CHATTERJEE and HADI Regression Analysis by Example, Fourth Edition
CHATTERJEE and HADI Sensitivity Analysis in Linear Regression
CHATTERJEE and PRICE Regression Analysis by Example, Third Edition
CHERNICK Bootstrap Methods: A Practitioner's Guide
CHERNICK and FRIIS Introductory Biostatistics for the Health Sciences
CHILÈS and DELFINER Geostatistics: Modeling Spatial Uncertainty
CHOW and LIU Design and Analysis of Clinical Trials: Concepts and Methodologies, Second Edition
CLARKE and DISNEY Probability and Random Processes: A First Course with Applications, Second Edition
COCHRAN and COX Experimental Designs, Second Edition

* Now available in a lower priced paperback edition in the Wiley Classics Library.

CONGDON Applied Bayesian Modelling
CONGDON Bayesian Models for Categorical Data
CONGDON Bayesian Statistical Modelling
CONGDON Bayesian Statistical Modelling, Second Edition
CONOVER Practical Nonparametric Statistics, Second Edition
COOK Regression Graphics
COOK and WEISBERG An Introduction to Regression Graphics
COOK and WEISBERG Applied Regression Including Computing and Graphics
CORNELL Experiments with Mixtures, Designs, Models, and the Analysis of Mixture Data, Third Edition
COVER and THOMAS Elements of Information Theory
COX A Handbook of Introductory Statistical Methods
COX Planning of Experiments
CRESSIE Statistics for Spatial Data, Revised Edition
CSÖRGŐ and HORVÁTH Limit Theorems in Change Point Analysis
DANIEL Applications of Statistics to Industrial Experimentation
DANIEL Biostatistics: A Foundation for Analysis in the Health Sciences, Sixth Edition
DANIEL Fitting Equations to Data: Computer Analysis of Multifactor Data, Second Edition
DASU and JOHNSON Exploratory Data Mining and Data Cleaning
DAVID and NAGARAJA Order Statistics, Third Edition
DEGROOT, FIENBERG, and KADANE Statistics and the Law
DEL CASTILLO Statistical Process Adjustment for Quality Control
DEMARIS Regression with Social Data: Modeling Continuous and Limited Response Variables
DEMIDENKO Mixed Models: Theory and Applications
DENISON, HOLMES, MALLICK, and SMITH Bayesian Methods for Nonlinear Classification and Regression
DETTE and STUDDEN The Theory of Canonical Moments with Applications in Statistics, Probability, and Analysis
DEY and MUKERJEE Fractional Factorial Plans
DILLON and GOLDSTEIN Multivariate Analysis: Methods and Applications
DODGE Alternative Methods of Regression
DODGE and ROMIG Sampling Inspection Tables, Second Edition
DOOB Stochastic Processes
DOWDY, WEARDEN, and CHILKO Statistics for Research, Third Edition
DRAPER and SMITH Applied Regression Analysis, Third Edition
DRYDEN and MARDIA Statistical Shape Analysis
DUDEWICZ and MISHRA Modern Mathematical Statistics
DUNN and CLARK Applied Statistics: Analysis of Variance and Regression, Second Edition
DUNN and CLARK Basic Statistics: A Primer for the Biomedical Sciences, Third Edition
DUPUIS and ELLIS A Weak Convergence Approach to the Theory of Large Deviations
EDLER and KITSOS (editors) Recent Advances in Quantitative Methods in Cancer and Human Health Risk Assessment
ELANDT-JOHNSON and JOHNSON Survival Models and Data Analysis
ENDERS Applied Econometric Time Series
ETHIER and KURTZ Markov Processes: Characterization and Convergence
EVANS, HASTINGS, and PEACOCK Statistical Distribution, Third Edition
FELLER An Introduction to Probability Theory and Its Applications, Volume I, Third Edition, Revised; Volume II, Second Edition
FISHER and VAN BELLE Biostatistics: A Methodology for the Health Sciences

* Now available in a lower priced paperback edition in the Wiley Classics Library.

FITZMAURICE, LAIRD, and WARE Applied Longitudinal Analysis
FLEISS The Design and Analysis of Clinical Experiments
FLEISS Statistical Methods for Rates and Proportions, Second Edition
FLEMING and HARRINGTON Counting Processes and Survival Analysis
FULLER Introduction to Statistical Time Series, Second Edition
FULLER Measurement Error Models
GALLANT Nonlinear Statistical Models.
GEISSER Modes of Parametric Statistical Inference
GELMAN and MENG (editors) Applied Bayesian Modeling and Casual Inference from Incomplete-data Perspectives
GEWEKE Contemporary Bayesian Econometrics and Statistics
GHOSH, MUKHOPADHYAY, and SEN Sequential Estimation
GIESBRECHT and GUMPERTZ Planning, Construction, and Statistical Analysis of Comparative Experiments
GIFI Nonlinear Multivariate Analysis
GIVENS and HOETING Computational Statistics
GLASSERMAN and YAO Monotone Structure in Discrete-Event Systems
GNANADESIKAN Methods for Statistical Data Analysis of Multivariate Observations, Second Edition
GOLDSTEIN and LEWIS Assessment: Problems, Development, and Statistical Issues
GREENWOOD and NIKULIN A Guide to Chi-Squared Testing
GROSS and HARRIS Fundamentals of Queueing Theory, Third Edition
HAHN and SHAPIRO Statistical Models in Engineering
HAHN and MEEKER Statistical Intervals: A Guide for Practitioners
HALD A History of Probability and Statistics and their Applications Before 1750
HALD A History of Mathematical Statistics from 1750 to 1930
HAMPEL Robust Statistics: The Approach Based on Influence Functions
HANNAN and DEISTLER The Statistical Theory of Linear Systems
HEIBERGER Computation for the Analysis of Designed Experiments
HEDAYAT and SINHA Design and Inference in Finite Population Sampling
HEDEKER and GIBBONS Longitudinal Data Analysis
HELLER MACSYMA for Statisticians
HINKELMANN and KEMPTHORNE Design and Analysis of Experiments, Volume 1: Introduction to Experimental Design
HINKELMANN and KEMPTHORNE Design and analysis of experiments, Volume 2: Advanced Experimental Design
HOAGLIN, MOSTELLER, and TUKEY Exploratory Approach to Analysis of Variance
HOAGLIN, MOSTELLER, and TUKEY Exploring Data Tables, Trends and Shapes
HOAGLIN, MOSTELLER, and TUKEY Understanding Robust and Exploratory Data Analysis
HOCHBERG and TAMHANE Multiple Comparison Procedures
HOCKING Methods and Applications of Linear Models: Regression and the Analysis of Variance, Second Edition
HOEL Introduction to Mathematical Statistics, Fifth Edition
HOGG and KLUGMAN Loss Distributions
HOLLANDER and WOLFE Nonparametric Statistical Methods, Second Edition
HOSMER and LEMESHOW Applied Logistic Regression, Second Edition
HOSMER and LEMESHOW Applied Survival Analysis: Regression Modeling of Time to Event Data
HUBER Robust Statistics
HUBERTY Applied Discriminant Analysis
HUNT and KENNEDY Financial Derivatives in Theory and Practice, Revised Edition
HUSKOVA, BERAN, and DUPAC Collected Works of Jaroslav Hajek—with Commentary
HUZURBAZAR Flowgraph Models for Multistate Time-to-Event Data

IMAN and CONOVER A Modern Approach to Statistics
JACKSON A User's Guide to Principle Components
JOHN Statistical Methods in Engineering and Quality Assurance
JOHNSON Multivariate Statistical Simulation
JOHNSON and BALAKRISHNAN Advances in the Theory and Practice of Statistics: A Volume in Honor of Samuel Kotz
JOHNSON and BHATTACHARYYA Statistics: Principles and Methods, Fifth Edition
JUDGE, GRIFFITHS, HILL, LU TKEPOHL, and LEE The Theory and Practice of Econometrics, Second Edition
JOHNSON and KOTZ Distributions in Statistics
JOHNSON and KOTZ (editors) Leading Personalities in Statistical Sciences: From the Seventeenth Century to the Present
JOHNSON, KOTZ, and BALAKRISHNAN Continuous Univariate Distributions, Volume 1, Second Edition
JOHNSON, KOTZ, and BALAKRISHNAN Continuous Univariate Distributions, Volume 2, Second Edition
JOHNSON, KOTZ, and BALAKRISHNAN Discrete Multivariate Distributions
JOHNSON, KOTZ, and KEMP Univariate Discrete Distributions, Second Edition
JURECKOVA and SEN Robust Statistical Procedures: Asymptotics and Interrelations
JUREK and MASON Operator-Limit Distributions in Probability Theory
KADANE Bayesian Methods and Ethics in a Clinical Trial Design
KADANE and SCHUM A Probabilistic Analysis of the Sacco and Vanzetti Evidence
KALBFLEISCH and PRENTICE The Statistical Analysis of Failure Time Data, Second Edition
KARIYA and KURATA Generalized Least Squares
KASS and VOS Geometrical Foundations of Asymptotic Inference
KAUFMAN and ROUSSEEUW Finding Groups in Data: An Introduction to Cluster Analysis
KEDEM and FOKIANOS Regression Models for Time Series Analysis
KENDALL, BARDEN, CARNE, and LE Shape and Shape Theory
KHURI Advanced Calculus with Applications in Statistics, Second Edition
KHURI, MATHEW, and SINHA Statistical Tests for Mixed Linear Models
KISH Statistical Design for Research
KLEIBER and KOTZ Statistical Size Distributions in Economics and Actuarial Sciences
KLUGMAN, PANJER, and WILLMOT Loss Models: From Data to Decisions
KLUGMAN, PANJER, and WILLMOT Solutions Manual to Accompany Loss Models: >From Data to Decisions
KOTZ, BALAKRISHNAN, and JOHNSON Continuous Multivariate Distributions, Volume 1, Second Edition
KOTZ and JOHNSON (editors) Encyclopedia of Statistical Sciences: Volumes 1 to 9 with Index
KOTZ and JOHNSON (editors) Encyclopedia of Statistical Sciences: Supplement Volume
KOTZ, READ, and BANKS (editors) Encyclopedia of Statistical Sciences: Update Volume 1
KOTZ, READ, and BANKS (editors) Encyclopedia of Statistical Sciences: Update Volume 2
KOVALENKO, KUZNETZOV, and PEGG Mathematical Theory of Reliability of Time-Dependent Systems with Practical Applications
KUROWICKA and COOKE Uncertainty Analysis with High Dimensional Dependence Modelling
LACHIN Biostatistical Methods: The Assessment of Relative Risks
LAD Operational Subjective Statistical Methods: A Mathematical, Philosophical, and Historical Introduction
LAMPERTI Probability: A Survey of the Mathematical Theory, Second Edition

* Now available in a lower priced paperback edition in the Wiley Classics Library.
* Now available in a lower priced paperback edition in the Wiley - Interscience Paperback Series.

LANGE, RYAN, BILLARD, BRILLINGER, CONQUEST, and GREENHOUSE Case Studies in Biometry
LARSON Introduction to Probability Theory and Statistical Inference, Third Edition
LAWLESS Statistical Models and Methods for Lifetime Data, Second Edition
LAWSON Statistical Methods in Spatial Epidemiology, Second Edition
LE Applied Categorical Data Analysis
LE Applied Survival Analysis
LEE Structural Equation Modeling: A Bayesian Approach
LEE and WANG Statistical Methods for Survival Data Analysis, Third Edition
LEPAGE and BILLARD Exploring the Limits of Bootstrap
LEYLAND and GOLDSTEIN (editors) Multilevel Modelling of Health Statistics
LIAO Statistical Group Comparison
LINDVALL Lectures on the Coupling Method
LINHART and ZUCCHINI Model Selection
LITTLE and RUBIN Statistical Analysis with Missing Data, Second Edition
LLOYD The Statistical Analysis of Categorical Data
LOWEN and TEICH Fractal-Based Point Processes
MAGNUS and NEUDECKER Matrix Differential Calculus with Applications in Statistics and Econometrics, Revised Edition
MALLER and ZHOU Survival Analysis with Long Term Survivors
MALLOWS Design, Data, and Analysis by Some Friends of Cuthbert Daniel
MANN, SCHAFER, and SINGPURWALLA Methods for Statistical Analysis of Reliability and Life Data
MANTON, WOODBURY, and TOLLEY Statistical Applications Using Fuzzy Sets
MARCHETTE Random Graphs for Statistical Pattern Recognition
MARDIA and JUPP Directional Statistics
MARONNA, MARTIN, and YOHAI Robust Statistics: Theory and Methods
MASON, GUNST, and HESS Statistical Design and Analysis of Experiments with Applications to Engineering and Science, Second Edition
MCCULLOCH and SERLE Generalized, Linear, and Mixed Models
MCFADDEN Management of Data in Clinical Trials
MCLACHLAN Discriminant Analysis and Statistical Pattern Recognition
MCLACHLAN, DO, and AMBROISE Analyzing Microarray Gene Expression Data
MCLACHLAN and KRISHNAN The EM Algorithm and Extensions
MCLACHLAN and PEEL Finite Mixture Models
MCNEIL Epidemiological Research Methods
MEEKER and ESCOBAR Statistical Methods for Reliability Data
MEERSCHAERT and SCHEFFLER Limit Distributions for Sums of Independent Random Vectors: Heavy Tails in Theory and Practice
MICKEY, DUNN, and CLARK Applied Statistics: Analysis of Variance and Regression, Third Edition
MILLER Survival Analysis, Second Edition
MONTGOMERY, PECK, and VINING Introduction to Linear Regression Analysis, Fourth Edition
MORGENTHALER and TUKEY Configural Polysampling: A Route to Practical Robustness
MUIRHEAD Aspects of Multivariate Statistical Theory
MULLER and STEWART Linear Model Theory: Univariate, Multivariate, and Mixed Models
MURRAY X-STAT 2.0 Statistical Experimentation, Design Data Analysis, and Nonlinear Optimization
MURTHY, XIE, and JIANG Weibull Models
MYERS and MONTGOMERY Response Surface Methodology: Process and Product Optimization Using Designed Experiments, Second Edition

MYERS, MONTGOMERY, and VINING Generalized Linear Models. With Applications in Engineering and the Sciences
NELSON Accelerated Testing, Statistical Models, Test Plans, and Data Analysis
NELSON Applied Life Data Analysis
NEWMAN Biostatistical Methods in Epidemiology
OCHI Applied Probability and Stochastic Processes in Engineering and Physical Sciences
OKABE, BOOTS, SUGIHARA, and CHIU Spatial Tesselations: Concepts and Applications of Voronoi Diagrams, Second Edition
OLIVER and SMITH Influence Diagrams, Belief Nets and Decision Analysis
PALTA Quantitative Methods in Population Health: Extentions of Ordinary Regression
PANJER Operational Risks: Modeling Analytics
PANKRATZ Forecasting with Dynamic Regression Models
PANKRATZ Forecasting with Univariate Box-Jenkins Models: Concepts and Cases
PARZEN Modern Probability Theory and Its Applications
PENA, TIAO, and TSAY A Course in Time Series Analysis
PIANTADOSI Clinical Trials: A Methodologic Perspective
PORT Theoretical Probability for Applications
POURAHMADI Foundations of Time Series Analysis and Prediction Theory
PRESS Bayesian Statistics: Principles, Models, and Applications
PRESS Subjective and Objective Bayesian Statistics, Second Edition
PRESS and TANUR The Subjectivity of Scientists and the Bayesian Approach
PUKELSHEIM Optimal Experimental Design
PURI, VILAPLANA, and WERTZ New Perspectives in Theoretical and Applied Statistics
PUTERMAN Markov Decision Processes: Discrete Stochastic Dynamic Programming
QIU Image Processing and Jump Regression Analysis
RAO Linear Statistical Inference and its Applications, Second Edition
RAUSAND and HOYLAND System Reliability Theory: Models, Statistical Methods and Applications, Second Edition
RENCHER Linear Models in Statistics
RENCHER Methods of Multivariate Analysis, Second Edition
RENCHER Multivariate Statistical Inference with Applications
RIPLEY Spatial Statistics
RIPLEY Stochastic Simulation
ROBINSON Practical Strategies for Experimenting
ROHATGI and SALEH An Introduction to Probability and Statistics, Second Edition
ROLSKI, SCHMIDLI, SCHMIDT, and TEUGELS Stochastic Processes for Insurance and Finance
ROSENBERGER and LACHIN Randomization in Clinical Trials: Theory and Practice
ROSS Introduction to Probability and Statistics for Engineers and Scientists
ROSSI, ALLENBY, and MCCULLOCH Bayesian Statistics and Marketing
ROUSSEEUW and LEROY Robust Regression and Outline Detection
RUBIN Multiple Imputation for Nonresponse in Surveys
RUBINSTEIN Simulation and the Monte Carlo Method
RUBINSTEIN and MELAMED Modern Simulation and Modeling
RYAN Modern Regression Methods
RYAN Statistical Methods for Quality Improvement, Second Edition
SALEH Theory of Preliminary Test and Stein-Type Estimation with Applications
SALTELLI, CHAN, and SCOTT (editors) Sensitivity Analysis
SCHEFFE The Analysis of Variance

* Now available in a lower priced paperback edition in the Wiley Classics Library.
* Now available in a lower priced paperback edition in the Wiley - Interscience Paperback Series.

SCHIMEK Smoothing and Regression: Approaches, Computation, and Application
SCHOTT Matrix Analysis for Statistics
SCHOUTENS Levy Processes in Finance: Pricing Financial Derivatives
SCHUSS Theory and Applications of Stochastic Differential Equations
SCOTT Multivariate Density Estimation: Theory, Practice, and Visualization
SEARLE Linear Models
SEARLE Linear Models for Unbalanced Data
SEARLE Matrix Algebra Useful for Statistics
SEARLE and WILLETT Matrix Algebra for Applied Economics
SEBER Multivariate Observations
SEBER and LEE Linear Regression Analysis, Second Edition
SEBER and WILD Nonlinear Regression
SENNOTT Stochastic Dynamic Programming and the Control of Queueing Systems
SERFLING Approximation Theorems of Mathematical Statistics
SHAFER and VOVK Probability and Finance: Its Only a Game!
SILVAPULLE and SEN Constrained Statistical Inference: Inequality, Order, and Shape Restrictions
SINGPURWALLA Reliability and Risk: A Bayesian Perspective
SMALL and MCLEISH Hilbert Space Methods in Probability and Statistical Inference
SRIVASTAVA Methods of Multivariate Statistics
STAPLETON Linear Statistical Models
STAUDTE and SHEATHER Robust Estimation and Testing
STOYAN, KENDALL, and MECKE Stochastic Geometry and Its Applications, Second Edition
STOYAN and STOYAN Fractals, Random and Point Fields: Methods of Geometrical Statistics
STYAN The Collected Papers of T. W. Anderson: 1943–1985
SUTTON, ABRAMS, JONES, SHELDON, and SONG Methods for Meta-Analysis in Medical Research
TANAKA Time Series Analysis: Nonstationary and Noninvertible Distribution Theory
THOMPSON Empirical Model Building
THOMPSON Sampling, Second Edition
THOMPSON Simulation: A Modeler's Approach
THOMPSON and SEBER Adaptive Sampling
THOMPSON, WILLIAMS, and FINDLAY Models for Investors in Real World Markets
TIAO, BISGAARD, HILL, PENA, and STIGLER (editors) Box on Quality and Discovery: with Design, Control, and Robustness
TIERNEY LISP-STAT: An Object-Oriented Environment for Statistical Computing and Dynamic Graphics
TSAY Analysis of Financial Time Series
UPTON and FINGLETON Spatial Data Analysis by Example, Volume II: Categorical and Directional Data
VAN BELLE Statistical Rules of Thumb
VAN BELLE, FISHER, HEAGERTY, and LUMLEY Biostatistics: A Methodology for the Health Sciences, Second Edition
VESTRUP The Theory of Measures and Integration
VIDAKOVIC Statistical Modeling by Wavelets
VINOD and REAGLE Preparing for the Worst: Incorporating Downside Risk in Stock Market Investments
WALLER and GOTWAY Applied Spatial Statistics for Public Health Data
WEERAHANDI Generalized Inference in Repeated Measures: Exact Methods in MANOVA and Mixed Models

*Now available in a lower priced paperback edition in the Wiley Classics Library.

WEISBERG Applied Linear Regression, Second Edition
WELISH Aspects of Statistical Inference
WESTFALL and YOUNG Resampling-Based Multiple Testing: Examples and Methods for p-Value Adjustment
WHITTAKER Graphical Models in Applied Multivariate Statistics
WINKER Optimization Heuristics in Economics: Applications of Threshold Accepting
WONNACOTT and WONNACOTT Econometrics, Second Edition
WOODING Planning Pharmaceutical Clinical Trials: Basic Statistical Principles
WOOLSON and CLARKE Statistical Methods for the Analysis of Biomedical Data, Second Edition
WU and HAMADA Experiments: Planning, Analysis, and Parameter Design Optimization
WU and ZHANG Nonparametric Regression Methods for Longitudinal Data Analysis: Mixed-Effects Modeling Approaches
YANG The Construction Theory of Denumerable Markov Processes
YOUNG, VALERO-MORA, and FRIENDLY Visual Statistics: Seeing Data with Dynamic Interactive Graphics
ZELLNER An Introduction to Bayesian Inference in Econometrics
ZELTERMAN Discrete Distributions: Applications in the Health Sciences
ZHOU, OBUCHOWSKI, and McCLISH Statistical Methods in Diagnostic Medicine

* Now available in a lower priced paperback edition in the Wiley Classics Library.

Printed and bound by CPI Group (UK) Ltd, Croydon, CR0 4YY

16/04/2025

14658497-0002